John Dover · Martin Warren · Tim Shreeve
Editors

Lepidoptera Conservation in a Changing World

Previously published in the *Journal of Insect Conservation* Volume 15,
Issues 1 and 2, 2011

Springer

Editors
John Dover
Staffordshire University
Faculty of Sciences
College Road
Stoke-on-Trent ST4 2DE
Staffordshire
United Kingdom

Martin Warren
Butterfly Conservation
Manor Yard
East Lulworth BH20 5QP
Dorset
United Kingdom

Tim Shreeve
Oxford Brookes University
School of Life Sciences
Gipsy Lane
Oxford OX3 0BP
Oxfordshire
United Kingdom

ISBN: 978-94-007-1441-0

Library of Congress Control Number: 2011924453

Life logo on front cover: The Symposium and its proceedings were part-funded by DG Environment. The opinions expressed are those of the authors and do not necessarily reflect those of the funders.

Printed on acid-free paper.

springer.com

Contents

Foreword
Sir David Attenborough

EDITORIAL NOTE

2010 and beyond for lepidoptera
J.W. Dover · M.S. Warren · T.G. Shreeve **1**

2010 ASSESSMENTS FOR LEPIDOPTERA

British butterfly distributions and the 2010 target
J. Asher · R. Fox · M.S. Warren **5**

The development of butterfly indicators in the United Kingdom and assessments in 2010
T. Brereton · D.B. Roy · I. Middlebrook · M. Botham · M. Warren **15**

Moths count: recording moths for conservation in the UK
R. Fox · Z. Randle · L. Hill · S. Anders · L. Wiffen · M.S. Parsons **29**

The state of the Dutch larger moth fauna
D. Groenendijk · W.N. Ellis **43**

Recent trends in butterfly populations from north-east Spain and Andorra in the light of habitat and climate change
C. Stefanescu · I. Torre · J. Jubany · F. Páramo **51**

Andorran butterfly monitoring scheme
R. Caritg · M. Domènech · J. Dantart · J. Jubany **63**

Assessing conservation status and trends for the world's butterflies: the Sampled Red List Index approach
O.T. Lewis · M.J.M. Senior **67**

Butterfly community recovery in degraded rainforest habitats in the Upper Guinean Forest Zone (Kakum forest, Ghana)
Sz. Sáfián · G. Csontos · D. Winkler **75**

Conservation of butterflies in Japan: status, actions and strategy
Y. Nakamura **85**

CLIMATE CHANGE IMPACTS AND ADAPTATION

Recent evidence for the climate change threat to Lepidoptera and other insects
R.J. Wilson · I.M.D. Maclean **103**

Synchronisation of egg hatching of brown hairstreak (*Thecla betulae*) and budburst of blackthorn (*Prunus spinosa*) in a warmer future
H.H. de Vries · S.H. Ens · G. de Graaf · L. Teunissen · R. van der Velde · L. Vogelaar · A. Winterink · M.E. Visser **113**

THE SCIENCE OF CONSERVATION MANAGEMENT

Evidence based conservation of butterflies
J.A. Thomas · D.J. Simcox · T. Hovestadt **123**

Butterfly abundance in a warming climate: patterns in space and time are not congruent
N.J.B. Isaac · M. Girardello · T.M. Brereton · D.B. Roy **141**

Declines of prairie butterflies in the midwestern USA
S.R. Swengel · D. Schlicht · F. Olsen · A.B. Swengel **149**

Habitat preferences of *Maculinea arion* and its *Myrmica* host ants: implications for habitat management in Italian Alps
L.P. Casacci · M. Witek · F. Barbero · D. Patricelli · G. Solazzo · E. Balletto · S. Bonelli **163**

Not only the butterflies: managing ants on road verges to benefit *Phengaris* (*Maculinea*) butterflies
I. Wynhoff · R. van Gestel · C. van Swaay · F. van Langevelde **171**

Butterflies on the brink: habitat requirements for declining populations of the marsh fritillary (*Euphydryas aurinia*) in SW England
M. Smee · W. Smyth · M. Tunmore · R. ffrench-Constant · D. Hodgson **189**

The effects of habitat fragmentation on niche requirements of the marsh fritillary, *Euphydryas aurinia*, (Rottemburg, 1775) on calcareous grasslands in southern UK
M.S. Botham · D. Ash · N. Aspey · N.A.D. Bourn · C.R. Bulman · D.B. Roy · J. Swain · A. Zannese · R.F. Pywell **201**

***Leptidea sinapis* (Wood White butterfly) egg-laying habitat and adult dispersal studies in Herefordshire**
S.A. Clarke · D.G. Green · J. Joy · K. Wollen · I. Butler **211**

New tools to boost butterfly habitat quality in existing grass buffer strips
R.J. Blake · B.A. Woodcock · D.B. Westbury · P. Sutton · S.G. Potts **225**

What can sown wildflower strips contribute to butterfly conservation?: an example from a Swiss lowland agricultural landscape
C. Haaland · L.-F. Bersier **237**

Land-use, environment, and their impact on butterfly populations in a mountainous pastoral landscape: individual species distribution and abundance
J.W. Dover · A. Rescia · S. Fungariño · J. Fairburn · P. Carey · P. Lunt · C. Arnot · R.L.H. Dennis · C.J. Dover **247**

LANDSCAPE-SCALE CONSERVATION

Landscape scale conservation: resources, behaviour, the matrix and opportunities
T.G. Shreeve · R.L.H. Dennis **261**

Habitat resources, remnant vegetation condition and area determine distribution patterns and abundance of butterflies and day-flying moths in a fragmented urban landscape, south-west Western Australia
M.R. Williams **271**

High and dry or sunk and dunked: lessons for tallgrass prairies from quaking bogs
A.B. Swengel · S.R. Swengel **289**

Landscape-scale conservation in practice: lessons from northern England, UK
S. Ellis · D. Wainwright · F. Berney · C. Bulman · N. Bourn **303**

Securing viable metapopulations of the Marsh Fritillary butterfly, *Euphydryas aurinia*, (Lepidoptera: Nymphalidae) in Northern England
K. Porter · S. Ellis **317**

Is the Natura 2000 network sufficient for conservation of butterfly diversity? A case study in Slovenia
R. Verovnik · M. Govedič · A. Šalamun **327**

Status and conservation of *Asclepiadaceae* and *Danaus* in southern Spain
J.F. Haeger · D. Jordano · M.L. Meléndez **333**

FUTURE CHALLENGES

Grassland butterflies and low intensity farming in Europe
J.W. Dover · S. Spencer · S. Collins · I. Hadjigeorgiou · A. Rescia **339**

Developing and launching a wider countryside butterfly survey across the United Kingdom
T.M. Brereton · K.L. Cruickshanks · K. Risely · D.G. Noble · D.B. Roy **349**

Ten challenges for 2010 and beyond to conserve Lepidoptera in Europe
M.S. Warren · N.A.D. Bourn **361**

Subject Index 367

Butterfly Conservation Sixth International Symposium Reading, 25–28 March 2010

Opening address by Sir David Attenborough President, Butterfly Conservation

"I believe this is a record-breaking gathering with nearly 300 delegates and many more who would like to be here but were not able to attend. As I understand it that makes this one of the biggest ever international gatherings of butterfly scientists and conservationists.

I will be surprised if there is anyone here who didn't first come to butterflies because they thought they are wonderful, extraordinary, heart lifting creatures – visions of beauty and visions of summer. That's what made me interested in butterflies.

Of course we now realise that we have a huge responsibility on our shoulders. The knowledge, information and painstaking research which that love of butterflies brought about has made you the guardians of great wisdom. You know how to conserve butterflies and the landscape - and the biodiversity in that landscape.

We came here through love of nature and you now have in your knowledge, your wisdom, the skills and experience which enable you to help others conserve the landscape. The world has never been in greater need of people like you. You know from your research on butterflies that they are in real danger and 10% of all European species are at serious risk of extinction.

What delegates at this conference have already begun to discover is that similar threats to wildlife are found across the world. However, as love of butterflies spreads across the world, with it spreads the knowledge of what has to be done. Butterflies are the canaries of the coalmine, they are the indicators that show that the whole of that biodiversity is in danger - so what you do here is of great importance for natural history, for wildlife, for the biodiversity of the world.

I count myself privileged to be involved in Butterfly Conservation and I wish you heartfelt success on your deliberations. What you discover here is of great importance to the natural world everywhere.

I wish you every success."

David. Attenborough

J Insect Conserv (2011) 15:1–3
DOI 10.1007/s10841-010-9368-1

EDITORIAL NOTE

2010 and beyond for Lepidoptera

J. W. Dover · M. S. Warren · T. G. Shreeve

Received: 21 October 2010 / Accepted: 10 November 2010 / Published online: 26 November 2010

The Convention on Biological Diversity (CBD) was an outcome of the 'Earth Summit' held in Rio de Janeiro in 1992. In 2002 the Convention committed the European Union (EU) and (currently) 192 other countries "to achieve by 2010 a significant reduction of the current rate of biodiversity loss at the global, regional and national level" (Anonymous 2010). The EU set an even more ambitious target: "…to protect and restore habitats and natural systems and halt the loss of biodiversity by 2010…" (Anonymous 2001). In March 2010, Butterfly Conservation held its 6th International Symposium on the topic, 'The 2010 Target and Beyond for Lepidoptera', to assess progress towards the target, report advances in conservation science, and look to the future. This volume presents a collection of papers from that meeting.

Evidence presented at the Symposium from across the world was bleak, with no evidence that the EU and CBD target had been met (Asher et al. 2011; Brereton et al. 2011; Caritg et al. 2011; Fox et al. 2011; Groenendijk and Ellis 2011; Lewis and Senior 2011; Nakamura 2011; Sáfián et al. 2011; Stefanescu et al. 2011). At the global scale the general trend for Lepidoptera is one of decline, and the overall trend for biodiversity is of a continuous and accelerating decline. In May 2010 the CBD published its overall assessment of progress towards the 2010 target and reported failure (Anonymous 2010). The 2002 CBD and EU targets were ambitious, and, despite not being met, they did manage to place conservation and biodiversity firmly on global and national political and development agendas and provided mechanisms for raising public awareness.

Failure can be a powerful stimulus to renewed action, especially if the reasons for failure are identified. New CBD 2020 targets are now being formulated (Djoghlaf 2010) alongside a vision for 2050. Initial progress suggests there will be a much closer integration of biodiversity with climate change adaptation/mitigation, ecosystem services, human health and the economic benefits of biodiversity (Djoghlaf 2010). Papers in this collection examine some of the likely impacts of climate change on Lepidoptera (e.g. de Vries et al. 2011; Wilson and Maclean 2011). Whilst climate change is an issue and will become an even bigger issue in the future, biodiversity loss is already a problem and likely to become worse unless measures are taken now. For some Lepidoptera there is a real and imminent threat of local, regional and global extinction. For such species, urgent action is needed now and climate change is a distant threat. The papers here outline the issues, and some demonstrate how approaches based on sound conservation science can improve the status of species, both preventing imminent extinction and making populations more resilient to future climate change (e.g. Ellis et al. 2011; Porter and Ellis 2011).

Meeting the new CBD targets, will require well researched solutions and this volume contains contributions to the challenges ahead under the broad headings of the Science of Conservation Management and Landscape-scale Conservation. Jeremy Thomas, for example, distills a lifetime of experience working with endangered species to identify key factors of importance in butterfly conservation

J. W. Dover (✉)
Institute for Environment, Sustainability and Regeneration, Staffordshire University, College Road, Stoke-on-Trent ST4 2DE, UK
e-mail: j.w.dover@staffs.ac.uk

M. S. Warren
Butterfly Conservation, Manor Yard, Wareham BH20 5QP, UK

T. G. Shreeve
School of Life Sciences, Oxford Brookes University, Headington, Oxford OX3 0BP, UK

(Thomas et al. 2011). Many contributions are species-specific in their approach e.g. Botham et al. (2011) whilst other papers examine how we can manipulate the agricultural environment to soften the impacts of intensive agriculture on a broad range of 'wider countryside' species (e.g. Haaland and Bersier 2011).

In 2002 when CBD participants committed themselves to the 2010 target, it was increasingly recognised that, whilst the optimal management of individual habitat patches will always underpin conservation action, long-term persistence of species in a given area was a landscape-scale issue. The spatial organisation of resources within a landscape is now recognised as a prime determinant of species' distributions and persistence (e.g. Williams 2011) and targeted action on the ground is demonstrating that such approaches can yield excellent results (e.g. Ellis et al. 2011). Whilst much landscape-scale conservation is aimed at repairing anthropogenic impacts that created extremely fragmented habitats, it is encouraging that naturally fragmented systems are being examined for cues to understand species responses to such systems (Swengel and Swengel 2011). Understanding species' responses to fragmented landscapes is exceptionally difficult, as experimentation at such a scale is often impossible, and even intensive agricultural landscapes rarely correspond to the binary patch/matrix simplicity of early modelling (Dennis et al. 2003; Dover and Settele 2009). In the absence of detailed data on species' behavioural responses in the 'matrix' between habitat patches and recognising that this is a complex issue to address, Shreeve and Dennis (2011) suggest that using agri-environment schemes to increase landscape heterogeneity and disturbance may be a pragmatic approach to improve dispersal and access to resources.

Agri-environment schemes have not always been well thought-out (Warren et al. 2005), and can have negative consequences (e.g. Konvička et al. 2008) although they are probably our best tool to work with in intensively managed landscapes (Settele et al. 2009). It is promising therefore that Brereton et al. (2011) reports clear evidence that targeted Higher Level agri-environment schemes are slowing, and in some cases reversing, the decline of threatened butterflies in the UK Schemes must be designed and implemented on the basis of science if they are to be effective in conserving biodiversity. In intensive agricultural areas the pressure is to maintain production at the expense of agri-environment schemes and in the EU they remain vulnerable to changes in the Common Agricultural Policy (CAP). In landscapes that are less intensively managed (extensive or cultural landscapes) there is a different problem: supporting existing agricultural systems that currently deliver some of the highest quality areas in terms of biodiversity (Dover et al. 2011a). This is not an issue of developing a new margin mixture or management prescription; in Europe it is about radically reforming the (CAP) and developing support measures that can underpin existing High Nature Value Farmland—it is a matter of socioeconomics and social justice (see Dover et al. 2011b).

Reforming the CAP is just one of the ten challenges that Warren and Bourn (2011) have identified as needing to be addressed if the new CBD 2020 targets and 2050 vision are to be realised. The challenges are undeniably tough; but the global community must rise to them, and we as conservationists need to fulfil the faith that is held in us and our ability to help effect change (see Attenborough 2011). We have much of the science to know what will be effective; a crucial issue now is how we apply that science to be effective against one of the greatest threats facing the planet: the rapid loss of biodiversity.

Acknowledgments The Symposium and its proceedings were part-funded by DG Environment. The opinions expressed are those of the authors and do not necessarily reflect those of the funders.

References

Anonymous (2001) Presidency conclusions. Götegorg European Council 15 and 16 June 2010 SN 200/1/01 REV 1. EU, Brussels

Anonymous (2010) Global biodiversity outlook 3—executive summary. Secretariat of the Convention on Biological Diversity, Montréal

Asher J, Fox R, Warren MS (2011) British butterfly distributions and the 2010 target. J Insect Conserv. doi: 10.1007/s10841-010-9346-7

Attenborough D (2011) Opening address by Sir David Attenborough, president of butterfly conservation, to the 6th international symposium reading, 25–28 March 2010. J Insect Conserv

Botham MS, Ash D, Aspey N, Bourn NAD, Bulman CR, Roy DB, Swain J, Zannese A, Pywell RF (2011) The effects of habitat fragmentation on niche requirements of the marsh fritillary, *Euphydryas aurinia*, (Rottemburg, 1775) on calcareous grasslands in southern UK. J Insect Conserv. doi: 10.1007/s10841-010-9344-9

Brereton TM, Roy DB, Middlebrook I, Botham M, Warren M (2011) The development of butterfly indicators in the United Kingdom and assessments on 2010. J Insect Conserv. doi: 10.1007/s10841-010-9333-z

Caritg R, Domènech M, Dantart J, Jubany J (2011) Andorran butterfly monitoring scheme. doi: 10.1007/s10841-010-9352-9

de Vries HH, Ens SH, de Graaf G, Teunissen L, van der Velde R, Vogelaar L, Winterink A, Visser ME (2011) Synchronisation of egg hatching of brown hairstreak (*Thecla betulae*) and budburst of blackthorn (*Prunus spinosa*) in a warmer future. J Insect Conserv. doi: 10.1007/s10841-010-9355-6

Dennis RLH, Shreeve TG, Van DycK H (2003) Towards a functional resource-based concept for a habitat: a butterfly biology viewpoint. Oikos 102:417–426

Djoghlaf A (2010) Statement by Mr Ahmed Djoghlaf, executive secretary of the convention on biological diversity on the occasion of the informal expert workshop on the updating of the strategic plan of the convention for the post-2010 period, London 18 January 2010. Secretariat of the Convention on Biological Diversity, UNEP, Montreal

Dover J, Settele J (2009) The influences of landscape structure on butterfly distribution and movement: a review. J Insect Conserv 13:3–27

Dover JW, Rescia A, Fungariño S, Fairburn J, Carey P, Lunt P, Arnot C, Dennis RLH, Dover CJ (2011a) Land-use, environment, and their impact on butterfly populations in a mountainous pastoral landscape: individual species distribution and abundance. J Insect Conserv. doi: 10.1007/s10841-010-9338-7

Dover JW, Spencer S, Collins S, Hadjigeorgiou I, Rescia A (2011b) Grassland butterflies and low intensity farming in Europe

Ellis S, Wainwright D, Berney F, Bulman C, Bourn N (2011) Landscape-scale conservation in practice: lessons from northern England, UK. J Insect Conserv. doi: 10.1007/s10841-010-9324-0

Fox R, Randle Z, Hill L, Anders S, Wiffen L, Parsons MS (2011) Moths count: recording moths for conservation in the UK. J Insect Conserv. doi: 10.1007/s10841-010-9309-z

Groenendijk D, Ellis WN (2011) The state of the Dutch larger moth fauna. J Insect Conserv. doi: 10.1007/s10841-010-9326-y

Haaland C, Bersier LF (2011) What can sown wildflower strips contribute to butterfly conservation?: an example from a Swiss lowland agricultural landscape. J Insect Conserv. doi: 10.1007/s10841-010-9353-8

Konvička M, Beneš J, Cizek O, Kopecek F, Konvicka O, Vitaz L (2008) How too much care kills species: grassland reserves, agri-environmental schemes and extinction of *Colias myrmidone* (Lepidoptera: Pieridae) from its former stronghold. J Insect Conserv 12:519–525

Lewis O, Senior MJM (2011) Assessing conservation status and trends for the world's butterflies: the sampled red list index approach. J Insect Conserv. doi: 10.1007/s10841-010-9329-8

Nakamura Y (2011) Conservation of butterflies in Japan: status, actions and strategy. J Insect Conserv. doi: 10.1007/s10841-010-9299-x

Porter K, Ellis S (2011) Securing viable metapopulations of the Marsh Fritillary butterfly, *Euphydryas aurinia*, (Lepidoptera: Nymphalidae) in Northern England. J Insect Conserv. doi: 10.1007/s10841-010-9328-9

Sáfián S, Csontos G, Winkler D (2011) Butterfly community recovery in degraded rainforest habitats in the upper Guinean forest zone (Kakum Forest, Ghana). J Insect Conserv. doi: 10.1007/s10841-010-9343-x

Settele J, Dover J, Dolek M, Konvička M (2009) Butterflies of European ecosystems: impact of land use and options for conservation management. In: Settele J, Shreeve T, Konvička M, Van Dyck H (eds) Ecology of butterflies in Europe. CUP, Cambridge, pp 353–370

Shreeve TG, Dennis RLH (2011) Landscape scale conservation: resources, behaviour, the matrix and opportunities. J Insect Conserv. doi: 10.1007/s10841-010-9338-7

Stefanescu C, Torres I, Jubany J, Páramo F (2011) Recent declines of butterflies in north-east Spain following habitat and climate change. J Insect Conserv. doi: 10.1007/s10841-010-9325-z

Swengel AB, Swengel SR (2011) High & dry or sunk & dunked: lessons for tallgrass prairies from quaking bogs. J Insect Conserv. doi: 10.1007/s10841-010-9335-x

Thomas JA, Simcox DJ, Hovestadt T (2011) Evidence based conservation of butterflies. J Insect Conserv. doi: 10.1007/s10841-010-9341-z

Warren MS, Bourn N (2011) Ten conservation challenges for 2010 and beyond to conserve Lepidoptera in Europe. J Insect Conserv. doi: 10.1007/s10841-010-9356-5

Warren M, Brereton T, Wigglesworth T (2005) Do agri-environment schemes help butterflies? Experience from the UK. In: Kűhn E, Feldmann R, Thomas JA, Settele J (eds) Studies on the ecology and conservation of butterflies in Europe, Vol 1: general concepts and case studies. Moscow, Pensoft, pp 121–123

Williams MR (2011) Habitat resources, remnant vegetation condition and area determine distribution patterns and abundance of butterflies and day-flying moths in a fragmented urban landscape, south-west Western Australia. J Insect Conserv. doi: 10.1007/s10841-010-9307-1

Wilson RJ, Maclean IMD (2011) Recent evidence for the climate threat to Lepidoptera and other insects. J Insect Conserv. doi: 10.1007/s10841-010-9342-y

J Insect Conserv (2011) 15:291–299
DOI 10.1007/s10841-010-9346-7

ORIGINAL PAPER

British butterfly distributions and the 2010 target

Jim Asher · Richard Fox · Martin S. Warren

Received: 7 April 2010 / Accepted: 6 September 2010 / Published online: 26 September 2010

Abstract Butterfly Conservation, in partnership with the Centre for Ecology and Hydrology and Dublin Naturalists' Field Club, has been coordinating the collation of detailed geographical records of butterfly sightings across Britain and Ireland, mainly made by volunteers, for a continuous period of 15 years since 1995. This has generated a dataset of over 7.5 million records, supplemented by collated historical records. Results at the end of the first five-year period (1995–9) and at the end of the second five-year period (2000–4) indicated significant range declines in many species over a 30-year period (since baseline distributions recorded in the 1970s); for other species, there has been a significant extension of range, principally towards more northerly latitudes. This paper shows the results of a preliminary analysis for the end of the third five-year period (2005–9), to assess progress towards the EU target of halting biodiversity loss by 2010. The findings indicate that populations are continuing to be lost at a similar rate to previous periods, contrary to the 2010 target. In contrast, major northward extensions of range are reported for some mobile generalist species. Conservation successes are also reported for a few species whose declines have been halted and in some cases, reversed. Changes in range and species diversity remain consistent with the primary hypotheses for the causes of change: decreases caused by destruction, deterioration and fragmentation of biotopes and range increases in response to climate warming.

Keywords Butterfly distributions · Biodiversity · Range changes · Climate change · 2010 Target

J. Asher (✉) · R. Fox · M. S. Warren
Butterfly Conservation, Manor Yard, East Lulworth, Dorset BH20 5QP, UK
e-mail: jim.asher@btinternet.com

Introduction

Since 1995, Butterfly Conservation, in partnership with the Centre for Ecology and Hydrology and the Dublin Naturalists' Field Club, has been collating butterfly records, collected across Britain and Ireland (the UK, Republic of Ireland, Isle of Man and Channel Islands). The vast majority of records have been submitted by volunteers. In addition, the project, called Butterflies for the New Millennium (BNM), has been a vehicle for the collation of historical records of butterflies which date back to 1690.

The BNM project now has over 7.5 million butterfly records, most of which arise from opportunistic rather than systematic or randomised recording. Although early records were very varied in their completeness, most records collected since 1995 have been gathered to minimum standards for spatial reference (mainly at a resolution of at least 1 km^2), date of sighting and indication of numbers of each species seen. This massive database has proved to be a valuable resource for research and for conservation priority-setting. Regular feedback to recorders has stimulated continued recording, whilst detailed records, after full validation, are posted on the National Biodiversity Network Gateway (www.nbn-gateway.org.uk).

The BNM data were used to provide maps in Asher et al. (2001) in which species data from 1995 to 1999 were compared to those in the first comprehensive atlas of butterflies in Britain and Ireland (Heath et al. 1984) covering records from 1970 to 1982. These showed significant changes in the range and distribution of many species over a 25-year period (Warren et al. 2001; Thomas et al. 2004). A further publication, Fox et al. (2006), based on a further 5 years of intensive recording (2000–4), which also made use of in the results of the UK Butterfly Monitoring Scheme (www.ukbms.org.uk), showed that substantial changes had continued, underlining the need for and the benefit of continued recording. 76% of species (excluding migrants that do not maintain permanent populations in Britain) had decreased in occurrence once variation in sampling effort had been accounted for (see Fox et al. 2006). Based on analysis of the BNM data, Warren et al. (2001) highlighted the opposing forces of habitat loss and climate change on species distributions, and Thomas et al. (2004) showed that declines in butterfly species distributions were progressing more rapidly than those for birds and for plants.

Previously published comparisons looked at medium-term changes between recent records and those from the 1970s (Asher et al. 2001; Fox et al. 2006). In making this comparison, a major challenge arose in quantifying the level of distribution change to account for the very different sampling intensity and resolution of the recent (Fox et al. 2006) and historic (Heath et al. 1984) data sets. Here we present, for an initial selection of species, an assessment of change over a much shorter (10-year) timescale, comparing periods with very similar sampling intensity.

Questions particularly relevant to this new analysis are:

1. Have butterfly distributions changed significantly within the past decade?
2. How does this impact on butterfly species diversity?
3. Have we achieved the EU target of halting biodiversity loss in UK by 2010?

The major data resource we now have provides an opportunity to compare two 5-year periods 10 years apart: 1995–9 and 2005–9, where the two datasets are much more closely comparable in their extent and nature than was previously possible, to address these urgent questions. This paper presents the first analyses of butterfly records collated from 2005 to 2009 with those collated from 1995 to 1999 in this context.

Methods

Data collection and collation

Butterfly records were collected in the field, mainly by volunteers, using forms or other recording instructions consistent with the standards set by the scheme (Asher et al. 2001). Records were submitted to a network of local/regional recording co-ordinators, many of whom are also volunteers. Records were checked and validated at local/regional level and completed datasets submitted typically once per year for collation at Britain and Ireland scale. A further stage of checking and validation at national scale ensures a high level of accuracy in the database.

A cycle of feedback at a local level to recorders provided a mechanism to improve recording in under-visited areas with the aim of achieving complete geographical coverage at 10 km × 10 km grid square resolution at a national scale; in many local areas, recorders were encouraged to achieve coverage at a finer resolution, typically over a 2 km × 2 km grid. A further cycle of feedback at a national level has been provided through project newsletters and an annual national butterfly recorders' meeting.

Analysis

We assessed proportional change in species' national distributions for individual species over the 10-year period (i.e. between 1995–9 and 2005–9).

Assessing range expansions and contractions

Analyses from the two time periods, 1995–9 and 2005–9, were restricted to those 10 km squares in Britain that have been visited in both periods, and included calculation of a distribution range (species occupancy), in terms of the number of 10 km squares visited in both periods in which each species had been recorded. The change in range for each species was then calculated as the percentage change in species occupancy between the two time periods.

Assessing range shifts within Britain

For four of the most rapidly expanding widespread species, the latitude shift of the northern margin in Britain has been measured by taking the mean OS grid north value of the ten most northerly occupied 10 km squares for each five-year time period (following Hill et al. 2002; Hickling et al. 2006; Pöyry et al. 2009; see also Franco et al. 2006). We have also taken the mean grid north value of all 10 km squares occupied by each species across the whole set of data for each time period in Britain, to investigate species showing no clear change in the northern range margin, but a strong indication of a change of occupancy in northern latitudes. We note that none of these species has a southern range margin in Britain.

Assessing changes in species diversity

In order to provide an answer to the question of whether we have achieved the EU target of halting biodiversity loss by 2010, we have analysed the changes in the number of species recorded in grid squares across Britain during each of the five-year periods 1995–9 and 2005–9. This follows a broadly similar approach to analysis of records from earlier periods by Menéndez et al. (2006). In this case, we have averaged the change in the number of species recorded in each 10 km square across 100 km squares to provide a simple picture of change. The analysis includes all butterfly species with the exception of three long-range migrants, *Colias croceus, Vanessa atalanta and V. cardui*; it has also been restricted to include only those 10 km squares visited in both periods and visited at least 10 times in each period, so as to reduce bias that may arise due to insufficient levels of recording in individual 10 km squares.

This paper aims to assess changes in overall butterfly species diversity in recent years, at an early stage in the data analysis. We have therefore limited detailed species-level results and interpretation to: the five most severely declining species; two localised species that appear to be more stable or increasing in range; and four of the most rapidly expanding widespread species. These provide examples of the types of change observed.

Results

The map in Fig. 1 indicates the intensity of recording effort in terms of total number of visits per 10 km square over the 15-year period from 1995 to 2009. There is a clear bias towards the more densely populated parts of England and central Scotland; recording intensity is much lower in remote and highland areas and across much of Ireland. Nevertheless, recording coverage reaches almost 100% at the 10 km grid square resolution in these areas.

The recording levels achieved in the BNM project are illustrated by Fig. 2, which shows the number of 'visits' (unique combinations of recorder, date and grid reference) and of species records collected between 1995 and 2009. The number of records collected each year does vary (for example a low point in 2001 occurred because access to the countryside was restricted due to widespread outbreaks of foot-and-mouth disease). The average level of recording has increased: ca. 39,000 records/year, ca. 39,000 records/year and ca. 47,000 records/year over 1995–9, 2000–4 and 2005–9, respectively.

The map in Fig. 3 illustrates the number of butterfly species recorded in each 10 km square across Britain and Ireland during the 15-year period 1995–2009. Two key aspects are evident from this map: (1) the gradient of decreasing species richness from the south and east towards the north and west; and (2) the localised hotspots of higher species richness in a number of specific geographical areas, most notably along the areas of calcareous surface geology in England.

A preliminary analysis of the percentage change in the number of occupied 10 km squares for each species (limited to 10 km squares in Britain visited in both periods),

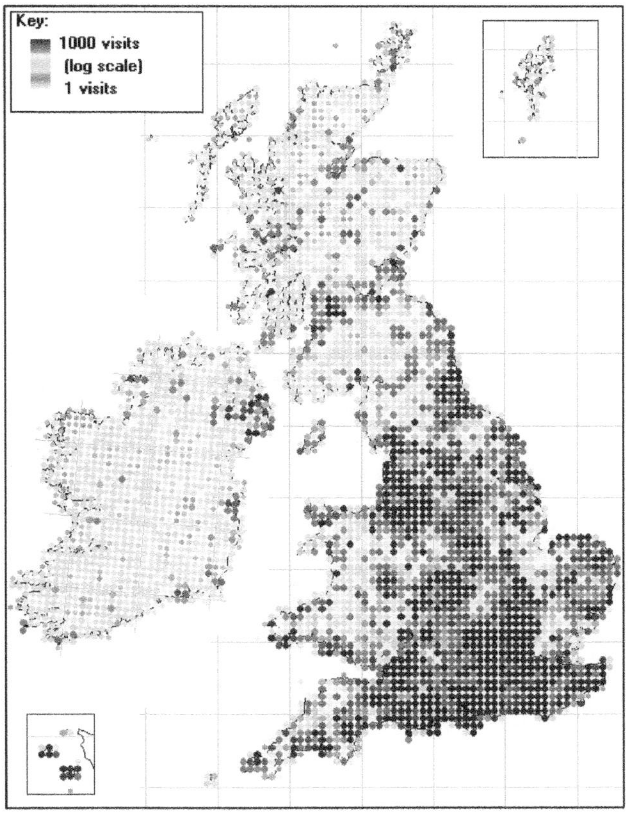

Fig. 1 Distribution of recording effort over the 15-year period 1995–2009

shows five species with losses of 20% or more between 1995–9 and 2005–9 (see Table 1). Figure 4 shows distribution maps for four of these five species; three of these are among our most threatened and localised butterfly species, High Brown Fritillary *Argynnis adippe,* Pearl-bordered Fritillary *Boloria euphrosyne* and Duke of Burgundy *Hamearis lucina;* one of these species has been regarded as a widespread generalist, Wall *Lasiommata megera.* These species show declines in the number of occupied squares of 49%, 43%, 28% and 20%, respectively between 1995–9 and 2005–9 (Table 1).

In contrast, for some of our threatened species, their range as measured by 10 km square occupancy is relatively stable, for example, the recorded change for the Heath Fritillary *Melitaea athalia* over the 10-year period is a decline of 2%, whilst for Silver-spotted Skipper *Hesperia comma,* there is an increase of 24%. In the case of *M. athalia,* this is attributed to successes of intensive conservation management on its few remaining sites (see also Brook et al. 2007 and Hodgson et al. 2009); while the increased range of *H. comma* is attributed to its ability to use a wider range of microhabitat niches in grazed calcareous grassland under the beneficial effects of climate warming (Davies et al. 2005 and 2006).

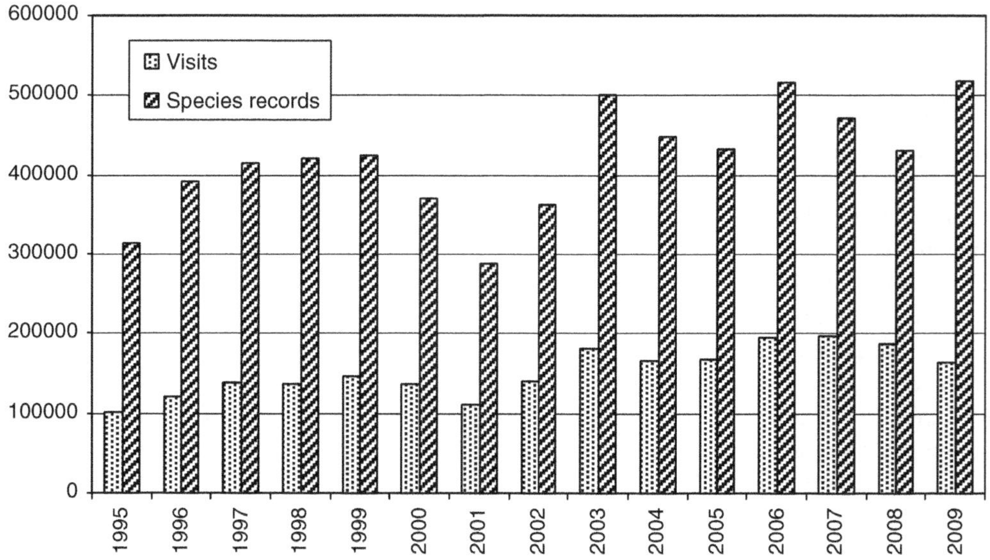

Fig. 2 Number of 'visits' and species records collected over the period 1995–2009

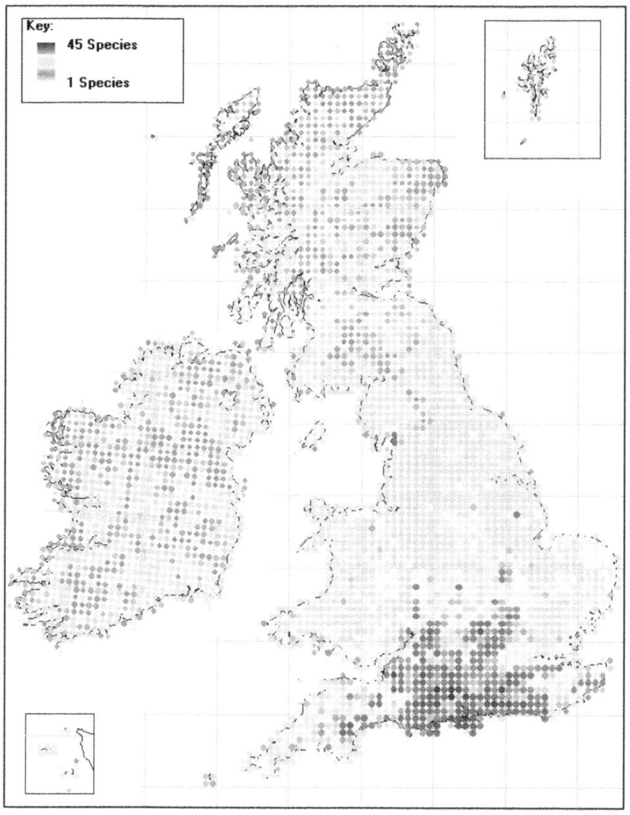

Fig. 3 Species density recorded from records collected over the period 1995–2009

Note that the recorded decline of *L. megera* has occurred across the very well-recorded areas apparent from Fig. 1, so absence of records from these areas is unlikely to be attributable to a lack of recording.

In contrast, Fig. 5 shows distribution maps for four generalist species that showed range expansion over the earlier 30-year comparison, and demonstrate further range expansion over the past 10 years: Comma *Polygonia c-album*, Ringlet *Aphantopus hyperantus*, Peacock *Inachis io*, and Speckled Wood *Pararge aegeria*.

Table 1 Change in 10 km square occupancy recorded over the 30-year period between 1970–82 and 1995–2004 (Fox et al. 2006, using sub-sampling) and the declines seen over the 10-year period between 1995–9 and 2005–9

Species	Decline in 10 km occupancy (%)	
	30 year period (%)	10 year period (%)
A. adippe	−79	−49
B. euphrosyne	−61	−43
H. lucina	−52	−28
L. sinapis	−65	−23
L. megera	−38	−20

These five species display the most serious declines over the 10-year period

The results of the analysis of range expansion for these species are shown in Table 2.

We note also that the northern range margin of Wall *L. megera* in Britain, measured in the same way, has also increased by 31 ± 8 km in the same 10-year period. This expansion to the north contrasts with its massive loss of range within the central part of its distribution in Britain.

The results of the analysis of changes in species diversity are shown in Fig. 6. In summary, the map indicates that declines in species diversity are occurring across most of the southern half of Britain, whilst in the north, species diversity appears to be increasing. This appears to differ from the results reported by Menéndez et al. (2006) which

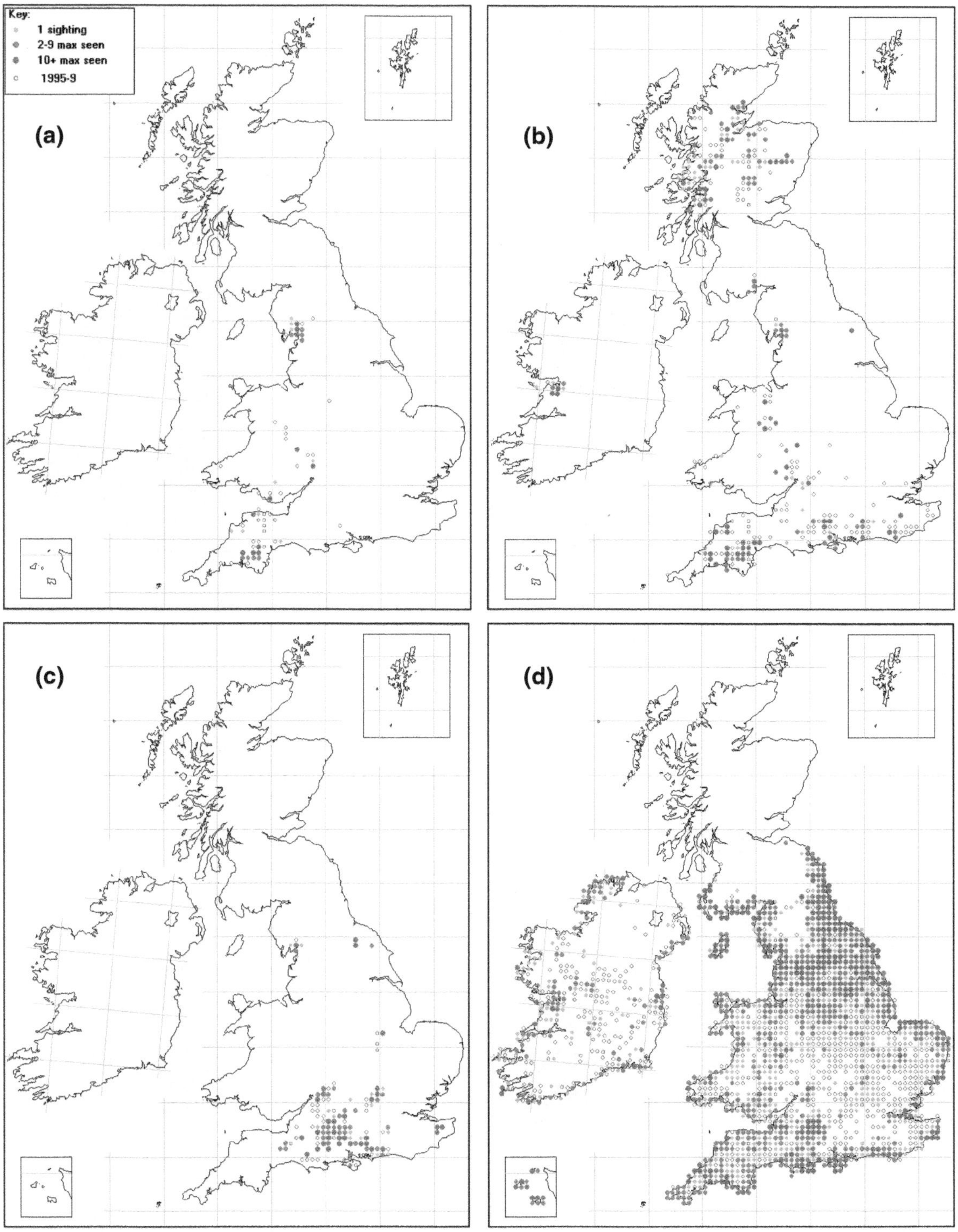

Fig. 4 Maps showing losses of populations occurring between 1995–9 and 2005–9 for (**a**) High Brown Fritillary *A. adippe,* (**b**) Pearl-bordered Fritillary *B. euphrosyne,* (**c**) Duke of Burgundy *H. lucina* and (**d**) Wall *L. megera*. *Solid dots* indicate populations extant in 2005–9; *open circles* indicate population extinctions since 1995–9

found broadly that a general increase in species richness was observed across Britain between 1970–82 and 1995–9. The observed reduction reported in this paper in central southern England is at least partly attributable to the extensive decline of the formerly widespread *L. megera* which appears to have accelerated over the last decade. The high levels of change in west Wales are partly attributed to under-recording in that area in 2005–9.

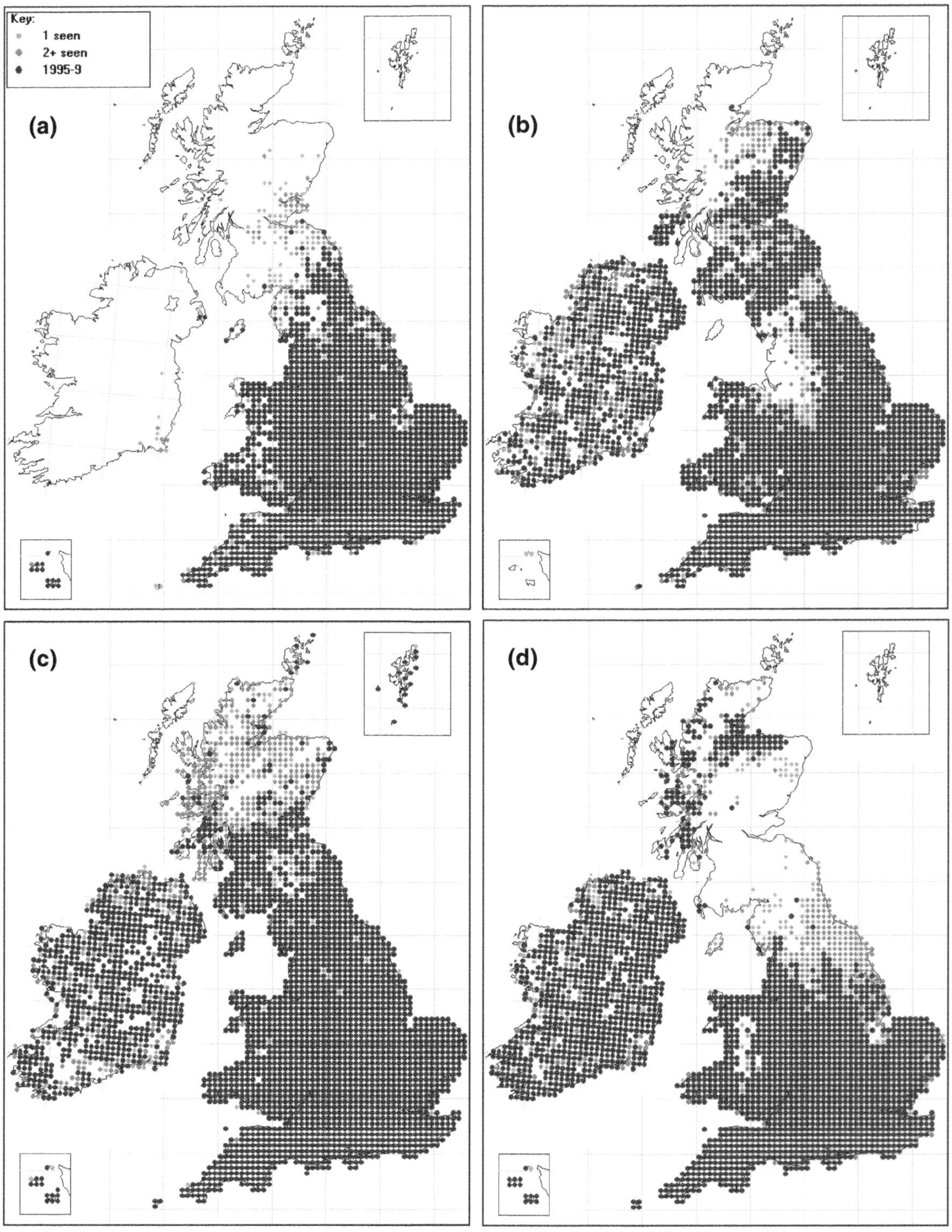

Fig. 5 Distribution maps of four butterfly species that have expanded their range between 1995–9 and 2005–9: Comma *P. c-album*, Ringlet *A. hyperantus*, Peacock *I. io*, and Speckled Wood *P. aegeria*. The *darkest dots* indicate populations recorded in 1995–9; *lighter dots* indicate populations newly recorded in 2005–9

Discussion

The decline of our most threatened species continues the trend noted in previous publications (Asher et al. 2001; Warren et al. 2001; Fox et al. 2006). For each of the five species shown in Table 1, the declines in 10 km square occupancy recorded over the 30-year period between 1970–82 and 1995–2004 (Fox et al. 2006) have continued

Table 2 Change in 10 km square occupancy (%), northerly range margin shift NLS (km) and shift in the mean latitude (MLS) of the whole record set in Britain (MLS) between 1995–9 and 2005–9

Species	10 km occupancy (%)	NLS (km)	MLS (km)
I. io	18 ± 3	8 ± 9	86 ± 7
P. c-album	7 ± 4	141 ± 12	86 ± 5
P. aegeria	22 ± 4	106 ± 12	77 ± 8
A. hyperantus	8 ± 3	15 ± 6	69 ± 7

In this analysis, grid north is used as a proxy for latitude and the quoted uncertainties are Standard Deviations

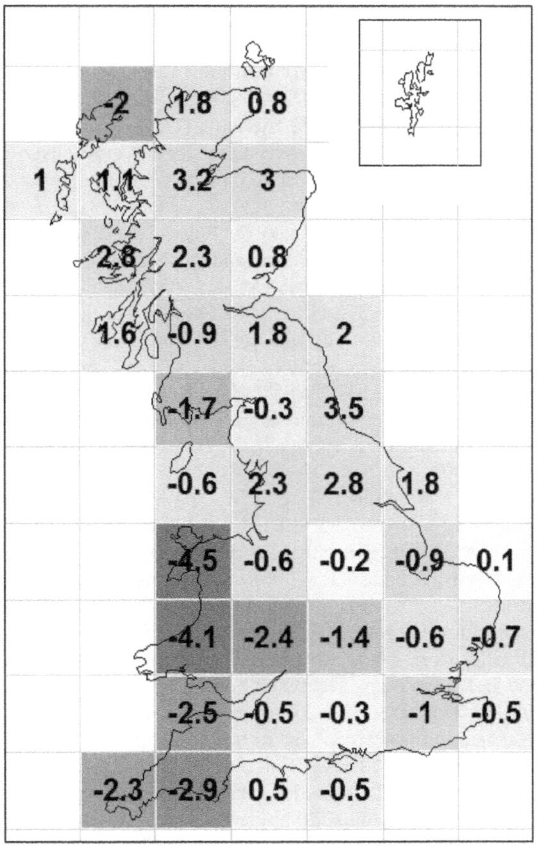

Fig. 6 Change in the number of species recorded in Britain per 10 km square, between 1995–9 and 2005–9, averaged across 100 km squares

over the last 10-year period (1995–9 and 2005–9). In each case, the indication is that the rate of decline may even be increasing. Note that the 30-year comparison made use of sub-sampling methodology (Fox et al. 2006) to compensate for very different levels of recording; the 10-year comparison reported here is a direct comparison of similar datasets, with no sub-sampling; in both comparisons, gains and losses were calculated only within 10 km squares recorded in both periods.

In the first three cases (*A. adippe, B. euphrosyne and H. lucina*), the declines are broadly consistent with the impacts of loss of specialist habitats in the modern landscape (Warren et al. 2001). Restoring populations of these species is a massive conservation challenge. In the case of *L. megera*, the cause of decline of this butterfly, once common across central southern England, remains unexplained. The population decline of the species on monitored sites is also very high (80% since 1976, Botham et al. 2008), mirroring declines across north-west Europe (Van Dyck et al. 2009). In contrast to these ongoing declines, there have been some conservation successes, notably the halting of the decline of *M. athalia*, one of the most threatened species in the 1995–9 period, as well as expansions of *H. comma*.

For the four examples of widespread generalist species that have shown greater range expansions, these are mainly accompanied by a shift in their northerly range margins. This is consistent with previous publications on British species (Hill et al. 1999; Warren et al. 2001; Hill et al. 2002; Hickling et al. 2006), and studies on butterflies in other countries (Parmesan et al. 1999; Pöyry et al. 2009). However, the interpretation of the data is not entirely straightforward. For *P. c-album* and *P. aegeria*, there is clear evidence for a northward shift in both the northern range margin and the mean of the full range, consistent with a simple latitudinal extension of the range (range centroid shifts have been used to assess changes in bird distributions by Huntley et al. 2008). In contrast, the results for Ringlet *A. hyperantus* and Peacock *I. io* show little change in the northern range margin, but a significant change in the mean of the full range. The expansion maps in Fig. 5 show that these two species appear to be filling in range in more northerly areas. *I. io* has been recorded for some time in Shetland, for example, but the number and density of records in northern mainland Scotland have apparently increased by a dramatic extent.

Ecological traits have been shown to correlate to distribution change and extinction risk (Cowley et al. 1999; Dennis et al. 2000, 2004, 2005; Mattila et al. 2008). Generalist butterflies tended to respond to climatic factors rather than ecological factors, such as trophic relationships and habitat heterogeneity (Menéndez et al. 2007). The results tend to show that species that have become highly specialised in our contemporary landscapes remain under threat and may respond negatively to further climate change; a number of generalist species are showing the ability to extend their range northward to a greater or lesser degree.

In general, the results here are consistent with these factors—four of the five most rapidly declining species are recognised specialists, whose isolated and fragmented habitat niches are most difficult to maintain in modern landscapes. The expansion of a number of widespread species towards unoccupied areas (generally northwards) is consistent with an ability to respond positively to climate change.

Conclusions

The European Union set a target of halting biodiversity loss by 2010, to focus attention on the need to take action at a time when evidence indicated an accelerating rate of decline. The prime purpose of the results presented here is to test whether this target has been achieved for butterfly species in UK.

Our overall conclusions from this preliminary analysis of recent data on butterfly distributions in Britain from 2005 to 9 with those a decade earlier (1995–9) are that:

(a) for five of the most threatened species, previously recorded declines have continued and may even have increased in the last decade. This is principally attributed to the continuing loss and deterioration of their habitats, combined with the effects of fragmentation and isolation. For these important indicator species, the data show that the UK will not achieve the 2010 target of halting biodiversity loss.
(b) for two threatened species, ranges may have stabilised. This is attributed in part to pro-active conservation projects focussed on their conservation, and partly to the beneficial effects of climate warming;
(c) for the four widespread species analysed here, ranges are increasing, attributed to climate warming and the ability of generalist species to move rapidly through the countryside. These factors appear to be contributing to a shift in their northerly range margin and/or greater occupancy within range towards the north of their previous range;
(d) recorded species diversity for all butterfly species (excepting three long-range migrant species) appears to have declined in the southern half of Britain and to have increased in the northern half of Britain. This is consistent with the opposing forces of habitat degradation and the beneficial impacts of climate warming.

Further analysis will be carried out to provide a more detailed picture across the whole of the British butterfly fauna, but this is unlikely to alter the main conclusions reached here.

Acknowledgments The authors are grateful to their project partners, the Centre for Hydrology and Ecology and the Dublin Naturalists Field Club; and to The John Ellerman Foundation, the Countryside Council for Wales, Natural England, Northern Ireland Environment Agency and Scottish Natural Heritage for grant contributions to the project. Most vitally of all, the authors are grateful to the many thousands of volunteers who have contributed records to the project over the years, and to the local and regional coordinators for collating and validating the records.

References

Asher J, Warren MS, Fox R, Harding PT, Jeffcoate G, Jeffcoate S (2001) The millennium atlas of butterflies of Britain and Ireland. Oxford University Press, Oxford

Botham MS, Brereton TM, Middlebrook I, Cruickshanks KL, Harrower C, Beckmann B, Roy DB (2008) United Kingdom butterfly monitoring scheme report for 2008. CEH Wallingford

Brook S, McCracken M, Bulman CR, Camp P, Bourn NAD (2007) Post-burn bracken *Pteridium aquilinum* control to manage habitat for the heath fritillary butterfly *Mellicta athalia* on Exmoor, Somerset, England. Conserv Evid 4:81–87

Cowley MJR, Thomas CD, Thomas JA, Warren MS (1999) Flight areas of British butterflies: assessing species status and decline. Proc R Soc 266:1587–1592

Davies ZG, Wilson RJ, Brereton TM, Thomas CD (2005) The re-expansion and improving status of the silver-spotted skipper butterfly (*Hesperia comma*) in Britain: a metapopulation success story. Biol Conserv 124:189–198

Davies ZG, Wilson RJ, Coles S, Thomas CD (2006) Changing habitat associations of a thermally constrained species, the silver-spotted skipper butterfly, in response to climate warming. J Animal Ecol 75:247–256

Dennis RLH, Sparks TH, Donato B, Pollard E (2000) Ecological correlates of island incidence and geographical range among British butterflies. Biodivers Conserv 9:343–359

Dennis RLH, Hodgson JG, Grenyer R, Shreeve TG, Roy DB (2004) Host plants and butterfly biology. Do host-plant strategies drive butterfly status? Ecol Ent 29:12–26

Dennis RLH, Shreeve TG, Arnold HR, Roy DB (2005) Does diet breadth control herbivorous insect distribution size? Life history and resource outlets for specialist butterflies. J Insect Conserv 9:187–200

Fox R, Asher J, Brereton T, Roy D, Warren MS (2006) The state of butterflies in Britain and Ireland. Pisces, Newbury

Franco AMA, Hill JK, Kitschke C, Collingham YC, Roy DB, Fox R, Huntley B, Thomas CD (2006) Impacts of climate warming and habitat loss on extinctions at species' low-altitude range boundaries. Glob Change Biol 12:1545–1553

Heath J, Pollard E, Thomas JA (1984) Atlas of butterflies in Britain and Ireland. Viking, Harmondsworth

Hickling R, Roy DB, Hill JK, Fox R, Thomas CD (2006) The distributions of a wide range of taxonomic groups are expanding polewards. Glob Change Biol 12:450–455

Hill JK, Thomas CD, Huntley B (1999) Climate and habitat availability determine 20th century changes in a butterfly's range margins. Proc R Soc B 266:1197–1206

Hill JK, Thomas CD, Fox R, Telfer MG, Willis SG, Asher J, Huntley B (2002) Responses of butterflies to 20th century climate warming: implications for future ranges. Proc R Soc B 269:2163–2171

Hodgson JA, Moilanen A, Bourn NAD, Bulman CR, Thomas CD (2009) Managing successional species: modelling the dependence of heath fritillary populations on the spatial distribution of woodland management. Biol Conserv 142:2743–2751

Huntley B, Collingham YC, Willis SG, Green RE (2008) Potential impacts of climatic change on European breeding birds. PLoS One. doi:10.1371/journal.pone.0001439

Mattila N, Kotiaho JS, Kaitala V, Komonen A (2008) The use of ecological traits in extinction risk assessments: a case study on geometrid moths. Biol Conserv 141:2322–2328

Menéndez R, González-Megías A, Hill JK, Braschler B, Willis SG, Collingham Y, Fox R, Roy DB, Thomas CD (2006) Species richness changes lag behind climate change. Proc R Soc B 273:1465–1470

Menéndez R, González-Megías A, Collingham Y, Fox R, Roy DB, Ohlemüller R, Thomas CD (2007) Direct and indirect effects of climate and habitat factors on specialist and generalist butterfly diversity. Ecology 88:605–611

Parmesan C, Ryrholm N, Stefanescu C, Hill JK, Thomas CD, Descimon H, Huntley B, Kaila L, Kullberg J, Tammaru T, Tennant J, Thomas JA, Warren M (1999) Polewards shifts in geographical ranges of butterfly species associated with regional warming. Nature 399:579–583

Pöyry J, Luoto M, Heikkinen RK, Kuussaari M, Saarinen K (2009) Species traits explain recent range shifts of Finnish butterflies. Glob Change Biol 15:732–743

Thomas JA, Telfer MG, Roy DB, Preston CD, Greenwood JJD, Asher J, Fox R, Clarke RT, Lawton JH (2004) Comparative losses of British butterflies, birds, and plants and the global extinction crisis. Science 303:1879–1881

Van Dyck H, van Strien AJ, Maes D, van Swaay CAM (2009) Declines in common, widespread butterflies in a landscape under intense human use. Conserv Biol 23:957–965

Warren MS, Hill JK, Thomas JA, Asher J, Fox R, Huntley B, Roy DB, Telfer MG, Jeffcoate S, Harding P, Jeffcoate G, Willis SG, Greatorex-Davies JN, Moss D, Thomas CD (2001) Rapid responses of British butterflies to opposing forces of climate and habitat change. Nature 414:65–69

J Insect Conserv (2011) 15:139–151
DOI 10.1007/s10841-010-9333-z

ORIGINAL PAPER

The development of butterfly indicators in the United Kingdom and assessments in 2010

T. Brereton · D. B. Roy · I. Middlebrook · M. Botham · M. Warren

Received: 26 April 2010 / Accepted: 21 August 2010 / Published online: 11 September 2010

Abstract The United Kingdom (UK) Government has national and international commitments to tackle the rate of biodiversity loss by 2010. Biodiversity indicators are used to measure and communicate progress in meeting these commitments. From 2005 onwards, butterflies have been adopted as Governmental biodiversity indicators in England, Scotland and for the UK as a whole. The indicators are compiled using butterfly abundance data collected through the UK Butterfly Monitoring Scheme, at a network of site established from 1976 onwards. The indicators show that butterfly numbers have fluctuated considerably from year-to-year, though analysis of the underlying smoothed multi-species trends for (habitat) 'specialist' species show significant long-term declines in each country since the 1970s. Trends in wider countryside 'generalist' species vary at the country-level from little or no overall change in Scotland and across the UK, to declines over selected years in England. Comparisons of changes in butterfly abundance before and after the 2010 target was set in 2002 suggest that the rate of decline at the UK-level is increasing for specialist species. In spite of large amounts of investment since 2000 to improve the habitat condition of protected areas, the trend for butterfly populations is no different in protected areas compared to elsewhere. Analysis by policy sector in England, shows that butterflies are declining rapidly in both forestry land and farmland, although in the latter habitat type, improvements are being seen on land entered into agri-environment schemes. We conclude by assessing the extent to which butterflies may represent broader biodiversity and help inform and evaluate conservation policy.

Keywords Butterfly monitoring · Biodiversity indicators · Abundance trends · 2010 target · Protected areas · Agri-environment schemes · UKBMS

T. Brereton (✉) · I. Middlebrook · M. Warren
Butterfly Conservation, Manor Yard, East Lulworth, Wareham, Dorset BH20 5QP, UK
e-mail: tbrereton@butterfly-conservation.org

D. B. Roy · M. Botham
Centre for Ecology & Hydrology, Maclean Building, Benson Lane, Crowmarsh Gifford, Wallingford, Oxfordshire OX10 8BB, UK

Introduction

Recent decades have seen commitments for action to address biodiversity loss agreed by nearly 200 countries through the 1993 Convention on Biological Diversity (CBD). The CBD is a legally binding global treaty that puts an obligation on individual governments to develop National Biodiversity Strategies and Action Plans for the conservation and sustainable use of wild animals and plants. The United Kingdom ratified the Convention in 1994, and in the same year published a 20-year UK Biodiversity Action Plan (UKBAP) (Anon 1994). Government responsibility for delivering biodiversity conservation within the UK has been devolved to constituent countries since 1998, with separate strategies produced for England (Defra 2002), Northern Ireland (DoE(NI) 2002), Scotland (Scottish Executive 2004) and Wales (WAG 2006a, b). For the UK as whole and at the country level, biodiversity conservation is tackled in a number of ways including through a network of protected areas, integration into relevant sectors of policy and decision-making, targeted action for declining species and habitats, and compliance with international laws and agreements (UK Biodiversity Partnership 2007). Agri-environment schemes, woodland grant schemes and achieving favourable condition of protected areas are the main land management mechanisms for conserving biodiversity in the UK.

In 2002, Parties to the CBD committed to achieve a significant reduction in the current rate of biodiversity loss at global, regional and national scales by 2010 (CBD 2006). The World Summit on Sustainable Development in 2002 endorsed the 2010 Biodiversity Target. Furthermore, the 2010 Target has been adopted by the European Union, with a more ambitious objective to halt biodiversity loss by 2010.

To measure and communicate progress in national biodiversity action plans, strategies and delivery of the 2010 target, biodiversity indicators are recommended (CBD 2004). For this purpose, the CBD adopted a framework in 2004 for the use of a range of biodiversity indicators at both national and International levels. Headline biodiversity indicators have been developed in a number of focal areas, related to environmental state, environmental pressures and policy responses (OECD 1994; UNEP 2003), whilst ecosystem services is an emerging category of indicators (Mace and Baillie 2007). The CBD and several authors have provided guidelines on the properties that a set of biodiversity indicators should have (e.g. Rees et al. 2008; Failing and Gregory 2003; Balmford et al. 2005; Buckland et al. 2005; Scholes and Biggs 2005). Criteria vary at the national level, but tend to focus on three key areas–(1) scientific credibility, (2) policy/social relevance and (3) practical monitoring and data requirements.

Within indicator frameworks, a key state indicator adopted at global, European, UK and country level scales is *status and trends in the components of biodiversity – the abundance and distribution of selected species*. Species trend indicators are considered a sensitive measure of biodiversity change (Balmford et al. 2003; Ten Brink et al. 1991; Ten Brink 2000, de Heer et al. 2005), although other measures that reflect the wider meaning of biodiversity are required (Feest et al. 2010). Butterflies have been proposed as species state indicators on scientific criteria due to their rapid and sensitive responses to subtle habitat and climatic changes and as representatives for the diversity and responses of other wildlife, especially insects (Rosenberg et al. 1986, Erhardt and Thomas 1991; Kremen 1992; New 1997; Beccaloni and Gaston 1995; Oostermeijer and van Swaay 1998; Fleishman et al. 2000; Maes and Van Dyck 2001; Ehrlich 2001, 2003; Parmesan 2003; Thomas 2005). Correlations between changes in ecosystem condition have already been demonstrated in a range of northern temperate habitats (Thomas and Harrison 1992; Pullin 1996; Sparks et al. 1996; Pollard et al. 1998; Schultz 1998; Swengel 1998; Pywell et al. 2004; Parmesan and Yohe 2003; Davies et al. 2005; Brereton et al. 2007). The loss, degradation and fragmentation of semi-natural habitats and the consequent effects on butterflies in the UK have been widely documented (Warren 1992; Asher et al. 2001; Fox et al. 2006). Studies have also identified patterns of change in butterflies and other taxa that are consistent with a climate explanation at global (Root et al. 2003; Parmesan et al. 1999), European (Thomas et al. 1999) and British (Thomas et al. 1999; Hill et al. 2002; Berry et al. 2002; Heath et al. 1984, Thackeray et al. 2010) scales.

Butterflies are typical insects with high reproductive rates, short life cycles and low trophic levels which allow them to respond rapidly to environmental change (Thomas et al. 2004). Representation for insects is extremely important as they comprise 56% of known species (Groombridge 1992) and an estimated 80% of global species (Stork 1997). In the UK, there are 28,500 insect species compared to 1,400 species of flowering plant, 210 species of breeding bird and 48 species of mammal (Anon 1994). Although there are only 58 regularly occurring butterfly species in the UK (Fox et al. 2006), their immature stages occur in all main terrestrial habitat types (except for dead wood) (Thomas 2005). As with many other insects, a high proportion of butterfly species are restricted to specific micro-habitats in relatively small areas of semi-natural habitat (Thomas 1991, 1993). Subtle changes in these habitats may substantially diminish insect diversity, but may not significantly impact upon higher trophic levels (e.g. bird populations) used as Biodiversity Indicators (Thomas 1995). Recent analyses of distribution data have also shown that butterflies are declining more rapidly than either birds or plants in Britain, emphasising their potential role as indicators (Thomas et al. 2004).

In practical terms, butterflies are relatively easy to study; their taxonomy is understood (Blair 1999; Caro and O'Doherty 1999; Ricketts et al. 2002) and a large body of high quality monitoring data is available over a relatively long time-series (since 1976) (Brereton et al. 2006; van Swaay and Van Strien 2008). Butterflies are the only insect group for which it is currently possible to estimate rates of abundance decline in most countries (Thomas and Clarke 2004; de Heer et al. 2005; Thomas 2005). Methods to monitor butterfly abundance are well described, extensively tested and scientifically sound (Pollard 1977; Pollard and Yates 1993). Furthermore, the high popular appeal of butterflies makes trends in their status a potentially powerful communication tool for biodiversity conservation (Fox et al. 2006; Kuhn et al. 2008). In UK policy terms, butterflies are important, because 40% of species are identified as Priorities for action in the UKBAP with specific targets set within the revised (UK BAP 2007) for seven of the species (see www.ukbap.org.uk).

The first global assessment of biodiversity change and progress was made in 2006, with UK butterflies featuring in the assessment (Secretariat of the Convention on

Biological Diversity 2006), highlighting the potential interest in using butterflies as indicators. Biodiversity indicators in the UK have been developed since 2002. Within the current suite, butterfly indicators have been adopted at the UK level and at the country level for England and Scotland, whilst candidate indicators have been compiled for Wales and at the regional level for the south west of England. In this paper we provide a first review of the development of butterfly biodiversity indicators in the UK. We describe policy function, data sources, how the indicators have been compiled and document peer-reviewed assessments of their suitability. Furthermore we assess the strengths and weakness of butterfly population data as a biodiversity indicator, discussing briefly how well trends in butterflies may represent trends in other insects groups.

Finally, we will provide a timely assessment of the latest trends in the indicators and also assess the extent to which two of the most important land management mechanisms for conservation (1) protected areas (Sites of Special Scientific Interest – SSSIs) and (2) agri-environment schemes may have influenced butterfly trends, using England as a case study.

Methods

Data sources

Butterfly indicators were compiled using annual butterfly abundance data for species collected at 1726 sites within the UK Butterfly Monitoring Scheme (UKBMS) (Botham et al. 2008). More than 1700 sites have contributed data to the UKBMS (Fig. 1), with more than 1000 sites monitored in 2009. The majority of sites (~90%) in the UKBMS are monitored by butterfly transects (Pollard and Yates 1993). The transect method was devised for the Butterfly Monitoring Scheme (BMS) co-ordinated by CEH (formerly the Institute of Terrestrial Ecology) and ran from 1976 until 2005, when it was merged with 'independent' transect data collated by Butterfly Conservation to form the UKBMS. The transect method involves weekly butterfly counts for up to 26 weeks of the year along fixed routes through the season made under defined weather, recording area and time of day criteria when butterflies are active. At the remainder of sites (~ 10%), two lower effort methods are employed as an alternative to transect monitoring, chiefly for selected specialist species; adult timed counts are used to sample fritillary and other colonial habitat specialist species occupying relatively inaccessible habitats (Warren et al. 1984) and larval web counts (Lewis and Hurford 1997) are used to monitor several species including Marsh Fritillary *Euphydryas aurinia*, where the immature stages

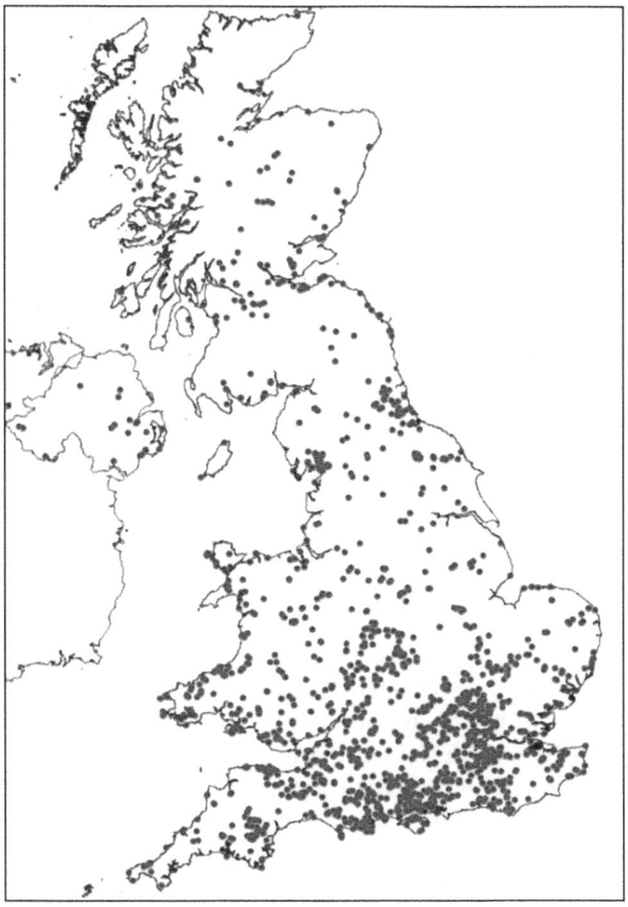

Fig. 1 Location of sites, where data is collected to compile the butterfly indicators

are generally easier to record. Both 'non-transect' methods involve timed searches and counts through a representative sample of the area occupied, with the counts converted to a robust index that accounts for both the size of the site and the time in the season when the count was made.

Compiling the indicators

Weekly transect counts for each species are summed to generate annual abundance indices. For sites with missing weekly counts, a Generalised Additive Model (GAM) is used to impute the missing values and to calculate the annual abundance index (Rothery and Roy 2001). In both timed and larval web counts, systematic recording is made on single days in suitable weather (UKBMS recording criteria met), with the counts converted to an *annual index* that accounts for both the size of the colony and the time in the season when the count was made.

Collated indices of abundance for each species at each site are derived from a log-linear model applied to the site totals, with effects for site and year that account for missing year values and adjustments for overdispersion and serial correlation (ter Braak et al. 1994; Pannekoek and Van Strien 2005). Transect and non-transect data are combined in the model, with all indices scaled comparatively (to 1 in the start year), as part of the Poisson regression algorithm. Previous research has confirmed that transect and non-transect data give similar regional trends and combining those data can increase the precision of trend estimates (Brereton et al. 2002).

Multi-species (composite) indices of butterfly abundance were calculated following methods developed for UK birds (Gregory et al. 2003, 2005) and as used for stock markets to monitor changes in the value of the top 100 companies (e.g. the FTSE 100). Multi-species indices were derived by calculating the geometric mean index across each species assemblage. In this, for each year separately, the log of each species index value was taken, then averaged across selected species and the exponential of the result calculated.

The following butterfly indicator groupings were made: (1) all species (up to 53 of the 58 resident UK species) (2) generalist (wider countryside) species - mobile species that occur in a wide range of habitats in the wider countryside and (3) (habitat) specialists - low mobility species chiefly restricted to semi-natural habitats and (4) regular migrants (Asher et al. 2001). The distinction between specialist and generalist species is somewhat arbitrary, as for example a specialist in one country may be a generalist in another, although the definitions have proven to be both useful, in for example highlighting agricultural policy issues in relation to farmland bird declines (Gregory et al. 2005). In England, butterfly indicators were broken down further by policy workstream, broadly following the indicator model already adopted for birds. A woodland butterfly indicator was compiled from monitored sites located within woodland, whilst a farmland butterfly indicator was compiled from open habitat sites subject to stock grazing, cultivation and other agricultural management practices.

Trends in the indicators

Trends and confidence intervals in multi-species butterfly indicators were assessed by structural time-series analysis and the Kalman Filter using the program Trendspotter (Soldaat et al. 2007). Structural time series models are regression models in which the explanatory variables are functions of time and the parameters are time-varying. The Kalman filter is a recursive filter that estimates the state of a dynamic system from a series of incomplete and noisy measurements (Harvey 1989). Trendspotter emphasizes temporal variation as the main source of error and provides a test of the difference between the smoothed population number of the last year against each of the preceding years, thereby taking into account serial correlation between years (Visser 2004; Soldaat et al. 2007). We applied Trendspotter to multi-species butterfly indicators to visualise the smoothed trend and to provide a status assessment of key periods of change in the index (Table 1). To describe the change in the multi-species index over the long-term (1976/9-2009) and short-term (2000–2009) we estimated the percentage change in abundance by ordinary linear regression of the log-transformed smoothed indices against year.

Policy responses—impacts of protected area management and agri-environment schemes on butterfly trends

SSSIs and agri-environment schemes are important mechanisms for conserving biodiversity in the UK. For example, SSSIs cover 8% of the land surface of England and 12% each in Scotland and Wales. Between 2000 and 2007 Government expenditure in managing the SSSI series amounted to £395 million and by 2011 is expected to exceed £100 million per annum (National Audit Office 2008. A driver for this is a Government target to achieve favourable condition on 95% of SSSIs by 2010 (National Audit Office 2008).

In England, funding for conservation management of SSSIs and other areas with wildlife interest comes from two main sources (1) Environmental Stewardship (agri-environment schemes) and (2) Woodland Grant Schemes. In England, 54 per cent of agricultural land is managed under an entry level agreement, whilst a further 13% is managed under a higher level or targeted scheme (Defra 2009). Agri-environment schemes are particularly important as more than 90% Britain's resident butterflies breed on farmland over at least part of their range (Brereton 2004). Management for UKBAP Priority and other specialist butterflies is a regular component of higher level agri-environment scheme agreements and the presence of specialist butterflies has been used to target schemes (Thompson pers. comm.). Over the period 2003–2007, more than 800 butterfly monitored sites were located on land entered into higher-level agri-environment scheme agreements (UKBMS unpublished data).

UKBMS monitoring sites were classified by their SSSI status (transect within or outside a SSSI) and whether the habitats sampled along the transect route were managed under a higher level agri-environment scheme (Environmentally Sensitive Areas, Countryside Stewardship and Higher Level Environmental Stewardship). Collated indices of abundance were calculated for each species for each group of sites (SSSI, non-SSSI, AgriEnvironment, non-AgriEnvironment). Analysis of Covariance (ANCOVA) was used to test for differences in the linear trend in index values over time, compared to paired site groupings

Table 1 Long-term and short-term trends in UK butterfly indicators

Indicator	No. spp.	Time period	% change over smoothed series	Status assessment (2009 versus other years)
Long-term trends				
England all butterflies	50	1976–2009	−19	Moderate decline 1976–2005
England specialist butterflies	26	1976–2009	−45	Moderate decline 1976–2007
England generalist butterflies	24	1976–2009	−12	Stable 1976–82, moderate decline 1983-98
England all farmland butterflies	48	1990–2009	−42	Moderate decline 1990–96, 98–2000
England specialist farmland butterflies	23	1990–2009	−36	Moderate decline 1990–2007
England generalist farmland butterflies	25	1990–2009	−47	Moderate decline 1990–94, Steep decline 1995–2008
England all woodland butterflies	42	1990–2009	−65	Steep decline 1990–2005 (exc. Moderate decline 1993-4, 98-01)
England specialist woodland butterflies	16	1990–2009	−56	Moderate decline 1990–2008
England generalist woodland butterflies	26	1990–2009	−67	Steep decline 1990–2007 (exc. Moderate decline 1993)
Scotland all butterflies	20	1979–2009	+23	Stable
Scotland specialist butterflies	6	1979–2009	−44	Moderate decline 1979–85
Scotland generalist butterflies	14	1979–2009	+22	Stable
UK specialist butterflies	25	1976–2009	−34	Moderate decline 1976–2006, exc. 1977, 82-87 (Deteriorating)
UK generalist butterflies	24	1976–2009	−6	Stable (exc. moderate decline 1988–97)
Short-term trends				
England all butterflies	50	2000–2009	−22	Moderate decline 2000–05
England specialist butterflies	26	2000–2009	−34	Moderate decline 2000–07
England generalist butterflies	24	2000–2009	−19	Uncertain
England all farmland butterflies	48	2000–2009	−47	Moderate decline 2000, steep decline 2001-07
England specialist farmland butterflies	23	2000–2009	−31	Moderate decline 2000–07
England generalist farmland butterflies	25	2000–2009	−58	Steep decline 2000–08
England all woodland butterflies	42	2000–2009	−59	Moderate decline 2000–06 (exc. steep decline 2002-05)
England specialist woodland butterflies	16	2000–2009	−38	Moderate decline 2000–08
England generalist woodland butterflies	26	2000–2009	−55	Steep decline 2000–07
Scotland all butterflies	20	2000–2009	+6	Stable
Scotland specialist butterflies	6	2000–2009	+7	Stable
Scotland generalist butterflies	14	2000–2009	+6	Stable
UK specialist butterflies	25	2000–2009	−22	Moderate decline 2000–06
UK generalist butterflies	24	2000–2009	−17	Stable

(e.g SSSI vs non-SSSI). In all cases the two series compared were for the same number (n) of years of annual data.

Sampling bias in the UKBMS and impacts on the quality of butterfly indicators

There has been a large increase in the number of UKBMS sites monitored annually, from 38 in 1976 to 275 by 1990, 703 by 2000 and 1009 by 2009. In the UKBMS, sites have been established by volunteer recorder 'free choice', rather than by allocation through a random sampling framework. This has led to biases in the sample, including oversampling of (1) species-rich butterfly habitats such as semi-natural grasslands, (2) protected land and (3) regions with higher butterfly recorder densities. For example, ~50% of sites are in protected areas (SSSIs and National Nature Reserves) yet protected areas occupy only ~10% of the UK land surface (National Audit Office 2008). 87% of sites are located within semi-natural habitats, which occupy less than a third of the UK land surface. Intensive farmland covers 46% of the UK land surface, yet is the primary habitat on only 6% of transects. In contrast, agriculturally unimproved calcareous grassland occupies less than 1% of the land surface, yet is the primary habitat for more than 20% of transects. The majority of sites are located in central-southern Britain, where recorder density is highest and the majority of species-rich calcareous grassland and ancient woodland butterfly habitats are found (Asher et al. 2001). Because site selection is strongly biased towards higher quality habitat, sample coverage of specialist species is generally good (Butterfly Conservation and CEH unpublished data). However, a concern is that the trend for generalist species may not be representative because the majority of their populations occur in intensive or lower quality habitats in the under-sampled wider countryside, leading in turn to biased butterfly indicators.

The effects of sampling biases were tested by comparing generalist species trends (1) at original BMS sites versus more recently established ('independent') non-BMS sites (2) on protected versus non protected land and (3) in semi-natural habitats versus the wider countryside using ANCOVA.

Results

Adoption of butterflies as governmental biodiversity indicators in the UK

In November 2005, a suite of butterfly indicators were published for England by Defra in an indicator review report documenting progress in implementing the England Biodiversity Strategy (Defra 2006). These included a top level ('headline') *Populations of Butterflies* indicator for all species, disaggregated into habitat specialists and wider countryside generalist species. Similar structured indicators were developed for two policy sectors, with a farmland butterfly for agriculture and a woodland butterfly indicator for forestry, with the base year for each indicator being 1990. The indicators are assessed on trends in the specialist species, as the majority are: (1) in long-term decline (Fox et al. 2006) and (2) identified as UK BAP priority species. For each indicator, there is a policy objective to reverse the long-term population declines.

In March 2007, Defra on behalf of the UK Biodiversity Partnership published a first set of biodiversity indicators for UK and international reporting of biodiversity conservation progress (Defra 2007). The set comprised 18 indicators, including a *Trends in populations of selected species (butterflies) indicator*, with separate grouping for (habitat) specialists and wider countryside (generalist) species.

In Scotland, a Butterfly Indicator was compiled for Scottish Natural Heritage (SNH) in November 2006. The indicator was developed in the same way as for England, with separate trends for all-species, habitat specialists and wider countryside (generalists) species, covering all years since 1979.

UK and England indicators include the assessment of both long-term (since the start of the series) and short-term (since 2000) trends. The Scottish Indicator is based on trend over the whole series.

Biodiversity indicators for Wales are still in development (Hockley et al. 2010), though as part of the process a test/candidate indicator of *Populations of Butterflies in Wales*, with separate trends for all-species, habitat specialists and wider countryside (generalists) species, covering the all years since 1976 has been compiled from UKBMS data.

UKBMS data has also been used in the development of a European Butterfly Indicator, as part of a European initiative co-ordinated by the European Environment Agency (EAA) launched to develop a First European Set of Biodiversity Indicators. As part of this process, Butterfly Conservation Europe produced a preliminary *Grassland Butterfly Indicator* in 2005, and the indicator was formally adopted by the European Union in 2007.

Trends in the indicators

The indicators for England, Scotland and the UK (Table 1) show that butterfly numbers have fluctuated considerably from year-to-year, most likely attributable to variations in annual weather conditions and other natural events (Pollard 1988; Roy et al. 2001). However, a general pattern is apparent in the multi-species trend for specialist species, with significant long-term declines in each country since the 1970s. Trends in generalist species vary at the country-level from showing little or no overall change in Scotland and across the UK, to declines over some years in England. In Wales, there has been a significant decline (by 87%) in the composite measure for specialist species since 1976, whilst the generalist multi-species trend is classed as stable (Butterfly Conservation and CEH unpublished data). Scotland is the only country for which there is a positive change in abundance over the whole time series since the 1970s (linear trend of index against year, for all-species and for generalists), although there is no evidence for statistically significant changes in the index (Trendspotter Status Assessment, Table 1). Analysis by policy sector in England since 1990 shows that at farmland and woodland sites, there have been significant declines in abundance in the composite indices for specialist, generalist and all-species combined.

Short-term trends (since 2000) show moderate to steep declines in all butterfly indicators, except for Scotland where trends are classed as stable and for generalist species in England, where the trend is uncertain. Progress in conserving butterflies through the 2010 target was assessed by comparing the percentage change in abundance in butterfly indicators since adoption of the target (2002–2009) with a comparable period prior to adoption (1994–2001). The time series was too short to assess whether the difference in butterfly indicator trends between each time period were

Table 2 Trends in UK butterfly indicators, before (1994–2001) and following (2002–2009) adoption of the 2010 target

Indicator	No. spp.	% change 1994–2001	% change 2002–2009
Long-term trends			
England all butterflies	50	−8.9	−19.4
England specialist butterflies	26	−10.5	−36.8
England generalist butterflies	24	−10.0	−15.5
England all farmland butterflies	48	−14.07	−56.1
England specialist farmland butterflies	23	−10.9	−28.5
England generalist farmland butterflies	25	−7.5	−69.1
England all woodland butterflies	42	−12.3	−53.3
England specialist woodland butterflies	16	−23.4	−31.9
England generalist woodland butterflies	26	−23.6	−62.3
Scotland all butterflies	20	+4.8	+4.5
Scotland specialist butterflies	6	+4.6	+4.4
Scotland generalist butterflies	14	−9.3	+7.3
UK specialist butterflies	25	−7.7	−13.9
UK generalist butterflies	24	−8.6	−25.0

significantly different, due to high annual variability in abundance giving low statistical power. With the exception of Scotland, the percentage change in decline in all butterfly indicators has accelerated since adoption of the 2012 target, with the largest negative changes noted for farmland sites in England (Table 2). In Scotland, the most favourable trend was for generalist species, where there was a 7.3% increase post-2001, compared to a 9.3% decrease pre-2001.

Impacts of land management and agri-environment schemes

For a sample of specialist species (n = 11) in England over the period 1990–2009, there has been little or no overall change in abundance and no difference in the trend ($F_{1,37} = 0.78$, $P = 0.38$) between protected areas (SSSIs) (n = 360 sites) and non-protected areas (n = 228 sites). At agriculturally managed sites in England, the composite measure of abundance for specialist species (n = 12) showed little or no overall change from 1994–2009 at sites entered into higher level agri-environment scheme sites, whilst there was a 35% (non-significant) decline in specialists at non-scheme sites. Figure 2 shows that from 1995–2001, the composite index for specialists was lower in every year (except 1999) at scheme sites, whilst from 2002–2008 the reverse was true.

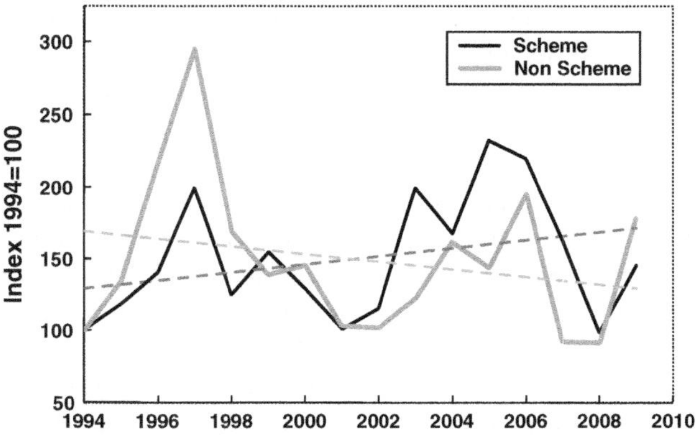

Fig. 2 Trends in specialist butterflies (n = 12 species) at higher level agri-environment scheme sites (n = 239) in England compared to non scheme sites (n = 233) 1994–2009

Impacts of sampling bias on the quality of butterfly indicators

Although there has been a sharp increase in the number of sites monitored over time, particularly since the establishment of independent transects in the early 1980s, there is no difference in the long-term abundance trend (1990–2009) for generalists between original BMS sites (−56% highly significant decline, $P = 0.001$, n = 177 sites) and non-BMS sites (−60% highly significant decline, $P = 0.001$, n = 1551 sites) ($F_{1,37} = 0.112$, $P = 0.74$). Similarly, there is no difference in the long-term abundance trend (1990–2009) in generalists between (1) protected areas (SSSIs) (−60% highly significant decline, $P = 0.001$, n = 475 sites) and non protected areas (−63% highly significant decline, $P < 0.001$, n = 489 sites) ($F_{1,37} = 0.061$, $P = 0.80$) and (2) semi-natural (−61% highly significant decline, $P < 0.001$, n = 570 sites) and wider countryside habitats (−35% n.s. decline, $P = 0.05$, n = 104 sites) ($F_{1,37} = 1.30$, $P = 0.26$).

Discussion

Trends in the indicators and causes of change

UK butterfly indicators show a long-term decline in the composite measures of abundance for specialist species but little or no overall change for generalist species. The indicators provide clear evidence that UK Government commitments to halt the long-term decline in biodiversity have not been met with respect to specialist butterflies. The very sharp decline in specialist species is considered to have been triggered by the 1976 drought (Pollard and Yates 1993) and their subsequent failure to recover appears linked to a range of factors including intensification of land management, abandonment/habitat neglect, inappropriate conservation management, climate change and habitat fragmentation (see Fox et al. 2006). The decline in abundance of specialist species in the UK parallels recent historical declines in range documented for the majority of these species (Asher et al. 2001; Fox et al. 2006). This is not just a UK problem, as specialist butterflies are declining in range and abundance across many parts of Europe (Van Swaay and Warren 1999, 2005; Van Swaay et al. 2010).

Generalists in the UK have fared better, and this is also demonstrated by distribution data, where a number of species are expanding their ranges northwards (Fox et al. 2006), consistent with predicted responses to higher summer temperatures over the period (Roy et al. 2001). That generalists show little or no overall long-term change, rather than a significant increase, may possibly suggest that the predicted beneficial impacts of climate change are masking regional declines due to intensification of land management and/or that climate change is having a negative impact on some of the species. The majority of specialist species have not responded in the same way to the warmer summers because of their lower mobility, which limits their potential to find the remaining fragments of suitable semi-natural habitat (Warren et al. 2001).

The UK Government has commitments under the CBD, to significantly reduce the rate of biodiversity loss by 2010. Our latest assessment shows that this will not be met for butterflies and thus for the other components of biodiversity which may be represented through them. Comparisons of changes in butterfly abundance pre-CBD target (1994–2001) and post-CBD target establishment (2002–2009) suggest that if anything the rate of decline at the UK-level is increasing for specialist species. However, caution is needed here, as the differences are not subject to statistical testing due to the short time series. Furthermore, short-term trends are more likely to be influenced by annual weather events and recent years have seen a run of poor summers, which have exacerbated long-term declines in specialist species (Botham et al. 2008). In the short-term, butterflies have fared better in Scotland (northern UK), where trends since 2000 are stable and there are more favourable trends for both specialists and generalists since the CBD target was set. However, there have still been substantial declines of some Priority species which are compensated for by increases of others.

Our analyses show that in spite of a large amount of investment to improve the condition of protected areas for biodiversity through positive conservation management, composite measures for specialist and generalist butterflies do not show improved trends on protected land compared to elsewhere. Although other studies have shown that selected butterflies have benefitted from SSSI management through improvements made in habitat condition (Brereton et al. 2002, 2005), there is a concern that uniform management on some sites is negatively impacting on species requiring a diverse habitat structure (Davies et al. 2007).

The butterfly indicators show that specialist butterflies in England face severe problems in both farmland and woodland habitats. The sharp decline in woodland butterflies is thought to be largely attributable to a corresponding loss of open habitats within woodland, due to management changes especially the decline in traditional woodland management (e.g. coppicing), the reduction in felling areas and the shading of rides and glades (Fox et al. 2006). Other factors including local increases in deer numbers, habitat fragmentation and climate change may also be important, but direct evidence is lacking. The decline suggests that the scale of management being delivered through Woodland Grant Schemes in England is insufficient and that more wide-ranging policy and land management measures are needed to reverse the fortunes of woodland butterflies. On farmland, although butterflies have been declining since the start of assessment in 1990, there are more encouraging signs that policy measures are beginning to have a beneficial impact. Specialist species performed significantly better at sites entered into higher level agri-environment scheme agreements, compared to elsewhere. The difference in trends indicates that site-specific conservation work carefully targeted at butterflies and other interest features, which is a feature of scheme management is beginning to make a difference in conservation terms.

Sampling bias in the UKBMS and the quality of butterfly indicators

We found that trends for generalist species were very similar at (1) original BMS sites and more recently established non-BMS sites (2) on protected and non protected land and (3) in semi-natural and wider countryside habitats. Thus our analyses suggest that the increase in the number of sites in the UKBMS over time and the heavy bias towards sampling of higher quality habitat are having little impact in biasing the direction and magnitude of the composite trend for generalist species. Studies in the Netherlands have produced similar results (Brereton et al. in press).

Function and evaluation of butterflies as governmental biodiversity indicators

The UK Government now recognises butterflies as "good indicators of the broad state of wildlife and the countryside" (Defra 2007). Reasons highlighted (within indicator fiches) as to why butterflies make good indicators include, their (1) rapid response to change, (2) wide range of habitat choices, (3) popular appeal, and (4) long-term data availability. Assessment roles for the butterfly indictors in policy terms include (1) the potential to assess anthropogenic impacts (climate change, agro-chemical, air pollution and habitat fragmentation), (2) to determine whether habitat quality and diversity is being maintained at fine scales through to landscape scales, (3) to measure lower trophic level food availability and (4) to provide biodiversity representation for changes in the abundance of other insects.

The indicators have been the subject of peer-review in England (Defra unpublished data), Scotland (SNH unpublished data) and at the European level (European Environment Agency, unpublished data), whilst there is likely to be a need to review indicators at the UK and England level in 2011 to reflect new international and domestic commitments (M. Stevenson, Defra pers. comm.). Each of the evaluations have been completed in a broadly similar way, using methods adapted from CBD core principles criteria (see http://www.cbd.int/2010-target/framework/indicators.shtml). Criteria include policy relevance, biodiversity relevance, scientifically sound and well founded methodologically, and several practical criteria including broad acceptance and understandability, affordable monitoring, available and routinely collected data. In each of the reviews, the indicators have tended to score highly in terms of data quality, and moderate to highly in terms of policy relevance.

Although there has been strong support within Government for the use of butterflies as indicators, it is further necessary to document the strengths, weaknesses and overall suitability of butterflies as biodiversity indicators, both for assessing broader biodiversity change and in helping to evaluate and inform conservation policies.

A crucial issue is the extent to which trends in butterflies represent trends in other biodiversity, particularly as butterfly species comprise only 0.2% of UK insect species (Anon 1994). The distribution of butterflies has been found to be a good predictor of areas of high biodiversity, species richness and or habitat quality in the majority, though not all studies (Beccaloni and Gaston 1995; Brown 1991; Brown and Freitas 2000; Simonson et al. 2001; Fleishman et al. 2005; Grill et al. 2005; Kerr et al. 2000; Ricketts et al. 2002; Kremen et al. 2003; Thomas and Clarke 2004; Maes and van Dyck 2005; Maes et al. 2005). There is only limited evidence to indicate that changes in butterfly abundance, species-richness and distribution mirror changes in other taxa (Blair 1999; Swengel and Swengel 1999; Brown and Freitas 2000; Conrad et al. 2004; Thomas and Clarke 2004; Thomas et al. 2004; Hickling et al. 2006). However such studies are unlikely to be fully conclusive as they may be dependent on the taxa and the spatial scales considered (Ricketts et al. 2002).

A general problem is a lack of available data on trends in the abundance of other insects for comparison. In the UK, the best available long-term dataset is for moths, through the Rothamsted Insect Survey (Woiwod and Harrington 1994; Conrad et al. 2004; Conrad et al. 2006), with trends in moth abundance significantly correlated with trends in butterfly abundance (Brereton et al. in press). Based on a comprehensive review of studies into their life-history traits, biology, relative sensitivity to climate change and adjusted extinction rates, recent reviews (Ehrlich 1994, 2001; Thomas 2005) have concluded that butterflies may be considered reasonable, albeit imperfect representative indicators of trends observed in the majority of other terrestrial insects (excluding for example invertebrate groups that are predominantly predators and parasitoids).

In policy terms, species biodiversity indicators can be assigned to one of four broad categories, according to their ability to generalise findings and their strength of relationship with environmental factors (Gregory et al. 2005). Type 1 indicators measure how specific species or species groups are doing, for example species of conservation concern. Type 2 indicators ask how biodiversity is doing more generally, with no direct aim to link these indicators with specific environmental factors. Type 3 indicators are used to show how specific species or groups of species are responding to environmental pressures, for example lichens to air pollution, and Type 4 indicators show how biodiversity is responding to environmental factors in general, for example farmland biodiversity to agricultural practices.

Headline all-species butterfly indicators are chiefly Type 2 indicators, especially in relation to providing the only direct representation for insects. The Butterfly indicators in the UK are disaggregated into specialist and generalist species. The specialist indicators are likely to represent a large amount of biodiversity as specialist butterflies are largely restricted to semi-natural habitats (Asher et. al. 2001), which are among the most species-rich (for both insects and plant) habitats and are also critically important for rare insect species (Fry and Lonsdale 1991; Thomas 2005). The specialist indictor also conforms to a Type 2 indicator, as the majority of specialist species are Priority Species in the UK BAP. The indicators can be broken down further and this is shown in England where butterfly indicators for agriculture and forestry sectors are compiled. These are more likely to directly assess land management impacts and to a degree they can be considered as Type 4 indicators. However, a policy limitation with butterflies and other species indicators is that species trends can rarely be explained by single policy drivers. This problem is exacerbated because indicators show average change across species, whilst individual species may be reacting differently to change.

The generalist indicator is a Type 2 indictor that represents the state of a smaller suite of biodiversity, but one which extends over much of the landscape (common species in the wider countryside). Due to sampling biases, here we largely infer trends in these species from apparent synchrony in trends on protected areas and semi-natural sites. To this end, we recommend improvements in wider countryside monitoring and developments are already in progress, with a new scheme established in 2009 to monitor butterflies in the wider countryside through a random sampling framework (Roy et al. 2007, Cruickshanks et al. this volume).

Conclusions

In spite of their imperfections, we conclude that the butterfly indicators have played a valuable role in documenting changes in biodiversity in the UK. In particular, the strong signals of decline in specialist species point to a wider biodiversity crisis on the remaining fragments of higher quality habitat, which require an urgent policy response.

Butterfly monitoring in the UK continues to grow, and with it the potential to develop a wider suite of butterfly indicators, both at the country and regional level and in relation to other environmental pressures such as climate change. The UK has also played an important role in helping to develop butterfly monitoring across Europe, which opens up further possibilities in the use of butterflies as indicators.

Acknowledgments We are indebted to all the volunteer recorders and local co-ordinators who have contributed to butterfly monitoring in general and the UKBMS specifically since 1976. The UKBMS is a partnership between Butterfly Conservation and the Natural Environment Research Council, Centre for Ecology & Hydrology. The UKBMS is funded by a multi-agency consortium led by Defra, and including the Countryside Council for Wales, the Joint Nature Conservation Committee, Forestry Commission, Natural England, the Natural Environment Research Council, the Northern Ireland Environment Agency and Scottish Natural Heritage.

References

Anon (1994) Biodiversity: the UK action plan. HMSO, London

Asher J, Warren MS, Fox R, Harding P, Jeffcoate G, Jeffcoate S (2001) The millennium atlas of butterflies in Britain and Ireland. Oxford University Press, Oxford, UK

Balmford A, Green RE, Jenkins M (2003) Measuring the changing state of nature. Trends Ecol Evol 18(7):326–330

Balmford A, Bennun L, Ten Brink B, Cooper D, Côté IM, Crane P, Dobson AP, Dudley N, Dutton I, Green RE, Gregory RD, Harrison J, Kennedy ET, Kremen C, Leader Williams N, Lovejoy TE, Mace P, May RM, Mayaux P, Morling P, Phillips J, Redford K, Ricketts TH, Rodriguez JP, Sanjayan MA, Schei PJ, Van Jaarsveld AS, Walther BA (2005) The convention on biological diversity's 2010 target. Science 307:212–213

Beccaloni GW, Gaston KJ (1995) Predicting species richness of neotropical forest butterflies: Ithominae (Lepidoptera: Nymphalidae) as indicators. Biol Indic 71:77–86

Berry PM, Dawson TP, Harrison PA, Pearson RG (2002) Modelling potential impacts of climate change on the bioclimatic envelope of species in Britain and Ireland. Glob Ecol Biogeogr 11:453–462

Blair RB (1999) Birds and butterflies along an urban gradient: surrogate taxa for assessing biodiversity? Ecol Appl 9: 164–170

Botham MS, Brereton TM, Middlebrook I, Cruickshank KL, Harrower C, Beckmann B, Roy DB (2008). United Kingdom Butterfly Monitoring Scheme report for 2008. CEH Wallingford

Brereton T (2004) Farming and butterflies. The Biologist 51: 32–36

Brereton T, Warren M, Stewart K (2002) BD1427: developing a system for assessing the effect of agri-environmentschemes on butterfly populations. Butterfly Conservation, Wareham

Brereton T, Wigglesworth T, Warren MS, Stewart K (2005) BD1446: Agri-environment schemes and butterflies: re-assessing the impacts and improving delivery of BAP targets. Butterfly Conservation, Wareham

Brereton T, Roy D, Greatorex-Davies N (2006) 30 years and counting. The contribution to conservation and ecology of butterfly monitoring in the UK. Br Widl 17:162–170

Brereton TM, Warren MS, Roy DB, Stewart K (2007) The changing status of the Chalkhill Blue butterfly *Polyommatus coridon* in the UK: the impacts of conservation policies and environmental factors. J Insect Conserv 12:629–638

Brereton TM, van Swaay C, van Strien A (in press) Developing a butterfly indicator to assess changes in Europe's biodiversity. Proceedings of the 17th International Conference of the European Bird Census Council, Bird Numbers 2007. Chiavenna, Italy

Brown KS (1991) Conservation of neotropical environments: insects as indicators. In: Collins NM, Thomas JA (eds) The conservation of insects and their habitats. Academic Press, London, pp 350–404

Brown KS, Freitas AVL (2000) Atlantic forest butterflies: indicators for landscape conservation. Biotropica 32:934–956

Buckland ST, Magurran AE, Green RE, Fewster RM (2005) Monitoring change in biodiversity through composite indices. Philos Trans R Soc B 360:243–254

Caro TM, O'Doherty G (1999) On the use of surrogate species in conservation biology. Conserv Biol 13:805–814

CBD (2004) Decision VII/30 of the Seventh Conference of the Parties to the Convention on Biological Diversity (CBD/COP7) "Strategic Plan: future evaluation of progress". http://www.biodiv.org/decisions/default.aspx?dec=VII/30

CBD (2006) Global Biodiversity Outlook 2. Secretariat of the. convention on biological diversity, Montreal, Canada

Conrad KF, Woiwod IP, Parsons M, Fox R, Warren MS (2004) Long-term population trends in widespread British moths. J Insect Conserv 8:119–136

Conrad KF, Warren MS, Fox R, Parsons MS, Woiwod IP (2006) Rapid declines of common, widespread British moths provide evidence of an insect biodiversity crisis. Biol Conserv 132:279–291

Davies ZG, Wilson RJ, Brereton TM, Thomas CD (2005) The re-expansion and improving status of the silver-spotted skipper butterfly (Hesperia comma) in Britain: a metapopulation success story. Biol Conserv 124:189–198

Davies H, Brereton T, Roy DBR, Fox R (2007) Government targets for protected area management: will threatened butterflies benefit? Biodivers Conserv 16(13):3719–3736

De Heer M, Kapos V, Ten Brink BJE (2005) Biodiversity trends in Europe: development and testing of a species trend indicator for evaluating progress towards the 2010 target. Philos Trans R Soc B 360:297–308

Defra (2002) Working with the grain of nature: a biodiversity strategy for England. Defra, London

Defra (2006) Working with the grain of nature–taking it forward: Volume II. Measuring progress on the England Biodiversity Strategy: 2006 assessment. Defra, London

Defra (2007) UK biodiversity indicators in your pocket 2007. Published by Defra on behalf of the UK Biodiversity Partnership. Defra, London

Defra (2009) UK biodiversity indicators in your pocket 2009. Published by Defra on behalf of the UK Biodiversity Partnership. Defra, London

DoE(NI) (2002). Northern Ireland Biodiversity Strategy. Department of the environment Northern Ireland (environment and heritage service)

Ehrlich PR (1994) Energy use and biodiversity loss. Philos Trans R Soc B 344:99–104

Ehrlich PR (2001) Tropical butterflies: a key model group that can be "completed". Lepid News 2(1):10–12

Ehrlich PR (2003) Introduction: butterflies, test systems, and biodiversity. In: Boggs L, Watt WB, Ehrlich PR (eds) Butterflies: ecology and evolution taking flight. University of Chicago Press, Chicago, pp 1–6

Erhardt A, Thomas JA (1991) Lepidoptera as indicators of change in the semi-natural grasslands of lowland and upland Europe. In: Collins NM, Thomas JA (eds) The conservation of insects and their habitats. Academic Press, London, pp 213–236

Failing L, Gregory R (2003) Ten common mistakes in designing biodiversity indicators for forest policy. J Environ Manage 68:121–132

Feest A, Aldreda TD, Jedamzika K (2010) Biodiversity quality: a paradigm for biodiversity. Ecological Indicators. Ecol Indic 10(6):1077–1082

Fleishman E, Fay JP, Murphy DD (2000) Upsides and downsides: contrasting topographic gradients in species richness and associated scenarios for climate change. J Biogeogr 27:1209–1219

Fleishman E, Thomson JR, MacNally R, Murphy DD, Fay JP (2005) Using indicator species to predict richness of multiple taxonomic groups. Conserv Biol 19:1125–1137

Fox R, Asher J, Brereton T, Roy D, Warren M (2006) The state of butterflies in Britain and Ireland. Pisces Publications, Newbury

Fry R, Lonsdale D (1991) Habitat conservation for Insects—a neglected green Issue. Amateur Entomological Society, London

Gregory RD, Noble D, Field R, Marchant JH, Raven M, Gibbons DW (2003) Using birds as indicators of biodiversity. Ornis Hung 12–13:11–24

Gregory RD, Van Strien AJ, Vorisek P, Gmelig Meyling AW, Noble DG, Foppen RPB, Gibbons DW (2005) Developing indicators for European birds. Philos Trans R Soc B 360:269–288

Grill A, Knoflach B, Cleary DFR, Kati V (2005) Butterfly, spider, and plant communities in different land-use types in Sardinia, Italy. Biodivers Conserv 14:1281–1300

Groombridge B (1992) Global biodiversity: status of the Earth's living resources. Chapman & Hall, London

Harvey AC (1989) Forecasting structural time series models and the kalman filter. Cambridge University Press, London

Heath J, Pollard E, Thomas JA (1984) Atlas of butterflies in Britain and Ireland. Harmondsworth, Viking

Hickling R, Roy DB, Hill JK, Fox R, Thomas CD (2006) The distributions of a wide range of taxonomic groups are expanding polewards. Glob Chang Biol 12:450–455

Hill JK, Thomas CD, Fox R, Telfer MG, Willis SG, Asher J, Huntley B (2002) Responses of butterflies to 20th century climate warming: implications for future ranges. Proceedings of the Royal Society B 269: 2163–2171

Hockley N, Farrar J, Gibbons JM, Hearn SM, Jones JPG, Kaiser M (2010) Biodiversity indicators for Wales—a report to the Welsh Assembly Government. University of Bangor, Bangor

Kerr JT, Sugar A, Packer L (2000) Indicator taxa, rapid biodiversity assessment, and nestedness in an endangered ecosystem. Conserv Biol 14:1726–1734

Kremen C (1992) Assessing the indicator properties of species assemblages for natural areas monitoring. Ecol Appl 2:203–217

Kűhn E, Feldmann R, Harpke A, Hirneisen N, Musche M, Leopold P, Settele J (2008) Getting the public involved into butterfly conservation—lessons learned from a new monitoring scheme in Germany. Isr J Ecol Evol 54:89–103

Lewis OT, Hurford C (1997) Assessing the status of the marsh fritillary butterfly (*Eurodryas aurinia*): an example from Glamorgan, UK. J Insect Conserv 1(3):159–166

Mace GM, Baillie JEM (2007) The 2010 biodiversity indicators: challenges for science and policy. Conserv Biol 21(6):1406–1413

Maes D, Van Dyck H (2001) Butterfly diversity loss in Flanders (north Belgium): Europe's worst case scenario? Biol Conserv 99:263–276

Maes D, Bauwens D, de Bruyn L, Anselin A, Vermeersch G, van Landuyt W, de Knijf G, Gilbert M (2005) Species richness coincidence: conservation strategies based on predictive modelling. Biodivers Conserv 14:1345–1364

National Audit Office (2008) Natural England's role in improving sites of special scientific interest. The Stationery Office, London

New TR (1997) Are Lepidoptera an effective 'umbrella group' for biodiversity conservation? J Insect Conserv 1:5–12

OECD (1994) Environmental indicators. OECD core sets, OECD, Paris

Oostermeijer JGB, van Swaay CAM (1998) The relationship between butterflies and environmental indicator values: a tool for conservation in a changing landscape. Biol Conserv 86:271–280

Pannekoek J, Van Strien AJ (2005) TRIM 3 manual. Trends and indices for monitoring data. CBS. Statistics Netherlands, Voorburg, Netherlands

Parmesan C (2003) Butterflies as bioindicators for climate change effects. In: Boggs CL, Watt WB, Ehrlich PR (eds) Butterflies—ecology and evolution taking flight. University of Chicago Press, Chicago, pp 541–560

Parmesan C, Yohe G (2003) A globally coherent fingerprint of climate change impacts across natural systems. Nature 421:37–42

Parmesan C, Ryrholm N, Stefanescu C, Hill JK, Thomas CD, Descimon H, Huntley B, Laila L, Kullberg J, Tammaru T, Tennent WJ, Thomas JA, Warren MS (1999) Poleward shifts in geographical ranges of butterfly species associated with regional warming. Nature 399:579–583

Pollard E (1977) A method of assessing changes in the abundance of butterflies. Biol Conserv 12:115–134

Pollard E (1988) Temperature, rainfall and butterfly numbers. J Appl Ecol 25:819–828
Pollard E, Yates TJ (1993) Monitoring butterflies for ecology and conservation. Chapman and Hall, London
Pollard E, Woiwod IP, Greatorex-Davies JN, Yates TJ, Welch RC (1998) The spread of coarse grasses and changes in numbers of Lepidoptera in a woodland nature reserve. Biol Conserv 84: 17–24
Pullin AS (1996) Restoration of butterfly populations in Britain. Restor Ecol 4:71–81
Pywell RF, Warman EA, Sparks TH, Greatorex-Davis JN, Walker KJ, Meek WR, Carvell C, Petit S, Firbank LG (2004) Assessing habitat quality for butterflies on intensively managed arable farmland. Biol Conserv 118:313–325
Rees HL, Hyland JL, Hylland K, Mercer Clarke CSL, Roff JC, Ware S (2008) Environmental indicators: utility in meeting regulatory needs an Overview. ICES J Mar Sci 65:1381–1386
Ricketts TH, Daily GD, Ehrlich PR (2002) Does butterfly diversity predict moth diversity? testing a popular indicator taxon at local scales. Biol Conserv 103:361–370
Root TL, Price JT, Hall KR, Schneider SH, Rosenzweig C, Pounds JA (2003) Fingerprints of global warming on wild animals and plants. Nature 421:57–60
Rosenberg DM, Danks HV, Lehmkuhl DM (1986) Importance of insects in environmental impact assessment. Environ Manage 10:773–783
Rothery P, Roy DB (2001) Application of generalized additive models to butterfly transect count data. J Appl Stat 28:897–909
Roy DB, Rothery P, Moss D, Pollard E, Thomas JA (2001) Butterfly numbers and weather: predicting historical trends in abundance and the future effects of climate change. J Anim Ecol 70:201–217
Roy DB, Rothery P, Brereton T (2007) Reduced-effort schemes for monitoring butterfly populations. J Appl Ecol 44:993–1000
Scholes RJ, Biggs R (2005) A biodiversity intactness index. Nature 434:45–49
Schultz CB (1998) Dispersal behaviour and its implications for reserve design in a rare Oregan butterfly. Conserv Biol 12:248–292
Scottish Executive (2004) Scotland 's Biodiversity—It's in Your Hands. A strategy for the conservation and enhancement of biodiversity in Scotland. Scottish Executive Environment and Rural Affairs Department, Edinburgh
Secretariat of the Convention on Biological Diversity (2006) Global biodiversity outlook 2. CBD, Montreal
Simonson SE, Opler PA, Stohlgren TJ, Chong GW (2001) Rapid assessment of butterfly diversity in a montane landscape. Biodivers Conserv 10:1369–1386
Soldaat L, Visser H, Van Roomen M, Van Strien AJ (2007) Smoothing and trend detection in waterbird monitoring data using structural time-series analysis and the Kalman filter. J Ornithol 148(suppl 2):351–357
Sparks TH, Greatorex-Davies JN, Mountford JO, Hall ML, Marrs RH (1996) The effects of shade on the plant communities of rides in plantation woodland and implications for butterfly conservation. For Ecol Manage 80:197–207
Stork N (1997) How many species are there? Biodivers Conserv 2:215–232
Swengel AB (1998) Effects of management on butterfly abundance in tallgrass prairie and pine barrens. Biol Conserv 83:77–89
Ten Brink BJE (2000) Biodiversity indicators for the OECD environmental outlook and strategy; a feasibility study. RIVM report 402001014, Bilthoven
Ten Brink BJE, Hosper SH, Colijn F (1991) A quantitative methods for description and assessment of ecosystems: the AMEOBA approach. Mar Poll Bull 23:265–270
Ter Braak CJF, van Strien AJ, Meijer R, Verstrael TJ (1994) Analysis of monitoring data with many missing values: which method? In: W. Hagemeijer & T. Verstrael, Eds. Bird Numbers 1992. Distribution, Monitoring and Ecological Aspects. Proceedings of the 12 th International Conference of the International Bird Census Committee and European Ornithological Atlas Committee. SOVON, Beek-Ubbergen, The Netherlands
Thackeray SJ, Sparks TH, Frederiksen M, Burthe S, Bacon, PJ Bell, JR Botham MS, Brereton TM, Bright PW, Carvalho L, Clutton-Brock T, Dawson A, Edwards M, Elliott JM, Harrington R, Johns D, Jones ID, Jones JT, Leech DI, Roy DB, Scott WA, Smith M, Smithers RJ, Winfield IJ, Wanless S (2010) Trophic level asynchrony in rates of phenological change for marine, freshwater and terrestrial environments. Glob Chang Biol. doi: 10.1111/j.1365-2486.2010.02165.x
Thomas JA (1991) Rare species conservation: case studies of European butterflies. In: Spellerberg IF, Goldsmith FB, Morris MG (eds) The scientific management of temperate communities for conservation. Blackwell's Scientific, Oxford, pp 149–197
Thomas JA (1993) Holocene climate changes and warm man-made refugia may explain why a sixth of British butterflies possess unnatural early successional habitats. Ecography 16:278–284
Thomas JA (2005) Monitoring change in the abundance and distribution of insects using butterflies and other indicator groups. Philos Trans R Soc B 360:339–357
Thomas JA, Clarke RT (2004) Extinction rates and butterflies–response. Science 305:1563–1564
Thomas CD, Harrison S (1992) Spatial dynamics of a patchily distributed butterfly species. J Anim Ecol 61:437–446
Thomas JA, Rose RJ, Clarke RT, Thomas CD, Webb NR (1999) Intraspecific variation in habitat availability among ectothermic animals near their climatic limits and their centres of range. Funct Ecol 13:55–64
Thomas JA, Telfer MG, Roy DB, Preston C, Greenwood JJD, Willis SG, Asher J, Fox R, Clarke RT, Lawton JH (2004) Comparative losses of British butterflies, birds, and plants and the global extinction crisis. Science 303:1879–1881
UK BAP (2007) UK List of Priority Species and Habitats. Available at http://www.ukbap.org.uk/newprioritylist.aspx
UK Biodiversity Partnership (2007) Conserving biodiversity–the UK approach. Defra, London
UNEP (2003) Report of expert meeting on indicators of biological diversity including indicators for rapid assessment of Inland water ecosystems. 10–14 November 2003, Montreal
Van Swaay CAM, Van Strien AJ (2005) Using butterfly monitoring data to develop a European grassland butterfly indicator. In: Kuehn E, Thomas J, Feldmann R, Settele J (eds.) 2005. Studies on the Ecology and Conservation of Butterflies in Europe. Proceedings of the Conference held in UFZ Leipzig, 5–9th of December, 2005
Van Swaay CAM, Van Strien AJ (2008) The European butterfly indicator for grassland species 1990–2007. Report VS2008.022, De Vlinderstichting, Wageningen
Van Swaay C, Warren MS (1999) Red data book of European butterflies (Rhopalocera). Council of Europe, Strasbourg (Nature and Environment Series No. 99)
Van Swaay CAM, Nowicki P, Settele J, van Strien AJ (2008) Butterfly monitoring in Europe: methods, applications and perspectives. Biodiver Conserv 17:3455–3469
Van Swaay C, Cuttelod A, Collins S, Maes D, López Munguira M, Šašić M, Settele J, Verovnik R, Verstrael T, Warren M, Wiemers M, Wynhof I (2010) European red list of butterflies. IUCN and butterfly conservation Europe. Publications Office of the European Union, Luxembourg
Visser H (2004) Estimation and detection of flexible trends. Atmos Environ 38:4135–4145

Warren MS (1992) The conservation of British butterflies. In: Dennis RLH (ed) The ecology of butterflies inBritain. Oxford University Press, Oxford, pp 246–274

Warren MS, Thomas CD, Thomas JA (1984) The status of the heath fritillary butterfly Mellicta athalia Rott. in Britain. Biol Conserv 29:287–305

Warren MS, Hill JK, Thomas JA, Asher J, Fox R, Huntley B, Roy DB, Telfer MG, Jeffcoate S, Harding P, Jeffcoate G, Willis SG, Greatorex-Davies JN, Moss D, Thomas CD (2001) Rapid responses of British butterflies to opposing forces of climate and habitat change. Nature 414:65–69

Welsh Assembly Government (2006a) Environment strategy for wales. Welsh Assembly Government, Cardiff

Welsh Assembly Government (2006b) Environment strategy for wales: report on progress—September 2007. Welsh Assembly Government, Cardiff

Woiwod IP, Harrington R (1994) Flying in the face of change: the Rothamsted Insect Survey. In: Leigh RA, Johnson AE (eds) Long-term experiments in agricultural and ecological sciences. CAB International, Wallingford, pp 321–342

J Insect Conserv (2011) 15:55–68
DOI 10.1007/s10841-010-9309-z

ORIGINAL PAPER

Moths count: recording moths for conservation in the UK

**Richard Fox · Zoë Randle · Les Hill ·
Susan Anders · Laura Wiffen · Mark S. Parsons**

Received: 31 March 2010 / Accepted: 30 June 2010 / Published online: 15 July 2010

Abstract Insects are under-represented in current assessments of biodiversity loss at global and national scales. Butterflies, and a few other insect taxa, have been used as indicators of biodiversity change and as flagships for conservation, especially in temperate zones, but these groups are typically species-poor and may not be representative of insects as a whole. Macro-moths, on the other hand, are an important component of UK biodiversity, in terms of both species richness and ecosystem function. Moth abundance has decreased significantly in the UK and there is rapidly growing interest in the surveillance and monitoring of macro-moths among amateur and professional scientists. The Moths Count project was instigated to raise awareness of moths, increase moth recording and set up a long-term National Moth Recording Scheme for c.900 species of macro-moth in the UK. Since 2006, Moths Count has promoted moths to millions of people, trained and engaged thousands in recording and collated over eight million species distribution records into the National Moth Recording Scheme. These data, though incomplete at present, are already contributing to assessments of biodiversity loss and to conservation initiatives. We present initial results which show that substantial changes are already evident for macro-moths in the UK. The northern range margins of a sample of 12 species have shifted northwards by 194.8 km on average between 1982 and 2009 (7.8 km year^{-1} over the 25 years between survey mid-points). Provisional distribution trends (not corrected for variation in recording effort) were calculated for 43 species and the results corroborated both positive and negative population-level changes assessed previously.

R. Fox (✉) · Z. Randle · L. Hill · S. Anders · L. Wiffen · M. S. Parsons
Butterfly Conservation, Wareham, UK
e-mail: rfox@butterfly-conservation.org

Keywords Macro-moths · Lepidoptera · Distribution trends · Range margins · Recording · Public involvement

Introduction

Biodiversity loss in the form of species extinction and habitat destruction is now a major global issue. Modern extinction rates among well-documented taxa are several orders of magnitude greater than the long-term average (Pimm et al. 1995; May 2010) and, in conjunction with regional and national biodiversity declines, have led to widespread concern about the sustainability of our planet's ecosystems and the essential 'services' they provide to humankind (Balmford and Bond 2005; Millennium Ecosystem Assessment 2005; Mooney 2010). Political recognition of the threats led to various international governmental targets to halt, or reduce the rate of, biodiversity loss by 2010.

In this context, insects occupy an important but problematic position. Insects form the most diverse animal taxon on earth and provide numerous, critical functions within ecosystems (Kremen and Chaplin-Kramer 2007). Knowledge of extinction rates and changing species richness among insects is, therefore, paramount to an understanding of global biodiversity loss and its implications. However, basic information from which to derive such knowledge is inadequate. The number of insect species is unknown and as many as 90% are yet to be discovered (Thomas 2005). Among described species, baseline information about extent of distribution, abundance and ecological requirements is scant and biased heavily towards temperate biomes. Only a tiny proportion of known insect species have been evaluated to assess their risk of extinction (Baillie et al. 2004) and few insect extinctions have been documented, although many are suspected (Dunn 2005). Global assessments of the biodiversity impacts of climate change have been undertaken with little reference to insects (Parmesan and Yohe 2003; Thomas et al. 2004a).

In spite of these substantial impediments, considerable efforts have been made in many countries, particularly in Europe, to develop surveillance and monitoring of insects (de Heer et al. 2005; van Swaay et al. 2008). In a few cases, notably in the United Kingdom, longer time-series of data are available (Thomas 2005; Conrad et al. 2007). The UK Butterfly Monitoring Scheme (Pollard and Yates 1993; Roy et al. 2007; Oliver et al. 2010), Rothamsted Insect Survey light-trap network (Woiwod and Harrington 1994; Conrad et al. 2004) and the Rothamsted Insect Survey suction-trap network (Benton et al. 2002; Shortall et al. 2009) have provided important results from intensive, fixed-point monitoring of insect populations. In Britain and Ireland, species surveillance, or biological recording, is a long-standing popular pastime with many thousands of members of the public engaged in a wide variety of local and national schemes run mainly by non-governmental organisations. Thus, considerable UK datasets exist for certain 'charismatic' insect taxa, such as butterflies, dragonflies, hoverflies and ladybirds. In the best case scenario, that of butterflies, these data have been utilised to calculate temporal trends in range size (Warren et al. 2001; Fox et al. 2006a), population size (Pollard and Yates 1993), phenology (Roy and Sparks 2000), shifts in range margins (Hill et al. 2002) and patterns of species richness (Menéndez et al. 2006) and community composition (González-Megías et al. 2008). Perhaps most significantly, it has proved possible to directly compare biodiversity loss of an insect group (butterflies) with vertebrates (birds) and vascular plants at a national scale (Thomas et al. 2004b). This study revealed significantly greater losses (population extinctions) and national extinctions among butterflies than in the other taxa, adding to concern that, without including insects, current assessments severely underestimate biodiversity loss.

Almost all insect taxa that have been monitored effectively have relatively low species richness. It is clearly advantageous to have trends on a few insect taxa rather than none, but are the declines of butterflies (Warren et al. 2001) or the northward spread of dragonflies (Hickling et al. 2005) in Britain representative of other insects, at the national or global scale? A considerable recent literature exists on the use of butterflies and other taxa as indicators, but little consensus has emerged (e.g. Ricketts et al. 2002; Kremen et al. 2003; Maes and Van Dyck 2005; Thomas 2005; Fleishman and Murphy 2009). A preferable approach would be to directly assess larger numbers of insect species.

There is one species-rich insect taxon, the moths, with potential for comprehensive surveillance and monitoring in the UK and elsewhere, in order to assess biodiversity change beyond estimates based on small 'charismatic' insect groups and vertebrates. Here we report on the development of national surveillance for the macro-moths in the UK, undertaken as part of the Moths Count project run by Butterfly Conservation. We provide some initial findings relating to increases and decreases in species occurrence and the first quantitative assessments of northward range margin shifts for macro-moths.

Macro-moths are uniquely placed for surveillance to better understand biodiversity change. The macro-moths comprise a relatively large and ecologically diverse group of insects in the UK (some 900 species in 19 lepidopteran families: Hepialidae, Cossidae, Zygaenidae, Limacodidae, Sesiidae, Lasiocampidae, Saturniidae, Endromidae, Drepanidae, Thyatiridae, Geometridae, Sphingidae, Notodontidae, Thaumetopoeidae, Lymantriidae, Arctiidae, Ctenuchidae, Nolidae and Noctuidae), an order of magnitude increase in species richness over the butterflies or dragonflies. Macro-moths are representative of almost all terrestrial biotopes, have a relatively stable taxonomy, and most are readily identified. There is a large UK workforce of skilled amateur moth recorders, at least 2,000 strong, and evidence that this number is increasing rapidly (Fox et al. 2005). A questionnaire of 1,032 moth recorders, carried out in 2003/04, found that 36% of respondents had been recording moths for less than 4 years (Spalding et al. 2005). The UK's moth recorders and their predecessors have generated a substantial dataset of historical information that, if collated and computerised, could provide a baseline against which to assess distribution change for each species.

There is already a long-standing national monitoring network for macro-moths, in the form of the Rothamsted Insect Survey light-trap network (Woiwod and Harrington 1994). Recent analysis of 35 year (1968–2002) population trends from the Rothamsted Insect Survey highlighted significant declines in the overall abundance of nocturnal moths in Britain and found that 66% of 337 widespread and common macro-moth species studied had declined (Conrad et al. 2006). Declines have also been reported from The Netherlands (Groenendijk and Ellis in press). Evidence of such widespread declines, as well as national-level extinctions (Parsons 2010) and highly threatened species (Parsons 2004), demands a conservation response. A national distribution recording scheme for macro-moths will provide essential knowledge of species occurrence in the landscape, underpinning almost all initiatives to reverse biodiversity loss or conserve species. It will also enable assessment of hundreds of less common macro-moth species not included in the Rothamsted population trend analysis. This paper describes the development of such a scheme and some initial results.

Methods

The Moths Count project

The use of volunteers to collect biodiversity surveillance and monitoring data has a long history in the UK. This approach has been validated by the numerous research and policy outputs emanating from such schemes, but also by recent research (e.g. Newman et al. 2003; Schmeller et al. 2009). Butterfly Conservation devised a four-year project, Moths Count, to develop a national surveillance system for macro-moths in the UK, Isle of Man and Channel Islands (the National Moth Recording Scheme, NMRS) that would harness the skills and effort of existing volunteer moth recorders, collate historical data sources, and recruit and train new volunteers. The willingness of the existing moth recorders to support the project was determined in advance by use of questionnaires, focus group meetings and promotional articles. The Moths Count project was led by Butterfly Conservation, but was an active partnership of many national and local organisations (including government agencies), businesses and individuals. Funding to run the project was secured from a variety of public and charitable sector sources, with the majority coming from the Heritage Lottery Fund, as well as private and business sector donations. The total project budget was over £1 million, with a team of five full-time staff over 4 years.

Promotion, feedback and training of volunteer scientists

The experience of organisations and professionals co-ordinating volunteer-based surveillance and monitoring programmes is that feedback to participants is vital to maintain motivation. This has also been shown in research, which demonstrated the high value placed by volunteers on the social aspects of such work (Bell et al. 2008; Kühn et al. 2008). The Moths Count project invested a considerable portion of staff time and budget expenditure in outreach work, designed to promote awareness of and interest in moths and moth recording, training and feedback.

Specific targets of the outreach work included holding 90 'public moth events', informal opportunities for the public to experience moth recording in the company of skilled volunteers, improving the profile of moths in the media, producing accessible education materials (leaflets and websites) and developing simple 'citizen science' surveys that could be undertaken by the public with no training or equipment. As volunteer involvement in the study and recording of moths prior to the Moths Count project appeared to be heavily biased toward older men, there was also an intention to engage more women and young people.

The project objective for training was to provide a framework of learning opportunities to help members of the public progress from complete beginners through to knowledgeable moth recorders, able to contribute high quality data to the NMRS in order to inform biodiversity conservation policy and practice. A target programme of 100 training courses was set, each to be led by a professional tutor and with continual monitoring of the quality of training provision (via feedback questionnaires completed by course participants). Increasing the number of active moth recorders and their level of expertise provides a direct benefit to the NMRS by increasing surveillance coverage at both national and local levels, but also leads to the establishment of sustainable, locally-based volunteer networks to continue recording in the future.

Substantial project resources were allocated to feedback for actual or potential participants in the NMRS. Volunteers study moths for a variety of personal reasons including enjoyment, relaxation and intellectual stimulation. However, they are generally willing to participate in surveillance and monitoring schemes, such as the NMRS, if the rationale behind the activity is made clear. Indeed, many gain additional motivation and satisfaction from helping to produce utilitarian benefits such as biodiversity conservation (Bell et al. 2008). The Moths Count project held regular moth recorders' conferences in each of the four UK nations, providing opportunities for two-way, face-to-face feedback between project staff, partner organisations and volunteer recorders, as well as informal social networking between individuals. A project website, email newsletter and magazine-style annual project newsletter were also used to provide feedback, in addition to direct contact between project staff and volunteers. This wide variety of media and approaches ensured that volunteers were involved in the project, received accurate and timely information about progress and understood the linkages between the NMRS and wider policy and conservation initiatives.

The National Moth Recording Scheme

The main aim of the Moths Count project was to develop a National Moth Recording Scheme through which spatially and temporally referenced macro-moth distribution records, both past and future, could be collated into a comprehensive, accurate and accessible database. The NMRS was developed to cover the c.900 macro-moth species present within the geographical area of the UK, Isle of Man and Channel Islands. The NMRS contributes to the long-term conservation of moths and their habitats, by making available the data needed to implement effective conservation policies and action at the local, regional, national and international levels by members of the public, communities, and public, private and charitable sector organisations (e.g. by shaping policy, informing decisions and contributing to public awareness and scientific knowledge).

The project planned to mobilize the experience and expertise of the existing moth recording community to create a successful and highly cost-effective recording network that could be sustained in the long-term. The NMRS was developed, therefore, using a hierarchical approach, which has worked successfully for other national surveillance and monitoring schemes: members of the public and volunteer recorders pass their moth observations to volunteer county moth recorders and, in due course, collated, computerised and verified county datasets of moth records are passed to the NMRS database. Feedback was disseminated effectively in the opposite direction, from the Moths Count project team to county moth recorders and then to volunteer recorders, local communities and wider audiences.

Commencing in 2007, a network of county moth recorders was established to form the backbone of the NMRS. Some of these highly-skilled volunteers were already in place at the outset of the Moths Count project, others had to be recruited. A national database was established by Butterfly Conservation and guidance provided to county moth recorders. The first county dataset of macro-moth records was incorporated into the NMRS database at the end of 2007 and it has grown rapidly since then (see Results). It is not possible at this early stage in the NMRS to undertake full analysis of every macro-moth species. However, here we present some preliminary analyses utilising the NMRS database, and of relevance to issues of biodiversity loss and climate change.

Northward shifts in range margins of British macro-moths

With much recent evidence of poleward range margin shifts among terrestrial organisms (e.g. Parmesan et al. 1999; Hickling et al. 2006; Hitch and Leberg 2007; Huntley et al. 2008), we examined the distribution data in the NMRS for 12 species of macro-moth, each with a southern distribution in Great Britain. These species were not chosen at random or systematically. Anecdotal evidence from the 'grey' literature, suggested that each of these species had moved northwards, but change in the northern range margins of these moths had not been assessed quantitatively. Eight of the macro-moths selected were long-term resident species in Great Britain, with populations supported by native host plants. Four of the species, in contrast, colonised Britain during the twentieth century and are wholly or largely reliant on alien host plant species. Blair's shoulder-knot *Lithophane leautieri* and cypress pug *Eupithecia phoeniceata* use non-native cypress trees and shrubs in the family Cupressaceae as larval host plants, varied coronet *Hadena compta* primarily utilises sweet William *Dianthus barbatus* and toadflax brocade *Calophasia lunula* uses the naturalised garden-escape purple toadflax *Linaria purpurea* (as well as native plants, including common toadflax *L. vulgaris* and pale toadflax *L. repens*).

We calculated the position of the species' northern range margin in Great Britain (i.e. excluding Northern Ireland, the Isle of Man and Channel Islands) in two separate time periods (1960–1982 and 1983–2009), corresponding to previous survey activity. In each time period, the range margin was determined as the average latitude of the 10 most northerly occupied 10 km × 10 km grid squares of the Ordnance Survey National Grid of Great Britain (following the method developed by Hill et al. 2002). The shift in range margin for each species was calculated as the difference (in km) between the locations of the range margin in the two time periods.

The NMRS database was used to determine range margins in the 1983–2009 period. For the earlier period the maps published in Heath and Emmet (1983a, b) were used, where possible. However, for the three Geometridae species selected (*Eupithecia phoeniceata*, dwarf cream wave *Idaea fuscovenosa* and least carpet *I. rusticata*) no such maps have been published, so recently-computerised, historical data from the Biological Records Centre were used via www.nbn.org.uk.

Distribution change of UK macro-moths

A key aim of the NMRS is to determine distribution trends over time for all macro-moth species. It is premature to undertake this large-scale analysis at present as the spatial and temporal coverage of the NMRS database is insufficient. Datasets of this kind are also subject to a variety of biases stemming from the opportunistic nature of recording and substantial variation in observer effort over time and space (see Discussion). These need to be accounted for in a full analysis of macro-moth distribution trends over time. However, with these caveats, the data gathered to date can be used to examine preliminary trends.

We selected 40 macro-moth species to examine distribution change from the provisional NMRS data. These were chosen from the species showing the greatest increases and decreases in population index based on the analysis of Rothamsted Insect Survey data in Conrad et al. (2006). Strong positive correlation has been shown for UK butterflies between population trends and distribution trends (Warren et al. 2001), and we hypothesise that the same relationship will hold for macro-moths. We cannot yet test this hypothesis, but examining patterns of distribution change at the extremes of recorded population trends may be indicative. We also assessed distribution change for three species that were considered stable. Two of these, scorched wing *Plagodis dolabraria* and flame shoulder *Ochropleura plecta*, had extremely low annual change rates in the 35 year population trends calculated by Conrad et al. (2006). The third 'stable' species, large yellow underwing *Noctua pronuba*, is a ubiquitous species found commonly throughout Britain and Ireland. The actual distribution of this species cannot increase substantially (as it already occurs almost everywhere), so any change in the recorded distribution will give information about bias in the NMRS database.

Distribution change was examined from the NMRS database. Extent of distribution for each of the 43 species was determined as the number of 10 km × 10 km grid squares in which the species was recorded at least once in a given time period. Two time periods were used, pre-2000 and 2000 onwards. These periods were chosen with reference to over-representation of recent moth records in the NMRS (see Results). The proportional change in distribution was calculated as the number of occupied grid squares 2000 onwards minus the number of pre-2000 occupied squares, divided by the number of pre-2000 occupied squares.

Results

Promotion, feedback and training of volunteer scientists

These aspects of the project were successful. The UK media, which had previously reflected the generally negative attitude of the public towards moths (especially in contrast to butterflies, New 2004), took interest in Moths Count from the outset. The project achieved seven appearances on national television and 18 on national radio, as well as 32 articles in national newspapers and over 60 articles in popular magazines. In one memorable newspaper headline, moths were described as the "Glory of our gardens" (Daily Mail 21 June 2008) to over two million readers. Journalists and producers now actively seek moth-related news and feature items, a sea change in attitude for which the Moths Count project can take some credit.

The public was also actively engaged. Ninety-five public moth events were held across the UK (2007–2009) attended by a total of 2,010 people and involving many partner organisations. At least 30% of participants were women and over a quarter were under 16. Over 100,000 leaflets were distributed and the project website received 17,500 hits in its first 8 months of operation (July 2009–March 2010). Simple online surveys of widespread and easily identified moths (e.g. humming-bird hawk-moth *Macroglossum stellatarum*) also proved effective. In 2009, over 2,300 people contributed to these surveys (www.butterfly-conservation.org).

Training courses were run throughout the UK and exceeded the project target. By the end of 2009, 110 training events had been held as part of the Moths Count project, attended by 1,481 people. The quality and appropriateness of the training provided was highly rated by participants. Feedback questionnaires revealed a mean annual satisfaction score of 95%. Many training courses focused on UK Biodiversity Action Plan priority species, including slender scotch burnet *Zygaena loti*, chalk carpet *Scotopteryx bipunctaria*, straw belle *Aspitates gilvaria*, narrow-bordered bee hawk-moth *Hemaris tityus*, marsh moth *Athetis pallustris* and four-spotted *Tyta luctuosa* among others, thus contributing towards conservation targets for these threatened species.

The National Moth Recording Scheme

The first local dataset of approximately 161,000 records (unique combinations of species occurrence, spatial location and date) was imported into the NMRS database in December 2007. Just over 2 years later, by March 2010, the NMRS comprised 8.2 million distribution records of macro-moths, making it the largest dataset for any animal taxon on the UK's National Biodiversity Network Gateway (www.nbn.org.uk). Ninety-three percent of counties had contributed datasets to the NMRS and the project aims to achieve complete county-level coverage during the first half of 2010.

At this stage the NMRS database shows a strong bias towards recent records (Fig. 1). The records spanned the period 1769 (a record of death's head hawk-moth *Acherontia atropos*) to 2009, but 61% of records were from the period 2000–2009 and only 9% were pre-1980. The relative paucity of data for 2008 and 2009 in the NMRS reflects time lags in the flow of records from field recorder, via county moth recorder, to the national scheme.

The bias towards modern records is in part a reflection of increasing recording effort over time, but is also an artefact of NMRS policy. Guidance given to county moth recorders prioritised recent records, recommending that they worked through moth records progressively backwards in time. The representation of historical records in the NMRS will increase (although the relative proportion may not), as county recorders alleviate backlogs of recent data and move onto older records. Furthermore, substantial centralised datasets of historical records were repatriated to county recorders by the Moths Count project during 2009 and 2010, and these will enter the NMRS database in due course.

Overall geographical coverage in the NMRS database was already comprehensive by March 2010 (Fig. 2). The aim of any such scheme is to determine the actual distribution of species in the landscape rather than simply mapping the location of active recorders. The widespread coverage of data collated into the NMRS in the short period of time since the recording scheme commenced, gives

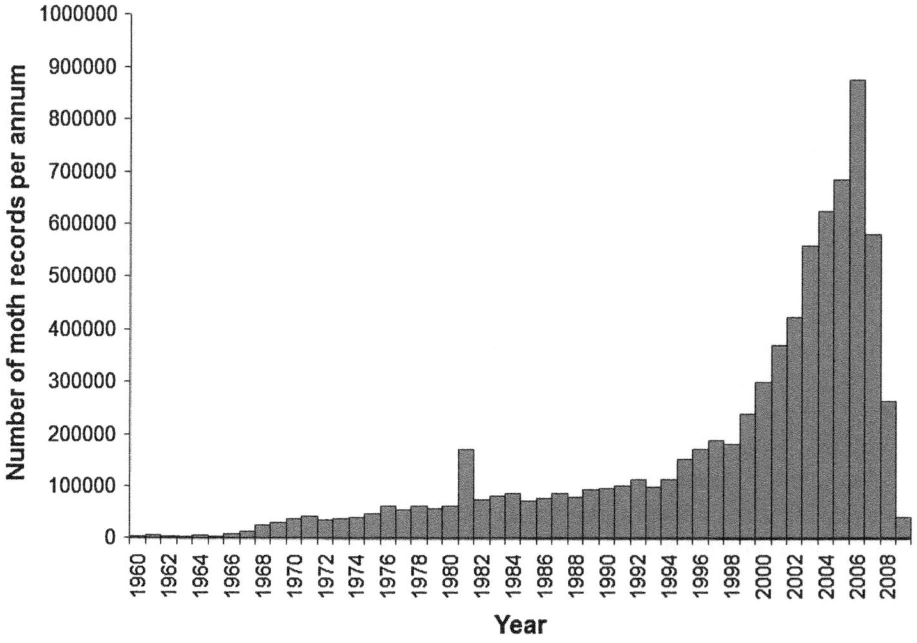

Fig. 1 The number of macro-moth distribution records per year (1960–2009) in the National Moth Recording Scheme database at March 2010. The relative paucity of records from 2008 and 2009 reflects a time lag in data flow through the recording scheme

confidence that robust assessments of species' status and trend will be possible in the future.

Northward shifts in range margins of British macro-moths

Each of the 12 macro-moth species selected for assessment of range margin change showed a substantial northward shift (mean = 194.8 km, $n = 12$) between the two date periods (25 years between mid-points) (Table 1). This equates to a mean annual shift of 7.8 km year^{-1} over the 25 years between survey mid-points. The mean range margin shift for macro-moths utilising native host plants (198.8 km, 8.0 km year^{-1}, $n = 8$) was not significantly different from that for those using alien host plants (mean = 187.0 km, 7.5 km year^{-1}, $n = 4$) (*T*-test, $t = 0.28$, $P = 0.78$). Maps showing the northward spread of some of these species are given in (Fig. 3).

Distribution change of UK macro-moths

A simple assessment of species distribution trend from the NMRS database revealed large-scale changes between the pre-2000 and 2000 onwards recording periods (Table 2). Each of the 20 macro-moth species selected because of large population index increases in Conrad et al. (2006) showed a positive distribution trend (mean = 64%, $n = 20$). The three species expected to indicate stability also showed substantial distribution increases (mean = 46%, $n = 3$). In contrast, the 20 species selected as examples of macro-moths with the greatest population declines in Conrad et al. (2006), had an overall decrease in distribution

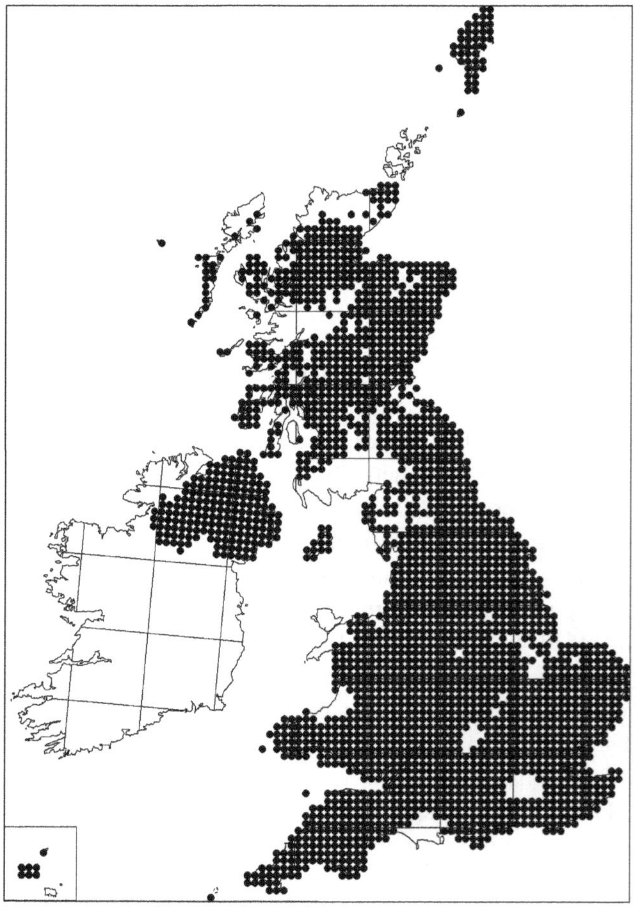

Fig. 2 Map of Britain and Ireland showing coverage of all macro-moth records (1769–2009), at 10 km grid square resolution, in the National Moth Recording Scheme database at March 2010. Major gaps represent local datasets that have yet to be incorporated into the database. The Republic of Ireland is not included in the NMRS

Table 1 Northward range margin shifts of 12 macro-moth species in Great Britain between 1960–1982 and 1983–2009

Species	Northwards range margin shift (km)
(a) Species using native host plants	
Red-necked footman *Atolmis rubricollis*	393
Shuttle-shaped dart *Agrotis puta*	262
Vine's rustic *Hoplodrina ambigua*	220
Buff footman *Eilema depressa*	209
Red underwing *Catocala nupta*	165
Dingy footman *Eilema griseola*	125
Dwarf cream wave *Idaea fuscovenosa*	122
Least carpet *Idaea rusticata*	94
(b) Species using alien host plants	
Blair's shoulder-knot *Lithophane leautieri*	365
Toadflax brocade *Calophasia lunula*	136
Cypress pug *Eupithecia phoeniceata*	130
Varied coronet *Hadena compta*	117

Species are listed in order of magnitude of range shift within groups utilising native and alien larval host plants

(mean = -14%, $n = 20$). Maps illustrating the distribution decreases for double dart *Graphiphora augur* and v-moth *Macaria wauaria* are shown in Fig. 4. Nevertheless, five species in this group had increased in distribution (but see discussion).

Discussion

Moths are often regarded as the poor cousins of butterflies. This attitude is especially prevalent among the public and in the media, but is also encountered among scientists and policy makers, where butterflies are firmly established as model organisms for research (Watt and Boggs 2003) and as flagships and indicators for biodiversity conservation (Samways 2005; van Swaay et al. 2008). Yet moths, because of their diversity and abundance, are of much greater functional significance in ecosystems (Proctor et al. 1996; Vaughan 1997; Wilson et al. 1999; Wickramasinghe et al. 2004; Fox et al. 2006b).

We have shown, through the Moths Count project, the potential to challenge negative attitudes successfully in the media and engage large numbers of the public in the study of macro-moths. We have also, in a short space of time, collated a substantial time-series of species occurrence data at the national scale in the UK. Despite the existence of a population monitoring scheme (the Rothamsted Insect Survey) covering macro-moths, these new data, collated from thousands of volunteers through the NMRS, are of great importance for biodiversity conservation. Robust quantitative trends in distribution over time have not been published for any species-rich terrestrial invertebrate taxon in the UK (although they have for marine zooplankton e.g. Beaugrand and Reid 2003; Beaugrand et al. 2009), but the NMRS will enable this to be done in the near future for c.900 species of macro-moth. Furthermore, the detailed spatial information in the NMRS can be applied in numerous ways to target and implement measures to reduce and reverse biodiversity loss.

Each of the 12 macro-moth species we examined for range margin change showed a strong northward shift. This was to be expected since we had selected them, from a pool of some 900 species, specifically because northward colonisation had been reported in personal communication, the 'grey' literature and recent field guides (e.g. Waring et al. 2009). Nevertheless, these range margin shifts have now been measured quantitatively and they add to and support the growing body of evidence for poleward shifts in species' distributions during a period of climatic warming (Parmesan and Yohe 2003). Butterflies have been particularly well-studied in this context both in Britain (Hill et al. 1999; Warren et al. 2001; Hill et al. 2002; Willis et al. 2009) and elsewhere (Parmesan et al. 1999; Crozier 2003, 2004; Mitikka et al. 2008; Pöyry et al. 2009). However, the total numbers of species assessed remain relatively low. The most inclusive of these analyses identified northward shifts in the range margins (either northern or southern limits of range) for only 36 species of butterflies (Parmesan et al. 1999). In Britain, Hill et al. (2002) found that 11 of 46 southerly distributed butterflies had extended their range northwards. Thus, in our provisional, selective and incomplete assessment of macro-moth range margins, we have identified more northward shifts than have been previously identified for butterflies in Britain. Hickling et al. (2006) conducted a range margin assessment for 329 species in 16 taxonomic groups in Britain, including 177 insect species in eight taxa (not including moths). The northern range margins of a total of 275 species moved northwards, including 151 insects (CD Thomas pers. comm.). When, in due course, we analyse range margin shifts for all southerly distributed macro-moths in Britain, the results may be expected to add significantly to the total number of insect examples currently in the literature.

The range margin shifts recorded for macro-moths in this study are towards the upper end of measurements from British butterfly studies. The 170 km northward shift (8.1 km year^{-1}) of the comma butterfly *Polygonia c-album* (between the periods 1970–1982 and 1995–1999) was the only species to exceed 100 km in the study by Hill et al. (2002). The common darter dragonfly *Sympetrum striolatum*, on the other hand, had shifted 346 km (13.8 km year^{-1}) northwards in Britain (between 1960–1970 and 1985–1995), and two other Odonata species by almost 200 km, rivalling the range extensions we recorded for

Fig. 3 Maps of selected macro-moth species that show substantial northward shifts of range margin (Table 1). *Black dots* indicate all 10 km grid squares with records of the species 1960–1982; *grey dots* indicate 10 km grid squares with only 1983–2009 records of the species in the NMRS database. Species are as follows: **a** red-necked footman *Atolmis rubricollis*, **b** shuttle-shaped dart *Agrotis puta*, **c** vine's rustic *Hoplodrina ambigua*, **d** buff footman *Eilema depressa*, **e** red underwing *Catocala nupta*, **f** dingy footman *Eilema griseola*, **g** Blair's shoulder-knot *Lithophane leautieri*, **h** toadflax brocade *Calophasia lunula*, **i** varied coronet *Hadena compta*. Species **a**–**f** utilise native larval host plants, while species **g**–**i** wholly or partially utilise alien host plant species. It should be noted that the data shown on these maps, drawn from the NMRS database, are not identical to those used to assess range margin shift, taken from Heath and Emmet (1983a, b)

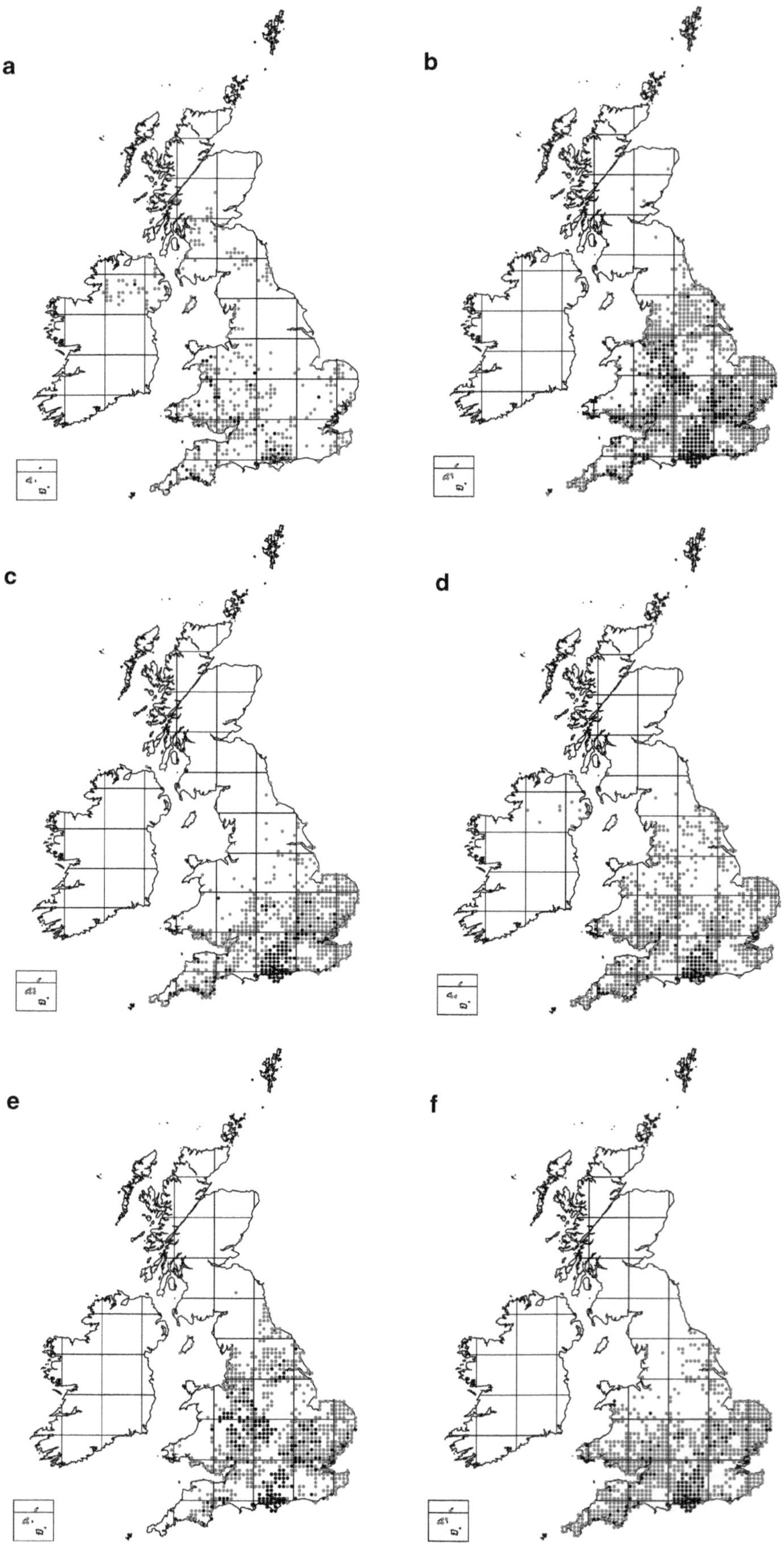

Fig. 3 continued

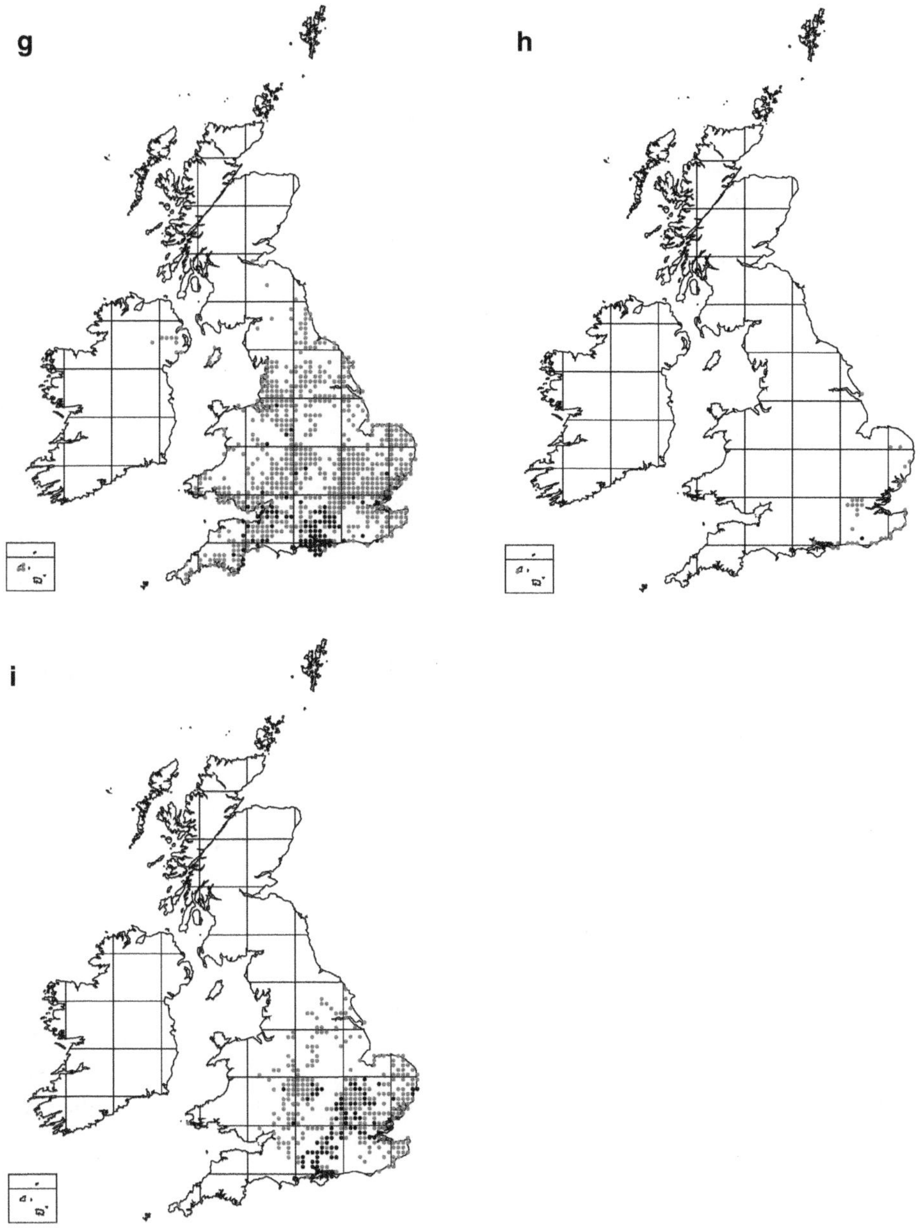

Atolmis rubricollis (15.7 km year^{-1}) and *Lithophane leautieri* (14.6 km year^{-1}) (Hickling et al. 2005).

Distribution change over the 25 year period between the mid-points of sampling periods 1960–1982 and 1983–2009, was assessed for 40 macro-moths likely to have shown substantial change and three further species expected to have changed little. Measuring distribution change over time from recording scheme datasets is problematic. In most such schemes, including the butterfly recording scheme for Britain and Ireland (Fox et al. 2006a), species occurrence records are collected opportunistically by volunteer recorders, with little direction and no random or systematic sampling of the landscape. It has been established that such data may contain bias emanating from uneven geographical recording effort (e.g. as a result of access and favoured 'hot spot' areas) (Dennis and Thomas 2000), differences in the apparency of species due to size, colouration and behaviour (Dennis et al. 2006), uneven recording over time (Dennis et al. 1999; Dennis 2001) and issues of scale (Thomas and Abery 1995). These causes of bias have not been investigated in detail in the NMRS database, although the issue of uneven recording over time is clear (Fig. 1).

In light of this, the distribution change analyses reported in this paper must be viewed tentatively, as early indications of patterns of change from a dataset that is still being developed (particularly with regard to the historical baseline). We would expect species to show increases in distribution over time simply as a result of the strong bias of records towards the modern period, which has not been

Table 2 Distribution change for 43 species assessed by comparison of the number of occupied 10 km grid squares in the NMRS database pre-2000 and 2000–2009

Species	% Distribution change
(a) Species with greatest increases in population trend	
Buff footman *Eilema depressa*	175
Red-green carpet *Chloroclysta siterata*	122
Least carpet *Idaea rusticata*	110
Dingy footman *Eilema griseola*	100
Spruce carpet *Thera britannica*	88
Blair's shoulder-knot *Lithophane leautieri*	74
Vine's rustic *Hoplodrina ambigua*	73
Grey shoulder-knot *Lithophane ornitopus*	69
Least black arches *Nola confusalis*	60
Pine beauty *Panolis flammea*	55
Satin beauty *Deileptenia ribeata*	51
Broad-bordered yellow underwing *Noctua fimbriata*	50
Treble brown spot *Idaea trigeminata*	45
Scarce footman *Eilema complana*	44
Dwarf cream wave *Idaea fuscovenosa*	32
Dotted carpet *Alcis jubata*	31
Marbled beauty *Cryphia domestica*	31
Juniper carpet *Thera juniperata*	28
Devon carpet *Lampropteryx otregiata*	25
Blue-bordered carpet *Plemyria rubiginata*	11
(b) 'Stable' species	
Scorched wing *Plagodis dolabraria*	50
Flame shoulder *Ochropleura plecta*	45
Large yellow underwing *Noctua pronuba*	41
(c) Species with greatest decreases in population trend	
Dusky thorn *Ennomos fuscantaria*	48
Autumnal rustic *Eugnorisma glareosa*	33
Beaded chestnut *Agrochola lychnidis*	24
Dark-barred twin-spot carpet *Xanthorhoe ferrugata*	21
Deep-brown dart *Aporophyla lutulenta*	5
Feathered gothic *Tholera decimalis*	−6
Grass rivulet *Perizoma albulata*	−10
Lackey *Malacosoma neustria*	−10
White-line dart *Euxoa tritici*	−12
Spinach *Eulithis mellinata*	−19
Dark spinach *Pelurga comitata*	−20
Anomalous *Stilbia anomala*	−21
Hedge rustic *Tholera cespitis*	−22
September thorn *Ennomos erosaria*	−24
Figure of eight *Diloba caeruleocephala*	−30
Dusky-lemon sallow *Xanthia gilvago*	−33
Garden dart *Euxoa nigricans*	−36
Brindled ochre *Dasypolia templi*	−46
Double dart *Graphiphora augur*	−48
V-moth *Macaria wauaria*	−77

Species are shown in order of magnitude of distribution change (from increase to decrease) within categories selected for study

accounted for in our analysis. The positive distribution trends (mean = 46% increase) measured for the 'stable' species *Plagodis dolabraria, Ochropleura plecta* and *Noctua pronuba* support this hypothesis and are indicative of what might be expected for species whose distributions have not actually changed over time. This, therefore, gives greater confidence that the distribution declines presented here are qualitatively correct and likely to be conservative estimates of change. All but one of the 20 species with large population declines (as well as eight of the species with large population increases) had distribution changes below this 46% threshold and are likely to have declined in real terms.

Despite the provisional nature of this analysis, it is clear that some macro-moth species have undergone substantial changes in distribution, both positive and negative, within the UK. Caution must be exercised in interpreting the results, but the distribution increases and decreases of macro-moths assessed here are in keeping with trends found in other studies in the UK and elsewhere. All of the moths for which distribution change was calculated are considered widespread and common species (a prerequisite of inclusion in the Rothamsted population trend study, Conrad et al. 2004). The greatest decreases measured here are comparable with the most rapid declines recorded among highly threatened butterflies species in Britain (Fox et al. 2006a). *Macaria wauaria* (77% decrease) compares to the 79% decline of high brown fritillary *Argynnis adippe* over a similar period, while *Graphiphora augur* (48% decrease) and *Dasypolia templi* (46% decrease) compare to the 46% decline of marsh fritillary *Euphydryas aurinia*. Both these butterflies are listed as priority species in the UK Biodiversity Action Plan and assessed as Critically Endangered and Vulnerable, respectively, in the Red List of British butterflies (Fox et al. 2010).

Conrad et al. (2006) reported population declines in 66% of a sample of 337 widespread macro-moth species in Britain, with 21% (71 species) decreasing rapidly (>30% 10 $year^{-1}$). Overall decreases in moth biodiversity have also been reported from The Netherlands (Groenendijk and van der Meulen 2004; Groenendijk and Ellis in press) and Finland (Mattila et al. 2006, 2008). Warren et al. (2001) found that the distributions of 74% of 46 butterfly species had declined in Britain since the 1970 s, while Fox et al.

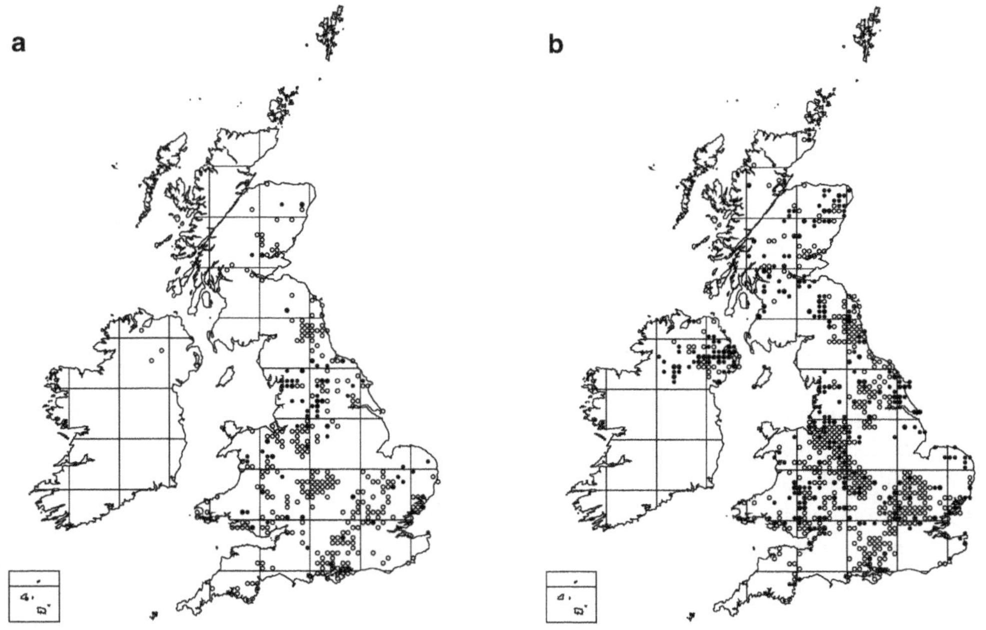

Fig. 4 Maps showing distribution decreases of two macro-moth species in the UK: **a** V-moth *Macaria wauaria*, 77% decrease in occupied squares, **b** double dart *Graphiphora augur*, 48% decrease. *Solid dots* indicate all 10 km grid squares with 2000–2009 records of the species in the NMRS database; *open dots* indicate 10 km grid squares with only pre-2000 occurrence records of the species

(2006a) reported decreases for 76% of 54 resident British butterflies, 14 of which had lost >40% of their distribution since the 1970 s. The butterfly faunas of Flanders (northern Belgium) and The Netherlands appear to have fared even worse (Maes and Van Dyck 2001; Van Dyck et al. 2009), while van Swaay et al. (2006, 2010) reported overall decreases of butterfly distributions at the European scale.

Simultaneously, other species have increased their distributions, often, but not exclusively, including a shift in range margin. The buff footman *Eilema depressa*, least carpet *Idaea rusticata* and dingy footman *E. griseola*, for example, have each doubled their recorded distribution and spread northwards by approximately 100 km. Whilst the distribution trends presented in this paper are by no means an exhaustive treatment, they do provide tentative support for a correlation between changes at the population level and distribution level, as has been shown previously for British butterflies (Warren et al. 2001).

The development of analytical methods for assessing change from biased datasets of species occurrence records is an area of intense research activity at present. Approaches dealing with static 'snap-shot' distribution data (e.g. Wilson et al. 2004) and time-series data (e.g. Maes and van Swaay 1997; Telfer et al. 2002; Dennis and Shreeve 2003; Thomas et al. 2004b; Fox et al. 2006a; Menéndez et al. 2006) provide a rich resource for full analysis of UK macro-moths once gross geographical and temporal coverage issues have been resolved.

Little research has been undertaken as yet into the causes of change in the extent and range margins of macro-moths in Britain. Climate has been implicated in the distribution decline of the garden tiger *Arctia caja* in Britain (Conrad et al. 2002, 2003), in the pattern of immigration to Britain (Sparks et al. 2005) and in altitudinal shifts of tropical moths in Borneo (Chen et al. 2009). Both the range margin shifts of the eight macro-moths that use native host plants and some of the distribution increases examined here are consistent with responses to recent climate change. However, there may be other causal factors. Four of the species examined for range margin shift utilise alien host plants exclusively or primarily. *Lithophane leautieri*, *Calophasia lunula*, *Eupithecia phoeniceata* and *Hadena compta* have colonised Great Britain and extended their range margins as a result of the widespread cultivation of their host plants, principally in domestic gardens. The potential influence of human planting of alien host plants can also be seen among the group of moths showing substantial increases in distribution size. For example, *Thera britannica*, *Panolis flammea*, *Deileptenia ribeata* and *T. juniperata* each make use of alien conifer species, which have been widely planted in commercial forests and domestic gardens.

Habitat loss or degradation is a likely causal factor in distribution decreases, as has been found for biodiversity worldwide. For example, Merckx et al. (2009) found positive effects of landscape features including hedgerow trees and wide, uncultivated field margins on macro-moth abundance and diversity in British farmland. Macro-moth population data from the Rothamsted Insect Survey also provide evidence of declines in response to agricultural improvement (Woiwod and Harrington 1994).

Furthermore, the trends we present here hint at causal factors relating to pollution. Several species expanding their distributions and range margins (e.g. *Eilema depressa*, *E. griseola* and *Atolmis rubricollis*) use lichens and algae as larval hosts, raising the potential of air quality as a driver of recent increases. In contrast, decreasing species such as *Macaria wauaria* and *Eulithis mellinata* utilise currants *Ribes* spp. and may be impacted by increased use of pesticides. Finally, light pollution should not be ruled out as a potential cause of biodiversity loss (Eisenbeis 2006; Frank 2006), although research in this area is urgently required (Sutherland et al. 2006).

Instigated by the Moths Count project, the National Moth Recording Scheme has, in just 27 months, collated one of the UK's largest datasets of species' occurrence records and one that will continue to grow rapidly. These data will be critical to understand the rapid changes taking place in populations, ranges and ecological communities, and to halt and reverse the biodiversity losses that increasingly threaten human welfare. Political, as well as scientific, commitment is needed and, therefore, public concern and support is important. The Moths Count project has engaged successfully with both politicians and the public, raising awareness of the issues among millions of people via the media and involving thousands directly in moth recording for conservation in the UK.

Acknowledgments We are very grateful to the thousands of volunteers past and present, especially the county moth recorders, for collecting the distribution records now collated into the NMRS. We would also like to thank the many partner organisations and individuals for their support of the Moths Count project. Principal funders included the Heritage Lottery Fund, Butterfly Conservation, Environment Agency, Natural England, Countryside Council for Wales, City Bridge Trust, Northern Ireland Environment Agency, Royal Entomological Society, Scottish Natural Heritage, Royal Society for the Protection of Birds, D'Oyly Carte Charitable Trust, The 1989 Willan Charitable Trust, John Spedan Lewis Foundation and British Entomological and Natural History Society. Many other organisations are involved, providing support and helping to host events. A full list of partners is provided at www.mothscount.org. We thank Martin Warren for guidance, Roger Dennis for statistical advice and two anonymous referees for constructive comments. RF, ZR, LH, SA and LW were funded by Heritage Lottery Fund project HG-05-00610.

References

Baillie JEM, Hilton-Taylor C, Stuart SN (2004) 2004 IUCN Red List of threatened species. A global species assessment. IUCN, Gland and Cambridge

Balmford A, Bond W (2005) Trends in the state of nature and their implications for human well being. Ecol Lett 8:1218–1234

Beaugrand G, Reid PC (2003) Long-term changes in phytoplankton, zooplankton and salmon related to climate. Glob Change Biol 9:801–817

Beaugrand G, Luczak C, Edwards M (2009) Rapid biogeographical plankton shifts in the North Atlantic Ocean. Glob Change Biol 15:1790–1803

Bell S, Marzano M, Cent J, Kobierska H, Podjed D, Vandzinskaite D, Reinert H, Armaitiene A, Grodzińska-Jurczak M, Muršič R (2008) What counts? Volunteers and their organisations in the recording and monitoring of biodiversity. Biodivers Conserv 17:3443–3454

Benton TG, Bryant DM, Cole L, Crick HQP (2002) Linking agricultural practice to insect and bird populations: a historical study over three decades. J Appl Ecol 39:673–687

Chen I-C, Shiu H-J, Benedick S, Holloway JD, Khen Chey V, Barlow HS, Hill JK, Thomas CD (2009) Elevation increases in moth assemblages over 42 years on a tropical mountain. Proc Nat Acad Sci 106:1479–1483

Conrad KF, Woiwod IP, Perry JN (2002) Long-term decline in abundance and distribution of the garden tiger moth (*Arctia caja*) in Great Britain. Biol Conserv 106:329–337

Conrad KF, Woiwod IP, Perry JN (2003) East Atlantic teleconnection pattern and the decline of a common arctiid moth. Glob Change Biol 9:13–125

Conrad KF, Woiwod IP, Parsons M, Fox R, Warren M (2004) Long-term population trends in widespread British moths. J Insect Conserv 8:119–136

Conrad KF, Warren M, Fox R, Parsons M, Woiwod IP (2006) Rapid declines of common, widespread British moths provide evidence of an insect biodiversity crisis. Biol Conserv 132:279–291

Conrad KF, Fox R, Woiwod IP (2007) Monitoring biodiversity: measuring long-term changes in insect abundance. In: Stewart AJA, New TR, Lewis OT (eds) Insect conservation biology. Royal Entomological Society, London, pp 203–225

Crozier L (2003) Winter warming facilitates range expansion: cold tolerance of the butterfly *Atalopedes campestris*. Oecologia 135:648–656

Crozier LG (2004) Field transplants reveal summer constraints on a butterfly range expansion. Oecologia 14:148–157

De Heer M, Kapos V, Ten Brink BJE (2005) Biodiversity trends in Europe: development and testing of a species trend indicator for evaluating progress towards the 2010 target. Phil Trans R Soc B 360:297–308

Dennis RLH (2001) Progressive bias in species status is symptomatic of fine-grained mapping units subject to repeated sampling. Biodivers Conserv 10:483–494

Dennis RLH, Shreeve TG (2003) Gains and losses of French butterflies: tests of predictions, under-recording and regional extinction from data in a new atlas. Biol Conserv 110:131–139

Dennis RLH, Thomas CD (2000) Bias in butterfly distribution maps: the influence of hot spots and access. J Insect Conserv 4:73–77

Dennis RLH, Sparks TH, Hardy PB (1999) Bias in butterfly distribution maps: the effects of sampling effort. J Insect Conserv 3:33–42

Dennis RLH, Shreeve TG, Isaac NJB, Roy DB, Hardy PB, Fox R, Asher J (2006) The effects of visual appearance on bias in butterfly recording and monitoring. Biol Conserv 128:486–492

Dunn RR (2005) Modern insect extinctions, the neglected majority. Conserv Biol 19:1030–1036

Eisenbeis G (2006) Artifical night lighting and insects: attraction of insects to streetlamps in a rural setting in Germany. In: Rich C, Longcore T (eds) Ecological consequences of artificial night lighting. Island Press, Washington, pp 281–304

Fleishman E, Murphy DD (2009) A realistic assessment of the indicator potential of butterflies and other charismatic taxonomic groups. Conserv Biol 23:1109–1116

Fox R, Spalding A, Tunmore M, Parsons M (2005) Planning a new national macro-moth recording scheme. Br J Ent Nat Hist 18:26–36

Fox R, Asher J, Brereton T, Roy D, Warren M (2006a) The state of butterflies in Britain and Ireland. Pisces Publications, Newbury

Fox R, Conrad KF, Parsons MS, Warren MS, Woiwod IP (2006b) The state of Britain's larger moths. Butterfly Conservation and Rothamsted Research, Wareham

Fox R, Warren MS, Brereton TM (2010) A new red list of British butterflies. Species status no. 12. Joint Nature Conservation Committee, Peterborough

Frank KD (2006) Effects of artificial night lighting on moths. In: Rich C, Longcore T (eds) Ecological consequences of artificial night lighting. Island Press, Washington, pp 305–344
González-Megías A, Menéndez R, Roy D, Brereton T, Thomas CD (2008) Changes in the composition of British butterfly assemblages over two decades. Glob Change Biol 14:1464–1474
Groenendijk D, Ellis WN (in press) The state of the Dutch larger moth fauna. J Insect Conserv
Groenendijk D, van der Meulen J (2004) Conservation of moths in The Netherlands: population trends, distribution patterns and monitoring techniques of day-flying moths. J Insect Conserv 8:109–118
Heath J, Emmet AM (1983a) The moths and butterflies of Great Britain and Ireland volume 9 Sphingidae—Noctuidae (part I). Harley Books, Colchester
Heath J, Emmet AM (1983b) The moths and butterflies of Great Britain and Ireland volume 10 Noctuidae (part II) and Agaristidae. Harley Books, Colchester
Hickling R, Roy DB, Hill JK, Thomas CD (2005) A northward shift of range margins in British Odonata. Glob Change Biol 11:502–506
Hickling R, Roy DB, Hill JK, Fox R, Thomas CD (2006) The distributions of a wide range of taxonomic groups are expanding polewards. Glob Change Biol 12:450–455
Hill JK, Thomas CD, Huntley B (1999) Climate and habitat availability determine 20th century changes in a butterfly's range margins. Proc R Soc B 266:1197–1206
Hill JK, Thomas CD, Fox R, Telfer MG, Willis SG, Asher J, Huntley B (2002) Responses of butterflies to 20th century climate warming: implications for future ranges. Proc R Soc B 269:2163–2171
Hitch AT, Leberg PL (2007) Breeding distributions of North American bird species moving north as a result of climate change. Conserv Biol 21:534–539
Huntley B, Collingham YC, Willis SG, Green RE (2008) Potential impacts of climatic change on European breeding birds. PLoS One. doi:10.1371/journal.pone.0001439
Kremen C, Chaplin-Kramer R (2007) Insects as providers of ecosystem services: crop pollination and pest control. In: Stewart AJA, New TR, Lewis OT (eds) Insect conservation biology. Royal Entomological Society, London, pp 349–382
Kremen C, Lees DC, Fay JP (2003) Butterflies and conservation planning in Madagascar: from pattern to practice. In: Boggs CL, Watt WB, Ehrlich PR (eds) Butterflies - ecology and evolution taking flight. University of Chicago Press, Chicago, pp 517–540
Kühn E, Feldmann R, Harpke A, Hirneisen N, Musche M, Leopold P, Settele J (2008) Getting the public involved in butterfly conservation: Lessons learned from a new monitoring scheme in Germany. Israel J Ecol Evol 54:89–103
Maes D, Van Dyck H (2001) Butterfly diversity loss in Flanders (North Belgium): Europe's worst case scenario? Biol Conserv 99:263–276
Maes D, Van Dyck H (2005) Habitat quality and biodiversity indicator performances of a threatened butterfly versus a multispecies group for wet heathland in Belgium. Biol Conserv 123:177–187
Maes D, Van Swaay CAM (1997) A new methodology for compiling national Red Lists applied to butterflies (Lepidoptera, Rhopalocera) in Flanders (N-Belgium) and the Netherlands. J Insect Conserv 1:113–124
Mattila N, Kaitala V, Komonen A, Kotiaho JS, Päivinen J (2006) Ecological determinants of distribution decline and risk of extinction in moths. Conserv Biol 20:1161–1168
Mattila N, Kotiaho JS, Kaitala V, Komonen A (2008) The use of ecological traits in extinction risk assessments: a case study on geometrid moths. Biol Conserv 141:2322–2328
May RM (2010) Ecological science and tomorrow's world. Phil Trans R Soc B 365:41–47
Menéndez R, González-Megías A, Hill JK, Braschler B, Willis SG, Collingham Y, Fox R, Roy DB, Thomas CD (2006) Species richness changes lag behind climate change. Proc R Soc B 273:1465–1470
Merckx T, Feber RE, Riordan P, Townsend MC, Bourn NAD, Parsons MS, Macdonald DW (2009) Optimizing the biodiversity gain from agri-environment schemes. Agric Ecosyst Environ 130:177–182
Millennium Ecosystem Assessment (2005) Ecosystems and human well-being: synthesis. Island Press, Washington DC
Mitikka V, Heikkinen RK, Luoto M, Araújo MB, Saarinen K, Pöyry J, Fronzek S (2008) Predicting range expansion of the map butterfly in Northern Europe using bioclimatic models. Biodivers Conserv 17:623–641
Mooney HA (2010) The ecosystem-service chain and the biological diversity crisis. Phil Trans R Soc B 365:31–39
New TR (2004) Moths (Insecta: Lepidoptera) and conservation: background and perspective. J Insect Conserv 8:79–94
Newman C, Buesching CD, Macdonald DW (2003) Validating mammal monitoring methods and assessing the performance of volunteers in wildlife conservation—"Sed quis custodiet ipsos custodies?". Biol Conserv 113:189–197
Oliver TH, Roy DB, Hill JK, Brereton T, Thomas CD (2010) Heterogeneous landscapes promote population stability. Ecol Lett 13:473–484
Parmesan C, Yohe G (2003) A globally coherent fingerprint of climate change impacts across. Natural systems. Nature 421:37–42
Parmesan C, Ryrholm N, Stefanescu C, Hill JK, Thomas CD, Descimon H, Huntley B, Kaila L, Kullberg J, Tammaru T, Tennant J, Thomas JA, Warren M (1999) Polewards shifts in geographical ranges of butterfly species associated with regional warming. Nature 399:579–583
Parsons MS (2004) The United Kingdom biodiversity action plan moths—selection, status and progress on conservation. J Insect Conserv 8:95–107
Parsons M (2010) The changing moth and butterfly fauna of Britain–the first decade of the twenty-first century (2000–2009). Entomologist's Rec J Var 122:13–22
Pimm SL, Russell GJ, Gittleman JL, Brooks TM (1995) The future of biodiversity. Science 269:347–350
Pollard E, Yates TJ (1993) Monitoring butterflies for ecology and conservation. Chapman & Hall, London
Pöyry J, Luoto M, Heikkinen RK, Kuussaari M, Saarinen K (2009) Species traits explain recent range shifts of Finnish butterflies. Glob Change Biol 15:732–743
Proctor M, Yeo P, Lack A (1996) The natural history of pollination. Harper Collins, London
Ricketts TH, Daily GC, Ehrlich PR (2002) Does butterfly diversity predict moth diversity? Testing a popular indicator taxon at local scales. Biol Conserv 103:361–370
Roy DB, Sparks TH (2000) Phenology of British butterflies and climate change. Glob Change Biol 6:407–416
Roy DB, Rothery P, Brereton T (2007) Reduced-effort schemes for monitoring butterfly populations. J Appl Ecol 44:993–1000
Samways MJ (2005) Insect diversity conservation. Cambridge University Press, Cambridge
Schmeller DS, Henry P-Y, Julliard R, Gruber B, Clobert J, Dziock F, Lengyel S, Nowicki P, Déri E, Budrys E, Kull T, Tali K, Bauch B, Settele J, van Swaay C, Kobler A, Babij V, Papastergiadou E, Henle K (2009) Advantages of volunteer-based biodiversity monitoring in Europe. Conserv Biol 23:307–316
Shortall CR, Moore A, Smith E, Hall MJ, Woiwod IP, Harrington R (2009) Long-term changes in the abundance of flying insects. Insect Conserv Divers 2:251–260

Spalding A, Tunmore M, Parsons M, Fox R (2005) The state of moth recording in Britain: the results of the national macro-moth recording scheme consultation questionnaire. Atropos 24:9–19
Sparks TH, Roy DB, Dennis RLH (2005) The influence of temperature on migration of Lepidoptera into Britain. Glob Change Biol 11:507–514
Sutherland WJ, Armstrong-Brown S, Armsworth PR, Brereton T, Brickland J, Campbell CD, Chamberlain DE, Cooke AI, Dulvy NK, Dusic NR et al (2006) The identification of 100 ecological questions of high policy relevance in the UK. J Appl Ecol 43:617–627
Telfer MG, Preston CD, Rothery P (2002) A general method for measuring relative change in range size from biological atlas data. Biol Conserv 107:99–109
Thomas JA (2005) Monitoring change in the abundance and distribution of insects using butterflies and other indicator groups. Phil Trans R Soc B 360:339–357
Thomas CD, Abery JCG (1995) Estimating rates of butterfly decline from distribution maps: the effect of scale. Biol Conserv 73:59–65
Thomas CD, Cameron A, Green RE, Bakkenes M, Beamont LJ, Collingham YC, Erasmus BFN, Ferreira de Siqueira M, Grainger A, Hannah L, Hughes L, Huntley B, van Jaarsveld AS, Midgley GF, Miles L, Ortega-Huerta MA, Peterson AT, Phillips OL, Williams SE (2004a) Extinction risk from climate change. Nature 427:145–148
Thomas JA, Telfer MG, Roy DB, Preston CD, Greenwood JJD, Asher J, Fox R, Clarke RT, Lawton JH (2004b) Comparative losses of British butterflies, birds, and plants and the global extinction crisis. Science 303:1879–1881
Van Dyck H, van Strien AJ, Maes D, van Swaay CAM (2009) Declines in common, widespread butterflies in a landscape under intense human use. Conserv Biol 23:957–965
Van Swaay C, Warren MS, Loïs G (2006) Biotope use and trends of European butterflies. J Insect Conserv 10:189–209
Van Swaay CAM, Nowicki P, Settele J, van Strien AJ (2008) Butterfly monitoring in Europe: methods, applications and perspectives. Biodivers Conserv 17:3455–3469
Van Swaay C, Cuttelod A, Collins S, Maes D, Lopez Munguira M, Šašić M, Settele J, Verovnik R, Verstrael T, Warren M, Wiemers M, Wynhof I (2010) European red list of butterflies. Publications Office of the European Union, Luxembourg
Vaughan N (1997) The diets of British bats (Chiroptera). Mamm Rev 27:77–94
Waring P, Townsend M, Lewington R (2009) Field guide to the moths of Great Britain and Ireland, 2nd edn. British Wildlife Publishing, Gillingham
Warren MS, Hill JK, Thomas JA, Asher J, Fox R, Huntley B, Roy DB, Telfer MG, Jeffcoate S, Harding P, Jeffcoate G, Willis SG, Greatorex-Davies JN, Moss D, Thomas CD (2001) Rapid responses of British butterflies to opposing forces of climate and habitat change. Nature 414:65–69
Watt WB, Boggs CL (2003) Synthesis: butterflies as model systems in ecology and evolution—present and future. In: Boggs CL, Watt WB, Ehrlich PR (eds) Butterflies—ecology and evolution taking flight. University of Chicago Press, Chicago, pp 603–613
Wickramasinghe LP, Harris S, Jones G, Jennings N (2004) Abundance and species richness of nocturnal insects on organic and conventional farms: effects of agricultural intensification on bat foraging. Conserv Biol 18:1283–1292
Willis SG, Thomas CD, Hill JK, Collingham YC, Telfer MG, Fox R, Huntley B (2009) Dynamic distribution modelling: predicting the present from the past. Ecography 32:5–12
Wilson JD, Morris AJ, Arroyo BE, Clark SC, Bradbury RB (1999) A review of the abundance and diversity of invertebrate and plant foods of granivorous birds in northern Europe in relation to agricultural change. Agric Ecosyst Environ 75:13–30
Wilson RJ, Thomas CD, Fox R, Roy DB, Kunin WE (2004) Spatial patterns in species distributions reveal biodiversity change. Nature 432:393–396
Woiwod IP, Harrington R (1994) Flying in the face of change: the rothamsted insect survey. In: Leigh R, Johnston A (eds) Long-term experiments in agricultural and ecological sciences. CAB International, Wallingford, pp 321–342

J Insect Conserv (2011) 15:95–101
DOI 10.1007/s10841-010-9326-y

ORIGINAL PAPER

The state of the Dutch larger moth fauna

D. Groenendijk · W. N. Ellis

Received: 24 March 2010 / Accepted: 3 August 2010 / Published online: 22 August 2010

Abstract Moths make up a significant part of the biodiversity both in northwestern Europe and globally. In the Netherlands for instance, 766 species of larger moths are considered native. The interest of amateur recorders in this group in the Netherlands is rapidly growing. This has led to a strong increase in the number of records in the central database, allowing better and more precise calculations on the local trends of moths. About one-third of the species in our analysis are decreasing and could be listed as threatened. On the other hand, some species are strongly increasing. All species together show a significant, declining, trend in abundance. At the same time there exist strong year to year fluctuations in this trend which are due to weather conditions. We conclude, therefore, that the changing climate is having a strong impact on our moth fauna. The absolute number of individuals of the common species has dropped by one-third. We argue that this had a negative impact on plant biodiversity. The need for better protection of moths in northwestern Europe is also implicated by the fact that many species of birds, bats and other groups depend on moths (or their larvae) as main food sources.

Keywords Lepidoptera · The Netherlands · Numerical decline · Climate effect · Performance · Abundance · Insectivorous birds · Phytophagous insects

D. Groenendijk (✉)
Dutch Butterfly Conservation, P.O. Box 506, 6700 AM Wageningen, The Netherlands
e-mail: dick.groenendijk@vlinderstichting.nl

W. N. Ellis
Working Group Lepidoptera Faunistics, Zoological Museum, Section Entomology, Plantage Middenlaan 64, 1018 DH Amsterdam, The Netherlands

Introduction

Moths make up a significant part of the biodiversity both in northwestern Europe and globally (Young 1997). They represent one of the largest insect groups and, therefore, are important elements in many ecosystems at different trophic levels. First of all, moths can act as pollinators for flowering plants (Boggs 1987). In some cases, adult moths are important components of the diets of birds. In addition, many species of bats feed on moths and in some species moths make up a substantial part of their diet (Vaughan 1997). The most important component in ecosystems, however, is the role of larvae in the food chain (Török et al. 2004; Visser et al. 2006). Many bird species eat larvae or feed their young with larvae of moths. In fact, most bird species are, at some time of the year, largely dependent on the presence of moth larvae for survival or successful reproduction (Hogstad 2005). Furthermore, their herbivorous habits often make them effective regulators of vegetation (Bernays et al. 2004; Schowalter 1981).

At the same time, moths have a strong aesthetic appeal, and form one of the insect groups that have attracted most interest among free-time entomologists. Interest among amateur entomologists in the Netherlands in this group is growing rapidly. This is partly due to the publication of field guides (e.g. a Dutch version of Waring and Townsend (2003) was the first available field guide on moths in the Netherlands and was published in 2006) and information on internet sites (e.g. www.vlindernet.nl). These activities are strongly facilitated by activities of Dutch Butterfly Conservation and the Working Group Lepidoptera Faunistics (Groenendijk and van der Meulen 2004). This in turn has resulted in a strong increase in the number of records in our database, allowing better and more precise calculations on trends of moths in the Netherlands. The name of the database is Noctua and this name is used throughout the paper. Noctua is managed by both the Working Group Lepidoptera Faunistics and Dutch Butterfly Conservation. Although Noctua covers faunistical information of all Lepidoptera, this paper focuses on the larger (macro) moths (defined as the species belonging to the families covered in Waring and Townsend (2003)), of which 766 species are considered as native in the Netherlands. Analysis of the data set available through Noctua has generated national population trends for hundreds of larger moths in the Netherlands, which have never been published before.

The aim of this paper is to present the current state of moth biodiversity trends in the Netherlands. We present (1) figures on trends of larger moths in the Netherlands and (2) a comparison with the state of Britain's larger moths as described in Conrad et al. (2006) and Fox et al. (2006).

Methods

The database Noctua, at the time of writing, contains about 2.3 million records, covering faunistic information of all Lepidoptera groups that have occurred in the Netherlands, from about 1800 till the present day. Table 1 summarizes the composition of Noctua. Sources include publications, unpublished records and notebooks, and information provided either directly, or through data collecting websites, in particular www.waarneming.nl. Data are validated in two steps. A first step automatically compares a record with the known flight period and with the national, regional, and local abundance of the species under consideration. Records that do not pass this test are further checked by specialists before inclusion in the database. Note that for the present publication over 1.5 million records on 766 species of larger moths (macrolepidoptera) are used (Table 1).

Table 1 Summary of the composition of the database Noctua

	Numbers of records (*1.000)			
Total number of records	2.270			
Not yet validated, rejected, or 'migrants_only'		230		
Available		2.040		
Microlepidoptera			401	
Rhopalocera			127	
Macrolepidoptera			1.512	
1850–1949 (100 years)				83
1950–1979 (30 years)				308
1980–1989 (10 years)				146
1990–1999 (10 years)				304
2000–2009 (10 years)				671

Note that for this publication only data on larger (macro) moths (macrolepidoptera) are used

It is important to stress that the moth data in Noctua are not the result of any monitoring scheme (cf Woiwod and Harrington 1994). Some volunteers that provide us with data do collect on an almost daily basis, but most data are the results of random excursions, holidays and so on. The calculated trends, therefore, may suffer from differences in the geographical coverage (cf Telfer et al. 2002). However, we believe that the Noctua database is sufficiently large to draw sensible conclusions for particular species, provided that the temporal or spatial scale is not too restricted. This conviction is reinforced by examination of the rank distribution of log-abundance values which does not show a disproportionally high number of very common or very rare species (Fig. 1).

Abundance, nationwide or in a particular region, and always for a specified time period, is calculated by obtaining the sum of the log-transformed number of individuals divided by the number of collection events. In this way the intensity of recording effort is factored out (cf Telfer et al. 2002). A collection event is defined as the event that a person observed one or more moth species at one date at one kilometer square. The calculations are described in detail at www.vlindernet.nl. In the present paper all calculations were limited to the period from 1980 onwards because around 1980 personal computers became available, and field entomologists started in earnest to store their observations in a structured way. Migrants, species

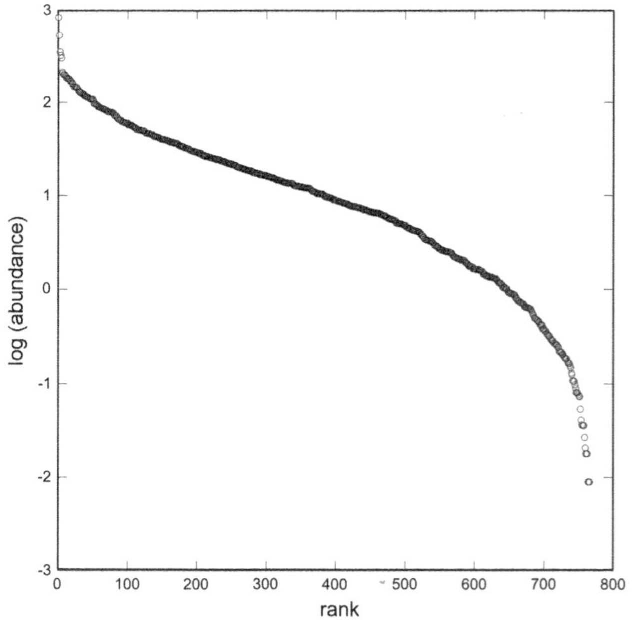

Fig. 1 Distribution of the log-abundance values of all native larger moth species in the Netherlands ($n = 766$)

that periodically moves from one geographical region to another, were left out of the calculations. In these non-resident species their population fluctuations are mostly determined elsewhere.

For those species that were not excessively rare (abundance > 1, $n = 626$), we then calculated their yearly 'performance' defined as the difference of their yearly abundance minus their average abundance, divided by the standard deviation. The phrase performance thus informs how well a species is doing in a given year. Moreover, its value is comparable between species. The slope of the linear regression of a species' performance on year (1980–2009) gives an indication of its rise or decline in that period, its 'population trend'.

A species' population trend and its abundance are the two key components of the degree in which it is under threat. We numerically estimated this threat by applying the rule:

if slope < 0 then threat $=$ slope/abundance else threat $=$ slope $\times$ abundance.

We also calculated the yearly average of the number of adult individuals collected or seen per collection event, for all common (abundance ≥ 100) macro-moths. Collection events in which less than five common species were recorded were disregarded, to exclude reports of only casual, potentially incomplete, observations.

Results

Trends and numbers

Figure 2 shows four examples of individual moth species with decreasing and increasing trends. Both Dark Spinach (*Pelurga comitata*) and Figure of Eight (*Diloba caeruleocephala*) show strong decreasing trends, whereas Orange Footman (*Eilema sororcula*) and Cinnabar (*Tyria jacobaeae*) are increasing.

Figure 3 shows the frequency distribution of all the species specific threat values. The number of threatened

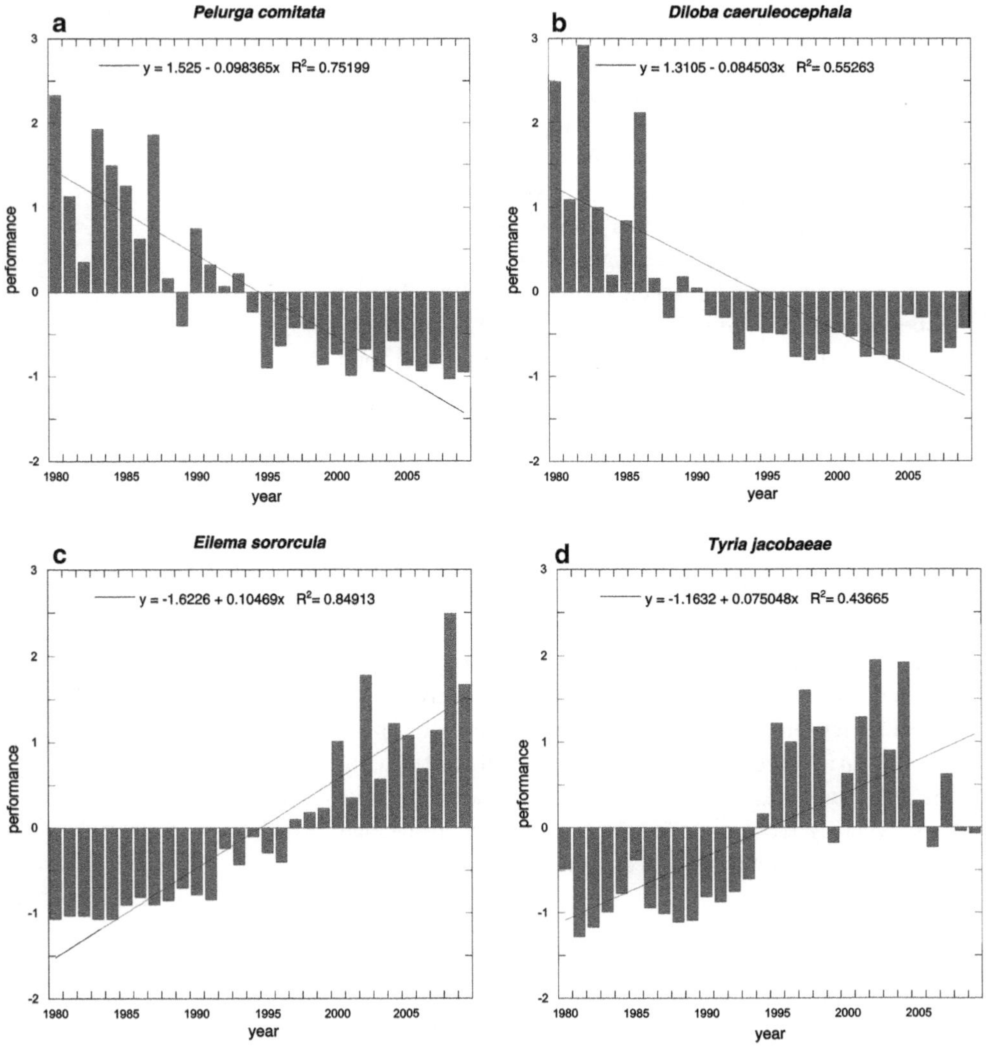

Fig. 2 Yearly performance (1980–2009), defined as the difference of the yearly abundance minus average abundance, divided by the standard deviation of *Pelurga comitata*, *Diloba caeruleocephala*, *Eilema sororcula* and *Tyria jacobaeae* in the Netherlands. The linear regression and its equation is also given

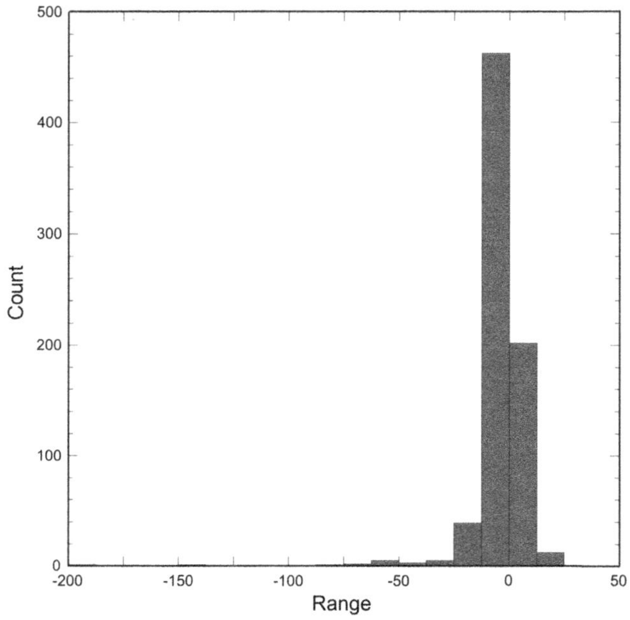

Fig. 3 Frequency distribution of threat values of all species of larger moths in the Netherlands for which these could be calculated ($n = 733$)

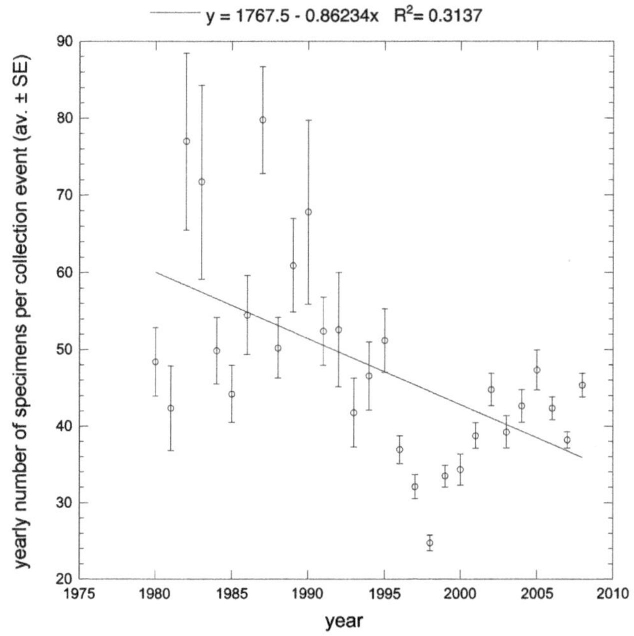

Fig. 4 Average number of individuals per observation per year. *Error bars* present standard errors. The linear regression and its equation is also given

species is about two-third compared with the not threatened ones. Out of a total of 733 species extant in the period 1980-present, 519 show a negative slope. Against 43 species strongly decreasing with a slope less than -0.08 there are only 13 with a strongly increasing slope > 0.08, or in terms of threat, against 91 species with a threat value below -10 there are 19 with a score higher than $+10$ (Fig. 3).

The negative trend in species also holds true for the number of observed individuals. Although subject to yearly fluctuations, the results strongly indicate that the total number of individual larger moths per collection event has dropped from 59 to 37 individuals, a decline of 37% in the period from 1980 onwards (Fig. 4).

Effect of life cycle and abundance

The rate of decline differs among species groups differing in hibernation strategies (Fig. 5). The average slope of all species hibernating as eggs is the lowest, and the average slope of the species overwintering as adults is slightly positive.

Another tendency in the results is the influence of abundance. When all larger moth species are grouped in six equally sized abundance classes, the rarest species show the greatest losses, whereas the most common species are on average stable or show only a slightly decreasing trend (Fig. 6).

Effect of climate change

One could assume, as a nul-hypothesis, that the yearly perfomance of each species is independant of that of any

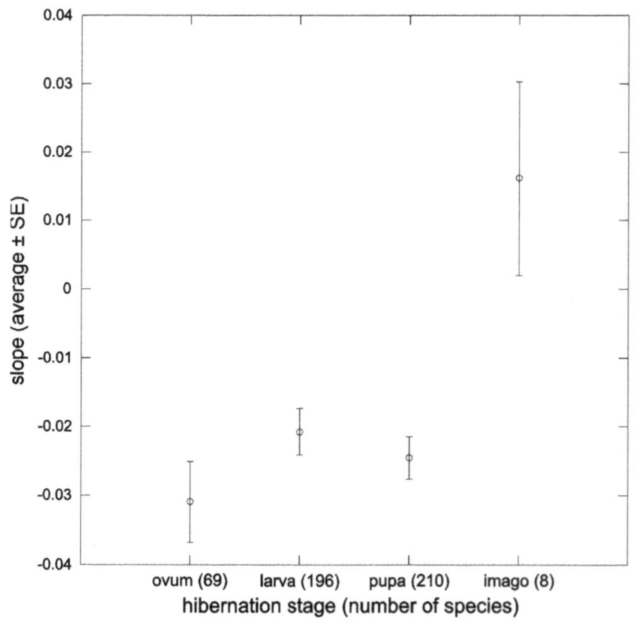

Fig. 5 Relationship between the average of the slopes of the linear regression of the performance on year, and hibernation stage. The number of larger moth species in each category is given in *brackets*. *Error bars* present standard errors

other. In that case the average of the yearly perfomances of all species would be roughly similar. Given the declining trend that has been documented above, one would expect that the course of the averages from 1980 on would show as a smoothly declining line. Contrary to this expectation,

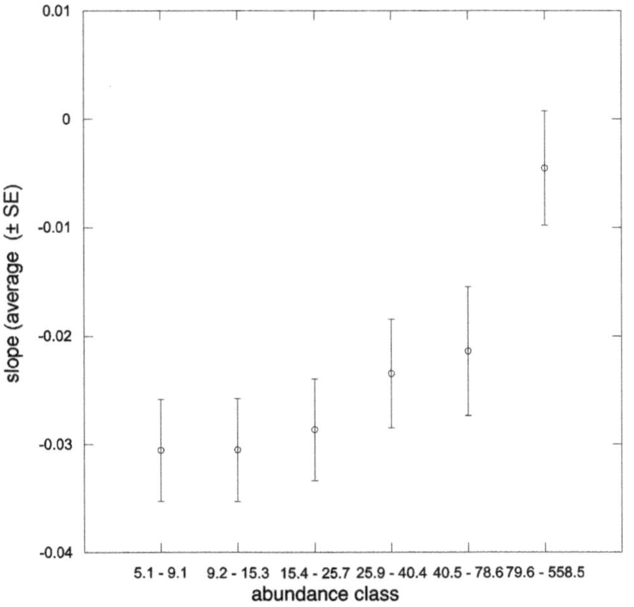

Fig. 6 Relationship between the average of the slopes of the linear regression of the performance on year, and six equally sized abundance classes. The *class limits* are given below. *Error bars* present standard errors

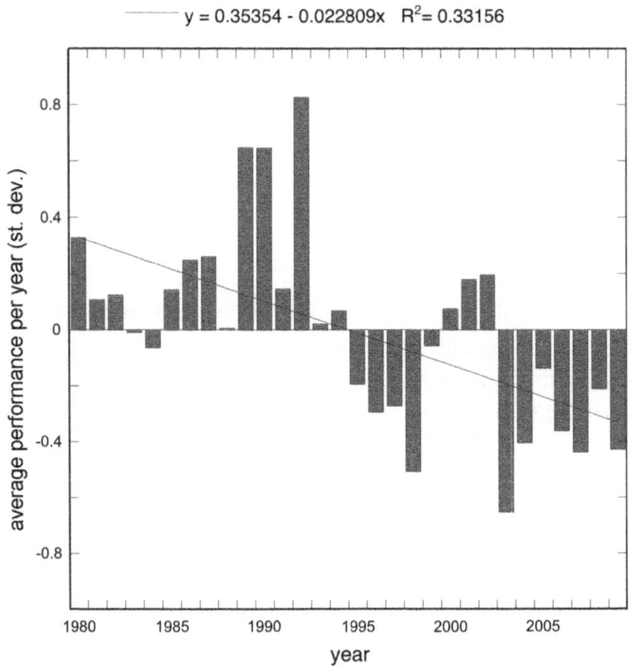

Fig. 7 Yearly averages of the performance of the macro-moths in the Netherlands (rarest species and migrants omitted; $n = 626$). The linear regression and its equation is also given

there is a very strong year-to-year fluctuation (Fig. 7), which is assumed to be largely influenced by climatic variables. Obviously, for the larger moths good and bad years are a reality (anova: $F = 86.78$; $P < 0.001$). As the regression equation shows, there is a highly significant declining trend. Moreover, note the strong match in year-to-year fluctuations with Fig. 4.

Discussion

Comparison with the British situation

Clearly, the bleak situation depicted by Fox et al. (2006) for Britain does hardly differ from that in the Netherlands. A comparison at the level of individual species, for instance, shows a close congruence. Species like Dark Spinach (*Pelurga comitata*) and Figure of Eight (*Diloba caeruleocephala*) are, like in Britain, among the most severely in decline in the Netherlands (Fig. 2). In addition, the increase of Orange Footman (*Eilema sororcula*) in the Netherlands is also supported by the positive trends of eight other footman species in Britain. Footman moths (subfamily Lithosiinae) are doing very well as a group. In Britain eight species are included in the analysis, all of which have increased (Fox et al. 2006). Larvae of Footman moths are algae and lichen feeders and lichens have been increasing in Northwestern Europe during recent decades (Gilbert 1992).

Comparing the species trends from Britain and the Netherlands, there are some interesting exceptions. The White Ermine (*Spilosoma lubricipeda*) for instance, which shows a decline of 77% in Britain over 35 years, is doing quite well in the Netherlands, with an abundance of 175 (average 1980-now) and a positive slope of 0.03, which equates to a 41% increase in 30 years. Most remarkable is the situation of Cinnabar (*Tyria jacobaeae*), that meets the criteria to be classified as vulnerable in Britain, whereas in the Netherlands it is one of the sharpest rising species. There are, however, some indications that the species is faring less well during the last few years (Fig. 2). Figure 2 gives a few examples of trends of moth species while the condition of each species of the Dutch larger moth fauna is shown on our website www.vlindernet.nl.

Not only at the species level the situation in Britain and the Netherlands is highly comparable, other patterns also show a striking similarity. The relation between moth species population trends and overwintering stage (Fig. 5) is very similar (Fox et al. 2006) and also the distribution of threat values (Fig. 3), are highly comparable with the figures presented in Fox et al. (2006). We conclude, therefore, that, although there are differences in the way of calculations and underlying data between the present study and the long term study based on the Rothamsted Insect Survey (Woiwod and Harrington 1994; Conrad et al. 2004), the situation in the Netherlands hardly differs from that in

Britain and is most likely an illustration of the situation in much of Northwestern Europe.

Climatic effects

As was shown in Fig. 4, the yearly performance, averaged over all species, declined strongly. Many interacting factors will have contributed to this effect, like loss of habitat quality and connectivity, loss of biodiversity in the lower trophic level, lowering of water tables, use of pesticides, light pollution and other factors. However, the figure also shows that this declining trend has strong year to year variations. An inspection of individual species illustrates this dramatically. Even species, that in the long run are doing exceptionally poor, most often show a peak in their performance in, for instance, the year 1992, which was on average a good year (Fig. 7). We do not believe that the peaks and troughs observed can be attributed only to the deleterious factors mentioned above. On average, they are long term effects that change only slowly and unidirectionally over time. Natural population fluctuations are not a likely explanation either, because it would be difficult to understand why these would be so strongly synchronised to give this averaged pattern. One might think of an observational artefact (a nice summer, many observers afoot, high scores), but observation effort is factored out in our calculations. The only environmental factor of a sufficiently large geographical scale that remains is the weather. Indeed, we think that weather fluctuations have played a decisive effect on top of the general trend. We conclude that weather has a strong influence on a species' performance, and that on average all species in our fauna react in a similar way. Future changes in the long term pattern of weather conditions, i.e. climate, therefore, are expected to have a strong impact on our moth fauna.

Numerical decline

The drop in the number of individuals across all species is certainly dramatic and almost certainly it is real. The rise in the number of recorders and observations in itself can have no effect, because the figures are based on the number of observations. As a matter of fact, big changes have occurred in the period since 1980 in the population of observers and their methods of observing and reporting. The most influential change, in our view, the rising popularity of online databases where anyone can record a casual observation, has been eliminated from the calculation. What remains is the rise in the proportion of new and little trained observers, and changes in the technical equipment. One would guess, however, that both would result in higher, rather than lower, numbers of individuals of common moths. This means that the changes in our larger moth fauna not only implies a replacement of rare species by common ones. Rather we see that the entire moth fauna, and by implication the entire basis of our ecosystem is being jeopardized with the trophic levels below, and above, the moth fauna being most severely affected.

Implications

As a consequence, we hypothesize that moth declines will have a huge impact on biodiversity in the Netherlands, the United Kingdom (cf Conrad et al. 2004) and, therefore, most likely in the whole of Northwestern Europe. The ecological implications of this decline are, however, still poorly understood.

The larvae of the majority of species of larger moths have a rather generalized spectrum of host plants. The decline in their number will translate, therefore, in a general decline in herbivore pressure. The impact of insect herbivory on vegetation composition is not particularly well known, especially because exclusion experiments have strong side effects. Yet, a general theme is a change in competition patterns, resulting in a decline of plant diversity, following exclusion (Brown et al. 1988; Crawley 1989; Maron and Crone 2006; Olf and Ritchie 1998; Scherber et al. 2006).

The same effect, lusher plant growth and a declining botanical diversity, is also caused by an enhanced fertilizer deposition, and by a reduction in herbivory by vertebrates (e.g., rabbits, sheep). This has been an reason of much concern among conservationists in the Netherlands (Tamis et al. 2005; Tilman 1987). Our results indicate that also a reduction in insect herbivory must be taken into consideration.

In addition, it may well be that the severe decline in some bird species in Northwestern Europe, especially of those that inhabit agricultural landscapes (Chamberlain et al. 2000) like Corn Bunting (*Miliaria calandra*) and Eurasian Skylark (*Alauda arvensis*), is related to the decline in moth diversity. It can be hypothesized, that the most important trigger is the period in which these birds feed their young. As larvae of moths are one of the most important prey items in these farmland birds, a decline in moths can be responsible for lower reproduction rates and, consequently, declining trends of these bird species.

Another example is found in research on European Nightjar (*Caprimulgus europaeus*), a migratory bird that is present in northwestern Europe from late April to early September. During its presence in northwestern Europe the birds mainly feed on moths (Schlegel 1994; Sierro et al. 2001). In addition, recent research showed that about 80% of the prey caught by adult birds to feed their young consist of moths. Therefore, we predict that the widespread declines of larger moths in the Netherlands and the United

Kingdom will be having strong effects on the birds, bats and mammals, which depend on them for food. This emphasizes the urgent need for the protection and restoration of moth habitats.

Acknowledgments We are grateful to all observers who have entered their moth records to the Noctua database, making this study possible. In addition, two anonymous reviewers and Dr. Martin Warren are acknowledged for their time and efforts to improve the manuscript.

References

Bernays EA, Singer MS, Rodrigues D (2004) Foraging in nature: foraging efficiency and attentiveness in caterpillars with different diet breadths. Ecol Entomol 29:389–397

Boggs CL (1987) Ecology of nectar and pollen feeding in Lepidoptera. In: Slansky F, Rodriguez JG (eds) Nutritional ecology of insects, mites, spiders, related invertebrates. Wiley, New York, pp 369–391

Brown VK, Jepsen M, Gibson CWD (1988) Insect herbivory: effects on early old field succession demonstrated by chemical exclusion methods. Oikos 52:293–302

Chamberlain DE, Fuller RJ, Bunce RGH, Duckworth JC, Shrubb M (2000) Changes in the abundance of farmland birds in relation to the timing of agricultural intensification in England and Wales. J Appl Ecol 37:771–788

Conrad KF, Woiwod IP, Parsons M, Fox R, Warren M (2004) Long-term population trends in widespread British moths. J Insect Conserv 8:119–136

Conrad KF, Warren M, Fox R, Parsons M, Woiwod IP (2006) Rapid declines of common, widespread British moths provide evidence of an insect biodiversity crisis. Biol Conserv 132:279–291

Crawley MJ (1989) Insect herbivores and plant population dynamics. Ann Rev Entomol 34:531–564

Fox R, Conrad KF, Parsons MS, Warren MS, Woiwod IP (2006) The state of Britain's larger moths. Butterfly Conservation and Rothamsted Research, Wareham, Dorset

Gilbert OL (1992) Lichen reinvasion with declining air pollution. In: Bates JW, Farmer AM (eds) Bryophytes and Lichens in a changing environment. Clarendon Press, Oxford, pp 159–177

Groenendijk D, van der Meulen J (2004) Conservation of moths in the Netherlands: population trends, distribution patterns and monitoring techniques of day-flying moths. J Insect Conserv 8:109–118

Hogstad O (2005) Numerical and functional responses of breeding passerine species to mass occurrence of geometrid caterpillars in a subalpine birch forest: a 30-year study. Ibis 147:77–91

Maron JL, Crone E (2006) Herbivory: effects on plant abundance, distribution and population growth. Proc R Soc B 273:2575–2584

Olf H, Ritchie ME (1998) Effects of herbivores on grassland plant diversity. TREE 13:261–265

Scherber C, Milcu A, Partsch A, Scheu S, Weisser WW (2006) The effects of plant diversity and insect herbivory on performance of individual plant species in experimental grassland. J Ecol 94:921–931

Schlegel R (1994) Der ziegenmelker. Die Neue Brehm-Bücherei, Westarp Wissenschaften

Schowalter TD (1981) Insect herbivore relationship to the state of the host plant: biotic regulation of ecosystem nutrient cycling through ecological succession. Oikos 37:126–130

Sierro A, Arlettaz R, Naef-Danzer B, Strebel S, Zbinden N (2001) Habitat use and foraging ecology of the nightjar (*Caprimulgus europaeus*) in the Swiss Alps: towards a conservation scheme. Biol Conserv 98:325–331

Tamis W, van 't ZelfdeM, van der Meijden R, Udo de Haes HA (2005) Changes in vascular plant biodiversity in the Netherlands in the 20th century explained by their climatic and other environmental characteristics. Clim Change 72:37–56

Telfer MG, Preston CD, Rothery P (2002) A general method for measuring relative change in range size from biological atlas data. Biol Conserv 107:99–107

Tilman D (1987) Secondary succession and the pattern of plant dominance along experimental nitrogen gradients. Ecol Monogr 57:189–214

Török J, Hegyi G, Tóth L, Könczey R (2004) Unpredictable food supply modifies costs of reproduction and hampers individual optimization. Oecologia 141:432–443

Vaughan N (1997) The diets of British bats (Chiroptera). Mammal Rev 27:77–94

Visser ME, Holleman LJM, Gienapp P (2006) Shifts in caterpillar biomass phenology due to climate change and its impact on the breeding biology of an insectivorous bird. Oecologia 147:164–172

Waring P, Townsend M (2003) Field Guide to the Moths of Great Britain and Ireland. Br Wildl Publ, Hook

Woiwod IP, Harrington R (1994) Flying in the face of change: the Rothamsted Insect Survey. In: Leigh R, Johnston A (eds) Long-term experiments in agricultural and ecological sciences. CAB International, UK, pp 321–342

Young M (1997) The natural history of moths. T & AD Poyser, London

J Insect Conserv (2011) 15:83–93
DOI 10.1007/s10841-010-9325-z

ORIGINAL PAPER

Recent trends in butterfly populations from north-east Spain and Andorra in the light of habitat and climate change

Constantí Stefanescu · Ignasi Torre · Jordi Jubany · Ferran Páramo

Received: 24 March 2010 / Accepted: 3 August 2010 / Published online: 25 August 2010

Abstract Although butterfly declines have been reported across Europe, no assessment based on detailed quantitative data has ever been made for any extensive area in the Mediterranean Basin. In 1994, a Butterfly Monitoring Scheme was launched in Catalonia (NE Spain), and in 2005 a similar, albeit much smaller, scheme started in the neighbouring Pyrenean country of Andorra. Here we provide a first thorough assessment of butterfly trends in both areas for the last 15 years. Several patterns emerged, above all a worrying decline of a substantial part of the fauna. It was also evident that habitat specialists are experiencing greater declines than habitat generalists, thereby butterfly communities becoming progressively dominated by common species. However, habitat indicators based on characteristic species also revealed that trends are actually associated with habitat types, grassland and scrub specialists declining strongly but woodland specialists showing a marginal increase. These differences are certainly related to profound landscape changes, mainly a dramatic reduction of semi-natural grasslands and open Mediterranean scrub, and a major increase in woodlands. The general effect of climatic warming on butterfly populations was investigated by using the temperature community index (CTI) approach. The thermal structure of butterfly communities remained very stable over time, except in one case where, contrary to the expectations, a significant negative trend in the CTI was noted. However, this surprising result can be explained by taking into account the above-reported pattern of butterfly communities becoming dominated by common species, characterized by low thermal indices in comparison with declining Mediterranean specialists.

Keywords Butterfly monitoring · Population trends · Land-use changes · Habitat indicators · Climatic warming · Mediterranean basin

Electronic supplementary material The online version of this article (doi:10.1007/978-94-007-1442-7_6) contains supplementary material, which is available to authorized users.

C. Stefanescu (✉) · I. Torre · J. Jubany · F. Páramo
Butterfly Monitoring Scheme, Museu Granollers-Ciències Naturals, Francesc Macià 51, 08402 Granollers (Barcelona), Spain
e-mail: canliro@teleline.es

Introduction

Following the seminal studies by Thomas (1984, 1991) in the UK, in recent decades interest in the ecology and conservation of butterflies has grown throughout Europe. Worrying trends, with range contractions and population declines, have been reported for many species and habitats in a number of countries, in most cases as a consequence of habitat destruction and fragmentation (van Swaay et al. 2006).

A commonly reported pattern is that of greater declines in habitat specialists than in habitat generalists (e.g. Warren et al. 2001; van Swaay et al. 2006). In the UK, for example, butterfly generalists have remained mostly stable or even experienced a slightly positive trend over the last 30 years (Botham et al. 2008). This phenomenon implies changes in the composition of butterfly communities, which gradually become dominated by common species with fewer habitat requirements to the detriment of species associated with more valuable habitats (González-Megías et al. 2008). Although this seems to be a well established pattern, geographical variability in population trends is also expected to occur because of differences in climate and land use between European countries. For instance, it has been predicted—and already seen to occur—that climate warming will positively affect butterfly species at the northern edges of their European ranges, but negatively affect those at the southern edges of their ranges in the Mediterranean basin (e.g. Warren et al. 2001; Stefanescu et al. 2004, 2010; Menéndez et al. 2006; Wilson et al. 2007; Merrill et al. 2008; Pöyry et al. 2009). However, at the northern margins the extent to which range expansions occur also depends on other factors such as landscape structure (Hill et al. 2001; Menéndez et al. 2006). Moreover, in some countries in central Europe human impact has been so high (e.g. in the form of levels of nitrogen pollution and changes in land use) that even the commonest and most widespread butterflies are undergoing serious declines (Van Dyck et al. 2009).

Unfortunately, the lack of comprehensive datasets on butterfly distribution and population abundance has generally prevented any attempt to go beyond speculation in the case of southern Europe (but see Wilson et al. 2005, 2007, for an exception). From the point of view of butterfly conservation this represents a serious flaw, as the Mediterranean basin is one of the continent's richest areas for butterflies and with the highest number of endemic species (Dennis and Schmitt 2009). In this paper, we focus on one of the few study systems in Southern Europe for which good butterfly population data exists. In 1994, a Butterfly Monitoring Scheme aimed at systematically recording butterfly populations over an extensive area was launched in Catalonia (NE Spain). This monitoring programme has remained active, and has even increased its geographical coverage and the number of studied butterfly sites; consequently, we are now in a position to perform a first thorough assessment of population trends over the last 15 years for a substantial part of the region's butterfly fauna. In addition to population trends, we examine the impacts of land-use change and climate change.

Materials and methods

Butterfly data

Data on butterfly abundance were collected between 1994 and 2008 at 95 sites as part of the Catalan Butterfly Monitoring Scheme or CBMS (recording period: 1994–2008, 89 sites) and the Andorran Butterfly Monitoring Scheme or BMSAnd (recording period: 2006–2008, 6 sites) (Fig. 1). Both programmes are based on standardized transect counts of adult butterflies and all use exactly the same methodology (for further details see: www.catalanbms.org/; www.bmsand.ad/) and thus for the purposes of the present analysis data from the CBMS and the BMSAnd were pooled into a single dataset. To ensure a more homogeneous geographical treatment, data from four sites in the Balearic Islands were excluded from the analysis.

The number of monitored transects increased steadily from 10 in 1994 to 62 in 2008 (Fig. S1). Even though there has been some turnover in the number of sites, an important fraction remained stable throughout the whole recording period (e.g. by 2008, data series of seven or more years were available for 37 transects; Fig. S1). Monitoring sites are situated at an altitudinal range of 0–2,275 m a.s.l. and represent a rich variety of biotopes, including arable farmland, different types of grasslands, shrubland and a variety of forests in coastal and upland areas (Table 1). At each site, counts are conducted along fixed routes (mean length ± SD: 1,675 ± 728 m; range: 728–4,297 m), typically stratified into shorter sections (9 ± 2.5 sections per transect; range: 5–16) to subsample major variations in habitat.

Whenever conditions were suitable for butterfly activity, counts were made on a weekly basis starting on March 1 and ending on September 26, a total of 30 recording weeks per year. All butterflies within 2.5 m on each side and 5 m in front of the recorder were counted, as per the standard procedure described by Pollard and Yates (1993). Whenever possible, each transect was always walked by the same person to minimize sampling errors.

Throughout the study period, a total of 170 butterfly species were recorded (see the complete list in Table S1). Of these, a subset of 78 species was used in the analysis of population trends (see below).

Indices of abundance and population trends

At the end of the recording season, indices of abundance were calculated for each of the commonest butterfly species, which were arbitrarily defined as those appearing in an average of seven or more transects per year (see Table S1). As an exception, we excluded a few species whose identification was problematic and often unreliable (e.g. Grizzled Skipper *Pyrgus malvoides*, Red Underwing Skipper *Spialia sertorius*, Chapman's Blue *Polyommatus thersites*, White Admiral *Limenitis camilla*, etc.; Table S1), but included a few uncommon species with locally strong populations (e.g. Provence Hairstreak *Tomares ballus* and Purple-shot Copper *Lycaena alciphron*) or considered to be of special conservation value (e.g. Large Blue *Maculinea arion* and Chestnut Heath *Coenonympha glycerion*; Table S1).

Because of complex phenology involving overlapping generations in most multivoltine species (e.g. Stefanescu et al. 2003), we calculated a single annual index of abundance, rather than separate indices for different generations. Each index included estimated values for missing counts (i.e. values estimated as the mean of the preceding

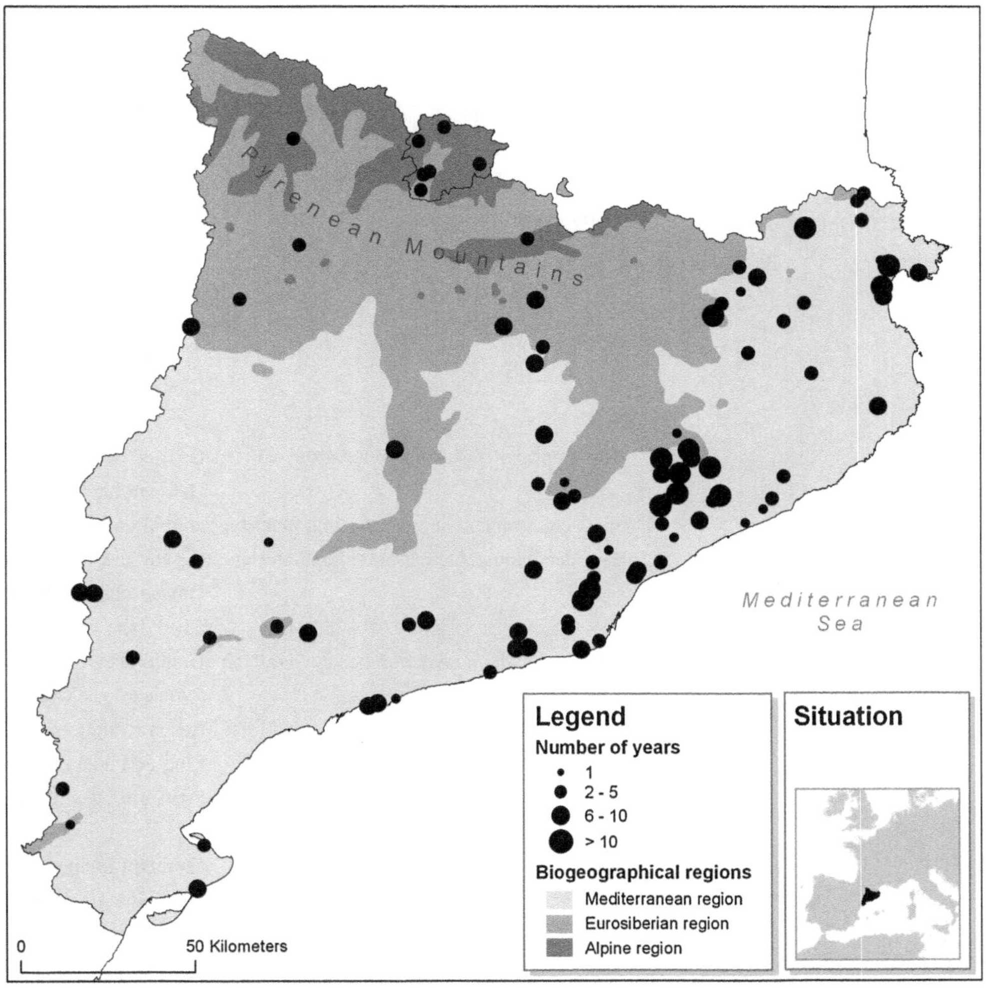

Fig. 1 Map of the study region showing the location of the 95 butterfly transects that were used in the calculation of population trends. The size of the circles is proportional to the number of recording years of each sampling site. Also shown are the generally accepted boundaries of the biogeographical regions present in Catalonia

and succeeding counts, as described by Pollard and Yates 1993). The number of missing counts was low throughout the recording period (3.99 ± 2.76 missing weeks per season out of 30 possible); transects with more than 15% of counts missing in any given year were excluded from the analyses.

For each species, indices for different sites were combined into a single national index of abundance using TRIM, a widely used freeware programme for the analysis of series of counts with missing data (e.g. monitoring data), based on loglinear regressions (Pannekoek and van Strien 2005). As is habitual in monitoring programmes based on volunteer work, recording sites were not randomly distributed and so certain areas were oversampled and others undersampled (see van Swaay et al. 2002 for a discussion of this problem). We thus used the weighting procedure implemented in TRIM to correct a sampling bias at county level (average surface area for each county or comarca = 767.35 ± 392.92 km^2, $n = 41$), given that at this level environmental conditions remain highly homogeneous. Each recording site was given a weight factor calculated as the quotient between the county's surface area and the number of recording sites in that county. All weights were expressed in relation to the lowest value, which was re-scaled to 1 (final weights ranged from 1 to 14.8). For TRIM requirements, the counties were nested into a covariable distinguishing between two strata (1-Mediterranean biogeographical region, 2-European, subalpine and alpine biogeographical regions).

Habitat specialisation and population trends

In order to test whether habitat specialists were undergoing more serious declines than habitat generalists, population trends were analysed in relation to the degree of habitat specialisation of each butterfly species. Following Julliard et al. (2006), an index of habitat specialisation (IHS) was calculated based on how the density of each species was partitioned across the habitats most widely represented in the CBMS. Density data were originally obtained from 314 transect sections representing 17 CORINE habitat types (Moss et al. 1990). To avoid pseudoreplication, this number was reduced to 173 independent units by pooling those sections representing the same habitat type in the same

Table 1 Main habitats and plant communities represented in the CBMS and BMSAnd in 1994–2008, with the number of sites at which they appear in

Habitat type and vegetation zone	Dominant plant community	Nr of sites
Mediterranean lowlands		
Holm oak woodland	Coastal holm oak woodland	43
	Upland holm oak woodland	4
	Continental holm oak woodland	6
Scrublands	Rosemary and *Erica* scrub	2
	Rosemary and white flax scrub	4
	Continental holly oak and *Rhamnus* garrigue	3
	Coastal holly oak and dwarf fan-palm garrigue	2
	Thyme and *Sideritis* scrub	1
	Continental gypsum and thyme scrub	1
Coast		
Riparian and freshwater vegetation	Coastal wetland communities	7
Coast	Halophilous communities	3
Sub-Mediterranean and central European-type humid uplands		
Dry deciduous oak and pine woodland	Downy oak and black pine woodland	1
	Downy oak woodland with box	4
	Black pine woodland	2
Humid deciduous oak and beech woodland	Humid deciduous oak and ash woodland	1
	Pyrenean oak woodland	1
	Beech woodland	2
	Ling and dog-violet scrub	1
Hazel and Scots pine woodland	Scots pine woodland	1
High mountain and subalpine habitats		
Subalpine	Common juniper scrub	1
	Subalpine grassland	5

transect route. For each species, the IHS was then calculated as the coefficient of variation of the average densities within the 17 CORINE habitat types. The IHS is closely related to the number of occupied habitat classes, with low values for species homogeneously distributed across all habitats and the opposite for species restricted to certain habitat types. To test whether or not population trends are dependent on the degree of habitat specialisation, the slopes of trends as calculated by TRIM were regressed against the IHS values: a significant negative relationship is predicted by the assumption of more serious declines in habitat specialists.

Habitat preferences of butterflies and habitat indicators

In a subsequent step, we investigated whether or not trends in biodiversity were linked to certain habitat types. Following the same methodology as developed for European birds (Gregory et al. 2005), habitat indicators were constructed based on the population trends of characteristic butterfly species. Annual indices rather than the abundances of characteristic species were averaged (using geometric means) to give each species an equal weight in the resulting indicator.

To identify habitat preferences we used species' density data across the 17 CORINE habitat types as detailed above. In this second analysis, habitat types were grouped into five broad habitat categories (scrub, grasslands, woodlands, wetlands and agricultural/ruderal habitats) and then the average abundance of each species was calculated for each category (expressed as the number of individuals/100 m per year). A species concentrating over 50% of its density in one particular habitat was considered as a specialist or as a characteristic species of that habitat; otherwise, it was considered as a generalist (Table 2). Knapweed Fritillary *Melitaea phoebe*, Dusky Heath *Coenonympha dorus*, Silver-washed Fritillary *Argynnis paphia* and Camberwell Beauty *Nymphalis antiopa* did not surpass the 50% density threshold in a specific habitat, but were considered as specialists (the first two of grasslands, the latter two of woodlands) following van Swaay et al. (2006) and our knowledge of their ecology in Catalonia.

No indicators were built for wetlands and agricultural/ruderal habitats, since the number of characteristic species was very low (none and four, respectively). On the other hand, an indicator of habitat generalists was developed to further test the hypothesis that population trends are dependent on the degree of habitat specialisation (see

Table 2 Characteristic species of the broad habitat types used for the habitat indicators

	IHS	Nr of sites
Scrub specialists		
Lycaena alciphron	2.073	13
Glaucopsyche melanops	2.067	43
Thymelicus sylvestris	2.013	34
Hipparchia semele	1.940	43
Melitaea cinxia	1.672	35
Issoria lathonia	1.638	55
Coenonympha arcania	1.497	51
Callophrys rubi	1.309	77
Pyronia tithonus	1.288	58
Grassland specialists		
Aglais urticae	3.127	32
Erynnis tages	2.299	34
Hesperia comma	2.180	23
Polyommatus coridon	2.090	25
Tomares ballus	2.003	28
Melanargia occitanica	1.992	21
Colias alfacariensis	1.779	54
Coenonympha dorus	1.716	34
Glaucopsyche alexis	1.705	40
Polyommatus bellargus	1.553	43
Melitaea didyma	1.352	58
Brintesia circe	1.173	69
Coenonympha pamphilus	1.064	50
Melitaea phoebe	1.052	64
Woodland specialists		
Neozephyrus quercus	1.543	43
Nymphalis antiopa	1.444	31
Pararge aegeria	1.423	90
Argynnis paphia	1.342	63
Polygonia c-album	1.299	58
Celastrina argiolus	1.245	83
Limenitis reducta	0.959	69
Generalist species		
Lasiommata megera	0.517	92
Gonepteryx cleopatra	0.635	86
Colias crocea	0.659	95
Thymelicus acteon	0.681	66
Pieris brassicae	0.748	95
Polyommatus icarus	0.750	92
Leptidea sinapis	0.802	79
Melanargia lachesis	0.806	79
Pieris rapae	0.852	95
Maniola jurtina	0.861	84
Lycaena phlaeas	0.893	89
Gonepteryx rhamni	0.921	77
Iphiclides podalirius	0.957	87

Also shown is their corresponding Index of Habitat Specialization (IHS, see text) and the number of sites where each species was recorded. Generalist species are those homogeneously distributed between habitat types

above). Only data for the most generalist species—here defined as those having IHS <1—were used (Table 2). The two long-distance migrants Red Admiral *Vanessa atalanta* (IHS = 0.489) and Painted Lady *Cynthia cardui* (IHS = 0.668) were excluded, as their population dynamics are totally dependent on the environmental situation in their areas of origin (Pollard et al. 1998; Stefanescu 2001).

Temporal trends of habitat indicators were tested by linear regression, with years as the independent variable.

Impact of climatic change on butterfly populations

To test for a general effect of climatic change on butterfly populations, the community approach developed by Devictor et al. (2008, in prep.) was used. Selected butterfly communities were characterized by their community temperature index (CTI) for each year from 1994 to 2008. The CTI has been defined as the average of the long-term mean temperature within the range of each species occurring in the assemblage, weighted by species abundance (Devictor et al. 2008). A direct response of butterfly species to climatic warming would lead to a predictable increase in the CTI, given that populations adjust their densities according to each species-specific temperature requirement. We tested for temporal trends in CTI for a subset of six monitored sites that had complete data from 1994 to 2008. These sites encompass an altitudinal range of 0–1,100 m and thus include a variety of environmental types (from coastal sites to lowland and upland sites dominated by different kinds of scrub and forests).

Results

A summary of the population trends recorded between 1994 and 2008 in Catalonia and Andorra is given in Table 3. Almost half of the studied species (37 out of 78 species, or 47.4%) showed significant population trends, with negative trends clearly predominating over positive trends (28 vs. 9 species). Furthermore, 10 trends were

Table 3 Population trends of 78 butterfly species in Catalonia and Andorra based on 15 years (1994–2008) of transect count data, as calculated with TRIM

Species	Overall trend
Stable trend	
Carcharodus alceae	−0.014
Thymelicus sylvestris	0.0051
Thymelicus acteon	−0.0089
Papilio machaon	−0.0176
Anthocharis cardamines	−0.0005
Euchloe crameri	0.0022
Pontia daplidice	0.0052
Callophrys rubi	−0.0164
Satyrium esculi	0.01
Lampides boeticus	−0.0091
Cupido alcetas	0.021
Celastrina argiolus	−0.0038
Inachis io	0.012
Polygonia c-album	0.0011
Limenitis reducta	−0.0047
Charaxes jasius	−0.0081
Pararge aegeria	−0.0046
Lasiommata megera	−0.0014
Coenonympha pamphilus	0.0081
Pyronia bathseba	−0.009
Maniola jurtina	0.0087
Melanargia lachesis	−0.0129
Uncertain trend	
Hesperia comma	−0.0316
Zerynthia rumina	−0.0416
Anthocharis euphenoides	−0.034
Aporia crataegi	0.0255
Neozephyrus quercus	0.0332
Tomares ballus	−0.0239
Cacyreus marshalli	−0.0584
Cupido minimus	−0.0784
Pseudophilotes panoptes	−0.0446
Glaucopsyche alexis	−0.0243
Glaucopsyche melanops	−0.0614
Aricia cramera	−0.0417
Polyommatus coridon	−0.014
Boloria dia	−0.073
Melitaea phoebe	0.0257
Melitaea didyma	0.0021
Melitaea deione	0.0481
Hipparchia statilinus	−0.0251
Hipparchia fidia	0.0356
Decline (Moderate)	
Erynnis tages	−0.0786
Ochlodes venatus	−0.0295
Iphiclides podalirius	−0.024
Leptidea sinapis	−0.0367
Pieris brassicae	−0.028
Pieris napi	−0.0437
Colias crocea	−0.0241
Colias alfacariensis	−0.0786
Lycaena phlaeas	−0.0342
Lycaena alciphron	−0.0501
Leptotes pirithous	−0.0334
Polyommatus semiargus	−0.0787
Polyommatus icarus	−0.0264
Polyommatus bellargus	−0.0805
Issoria lathonia	−0.0304
Vanessa atalanta	−0.012
Euphydryas aurinia	−0.0696
Pyronia tithonus	−0.0202
(Steep)	
Maculinea arion	−0.398
Polyommatus escheri	−0.1498
Cynthia cardui	−0.0792
Aglais urticae	−0.1041
Melitaea cinxia	−0.1061
Coenonympha arcania	−0.0825
Coenonympha glycerion	−0.1725
Coenonympha dorus	−0.103
Pyronia cecilia	−0.1241
Melanargia occitanica	−0.2259
Increase (Moderate)	
Pieris rapae	0.0389
Gonepteryx rhamni	0.0459
Argynnis paphia	0.0408
Hipparchia semele	0.048
Brintesia circe	0.0678
(Steep)	
Gonepteryx cleopatra	0.106
Libythea celtis	0.1952
Nymphalis antiopa	0.1921
Nymphalis polychloros	0.1355

The overall trend is the slope of the regression line over the whole study period, based on imputed indices with intercept. The 95% confidence interval of a trend estimate is computed by multiplying the standard error by 1.96. If this interval does not include the value 1, then the trend is statistically significant. Also shown is the classification of the trend estimate given by TRIM for a 20-year period. Declines or increases refer to significant trends, while stable or uncertain trends are non-significant

classified by TRIM as steep declines, compared to only four strong increases. Among the species with uncertain trends, negative values were also predominant (13 vs. 6 species), suggesting that a general pattern of population

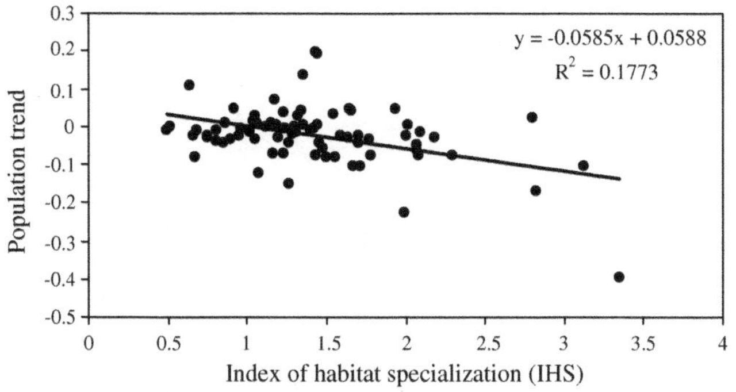

Fig. 2 The relationship between population trends and an index of habitat specialization (IHS) in the 78 studied butterfly species. The highly significant negative correlation between both variables ($r = -0.421$, $P < 0.0001$) reveals a trend for greater population declines in habitat specialists than in habitat generalists

declines in the region will be further confirmed as more data become available in future years. Only 22 species (28%) remained stable throughout the whole period.

There was a significant relationship between population trends and habitat specialization ($R^2 = 0.18$), with specialists (i.e. species with high IHS values) generally showing steeper declines than generalists (Fig. 2). However, the analysis of habitat indicators clearly indicated that not all habitat specialists were declining and that their fate was actually associated with certain habitat types (Fig. 3). Thus, although a grassland and a scrub indicator summarizing broad trends experienced by characteristic species revealed a strong decline between 1994 and 2008 (grassland indicator: $F = 11.41$, $P = 0.0049$; scrub indicator: $F = 14,66$, $P = 0.0021$), a woodland indicator showed a marginal increase ($F = 4.55$, $P = 0.053$). By 2008, the indices for grassland and scrub specialists were 67 and 50% lower than the base year in 1994, respectively, while the indicator for woodland specialists was 115% higher (Fig. 3). Interestingly, the overall abundance of generalist species (i.e. those with no clear habitat preferences) remained very stable ($F = 0.00$, $P = 0.98$), thereby confirming the pattern—that common and widespread butterflies are to some extent buffered against population declines—shown in Fig. 2.

On the other hand, no clear pattern emerged from the analysis of the CTI (Fig. 4). Only in one out of six butterfly communities analysed was there a significant negative trend in the CTI. In this case, contrary to expectations under a scenario of climatic warming, the community became progressively dominated by cold species. Apart from this exception, the thermal structure of butterfly communities remained very stable over time. The lack of significant trends cannot be attributed to a bias in the choice of the sampling sites since they included a wide range of altitudes and ecological conditions (Fig. 4).

Discussion

Our study is the first to provide a general quantitative assessment of recent butterfly population trends in a large region in the Mediterranean basin (NE Iberian Peninsula). On the basis of 15 years of transect data, we have shown that in nearly half of the species significant trends are occurring, consisting mainly of population declines. Although the assessment was based on a subset of the species known in the region (78 out of 200 species), our conclusions can be probably applied to the whole of the region's butterfly fauna. Indeed, the analysis was mostly focused on common butterflies widely distributed across the country, that is, species characterised by a relative ecological flexibility and thus more resistant to environmental

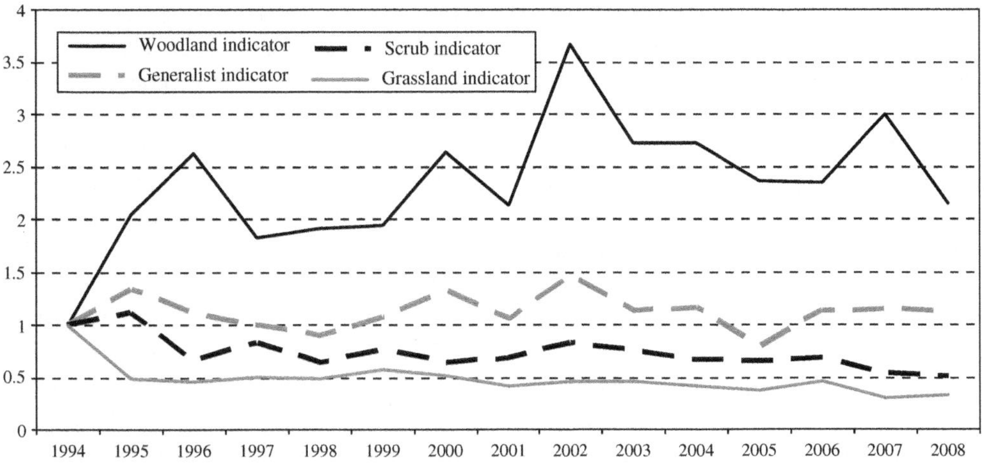

Fig. 3 Temporal trends in habitat indicators based on the characteristic species of broad habitat types (woodland: 7 species; scrub: 9 species; grassland: 14 species; generalist: 13 species). Grassland and scrub indicators declined significantly, while the woodland indicator marginally increased between 1994 and 2008. By contrast, the generalist indicator remained stable over the same period

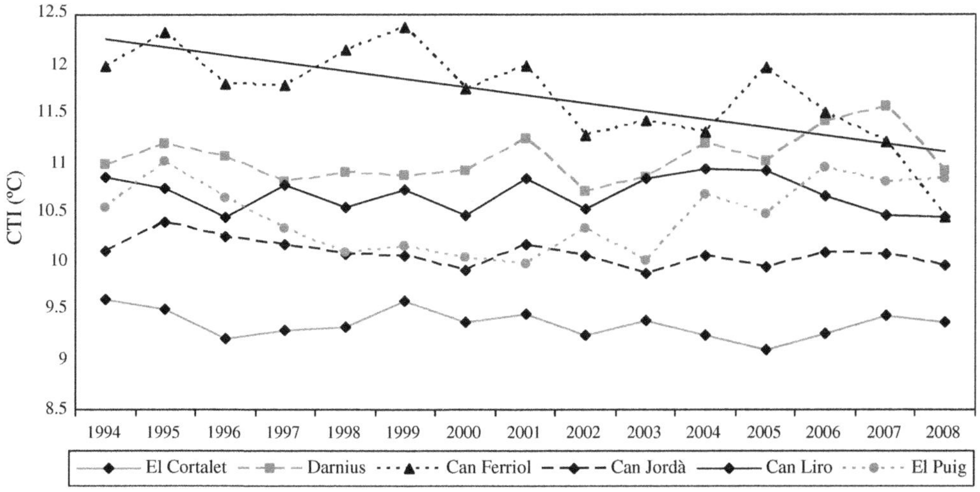

Fig. 4 Trends in the Community Temperature Index (CTI) of six butterfly communities between 1994 and 2008. Only Can Ferriol showed a significant negative trend ($b = -0.082$, $F = 14.94$, $P = 0.002$). The altitudes and main habitat types of each sampling site are as follows: El Cortalet, 0 m a.s.l.; Can Ferriol: 200 m; Darnius: 200 m; Can Liro: 330 m; Can Jordà: 600 m; El Puig: 1,100 m

change. Unfortunately, data for rare species are generally too scarce to be used in any comparable analysis. However, even with the available dataset, we obtained clear evidence that the species showing the most serious declines are true habitat specialists (e.g. the uncommon and locally distributed species) (Fig. 2). As an example, the steepest declines were recorded for *Maculinea arion*, Western Marbled White *Melanargia occitanica* and *Coenonympha glycerion*, all three scarce butterflies with narrow ecological requirements and very local populations in Catalonia and Andorra. Thus, if the whole pool of species, including even the rarest ones, could be analysed, it is highly likely that butterfly declines would appear to be even more widespread.

Moreover, our finding further confirms the general phenomenon whereby butterfly communities are becoming more dominated by generalist species (Polus et al. 2007; González-Megías et al. 2008), which has been described in countries such as Belgium and the UK subject to more severe processes of landscape intensification. This is a particularly worrying result in a region widely recognized as a world biodiversity hotspot (Blondel and Aronson 1999; Myers et al. 2000) in which a high concentration of endemic butterfly species occurs (Dennis and Schmitt 2009).

The analysis of habitat indicators provides some clues as to which factors are responsible for the above-reported pattern. Our data show very clearly that population trends differ widely depending on the habitat type under consideration: while the most characteristic species of open habitats such as different kinds of grasslands and scrub have undergone very rapid declines in recent years, an opposite, albeit weaker trend, has been observed for woodland species. This is certainly related to the profound changes in the landscape that have occurred in the country—and, more generally, in vast areas of the northern Mediterranean basin—in recent decades. Amongst the most evident are the abandonment of traditional land uses such as grazing and mowing in low productive upland areas (e.g. Preiss et al. 1997; Debussche et al. 1999) and pressure from increasing agricultural intensification and urbanization in fertile lowlands and coastal areas (e.g. Petit et al. 2001; Pino et al. 2009). Conjointly, these opposing forces are leading to a dramatic reduction of semi-natural grasslands and open Mediterranean scrub, and also to a major increase in woodlands, especially in mountain areas. In a mere 100 years woodland cover in Catalonia has more than doubled, increasing from 600,000 to 1,400,000 ha, while in the short term between 1987 and 1997 the increase has been estimated to be in the order of 15,000 ha (Brotons et al. 2004).

These large scale landscape changes have probably seriously affected the fauna and flora. We hypothesize that woodland butterflies benefit from the increase in forest cover, just as has been observed in Catalan birds (Brotons et al. 2004). The situation in open habitats is radically different and bears a close resemblance to that in central and northern Europe (e.g. Wallis de Vries et al. 2002; Krauss et al. 2010). For butterflies in particular, over the last 15 years there has been a strikingly similar decrease in the average abundance of grassland species in our region and in western Europe in general (van Swaay et al. 2008).

Aside from habitat preferences, other ecological correlates could also be influencing the direction of population trends, as has been shown by other Lepidopteran studies (e.g. Koh et al. 2004; Mattila et al. 2006; Pöyry et al. 2009). A formal analysis considering different life-history traits is beyond the scope of this paper, but could be the subject of a future study. However, it is interesting to note that population trends differed significantly between overwintering strategies ($F_{3,67} = 5.90$, $P = 0.0012$), with species overwintering as adults (except the Small Tortoishell *Aglais urticae*) showing comparatively stronger increases than the others. Undoubtedly, this remarkable finding deserves further investigation.

One may also expect climatic warming to play an important role in population trends, above all in view of the compelling evidence highlighting this factor as one of the main drivers of change in butterfly communities currently at work (e.g. Menéndez et al. 2006; Wilson et al. 2007; Pöyry et al. 2009). This possibility seems actually more likely given the theoretical work that predicts that climatic warming will lead to losses in butterfly species richness in the study region (Stefanescu et al. 2004, 2010). Recently, model projections of change in potential range extent have been used as a proxy for the impact of climate change on butterfly species at a continental scale (Heikkinen et al. 2010). Unfortunately, climatic envelope models based on high resolution data are not available for butterfly species at the scale of the study area and so such an approach in our case would be inapplicable. Instead, here we have used the temperature community approach developed by Devictor et al. (2008, submitted) to test for a general effect of climatic change on butterfly populations.

Quite surprisingly, none of the six communities for which data were available from 1994 to 2008 showed an increase in temperature. In fact, the only significant trend was in the opposite direction, that is, one community apparently became dominated by colder species during the study period. Although our results contradict previous findings for bird and butterfly communities in various central and northern European countries (Devictor et al. 2008, submitted), we believe that this apparent abnormal pattern can be explained by taking into account some confounding factors. In particular, land use changes may compensate for the expected increase in CTI by favouring butterfly generalists and, conversely, by being detrimental to butterfly specialists. The former species are typically more widespread in Europe, which means that their Species Temperature Index or STI (i.e. an index summarising the long-term average temperature over the species' entire European range) are usually much lower than the STI of Mediterranean specialists inhabiting dry meadows and sclerophyllous scrub in southern latitudes. The CTI of a butterfly community for a given year at a given site is the average of each STI weighted by the species' abundance. Therefore, the above-reported phenomenon of butterfly communities becoming progressively dominated by generalists would lower the CTI values in most cases and thus act in the opposite direction to climatic warming.

Interestingly, the only site where a significant trend in the CTI was recorded is situated in the metropolitan area of Barcelona in an area subject to high levels of human disturbance (e.g. human frequentation and habitat fragmentation). This continuous pressure results in the deterioration of natural habitats such as dry meadows and evergreen oak forest, and the expansion of ruderal habitats (e.g. Guirado et al. 2007). As a consequence, Mediterranean endemics such as Southern Gatekeeper *Pyronia cecilia*, *Melanargia occitanica* and Dusky Heath *Coenonympha dorus* have experienced dramatic declines in the last two decades, while generalist species such as Small White *Pieris rapae* and Speckled Wood *Pararge aegeria* have become considerably more common. These changes have led to a substantial decrease in the CTI of this butterfly community.

Although not revealed by our analysis, other data strongly suggest that climatic warming is indeed influencing Mediterranean butterfly communities. For instance, several highly mobile African species have expanded their ranges recently and have started to be increasingly recorded in our region (e.g. Plain Tiger *Danaus chrysippus*, Monarch *Danaus plexippus*, Desert Orange Tip *Colotis evagore* and Green-striped White *Euchloe belemia*), while remarkable advances in the flight periods of different species have also been recorded at a number of sites (Peñuelas et al. 2002; Stefanescu et al. 2003; Gordo and Sanz 2005). Moreover, detailed community and auto-ecological studies carried out in a mountain range in central Spain have convincingly shown altitudinal shifts in the predicted direction and also provide a number of mechanistic processes that help explain the observed patterns (Wilson et al. 2005, 2007; Merrill et al. 2008).

We can therefore conclude that even though the approach of Devictor et al. (2008, submitted) has been successfully used for detecting rapid responses of birds and butterflies to climatic warming, its usefulness may be constrained by other factors such as land use changes that have a potentially greater impact on the structure of natural communities.

Acknowledgments Thanks are due to all the recorders who contribute data to the Catalan Butterfly Monitoring Scheme and BMSAnd. The Butterfly Monitoring Scheme in Catalonia is funded by the Departament de Medi Ambient i Habitatge de la Generalitat de Catalunya. The Diputació de Barcelona, Patronat Metropolità Parc de Collserola and Fundació Caixa de Catalunya have also given financial support to this project. The BMSAnd project is supported by the Centre d'Estudis de la Neu i la Muntanya d'Andorra (CENMA), from the Institut d'Estudis Andorrans (IEA).

References

Blondel J, Aronson J (1999) Biology and wildlife of the mediterranean region. Cambridge University Press, Cambridge

Botham MS, Brereton TM, Middlebrook I, Cruickshanks KL, Harrower C, Beckmann B, Roy DB (2008) United Kingdom butterfly monitoring scheme report for 2008. CEH Wallingford

Brotons Ll, Herrando S, Estrada J, Pedrocchi V (2004) Patrons generals dels canvis en la distribució de les espècies i l'ús del sòl en el període entre els dos atles. In: Estrada J, Pedrocchi V, Brotons Ll, Herrando S (eds) Atles dels ocells nidificants de Catalunya 1999–2002. Institut Català d'Ornitologia/Lynx Edicions, Barcelona, pp 567–581

Debussche M, Lepart J, Dervieux A (1999) Mediterranean landscape changes: evidence from old postcards. Glob Ecol Biogeogr 8:3–15

Dennis RLH, Schmitt T (2009) Faunal structures, phylogeography and historical inference. In: Settele J, Shreeve T, Konvicka M, Van Dyck H (eds) Ecology of butterflies in Europe. Cambridge University Press, Cambridge, pp 250–280

Devictor V, Julliard R, Couvet D, Jiguet F (2008) Birds are tracking climate warming, but not fast enough. Proc R Soc Lond B 275:2743–2748

Devictor V, van Swaay CAM, Brereton T, Brotons Ll, Chamberlain D, Heliölä J, Herrando S, Julliard R, Kuussaari M, Lindström A, Reif J, Roy D, van Strien A, Settele J, Schweiger O, Stefanescu C, Vermouzek Z, van Turnhout C, Wallis de Vries M, Wynhoff I, Jiguet F (submitted) Climate change induces desynchronization between European birds and butterflies

González-Megías A, Menéndez R, Roy D, Brereton T, Thomas CD (2008) Changes in the composition of British butterfly assemblages over two decades. Glob Change Biol 14:1464–1474

Gordo O, Sanz JJ (2005) Phenology and climate change: a long-term study in a Mediterranean locality. Oecologia 146:484–495

Gregory RD, van Strien A, Vorisek P, Gmelig Meyling AW, Noble DG, Foppen RPB, Gibbons DW (2005) Developing indicators for European birds. Phil Trans R Soc B 360:269–288

Guirado M, Pino J, Rodà F (2007) Comparing the role of site disturbance and landscape properties on understory species richness in fragmented periurban Mediterranean forests. Landsc Ecol 22:117–129

Heikkinen RK, Luoto M, Leikola N, Pöyry J, Settele J, Kudrna O, Marmion M, Fronzek S, Thuiller W (2010) Assessing the vulnerability of European butterflies to climate change using multiple criteria. Biodivers Conserv 19:695–723

Hill JK, Collingham YC, Thomas CD, Blakeley DS, Fox R, Moss D, Huntley B (2001) Impacts of landscape structure on butterfly range expansion. Ecol Lett 4:313–321

Julliard R, Clavel J, Devictor V, Jiguet F, Couvet D (2006) Spatial segregation of specialists and generalists in bird communities. Ecol Lett 9:1237–1244

Koh LP, Sodhi NS, Brook BW (2004) Ecological correlates of extinction proneness in tropical butterflies. Conserv Biol 18: 571–1578

Krauss J, Bommarco R, Guardiola M, Heikkinen RK, Helm A, Kuussaari M, Lindborg R, Öckinger E, Pärtel M, Pino J, Pöyry J, Raatikainen RM, Sang A, Stefanescu C, Teder T, Zobel M, Steffan-Dewenter I (2010) Habitat fragmentation causes immediate and time-delayed biodiversity loss at different trophic levels. Ecol Lett 13:597–605

Mattila N, Kaitala V, Komonen A, Kotiaho JS, Päivinen J (2006) Ecological determinants of distribution decline and risk of extinction in moths. Conserv Biol 20:1161–1168

Menéndez R, González-Megías A, Hill JK, Braschler B, Willis SG, Collingham Y, Fox R, Roy DB, Thomas CD (2006) Species richness changes lag behind climate change. Proc R Soc Lond B 273:1465–1470

Merrill RM, Gutiérrez D, Lewis OT, Gutiérrez J, Díez SB, Wilson RJ (2008) Combined effects of climate and biotic interactions on the elevational range of a phytophagous insect. J Anim Ecol 77: 145–155

Moss D, Wyatt B, Cornaert MH, Roekaerts M (1990) CORINE Biotopes. The design, compilation and use of an inventory of sites of major importance for nature conservation in the European Community. Directorate-General Environment, Nuclear Safety and Civil Protection, Luxembourg

Myers N, Mittermeier RA, Mittermeier CG, da Fonseca GAB, Kent J (2000) Biodiversity hotspots for conservation priorities. Nature 403:853–858

Pannekoek J, van Strien AJ (2005) TRIM 3 manual. Trends and indices for monitoring data. CBS, Statistics Netherlands, Voorburg, Netherlands Freely available via www.ebcc.info

Peñuelas J, Filella I, Comas P (2002) Changed plant and animal life cycles from 1952 to 2000 in the Mediterranean region. Glob Change Biol 8:1–14

Petit S, Firbank L, Wyatt B, Howard D (2001) MIRABEL: models for integrated review and assessment of biodiversity in European landscapes. Ambio 30:81–88

Pino J, Rodà F, Basnou C, Guirado M (2009) El canvi ambiental a la Mediterrània: la perspectiva del paisatge. In: Barriocanal C, Varga D, Vila J (eds) Canvi ambiental global: una perspectiva multiescalar. Quaderns de Medi Ambient, 1. Documenta Universitaria, Girona, pp 91–102

Pollard E, Yates T (1993) Monitoring butterflies for ecology and conservation. Chapman and Hall, London

Pollard E, van Swaay CAM, Stefanescu C, Lundsten KE, Maes D, Greatorex-Davies JN (1998) Migration of the painted lady butterfly *Cynthia cardui* in Europe: evidence from monitoring. Divers Distrib 4:243–253

Polus E, Vandewoestijne S, Choutt J, Baguette M (2007) Tracking the effects of one century of habitat loss and fragmentation on calcareous grassland butterfly communities. Biodivers Conserv 16:3423–3436

Pöyry J, Luoto M, Heikkinen RK, Kuussaari M, Saarinen K (2009) Species traits explain recent range shifts of Finnish butterflies. Glob Change Biol 15:732–743

Preiss E, Martin JL, Debussche M (1997) Rural depopulation and recent landscape changes in a Mediterranean region: consequences to the breeding avifauna. Landsc Ecol 12:51–61

Stefanescu C (2001) The nature of migration in the red admiral butterfly *Vanessa atalanta*: evidence from the population ecology in its southern range. Ecol Entom 26:525–536

Stefanescu C, Peñuelas J, Filella I (2003) Effects of climatic change on the phenology of butterflies in the northwest Mediterranean basin. Glob Change Biol 9:1494–1506

Stefanescu C, Herrando S, Páramo F (2004) Butterfly species richness in the north-west Mediterranean basin: the role of natural and human-induced factors. J Biogeogr 31:905–915

Stefanescu C, Carnicer J, Peñuelas J (2010) Determinants of Mediterranean butterfly species richness: the negative synergistic forces of climate and habitat change. Ecography. doi:10.1111/j.1600-0587.2010.06264.x

Thomas JA (1984) The conservation of butterflies in temperate countries: past efforts and lessons for the future. In: Vane-Wright RI, Ackery P (eds) Biology of butterflies. Symposium of the royal entomological society 11. Academic Press, New York, pp 333–353

Thomas JA (1991) Rare species conservation: case studies of European butterflies. In: Spellerberg IF, Goldsmith FB, Morris MG (eds) The scientific management of temperate communities for conservation. BES Symposium (1989). Blackwell Scientific Publications, Oxford, pp 149–197

van Dyck H, van Strien AJ, Maes D, van Swaay CAM (2009) Declines in common, widespread butterflies in a landscape under intense human use. Conserv Biol 23:957–965

van Swaay CAM, Plate CL, van Strien AJ (2002) Monitoring butterflies in the Netherlands: how to get unbiased indices. Proc Exper Appl Entomol, NEV (Amsterdam) 13:21–27

van Swaay CAM, Warren M, Loïs G (2006) Biotope use and trends of European butterflies. J Insect Conserv 10:189–209

van Swaay CAM, Nowicki P, Settele J, van Strien AJ (2008) Butterfly monitoring in Europe: methods, applications and perspectives. Biodivers Conserv 17:3455–3469

Wallis de Vries MF, Poschlod P, Willems JH (2002) Challenges for the conservation of calcareous grasslands in northwestern

Europe: integrating the requirements of flora and fauna. Biol Conserv 104:265–273
Warren MS, Hill JK, Thomas JA, Asher J, Fox R, Huntley B, Roy DB, Telfer MG, Jeffcoate S, Harding P, Jeffcoate G, Willis SG, Greatorex-Davies JN, Moss D, Thomas CD (2001) Rapid responses of British butterflies to opposing forces of climate and habitat change. Nature 414:65–69
Wilson RJ, Gutiérrez D, Gutiérrez J, Martínez D, Agudo R, Montserrat VJ (2005) Changes to the elevational limits and extent of species ranges associated with climate change. Ecol Lett 8:1138–1146
Wilson RJ, Gutiérrez D, Gutiérrez J, Monserrat VJ (2007) An elevational shift in butterfly species richness and composition accompanying recent climate change. Glob Change Biol 13:1873–1887

J Insect Conserv (2011) 15:341–344
DOI 10.1007/s10841-010-9352-9

SHORT COMMUNICATION

Andorran butterfly monitoring scheme

**R. Caritg · M. Domènech · J. Dantart ·
J. Jubany**

Received: 25 March 2010 / Accepted: 7 September 2010 / Published online: 24 September 2010

Abstract In 2004 Andorra started its own Butterfly Monitoring Scheme, BMSAnd. At present, the network is made up of 6 sites distributed across the country which are widely representative of the habitats present in Andorra. Up until 2009, 129 species had been recorded on the transect routes, which represent 83% of the 156 species identified in Andorra. Of these, 39 are listed as Species of European Conservation Concern (SPEC) in the Red Data Book of European Butterflies, 5 of which are in the SPEC1-3 category and the other 34 are in the SPEC4a and 4b categories.

Keywords Monitoring · Scheme · Butterflies · Pyrenees · Andorra

Introduction

The Butterfly Monitoring Scheme in Andorra (BMSAnd) has been running since 2004. The aim of the project is to evaluate the diversity and abundance of butterflies in Andorra. The BMSAnd network is coordinated by the Snow and Mountain Research Centre in Andorra—CENMA (Centre d'Estudis de la Neu i de la Muntanya d'Andorra) under the supervision of the Catalan Butterfly Monitoring Scheme (CBMS). A team of collaborators form the body of the project and make it viable.

The objectives of the programme are not only to evaluate the diversity and abundance of butterflies but also include obtaining phenological information, improving knowledge on butterflies' environmental preferences, their migratory movements, the colonization of new territories, the extinction of some populations as well as learning about the conservation state of some species in Andorra.

R. Caritg · M. Domènech (✉)
Snow and Mountain Research Centre of Andorra (CENMA) in Andorran Studies Institute (IEA), Av. Rocafort 21-23, edifici Molí, 3r pis, Sant Julià de Lòria, Andorra
e-mail: mdomenech.cenma@iea.ad

J. Dantart · J. Jubany
Catalan Butterfly Monitoring Scheme (CBMS), Museu de Granollers, Ciències Naturals, Avinguda Francesc Macià, 51, 08402 Granollers Catalunya, Spain

Methods

The BMSAnd follows the same methodology developed in the United Kingdom by Pollard and Yates (1993) which was later adopted by Catalonia in 1994 and Andorra in 2004. All the butterflies seen on the transect route are recorded (those located 2.5 m either side of the route and 5 m ahead of the recorder) on a weekly basis for a period which runs between 1st March and 26th September. For the alpine transect routes, which are at higher altitudes, the scheme starts on 1st May, once the snow has melted.

Although the project started in 2004, the data available only dates from 2006. This is due to the fact that a site is not fully established until the data collected by the recorder is considered to be reliable. In most cases this tends to be after the first or second year of being in operation, depending on the richness of the species and the recorder's knowledge.

The BMSAnd network (Table 1; Fig. 1) is currently made up of 6 sites carefully chosen to provide a wide representation of the habitats present in Andorra quoted by Carreras et al. (2003). The data collected are added to and dealt together with the data obtained from the CBMS network sites thus increasing the representation of high, subalpine mountain environments.

Table 1 Description of BMSAnd sites (location, UTM, average height and length of transect route), adapted from Dantart and Jubany (2008)

Location	UTM	Average height (m)	Transect length (m)	Vegetation zone
Fontaneda	31TCH70	963	1311.6	Mediterranean influence, Holm oak area
Rec del Solà	31TCH70	1,092	1321.5	Rainy submediterranean and medioeuropean mountain, oak and dry pine area
Enclar	31TCH70	1,180	1557.8	Rainy submediterranean and medioeuropean mountain, oak and dry pine area
Comapedrosa natural park	31TCH71	1,781	1222.0	Rainy submediterranean and medioeuropean mountain, oak and dry pine area, hazel and Scotch pine area
Sorteny natural park	31TCH81, CH82	1,943	970.2	Subalpine high mountain, pastures and pine areas
Pessons	31TCH90	2,275	2207.7	Subalpine high mountain, pastures and pine areas

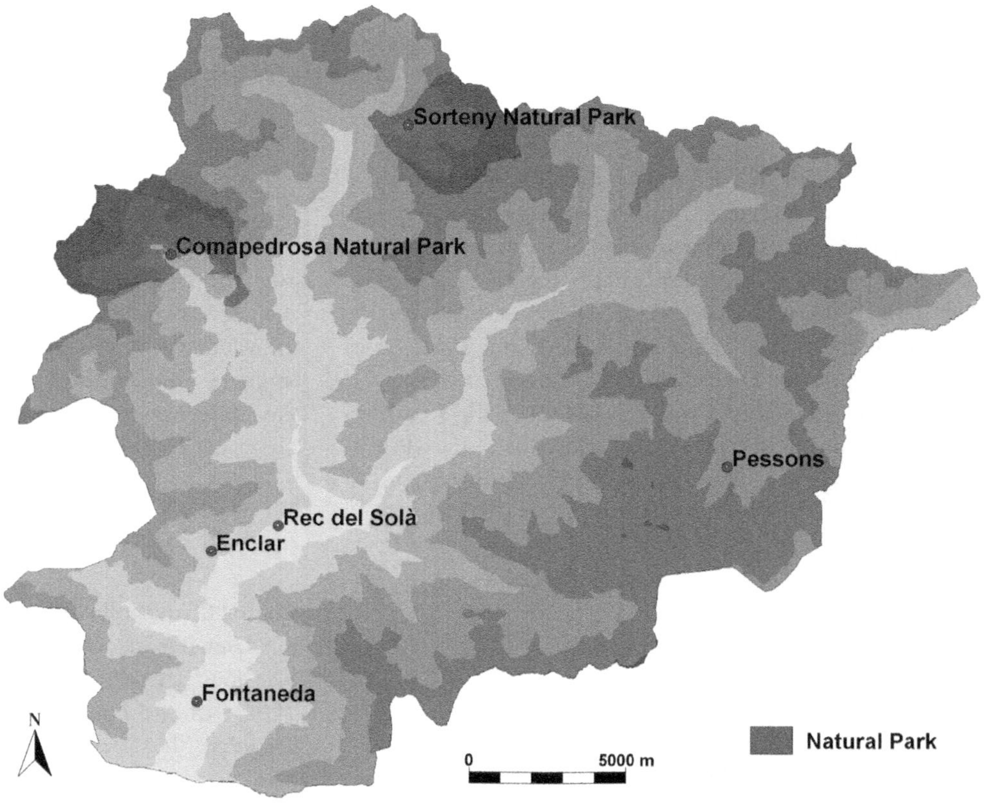

Fig. 1 Geographical situation of BMSAnd sites in Andorra

Results

Overall, data on 129 species (Table 2) have been gathered from the BMSAnd network sites (Dantart and Jubany 2010).

Fontaneda

Observations. The Fontaneda site is one of great interest in the Andorran network as it is located in one of the few environments with a significant Mediterranean influence. It is worth mentioning the presence of *Aristolochia pistolochia*, the host plant of *Zerynthia rumina* species which was later also recorded on the site. In all likelihood, this is one of the northernmost boundaries of the species' distribution area in the Iberian Peninsula. The most abundant species at this site are: *Pieris rapae* (13% of all individuals), *Cynthia cardui* (8%) and *Satyrium esculi* (6%).

Rec del Solà

Observations. This site is characterised by significant human influence due to its proximity to the country's capital. It is an area with a combination of vegetable plots and the pre-existing natural vegetation. The most

Table 2 Results per site: number of years with data, of species per year and of specimens per year. These values are the average of the active period for every site

Location	No. years with data	No. species/year	No. specimens/year
Fontaneda	3	59.3	1225.0
Rec del Solà	3	35.7	602.5
Enclar	4	57.0	1100.5
Comapedrosa natural Park	4	42.5	713.8
Sorteny natural park	4	49.8	583.4
Pessons	3	31.3	493.5

abundant species are: *Pieris rapae* (26% of all individuals), *Pieris brassicae* (16%) and *Cynthia cardui* (15%).

Enclar

Observations. There is a significant height difference within the Enclar site (around 350 m) as well as changes in aspect. There is, therefore a great variety of habitats and a considerable richness in species. Species worth noting include *Lycaena tityrus*, *Polyommatus amanda*, *Hipparchia statilinus*, *Plebicula dorylas* and *Boloria selene*. The most abundant species are: *Pieris rapae* (10% of all individuals), *Pararge aegeria* (10%) and *Pieris napi* (5%).

Comapedrosa natural park

Observations. The species to highlight in this site are *Parnassius apollo*, *Euchloe simplonia* and those of the *Erebia* genus, all of which fall into the SPEC categories in the European context. The most abundant species: *Pieris rapae* (10% of all individuals), *Cynthia cardui* (8%) and *Colias crocea* (7%).

Sorteny natural park

Observations. There are some species which have been recorded only on this site out of all the sites which make up both the BMSAnd and the CBMS network. These are: *Aricia eumedon*, *Aricia nicias* and *Boloria pales*. The most abundant species are: *Aglais urticae* (10% of all individuals), *Polyommatus coridon* (10%) and *Colias crocea* (9%).

Pessons

Observations. The Pessons site is of great interest for both Andorra and as a contribution to the CBMS network due to its subalpine character. It is a transect route in which high mountain species can be detected and one that allows the altitudinal limit for many other species to be established. It supports species of the *Erebia* genus which are found in high mountain areas, such as *E. epiphron* and *E. cassioides*, both of which are SPEC4a. The potential of this transect lies in the hope of finding high mountain species which have not been recorded in any of the CBMS or other BMSAnd sites to date. The most abundant species are: *Cynthia cardui* (23% of all individuals), *Erebeia cassioides* (15%) and *Colias crocea* (10%).

Conclusions

Overall, 129 species have been recorded on the BMSAnd network sites (Dantart and Jubany 2010). This figure represents 83% of the 156 species that have been identified in Andorra (Dantart 2004; Dantart and Jubany 2008, 2009, 2010). On the BMSAnd sites, 39 species are listed as Species of European Conservation Concern (SPEC) in the Red Data Book of European Butterflies (van Swaay and Warren, 1999), 5 of which are in SPEC1-3 and the remaining 34 are in SPEC4a and 4b categories. The species listed as SPEC1-3 are: *Pyrgus cirsii*, *Thymelicus acteon*, *Parnassius apollo*, *Euchloe simplonia* and *Maculinea arion*. At the same time, two of these species, *P. apollo* and *M. arion* are included in Annex II of the Bern Convention. The remaining 34 species are of certain interest on a European level as their world distribution is restricted to Europe, although they are not yet considered as threatened.

The regular nature of the fieldwork and the incorporation of new data over the years make this an interesting project which hopes to continue widening knowledge on butterfly fauna in Andorra and to continue involving more collaborators, the civilian population and other Andorran entities.

Acknowledgments To the BMSAnd collaborators: Jordi Dalmau, Lurdes Garet, Marta March, Montse Mases, Ann Matschke, Carles Mújica and Josep Palau (Parc natural de Sorteny), Jordi Nicolau (Parc natural de les Valls del Comapedrosa), Eric Sylvestre, Laura Trapero Helena Izquierdo and Anna Viaplana (Centre Andorra Sostenible). Finally we would like to thank the anonymous reviewers for their useful comments.

References

Carreras J, Carrillo, E, Ferré A, Ninot JM (2003) Mapa digital dels hàbitats d'Andorra escala 1:25.000. Centre de Recerca de Biodiversitat Vegetal—Universitat de Barcelona. Centre de Biodiversitat—IEA

Dantart J (2004) Diversitat de papallones diürnes d'Andorra. Hàbitats. Revista del Centre de Biodiversitat 8:20–33

Dantart J, Jubany J (2008) Actualització de les dades recollides a les estacions del butterfly monitoring scheme a Andorra (BMSAnd). Temporada 2007. Museu de Granollers, Ciències Naturals. Granollers 2008

Dantart J, Jubany J (2009) Actualització de les dades recollides a les estacions del butterfly monitoring scheme a Andorra (BMSAnd). Temporada 2008. Museu de Granollers, Ciències Naturals. Granollers 2009

Dantart J, Jubany J (2010) Actualització de les dades recollides a les estacions del butterfly monitoring scheme a Andorra (BMSAnd). Temporada 2009. Museu de Granollers, Ciències Naturals. Granollers 2010

Pollard E, Yates TJ (1993) Monitoring butterflies for ecology and conservation. Chapman Hall, London

van Swaay CAM, Warren MS (1999) Red data book of European butterflies (Rhopalocera). Nature and environment, vol 99. Council Europe Publishing, Strasbourg

J Insect Conserv (2011) 15:121–128
DOI 10.1007/s10841-010-9329-8

ORIGINAL PAPER

Assessing conservation status and trends for the world's butterflies: the Sampled Red List Index approach

Owen T. Lewis · Michael J. M. Senior

Received: 25 March 2010 / Accepted: 3 August 2010 / Published online: 15 August 2010

Abstract Red List Indices provide a method for assessing global trends in species' conservation status, and for monitoring progress towards achieving conservation targets (for example, commitments under the Convention on Biological Diversity). Red List Indices are based on categorization of taxa in terms of their threat status using information on, for example, current and projected abundances, distributions, and threats. Global assessments have now been undertaken for a suite of well-known vertebrate taxa. However, highly diverse invertebrate taxa are currently very poorly represented in such assessments, and there is a danger that their threats and their utility as biodiversity indicators will be overlooked. Unlike most invertebrates, butterflies are relatively well-known globally. We describe ongoing efforts to incorporate butterflies into the Red List Index process. Because of high species richness (approximately 15,000 Papilionoidea globally) a comprehensive assessment is not feasible. Instead, we apply a 'Sampled Red List Index' approach which draws on a subset of 1,500 focal taxa. We illustrate the process and the challenges (particularly taxonomic issues and issues of data deficiency) using a variety of case studies. The information provided should be relevant to other researchers seeking to apply the Red List Index approach to invertebrates and other diverse but poorly studied taxa.

Keywords Africa · Extinction · Global · Indicator · Lepidoptera · Red List Index · Monitoring

O. T. Lewis (✉) · M. J. M. Senior
Department of Zoology, University of Oxford, South Parks Road, Oxford OX1 3PS, UK
e-mail: owen.lewis@zoo.ox.ac.uk

Introduction

Approximately 10% of all described species are Lepidoptera: some 180,000 species in total. Of these, about 10% or 18,000 species are butterflies, members of the superfamilies Papilionoidea and Hesperioidea (Robbins and Opler 1997; Ackery et al. 1999). Unlike most Lepidoptera and indeed most non-vertebrates, butterflies are relatively well-studied globally, being perhaps the most charismatic and best-known terrestrial invertebrates. In temperate regions of Europe and North America their status and ecology is thoroughly understood, and individual species have been the focus of intensive—and often successful—single-species conservation efforts (e.g., New 1991; Thomas et al. 2009). Furthermore, butterflies have been widely adopted as ecological indicators for conservation assessments and as indicators of wider trends in the status of biodiversity (e.g., Thomas et al. 2004; Thomas 2005).

However, the great majority of butterfly species are restricted to the tropics: little is known about most of these species and their taxonomy is often uncertain or confused. While we might predict that major anthropogenic drivers such as habitat destruction and climate change are having detrimental effects on their abundance and distribution, no concrete multi-taxa data exist to document this. We do not know what fraction of butterflies globally is threatened with extinction; whether levels of threat vary geographically or taxonomically; what the major threatening processes are, and whether the nature of the threats to butterflies differ from those affecting other taxa.

In this paper, we describe a new and ongoing effort to rectify this situation and to bring butterflies into the mainstream of conservation assessment and planning globally, through targeted application of the Red List assessments and application of the Red List Index (RLI) approach. We describe the rationale of this approach and discuss the practicalities and difficulties of applying it to butterflies.

The Red List and Red List Indices

The International Union for the Conservation of Nature and National Resources (IUCN) provides an internationally-recognized categorization of species in terms of their risk of extinction. Species are allocated to one of nine categories (Table 1) based on factors such as rates of population decline, population sizes, area of geographic distribution, and the degree of population and distribution fragmentation. 'Threatened' species (those considered to be of high risk of extinction) are those in the categories Critically Endangered, Endangered and Vulnerable. The specific criteria used to allocate species to categories are summarised in Table 2. For example, under criterion B a species could be assessed as Vulnerable if its geographic range (known as the extent of occurrence) is less than 20,000 km^2, but only if two or more of the following apply: (a) the distribution is severely fragmented or the species occurs in few localities, (b) there is continuing decline in its distribution, or (c) there are extreme fluctuations in distribution. In this example, if the distribution is less than 5,000 km^2 then the species would be assessed as Endangered; less than 100 km^2 then it would be Critically Endangered (IUCN 2001). While the criteria are fully quantitative, precise and detailed information is not essential to apply them (Akçakaya et al. 2000; Mace et al. 2008): assessors are encouraged to use expert knowledge along with the best available information to estimate the relevant values, provided that these limitations are explicitly specified in the accompanying rationale (IUCN 2001). Because Red List categories are so broad, even if estimates of particular parameters used for the assessments (e.g., extent of occurrence) are imprecise, the process will normally still allocate species to a consistent category.

Such assessments are influential in their own right because they draw attention and sometimes funding to threatened taxa that are a high priority for conservation action (Possingham et al. 2002; Rodrigues et al. 2006). Systematic compilations of data on conservation status for multiple taxa have further potential: a snapshot survey of conservation status for a single taxonomic group (e.g., the approximately 9,500 species of birds) can allow conservationists to identify taxa, regions or habitats that are of particular conservation concern (Butchart et al. 2004). Furthermore, if such surveys are repeated regularly then trends through time can be measured. For example, Smith et al. (1993) used the speed at which species climb the ladder of endangerment through time to predict extinction rates.

Similar logic is applied in calculating Red List Indices, which have been adopted by the Convention on Biological Diversity as one of the methods for monitoring progress towards the target to achieve 'a significant reduction of the current rate of biodiversity loss at the global, regional and national levels' by 2010 (Secretariat of the Convention on Biological Diversity 2003). The RLI approach was first applied to birds, which have now been comprehensively assessed at a global level four times (Butchart et al. 2004). The RLI works by comparing the number of threatened species in repeated re-assessments. In the revised formulation of Butchart et al. (2007), the RLI value for a given set of assessments falls between 0 (all species are Least Concern) and 1 (all species are Extinct), and allows comparisons between sets of species in terms of their overall level of extinction risk, but also in terms of the rate at which this changes through time. In calculating the index, shifts between particular categories can be given equal weighting so that (for example), if 20 Near Threatened species shift down one category to Least Concern, and 20

Table 1 The IUCN Red List categories

Category	Abbreviation	Description
Extinct	EX	No individuals remaining
Extinct in the Wild	EW	Known only to survive in captivity, or as a naturalized population outside its historic range
Critically Endangered	CR	Extremely high risk of extinction in the wild
Endangered	EN	High risk of extinction in the wild
Vulnerable	VU	High risk of endangerment in the wild
Near threatened	NT	Likely to become threatened in the near future
Least Concern	LC	Lowest risk. Does not qualify for a more at risk category. Widespread and abundant taxa are included in this category
Data Deficient	DD	Insufficient data to make an assessment of its risk of extinction
Not Evaluated	NE	Has not yet been evaluated against the criteria

The term "threatened" describes taxa in the categories: Critically Endangered, Endangered, and Vulnerable

Table 2 A simplified overview of thresholds for the IUCN Red List criteria (modified from Butchart et al. 2005)

Criterion	Critically Endangered	Endangered	Vulnerable	Qualifiers and notes
A1: reduction in population size	≥90%	≥70%	≥50%	Over 10 years/3 generations in the past, where causes are reversible, understood and have ceased
A2–4: reduction in population size	≥80%	≥50%	≥30%	Over 10 years/3 generations in past, future or combination
B1: small range (extent of occurrence)	<100 km^2	<5,000 km^2	<20,000 km^2	Plus two of (a) severe fragmentation/few localities (1, ≤5, ≤10), (b) continuing decline, (c) extreme fluctuation
B2: small range (area of occupancy)	<10 km^2	<500 km^2	<2,000 km^2	Plus two of (a) severe fragmentation/few localities (1, ≤5, ≤10), (b) continuing decline, (c) extreme fluctuation
C: small and declining population	<250	<2,500	<10,000	Mature individuals. Continuing decline either (1) over specified rates and time periods or (2) with (a) specified population structure or (b) extreme fluctuation
D1: very small population	<50	<250	<1,000	Mature individuals
D2: very small range	N/A	N/A	<20 km^2 or ≤5 locations	Capable of becoming Critically Endangered or Extinct within a very short time
E: quantitative analysis	≥50% in 10 years/3 generations	≥20% in 20 years/5 generations	≥10% in 100 years	Estimated extinction-risk using quantitative models, e.g., population viability analyses

Critically Endangered species go Extinct, the index value does not change. Alternatively, weights can be applied to emphasise the importance of changes for highly endangered species. Red List Indices can be interpreted as "the proportion of species expected to remain extant in the near future in the absence of any conservation action" (Butchart et al. 2007).

One potential problem with this approach is that many shifts of status will be artifacts, generated because of changing levels of knowledge about individual species, changes to the specific criteria used to categorise species, or taxonomic changes. For example, only 20% of category changes for birds between 2000 and 2004 reflected genuine improvement or deterioration in status (Butchart et al. 2004). The IUCN Red List assessment procedure requires the reason for any category change to be recorded, allowing these 'false' category changes to be excluded from calculation of the index. A further criticism of the SRLI approach is that quite substantial changes to a species' distribution or abundance may not be detected because the categories are so broad; but broad categories seem essential to allow any assessments to take place for less well know taxa such as butterflies. While the 'headline' values for indices should reflect global trends, indices can be usefully 'disaggregated' to provide more focused and comparable information for particular biogeographic realms, habitat types and taxa. For example, while overall Red List Indices for birds have a value around 0.92 and show a gradual decline through time, particularly steep declines in 'disaggregated' Red List Indices have been documented for Indo-Malayan birds, driven by severe deforestation, and for petrels and albatrosses, driven by incidental mortality associated with longline fisheries (Butchart et al. 2007). Such results can clearly be useful in identifying urgent conservation priorities.

Sampled Red List Index (SRLI) approaches

High profile global Red List assessments have now been undertaken for a suite of vertebrate taxa (birds, amphibians and mammals). However, highly diverse invertebrate taxa are currently very poorly represented in such assessments, and there is a danger that their threats and utility as biodiversity indicators will be overlooked. The fact that Red List assessments are likely to be challenging for many invertebrates, including butterflies, should not deter us (Collen et al. 2008). Clearly, there are too many butterfly species in relation to the number of butterfly experts for us to assess the conservation status of each of the earth's species. Instead, we select a random subset of these species for assessment. If the sub-sample is genuinely random, then it should be representative of butterflies more widely in terms of the geographic and taxonomic spread of species as well as their conservation status. Simulations have demonstrated that an initial sample of 1,500 species is sufficient for these purposes, even if a substantial fraction of these species prove to be Data Deficient (Baillie et al. 2008). As for the Red List Index, periodic reassessments of these species should allow us to monitor long-term trends in species' status. The SRLI project for butterflies, supported by the Zoological Society of London, and with funding from the Esmee Fairbairn Foundation in currently underway and should be completed by the end of 2010. Because of particular problems with synonymy and likely high levels of data deficiency, early on the decision was taken to restrict the butterfly Red List Index process to the 'true butterflies' (superfamily Papilionoidea, approximately 15,000 species) and not the skippers (superfamily Hesperioidea, approximately 3,000 species). The project to date has involved data compilation by a network of approximately 60 butterfly experts across the globe. The process should be complete by late 2010 and will be the first global assessment of the status of world's butterflies. The intention is to carry out repeat assessments for the same sample of species approximately every 10 years.

Butterflies on the Red List

Red List assessments of butterflies have been carried out both internationally (taxonomically restricted, e.g., the swallowtail red data book: Collins and Morris 1985) and more locally, e.g., for Europe (van Swaay and Warren 1999; van Swaay et al. 2010) and South Africa (Henning 1989; Henning et al. 2009). At the time of writing (22 March 2010), just 425 Lepidoptera species have global Red List assessments, of which 372 are Papilionoidea (IUCN Red List 2010). This compares very unfavourably with other non-vertebrate taxa, such as plants (Brummitt et al. 2008). Since few of the species included on the SRLI have already been assessed, the SRLI project will increase this total by almost 1,500, in a representative manner. The existing coverage of butterfly assessments is very patchy and idiosyncratic, and a large fraction of existing butterfly assessments appear either to apply the criteria incorrectly, or are out of date because they are based on earlier definitions of categories and criteria which were more subjective and less quantitative than those currently applied. The taxonomic and geographic coverage of existing Red List butterflies is also very uneven, and biased towards 'popular' taxa and regions. Furthermore, of those species currently listed as threatened, a large fraction appears to have been accorded a threatened status that exaggerates the risk of extinction. Understandably, many international butterfly experts approached to assist with the SRLI project were initially very skeptical of the idea of applying the Red List Criteria to butterflies. Clearly, a general understanding of the status of butterflies globally is unlikely to emerge from an ad hoc approach to Red List assessment, and the more systematic approach being applied through the SRLI should help in this regard. An additional, beneficial side-effect of the SRLI project is that it will greatly broaden the geographic and taxonomic spread of butterflies represented on the Red List, as well as greatly increasing the total number of species included. The SRLI approach has already been applied successfully to other invertebrate taxa, including Odonata (Clausnitzer et al. 2009) and freshwater crabs (Cumberlidge et al. 2009).

Practicalities of the butterfly SRLI

Generating a species list for the assessments

Surprisingly, considering that butterflies are generally thought to be among the best-studied insects in terms of their taxonomy and ecology, there is still no global checklist of butterfly species names. This provides a major challenge for the sampled approach to Red Listing, since the random subset of species for assessment purposes must be drawn from a global list. Considerable effort was made at the start of the Red List Index project to make our species selection as unbiased as possible. Initially, our list was based on LepIndex (Beccaloni et al. 2003), a catalogue maintained by the Natural History Museum, London. LepIndex is the only source which contains (in theory) all the global Lepidoptera names, although species described since about 1980 are not included. The problem with LepIndex is that most of these names are "species group names" that do not necessarily correspond to species: 75% of the names on the list represent invalid junior homonyms or subspecies names (G. Beccaloni, personal communication).

We used LepIndex to generate information on the fraction of species to be drawn from each biogeographic realm; these values were consistent with information in other sources (Table 3). In the case of Afrotropical, Australasian and Neotropical faunas we were then able to use new and up-to-date data collated under the Butterfly

Table 3 Representation of butterflies on the Red List Index by Biogeographic Realm

Biogeographic realm	Estimated % of global fauna	Species to be assessed
Neotropical	42	630
Afrotropical	21	315
Oriental	14	210
Australasian	8	120
Palaearctic	11	165
Nearctic	4	60
Total	100	1,500

Estimates of species numbers exclude Hesperioidea and are based on a consensus assessment using Heppner (1991), Robbins and Opler (1997), and Larsen (2005a) in combination with the standardised random draw of taxon names from LepIndex

Taxome Project (Mallet 2007) to build species lists. Thus, for these regions the strategy chosen has been to ignore the names generated using the random draw from LepIndex and instead select directly from the up-to-date list appropriate numbers of taxa to reflect the proportion of the global butterfly fauna represented in each biogeographical realm (Table 3). If the random names picked in LepIndex came from biogeographic realms not currently covered by the Butterfly Taxome Project (in particular the Oriental and Palaearctic realms), our strategy was to search using printed or online information to pin down an appropriate classification. If we were unable to determine the name's status (and species), then we dropped that species from the list and picked the next species off a 'reserve list' for the same biogeographic realm. Although different approaches were necessary for different biogeographic realms, we emphasise that the species selection is genuinely random and is not for example stratified taxonomically. The aim is to represent butterflies as a taxon, and therefore the randomly-selected species should be representative of all butterflies in terms of the fraction of species included from each biogeographic realm, habitat and family.

Applying the Red List categories and criteria to butterflies

The first hurdle to overcome in assessing the Red List status of butterflies is that the original IUCN Red List categories and criteria were designed essentially for vertebrates (IUCN 1966; Mace et al. 2008). More recently, extensive efforts have been made to increase the applicability of the criteria to other taxa (IUCN 2001), but there are still limitations in their applicability for groups like butterflies. This is as much to do with data availability as it is to do with the specification of the criteria.

For butterflies, criterion B is the primary criterion used for assessments, as for most other invertebrate and plant groups. Criterion B categorizes species according to the extent and rate of decline of their geographic range, measured as extent of occurrence (EOO) and/or area of occupancy (AOO) (IUCN 2001). The EOO is the smallest area within a boundary covering the total extent of observed, projected or inferred species records (Gaston 1991). The AOO includes solely the area within the EOO that the species actually occupies. For instance, an EOO will include areas of unsuitable habitat or uncolonized areas, but the AOO records only the occupied area within the EOO (Gaston 1991). For most butterfly species, the EOO is straightforward to calculate as a convex polygon based on locality records. However, measurement of AOO is more complicated as it requires comprehensive species or habitat records collected at an appropriate spatial scale (IUCN Standards and Petitions Subcommittee 2010).

Criterion D2 is also regularly used for butterfly assessment; it is applied if a species faces current or future threats and has a very restricted AOO, or is found in fewer than five locations (IUCN 2001). Only on rare occasions are criteria other than B and D2 applied to butterfly assessments. The population data required to apply criterion A are rarely available for butterfly species, although they have been used widely in recent Red List Assessments for European species, where transect-based monitoring data exist for some countries (van Swaay et al. 2010). In a few cases, criteria A2–4 can be applied based on inferred population declines resulting from habitat loss or degradation.

Data deficiency

Butterflies are a well known taxon when compared to other invertebrates but are extremely poorly known when compared to many vertebrates. Our knowledge of the world's butterflies is highly variable, both taxonomically and geographically. In terms of the challenges posed by different butterfly families, the Riodinidae and Lycaenidae are most problematic. These families have much in common, often exhibiting highly specialised ecological relationships such as myrmecophily (Cottrell 1984), and having localised distributions (New 1993). While the Riodinidae are mostly Neotropical, the Lycaenidae are globally distributed; this is the most speciose butterfly family with approximately 6,000 species. Many are known from a handful of records, and some from the type specimen alone. Geographically, the problems are compounded by the fact that the highest species richness of butterflies is in tropical forests, some of the most data poor and most difficult to study areas. For example, around 20% of the world's butterfly species occur in Colombia, a country that makes up <1% of the earth's land area (Le Crom et al. 2002; Le Crom et al. 2004).

Because of these problems, a large proportion of butterflies selected for the SRLI have been categorized as Data Deficient (DD). A listing of DD indicates that too little is known about the species' distribution, status or threats to make an assessment of its extinction risk (IUCN 2001). Initial results from the Afrotropical species included in the SRLI suggest that around 15% are classed as DD and this seems likely to be representative of the global trend. To put this in perspective, on the current Red Lists only 1% of bird species, 15% of mammals, 25% of amphibians and 17% of reef-building corals are listed as DD (IUCN 2010). Preliminary indications from parallel SRLI exercises on Odonata (dragonflies) and Scarabaeidae (dung beetles) indicate that 35–40% of species in those taxa are likely to be categorised as DD (Clausnitzer et al. 2009; S. Spector, personal communication 2010). This suggests a lower fraction of DD butterflies compared to other invertebrates.

Alternatively, it may be that butterfly assessors are more willing to assess species based on low levels of data and greater uncertainty. Under IUCN guidelines, two key concepts in this context are dispute tolerance and risk tolerance (IUCN Standards and Petitions Subcommittee 2010). Dispute tolerance is the extent to which an assessor excludes extreme values from an assessment. Low dispute tolerance, including extreme values, increases the uncertainty of the assessment. Risk tolerance is the extent to which an assessor adopts an evidentiary (high risk tolerance) or precautionary (low risk tolerance) attitude towards risk (IUCN Standards and Petitions Subcommittee 2010). Precautionary assessors will assess a species as threatened unless it is highly unlikely that the species is not threatened, whilst an evidentiary assessor only classes a species as threatened if strong evidence supports that classification. The IUCN supports the general view that a moderate dispute tolerance and a low risk tolerance should be adopted (IUCN Standards and Petitions Subcommittee 2010). This should minimize the use of the DD category whilst maintaining accuracy in assessments, although it does introduce greater subjectivity.

In assessing butterflies for the SRLI we have tried to adhere closely to the IUCN's view on risk and dispute tolerance. However, the limited data available have led to greater projection and inference than is likely to be typical for better-known taxa. The majority of our assessments are thus characterized by low dispute tolerance and an inevitable low risk tolerance, as strong evidence is not available for many species and we try to use whatever limited, but reliable, data are available. Assessors working on groups such as mammals and birds are likely to adopt more of an evidentiary attitude whilst still maintaining a low risk tolerance, simply through having access to a greater volume of data.

Taxonomic uncertainty

Although butterflies are relatively well characterized taxonomically compared to other invertebrates (Wahlberg et al. 2005), we have still encountered widespread problems with synonymy, unresolved taxa and incorrect classification. It is not easy to assess the conservation status of a species when you don't know what that species is. Some of these problems are illustrated by the genus *Melinaea* (Nymphalidae: Ithomiini) in the Neotropics. Ithomiines include numerous Müllerian mimics which make their identification difficult. Furthermore, Müllerian mimicry rings involve multiple genera and even multiple families (Joron and Mallet 1998), so *Melinaea* spp. can also be confused with more distantly related species. Intraspecific polymorphism within *Melinaea* spp. and between subspecies further complicates matters, and molecular studies have had varying success in differentiating morphological species within the genus (Whinnett et al. 2005; Elias et al. 2007; Dasmahapatra et al. 2009). Mitochondrial DNA barcoding failed to identify any distinct groups, whilst AFLP markers had some success in dividing the genus into more distinct morphological groups (Dasmahapatra et al. 2009). Until these problems are resolved, most *Melinaea* spp. are likely to be considered Data Deficient.

Case studies

Even if we can overcome taxonomic problems, challenges in applying the Red List criteria remain. We will illustrate some of these problems using two Afrotropical case studies, from a part of the world where the extent of deforestation has been spectacular, but few if any butterfly extinctions have been documented (Larsen 2008).

Liptena ilaro

This lycaenid was described in 1974 from the Ilaro Forest in Nigeria but has not been collected since the original type series in 1967 (Larsen 2005b). Ilaro is close to the Nigerian capital Lagos, in the most densely populated area of Nigeria. The Ilaro Forest was totally cleared in the 1980s by illegal logging. No forest remains in the Ilaro area, and little more than 100 km^2 of severely fragmented forest is left in the nearby Ibadan area (Manu et al. 2005). The extent of occurrence (EOO) of this butterfly appears to be less than 5,000 km^2 and the area of occupancy (AOO) less than 100 km^2. With ongoing decline in the butterfly's EOO, AOO and habitat, and severe fragmentation of its distribution, the species seems likely to qualify as Endangered under the B criteria.

However, there are arguments for both a lower and higher threat classification. The species has not been recorded for over 40 years, its rare forest habitat has now been largely destroyed and no detailed data are available on forest cover, AOO or EOO. Therefore, it is possible that the species is Critically Endangered or perhaps even Extinct in the Wild. On the other hand, butterflies from the tribe Liptenini are able to persist in small forest fragments, this species is only readily identified by experts and there has been little recent collecting in the area of Nigeria where this butterfly occurs (T. B. Larsen, personal communication 2008). Thus, there is a slight possibility that *L. ilaro* could be Vulnerable or Near Threatened.

In this situation, we return to the concept of risk tolerance. A purely evidentiary (high risk tolerance) approach would classify the species as Vulnerable or even Data Deficient, whilst a precautionary approach would categorise the species as Critically Endangered. Extinct seems less appropriate for *L. ilaro*. The IUCN encourages an evidentiary approach when classifying extinctions, in order to promote ongoing conservation measures, and requires exhaustive surveys in all possible habitat (IUCN Standards and Petitions Subcommittee 2010). The IUCN suggests that a tag (not category) of possibly extinct is used in certain situations, for instance when severe threatening processes have occurred (IUCN Standards and Petitions Subcommittee 2010). In this case, a classification of Critically Endangered (possibly extinct) may be appropriate for *L. ilaro*.

Pseudaletis arrhon

This Afrotropical lycaenid has been given a preliminary classification of Data Deficient and is typical of the numerous DD species encountered whilst carrying out the SRLI. It is only known from its male type specimen collected from Bitje in southeastern Cameroon around 100 years ago (Libert 2007). Theclinae of this genus are normally extremely localised and elusive as a result of obligate associations with *Crematogaster* sp. ant nests found in particular trees (New 1993). Recent targeted collecting by the African Butterfly Research Institute around *Crematogaster* nests has resulted in a huge increase in Theclinae specimens, but no further specimens of this species. This suggests that the species could be threatened; however, there are still large areas of forest remaining in southeastern Cameroon (FAO 2005). Although we aim to take a precautionary approach and the area has been reasonably well researched by lepidopterists, until the area has been more thoroughly surveyed it would be inappropriate to consider any classification other than DD for a species which is known from just a single specimen, and which is likely to be naturally local in its distribution and elusive in terms of its behaviour.

Conclusions

We believe that the Red List Index approach has considerable potential for informing conservation policy and practice, and that butterflies, as 'flagship' terrestrial invertebrates (New 1991), should be part of this process. Other approaches to monitoring trends (e.g., the Living Planet Index) rely on long-term population data, which are very rarely available for butterflies and other invertebrates (Butchart et al. 2010). Here we have discussed the role of butterflies in global conservation assessments, but equivalent processes can and should be applied at regional scales. National level Red List assessments are now increasingly including butterflies, for example the recent Brazilian and Venezuelan red data books (Machado et al. 2005; Rodríguez and Rojas-Suárez 2008). While we have focused on the Red List Index approach, the wider benefits of more systematic approaches to Red Listing of butterflies (and other invertebrate taxa) are worth emphasizing. Only with standardization can we hope to know what fraction of butterflies globally is threatened with extinction, whether this is representative of other terrestrial taxa, and where the most urgent areas for conservation action are located.

Acknowledgments Ben Collen, Torben Larsen and two anonymous reviewers provided helpful comments on the manuscript. Mala Ram and Nadia Dewhurst have provided help and support throughout the SRLI process. We are particularly grateful to Torben Larsen whose encyclopedic knowledge of African (and other) butterflies has made the SRLI process much less daunting, and whose data were used for the case studies illustrated here. We also thank the numerous lepidopterists who have contributed to initiating the butterfly SRLI, in particular George Beccaloni, Blanca Huertas, Jim Mallet, Chris Muller, John Tennent and Keith Willmott. Helen Chadburn, Chris Gröbler, Chris Jeffs, Emily Lloyd, Ed Parker, Eleanor Slade, Chun Siew, Bee Smith and Chun Yuen Wong provided invaluable research assistance.

References

Ackery PR, de Jong R, Vane-Wright RI (1999) The butterflies: Hedyloidea, Hesperioidea and Papilionoidea. In: Kristensen NP (ed) Lepidoptera, moths and butterflies, vol 1: evolution, systematics, and biogeography. Handbuch der zoologie part 35. Walter de Gruyter, Berlin, pp 263–300

Akçakaya HR, Ferson S, Burgman MA, Keith DA, Mace GM, Todd CA (2000) Making consistent IUCN classifications under uncertainty. Conserv Biol 14:1001–1013

Baillie JEM, Collen B, Amin R, Akçakaya HR et al (2008) Toward monitoring global biodiversity. Conserv Lett 1:18–26

Beccaloni GW, Scoble MJ, Robinson GS, Pitkin B (eds) (2003) The global lepidoptera names index (LepIndex). http://www.nhm.ac.uk/entomology/lepindex. Accessed 20 Mar 2010

Brummitt N, Bachman SP, Moat J (2008) Applications of the IUCN Red List: towards a global barometer for plant diversity. Conserv Lett 6:127–135

Butchart SH, Stattersfield AJ, Bennun LA et al (2004) Measuring global trends in the status of biodiversity: Red List Indices for birds. PLoS Biol 2:e383. doi:10.1371/journal.pbio.0020383

Butchart SH, Stattersfield AJ, Baillie J et al (2005) Using Red List Indices to measure progress towards the 2010 target and beyond. Philos Trans R Soc Lond B Biol Sci 360:255–268

Butchart SH, Akçakaya HR, Chanson J et al (2007) Improvements to the Red List Index. PLoS ONE 2:e140

Butchart SHM, Walpole M, Collen B, van Strien A, Scharleman JPW, Almond REA, Baillie JEM, Bomhard B, Brown C, Bruno J, Carpenter KE, Carr GM, Chanson J, Chenery A, Csirke J, Davidson NC, Dentener F, Foster M, Galli A, Galloway JN, Genovesi P, Gregory R, Hockings M, Kapos V, Lamarque J-F, Leverington F, Loh J, McGeoch MA, McRae L, Minasyan A, Hernández Morcillo M, Oldfield T, Pauly D, Quader S, Revenga C, Sauer J, Skolnik B, Spear D, Stanwell-Smith D, Stuart SN, Symes A, Tierney M, Tyrrell TR, Vié J-C, Watson R (2010) Global biodiversity: indicators of recent declines. Science 328:1164–1168

Clausnitzer V, Kalkman VJ, Ram M, Collen B, Baillie JEM, Bedjanič M, Darwall WRT, Dijkstra KB, Dow R, Hawking J, Karube H,

Malikova E, Paulson D, Schüte K, Suhling F, Villanueva RJ, von Ellenrieder N, Wilson K (2009) Odonata enter the biodiversity crisis debate: the first global assessment of an insect group. Biol Conserv 142:1864–1869
Collen B, Ram M, Dewhurst N, Clausnitzer V, Kalkman V, Cumberlidge N, Baillie JEM (2008) Broadening the coverage of biodiversity assessments. In: Vié J-C, Hilton-Taylor C, Stuart SN (eds) The 2008 review of the IUCN Red List of threatened species. IUCN, Gland
Collins NM, Morris MG (1985) Threatened swallowtail butterflies of the world. IUCN, Gland
Cottrell CB (1984) Aphytophagy in butterflies—its relationship to myrmecophily. Zool J Linn Soc 80:1–57
Cumberlidge N, Ng PK, Yeo DCJ, Magalhaes C, Campos MR, Alvarez F, Naruse T, Daniel SR, Esser LJ, Attipoe FYK, Clotilde-Ba F-L, Darwall W, Mcivor A, Ram M, Baillie JEM, Collen B (2009) Freshwater crabs and the biodiversity crisis: importance, threats, status, and conservation challenges. Biol Conserv 142:1665–1673
Dasmahapatra KK, Elias M, Hill RI, Hoffman JI, Mallet J (2009) Mitochondrial DNA barcoding detects some species that are real, and some that are not. Mol Ecol Res 10:264–273
Elias M, Hill RI, Willmott KR et al (2007) Limited performance of DNA barcoding in a diverse community of tropical butterflies. Proc Roy Soc Lond Ser B Biol Sci 274:2881–2889
Food and Agriculture Organization (FAO) (2005) Global Forest Resources Assessment 2005. FAO, Rome. Downloadable at: http://www.fao.org/forestry/fra/fra2005/en/
Gaston KJ (1991) How large is a species' geographic range? Oikos 61:434–438
Henning SF, Henning GA (1989) South African red data book: butterflies. Council for Scientific and Industrial Research, National Scientific Programmes Unit, Pretoria (South African National Scientific Programmes report no. 158)
Henning GA, Terblanche RF, Ball JB (eds) (2009) South African red data book: butterflies. SANBI biodiversity series 13. South African National Biodiversity Institute, Pretoria
Heppner JB (1991) Faunal regions and the diversity of Lepidoptera. Trop Lepid 2(Suppl):1–84
IUCN (1966) Survival service commision. Red data book, vol 1–4. Morges, Switzerland
IUCN (2001) IUCN Red List categories and criteria: version 3.1. IUCN Species Survival Commission, Gland
IUCN (2010) IUCN Red List of threatened species. Version 2010.1. <www.iucnredlist.org>. Downloaded on 24 Mar 2010
IUCN Standards and Petitions Subcommittee (2010) Guidelines for using the IUCN Red List categories and criteria. Version 8.0. Prepared by the standards and petitions subcommittee in March 2010. http://intranet.iucn.org/webfiles/doc/SSC/RedList/RedList Guidelines.pdf. Accessed 22 Feb 2009
Joron M, Mallet J (1998) Diversity in mimicry: paradox or paradigm? Trends Ecol Evol 13:461–466
Larsen TB (2005a) Butterflies of West Africa. Apollo Books, Denmark
Larsen TB (2005b) Hazards of butterfly collecting—Fossil on a pin, UK. Ent Rec J Var 117:109–111
Larsen TB (2008) Forest butterflies in West Africa have resisted extinction… so far (Lepidoptera: Papilionoidea and Hesperioidea). Biodivers Conserv 17:2833–2847
Le Crom JF, Constantino LM, Salazar JA (2002) Mariposas de Colombia. Tomo I: papilionidae. Carlec Ltda, Bogotá
Le Crom JF, Llorente-Bousquets J, Constantino LM, Salazar JA (2004) Mariposas de Colombia. Tomo II: pieridae. Carlec Ltda, Bogotá
Libert M (2007) Révision du genre *Pseudaletis* Druce (Lepidoptera, Lycaenidae) (Format revue). ABRI, Nairobi & Lambillionea, Belgium
Mace GM, Collar NJ, Gaston KJ, Hilton-Taylor C, Akcakaya HR, Leader-Williams N, Milner-Gulland EJ, Stuart SN (2008) Quantification of extinction risk: IUCN's system for classifying threatened species. Conserv Biol 22:1424–1442
Machado ABM, Martins CS, Drummond GM (2005) Lista da fauna Brasileira ameaçada de extinção: incluindo as espécies quase ameaçadas e deficientes em dados. Fundação Biodiversitas, Belo Horizonte
Mallet J (2007) The butterfly taxome project. http://www.ucl.ac.uk/taxome/intro.html. Accessed 22 Mar 2010
Manu S, Peach W, Bowden C, Cresswell W (2005) The effects of forest fragmentation on the population density and distribution of the globally endangered Ibadan Malimbe *Malimbus ibadanensis* and other *Malimbe* species. Bird Conserv Int 15:275–285
New TR (1991) Butterfly conservation. Oxford University Press, Oxford
New TR (ed) (1993) Conservation Biology of the Lycaenidae (Butterflies). Occasional Paper of the International Union for Conservation of Nature and Natural Resources Species Survival Commission no. 8, Gland
Possingham HP, Andelman SJ, Burgman MA, Medellín RA, Master LL, Keith DA (2002) Limits to the use of threatened species lists. Trends Ecol Evol 17:503–507
Robbins RK, Opler PA (1997) Butterfly diversity and a preliminary comparison with bird and mammal diversity. In: Wilson DE, Reaka-Kudla ML, Wilson EO (eds) Biodiversity II, understanding and protecting our biological resources. Joseph Henry Press, Washington, pp 69–82
Rodrigues ASL, Pilgrim JD, Lamoreux JF, Hoffmann M, Brooks TM (2006) The value of the IUCN Red List for conservation. Trends Ecol Evol 21:71–76
Rodríguez JP, Rojas-Suárez F (eds) (2008) Libro Rojo de la Fauna Venezolana, Tercera Edición. Provita y Shell Venezuela, S.A, Caracas
Secretariat of the Convention of Biological Diversity (2003) Handbook of the convention on biological diversity, 2nd edn. Earthscan, London
Smith FDM, May RM, Pellew R, Johnson TH, Walter KS (1993) Estimating extinction rates. Nature 364:494–496
Thomas JA (2005) Monitoring change in the abundance and distribution of insects using butterflies and other indicator groups. Phil Trans R Soc B 360:339–357
Thomas JA, Telfer MG, Roy DB, Preston CD, Greenwood JJD, Asher J, Fox R, Clarke RT, Lawton JH (2004) Comparative losses of British butterflies, birds and plants and the global extinction crisis. Science 303:1879–1881
Thomas JA, Simcox DJ, Clarke RT (2009) Successful conservation of a threatened *Maculinea* butterfly. Science 325:80–83
van Swaay CAM, Warren MS (1999) Red data book of European butterflies. (Rhopalocera). Nature and environment, No. 99. Council of Europe Publishing, Strasbourg
van Swaay C, Cuttelod A, Collins S, Maes D, Lopez Munguira M, Šašić M, Settele J, Verovnik R, Verstrael T, Warren M, Wiemers M, Wynhof I (2010) European Red List of Butterfies. Publications Office of the European Union, Luxembourg. http://ec.europa.eu/environment/nature/conservation/species/redlist/index_en.htm (Accessed 23 Mar 2010)
Wahlberg N, Braby M, Brower A, de Jong R, Lee M, Nylin S, Pierce N, Sperling F, Vila R, Warren A, Zakharov E (2005) Synergistic effects of combining morphological and molecular data in resolving the phylogeny of butterflies and skippers. Proc R Soc B Biol Sci 272:1577–1586
Whinnett A, Zimmermann M, Willmott KR, Herrera N, Mallarino R, Simpson F, Joron M, Lamas G, Mallet J (2005) Strikingly variable divergence times inferred across an Amazonian butterfly 'suture zone'. Proc R Soc B 272:2525–2533. doi:10.1098/rspb.2005.3247

J Insect Conserv (2011) 15:351–359
DOI 10.1007/s10841-010-9343-x

SHORT COMMUNICATION

Butterfly community recovery in degraded rainforest habitats in the Upper Guinean Forest Zone (Kakum forest, Ghana)

Szabolcs Sáfián · Gábor Csontos · Dániel Winkler

Received: 31 March 2010 / Accepted: 3 September 2010 / Published online: 22 September 2010

Abstract Successful regeneration of secondary tropical forest might be crucial in the conservation of rainforests, since large areas of primary forest have been destroyed or degraded. Animal communities might play an important role in restoration of biological diversity in these secondary habitats, since some groups have high mobility and capacity for dispersal. Fruit-feeding butterflies were trapped to measure differences between butterfly communities in primary rainforest and disturbed forest habitats of different stage of regeneration including clear-cut, abandoned farmland, newly planted forest and middle-aged secondary growth. 3465 specimens representing 114 species were identified from 56 traps operated for 36 days. Extremely high values of rarefied species richness were estimated in the clear-cut habitat, due to the high number of singletons and doubletons. This was caused by a gap-effect that allowed penetration of canopy and open area species after disturbance. The differences between butterfly communities were best demonstrated through ecological composition, richness and abundance of indicator groups and habitat similarity based on Jaccard's similarity index. The results show a clear ability of butterfly communities in degraded forest habitats to regenerate in 50–60 years after clearance.

Keywords Butterfly community recovery · Fruit-feeding butterflies · Rainforest regeneration · Kakum forest · Ghana

Sz. Sáfián (✉)
Institute of Silviculture and Forest Protection, University of West Hungary, Bajcsy-Zsilinszky u. 4, 9400 Sopron, Hungary
e-mail: safian@bcghana.org

G. Csontos
Georgikon Faculty of Agriculture, University of Pannonia, Deák Ferenc u. 16, 8360 Keszthely, Hungary

D. Winkler
Institute of Wildlife Management and Vertebrate Zoology, University of West Hungary, Bajcsy-Zsilinszky u. 4, 9400 Sopron, Hungary

Introduction

Forest regeneration in tropical areas has received high publicity in the recent years (Wunderle 1997; Parrotta et al. 1997; Aide et al. 2000), since tropical rainforests are known as one of the most important mitigators of the effects of global climate change through fixing and storing a considerable amount of carbon stock (Foody et al. 1996; Chiti et al. 2010). Protection of the remaining rainforest areas in West Africa is also becoming increasingly important to conservation bodies as the Upper Guinean Forest Zone is known as one of the twenty-five biodiversity hotspots in the world (Myers et al. 2000) and the loss of the original forest cover reaches over 90% in several West African countries (UNEP/GRID-Arendal 2002). The recognition of the excessive need for wood products in these countries has led to the initiation of restoration programs, although the new forest plantations mostly consist of fast-growing non-indigenous tree species such as teak and eucalyptus (Tiarks et al. 1998). Although studies on the plant composition have been carried out in regenerating rainforests (Otsamo 1999; Kassi N'Dja 2008), little information has been published on the regeneration of animal communities in rainforests, especially in the Upper Guinean forest zone (Dunk 2004). Butterfly species or communities are well known indicators of different ecological factors since they usually react quickly to negative changes, but their flying ability also allows them to return if conditions become favourable (Lawton et al. 1998). Fruit-feeding butterfly communities are frequently used to measure these factors including their reaction to edge-effect, stratification, fragmentation or habitat loss (Fermon et al. 2000, 2003; Vanclay 2004; Bobo et al. 2006; Bossart et al. 2006; Bossart and Opuni-Frimpong 2009; Elbers and Bossart 2009). In this paper analysis of quantified samples collected in forests at different stages of degradation-regeneration (from clear-cut to primary forest through abandoned farmland, young planted indigenous forest and middle-aged secondary growth) reveals evidence that rainforest associated fruit-feeding butterfly communities are able to recover.

Study area

Kakum Forest has become a famous tourist attraction in Ghana's Central Region since the completion of the 'canopy walkway', a series of rope-bridges in the canopy level of the rainforest. Besides playing an important role in eco-tourism in the country, Kakum is also a biodiversity hot-spot. Covering more than 360 km^2 Kakum is one of the largest protected blocks of intact moist evergreen and semi-deciduous forest (with the adjacent Assin Attandanso Reserve) in Ghana, which hosts a viable population of forest elephant, a good number of primates, hundreds of birds and more than 600 butterfly species (BirdLife International 2009, Larsen 2006a, Riley and Riley 2005). Our sampling sites were selected in a 40 hectare area, which lies on the boundary of Kakum National Park (5°21′35.77″N; 1°22′8.49″W) (Fig. 1). It proved ideal for a comparative study of butterfly communities since the area was originally covered by primary rainforest.

However, during the last four decades parts were used by local farmers who created a sort of mosaic type vegetation structure including patches of untouched primary forest, formerly logged middle-aged secondary growth, oil-palm, cocoa and cassava plantations. The area has now been abandoned for over ten years, and some formerly utilized plantations were replanted with indigenous tree species such as *Ceiba pentandra*, *Chlorophora excelsa*, *Terminalia ivoriensis* and *Khaya ivoriensis* nine years ago by the present owner. An illegal clear-felling of a primary rainforest area within the study site created an extreme disturbance of forest habitat, leaving it practically without forest cover. A full set of disturbed forest habitats in different stages of regeneration were available for a comparative study of the butterfly communities.

The habitats studied were in close proximity and were interconnected by natural corridors that allowed dispersal of butterflies, and therefore extinction and re-colonisation depended mostly on ecological factors but not geographical isolation. The mean distance between sampling sites was approximately 250 m. Two control transects in the primary upland evergreen rainforest of the Atewa Range (6°14′54.83″N; 0°33′53.03″W) were selected (Fig. 1) to be compared with the samples collected in Kakum. With an estimated 700 species, Atewa Range is recognized for hosting the highest diversity of butterflies in West Africa west of the Dahomey Gap (Larsen 2007).

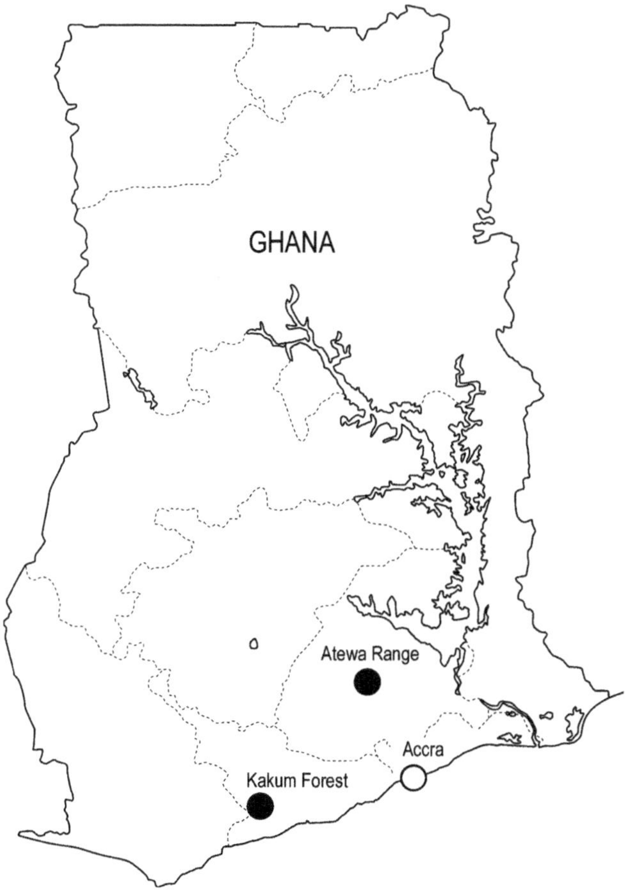

Fig. 1 Map of sampling areas

Sampled habitats:

Primary Forest (PF): Old primary rainforest with many mature trees, natural forest structure and stratification, including a very dense shrub-layer. Canopy cover is 100%.

Secondary Forest (SF): Middle-aged secondary growth with a few mature trees and dense understorey vegetation. The canopy cover is almost 100%.

Young Planted Forest (NF): Even-aged young forest, planted 9 years ago with indigenous trees. It has 100% closed canopy cover during wet season, but the canopy is more open in the dry period. Large trees and the shrub-layer are completely missing.

Farmland (FL): Semi-open area with middle-aged oil palms and 6–12 year-old trees and thickets. The canopy cover is less than 60%.

Clear-cut (CC): Open area of a few hectares, illegally clear-felled in December 2006 in the forest. There was virtually no arboraceous vegetation only thick coppices, tall grassy vegetation and invasive shrubs in 2007 when the sampling began.

Control 1 (K1): Wet primary evergreen forest with slight disturbance in a valley in the Atewa Range.

Control 2 (K2): Wet primary evergreen forest with slight disturbance on a hilltop in the Atewa Range.

Methods

Trap data

Fruit-feeding butterflies (Nymphalidae: Satyrinae, Charaxinae, Nymphalinae and Limenitinae) were attracted by mashed banana bait and captured by net traps, modified from IKEA's children's toy storage net "MINIFÅNGST" by the senior author (http://www.independent.co.uk/life-style/house-and-home/interiors/the-50-best-storage-ideas-96 4981.html?action=Popup&ino=17) (dimensions: $h = 75$ cm, $d = 23$ cm). The original structure was altered by cutting out the separations between compartments and covering all holes on the side of the net. A new, horizontal entrance gap was opened on the side almost at the bottom to ensure that butterflies are able to enter the trap when seeking the bait, which is placed in a plastic plate at the bottom of the trap. They nearly always fly upwards when alarmed and therefore they get trapped in the top section of the net. The banana bait was changed on every fourth day although the bait was refreshed by stirring every day after emptying the traps. Butterflies were collected from the traps once a day on a regular basis.

Eight traps were set along linear transects in each sampled habitats (56 traps in total). The bait was placed just above ground level (not more than 10 cm from the ground), which allowed sampling of fruit-feeding butterflies that are normally restricted to the forest floor and the dense undergrowth. The traps were operated for twelve consecutive days allowing one sample in each habitat. The data included comprise the results of three sampling periods (36 cumulative days/transects). The distance between traps was in each case 30 m—the same as the distance of the first and last traps from the habitat boundaries, to reduce the bias of sampling and edge effect as much as possible. The uniqueness of this kind of habitat complex and the overall size of each habitat patches did not allow replication of sampling under similar conditions. Sampling of the two control transects was carried out immediately after the Kakum transects to ensure that no seasonal differences would occur in the samples. All the butterflies were indentified by the senior author, with the help of his reference collection of Ghanaian butterflies and the comprehensive work of Larsen (2005) in case of particularly difficult specimens.

Analysis

The attributes of butterfly communities in the sampled habitats are presented via comparison of species richness, ecological composition, richness and abundance of indicator groups and diversity indices. In addition to the observed species richness nonparametric richness estimators (abundance-based estimators ACE and Chao1 and incidence-based estimators ICE and Chao2) were evaluated using the Species Richness Estimators v2.1 module of www.eco-tools.net. Singletons and doubletons (number of species represented by one or two individuals) were also verified. Three measures of species α diversity were calculated for each habitat: the Shannon index ($H' = -\sum p_i \ln p_i$), equitability ($J = H'/\ln S$—where S is species richness) and Fisher's alpha ($S = \alpha \ln (1 + N/\alpha)$. Community structure comparison between the different forest habitats was estimated using single linkage cluster analysis based on *Jaccard's* similarity index.

Results

Species richness and diversity

3,465 specimens belonging to 114 species were collected during three sampling periods (2,016 cumulative trapping days with an average 1, 7 specimens/trap/day) in 2007 and 2009 using 56 traps (Appendix 1). *Bicyclus larseni*, *B. uniformis*, *B. safitza*, *B. silvicolus*, *B. istaris*, *Charaxes candiope*, *C. pollux*, *C. zelica* were recorded as new to Kakum National Park since the latest checklist (Larsen 2006a). The highest observed species richness was identified from the 'primary forest' transect (62) and the control transects (59, 59), while the secondary growth, farmland, clear-cut and new forest had almost evenly low number of species; 51, 52, 53 and 54, respectively (Table 1; Fig. 2). The estimated species richness was the highest in the clear-cut habitat spanning from 77 (incidence-based ICE) to 128 (abundance based Chao1) compared with the 53 actually collected. Chao1 estimator gave the highest estimate of species richness in the two most 'disturbed' habitats (CC and FL). The Shannon diversity index varied from 2.989 (NF) to 3.768 (PF) and Fisher α ranged from 14.54 (FL) to 21.98 (PF). Species evenness was the highest in the

Table 1 Estimated species richness and diversity indices of the sampled habitats

Habitat	*S*	*N*	Singletons/ doubletons	Richness estimates		Number of presumed species	*H'*	*J*	α
				ACE/ICE	Chao1/Chao2				
PF	62	347	13/12	72/72	68/69	6–10	3.768	0.9130	21.98
SF	51	449	17/10	72/73	63/62	11–22	3.011	0.7658	14.81
NF	54	461	18/12	72/75	66/68	12–18	2.989	0.7494	15.87
FL	52	505	17/3	67/67	86/79	15–34	3.016	0.7632	14.54
CC	53	361	22/2	78/77	128/109	25–75	3.157	0.7952	17.13
K1	59	632	12/9	69/68	66/64	5–10	3.482	0.8539	15.92
K2	59	710	14/9	69/72	68/69	9–13	3.236	0.7936	15.29

S total number of species trapped, *N* number of individuals; *ACE*, *ICE*, *Chao1* and *Chao2* nonparametric richness estimators, *H'* Shannon's diversity index, *J* Pielou's evenness index, α Fisher's alpha diversity index

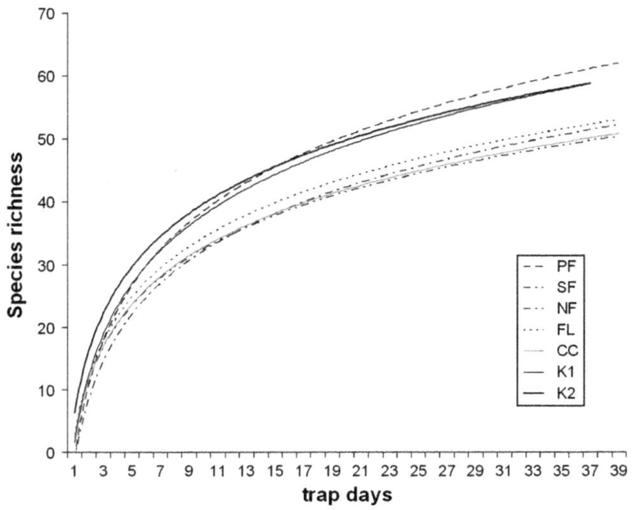

Fig. 2 Species accumulation curves show the actual number of recorded species against trapping days in each sampled habitat

'primary forest' ($J = 0.913$) followed by the 'clear cut' area ($J = 0.8539$).

Ecological composition

The analysis of ecological composition of butterflies based on Larsen (2006a) resulted in relatively small but clear differences between butterfly communities (Table 2). Generally all samples were dominated by forest butterflies; ubiquitous species and those centred on savannah land remained generally low in all samples: 0–5.6 and 3.2–5.8%, respectively. Practically no savannah butterflies were recorded from the PF, SF and NF samples. Mesophilous forest species represent almost one-third, while generalist forest species comprise about half of all recorded species. In the former group the PF reached the highest proportion with approximately 10% higher value of all other samples, in the latter, values in PF, K1 and K2 samples reached averagely 10% higher in all disturbed habitats. The differences are well pronounced in the wet forest samples where all forest samples shown relevantly higher proportion than disturbed habitats, reaching one-fifth of all species in the PF and K2.

Table 2 Number and proportion of butterfly species per ecological categories in the sampled habitats

	PF	SF	NF	FL	CC	K1	K2
DRF	0/0.0	1/2.0	2/3.7	3/5.8	1/1.9	1/1.7	3/5.1
MEF	23/37.1	13/25.5	15/27.8	13/25.0	15/27.8	17/28.8	15/25.4
WEF	12/19.4	6/11.8	7/13.0	5/9.6	3/5.6	10/16.9	12/20.3
ALF	25/40.3	29/56.9	27/50.0	27/51.9	30/55.6	27/45.8	25/42.4
GUI	0/0.0	0/0.0	0/0.0	1/1.9	3/5.6	2/3.4	2/3.4
UBQ	2/3.2	2/3.9	3/5.6	3/5.8	2/3.7	2/3.4	2/3.4

DRF dry forest species, *MEF* mesophilous forest species, *WEF* wet forest species, *ALF* forest species with wide ecological tolerance, *GUI* Guinea savannah species, *UBQ* ubiquitous species (based on Larsen 2006a)

Habitat quality indication

A comparison of species richness and abundance of three genera was used to demonstrate differences between butterfly communities and as an indicator of habitat quality in the sampled habitats (Fig. 3). Species belonging to *Euphaedra* and *Euriphene* were found restricted to intact forest habitats, showing high richness in PF, SF and the control transects. Samples from strongly disturbed habitats contain many fewer species of these groups. The proportion of *Euriphene* is 0% in both CC and FL samples, while the values of *Euphaedra* reach 7.6% (CC) and 38% (FL) of the highest recorded in the PF. The contrast is even more pronounced when abundance is examined. The highest number of individuals of both *Euriphene* and *Euphaedra* were captured in PF, as was the case in the control transects. Abundance drops significantly in disturbed habitats, showing zero values in CC and FL.

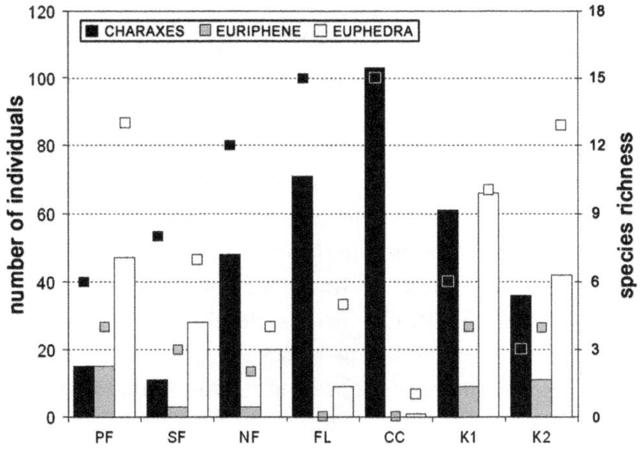

Fig. 3 Species richness (squares) and abundance (columns) of the genera *Euphaedra* (white), *Euriphene* (grey) and *Charaxes* (black) in the sampled habitats. The highest values of *Euphaedra* and *Euriphene* indicate intact forest, while the highest species richness and abundance of the genus *Charaxes* appeared to be in the most disturbed (open) habitats

The differences between the highest and lowest abundance values are 46 specimens for *Euphaedra* and 15 for *Euriphene* (if the controls are left out of the comparison). *Charaxes* reached significantly higher species richness in more disturbed habitats. The number of *Charaxes* species was 60% higher in FL and CC than in PF and 50% higher than in SF (the lowest species richness of *Charaxes* actually occurred in K1 and K2). The abundance values also show significant differences between disturbed habitats and forests. The lowest abundance values were found in the PF and SF, while outstandingly high values were recorded in the FL and CC where no real canopy cover exists. The difference between CC and PF was 88 specimens (85%), while it is even higher comparing to the SF: 92 specimens (90%). Abundance of *Charaxes* in the control transects appears to be higher than in PF and SF, they remain still 35 and 59% lower than the highest value, quite similarly to NF.

Habitat similarity

The agglomerative cluster analysis based on Jaccard's single linear similarity index clearly demonstrate the differences and similarities between butterfly communities in different habitat types. Cluster analysis separated the sampled habitats into two main groups: there is a complete separation between the totally open clear-cut habitat (CC) and the forest habitats. This second group was further subdivided into two subgroups, where transects K1 and K2 (which represent the highest similarity) cluster only with PF, while all disturbed forest communities are in a distinct subgroup with relevant differences (Fig. 4).

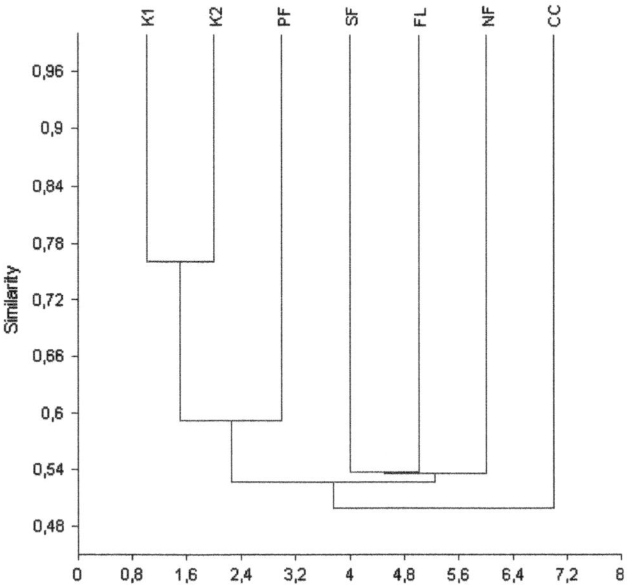

Fig. 4 Dendrogram based on cluster analysis using *Jaccard's* coefficient of similarity on the butterfly communities of the sampled habitats

Discussion

Species richness and diversity

Tropical forest butterfly communities are usually characterized by their high species richness and diversity. Although several studies from the Afrotropical region (Rogo and Odulaja 2001; Stork et al. 2003; Bobo et al. 2006; Bossart et al. 2006; Akite 2008) showed evidence that creation of gaps in forest habitats by logging, clearing for farming or complete clear-felling have a significant negative effect on butterfly species richness and diversity, this trend was the opposite in some Neotropical case studies (Wood and Gillman 1998; Horner-Devine et al. 2003). The increase in species richness and diversity can particularly be observed at a relatively small spatial scale (Hill and Hamer 2004; Bonebrake et al. 2010). It was therefore no surprise that estimated species richness was relatively high in the 'clear-cut' transect, compared to the difference between the estimated and observed richness in the 'farmland' (estimates of total richness represents a 22–40% increase).

In both habitats, maximum species richness was predicted by Chao1 which uses the relative proportions of singletons and doubletons for calculating estimates. Percentage of singletons was especially high in these two habitats (33% in FL and 42% in CC). The results can be explained by the nature of temporary gap habitats in rainforests, where more mobile species can occupy the open area, while a number of forest species are still able to utilize or disperse through open gaps in the forest in low numbers (Hill et al. 2001). Elbers and Bossart (2009) captured several real forest butterflies in more open farmland habitat close to the forest boundary. This gap effect is only partially reflected in the butterfly diversity values, since both Shannon and Fisher α indicate the highest diversity in primary forest habitat with evenly lower values in all other samples.

Ecological composition

The recorded differences between ecological composition of butterflies in the samples are not necessarily emphasised by the penetration of species characteristic of more open country into the degraded habitats but is best viewed through the richness of the 'wet forest species' (WEF). The high values in the PF and the control samples and the opposite in the disturbed forest samples show evidence that butterflies centred in wetter types of forests are sensitive enough to disappear from forest habitats upon disturbance. This is also well supported well by Sáfián et al. (2009) who found WEF almost completely missing from samples collected in heavily degraded forest habitats in the Togo Mountains (Togo), while Larsen (2006b) records several wet forest specialists from the better conserved forest in the same type of area in Ghana (Wli Falls). The generally low number of ubiquitous species (UBQ) and those centred in savannah habitats (GUI) indicates that temporal gaps in forest areas created by smaller scale disturbance might not necessarily establish favourable conditions for colonisation of butterflies inhabiting open areas. Habitat disturbance on a larger scale however caused drastic changes in the butterfly composition in Sierra Leone, increasing the proportion of UBQ and GUI species significantly (Sundufu and Dumbuya 2008; Sáfián et al. 2009). Minor but interesting differences are shown by the value of mesophilous forest species (MEF) in the PF. This 10% higher proportion could be explained by the disappearance of more sensitive species upon disturbance, just like the opposite in the generalist forest species (ALF), where butterflies with higher tolerance for habitat disturbance were able to appear in the SF, NF, FL and CC samples.

Habitat quality indication

The genera *Euriphene* and *Euphaedra* are known as forest interior species in the rainforest belt of tropical Africa that have strong fidelity to their habitats (Larsen 2005). Fermon et al. (2000) found *Euriphene* and *Euphaedra* almost exclusively in forest samples in Ivory Coast, while they were almost absent from plantation forest. Elbers and Bossart (2009) recorded only a few *Euphaedra* specimens and none of the *Euriphene* from their 'matrix' non-forest samples. This corresponds well with our results, strengthening the hypothesis that higher species richness and abundance of *Euphaedra* and *Euriphene* indicate more intact rainforest ecosystem in the Afrotropics. Although Molleman et al. (2006) recorded several *Charaxes* species both from understorey and canopy traps, other surveys and observations showed that the majority of forest-dwelling *Charaxes* are usually canopy (or middle canopy) butterflies that avoid entering dense undergrowth of forest but are able to appear quickly in open areas in forested habitats (Larsen 2005; Fermon et al. 2003). Our results clearly indicate that disturbance in forest habitats increases the species richness and abundance of *Charaxes* appearing in fruit-baited understorey traps. The abundance values in the control transects were actually caused almost exclusively by a single species *Charaxes fulvescens*, which is among the few *Charaxes* that regularly penetrate forest interior (pers. obs., Molleman et al. 2006).

Conclusions

The sampled clear-cut habitat has a low degree of similarity with those of forest habitats, based on the ecological analyses of butterfly community composition. However, the recovery of the primary forest butterfly community was observed from newly planted young forest and farm land towards middle-aged secondary growth. Diversity and rarefied species richness indices alone are not necessarily adaptable to evaluate butterfly community collapse and recovery in rainforest habitats, especially when the disturbed area is surrounded by habitats of more natural type, since disturbance allow a wide range of species to penetrate degraded habitats biasing diversity data. The genera, *Euriphene* and *Euphaedra* have proven to be indicators of forest interior in good condition, because they appeared to demonstrate very strong fidelity to the middle-aged and mature forest habitats; their high species richness and abundance in the secondary forest shown clear evidence of regeneration of the forest butterfly community. The presence of canopy dwelling *Charaxes* in large quantities indicated considerable habitat disturbance—extreme abundance was recorded only from the clear-cut with complete absence of forest cover. These results might prove important in further evaluations of the success of forest conservation and regeneration projects especially in afro-tropical areas.

Acknowledgments We thank all field assistants, namely Charles Owusu, Erika Zakar, Janka Bodorik, Zsuzsanna Kúti, Bence Farkas, Előd Kondorosy, Nikolett Rendes. Szabolcs Sáfián and Gábor Csontos are especially grateful to Simon Meadow-Smith who develops the studied area for conservation and eco-tourism, in which butterflies play an important role. Our research could not have been carried out without the various types of support from László Tóth (Minerex Drilling Ltd., Accra), Kati Torda (Sundtrade Ltd., Accra) and Nana Mike Adu-Nsiah (Wildlife Division of Forestry Commission, Ghana). Torben Larsen, David Dennis, Martin Warren and an anonymous reviewer helped to improve the manuscript with their comments and corrections.

Appendix I

See Table 3.

Table 3 Recorded butterfly species and abundance values from the sampled habitats

	PF	SF	NF	FL	CC	K1	K2
Family Lycaenidae							
Subfamily Polyommatinae							
Anthene larydas (Cramer, 1780)	0	0	0	0	1	0	0
Family Nymphalidae							
Subfamily Libytheinae							
Libythea labdaca Westwood, 1851	0	0	0	0	11	0	0
Subfamily Satyrinae							
Gnophodes betsimena (Boisduval, 1833)	5	14	16	11	4	27	36
Gnophodes chelys (Fabricius, 1793)	10	0	0	0	4	18	28
Melanitis leda (Linnaeus, 1758)	2	2	1	5	1	17	8
Melanitis libya Distant, 1882	3	17	6	8	8	1	1
Elymniopsis bammakoo (Westwood, [1851])	0	35	44	23	10	0	0
Bicyclus xeneas (Hewitson, 1866)	13	9	2	4	3	2	2
Bicyclus evadne (Cramer, 1779)	14	7	2	5	6	2	4
Bicyclus ephorus Weymer, 1892	3	0	1	0	0	2	3
Bicyclus zinebi (Butler, 1869)	7	3	2	0	1	27	23
Bicyclus uniformis (Bethune-Baker, 1908)	1	0	0	0	0	27	29
Bicyclus procora (Karsch, 1893)	12	13	0	3	0	16	27
Bicyclus larseni vande Weghe, 2009	3	1	1	1	0	1	2
Bicyclus taenias (Hewitson, 1877)	4	49	34	49	19	5	16
Bicyclus vulgaris (Butler, 1868)	0	5	8	15	11	5	1
Bicyclus dorothea (Cramer, 1779)	1	3	23	55	8	8	4
Bicyclus sandace (Hewitson, 1877)	6	14	63	83	30	9	5
Bicyclus sambulos (Hewitson, 1877)	12	12	6	1	0	42	43
Bicyclus sangmelinae Condamin, 1963	12	14	12	13	5	11	19
Bicyclus auricruda (Butler, 1868)	0	0	0	0	0	13	14
Bicyclus sylvicolus Condamin, 1965	6	0	0	0	0	16	17
Bicyclus abnormis (Dudgeon, 1909)	1	0	0	0	0	2	1
Bicyclus safitza (Westwood, 1850)	0	0	0	0	3	0	1

Table 3 continued

	PF	SF	NF	FL	CC	K1	K2
Bicyclus funebris (Guérin-Méneville, 1844)	0	0	0	0	0	21	3
Bicyclus istaris (Plötz, 1880)	2	0	0	0	0	0	0
Bicyclus madetes (Hewitson, 1874)	15	41	20	14	10	36	59
Bicyclus martius (Fabricius, 1793)	21	98	97	80	87	32	39
Hallelesis halyma (Fabricius, 1793)	2	13	0	1	1	3	6
Henotesia decira (Plötz, 1880)	0	0	0	0	0	26	29
Subfamily Charaxinae							
Charaxes varanes (Cramer, 1777)	0	0	0	0	1	0	0
Charaxes fulvescens (Aurivillius, 1891)	4	1	2	0	1	29	19
Charaxes candiope (Godart, 1824)	0	0	0	0	1	1	0
Charaxes protoclea Feisthamel, 1850	3	2	0	1	4	20	14
Charaxes boueti Feisthamel, 1850	0	0	3	1	5	0	0
Charaxes cynthia Butler, 1866	1	1	6	5	8	0	0
Charaxes lucretius Cramer, [1775]	0	0	1	4	5	0	0
Charaxes castor (Cramer, 1775)	0	0	2	1	0	0	0
Charaxes brutus (Cramer, 1779)	0	1	11	20	14	0	0
Charaxes pollux (Cramer, 1775)	0	0	0	1	0	0	0
Charaxes tiridates (Cramer, 1777)	0	2	5	12	22	1	0
Charaxes bipunctatus Rothschild, 1894	0	0	2	1	2	0	0
Charaxes numenes (Hewitson, 1859)	1	0	6	3	19	4	0
Charaxes zingha (Stoll, 1780)	6	1	5	3	9	6	3
Charaxes eupale (Drury, 1782)	0	1	0	6	1	0	0
Charaxes anticlea (Drury, 1782)	0	0	4	7	4	0	0
Charaxes etheocles agg. (Cramer, 1777)	0	0	0	4	7	0	0
Charaxes pleione (Godart, 1824)	0	0	0	2	0	0	0
Charaxes paphianus Ward, 1871	0	2	0	0	0	0	0
Charaxes zelica Butler, 1869	0	0	1	0	0	0	0
Charaxes eurinome (Cramer, 1775)	0	0	2	0	0	0	0
Palla violinitens (Crowley, 1890)	2	0	2	0	5	1	0
Palla decius (Cramer, 1777)	0	1	1	1	1	0	0
Palla ussheri (Butler, 1870)	4	1	0	4	6	6	2
Palla publius Staudinger, 1892	0	0	0	2	1	0	1
Subfamily Nymphalinae							
Kallimoides rumia (Doubleday, 1849)	9	3	1	2	1	75	142
Antanartia delius (Drury, 1782)	0	0	0	0	0	1	0
Hypolimnas anthedon (Doubleday, 1845)	0	0	0	0	1	0	0
Hypolimnas salmacis (Drury, 1773)	0	0	0	1	1	0	0
Protogoniomorpha parhassus (Drury, 1782)	0	0	1	0	1	0	1

Table 3 continued

	PF	SF	NF	FL	CC	K1	K2
Junonia stygia (Aurivillius, 1894)	0	0	0	0	1	0	0
Subfamily Biblidinae							
Byblia anvatara (Boisduval, 1833)	0	0	1	4	0	0	0
Ariadne enotrea (Cramer, 1779)	0	0	0	0	5	0	0
Subfamily Limenitinae							
Cymothoe fumana (Westwood, 1850)	8	2	0	0	2	2	6
Cymothoe egesta (Cramer, 1775)	6	0	2	1	0	2	2
Cymothoe aubergeri Plantrou, 1977	1	0	0	0	1	0	0
Cymothoe druryi van Velzen & Larsen, 2009	0	1	3	3	0	0	0
Cymothoe althea (Cramer, 1776)	0	0	1	0	0	0	0
Cymothoe jodutta (Westwood, 1850)	0	0	0	0	1	1	0
Cymothoe mabillei Overlaet, 1944	2	2	1	0	0	0	0
Cymothoe sangaris (Godart, 1824)	0	0	0	0	0	1	5
Catuna oberthueri Karsch, 1894	0	1	0	0	0	0	0
Catuna angustatum (Felder & Felder, 1867)	0	0	0	0	0	1	0
Euryphura chalcis (Felder & Felder, 1860)	0	0	0	1	1	0	0
Hamanumida daedalus (Fabricius, 1775)	0	0	0	1	0	0	0
Aterica galene (Brown, 1776)	11	7	1	8	1	6	12
Euriphene barombina (Aurivillius, 1894)	2	1	0	0	0	1	3
Euriphene simplex (Staudinger, 1891)	0	0	0	0	0	0	1
Euriphene aridatha (Hewitson, 1866)	2	1	0	0	0	0	0
Euriphene gambiae Feisthamel, 1850	10	1	2	0	0	5	5
Euriphene ampedusa (Hewitson, 1866)	0	0	1	0	0	2	2
Euriphene atossa (Hewitson, 1865)	1	0	0	0	0	0	0
Euriphene doriclea (Drury, 1782)	0	0	0	0	0	1	0
Bebearia tentyris (Hewitson, 1866)	2	0	0	0	0	0	0
Bebearia absolon (Fabricius, 1793)	0	1	1	0	0	0	0
Bebearia zonara (Butler, 1871)	2	0	0	0	0	0	0
Bebearia mandinga (Felder & Felder, 1860)	0	1	0	0	0	0	0
Bebearia abesa (Hewitson, 1869)	1	0	0	0	0	0	0
Bebearia mardania (Fabricius, 1793)	15	22	23	17	3	3	5
Bebearia cocalia (Fabricius, 1793)	6	6	4	3	1	2	1
Bebearia paludicola Holmes, 2001	5	3	1	5	0	0	0
Bebearia sophus (Fabricius, 1793)	14	4	3	0	0	3	2

Table 3 continued

	PF	SF	NF	FL	CC	K1	K2
Bebearia phantasina (Staudinger, 1891)	16	2	2	1	0	23	21
Bebearia cutteri (Hewitson, 1865)	1	0	0	0	0	0	0
Euphaedra medon (Linnaeus, 1763)	14	17	14	3	0	10	6
Euphaedra gausape (Butler, 1866)	1	0	0	0	0	0	0
Euphaedra xypete (Hewitson, 1865)	1	2	0	0	0	3	2
Euphaedra hebes Hecq, 1980	3	0	0	0	0	0	0
Euphaedra diffusa Gaede, 1916	0	0	0	0	0	0	1
Euphaedra crockeri (Butler, 1869)	2	0	0	0	0	8	6
Euphaedra cyparissa (Cramer, 1775)	0	0	0	0	0	0	1
Euphaedra themis (Hübner, 1807)	0	1	0	1	0	1	1
Euphaedra minuta Hecq, 1982	1	0	0	0	0	0	0
Euphaedra janetta (Butler, 1871)	3	0	0	1	0	3	2
Euphaedra ceres (Fabricius, 1775)	8	3	3	3	0	12	5
Euphaedra phaethusa (Butler, 1866)	6	2	1	0	1	9	8
Euphaedra inanum (Butler, 1873)	1	0	0	0	0	0	0
Euphaedra francina (Godart, 1824)	0	0	0	0	0	2	1
Euphaedra zampa (Westwood, 1850)	2	0	0	0	0	15	6
Euphaedra perseis (Drury, 1773)	0	1	0	0	0	0	1
Euphaedra harpalyce (Cramer, 1777)	3	2	0	1	0	3	2
Euphaedra eupalus (Fabricius, 1781)	2	0	2	0	0	0	0
Subfamily Heliconiinae							
Lachnoptera anticlia (Hübner, 1819)	0	0	1	0	0	0	1
Family Hesperiidae							
Subfamily Pyrginae							
Celaenorrhinus galenus (Fabricius, 1793)	0	0	1	0	1	0	0

PF Primary Forest, *SF* Secondary Forest, *NF* Young Planted Forest, *FL* Farmland, *CC* Clear-cut, *K1* Control 1, *K2* Control 2. Systematics and nomenclature follow Larsen (2005) complemented by the latest taxonomical updates (Libert 2006, Aduse-Poku et al. 2009, van Velzen and Larsen 2009, vande Weghe 2009)

References

Aduse-Poku K, Vingehoedt E, Wahlberg N (2009) Out-of-Africa again: a phylogenetic hypothesis of the genus *Charaxes* (Lepidoptera: Nymphalidae) based on five gene regions. Molec Phylogen Evol 53(2):463–478

Aide TM, Zimmerman JK, Pascarella JB, Rivera L, Marcano-Vega H (2000) Forest regeneration in a chronosequence of tropical abandoned pastures: implications for restoration ecology. Rest Ecol 8:328–338

Akite P (2008) Effects of anthropogenic disturbances on the diversity and composition of the butterfly fauna of sites in the Sango Bay and Iriiri areas, Uganda: implications for conservation. Afr J Ecol 46:3–13

BirdLife International (2009) Important bird area factsheet: kakum national park—Assin attandaso resource reserve, Ghana. Data zone. http://www.birdlife.org. Accessed 25 July 2010

Bobo KS, Waltert M, Fermon H, Njokagbor J, Mühlenberg M (2006) From forest to farmland: butterfly diversity and habitat associations along a gradient of forest conversion in Southwestern Cameroon. J Insect Conserv 10:29–42

Bonebrake TC, Ponisio LC, Boggs CL, Ehrlich PR (2010) More than just indicators: a review of tropical butterfly ecology and conservation. Biol Cons 143:1831–1841

Bossart JL, Opuni-Frimpong E (2009) Distance from edge determines fruit-feeding butterfly community diversity in afrotropical forest fragments. Environ Entomol 38(1):43–52

Bossart JL, Opuni-Frimpong E, Kuudaar S, Nkrumah E (2006) Richness, abundance, and complementarity of fruit-feeding butterfly species in relict sacred forests and forest reserves of Ghana. Biodivers Conserv 15(1):333–359

Chiti T, Certini G, Grieco E, Valentini R (2010) The role of soil in storing carbon in tropical rainforests: the case of ankasa park, Ghana. Plant soil. doi: 10.1007/s11104-009-0265-x

Dunk RR (2004) Recovery of faunal communities during tropical forest regeneration. Conserv Biol 18:302–309

Elbers JP, Bossart JL (2009) Occurrences of forest butterflies in the farm bush savannah outside a forest reserve in Ghana, West Africa. Int J Trop Insect Sci 29(3):141–150

Fermon H, Waltert M, Larsen TB, Dall'Asta U, Mühlenberg M (2000) Effects of forest management on diversity and abundance of fruit-feeding nymphalid butterflies in south-eastern Cote d'Ivoire. J Insect Conserv 4(3):173–189

Fermon H, Waltert M, Mühlenberg M (2003) Movement and vertical stratification of fruit-feeding butterflies in a managed West African rainforest. J Insect Conserv 7(1):7–19

Foody GM, Palubinskas G, Lucas RM, Curran PJ, Honzak M (1996) Identifying terrestrial carbon sinks: classification of successional stages in regenerating tropical forest from Landsat TM data. Remote Sens Environ 55:205–216

Hill JK, Hamer KC (2004) Determining impacts of habitat modification on diversity of tropical forest fauna: the importance of spatial scale. J Appl Ecol 41:744–754

Hill JK, Hamer KC, Tangah J, Dawood M (2001) Ecology of tropical butterflies in rainforest gaps. Oecologia 128:294–302

Horner-Devine MC, Daily GC, Ehrlich PR, Boggs CL (2003) Countryside biogeography of tropical butterflies. Conserv Biol 17:168–177

Kassi N'Dja JK (2008) Successional patterns of plant species and community diversity in a semi-deciduous tropical forest under shifting cultivation. J Veg Sci 19(6):809–820

Larsen TB (2005) The butterflies of West Africa. Apollo Books, Svendborg, Denmark

Larsen TB (2006a) The Ghana butterfly fauna and its contribution to the objectives of the protected areas system. WDSP report no. 63. Wildlife division (Forestry commission) & IUCN (World Conservation Union)

Larsen TB (2006b) The butterflies of kyabobo national park, Ghana, and those of the Volta region. WDSP report no. 64. Wildlife Division (Forestry Commission) & IUCN (World Conservation Union)

Larsen TB (2007) Additional comments on butterflies of the upland evergreen forest of the atewa range forest reserve, Ghana. In: McCullough J, Alonso LE, Naskrecki P, Wright HE, Osei-Owusu Y (eds) A rapid biological assessment of the atewa range forest reserve, Eastern Ghana. RAP bulletin of biological assessment, vol 47. Conservation International, Arlington, pp 61–62

Lawton JH, Bignell DE, Bolton B, Bloemers GF, Eggleton P, Hammond PM, Hodda M, Holt RD, Larsen TB, Mawdsley NA, Stork NE, Srivastava DS, Watt AD (1998) Biodiversity inventories, indicator taxa and effects of habitat modification in tropical forest. Nature 391:72–76

Libert M (2006) Note sur le genre Henotesia Butler (Lepidoptera, Satyridae). Lambillionea 106(1):1–14

Molleman F, Kop A, Brakefield PM, De Vries PJ, Zwaan BJ (2006) Vertical and temporal patterns of biodiversity of fruit-feeding butterflies in a tropical forest in Uganda. Biodiv Conserv 15(1):107–121

Myers N, Mittermeier RA, Mittermeier CG, da Fonesca GAB, Kent J (2000) Biodiversity hotspots for conservation priorities. Nature 403:853–858

Otsamo R (1999) Secondary forest regeneration under fast-growing forest plantations on degraded *Imperata cylindrica* grasslands. New Forest 19:69–93

Parrotta JA, Turnbull JW, Jones N (1997) Catalyzing native forest regeneration on degraded tropical lands. Forest Ecol Manag 99:1–7

Riley L, Riley W (2005) Nature's strongholds: the world's great wildlife reserves. Princeton University Press, Princeton and Oxford

Rogo L, Odulaja A (2001) Butterfly populations in two forest fragments at the Kenya coast. Afr J Ecol 39:266–275

Sáfián Sz, Csontos G, Kormos B (2009) Butterfly (Lepidoptera: Papilionoidea & Hesperioidea) records from Mount Agou and Klouto Region (near Kpalimé), Togo Mountains, Togo (with notes on the species composition in highly degraded forest habitats). Metamorph J Lep Soc Afr 20(4):115–129

Stork NE, Srivastava DS, Watt AD, Larsen TB (2003) Butterfly diversity and silvicultural practice in lowland rainforests of Cameroon. Biodivers Conserv 12:387–410

Sundufu AJ, Dumbuya R (2008) Habitat preferences of butterflies in the bumbuna forest, Northern Sierra Leone. J Insect Sci 8:64

Tiarks A, Nambiar EKS, Cossalter C (1998) Site Management and Productivity in Tropical Forest Plantations. CIFOR Occasional Paper No. 16:1–10

UNEP/GRID-Arendal (2002) Deforestation in West Africa: case Cote-d'Ivoire. In: UNEP/GRID-Arendal maps and graphics library. http://maps.grida.no/go/graphic/deforestation_in_west_africa_case_cote_d_ivoire(Accessed March 15, 2010)

van Velzen R, Larsen TB (2009) A new hidden species of the Cymothoe caenis-complex (Lepidoptera: Nymphalidae) from western Africa. Zootaxa 2197:53–63

Vanclay KJ (2004) Indicator groups and faunal richness. FBMIS Volume 1. http://www.fbmis.info/. Accessed 15 March 2010

vande Weghe G (2009) Description de nouveaux taxons et contribution à l'étude des Lépidoptères afrotropicaux (Lepidoptera: Nymphalidae, Limenitidinae, Satyrinae; Hesperiidae, Hesperiinae; Lycaenidae, Theclinae). Entomol Afric 14(1):1–24

Wood B, Gillman MP (1998) The effects of disturbance on forest butterflies using two methods of sampling in Trinidad. Biodivers Conserv 7:597–616

Wunderle JM Jr (1997) The role of animal seed dispersal in accelerating native forest regeneration on degraded tropical lands. Forest Ecol Manag 99(1–2):223–235

J Insect Conserv (2011) 15:5–22
DOI 10.1007/s10841-010-9299-x

ORIGINAL PAPER

Conservation of butterflies in Japan: status, actions and strategy

Yasuhiro Nakamura

Received: 23 March 2010 / Accepted: 1 June 2010 / Published online: 1 October 2010

Abstract There are about 240 butterfly species in Japan of which 15% are endangered. Grassland butterflies are the most threatened and have declined most widely, some with an extremely high rate of decline of over 80% in 40 years. The main cause is the change of "Satoyama" ecosystems, a traditional landscape including secondary woodlands, grasslands, paddy field and other habitats. However, most "Satoyama" ecosystems have been abandoned or destroyed as the landscape has been industrialized. This process has accelerated in recent years, leading to even greater impacts on butterflies. To halt this decline, the Japan Butterfly Conservation Society (JBCS) was founded in 2004 and is doing various crucial activities. Local groups have been formed and numbers have been growing steadily. Also, public awareness toward conservation of the natural environment has been increasing and conservation groups related to "Satoyama" exceed 1000. Government policy on biodiversity has developed since The National Biodiversity Strategy of Japan was published in 1995. JBCS has set targets to prevent threatened butterflies from becoming extinct at national and regional level. However, many difficulties remain and a greater effort is needed to develop a social system to maintain "Satoyama" as an industry. Although interest in the conservation of Japanese butterflies began several decades ago, full scale conservation activities have only started recently and future progress is expected.

Guest Editor: Dr. M.S. Warren.

Y. Nakamura (✉)
Japan Butterfly Conservation Society (JBCS), 1-36-1-301 Ohi, Shinagawa-ku, Tokyo 140-0014, Japan
e-mail: ynakamura@japan-inter.net

Keywords Butterflies · Conservation · Threatened species · Satoyama · Red Data Book · Japan

Introduction

While large numbers of wildlife are endangered on a global basis (Hilton-Taylor 2000), many species of butterflies are also threatened with extinction at national and regional level (e.g. Collins and Morris 1985; New 1993; Van Swaay and Warren 1999; Van Swaay et al. 2010).

Butterflies are a good indicator group, because their populations respond more rapidly to adverse environmental change than longer-lived organisms (Thomas et al. 2004) and several studies have related their changes to important environment issues such as global warming (e.g. Margaret 2008; Settele et al. 2008), agricultural intensification (e.g. Van Swaay et al. 2010), and the effect of air pollution (e.g. Corke 1999; Öckinger et al. 2006). Trends in butterflies can also help assess conservation measures such as agri-environment schemes (e.g. Rundlöf and Smith 2006), landscape scale conservation (e.g. Dennis 2010), environmental restoration (e.g. Lomov et al. 2006), and conservation planning (e.g. Cabeza et al. 2010). In addition, butterflies are one of the most popular wildlife elements and there are many thousands of amateur enthusiasts in Japan (Ishii 1996) and many more around the world.

Against this background, butterfly conservation has made considerable progress since the 1990s (New 1997) and efforts have spread around the world. Non-governmental organizations now play a considerable role and there are societies at national level dedicated to butterfly conservation in the following countries; Europe: United Kingdom, the Netherlands, Czechoslovakia, Poland, Switzerland, Slovenia etc.; America: USA; Africa: Ghana;

Asia: Japan and Republic of China. There are even more groups operating at local level. In the UK, Butterfly Conservation was founded in 1968 and now has a membership exceeding 14,000. Butterflies have become one of the "flagships" in biodiversity conservation.

In Japan, the conservation of butterflies started mainly in the late 1960s but has become more active since the late 1980s (Shibatani 1989, 1990; Ishii 1996). The main aim of this paper is to review the recent situation of endangered butterflies in Japan and describe the causes of decline, historical background, practical conservation, problems, and future conservation strategy.

The taxonomic order and nomenclature of butterflies follows Yata (2007). There are two levels of local public authorities in Japan. The country is currently divided into 47 "prefectures" (8,041 ± 11,575 km^2; mean area ± SD) and 1,727 "municipalities" (216 ± 247 km^2; mean area ± SD, in 2009, not divided into "ward" unit). Japan is an arc-shaped archipelago consisting of four main islands, Hokkaido, Honshu, Shikoku, and Kyushu, from northeast to southwest. Detailed data of species distribution are available from the author.

The status of threatened butterflies in Japan

National level

There are about 240 butterfly species in Japan. The revised Red List of butterflies in 2007 shows about 15% of butterfly species are threatened either in the category "Critically Endangered (CR) and Endangered (EN)" or "Vulnerable (VU)" (Ministry of the Environment 2007; Table 1). The national Red List produced by the Ministry of the Environment includes subspecies as the selection unit and combines the categories of "Critically Endangered" and "Endangered" for all insects (Table 1).

In the most threatened category "CR + EN", 10 out of 12 species are grassland butterflies, exceptions being the locally restricted subspecies of *Antigius butleri kurinodakensis*, *Japonica onoi mizobei* and *Oeneis norna sugitanii*. The other two are forest butterflies, *Shijimia moorei* and *Celastrina ogasawaraensis*, whose distribution is originally highly restricted. The trend is similar for VU species and it is clear that grassland butterflies are the most threatened and have declined more widely in Japan (Nakamura 2003, 2005; Imura and Shi 2004; Fujii 2006, 2007).

As yet, no butterfly species has become extinct in Japan, but several are close to extinction. The decline of grassland butterflies of "CR + EN" exceeds 50%, and some grassland butterflies have extremely high rates of decline, for example *Melitaea scotosia* (98%) and *Fabriciana nerippe* (95%) (Table 1; Fig. 1). Two other forest butterflies are now extremely rare, including *Shijimia moorei* which is found at only two sites, and *Celastrina ogasawaraensis* which is at a very low population level (Nakamura 2009b). The decline of grassland butterflies in Japan has parallels with those described in Europe, where many grassland species are listed as threatened, including *Maculinea teleius* (VU) and *Coenonympha oedippus* (EN) which both also occur in Japan (Van Swaay et al. 2010).

The main causes of drastic decline of grassland butterflies are abandonment of and changes to traditional land use, especially the degradation of "Satoyama" (see below). Over 40 butterflies breed in actively managed coppice woodland in Japan, but only a few are listed as threatened and the rate of the decline remains at relatively low level (e.g. *Luehdorfia japonica*: ca. 22%). Their decline is also caused by the abandonment of this traditional management.

Some butterflies that are endangered in Japan are common in South Korea, including *Shijimiaeoides divinus*, *Plebejus subsolanus*, *Maculinea teleius*, *Melitaea scotosia*, *Fabriciana nerippe* (Kim 2002).

Regional level

At the regional level, a large number of species have become extinct (Table 2). Within the major island units, for example, four species had already disappeared in Shikoku; *Pyrgus maculatus*, *Everes lacturnus*, *Niphanda fusca*, *Fabriciana nerippe*. The total number of extinctions at main island level is around seven. At the level of prefecture units, the total number of "extinct" and "probably extinct" has reached around 260 populations (Table 2). Moreover, a further 160 regional populations are at a Critically Endangered level. If the conservation target is set to conserve regional populations at prefecture unit level, more extensive conservation measures are needed immediately (Nakamura 2005).

There are the regional differences in the decline of butterfly species (Nakamura 2003). For example, the percentage decline at municipality level of *Maculinea teleius* is fairly low (Table 1), but varies in different subspecies (*M. teleius ogumae*: 3%, *M. t. kazamoto*: 86%, *M. t. hosonoi*: 18%, *M. t. daisensis*: 46%). Similarly, the decline of *Shijimiaeoides divinus* is different in Honshu (*S. divinus barine*: 96%) compared to Kyushu (*S. d. asonis*: 20%).

Causes of decline

The causes of decline of each endangered butterflies are listed on prefecture level in the prefecture unit Red Data List (Yata and Ueda 1993; Sunose and Eda 2003; Mano and Fujii 2009) and prefecture unit Red Data Book (published by 47 prefectural authorities). Inoue (2005)

Table 1 Endangered butterflies in Japan

Species[a]	Red List[b]	Habitat	Geographical distribution[c]	No. prefectures with records[d]	Prefectures with current records (% decline)[e]	Decline at municipality level[f]	Main cause of decline[g]
Hesperiidae							
Aeromachus inachus	CR+EN	Grassland	Chinese type	28	21 (25.0%)	69.8%	Degradation of Satoyama
Pyrgus maculatus	CR+EN	Grassland	Chinese type	20	6 (70.0%)	75.0%	Degradation of Satoyama, Agricultural develop.
Lycaenidae							
Antigius butleri kurinodakensis (2)	CR+EN	Woodland	Ussuri type	1	1 (0%)	0%	Degradation of Satoyama
Celastrina ogasawaraensis	CR+EN	Forest	Endemic in Ogasawara Islands	1	1 (0%)	75.0%*	Alien species (predation and vegetation change)
Everes lacturnus (2)	CR+EN	Grassland	Pan-tropical type	12	6 (50.0%)	58.6%	Urban develop., Degradation of Satoyama
Niphanda fusca	CR+EN	Grassland/Open woodland	Chinese type	44	27 (38.6%)	76.3%	Degradation of Satoyama, Urban develop.
Shijimia moorei	CR+EN	Forest	Himalayan type	3	1 (66.7%)	57.1%	Forestry, Overcollecting
Shijimiaeoides divinus (2)	CR+EN	Grassland	Ussuri type	8	3 (62.5%)	82.5%	Agricultural develop., Degradation of Satoyama, Overcollecting
Zizina emelina	CR+EN	Grassland	Pan-tropical type	36	18 (50.0%)	62.1%	Waterside develop., Degradation of Satoyama
Nymphalidae							
Fabriciana nerippe	CR+EN	Grassland	Chinese type	43	7 (83.7%)	94.7%	Degradation of Satoyama
Melitaea protomedia	CR+EN	Grassland	Siberian type	5	4 (20.0%)	87.0%	Degradation of Satoyama
Melitaea scotosia	CR+EN	Grassland	Ussuri type	14	1 (92.9%)	97.8%	Degradation of Satoyama, Agricultural develop.
Coenonympha oedippus (2)	CR+EN	Grassland	Siberian type	17	9 (47.1%)	77.7%	Degradation of Satoyama, Urban develop.
Hesperiidae							
Carterocephalus palaemon akaishianus (2)	VU	Subalpine grassland	Siberian type	2	2 (0%)	0%	Global warming (vegetation change)?
Hesperia florinda	VU	Grassland	Siberian type	10	5 (50.0%)	64.8%	Degradation of Satoyama
Ochlodes subhyalina	VU	Bamboo grove	Chinese type	1	1 (0%)	0%	Other develop.**
Parnara ogasawarensis	VU	Grassland	Endemic in Ogasawara Islands	1	1 (0%)	14.3%*	Alien species (predation)?
Pyrgus malvae	VU	Alpine grassland	Siberian type	1	1 (0%)	0%	Global warming (vegetation change)?
Papilionidae							
Luehdorfia japonica	VU	Woodland / Forest	Japanese type	26	24 (7.7%)	21.5%	Degradation of Satoyama, Forestry, Overcollecting

Table 1 continued

Species[a]	Red List[b]	Habitat	Geographical distribution[c]	No. prefectures with records[d]	Prefectures with current records (% decline)[e]	Decline at municipality level[f]	Main cause of decline[g]
Pieridae							
Aporia hippia	VU	Open forest	Ussuri type	4	4 (0%)	48.5%	Forestry, Tourism develop., Agricultural develop.
Eurema laeta	VU	Grassland	Pan-tropical type	37	26 (29.7%)	>55%	Waterside develop., Agricultural develop., Degradation of Satoyama
Gonepteryx maxima	VU	Open woodland	Chinese type	11	5 (54.5%)	67.3%	Degradation of Satoyama, Agricultural develop.
Leptidea amurensis	VU	Grassland	Ussuri type	22	15 (31.8%)	63.2%	Degradation of Satoyama, Agricultural develop., Waterside develop.
Lycaenidae							
Coreana raphaelis	VU	Woodland	Ussuri type	3	3 (0%)	25.0%	Agricultural develop., Degradation of Satoyama
Japonica onoi onoi / *J. o. mizobei* (2)	VU/CR+EN	Woodland	Chinese type	6 / 2	5 (16.7%) / 2 (0%)	2.3% / 0%	Degradation of Satoyama, Urban develop., Agricultural develop.
Plebejus argyrognomon	VU	Grassland	Siberian type	16	9 (43.8%)	65.1%	Waterside develop., Degradation of Satoyama
Plebejus subsolanus (3)	VU	Grassland	Siberian type	12	8 (33.3%)	56.4%	Degradation of Satoyama, Forestry, Urban develop.
Maculinea teleius (4)	VU	Grassland	Siberian type	22	15 (31.8%)	38.4%	Degradation of Satoyama, Agricultural develop., Overcollecting
Panchala ganesa	VU	Forest	Himalayan type	14	9 (35.7%)	35.2%	Forestry
Zizeeria karsandra	VU	Grassland	Pan-tropical type	1	1 (0%)	>80%*	Alien species (vegetation change)
Nymphalidae							
Limenitis populi	VU	Forest	Siberian type	11	3 (72.7%)	41.4%	Forestry, Waterside develop.
Melitaea ambigua	VU	Grassland	Ussuri type	9	4 (55.6%)	62.8%	Degradation of Satoyama, Agricultural develop.
Lethe marginalis	VU	Woodland	Chinese type	28	23 (17.9%)	>55%	Degradation of Satoyama, Forestry
Oeneis norna asamana / *O. n. sugitanii* (2)	VU/CR+EN	Alpine grassland	Siberian type	4 / 1	4 (0%) / 1 (0%)	0% / 0%	Tread under foot **
Ypthima multistriata niphonica (2)	VU	Grassland	Chinese type	31	21 (32.3%)	>50%	Degradation of Satoyama, Agricultural develop.
Red List Status							
"Near threatened"[a]							

Hesperiidae: *Carterocephalus palaemon satakei* (2), *Leptalina unicolor*, *Parnara bada*, *Thymelicus leoninus* (2), Papilionidae: *Luehdorfia puziloi* (2), *Parnassius eversmanni*, Pieridae: *Anthocharis cardamines* (2), *Colias palaeno* (2), Lycaenidae: *Artipe eryx*, *Fixsenia iyonis* (3), *Maculinea arionides*, *Pithecops corvus*, *Pithecops fulgens*, *Plebejus argus micrargus* (2),

Table 1 continued

Species[a]	Red List[b]	Habitat	Geographical distribution[c]	No. prefectures with records[d]	Prefectures with current records (% decline)[e]	Decline at municipality level[f]	Main cause of decline[g]
Spindasis takanonis, *Tongeia fischeri* (3), *Albulina optilete*, Nymphalidae: *Argyronome laodice*, *Brenthis daphne* (2), *Clossiana freija*, *Clossiana iphigenia*, *Hestina assimilis shirakii* (2)[h], *Kallima inachus*, *Polyura eudamippus*, *Sasakia charonda*, *Coenonympha hero neoperseis* (2), *Erebia ligea* (2), *Erebia neriene niphonica* (2), *Kirinia fentoni*, *Lethe europa*, *Oeneis melissa*, *Ypthima masakii*, *Ypthima riukiuana*, *Ypthima yayeyamana*							

[a] Data from Ministry of the Environment (2007). The number in brackets shows total sub-species occurring in Japan

[b] Threat status: *CR+EN* Critically Endangered or Endangered, *VU* Vulnerable

[c] From Matsumoto (2006) Widely distributed in Eurasia north of Himalayas: Siberian type; Korean Peninsula to Ussuri: Ussuri type; Eastern China to Korean Peninsula: Chinese type; Endemic in Japan: Japanese type; China to Korean Peninsula, Himalaya: Himalayan type; Widely in Wallacea: Pan-tropical type. Geographical range of subspecies is species level

[d] Distribution by no. of prefectural authorities. Data from Ministry of the Environment (2006) and related publications

[e] Number of prefectural authorities where still recorded (% decline in brackets). Data from Ministry of the Environment (2006) updated with recent information

[f] % decline by municipality unit (not divided into "ward" unit; same below). The unit of municipality is taken in 1998 ($N = 3,239$, mean area 116 ± 140 km^2; not divided into "ward" unit), after which many municipalities were merged. All data are included, but most change occurred between the 1950s and 2000s. Three species * show the change by island unit. Data from Sunose and Eda (2003), Mano and Fujii (2009), Red Data Book of Prefecture, and related publications updated with recent information

[g] References: Inoue (2005), Yata and Ueda (1993), Sunose and Eda (2003), Imura and Shi (2004), Mano and Fujii (2009) etc. The classification of the cause is shown in Table 3 except **

[h] Other subspecies *Hestina assimilis assimilis* is alien species (See text)

classified these causes into 13 categories and showed that deforestation and the conversion of natural forests were the main causes of forest butterfly decline, whilst the recent cessation of traditional mowing has severely affected grassland butterflies. However, the Third National Biodiversity Strategy of Japan in 2007 lists four major threats for biodiversity of Japan. Table 3 summarises the principal causes of the decline of Japanese butterflies based on Shibatani (1989, 1990), Imura and Shi (2004), Inoue (2005), Fujii (2006, 2007), Ministry of the Environment (2006), JBCS (2007). Detailed explanations of each cause are as follows.

Habitat change and destruction

Various types of development and forestry have destroyed many of the best localities for butterflies since the 1950s. Within forests, the biggest impact has been from destruction of natural forest followed by afforestation. In the period of rapid economic growth around 1955, there was a big increase in the demand for timber and an Afforestation Policy was implemented in the mid1950s. The natural forest was replaced by conifer plantations such as *Cryptomeria japonica*, *Chamaecyparis obtusa*, and *Larix kaempferi*. Plantation forests now makes up around 40% of the forest area, compared to around 20% in 1950. This has led to the degradation of habitat for forest butterflies from suburban areas to mountain districts. Despite these changes, most forest butterflies have not yet become threatened with extinction because of their wide distribution and the extent of total forest area (ca. 66%). However, there have been a few exceptions, notably *Shijimia moorei* and *Panchala ganesa* (Table 1), which both had restricted distributions and are now threatened due to loss of their original forest habitat.

The rate of afforestation and related forest development has decreased in recent years because of reduced demand and the importation of timber, as well as growing public awareness of nature conservation. Since the 1980s, the area of artificial forests has remained the same. In addition, since 2000, the Forestry Agency has moved ahead with the conservation of biodiversity in forests. Inoue (2005) pointed out that the impact of deforestation on butterflies has been smaller in many districts in recent years. The impact of afforestation is thus likely to be much smaller in the future.

In contrast, the reduced level of forest clearance and afforestation has adversely affected grassland butterflies (Tashita 1989, 1993; Inoue 2005, 2007). At the latter half of peak period of afforestation, the total area of forest clearings for grassland butterflies might have reached 8–9% between 1950s and 1970s in Japan, however, these temporal habitats had mostly disappeared after this period

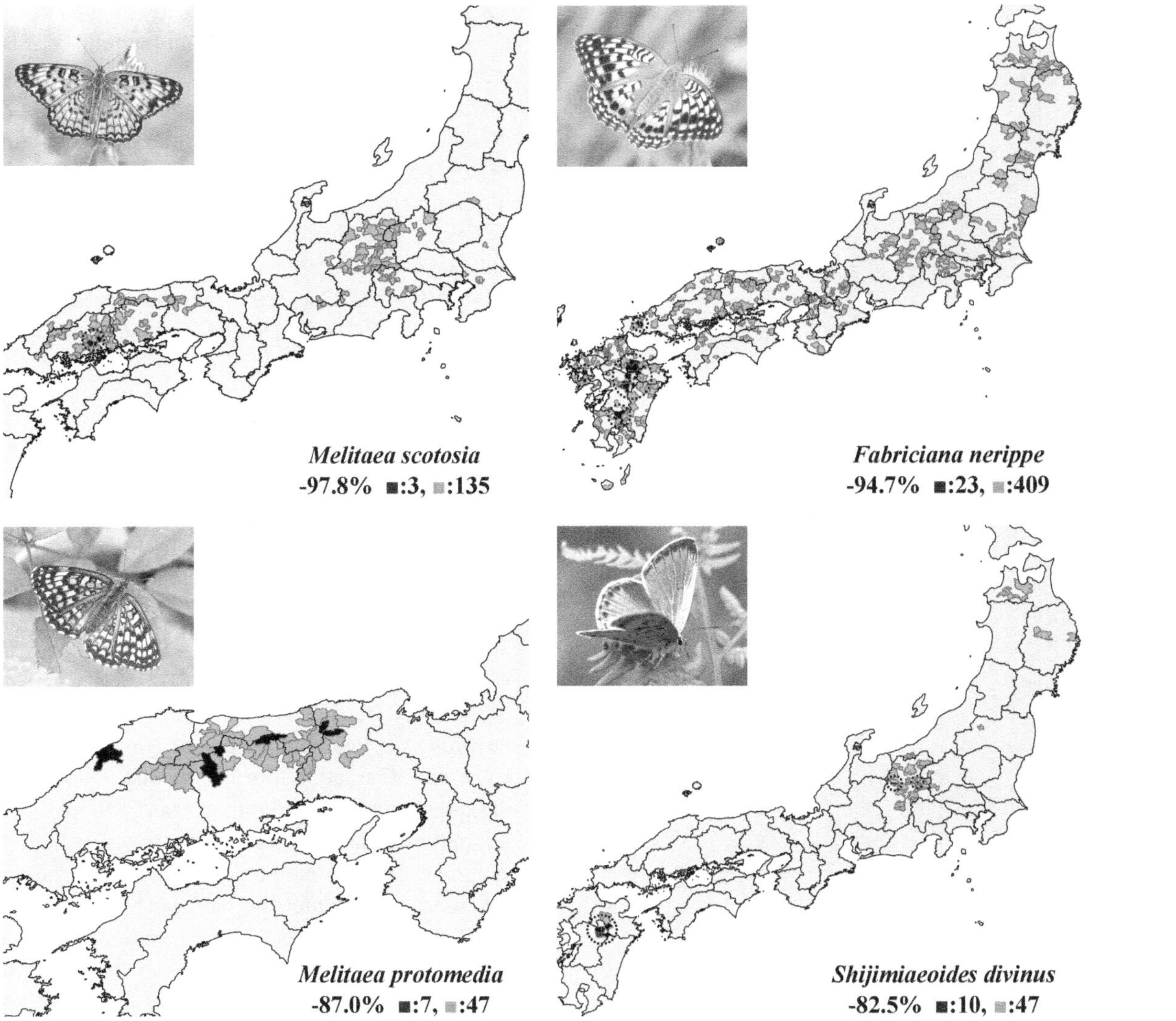

Fig. 1 Distribution change of four of the most rapidly declining grassland butterflies in Japan shown by municipality unit. The % declines are mainly from 1950s to 2000s (see Table 1). *Black colour*: Still present. *Grey colour*: Extinct. *Black lines* show boundaries of prefectural authorities, except for small islands. *Grey lines* show boundaries of municipal authorities as at 1998

Table 2 Status of butterflies at prefecture level and conservation action (modified from Nakamura 2005)

Status	Approximate No. populations[a]	No. populations with conservation action
Extinct	180	1
Probably extinct	80	2
Critically endangered: few habitats are known and conservation measures needed immediately	160	30
"Endangered" or "Vulnerable"	450	30–40

[a] Totals for all species within each prefecture unit. E.g. *F.nerippe*: Extinct = 36, *M. scotosia*: Extinct = 13 (See Table 1). Data from Sunose and Eda (2003), Mano and Fujii (2009), Red Data Book of Prefecture, and related publications updated with recent information

with a damaging impact on grassland species (Inoue 2005, 2007). For example, forest clearings remain suitable for *Pyrgus maculatus* for only five years and the butterfly soon declined following the growth of planted conifer trees (Tashita 1989).

Within agricultural areas, the main impact on butterflies is from farmland development and farm land consolidation. These developments have changed traditional small-scale farmland into more intensive large-scale operations. This destroyed not only the farmland but also the surrounding areas. A new agricultural complex was constructed in a non-farmland area, which introduced alien plants to the farmland. These changes impacted on almost all butterflies, including endangered butterflies like *Shijimiaeoides divinus* (Kobayashi 1989; Muroya 1989, 2007), *Plebejus argus* (Miyake 1993), and *Melitaea scotosia* (Miyagawa 1989).

Table 3 Cause of decline of Japanese butterflies

Biodiversity crisis factor[a]	Causes of decline for butterflies	Period mainly affected[b]
1 Species and habitat degradation due to excessive human activities	(1) Habitat change and destruction	
	a) Forestry: afforestation, deforestation, forest development, forest road construction, decrease of forest clearing (logging area), etc.	Afforestation: 1950s–1980s Forest road construction: 1970s-present Decrease of forest clearing: 1980s-present
	b) Agricultural development: farmland development, farm land consolidation, pasture, grassland improvement, land drainage, etc.	1970s-present
	c) Urban development: housing land development, road & highway construction, etc.	1970s-present
	d) Tourism development: construction of golf course, ski resort, scenic road and leisure home, etc.	Golf course: 1950s–early 1990s
	e) Waterside development: river improvement, dam construction, leisure of riverbed, etc.	Dam: 1960s-present
	(2) Over-collecting	1990s-present
2 Degradation of Satoyama[c] due to insufficient level of management	(3) Vegetation change	
	a) Abandonment and changing management of coppiced woodland	1980s-present
	b) Abandonment and changing management of semi-natural grassland	1950s-present
	(4) Effects for vegetation by sika deer and other mammals	Mid 1980s-present
3 Ecosystem disturbances caused by the introduced alien species and chemical contaminations	(5) Alien species	
	a) Vegetation change	1950s-present
	b) Predation by alien species	1980s-present
	c) Interspecific competition	Late 1970s-present
	(6) Chemicals: pesticide, insecticide, herbicide, aerial pollution	Pesticide etc.: 1950s-present
4 Global warming	(7) Climate and weather change	
	(8) Vegetation change	

[a] The Third National Biodiversity Strategy of Japan (Government of Japan 2007)

[b] The period is not a quantitative effect for butterflies except (2), (3) and (5b, c)

[c] Rural landscapes formed by sustainable use of natural resources

Fortunately, biodiversity conservation in farmland has become more prominent in recent years, following a revised Land Improvement Act in 2001 and the National Biodiversity Strategy of Japan.

In urban regions, butterflies have been affected by housing land development with related construction such as roads and parks (e.g. Iwano 1993). Large scale highway construction has had a big impact as well, especially for woodland and forest butterflies such as *Luehdorfia japonica*. This kind of development continues but more attention has to be paid to biodiversity using Environmental Impact Assessment (EIA).

Tourism development has caused many conservation problems in Japan since the 1980s. The "Resort Law" was enacted in 1987 against a background of increasing domestic demand. This led to large-scale destruction of habitat around newly constructed resorts. For example, the number of golf courses in 1957 was only 116, but by 1996 this number had increased by a factor of 20–2,340. The golf courses were located mainly in woodland areas around urban neighbourhoods in "Satoyama" landscapes, and adversely affected woodland butterflies. Since 1996, tourism development has decreased because economic conditions have deteriorated.

Waterside developments have also been damaging, for example large scale dam construction has had a big impact on the forest area. River banks and riverbeds provide good habitats for grassland butterflies, but have been destroyed or changed by river improvement and leisure utilization of the riverbed (Sei 1996). This change has had a big impact on butterflies, as have changes to the river ecosystem following the flood control measures (construction of dams, sand control dams and bunds). This has led to less frequent flooding and changes in the riverbed and surrounding vegetation, which has impacted on butterflies such as *Anthocharis cardamines, Plebejus subsolanus* and *Limenitis populi* in the upper reaches of rivers (Hama 1996) and *Eurema laeta, Zizina emelina* and *Plebejus argyrognomon* in middle to lower reaches of rivers (Ministry of the Environment 2006; Kato et al. 2007; Suda 2007).

Until the 1990s, conservation measures for wildlife were seldom carried out because similar habitats were perceived as common in surrounding areas. However, the Environmental Impact Assessment Law was enacted in June 1997 and now all larger projects are subject to the EIA procedure and require conservation measures for biodiversity.

Over-collecting

In Europe, collecting is a comparatively minor threat and a simple ban on collecting is not regarded as an effective way of conserving butterflies (Van Swaay and Warren 1999). In Japan, collecting is widely listed as a cause of decline, especially of endangered and popular butterflies such as *Pyrgus maculatus*, *Luehdorfia japonica*, *Shijimiaeoides divinus*, *Maculinea teleius*, *Coenonympha oedippus*, etc (Sunose and Eda 2003; Mano and Fujii 2009). Regarding threatened butterflies, most colonies are now small and isolated and the damage caused by collecting could be much higher than before. Over several hundred collectors are thought to be active and collectors sometimes take dozens of specimens of one species in a day at one place. For example, in 2009 over 90 individuals of *Maculinea teleius* were taken at a single site in a day, and over 45 specimens of *Luehdorfia japonica*. Although most of collectors collect responsibly, there are many other records of excessive collecting listed in the publication of amateur groups. Only a few endangered butterflies are banned from collecting by the law (e.g. "Law for Protection of Cultural Properties", "the Law for the Conservation of Endangered Species of Wild Fauna and Flora (LCES)"). Conflicts between collectors and local people have recently become a serious problem, especially for local nature conservation groups.

In 1970 and 1971, there have been symbolic examples of over-collecting. Oak trees were logged to get the eggs of *Chrysozephyrus hisamatsusanus* in an area, when the butterfly was extremely rare in those days and right after the discovery of hostplant of the butterfly (The Nature and Insects 1971). This destruction was reported in the media, and amateur entomologists were heavily criticized by conservation societies; lots of opinions were submitted, and all raised the concern about overexploitation of the butterflies (e.g. Hara et al. 1971). That was the first opportunity to think about the way of insect collecting seriously in Japan.

Against this background, collecting has been discussed widely by local entomological groups and the Lepidopterological Society of Japan. The Lepidopterological Society produced "guidelines and rules for collecting" in 1992 which states "exercise moderation in collecting specimens of butterflies and moths" (the Lepidopterological Society of Japan 1994). More recently, their conservation committee advocated to discourage the collecting of all endangered butterflies (Fujii and Nakamura 2000), but this has not necessarily changed the behaviour of collectors.

Degradation of Satoyama

Satoyama refers to traditionally managed woodlands such as oak coppices, pinewoods and bamboo groves, as well as grasslands managed for thatch, fodder and compost (Ministry of the Environment 2009). These secondary environments were maintained by the long-term sustainable use of essential natural resources. This traditional landscape includes various other rural environments, such as arable fields and orchards, rice paddies, irrigation ponds and ditches, and the villages and farmsteads themselves. The complex rural ecosystem formed by the combination of Satoyama and these other environments is called the "Satoyama Landscape" (Ministry of the Environment 2009). Technically speaking, the "Satoyama Landscape" and "Satochi" are far more extensive than Satoyama (Takeuchi 2001; Yamamoto 2001). However, the word Satoyama is more popular amongst the general public and has a similar meaning (Tabata 1997; Hiroki 2002).

The landscape mosaic of the Satoyama provides a variety of habitats for wild animals and plants (Washitani 2001) and is an important habitat for insects (Kato 2001) including butterflies (Ishii 2001; Nishinaka and Ishii 2007, Ohwaki et al. 2007). The Satoyama are found between urban areas and mountain areas, and the core Satoyama area covered about 20% of the total land area, and about 40% when surrounding agricultural lands are included (Government of Japan 2007). However, as their economic value has been lost to waves of industrialization and the westernization of Japanese society, most Satoyama ecosystems have been abandoned or destroyed, a tendency that has accelerated in the last 30–40 years (Washitani 2001).

The abandonment of management has threatened many plants and animals of the Satoyama including butterflies (e.g. Kobayashi 1972; Moriyama 1988; Ishii et al. 1993). The latter breed mainly in the semi-natural grassland, coppiced woodland and farmland in Satoyama. Mowing land and pasture areas are maintained as semi-natural grassland and used for animal feed, manure, roof cover, and fuel. Kitahara and Sei (2001) showed that the butterfly communities in semi-natural grassland had greater species richness and diversity than other types of human-modified grasslands. However, most grassland now has little or no economic value because of the decrease in demand for grass and an increasing use of imported feed and chemical fertilizers, and has deteriorated due to factors such as the cessation of grass cutting, grazing or fire-burning, destruction by various development, and afforestation. Other areas have been replaced by artificial grassland. The traditional method of grass cutting, which was done by hand and maintained the grass at a wide range of heights, has been changed and now the grass is usually cut extensively and uniformly by machine. This change has had an impact on butterflies (Sei 1996). In the early 1900s, grasslands spread in Japan and reached about 13% of the land area, while in the earlier Edo Period (1603–1868), it probably increased by over 30% because of the intensive use of grassland resources (Ogura 2006). However, the grassland area has since declined and is now about 0.9% (Ogura 2006).

The changes to Satoyama habitats have resulted in a serious decline of grassland butterflies (Ishii 1990, 1996; Nakamura 2005; Table 1; Figs. 1, 2). The regional extinction in prefectures of the two most threatened butterflies, *Fabriciana nerippe* and *Melitaea scotosia*, began at the end of the 1940s and their sharp decline continued until the end of the 1980s. So, their steep decline occurred within just a 40 year period. Today, good habitats for grassland butterflies are very limited and confined to areas still maintained by traditional Satoyama land use. Some of the largest maintained grassland areas are within military training lands, which are the best localities for grassland butterflies (Ishii 1994), especially *Fabriciana nerippe* (e.g. Iwasaki 1994).

In the temperate region of Japan, coppiced woodland in Satoyama is generally dominated by oak (e.g. *Quercus serrata*, *Q. acutissima*) which was harvested for use as firewood and charcoal, or for the cultivation of Japanese mushrooms. These trees were cut on a 10–30 year cycle, providing various stages of vegetation and a mosaic landscape that supported a high diversity of butterflies (Inoue 2003; Nishinaka and Ishii 2007). However, since the 1970s, the demand for wood fuel has dropped and many coppices have been abandoned, causing the woodland to revert to dense thicket or broad-leaved evergreen forest.

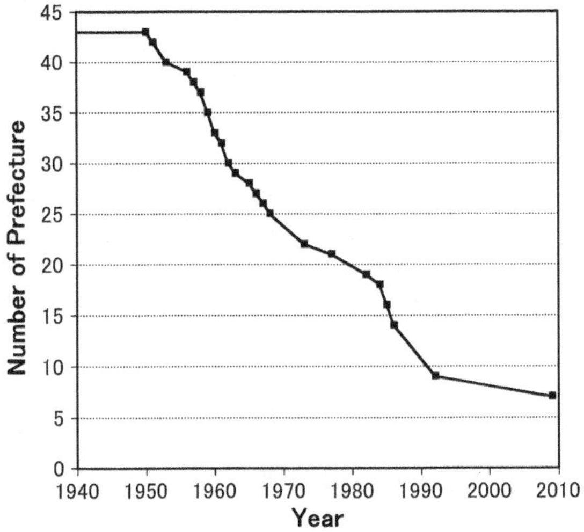

Fig. 2 The distribution change of *Fabriciana nerippe* between 1940 to 2009 in prefectural units. Data from Mano and Fujii (2009) and related publications, based on the last record for each prefecture

These vegetation changes and lack of early successional woodland stages are leading to the decline of many butterfly species.

Within the farmland of Satoyama, the area cultivated has declined since the 1970s, especially in paddy fields. The grassland around cultivated land has also become abandoned and strips of grassland have been lost around paddy fields, leading to the decline of grassland butterflies. With these changes and the decrease in hunting, animals such sika deer, boar, and monkey have increased, as has their impact on the vegetation (Miura 1999). The intensive feeding of the sika deer changed the vegetation, leading the extinction of *Melitaea ambigua* in one locality (Hasegawa 1994). The Ministry of the Environment has already taken conservation measures for biodiversity in some areas, but is facing difficulty in controlling the numbers of these mammals.

Another problem facing Satoyama areas is the phenomenon of underpopulation in rural areas as people move to towns and cities. If the trend continues humans may also disappear from these areas.

Alien species

The influence of alien species for butterfly decline in Japan is mainly divided into (1) vegetation change by alien plants; (2) predation by alien animals; and (3) interspecific competition by introduced alien butterflies.

(1) The number of alien plants in Japan has reached about 1,500 species (The Ecological Society of Japan 2002). Invasion by alien plants constitutes one of the most serious threats to biodiversity, through the displacement of native plants in urban areas, farmland, and riparian habitats (Miyawaki and Washitani 2004), as well as on oceanic Ogasawara Islands (e.g. Shimizu 2003). The spread of alien plants in riparian areas has been reported as the cause of decline of *Plebejus argyrognomon* (Takahashi 2004; Hagiwara et al. 2009) and *Zizina emelina* (Kato et al. 2007; Suda 2007).

(2) Threat to butterflies through predation by alien animals is reported only in *Celastrina ogasawaraensis* (see below).

(3) Interspecific competition from introduced alien butterflies is now a serious problem following releases of two introduced species/sub-species. *Sericinus montela* was first found in the field in 1978 within the Tokyo Prefecture and was considered to have been introduced from South Korea by butterfly enthusiasts (Fujii 2002). Since then, this species has been found in over 5 prefectures, some the result of additional releases. Similarly, *Hestina assimilis assimilis* was found in 1998 within Kanagawa Prefecture next to Tokyo Prefecture (Nakamura et al. 2003b).

Originally, the endemic subspecies *Hestina assimilis shirakii* was distributed in the Southwest Islands in Japan. However, *Hestina assimilis assimilis*, which inhabits the Korean Peninsula to China, was considered to be introduced by man-caused movement (Nakamura et al. 2003b). This subspecies has been spreading rapidly, mainly by natural dispersal within Kanagawa Prefecture and the surrounding prefectures (Nakamura et al. 2003b; Iwano and Sugai 2007).

The introduction of live butterflies is banned by the "Plant Protection Act", but this has not stopped some butterfly enthusiasts from breaking the law. Native butterflies already use the hostplant of each invasive species, and concerns over interspecific competition between native and non-native butterflies have been raised by the introduction of both *Sericinus montela* (Sakuratani and Kanno 2003; Inoue and Kon 2006) and *Hestina assimilis assimilis*. The man-caused movement and release of other wild animals and plants within Japan is a general problem for biodiversity as well (e.g. The Ecological Society of Japan 2002). For butterflies, the introduction of alien populations, especially of popular butterflies like *Luehdorfia japonica* and *Sasakia charonda*, has often been done without careful consideration and may cause genetic pollution. The reason of such easy release is solely a lack of specialist knowledge (Ito and Nanba 2007).

Chemicals

Pesticide and insecticides affect butterflies, mainly in farmland and forest areas. In farmland areas, spraying of insecticides in paddy field, dry fields and fruit farms may cause the death of a wide range of butterflies breeding in or near these fields (Yata and Ueda 1993; Sunose and Eda 2003). The toxicity of insecticides has recently been reduced but the effects are still a concern, especially around fruit farms.

In pine forests, aerial spraying of insecticides have been used widely since 1977 to kill the transmitting longicorn beetle *Monochamus alternatus* and defend against the pine wilt disease induced by the pinewood nematode *Bursaphelenchus xylophilus* (alien species from North America). This insecticides damaged most of insects (Obara and Yodoe 1994), and the disappearance of *Fabriciana nerippe* after spraying has been reported (Yodoe 1997). However, since human health is also adversely affected by the spraying of insecticides as well, the practice has decreased due to a citizens' opposition movement in many parts of Japan.

Global warming

There is evidence in other countries that many butterfly species are responding to climate warming by shifting their distributions polewards and/or to higher latitudes (e.g. Hill et al. 2009). In Japan, over 10 species may be changing their distribution to the northward, including *Papilio memnon, Narathura bazalus, Argyreus hyperbius, Cyrestis thyodamas, and Melanitis phedima* (Yoshio 2002; Kitahara 2008). Adverse effects are mainly due to two factors: "climate and weather change" and "vegetation change". Regarding the former, the problems relate to extraordinary and unseasonal events, such as intense rainfall, extreme low and high temperature. For example, in spring 2010, the adult number of *Luehdorfia japonica* was relatively low in most of the area, probably due to the unseasonal cool climate in April.

Regarding the latter, a large change in the distribution of the natural vegetation of Japan has been predicted (Tunekawa et al. 1996), which could have a strong adverse effect on butterflies. In beech (*Fagus crenata*) forests, suitable habitats are predicted to disappear in Kyushu, Shikoku and the Pacific Ocean side of Honshu, and will decrease greatly even in the Tohoku area of northern Honshu with the most extensive suitable habitats by 2100 (Tanaka et al. 2006). For *Sibataniozephyrus fujisanus*, the hostplants are *Fagus crenata* and *Fagus japonica*, and the butterfly is expected to decline gradually along with the shrinking of *Fagus* forest. Because *Fagus* forests are found at over 800 meter elevation in the western parts of Japan, it might be difficult for the species to shift its distribution in response to climate change.

There are also concerns over the change of alpine grassland in response to climate change. *Pyrgus malvae* occurs only around Mt. Apoi at altitudes of mainly 600–700 m in Hokkaido (Watanabe 1989). The alpine meadow habitat of this species is already changing to forest rapidly because of climate change (Watanabe 2005). Detailed monitoring is needed to assess the conservation needs of this butterfly.

Adverse vegetation change induced by sika deer has also been noticed in alpine vegetation (Motojima 2009). The range expansion of sika deer was considered to be caused by the decrease of snow accumulation.

In future decades, climate change may cause a broader range change in butterflies and could cause the large scale decline of butterflies in Europe (Settele et al. 2008). As Japan is an island nation, distribution changes in response to climate change will be limited for most of the wildlife, including butterflies, and it could have a more damaging impact on biodiversity.

Habitat isolation and fragmentation

This factor was not listed in Table 3, because it is a consequence of all other causes of decline. Habitat isolation and fragmentation induced by habitat loss causes the steady decline of endangered butterflies. Many local populations of endangered butterflies have disappeared without any obvious habitat change in recent years (Nakamura unpublished). Because it takes some time following habitat loss before extinction occurs (Hanski and Ovaskainen 2002), this factor could have a growing impact in the future.

Historical background to butterfly conservation

The actions for butterfly conservation in Japan can be classified into two broad types; actions by governments and actions within the private sector. The first action by national government was the designation of *Panchala ganesa* and its habitat in one area as "Tennenkinenbutsu", as a protected species in 1932, leading to the protection of its habitat and prohibition of collection. Further designations of protected species were made in and around 1960s, and prefectural and municipal authorities began to designate local protected species after the 1950s. Currently, seven species are protected nationally and around 70 kinds of butterfly populations are protected by prefectural and municipal authorities (Kamada and Fujii 2010).

The law, however, which protects not only valuable wild animals and plants but also historical sites or beauty spots was established in 1919 and revised as the "Law for Protection of Cultural Properties" in 1950. These used to designate mainly "rare" and "famous" butterflies rather than endangered ones, because the law stipulates the selection criteria as "academically valuable and also celebrating the nation's nature". In fact, the popular species *Luehdorfia japonica* and *Luehdorfia puziloi* account for 39% of designated protected butterfly species by municipal authorities. Many entomologists have since pointed out the ineffectiveness of this law because practical conservation such as habitat management nor monitoring has not been carried out for most designated butterflies (e.g. Shibatani 1989, 1990). On the plus side, the law may have played a role by drawing attention to development and the problems of over-collecting.

In 1991, the Japanese government published the first Red Data Book and in 1993 new legislation came into force to conserve these animals and plants (LCES). Two butterflies are designated under this law: *Shijimia moorei* (in 1996) and *Celastrina ogasawaraensis* (in 2008), and the government produced "Programmes for Rehabilitation of Natural Habitats and Maintenance of Viable Populations". Projects to conserve and restore habitats have since started. The first Red Data Book of butterflies in 1991 was not based on sufficient data and did not reflect the actual status. A revised Red List of 2000 revealed the more precise situation of threatened status, following work by the Conservation Committee of the Lepidopterological Society of Japan (Nakamura 2005; Ministry of the Environment 2006).

Influenced by the new trends, conservation actions spread across prefectures. From 1995, Red Data Books were produced by prefectures in order to recognise endangered species by regional units, and by 2005 they were published in all 47 prefectures. Also, over half of prefectures have developed regulations to conserve endangered animals and plants pursuant to the national LCES, and 21 kinds of butterfly populations were designated in seven prefectures by 2009 (Kamada and Fujii 2010). Municipal authorities have followed this lead and also conducted practical conservation projects on locally protected species.

Furthermore, various national biodiversity projects are making substantial progress. With the establishment of the National Biodiversity Strategy of Japan in 1995, conservation and restoration of biodiversity enterprises have been undertaken. These include the maintenance of secondary environmental protection in natural parks (a Revised Natural Parks Law came into force in 2003.); conservation of endangered animals in national parks, in which three butterflies, *Aporia hippia, Everes lacturnus, Melitaea protomedia*, were selected in 2006 (Nakamura 2008); projects to restore natural environments (the Law for the Promotion of Nature Restoration came into force in 2003.); actions to conserve Satoyama habitats (Satochi Satoyama conservation model projects have been enforced since 2004); and support for the citizen groups in Satoyama. Conservation actions aimed at the whole natural environment are relatively large-scaled and are connected with the conservation of butterflies at community level, and in some cases with the conservation of endangered species.

Actions to save butterflies have been developed at local level amongst the private sector, including actions to protect valuable wildlife by local citizens. This movement became active in the 1980s and local groups to conserve butterflies were set up in many regions. Targeted butterflies were mainly *Sasakia charonda* (e.g. Bandai 1996; Makibayashi 1996), *Luehdorfia japonica* and *L. puziloi* (Ihara 1996; Matsumura 2008), *Coreana raphaelis* (Kusakari 1996; Ogata 2009). *S. charonda* is famous as a "National Butterfly" and its largeness and beauty attract people, while *Luehdorfia* is also popular because of its beauty and ephemeral appearance in the spring, and *C. raphaelis* is recognized as very valuable because it inhabits a restricted area and was discovered relatively recently in 1951 (Nakahara and Oda 1953).

Specialist societies also became involved. In 1965 the Research Group of Nature Conservation (renamed as the Conservation Committee in 1975) was established within the Lepidopterological Society of Japan, and a Conservation Committee was established within the Entomological Society of Japan in 1966. The Lepidopterological Society of Japan took the initiative for butterfly conservation and their main initial activities were to announce statements of nature protection (in 1970) and to educate people by promoting controlled collecting. The committee has since started the following: seminars about butterflies and nature conservation since 1990, publications about butterfly conservation (Hama et al.1989; Yata and Ueda 1993; Ae et al. 1996; Tanaka and Arita 1996; Sunose and Eda 2003; Mano and Fujii 2009), organized a Red List of butterflies by prefecture (Yata and Ueda 1993; Sunose and Eda 2003; Mano and Fujii 2009), established a garden butterfly survey from 1997 to 2002 (Ishii 2003), submitted a petition for butterfly conservation, and implemented conservation activities for threatened species.

Despite these efforts, the conservation of endangered species was not sufficient and the decline of butterflies accelerated, and foundation of non-profit organization at national level for butterfly conservation was urgently needed (Nakamura 2003). In 2004, the Japan Butterfly Conservation Network was established in order conserve biodiversity using butterflies as a symbol (JBCN 2004). This was renamed as the Japan Butterfly Conservation Society (JBCS) in 2006. The society now has about 500 memberships and conducts the following main activities; (1) research on endangered species and its conservation, (2) publicity and educational activities, and (3) advice on butterfly conservation to administrative, citizen groups and general public. It has promoted monitoring and conservation activities for more than 15 endangered species in 30 localities. Regarding publicity and education, the society promotes the importance of conservation in general and butterflies as an indicator of the natural environment. It also holds an annual butterfly conservation symposium as well as other events for general public, and has published a general butterfly conservation guidebook (JBCS 2007) and conservation manuals for each endangered species (JBCN 2005; JBCS 2008b, 2009).

Meanwhile, general public awareness on the conservation of the natural environment has been increasing and activities to conserve Satoyama have become active since the 1990s. Conservation activities of Satoyama are not only aimed at biodiversity but for the improvement of the landscape and forests, recreation, and other purposes. There are now over 1000 groups related to Satoyama.

Regarding monitoring, The Nature Conservation Society of Japan started a wildlife monitoring project in Satoyama area during 2008 under the "Monitoring Sites 1000" project by the Ministry of the Environment. This includes butterfly transect counts on 50 sites.

Practical conservation

Practical conservation measures have been undertaken by the government (nationally or locally), research institutes, universities, non profit organizations (include JBCS), and local citizen groups. For threatened butterflies, conservation projects have been started for all of "CR + EN" and 12 out of 22 VU species.

Table 2 summarises efforts to conserve threatened species at prefectural level, most of them carried out by voluntary groups. Conservation has started for about 19% of CR species and in about 8% of EN or VU species. These rates are fairly low but new projects have started in recent years. Measures are divided into (1) scientific research, monitoring, (2) protection and management of habitat, (3) re-introduction, re-inforcement, (4) bans on collecting, (5) publicity and educational activities, lectures, symposia, campaigns, pamphlets, etc. (6) policy recommendations to the government. The following are examples for some endangered species:

Luehdorfia japonica and *Luehdorfia puziloi*

These are amongst the most popular butterflies in Japan and there are more than 20 citizen conservation groups in various regions (JBCS 2008a). Recently, the main causes of decline are changes to Satoyama and the main actions taken are the management of coppiced woodland (e.g. JBCS 2008a). These butterflies have become symbols or "flagship species" of Satoyama conservation. In addition, bans on collecting are spreading in some regions in response to growing concern about biodiversity conservation. JBCS has held the National Butterfly Conservation Symposium under the theme of the *Luehdorfia* three times from 2007 to 2008, each attended by about 300 people.

Aporia hippia

The habitats of *Aporia hippia* are distributed at high altitudes between 1400 and 2000 meters. Though butterflies inhabiting higher altitudes are relatively safe compared to lowlands, the butterfly has declined mainly due to afforestation and tourism development. Since the Ministry of the Environment started a detailed survey of the butterfly in 2006, active conservation measures have been taken by citizen groups, JBCS and the government. The main activities are the restoration of the habitats by logging the conifer plantation and monitoring.

Celastrina ogasawaraensis

The butterfly is endemic in the oceanic Ogasawara (Bonin) Islands. The Ogasawaras are separated by a vast sea from the Asian Continent and the mainland of Japan which were the sources of the islands' creatures. Due to their long isolation, about 30% of the islands' insects are endemic (Kato 1992). However, many foreign plants and animals were introduced or entered the Ogasawaras since the first settlement, and many have gone wild, leading a threat for native plants and animals (e.g. Shimizu 2003). Since late 1990s, various actions have been taken to conserve this wildlife. The butterfly was relatively common until the 1970s, but subsequently declined rapidly and it now inhabits only one island (Takakuwa and Suda 2004). The main causes of decline are predation by the alien species, Green Anole *Anolis carolinensis* from North America and decrease in the hostplant due to alien plants, mainly *Bischofia javanica* from South East Asia (Takakuwa and Suda 2004; Nakamura 2006, 2009b). Efforts are now being made to eradicate these alien species for the butterfly (Nakamura 2006, 2009b). Increases in egg numbers after the elimination of this plant have already been confirmed.

However, the estimated population of Green Anole is 6.4 million in the Ogasawaras and eradication is considered to be almost impossible, thus, in practice, a high priority for exterminating the lizard is given to some important areas (Toda et al. 2009). The adverse effects of Green Anole has been confirmed in other small-size, diurnal insects (Karube and Suda 2004; Karube 2005) and it has caused the serious decline on endemic dragonflies and other insects as well (Karube 2001, 2004, 2005). The population of the butterfly is considered at seriously low level and there is a high risk of imminent extinction (Nakamura 2009b).

Shijimia moorei

The butterfly inhabits in Kumamoto Prefecture (central Kyushu) and is now restricted to two areas. The habitat is virgin forest and the butterfly had declined due to deforestation as well as the poaching of larvae and their hostplant, *Lysionotus pauciflorus*. Conservation measures have been started by the Forestry Agency since it was protected by law. Its habitats are now preserved within a protected area, and efforts have started to monitor and expand the hostplant (Saigusa 1997).

Shijimiaeoides divinus

The butterfly now inhabits Nagano Prefecture (central Honshu) and central Kyushu (Fig. 1). In the former, the butterfly had disappeared from most of the area by the mid 1980s due to farm land consolidation, starting in the 1970s (Kobayashi 1989). Since 1990s, conservation measures have started in two areas, and monitoring, habitat management, restoration of habitat, reintroduction and educational activity have been implemented vigorously (Hama 2007; Tashita and Maruyama 2007; Shimizu 2009). Reintroduction programmes have been successful in one area (Shimizu 2009) and failed in the other due to the high rate of parasitism by an egg parasitoid (Hama 2007; Hirabayashi et al. 2009).

In central Kyushu, the habitat is pasture where the hostplant *Sophora flavescens* is found and where fields are burned off annually. The causes of decline are development, pesticide spraying, cessation of burning, and overcollecting (Murata and Nohara 1993). Continued burning protects the hostplant from other competitive species and is thus essential in maintaining the habitat (Murata and Nohara 1993; Murata et al 1998). The Ministry of the Environment has developed a project to maintain and expand the grasslands in this area, using the butterfly as one of the target species.

Melitaea scotosia

The butterfly has declined seriously and is now restricted to a narrow range (about 20 × 20 km) in the Hiroshima Prefecture (Fig. 1; JBCS 2008b). In this area, the habitat is natural marshy grassland or abandoned paddy fields which develop into marshy grassland. There are now only around 30 local populations remaining and the number of the larval nests in 2009 was only around 200 (Nakamura unpublished). The decline is continuing due to the isolation and fragmentation of habitat patches. The metapopulation structure of the butterfly is almost destroyed because the number of habitat patches has decreased continuously since 1980s, due to vegetation succession following agricultural decline as well as farmland development, farm land consolidation, and road construction. The situation for its conservation is thus extremely difficult (Nakamura 2007). The local conservation group for the butterfly was founded in 2001 and since then has conducted monitoring, habitat management and restoration, and educational activities (Nakamura et al. 2003a; Nakamura 2007; Iwami et al. 2006; Iwami 2009).

Melitaea protomedia

The distribution of the butterfly restricted in western parts of Honshu, and habitat is semi-natural grassland maintained mainly as pasture or meadow (Fig. 1). The butterfly has declined seriously due to the abandonment of management, and now inhabits only five areas. Conservation activities have started in all areas (Nakamura 2004), focussed on the maintenance or restoration of grassland management. In three areas, voluntary-based mowing has been assisting its conservation (Kawabe et al. 2005; Tanigami 2008; Nanba 2009). The optimal way of grassland management for the butterfly was studied, and mowing in June tended to produce the highest species diversity and abundance of the hostplant *Patrinia scanbiosaefolia* (Takahashi et al. 2008). The habitat is relatively extensive, so it might be difficult to continue the mowing on a voluntary basis.

A strategy for conserving butterflies in Japan

The Third National Biodiversity Strategy of Japan shows one of its three targets as "Conserve region-specific animals/plants and ecosystems in accordance with regional characteristics, and maintain and restore the country-level biodiversity by building an ecological network. In particular, maintain and restore the populations and habitat environments of species currently threatened with extinction, as well as take proper measures to prevent threatened species living in Japan from increasing". Local prefectural governments also speak publicly about the importance of conservation of threatened species. However, it is clear that existing effort is insufficient given the huge scale of the problems.

JBCS set targets to prevent threatened butterflies from extinction firstly at the national level, and secondly at the prefecture level. To improve the situation, we need to solve problems from both the scientific and social point of view as these are closely linked. Moreover, we need to improve the wider range of landscape for butterflies, not only threatened species. The measures needed are as follows;

Conserving and restoring the Satoyama

To conserve and restore the Satoyama, new modern values different from the past farming/life are needed to overcome the current biodiversity crisis and ecosystem deterioration (e.g. Washitani 2001). Changes to Satoyama have become the main threat for butterflies as a whole and the major problem to overcome for their conservation. Wildlife, including butterflies, had survived well with human activity in the Satoyama ecosystem. At present, the conservation of butterflies is focused on maintaining small number of habitats and does not cover the whole Satoyama ecosystem. This will enable the conservation of only small regional populations of threatened butterflies and butterfly communities.

To expand this effort, a new social system is needed to maintain the Satoyama landscape as an industry. For example, for grassland, an attempt is underway to use native grass as a biomass resource, though there are still some problems (Takahashi 2008). The awareness of grassland restoration is increasing and a National Grassland Restoration Network was founded in 2007. As for coppiced woodland, utilization as woody biomass is underway and could play a role in combating global warming (e.g. Tabata 2009). As interest in the conservation of Satoyama has increased, Ministry of the Environment has made efforts to conserve Satoyama landscapes for biodiversity. Further assistances including subsidies for local groups by the government are needed.

Agricultural reform

Agricultural developments are still ongoing and there are many cases where habitats are not paid attention to. More careful and intensive measures for conserving biodiversity are needed for these developments. Also, the improvement of agricultural area for biodiversity is needed. Ministry of Agriculture, Forestry and Fisheries (MAFF) has started the payment of agricultural subsidies for improvement of agricultural environment including biodiversity conservation since 2007. This helps conservation of threatened butterflies in some area.

Raise awareness of butterfly conservation

To conserve butterflies at the local level, more collaborators are needed. In Japan, the largest society for nature conservation is the Wild Bird Society of Japan, which was founded in 1934 and had a membership of around 43,000 in 2009. The second biggest is "WWF Japan", founded in 1971, with around 35,000 members in 2009 (NB the population in Japan is 127 million). Compared to these societies, the membership of JBCS is relatively low and we need more effort to expand membership. A larger membership could enable JBCS to promote bigger projects in Japan. Also, it is needed to make both the general public and policy makers more aware of the fate of butterflies and the threats they face, and let them understand the importance of butterflies as indicators of a healthy environment.

Promote scientific research and monitoring for butterflies

To understand the overall situation of butterflies, regular monitoring is needed. However, the monitoring of butterflies has just begun, and the number of sites is limited. We need to found a system of monitoring with participation of many citizens on a voluntary basis. Fund raising is also needed.

Strengthening of wildlife law and policy

Although the establishment of regulations on biodiversity is in progress, it is not sufficiently well linked to butterfly conservation. The LCES designates only two butterflies and many more species need to be designated and more sites protected. The habitats of endangered butterflies are not sufficiently contained in existing protected sites, because the habitats were not uncommon and had not been considered as important sites for wildlife. Though creating new protected areas around farmland and related areas presents many difficulties, we need to make new and effective systems.

Promotion of conservation in national parks

National parks and related protected areas are valuable sites for wildlife, which also contain key important habitats that are required for the conservation of butterflies as well. However, the monitoring of wildlife is rarely done, and so information on wildlife including butterflies is lacking. Also, the management of semi-natural environment for wildlife is not sufficient. The conservation effort for endangered butterflies in national parks has started since 2006 by the government, although the targets are limited now. Promotion of the monitoring and conservation of butterflies at community level is necessary to cope with ongoing extinction of local populations.

To halt the extinction of butterfly species

There are no butterfly species extinct in Japan. However, now two species, *Celastrina ogasawaraensis* and *Melitaea scotosia*, are extremely close to extinction followed by *Shijimia moorei* and *Melitaea protomedia*. It is important to set high priority to conserve these species, and to do anything to try to halt the extinctions. The funds by the government are insufficient, and there is no fund assistance by the government for *M. scotosia*. Therefore, more funding is needed as well as detailed monitoring and ecological research.

Acknowledgments The author heartily thanks Martin Warren for his kind help in editing the text, and Yoshiyuki Nagahata, Masahiko Kitahara, Minoru Ishii, Hidetoshi Tamate, Rashida Dumbuya, Matthew Oates and Nigel Bourn for valuable comments on the manuscript. Thanks are also due to K. Maeda, T. Kudo, T. Abe, A. Araya, Y. Nagahata, T. Saito, T. Mifune, M. Shiota, T. Kuzuya, T. Matsumura, R. Kaga, M. Takakuwa, T. Turufuji, S. Suda, S. Minohara, M. Yago, S. Watanuki, T. Shimonoya, T. Takeuchi, S. Umemura, S. Hayami, M. Kitahara, R. Kato, M. Ihara, A. Ono, K. Maruyama, Y. Ando, M. Takahashi, J. Togari, H. Fujii, S. Miwa, H. Takei, S. Morichi, O. Yamamoto, K. Kamada, S. Kondo, Y. Tateiwa, F. Ito, A. Tamura, K. Yodoe, S. Miyake, Y. Suemune, A. Kanayashiki, J. Iwami, K. Ohara, T. Dejima, K. Ohta, R. Arakawa, O. Fukuda, N. Nakazono, T. Mizobe, M. Kawano, T. Komatsu, Miyazaki Entomologists' Society, Y. Nakamine and K. Kanai for providing latest distribution data of threatened butterflies and helpful information, and J. Togari for providing photograph.

References

Ae SA, Hirowatari T, Ishii M, Brower LP (eds) (1996) Decline and conservation of butterflies in Japan III. The Lepidopterological Society of Japan, Osaka

Bandai K (1996) Preserving woodlands for butterflies—conservation of the giant purple, *Sasakia charonda*, in Nagasaka-cho, Yamanashi Prefecture. In: Ae SA, Hirowatari T, Ishii M, Brower LP (eds) Decline and conservation of butterflies in Japan III. The Lepidopterological Society of Japan, Osaka, pp 180–184

Cabeza M, Arponen A, Jäättelä L, Kujala H, van Teeffelen A, Hanski I (2010) Conservation planning with insects at three different spatial scales. Ecography 33:54–63

Collins NM, Morris MG (1985) Threatened swallowtail butterflies of the world: the IUCN Red Data Book. IUCN, Gland and Cambridge

Corke D (1999) Are honeydew/sap-feeding butterflies (Lepidoptera: Rhopalocera) affected by particulate air-pollution? J Insect Conserv 3:5–14

Dennis RLH (2010) A resource-based habitat view for conservation: butterflies in the British landscape. Wiley-Blackwell, West Sussex

Fujii H (2002) *Sericinus montela*. In: The Ecological Society of Japan (ed) Handbook of alien species in Japan. Chijinshokan, Tokyo, p 157 (In Japanese)

Fujii H (2006) Decline of grassland butterflies in Japan and conservation activities. Nat Insects 41(3):2–6 (In Japanese)

Fujii H (2007) Decline of butterflies in Japan and conservation activities. Biol Sci Genet 61(3):29–35 (In Japanese)

Fujii H, Nakamura Y (2000) The request for no collecting of the endangered butterflies. Jpn Insect Soc Newslett 43:37–38 (In Japanese)

Government of Japan (2007) The third national biodiversity strategy of Japan

Hagiwara Y, Yamashita M, Sawada H (2009) Invasion status of *Eragrostis curvula* in the *Lycaeides argyrognomon praeterinsularis* habitat along the Abe River, and the effects of its removal. Jpn J Conserv Ecol 14:25–35 (In Japanese with English abstract)

Hama E (1996) Thriving and survival of butterflies in the habitat on dry riverbed. In: Tanaka B, Arita Y (eds) Decline and conservation of butterflies in Japan IV. The Lepidopterological Society of Japan, Osaka, pp 157–172 (In Japanese with English abstract)

Hama E (2007) Conservation of the habitats of *Shijimiaeoides divinus barine*, liberated at Azuminoshi, Nagano. Nat Insects 42(7): 27–31 (In Japanese)

Hama E, Ishii M, Sibatani A (eds) (1989) Decline and conservation of butterflies in Japan I. The Lepidopterological Society of Japan, Osaka (In Japanese with English abstract)

Hanski I, Ovaskainen O (2002) Extinction debt at extinction threshold. Conserv Biol 16:666–673

Hara S, Ito T, Ishikawa K, Sibuya M, Shinkawa T, Suzuki H, Tajima S, Makibayashi I, Yamauchi H (1971) Open letter to Lepidopterological Society of Japan. Yadoriga 67:20 (In Japanese)

Hasegawa J (1994) Extinction of *Mellicta ambigua* by sika deer in Nikko. Gekkan Mushi 278:30–32 (In Japanese)

Hill JK, Ohlemüller R, Fox R, Thomas CD (2009) Climate warming and distribution changes in butterflies. In: Settele J, Shreeve T, Konvička M, Van Dyck H (eds) Ecology of butterflies in Europe. Cambridge University Press, Cambridge, pp 315–321

Hilton-Taylor C (Compiler) (2000) 2000 IUCN Red List of threatened species. IUCN Gland, Switzerland and Cambridge, UK

Hirabayashi J, Koda K, Nakamura H (2009) Life table of *Shijimiaeoides divinus barine* population introduced at pupal stage in a sanctuary of Alps Azumino National Government Park. J Fac Agric Shinshu Univ 45(1–2):21–30 (In Japanese with English abstract)

Hiroki S (ed) (2002) The ecology of Satoyama. The University of Nagoya Press, Nagoya (In Japanese)

Ihara M (1996) Conservation of *Luehdorfia japonica* in Iida City. In: Ae SA, Hirowatari T, Ishii M, Brower LP (eds) Decline and conservation of butterflies in Japan III. The Lepidopterological Society of Japan, Osaka, pp 172–173

Imura O, Shi K (2004) The problems of grassland biodiversity conservation from the view point of grassland butterflies. Agric Hortic 79:352–357 (In Japanese)

Inoue T (2003) Chronosequential change in a butterfly community after clear-cutting of deciduous forests in a cool temperate region of central Japan. Entomol Sci 6:151–163

Inoue T (2005) Causes of butterfly decline in Japan. Jpn J Ent 8:43–64 (In Japanese with English abstract)

Inoue T (2007) Conservation of butterflies in relation to the recent land use change in Japan. J Jpn Soc Grassland Sci 53:40–46 (In Japanese)

Inoue Y, Kon M (2006) Seasonal prevalence of an alien papilionid, *Sericinus montela*, and a native papilionid, *Atrophaneura alcinous*, on the bank of the Kizu River, Kyoto Prefecture. Nat Insects 41(13):33–37 (In Japanese)

Ishii M (1990) An elegy on the fall of Satoyama, coppices-broken relations between agriculture and butterflies in villages. Kagaku Asahi 50(11):22–27 (In Japanese)

Ishii M (1994) Grassland butterflies protected in exercise fields of the self-defence force. Nat Insects 29(6):2–3 (In Japanese)

Ishii M (1996) Decline and conservation of butterflies in Japan. In: Ae SA, Hirowatari T, Ishii M, Brower LP (eds) Decline and conservation of butterflies in Japan III. The Lepidopterological Society of Japan, Osaka, pp 157–167

Ishii M (2001) Studies on insects and their habitats in the Satoyama in a broader sense. Jpn J Environ Entomol Zool 12:187–193 (In Japanese with English abstract)

Ishii M (2003) Result of garden butterfly survey in 1997 to 2002. In: Sunose T, Eda K (eds) Decline and conservation of butterflies in Japan V. The Lepidopterological Society of Japan, Tokyo, pp 225–242 (In Japanese)

Ishii M, Ueda K, Shigematu T (1993) The nature conservation of Satoyama. Tsukiji-shokan, Tokyo (In Japanese)

Ito K, Nanba M (2007) Artificial distribution expansion of *Luehdorfia japonica* in Okayama Prefecture. Nat Insects 42(7):22–26 (In Japanese)

Iwami J (2009) Challenges and those overcoming in conservation activities of Scotosia checkerspot butterfly, *Melitaea scotosia*. Nat Insects 44(2):19–23 (In Japanese)

Iwami J, Kamada K, Shintani T, Nakamura Y, Nakamoto M (2006) Conservation action for the Scotosia Checkerspot, *Melitaea scotosia*. Nat Insects 41(3):10–15 (In Japanese)

Iwano H (1993) Decline of an *Artopoetes pryeri* population in Zama City, Kanagawa Prefecture. In: Yata O, Ueda K (eds) Decline and conservation of butterflies in Japan II. The Lepidopterological Society of Japan and The Nature Conservation Society of Japan, Osaka, pp 133–139 (In Japanese with English abstract)

Iwano H, Sugai T (2007) Expansion of distribution of a nymphalid butterfly, *Hestina assimilis*, in Kanagawa Prefecture. Nat Insects 42(7):18–21 (In Japanese)

Iwasaki I (1994) Population status of the nymphalid *Fabriciana nerippe* and the conservation practices in the Kirishima exercise field of the self-defence force in southern Kyushu. Nat Insects 29(6):9–12 (In Japanese)

JBCN (2004) The Japan butterfly conservation network newsletter No.1 (In Japanese)

JBCN (2005) A guide to conserving *Melitaea protomedia* and its habitats. Japan Butterfly Conservation Network, Kanagawa (In Japanese)

JBCS (2007) Butterfly conservation guide of Japan. Japan Butterfly Conservation Society, Kanagawa (In Japanese)

JBCS (2008a) The Japan butterfly conservation society newsletter No.7 (In Japanese)

JBCS (2008b) A guide to conserving *Melitaea scotosia* and its habitats. Japan Butterfly Conservation Society, Kanagawa (In Japanese)

JBCS (2009) A guide to conserving *Aporia hippia* and its habitats. Japan Butterfly Conservation Society, Tokyo (In Japanese)

Kamada K, Fujii H (2010) An explanatory guide to laws or ordinances for insect researchers and insect enthusiasts. Gekkan Mushi 469:23–34 (In Japanese)

Karube H (2001) The crisis of endemic dragonflies in the Ogasawara islands. Gekkan Mushi 369:22–32 (In Japanese)

Karube H (2004) The present situation of the endemic dragonflies in the Ogasawara islands-when and why have they declined? Res Rep Kanagawa Prefect Mus Nat Hist 12:31–45 (In Japanese)

Karube H (2005) The influence of the introduced lizard, *Anolis carolinensis* on the native insect fauna of the Ogasawara Islands. Bull Herpetol Soc Jpn 2005(2):163–168 (In Japanese)

Karube H, Suda S (2004) A preliminary report on influence of an introduced lizard, *Anolis carolinensis* on the native insect fauna of the Ogasawara islands. Res Rep Kanagawa Prefect Mus Nat Hist 12:21–30 (In Japanese)

Kato M (1992) List of insects in the Ogasawara Islands. WWFJ Sci Rep 1(Part 1):73–105 (In Japanese)

Kato M (2001) "Satoyama" and biodiversity conservation: "Satoyama" as important insect habitats. Global Environ Res 5: 135–149

Kato K, Suda S, Washitani I (2007) Conservation project of *Zizina emelina* in Kinu-gawa river basin, Tochigi Prefecture. Nat Insects 42(6):23–26 (In Japanese)

Kawabe S, Takasaki H, Nakamura Y, Miyake S (2005) Present condition of *Melitaea regama* Fruhstorfer (Endangered species I) in the Kusama, Niimi-shi area, and preservation and breeding of habitation environment. Bull Kurashiki Univ Sci Arts 10:93–103 (In Japanese with English abstract)

Kim Y (2002) Illustrated book of Korean butterflies in color. Kyo-Hak Publishing, Seoul

Kitahara M (2008) Poleward range shift of butterflies under global warming. Nat Insects 43(4):19–23 (In Japanese)

Kitahara M, Sei K (2001) A comparison of the diversity and structure of butterfly communities in semi-natural and human-modified grassland habitats at the foot of Mt. Fuji, central Japan. Biodivers Conserv 10:331–351

Kobayashi K (1972) Conservation of non-climax nature. Daisen 15:2 (In Japanese)

Kobayashi Y (1989) Decline of *Shijimiaeoides divinus* populations in Azumino, Nagano Prefecture. In: Hama E, Ishii M, Sibatani A (eds) Decline and conservation of butterflies in Japan I. The Lepidopterological Society of Japan, Osaka, pp 97–98 (In Japanese with English abstract)

Kusakari K (1996) Conservation of *Coreana raphaelis* in Kawanishi Town, Yamagata Prefecture, Japan. In: Ae SA, Hirowatari T, Ishii M, Brower LP (eds) Decline and conservation of butterflies in Japan III. The Lepidopterological Society of Japan, Osaka, pp 205–210

Li Y, Maruyama N, Koganezawa M, Kanzaki N (1996) Wintering range expansion and increase of sika deer in Nikko in relation to global warming. Wildlife Conservation Japan 2:23–35

Lomov B, Keith D, Britton D, Hochuli D (2006) Are butterflies and moths useful indicators for restoration monitoring? A pilot study in Sydney's Cumberland Plain Woodland. Ecol Manag Restor 7:204–210

Makibayashi I (1996) Recent progress on the conservation effort for *Sasakia charonda* (the Great Purple Emperor) in Ranzanmachi, Saitama Prefecture. In: Ae SA, Hirowatari T, Ishii M, Brower LP (eds) Decline and conservation of butterflies in Japan III. The Lepidopterological Society of Japan, Osaka, pp 176–179

Mano T, Fujii H (eds) (2009) Decline and conservation of butterflies and moths in Japan VI. The Lepidopterological Society of Japan, Tokyo (In Japanese)

Margaret V (2008) Butterflies as indicators of climate change. Sci Prog 91:193–201

Matsumoto K (2006) Butterfly fauna of Tama Forest Science Garden, Forestry and Forest Products Research Institute. Bull FFPRI 5:69–84 (In Japanese with English abstract)
Matsumura T (2008) Twenty years of *Luehdorfia puziloi* conservation at Mt. Akagi. Nat Insects 43(11):27–30 (In Japanese)
Ministry of the Environment (2006) Threatened wildlife of Japan—Red Data Book, 2nd ed, vol 5, Insecta. Japan Wildlife Research Center, Tokyo (In Japanese with English abstract)
Ministry of the Environment (2007) Revised Red List of mammals, brackish and fresh water fishes, insects, shells and plants (Press release document 3rd Aug 2007) (In Japanese)
Ministry of the Environment (2009) The Satoyama initiative. Ministry of the Environment, Government of Japan, Japan
Miura S (1999) Ecology of wild animals and damages of agriculture and forestry. National Forestry Extension Association in Japan, Tokyo (In Japanese)
Miyagawa K (1989) Changes in the habitat of *Melitaea scotosia* in Geihoku, Hiroshima Prefecture. In: Hama E, Ishii M, Sibatani A (eds) Decline and conservation of butterflies in Japan I. The Lepidopterological Society of Japan, Osaka, pp 127–129 (In Japanese with English abstract)
Miyake T (1993) Decline of *Plebejus argus* population in Kyushu (Oita Prefecture). In: Yata O, Ueda K (eds) Decline and conservation of butterflies in Japan II. The Lepidopterological Society of Japan and The Nature Conservation Society of Japan, Osaka, pp 169–173 (In Japanese with English abstract)
Miyawaki S, Washitani I (2004) Invasive alien plant species in riparian areas of Japan: the contribution of agricultural weeds, revegetation species and aquacultural species. Global Environ Res 8:89–101
Moriyama H (1988) What is the meaning of nature conservation? Rural Culture Association, Tokyo (In Japanese)
Motojima K (2009) Present situation of the vegetational damage by sika deer in South Alps. Sanrin 1503:64–69 (In Japanese)
Murata K, Nohara K (1993) Decline and conservation of *Shijimiaeoides divinus* (Matsumura) in Kumamoto Prefecture. In: Yata O, Ueda K (eds) Decline and conservation of butterflies in Japan II. The Lepidopterological Society of Japan and The Nature Conservation Society of Japan, Osaka, pp 151–159 (In Japanese with English abstract)
Murata K, Nohara K, Abe M (1998) Effect of routine fire-burning of the habitat on the emergence of the butterfly, *Shijimiaeoides divinus asonis* (Matsumura). Jpn J Ent 1:21–33 (In Japanese with English abstract)
Muroya Y (1989) Decline of *Shijimiaeoides divinus* populations in Aomori Prefecture. In: Hama E, Ishii M, Sibatani A (eds) Decline and conservation of butterflies in Japan I. The Lepidopterological Society of Japan, Osaka, pp 90–97 (In Japanese with English abstract)
Muroya Y (2007) Discovery and extinction of the large shijimi blue, *Shijimiaeoides divinus* (Fixsen, 1887) in north eastern Japan. Butterflies (S.fujisanus) 44:1–19 (In Japanese with English abstract)
Nakahara Y, Oda K (1953) An unrecorded Lycaenid butterfly from Japan. Tyō to Ga 4:20 (In Japanese)
Nakamura Y (2003) The present situation and future of butterflies in Japan. In: Sunose T, Eda K (eds) Decline and conservation of butterflies in Japan V. The Lepidopterological Society of Japan, Tokyo, pp 171–176 (In Japanese)
Nakamura Y (2004) Conservation activities of *Melitaea protomedia* started in various locations. Gekkan Mushi 401:36–39 (In Japanese)
Nakamura Y (2005) The revised red list of Japanese butterflies: their criteria, status and conservation. Nat Insects 40(14):10–14 (In Japanese)
Nakamura Y (2006) Recent situation of *Celastrina ogasawaraensis* and its conservation. Jpn Butterfly Conserv Soc Newslett 5:4–7 (In Japanese)
Nakamura Y (2007) Insect conservation by insect enthusiast (2); *Melitaea scotosia*. Gekkan Mushi 437:58–64 (In Japanese)
Nakamura Y (2008) Butterfly conservation in natural parks. Natl Parks 633:9–12 (In Japanese)
Nakamura Y (2009a) The present situation and the problems of butterfly conservation in Japan. Nat Insects 44(2):2–3 (In Japanese)
Nakamura Y (2009b) The present status of *Celastrina ogasawaraensis* and its conservation. Nat Insects 44(6):17–20 (In Japanese)
Nakamura Y, Iwami J, Kamada K, Miyazaki Y, Ikejiri S, Shintani T, Nakamoto M (2003a) Conservation of *Melitaea scotosia* in Sera and Kamo Plateau in Hiroshima Prefecture. In: Sunose T, Eda K (eds) Decline and conservation of butterflies in Japan V. The Lepidopterological Society of Japan, Tokyo, pp 199–205 (In Japanese)
Nakamura S, Sugai T, Kishi K (2003b) Mysterious occurrence of a nymphalid butterfly, *Hestina assimilis*, in Kanagawa Prefecture, central Japan. Gekkan Mushi 384:38–41 (In Japanese)
Nanba M (2009) Ecology and conservation of *Melitaea protomedia* in Ombara Highland, Okayama Prefecture. In: Mano T, Fujii H (eds) Decline and conservation of butterflies and moths in Japan VI. The Lepidopterological Society of Japan, Tokyo, pp 39–48 (In Japanese)
New TR (ed) (1993) Conservation biology of Lycaenidae (Butterflies). IUCN Gland, Switzerland
New TR (1997) Butterfly conservation 2nd ed. Oxford University Press, Melbourn
Nishinaka Y, Ishii M (2007) Mosaic of various seral stages of vegetation in the Satoyama, the traditional rural landscape of Japan as an important habitat for butterflies. Trans Lepid Soc Japan 58:69–90
Öckinger E, Hammarstedt O, Nilsson SG, Smith HG (2006) The relationship between local extinctions of grassland butterflies and increased soil nitrogen levels. Biol Conserv 128:564–573
Obara K, Yodoe K (1994) The problems of insecticide spraying of pine forest. Insects Nat 2:24–29
Ogata Y (2009) Recovery action of *Coreana raphaelis* in Iwate Pref. Nat Insects 44(2):15–18 (In Japanese)
Ogura J (2006) The changes of grassland area in Japan. J Kyoto Seika Univ 30:160–172 (In Japanese)
Ohwaki A, Nakamura K, Tanabe S (2007) Butterfly assemblages in a traditional agricultural landscape: importance of secondary forests for conserving diversity, life history specialists and endemics. Biodivers Conserv 16:1521–1539
Rundlöf M, Smith HG (2006) The effect of organic farming on butterfly diversity depends on landscape context. J Appl Ecol 43:1121–1127
Saigusa T (1997) Conservation of *Shijimia moorei* Leech (Lepidoptera, Lycaenidae). Nat Insects 32(8):8–12 (In Japanese)
Sakuratani Y, Kanno K (2003) Ecology of *Sericinus montela* in Kizu River bank, Kyoto Prefecture. In: Sunose T, Eda K (eds) Decline and conservation of butterflies in Japan V. The Lepidopterological Society of Japan, Tokyo, pp 181–184 (In Japanese)
Sei K (1996) The butterfly habitats on the riverside. In: Tanaka B, Arita Y (eds) Decline and conservation of butterflies in Japan IV. The Lepidopterological Society of Japan, Osaka, pp 113–117 (In Japanese with English abstract)
Settele J, Kudrna O, Harpke A et al (2008) Climatic risk atlas of European butterflies. Pensoft Publishers, Bulgaria. doi:10.3897/biorisk.1
Shibatani A (1989) Decline and conservation of butterflies in Japan. In: Hama E, Ishii M, Sibatani A (eds) Decline and conservation

of butterflies in Japan I. The Lepidopterological Society of Japan, Osaka, pp 16–22
Shibatani A (1990) Decline and conservation of butterflies in Japan. J Res Lepidoptera 29:305–315
Shimizu Y (2003) The nature of Ogasawara and its conservation. Global Environ Res 7:3–14
Shimizu T (2009) Recovery action of *Shijimiaeoides divinus barine* in Tomi City, Nagano Pref. Nat Insects 44(2):4–9 (In Japanese)
Suda S (2007) Conservation of nature and insects of gravelly floodplain. Biol Sci Genet 61(3):43–47 (In Japanese)
Sunose T, Eda K (2003) Decline and conservation of butterflies in Japan V. The Lepidopterological Society of Japan, Tokyo (In Japanese)
Tabata H (ed) (1997) Satoyama and its conservation. Hoikusha, Osaka (In Japanese)
Tabata H (2009) Woody biomass utilization and conservation of Satoyama forests. J Environ Conserv Eng 38:189–196 (In Japanese)
Takahashi M (2004) Decline of *Lycaeides argyrognomon* in the Abe river basin. Suruga no Konchu 206:5741–5746 (In Japanese)
Takahashi Y (2008) Possible utilization of native grass resources other than as a feed for livestock. J Jpn Soc Grassland Sci 53:318–325 (In Japanese)
Takahashi Y, Inoue M, Ondopa J (2008) Mowing and grazing management of *Micanthus*-type grasslands at the foot of Mount Sanbe in relation to the conservation of the *Patrinia scabiosaefolia* population. Bull Shimane Nat Museum Mt. Sanbe 6:1–6 (In Japanese with English abstract)
Takakuwa M, Suda S (2004) The decline of a Lycaenid butterfly *Celastrina ogasawaraensis*, with reference to its cause. Res Rep Kanagawa Prefect Mus Nat Hist 12:47–53 (In Japanese)
Takeuchi K (2001) Nature conservation strategies for the "Satoyama" and "Satochi", habitats for secondary nature in Japan. Global Environ Res 5:193–198
Tanaka B, Arita Y (1996) Decline and conservation of butterflies in Japan IV. The Lepidopterological Society of Japan, Osaka (In Japanese with English abstract)
Tanaka N, Matsui T, Yagihashi T, Taoda H (2006) Climatic controls on natural forest distribution and predicting the impact of climate warming: especially referring to Buna (*Fagus crenata*) forests. Global Environ Res 10:151–160
Tanigami M (2008) Tne conservation activities of Yodo Conservation Group for the *Melitaea protomedia*. Jpn Butterfly Conserv Newslett 8:12–13 (In Japanese)
Tashita M (1989) Decrease in the number of individuals of the Maculatus Skipper, *Pyrgus maculatus* Bremer & Grey (Lepidoptera, Hesperiidae) according to the secondary succession. Tyō to Ga 40:85–92 (In Japanese with English abstract)
Tashita M (1993) Degradation of a *Pyrgus maculatus* population caused by environmental alterations in Kiso, Nagano Prefecture. In: Yata O, Ueda K (eds) Decline and conservation of butterflies in Japan II. The Lepidopterological Society of Japan and The Nature Conservation Society of Japan, Osaka, pp 183–188 (In Japanese with English abstract)
Tashita M, Maruyama K (2007) Outline of the habitat of *Shijimiaeoides divinus barine* in Chubu district, central Japan. Butterflies (S. fujisanus) 44:24–31 (In Japanese with English abstract)
The Ecological Society of Japan (ed) (2002) Handbook of alien species in Japan. Chijinshokan, Tokyo (In Japanese)
The Lepidopterological Society of Japan (1994) Guidelines and rules for collecting. Tyō to Ga 44:258–259
The Nature and Insects (1971) Save *Chrysozephyrus hisamatsusanus*!. Nat Insects 6(3):1 (In Japanese)
Thomas JA, Telfer MG, Roy DB, Preston CD, Greenwood JJD, Asher J, Fox R, Clarke RT, Lawton JH (2004) Comparative losses of British butterflies, birds, and plants and the global extinction crisis. Science 303:1879–1881
Toda M, Nakagawa N, Sukigara N (2009) Ecology and control of green anole *Anolis carolinensis*, invasive alien species in the Ogasawara Islands. Global Environ Res 14:39–46 (In Japanese)
Tunekawa A, Ikeguchi H, Omasa K (1996) Prediction of Japanese potential vegetation distribution in response to climate change. In: Omasa K, Kai K, Taoda H, Uchijima Z, Yoshino M (eds) Climate change and plants in East Asia. Springer-Verlag, Tokyo, pp 57–65
Van Swaay CAM, Warren MS (1999) Red Data Book of European butterflies (Rhopalocera). Nature and Environment, No. 99. Council of Europe Publishing, Strasbourg
Van Swaay C, Cuttelod A, Collins S, Maes D, López Munguira M, Šašić M, Settele J, Verovnik R, Verstrael T, Warren M, Wiemers M, Wynhoff I (2010) European Red List of butterflies. Publications Office of the European Union, Luxembourg
Washitani I (2001) Traditional sustainable ecosystem "Satoyama" and biodiversity crisis in Japan: conservation ecological perspective. Global Environ Res 5:119–133
Watanabe Y (1989) *Pyrgus malvae* Linnaeus. In: Hama E, Ishii M, Sibatani A (eds) Decline and conservation of butterflies in Japan I. The Lepidopterological Society of Japan, Osaka, pp 23–25 (In Japanese)
Watanabe S (2005) Decline of ultrabasicosaxicolous flora from 1954 to 2003 on Mt. Apoi, Hidaka Province, Hokkaido, Japan. Jpn J Ecol 55:105–110 (In Japanese with English abstract)
Yamamoto S (2001) Studies on the effect of changes in rural landscape structure on secondary forest plants in Japanese rural areas. Bull Natl Inst Agro-environ Sci 20:1–105 (In Japanese with English abstract)
Yata O (2007) Iconographia insectorum Japonicorum colore naturali edita, vol I (Lepidoptera). Hokuryukan, Tokyo (In Japanese)
Yata O, Ueda K (1993) Decline and conservation of butterflies in Japan II. The Lepidopterological Society of Japan and The Nature Conservation Society of Japan, Osaka (In Japanese with English abstract)
Yodoe K (1997) *Fabriciana nerippe*. In: Shimane Prefecture (ed) Red Data Book of Shimane Prefecture, pp 172–173 (In Japanese)
Yoshio M (2002) Northward invasions of butterflies and climatic warming. Nat Insects 37(1):4–7 (In Japanese)

J Insect Conserv (2011) 15:259–268
DOI 10.1007/s10841-010-9342-y

ORIGINAL PAPER

Recent evidence for the climate change threat to Lepidoptera and other insects

Robert J. Wilson · Ilya M. D. Maclean

Received: 19 May 2010 / Accepted: 3 September 2010 / Published online: 19 September 2010

Abstract Climate change is now estimated by some biologists to be the main threat to biodiversity, but doubts persist regarding which species are most at risk, and how best to adapt conservation management. Insects are expected to be highly responsive to climate change, because they have short life cycles which are strongly influenced by temperature. Insects also constitute the most diverse taxonomic group, carrying out biotic interactions of importance for ecological functioning and ecosystem services, so their responses to climate change are likely to be of considerable wider ecological significance. However, a review of recent published evidence of observed and modelled effects of climate change in ten high-ranking journals shows that comparatively few such studies have focused on insects. The majority of these studies are on Lepidoptera, because of the existence of detailed contemporary and historical datasets. These biases in published information may influence conclusions regarding the threat of climate change to insect biodiversity. Assessment of the vulnerability of insect species protected by the Bern Convention on the Conservation of European Wildlife and Natural Habitats also emphasises that most information is available for the Lepidoptera. In the absence of the necessary data to carry out detailed assessments of the likely effects of climate change on most threatened insects, we consider how autecological studies may help to illuminate the potential vulnerability of species, and draw preliminary conclusions about the priorities for insect conservation and research in a changing climate.

Keywords Bioclimate models · Conservation priorities · Flagship species · Global change · Range shifts

R. J. Wilson (✉) · I. M. D. Maclean
Centre for Ecology and Conservation, University of Exeter, Cornwall Campus, Penryn TR10 9EZ, UK
e-mail: R.J.Wilson@exeter.ac.uk

I. M. D. Maclean
e-mail: I.M.D.Maclean@exeter.ac.uk

Introduction

Insects are expected to show rapid responses to climate change, because they are ectotherms with short life cycles and wide variation in population size over space and time (Bale et al. 2002). The insects are also a highly diverse taxonomic group, and their biotic interactions as natural enemies of plants and other animals, decomposers, and pollinators are critical for ecological functioning and ecosystem services (Ayres and Lombardero 2000; Bradford et al. 2002; Harrington 2002; Kremen et al. 2007). Therefore the effects of climate change on insect biodiversity have considerable applied ecological importance.

Some of the first clear evidence for ecological responses to climate change was drawn from studies on insects. Poleward shifts in the distributions of butterflies in Europe and North America represented clear signals of recent warming (Parmesan 1996; Parmesan et al. 1999), as did phenological advances towards earlier dates in the spring activity of temperate Lepidoptera (Roy and Sparks 2000; Peñuelas et al. 2002; Stefanescu et al. 2003). Subsequently, evidence has also been presented for range shifts (Hickling et al. 2005, 2006) and phenological shifts (Gordo and Sanz 2005, 2006; Hassall et al. 2007) in other insect groups. Now that the ecological effects of climate change have been widely documented (Walther et al. 2002; Parmesan and Yohe 2003), research has begun to focus more on climate change as a threat to biodiversity. In the most well-known study to recognise climate change as perhaps the major twenty-first Century threat to biodiversity, bioclimate models of the distributions of 69 butterfly species in Mexico, South Africa and Australia contributed to the estimates of extinction risk from climate change for 1,103 species in total (Thomas et al. 2004). The ecological importance and potential rapidity of insect responses imply that much published research since Thomas et al.'s (2004) paper would have focused on the effects of climate change on this taxonomic group, and on how to adapt conservation management in the light of the threat from climate change.

Here we review two sources of evidence for the threat of climate change to insect biodiversity. First, we conduct a comprehensive review of research published in ten high-ranking journals in the past 5 years: (1) to determine whether studies on insects make an important contribution to high profile studies of climate change effects; (2) to quantify the threat of climate change to insects, based on studies of observed and modelled responses; (3) to show whether there are taxonomic or regional biases in published research, and what the implications of any such biases may be for estimates of the threat of climate change. Second, given a lack of research into the observed or potential impacts of climate change on most insect groups, we review a wider range of literature to determine what information is available concerning the climate change risk to a group of threatened insects comprising representatives from a range of different taxa. We use insects listed by the Bern Convention on the Conservation of European Wildlife and Natural Habitats as our exemplar group. Where direct information for the effects of climate change on threatened species is not available, we consider how autecological data can be used to make preliminary recommendations for conservation or research needs for insects in a changing climate.

Methods

Evidence of climate change threat for insects

In this review we used the ISI Web of Science to find all papers published between January 2005 and December 2009 in ten leading journals which included "climate change" in the title, keywords or abstract. We reviewed journals of general science (*Nature*, *Proceedings of the National Academy of Science of the USA*, *Science*), general biology or ecology (*Ecology Letters*, *Proceedings of the Royal Society of London B*,), applied ecology (*Ecological Applications*, *Global Change Biology*, *Journal of Applied Ecology*), and conservation biology (*Biological Conservation*, *Conservation Biology*). We searched all papers with "climate change" and "biodiversity" in the title, abstract or keywords in *Global Change Biology* and *Proceedings of the National Academy of Sciences*, both of which have a very large number of studies on climate change. We widened the search to include any study with just "climate change" in the title, abstract or keywords for the remainder of journals, leading to a final total of 1,120 papers reviewed. In our quantitative analysis, we only included those papers in which (a) IUCN Red List categories were assigned to taxa; (b) a change in population size or range were reported; or (c) extinction risk was estimated; which greatly reduced the number of studies. We also only included those studies in which the response could be primarily attributed to climate change or where climate change impacts were separated from other drivers of change. We separated papers into those which provided empirical evidence of recent change ("observed") and those which predicted future change ("predicted"). Papers could include both observed and predicted estimates of change, and a number of papers included measurements for more than one taxonomic group.

The IUCN Red List criteria provided a means of assigning each taxon studied to a particular threat category on the basis of extinction risk or observed or projected changes in population size over set time periods (see Mace et al. 2008 for details). In the absence of better evidence we assumed that a reduction in range size led to a corresponding reduction in population size. We estimated the proportion of taxa that would be in each IUCN Red List category in 2100 (Table 3). To make this estimation, it is necessary to convert the estimates of extinction risk and population change for the time periods associated with each study to a standard of 90-year representing the period 2010–2100. We did this by assuming that rates of population changes and the frequency of extinction events remained constant through time. Overall population declines thus increase with time by an amount determined by geometric decay, and overall extinction risks increase with time by an amount determined by multiple-event probability theory. Annual rates of population decline (r) were estimated from each study as

$$r = \ln\left(\frac{1 - decline_t}{t}\right)$$

where $decline_t$ is the decline noted in the study and t is the time period of the study. Total decline for the period 2010–2100 (90 years) is then given by:

$$decline_{90} = 1 - \exp(r \times 90).$$

The extinction risk (E_{90}) for the period 2010–2100 (90 years) can be calculated from the extinction risk (E_t) associated with each study over the period (t) of each study as:

$$E_{90} = 1 - (1 - E_t)^{\frac{90}{t}}.$$

Climate change and protected insects

The 1979 Bern Convention on the Conservation of European Wildlife and Natural Habitats is an international legal agreement which aims to promote European cooperation for nature conservation. Appendix II of the Convention includes 56 strictly protected insect species, of which 26 are Lepidoptera (Table 1). We carried out literature searches in May 2009 using the scientific name of each of these species in ISI Web of Science, and in Google Scholar to include a wider range of potential sources.

A recently published "Climatic Risk Atlas" for European butterflies (Settele et al. 2008) includes bioclimate models of the future climatic ranges for 15 species of butterfly listed by the Bern Convention. The atlas uses three different climate change scenarios for the years 2050 and 2080, and models future potential ranges if species are able to fully occupy future modelled climate space ("Full Dispersal"), or only able to occupy those parts of the range which overlap with their current suitable climate space ("No dispersal"). The three scenarios were based on different trajectories of economic growth and environmental policy, ranging from sustainability (SEDG, IPCC scenario B1, +2.4°C by 2080), through a business as usual approach (BAMBU, IPCC scenario A2, +3.1°C by 2080) to development based on economic growth (GRAS, IPCC scenario A1F1, +4.1°C by 2080) (Nakicenovic et al. 2000; Spangenberg 2007; Settele et al. 2008). We use the published results of these models for our main quantitative data in this part of the review, asking two questions for the modelled effects of climate change on butterflies listed by the Bern Convention: (1) what are the differences in modelled future range size under different scenarios of climate warming (+2.4°C, +3.1°C, +4.1°C)? and (2) what are the differences in range size for full dispersal versus no dispersal? The two questions are relevant to determine the importance for conserving threatened taxa by (1) mitigating levels of climate change, and (2) by promoting range shifts by species into regions which become climatically favourable.

We encountered very few direct estimates of climate change effects on Bern Convention insects other than Lepidoptera, so our Discussion relating to this part of the review focuses on the evidence for sensitivity of the listed species to climate change, based on their ecological characteristics (e.g. dispersal propensity, range size, habitat specialism), and uses this information to make recommendations for research or conservation priorities for insects in a changing climate.

Table 1 Insects listed as strictly protected by Appendix II of the Bern Convention (Council of Europe 2010)

Orthoptera	**Ephemeroptera**
Baetica ustulata	*Palingenia longicauda*
Saga pedo	**Mantodea**
Odonata	*Apteromantis aptera*
Aeshna viridis	**Lepidoptera**
Brachythemis fuscopalliata	*Apatura metis*
Calopteryx syriaca	*Coenonympha hero*
Coenagrion freyi	*Coenonympha oedippus*
Coenagrion mercuriale	*Erebia calcaria*
Cordulegaster trinacriae	*Erebia christi*
Gomphus graslinii	*Erebia sudetica*
Leucorrhinia albifrons	*Eriogaster catax*
Leucorrhinia caudalis	*Euphydryas (Eurodryas) aurinia*
Leucorrhinia pectoralis	*Fabriciana elisa*
Lindenia tetraphylla	*Hyles hippophaes*
Macromia splendens	*Hypodryas maturna*
Ophiogomphus cecilia	*Lopinga achine*
Oxygastra curtisii	*Lycaena dispar*
Stylurus (= Gomphus) flavipes	*Maculinea arion*
Sympecma braueri	*Maculinea nausithous*
Coleoptera	*Maculinea teleius*
Buprestis splendens	*Melanargia arge*
Carabus bessarabicus	*Papilio alexanor*
Carabus hungaricus	*Papilio hospiton*
Carabus olympiae	*Parnassius apollo*
Cerambyx cerdo	*Parnassius mnemosyne*
Cucujus cinnaberinus	*Plebicula golgus*
Dytiscus latissimus	*Polyommatus galloi*
Graphoderus bilineatus	*Polyommatus humedasae*
Osmoderma eremita	*Proserpinus prosperpina*
Rosalia alpina	*Zerynthia polyxena*

Results

Evidence of climate change threat for insects

73 papers in the review included evidence of changes to extinction risk linked to climate change: 39 papers including observed changes, and 47 including predicted changes. Only nine of these papers dealt with insects (4 observed, 5 predicted), of which eight included evidence for the Lepidoptera (Table 2). All papers linking climate change to *observed* changes in extinction risk (usually range or population size) presented evidence for Lepidoptera (Table 2). Projecting observed or predicted changes to extinction risk over time did not lead to any estimates of decline over the period 2010–2100 which would classify the study taxa as Endangered or Critically Endangered, but 10% of taxa

Table 2 Papers in ten leading journals between 2005 and 2009 which linked climate change to observed or predicted changes in distribution or population size of insects

Reference	Taxa	Location
Observed		
Chen et al. (2009)	Lepidoptera (Geometridae + Others)	Borneo
Davies et al. (2005)	Lepidoptera (*Hesperia comma*)	England
WallisDeVries and Van Swaay (2006)	Lepidoptera (Rhopalocera)	Netherlands
Wilson et al. (2005)	Lepidoptera (Rhopalocera)	Spain
Predicted		
Colwell et al. (2008)	Lepidoptera (Geometridae) + Hymenoptera (Formicidae)	Costa Rica
Dangles et al. 2008	Lepidoptera (Gelechiidae)	Ecuador
Goulson et al. (2005)	Diptera (Calliphoridae, Muscidae)	England
Hoyle and James (2005)	Lepidoptera (*Pseudophilotes sinaicus*)	Egypt
Logan et al. (2007)	Lepidoptera (*Lymantria dispar*)	Utah (USA)

Table 3 IUCN criteria for population decline and extinction risk estimated for a period of 90 years (2010–2100) (see "Methods"), and the percentage of cases for observed and predicted effects of climate change on insect taxa which classify as each of the three threatened classes (CR, EN and VU)

	Critically Endangered	Endangered	Vulnerable
Population decline			
IUCN criteria/10 years	80%	50%	10%
Equivalent/90 years	99.99%	99.80%	95.96%
Extinction risk			
IUCN criteria	50%/10 years	20%/20 years	10%/100 years
Equivalent/90 years	99.80%	63.36%	9.05%
% observed	0	0	10
% predicted	0	0	9

Data sources are shown in Table 2

would be classified as Vulnerable based on observed changes (9% based on predicted changes) (Table 3).

Despite the small sample of studies of the climate change threat to insects, there were examples from a relatively wide range of geographical locations (Table 2). The nine papers included three from northern Europe (Britain and the Netherlands), one from north America (Utah), two from mountain regions in the Mediterranean (Spain and Egypt), and three from the Tropics (Borneo, Ecuador, Costa Rica), again with a focus on elevational gradients.

Climate change and protected insects

Nine of the 15 butterfly species whose climate niches were modelled by Settele et al. (2008) were categorised as having High, Very High, or Extremely High climate change risk, indicating that more than 70% of their current European distributions are predicted to be climatically unfavourable by 2080 in at least one climate change scenario (Table 4; Fig. 1). Four of the remaining species were categorised as experiencing some climate change risk, because more than 50% of currently suitable cells were predicted to be unsuitable by 2080 in one or more scenario. Only one species (the scarce fritillary *Hypodryas maturna*) did not show modelled losses of >50% in any 2080 scenario (Lower Climate Change Risk), while the current distribution of marsh fritillary (*Euphydryas aurinia*) was only related to climate conditions to a limited extent (AUC < 0.75).

Increases in temperature by 2.4°C (SEDG scenario), 3.1°C (BAMBU scenario) or 4.1°C (GRAS scenario) produced clear differences in predicted distribution sizes of Bern Convention butterfly species. In both 2050 and 2080 the highest warming scenario (GRAS) gave consistently lower modelled distribution sizes than both other scenarios. In 2080, the intermediate scenario (BAMBU) produced consistently reduced distribution sizes relative to the most conservative scenario (SEDG) (Fig. 1).

Future modelled climate space for all species included areas which were not modelled as suitable for the species in 2000. Hence the Full Dispersal scenario of distribution change always led to a less severe modelled decline in

Table 4 Modelled changes by 2080 to the climatically favourable range area in Europe for Bern Convention species of butterfly (after Settele et al. 2008), averaged for three scenarios of climate change

	Model accuracy (AUC)	Climate risk category+	Mean 2080 modelled change (%)*	
			Full dispersal	No dispersal
Apatura metis	0.98	HHR	+47	−71
Coenonympha hero	0.88	HHR	−44	−76
Coenonympha oedippus	0.95	R	+161	−50
Euphydryas aurinia	0.72	PR	−24	−41
Hypodryas maturna	0.79	LR	+46	−37
Lopinga achine	0.81	R	+58	−43
Lycaena dispar	0.88	R	+54	−41
Maculinea arion	0.77	R	−17	−52
Maculinea nausithous	0.91	HHHR	−45	−84
Maculinea teleius	0.84	HHR	−26	−74
Melanargia arge	0.98	HR	+119	−64
Papilio alexanor	0.94	HHR	−53	−77
Parnassius apollo	0.80	HR	−51	−66
Parnassius mnemosyne	0.77	HR	−18	−65
Zerynthia polyxena	0.85	HR	+80	−59

Results for % change are based either on "Full dispersal" (entire area in 2080) or "No dispersal" (area of overlap between area in 2000 and in 2080)

* Mean modelled change is shown as the mean of the three climate change scenarios (SEDG, BAMBU, GRAS; see "Methods"). Differences among scenarios are shown in Fig 1

+ Climate Risk Category based on maximum decline: >95% Extremely High Risk (HHHR), >85% Very High Risk (HHR), >70% High Risk (HR), >50% Risk (R), <50% Lower Risk (LR); AUC <0.75 Potential Risk (PR) (Settele et al. 2008)

potential distribution size than the No Dispersal scenario; and in seven out of 15 cases, the Full Dispersal scenario led on average (across the three climate change scenarios) to modelled increases in the size of the climate niche (Fig. 1; Table 4).

Discussion

Our review suggests that relatively few papers published in ten high-ranking journals in the past 5 years used insect systems to show evidence of the extinction threat from climate change (9 out of 73 papers showing a threat to biodiversity, or 12%). A further 253 published papers in the review related climate change to ecological change, but did not provide data which satisfied the criteria for inclusion in the review: 43 (17%) of these cases dealt with insects. The relatively low proportion of insect-based papers (relative to insect diversity) may reflect a wider tendency of insect research to be somewhat marginalised (Leather 2010). A wide range of journals now publish research in ecology, conservation and global change, as is evident from the reference list of the current article, and inclusion of these journals would have increased the sample size. Nevertheless, journals focusing more specifically on entomology or insect conservation are likely to play a key role in presenting the evidence for climate change effects on insects.

The few papers identified by the review illustrated some important points about evidence for the threat of climate change to insects. Evidence of *observed change* appeared to support estimates of threat from *predictions* of the effects of climate change: projecting the changes from the published papers over the period 2010–2100 predicted that no taxa would be categorised as Critically Endangered or Endangered, but that 10% would be classified as Vulnerable based on observed changes, or 9% based on predicted changes (Table 3). Lepidoptera played a key role in providing evidence of change, as would be expected, given the greater availability of current and historical information on species distributions and ecological requirements than for other taxa. A number of studies were conducted at temperate latitudes and on pest species (e.g. Goulson et al. 2005; Logan et al. 2007). The distributions of many ectothermic species including insects are limited by low temperatures at high latitudes, and so are expected (and observed) to expand their distributions in northern Europe because of climate change (e.g. Hickling et al. 2006), and widely dispersing generalists like many pests may also be able to expand their distributions relatively rapidly compared with more sedentary or habitat

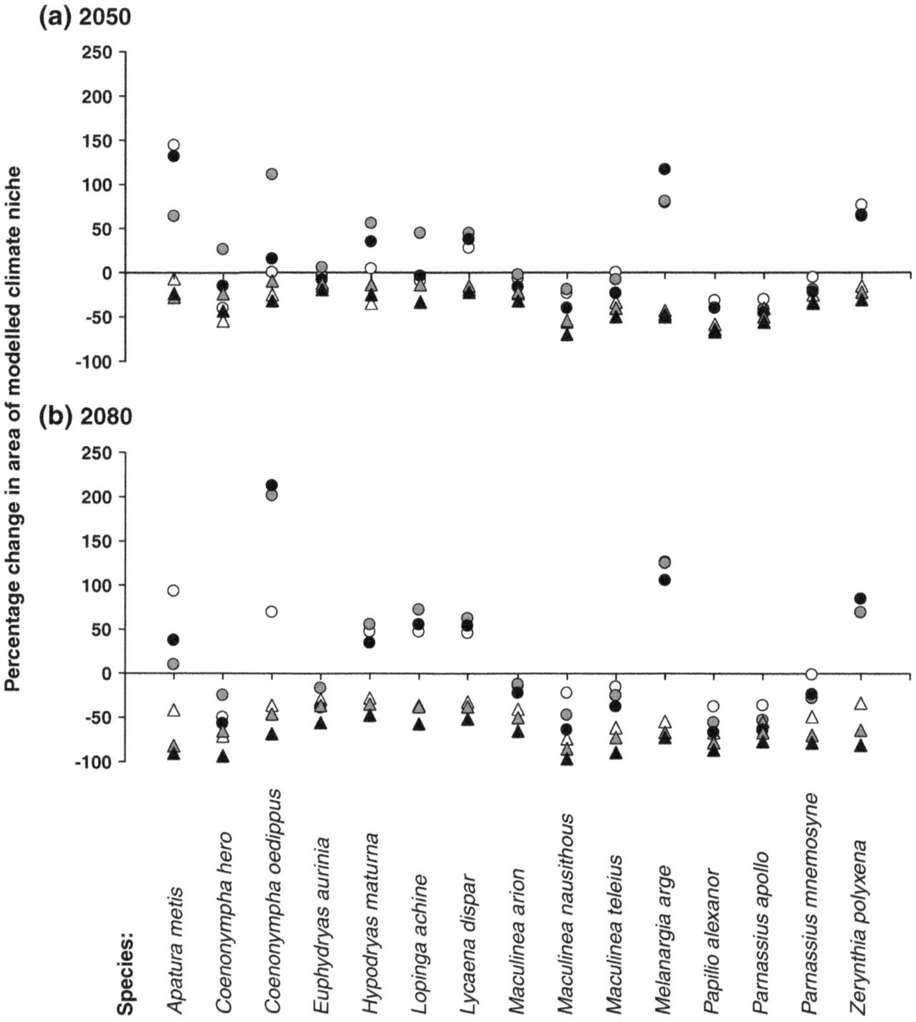

Fig. 1 Modelled changes to the European climate niche space for butterflies listed by the Bern Convention (adapted from Settele et al. 2008). Data show percentage change in modelled niche space between 2000 and *a* 2050, or *b* 2080, assuming either full dispersal (*circles*) or no dispersal (*triangles*), and are shown for three climate change scenarios, for 2.4°C increase (SEDG, *white symbols*), 3.1°C increase (BAMBU, *grey*), and 4.1°C increase (GRAS, *black*)

specialist species (Warren et al. 2001; Pöyry et al. 2009). Hence a preponderance of studies of insect responses to climate change at relatively high latitudes could lead to underestimates of the threat of climate change (but see Franco et al. 2006). Where studies have been carried out at lower latitudes, such as in Mediterranean mountains, species distributions have been observed to contract (Wilson et al. 2005) or predicted to do so (Hoyle and James 2005). Clearly evidence from lower latitudes is vital for identifying the real risk that climate change will erode species distributions, or whether species are able to adapt to local or regional changes in climate. Encouragingly, a few high profile studies have begun to gather evidence of climate change effects on tropical insects (e.g. Chen et al. 2009). In the Tropics, where latitudinal gradients in climate are negligible, elevational range shifts may play a major role in allowing species to track the availability of suitable climatic conditions, and work based to a large extent on insects has been important in illustrating this point (Colwell et al. 2008).

Quantitative information on expected effects of climate change on protected insects (those listed by the Bern Convention) is also most forthcoming for the Lepidoptera (Settele et al. 2008), and highlights two important points. First, more marked increases in temperature lead to more pronounced declines in the area of modelled climate space for threatened species. Second, if species are able to colonize areas which become climatically favourable, they are likely to experience less drastic declines in distribution area than if they are limited to future favourable areas which overlap with their current distributions. Since most Lepidoptera listed by the Bern Convention have highly specialist habitat requirements and therefore fragmented distributions of habitat, it is unlikely that most will be able to colonize regions that become climatically favourable (Warren et al. 2001; Menéndez et al. 2006). Hence estimates of future distribution sizes based on a No Dispersal scenario may be more realistic for threatened species than those assuming Full Dispersal, unless serious measures are put in place to promote range shifts by species. In addition, the current and future distributions of specific larval host plants will impose a further constraint on the distribution of many herbivorous insects (Araújo and Luoto 2007; Schweiger et al. 2008).

Protection and active management of current habitat and populations represent vital parts of conservation programmes designed to facilitate species range shifts, being necessary for the survival of species' current populations and the maintenance of future adaptive capacity. Habitat fragmentation is likely to constrain range expansions by species in many parts of the world, even when climate conditions become more favourable (Wilson et al. 2009), so the management of networks of habitat, or of landscapes in which favourable habitats are maintained, is necessary for species to be able to expand their distributions. Conservation interventions in the form of assisted colonization may become an appropriate technique for those species which are likely to suffer the most severe reductions in distribution size, and which have the least chance of reaching locations which become climatically suitable (Hoegh-Guldberg et al. 2008). Insects have a combination of characteristics (e.g. small size, relatively small area requirements, relatively large population sizes) which may make them appropriate for such techniques, and it is known that introductions of insects to climatically favourable regions may be successful (Menéndez et al. 2006; Willis et al. 2009). However, in situ conservation of extant populations should remain the bedrock of species conservation; and habitat management based on sound ecological science at introduction sites is crucial for the success of translocations, such as in the case of the reintroduction of Large Blue *Maculinea arion* to the UK (Thomas et al. 2009).

Research and conservation priorities for threatened Lepidoptera

Many Lepidoptera listed by the Bern Convention possess species traits which imply high climate change sensitivity, or reduced capacity for adaptation. All eight butterfly species listed by the Bern Convention but not modelled by Settele et al. (2008) are European endemics, generally with extremely restricted distributions to islands or mountain ranges. Given their often small distributions or population sizes, and their isolation to narrow suitable regions, these species have very little opportunity for latitudinal range shifts. For mountain-dwelling species, there may be some limited scope for elevational range shifts, or for survival under localised climate conditions provided by topographic heterogeneity (Loarie et al. 2009). Research into the habitat requirements and dispersal capacity of such species can be extremely valuable in identifying the factors limiting species distributions and their capacity to colonize locations which become suitable. For example, movement patterns have been quantified for *Erebia sudetica* in the Czech Republic (Kuras et al. 2003), and habitat requirements for *Erebia calcaria* have been modelled in Slovenia (de Groot et al. 2009). For several protected butterfly species, research into dispersal capacity suggests that distances of 10–50 km between habitat areas would almost certainly prevent natural colonizations (for *Parnassius apollo* see Brommer and Fred 1999; for *Parnassius mnemosyne* see Meglecz et al. 1999; Valimaki and Itamies 2003; for *Coenonympha hero* see Cassel-Lundhagen and Sjogren-Gulve 2007, 2008), again emphasising the importance of managing whole landscapes or habitat networks.

Research showing how distribution patterns can be predicted by a combination of the effects of climatic conditions and landcover can be helpful in predicting the locations and habitats which will be suitable for species in the future, but also in targeting sites for management or species searches in the present day. This approach has been successfully applied for *Parnassius mnemosyne* (Heikkinen et al. 2005, 2007; Luoto et al. 2007) and *Maculinea nausithous* (Jiménez-Valverde et al. 2008). At a finer scale, understanding how climate conditions influence resource availability for threatened species is vital to manage habitat successfully in a changing climate. The host ants of *Maculinea* species have fine-scale distributions which depend on local microclimate, and the habitat structure which provides suitable microclimates varies depending on prevailing regional climates (Thomas et al. 1998; Mouquet et al. 2005). For the reintroduction of *Maculinea arion* to the UK, management of sward structure in order to provide suitable microclimates for the larval host ants (*Myrmica sabuleti*) was an essential step (Thomas et al. 2009). Continued appropriate management has allowed reintroduced populations to survive in the UK and colonize additional suitable habitat nearby, in an exemplary case of how detailed ecological knowledge and habitat management can result in successful landscape-scale conservation.

Monitoring of habitat use and population responses to habitat management are vital because changes to climate conditions may mean that there are changes to the microhabitat types which satisfy the microclimate requirements of species. Larvae in both *Parnassius apollo* and *P. mnemosyne* appear to thermoregulate by moving between areas of litter, bare ground and more shaded vegetation (Valimaki and Itamies 2005; Ashton et al. 2009). In the case of *P. apollo*, populations at lower elevations in central Spain are associated with shrubbier, less-open conditions than populations at higher elevations, implying that habitat management can be adapted to local climates (Ashton et al. 2009). This mountain species has already suffered local extinctions linked to climate warming in France (Descimon et al. 2006) and Spain (Wilson et al. 2005; Ashton et al. 2009), with the low-elevation limits of the species shifting uphill in both regions. "False-spring events" in France appear to result in the early emergence of *P. apollo* larvae from winter diapause, leading to starvation when subsequent conditions are too cold for feeding activity (Descimon et al. 2006). The very small populations of *P. apollo* which remain in some parts of its distribution may have led to inbreeding depression which renders the species less able to adapt to changing conditions, although reintroductions for this species in Poland have had some measure of success (Adamski and Witkowski 2007).

Priorities for other threatened Insecta

In common with the butterflies listed by the Bern Convention, many narrow range European endemic species from other taxa are likely to have very limited scope for range expansion to future climatically favourable regions. For example, *Carabus olympiae* (Coleoptera) is known to have a restricted habitat distribution in the Italian alps, and increased conversion of beech woodland or scrub to pasture are likely to prevent the species from colonizing locations outside its current narrow distribution (Negro et al. 2007, 2008). There are also a few studies which point to potential climate effects on habitat associations. *Saga pedo* (Orthoptera) is known to be restricted to the warm microclimates of steep south-west facing slopes at its northern range limit (Kristin and Kanuch 2007), as is the case for many butterfly species (Thomas 1993). *Cerambyx cerdo* (Coleoptera) requires tree trunks receiving high levels of solar insolation in central Europe, so semi-open pasture landscapes favour the survival of this saproxylic beetle (Buse et al. 2007, 2008). If the habitat requirements of many insect species vary across their latitudinal or elevational ranges, then future habitat management may need to accommodate changes to habitat management to maintain suitable microclimates (Davies et al. 2006). A safety-first approach to management in the mean time would be to maintain heterogeneous conditions, which have been shown to stabilise insect population dynamics (Kindvall 1996; Sutcliffe et al. 1997; Oliver et al. 2010).

There is strong autecological evidence that the beetle *Osmoderma eremita* will be affected by climate change. This species is sensitive to increases in air temperature (Renault et al. 2005). It also has extremely limited dispersal (Hedin et al. 2008), so that persistence within woodlands is dependent on aggregations of old growth oak trees, and in fragmented habitats the species rarely moves between woodlands (Ranius 2000, 2007). Conservation of highly sedentary, habitat specialist species which are sensitive to climate represents a significant challenge under climate change, but the information on habitat requirements such as exists for *O. eremita* is a vital first step.

Species distribution models which incorporate the effects of climate and landcover may be a greater challenge for non-lepidopteran insect groups because less information on species distributions and habitat associations are generally available. However, where distribution data do exist, their use in developing such models can play a key role in identifying regions for targeted species searches, which can help to determine the conservation status of species, as well as strengthening evidence for the modelled effects of climate or habitat management. The Bern-listed Iberian endemic spider *Macrothele calpeiana* provides an excellent example of these points. The species' distribution appears to be determined mainly by climate variables, occurring in regions with high precipitation and high precipitation periodicity, but absent from regions where temperatures reach extremes (Jiménez-Valverde and Lobo 2006). A distribution model developed for *M. calpeiana* proved accurate at identifying a region of southern Portugal in which populations of the species occurred but had not previously been found (Jiménez-Valverde et al. 2007). The models also suggest that the species may require forest cover in order for populations to persist, or shift their range as the climate warms, hence identifying the key role that habitat management may play in allowing the adaptation of conservation management to climate change.

Conclusions

Strong evidence has been available for more than a decade that the distributions and phenology of insects have changed in response to climate change (Menéndez 2007; Wilson et al. 2007). Examples showing that the population or distribution sizes of insects have declined are generally more recent, and generally refer to Lepidoptera (Wilson et al. 2005; Franco et al. 2006; Chen et al. 2009; but see Parmesan 1996, 2006). In a review of ten high-ranking journals, insect examples represented only 12% of published cases directly relating a changed extinction risk to climate change. However, in these examples, predicted estimates of threat corresponded closely to those shown by empirical evidence for observed change. It is therefore appropriate for conservation biologists to consider climate change as an important threat to insects, and accordingly to include climate change adaptation measures in conservation programmes.

In a review of the climate change threat to an exemplar group of threatened insects (those listed by the Bern Convention) we only found direct predictions of threat for the butterflies (Settele et al. 2008), but autecological information for a number of other taxa suggested a strong sensitivity to climate change. Evidence for the influence of climate on habitat associations and population—and metapopulation-dynamics will be increasingly valuable for conserving these and other threatened insects as the climate changes.

Acknowledgments We were funded by the European Social Fund (Project "Climate Change and Landscape: Imagining the Future") and by the Council of Europe. RJW would like to thank The Bern Convention's Group of Experts on Biodiversity and Climate Change for assistance and discussion, particularly Deborah Procter, Vernon Heywood, Carolina Lasén Díaz and Véronique du Cussac.

References

Adamski P, Witkowski ZJ (2007) Effectiveness of population recovery projects based on captive breeding. Biol Conserv 140:1–7

Araújo MB, Luoto M (2007) The importance of biotic interactions for modelling species distributions under climate change. Global Ecol Biogeogr 16:743–753

Ashton S, Gutiérrez D, Wilson RJ (2009) Effects of temperature and elevation on habitat use by a rare mountain butterfly: implications for species responses to climate change. Ecol Entomol 34:437–446

Ayres MP, Lombardero MJ (2000) Assessing the consequences of global change for forest disturbance from herbivores and pathogens. Sci Total Environ 262:263–286

Bale JS, Masters GJ, Hodkinson ID et al (2002) Herbivory in global climate change research: direct effects of rising temperatures on insect herbivores. Global Change Biol 8:1–16

Bradford MA, Jones TH, Bardgett RD et al (2002) Impacts of soil faunal community composition on model grassland ecosystems. Science 298:615–617

Brommer JE, Fred MS (1999) Movement of the Apollo butterfly *Parnassius apollo* related to host plant and nectar plant patches. Ecol Entomol 24:125–131

Buse J, Schroder B, Assmann T (2007) Modelling habitat and spatial distribution of an endangered longhorn beetle—a case study for saproxylic insect conservation. Biol Conserv 137:372–381

Buse J, Ranius T, Assmann T (2008) An endangered longhorn beetle associated with old oaks and its possible role as an ecosystem engineer. Conserv Biol 22:329–337

Cassel-Lundhagen A, Sjogren-Gulve P (2007) Limited dispersal by the rare scarce heath butterfly—potential consequences for population persistence. J Insect Conserv 11:113–121

Cassel-Lundhagen A, Sjogren-Gulve P, Berglind SA (2008) Effects of patch characteristics and isolation on relative abundance of the scarce heath butterfly *Coenonympha hero* (Nymphalidae). J Insect Conserv 12:477–482

Chen I-C, Shiu H-J, Benedick S et al (2009) Elevation increases in moth assemblages over 42 years on a tropical mountain. Proc Nat Acad Sci USA 106:1479–1483

Colwell RK, Brehm G, Cardelús CL, Gilman AC, Longino JT (2008) Global warming, elevational range shifts, and lowland biotic attrition in the wet tropics. Science 322:258–261

Council of Europe (2010) Convention on the conservation of European wildlife and natural habitats Appendix II. http://conventions.coe.int/Treaty/FR/Treaties/Html/104-2.htm. Accessed 10 May 2010

Dangles O, Carpio C, Barragan AR, Zeddam JL, Silvain JF (2008) Temperature as a key driver of ecological sorting among invasive pest species in the tropical Andes. Ecol Appl 18:1795–1809

Davies ZG, Wilson RJ, Brereton TM, Thomas CD (2005) The re-expansion and improving status of the silver-spotted skipper butterfly (*Hesperia comma*) in Britain: a metapopulation success story. Biol Conserv 124:189–198

Davies ZG, Wilson RJ, Coles S, Thomas CD (2006) Changing habitat associations of a thermally constrained species, the silver-spotted skipper butterfly, in response to climate warming. J Anim Ecol 75:247–256

De Groot M, Rebeusek F, Grobelnik V, Govedic M, Salamun A, Verovnik R (2009) Distribution modelling as an approach to the conservation of a threatened alpine endemic butterfly (Lepidoptera: Satyridae). Eur J Entomol 106:77–84

Descimon H, Bachelard P, Boitier E, Pierrat V (2006) Decline and extinction of *Parnassius apollo* populations in France-continued. In: Kühn E, Feldmann R, Thomas JA, Settele J (eds) Studies on the ecology and conservation of butterflies in Europe, vol 1 General concepts and case studies. Pensoft, Sofia, pp 114–115

Franco AMA, Hill JK, Kitschke C, Collingham YC, Roy DB, Fox R, Huntley B, Thomas CD (2006) Impacts of climate warming and habitat loss on extinctions at species' low-latitude range boundaries. Global Change Biol 12:1545–1553

Gordo O, Sanz JJ (2005) Phenology and climate change: a long-term study in a Mediterranean locality. Oecologia 146:484–495

Gordo O, Sanz JJ (2006) Temporal trends in phenology of the honey bee *Apis mellifera* (L.) and the small white *Pieris rapae* (L.) in the Iberian Peninsula (1952–2004). Ecol Entomol 31:261–268

Goulson D, Derwent LC, Hanley ME, Dunn DW, Abolins SR (2005) Predicting calyptrate fly populations from the weather, and probable consequences of climate change. J Appl Ecol 42:795–804

Harrington R (2002) Insect pests and global environmental change. In: Douglas I (ed) Encyclopedia of global environmental change, vol. 3. Wiley, Chichester, pp 381–386

Hassall C, Thompson DJ, French GC, Harvey IF (2007) Historical changes in the phenology of British Odonata are related to climate. Global Change Biol 13:933–941

Hedin J, Ranius T, Nilsson SG, Smith HG (2008) Restricted dispersal in a flying beetle assessed by telemetry. Biodivers Conserv 17:675–684

Heikkinen RK, Luoto M, Kuussaari M, Poyry J (2005) New insights into butterfly-environment relationships using partitioning methods. Proc R Soc Lond B 272:2203–2210

Heikkinen RK, Luoto M, Kuussaari M, Toivonen T (2007) Modelling the spatial distribution of a threatened butterfly: impacts of scale and statistical technique. Landscape Urban Plan 79:347–357

Hickling R, Roy DB, Hill JK, Thomas CD (2005) A northward shift of range margins in British Odonata. Global Change Biol 11:502–506

Hickling R, Roy DB, Hill JK, Fox R, Thomas CD (2006) The distributions of a wide range of taxonomic groups are expanding polewards. Global Change Biol 12:450–455

Hoegh-Guldberg O, Hughes L, Mcintyre S et al (2008) Assisted colonization and rapid climate change. Science 321:345–346

Hoyle M, James M (2005) Global warming, human population pressure, and viability of the world's smallest butterfly. Conserv Biol 19:1113–1124

Jiménez-Valverde A, Lobo JM (2006) Distribution determinants of endangered Iberian spider *Macrothele calpeiana* (Araneae, Hexathelidae). Environ Entomol 35:1491–1499

Jiménez-Valverde A, García-Díez T, Bogaerts S (2007) First records of the endangered spider *Macrothele calpeiana* (Walckenaer, 1805) (Hexathelidae) in Portugal. Bol Soc Entomol Aragonesa 41:445–446

Jiménez-Valverde A, Gómez JF, Lobo JM, Baselga A, Hortal J (2008) Challenging species distribution models: the case of *Maculinea nausithous* in the Iberian Peninsula. Ann Zool Fenn 45:200–210

Kindvall O (1996) Habitat heterogeneity and survival in a bush cricket metapopulation. Ecology 77:207–214

Kremen C, Williams NM, Aizen MA et al (2007) Pollination and other ecosystem services produced by mobile organisms: a conceptual framework for the effects of land-use change. Ecol Lett 10:299–314

Kristin A, Kanuch P (2007) Population, ecology and morphology of *Saga pedo* (Orthoptera: Tettigoniidae) at the northern limit of its distribution. Eur J Entomol 104:73–79

Kuras T, Benes J, Fric Z, Konvicka M (2003) Dispersal patterns of endemic alpine butterflies with contrasting population structures: *Erebia epiphron* and *E. sudetica*. Popul Ecol 45:115–123
Leather SR (2010) Institutional vertebratism threatens UK food security. Trends Ecol Evol 24:413–414
Loarie SR, Duffy PB, Hamilton H, Asner GP, Field CB, Ackerley DD (2009) The velocity of climate change. Nature 462:1052–1055
Logan JA, Regniere J, Gray DR, Munson AS (2007) Risk assessment in the face of a changing environment: gypsy moth and climate change in Utah. Ecol Appl 17:101–117
Luoto M, Kuussaari M, Toivonen T (2007) Modelling butterfly distribution based on remote sensing data. J Biogeog 29:1027–1037
Mace GM, Collar NJ, Gaston KJ et al (2008) Quantification of extinction risk: the IUCN's system for classifying threatened species. Conserv Biol 22:1424–1442
Meglecz E, Neve G, Pecsenye K, Varga Z (1999) Genetic variations in space and time in *Parnassius mnemosyne* (L.) (Lepidoptera) populations in north-east Hungary: implications for conservation. Biol Conserv 89:251–259
Menéndez R (2007) How are insects responding to global warming? Tijdschr Entomol 150:355–365
Menéndez R, González Megías A, Hill JK et al (2006) Species richness changes lag behind climate change. Proc R Soc Lond B 273:1465–1470
Mouquet N, Thomas JA, Elmes GW, Clarke RT, Hochberg ME (2005) Population dynamics and conservation of a specialized predator: a case study of *Maculinea arion*. Ecol Monogr 75:525–542
Nakicenovic N et al (2000) Special report on emissions scenarios: a special report of working group III of the intergovernmental panel on climate change. Cambridge University Press, New York
Negro M, Casale A, Migliore L, Palestrini C, Rolando A (2007) The effect of local anthropogenic habitat heterogeneity on assemblages of carabids (Coleoptera, Caraboidea) endemic to the Alps. Biodiv Conserv 16:3919–3932
Negro M, Casale A, Migliore L, Palestrini C, Rolando A (2008) Habitat use and movement patterns in the endangered ground beetle species, *Carabus olympiae* (Coleoptera : Carabidae). Eur J Entomol 105:105–112
Oliver T, Roy DB, Hill JK, Brereton T, Thomas CD (2010) Heterogeneous landscapes promote population stability. Ecol Lett 13:473–484
Parmesan C (1996) Climate and species range. Nature 382:765–766
Parmesan C (2006) Ecological and evolutionary responses to recent climate change. Ann Rev Ecol Evol Syst 37:637–669
Parmesan C, Yohe G (2003) A globally coherent fingerprint of climate change impacts across natural systems. Nature 421:37–42
Parmesan C, Ryrholm N, Stefanescu C et al (1999) Poleward shifts in geographical ranges of butterfly species associated with regional warming. Nature 399:579–583
Peñuelas J, Filella I, Comas P (2002) Changed plant and animal life cycles from 1952 to 2000 in the Mediterranean region. Global Change Biol 8:531–544
Pöyry J, Luoto M, Heikkinen RK, Kuussaari M, Saarinen K (2009) Species traits explain recent range shifts of Finnish butterflies. Global Change Biol 15:732–743
Ranius T (2000) Minimum viable metapopulation size of a beetle, *Osmoderma eremita*, living in tree hollows. Anim Conserv 3:37–43
Ranius T (2007) Extinction risks in metapopulations of a beetle inhabiting hollow trees predicted from time series. Ecography 30:716–726
Renault D, Vernon P, Vannier G (2005) Critical thermal maximum and body water loss in first instar larvae of three Cetoniidae species (Coleoptera). J Thermal Biol 30:611–617
Roy DB, Sparks TH (2000) Phenology of British butterflies and climate change. Global Change Biol 6:407–416
Schweiger O, Settele J, Kudrna O, Klotz S, Kühn I (2008) Climate change can cause spatial mismatch of trophically interacting species. Ecology 89:3472–3479
Settele J, Kudrna O, Harpke A et al (2008) Climatic risk atlas of European butterflies. BioRisk 1 special issue. Pensoft, Sofia-Moscow
Spangenberg JH (2007) Integrated scenarios for assessing biodiversity risks. Sustain Dev 15:343–356
Stefanescu C, Peñuelas J, Filella I (2003) Effects of climatic change on the phenology of butterflies in the northwest Mediterranean Basin. Global Change Biol 9:1494–1506
Sutcliffe OL, Thomas CD, Yates TJ, Greatorex-Davies JN (1997) Correlated extinctions, colonizations and population fluctuations in a highly connected ringlet butterfly metapopulation. Oecologia 109:235–241
Thomas JA (1993) Holocene climate changes and warm man-made refugia may explain why a sixth of British butterflies possess unnatural early-successional habitats. Ecography 16:278–284
Thomas JA, Simcox DJ, Wardlaw JC, Elmes GW, Hochberg ME, Clarke RT (1998) Effects of latitude, altitude and climate on the habitat and conservation of the endangered butterfly *Maculinea arion* and its *Myrmica* ant hosts. J Insect Conserv 2:39–46
Thomas CD, Cameron A, Green RE et al (2004) Extinction risk from climate change. Nature 427:145–148
Thomas JA, Simcox DJ, Clarke RT (2009) Successful conservation of a threatened *Maculinea* butterfly. Science 325:80–83
Valimaki P, Itamies J (2003) Migration of the clouded Apollo butterfly *Parnassius mnemosyne* in a network of suitable habitats—effects of patch characteristics. Ecography 26:679–691
Valimaki P, Itamies J (2005) Effects of canopy coverage on the immature stages of the Clouded Apollo butterfly [*Parnassius mnemosyne* (L.)] with observations on larval behaviour. Entomol Fennica 16:117–123
WallisDeVries MF, Van Swaay CAM (2006) Global warming and excess nitrogen may induce butterfly decline by microclimatic cooling. Global Change Biol 12:1620–1626
Walther G-R, Post E, Convey P et al (2002) Ecological responses to recent climate change. Nature 416:389–395
Warren MS, Hill JK, Thomas JA et al (2001) Rapid responses of British butterflies to opposing forces of climate and habitat change. Nature 414:65–69
Willis SG, Hill JK, Thomas CD et al (2009) Assisted colonization in a changing climate: a test-study using two UK butterflies. Conserv Lett 2:45–51
Wilson RJ, Gutiérrez D, Gutiérrez J, Martínez D, Agudo R, Monserrat VJ (2005) Changes to the elevational limits and extent of species ranges associated with climate change. Ecol Lett 8:1138–1146
Wilson RJ, Davies ZG, Thomas CD (2007) Insects and climate change: processes, patterns and implications for conservation. In: Stewart AJA, New TR, Lewis OT (eds) Insect conservation biology. Proceedings of the royal entomological society's 22nd symposium. CABI Publishing, Wallingford, UK, pp 245–279
Wilson RJ, Davies ZG, Thomas CD (2009) Modelling the effect of habitat fragmentation on range expansion in a butterfly. Proc R Soc Lond B 276:1421–1427

J Insect Conserv (2011) 15:311–319
DOI 10.1007/s10841-010-9355-6

ORIGINAL PAPER

Synchronisation of egg hatching of brown hairstreak (*Thecla betulae*) and budburst of blackthorn (*Prunus spinosa*) in a warmer future

H. H. de Vries · S. H. Ens · G. de Graaf · L. Teunissen · R. van der Velde · L. Vogelaar · A. Winterink · M. E. Visser

Received: 24 March 2010 / Accepted: 9 September 2010 / Published online: 25 September 2010

Abstract Synchronisation of the phenology of insect herbivores and their larval food plant is essential for the herbivores' fitness. The monophagous brown hairstreak (*Thecla betulae*) lays its eggs during summer, hibernates as an egg, and hatches in April or May in the Netherlands. Its main larval food plant blackthorn (*Prunus spinosa*) flowers in early spring, just before the leaves appear. As soon as the Blackthorn opens its buds, and this varies with spring temperatures, food becomes available for the brown hairstreak. However, the suitability of the leaves as food for the young caterpillars is expected to decrease rapidly. Therefore, the timing of egg hatch is an important factor for larval growth. This study evaluates food availability for brown hairstreak at different temperatures. Egg hatch and budburst were monitored from 2004 to 2008 at different sites in the Netherlands. Results showed ample food availability at all monitored temperatures and sites but the degree of synchrony varied strongly with spring temperatures. To further study the effect of temperature on synchronisation, an experiment using normal temperatures of a reference year (T) and temperatures of $T + 5°C$ was carried out in climate chambers. At $T + 5°C$, both budburst and egg hatch took place about 20 days earlier and thus, on average, elevated temperature did not affect synchrony. However, the total period of budburst was 11 days longer, whereas the period of egg hatching was 3 days shorter. The implications for larval growth by the brown hairstreak under a warmer climate are considered.

Keywords Timing · *Thecla betulae* · *Prunus spinosa* · Climate change · Lepidoptera · Herbivory

H. H. de Vries (✉) · S. H. Ens · G. de Graaf · L. Teunissen · R. van der Velde · L. Vogelaar · A. Winterink
Dutch Butterfly Conservation, Mennonietenweg 10, 6700 AM Wageningen, The Netherlands
e-mail: henk.devries@vlinderstichting.nl

M. E. Visser
Netherlands Institute of Ecology (NIOO-KNAW), P.O. Box 40, 6666 ZG Heteren, The Netherlands

Introduction

Egg hatching is a crucial moment in the development of diapausing species (Tauber and Tauber 1976; Harrington et al. 1999). Adverse conditions in the lifetime of individuals are met by going into diapause until more favourable conditions are present. Univoltine species with prolonged periods of diapause are the result of selection against prolonged presence in juvenile or adult life forms. The timing of the irreversible ending of the diapause is an important moment in the life of an individual of such species; young individuals are often vulnerable to starvation, drought or predation. As environments change seasonally from harsh and hostile into friendly and food providing, accurate timing of emergence will ensure a higher chance of survival. Cues indicating these favourable conditions are generally based on day length or temperature, although cues linked more directly to the physiological status of the host plant could theoretically be the best (Harrington et al. 1999). Several papers deal with the importance of timing in lepidopteran species for ending the winter diapause (Dewar and Watt 1992; Buse and Good 1996; Van Dongen et al. 1997; Visser and Holleman 2001; Klemola et al. 2003; Van Asch and Visser 2007; Forkner et al. 2008; Both et al. 2009) showing that mistiming of the ending of diapause can result in high mortality rates. Disruption of the existing synchrony between food plant and butterfly has been observed as a result from climate change

(Thackeray et al. 2010). For instance, Visser and Holleman (2001) showed that the egg hatch of winter moth (*Operophtera brumata*) took place 25 days earlier during warm springs, which resulted in first instars being deprived of food due to closed oak buds. Delayed ending of the diapause can also have disadvantages. Late-hatched larvae will be confronted with ample presence of mature leaves, which can have a negative influence on larval growth (Scriber and Slansky 1981; Schroeder 1986; Raupp et al. 1988, Coley et al. 2006; Van Asch et al. 2007). Studies of the influence of temperature on the timing of egg hatch has shown that the effects can be variable and species specific. In some cases, butterfly species normally appear somewhat earlier than budburst (Winter Moth: Visser and Holleman 2001), in other cases, somewhat later (Myers 1992).

In this paper, we investigate the role of temperature on egg hatching in the brown hairstreak (*Thecla betulae, Lycaenidae, Lepidoptera)* and its possible synchronisation with the budburst of blackthorn (*Prunus spinosa*). In the Netherlands, *T. betulae* is a rare butterfly, its distribution being restricted to parts of the east and south of the country. The shrub *P. spinosa* is much more widely distributed, partly as a result of the widespread use of this plant in hedgerows and public parks. During recent years, *T. betulae* has been slowly declining in the Netherlands. This may be due to a disruption of the synchronisation of egg hatching and budburst during recent years. Larvae that hatch too early will be deprived of food, whereas larvae that hatch late can be confronted with food of reduced quality. Unpublished breeding attempts with *T. betulae* indicated low survival rates of first instars when only fed with mature leaves (pers. comm. Groenendijk). Therefore, we hypothesise that accurate timing of egg hatch is essential for creating conditions for optimal growth of first instars of *T. betulae*. Our main question is whether in a warm spring the larvae of *T. betulae* are confronted with closed buds of *P. spinosa* as found for the winter moth by Visser and Holleman (2001). The second question is whether the synchronisation between egg hatch and budburst is similar in different regions.

Materials and methods

Study organisms

In the Netherlands and neighbouring countries, brown hairstreak (*Thecla betulae*) is a rare and red listed species (Van Swaay 2006; Dudler et al. 1999; Maes and Van Dyck 1996). Since 1998, in the Netherlands this species has been monitored annually by egg counting surveys (Fig. 1). The species' recent decline (slope 0.947 ± 0.0235, $P < 0.05$) has led Dutch Butterfly Conservation to look for possible threats. Adult butterflies are often hidden in tree tops and difficult to observe. The females come down to lay their eggs at observable heights. *P. spinosa* is used for egg deposition and is the main larval food plant during spring and summer. Other *Prunus* species can be accepted as well (Stefanescu 2000; Bink and Moenen 2004; Koschuh et al. 2005). The distribution of *T. betulae* is limited to the south of the Netherlands (province of Limburg), the edge of the Veluwe, a sandy area (provincie of Gelderland), and some smaller areas in the Northeast (provinces of Overijssel and Drenthe). During summer, from July to September, female butterflies oviposit on young *P. spinosa* branches (Fartmann and Timmermann 2006; Merckx and Berwaerts 2010). The eggs are usually laid singly at heights of between 50 cm and 2 m. This is in accordance with observations in other countries (Thomas 1974; Fartmann and Timmermann 2006), but Fartmann and Timmermann (2006) concluded from a literature review that there is a slight shift towards a greater oviposition height from Western Europe to Southeast Austria. In the Netherlands, *P. spinosa* is one of the first shrubs to flower, starting in February–March (data www.natuurkalender.nl). Normally, flowering takes place a few weeks before budburst starts. Opening of the buds in April–May gives the young caterpillars the possibility of going inside the newly opened bud and eating the nutritious folded leaves inside. Hatching of the eggs generally starts in April and ends a few weeks later (personal observations). According to Thomas (1974), the first thing the hatched larva does is to search for an open bud and hide inside it. Later, the older instars rest under the leaf, feeding during the evening.

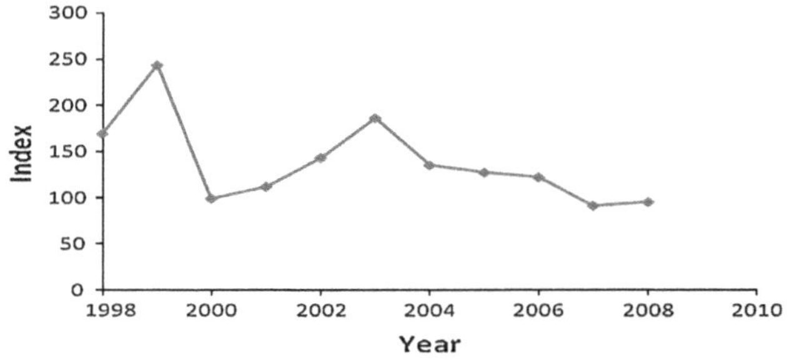

Fig. 1 The results of egg monitoring of *T. betulae* in the Netherlands (Dutch Butterfly Conservation/Statistics Netherlands). The index (Van Swaay et al. 2002) was fixed at 100 for the year 2000

Study areas

For this study, we observed ten populations of *T. betulae*, eight populations are in town parks, whereas the other two occur in nature reserves. Almost all areas are managed in such a way that the *P. spinosa* shrubs are cut on a regular basis, resulting in presence of young branches for the female *T. betulae* to deposit their eggs on. Without this kind of management many sites become rapidly unsuitable,

Fig. 2 Eggs were collected at ten locations in seven regions: *1* Steenwijk, *2* Zuidwolde, *3* Apeldoorn (4 locations), *4* Arnhem, *5* Wageningen, *6* Overlangbroek, *7* Zuid-Limburg. *Dots* represent findings of *T. betulae* in a 5 km grid (1995–2003), size and coloration indicate relative spatial distribution and number of observations respectively (Bos et al. 2006). Station De Bilt (*c*) was associated with populations 4, 5 and 6, Station Eelde (*a*) with 1 and 2, Station Twente (*b*) with 3, and Station Maastricht (*d*) with 7

because old branches become overgrown with lichens and algae. The latitudinal positions of the areas were measured using a Dutch kilometre grid (Amersfoort Coordinates). During a 5-year period, we collected data from eggs and plants, monitoring at ten locations in seven regions for 1–5 years (Fig. 2 and Table 1).

Field monitoring

The eggs are conspicuous white spots on the dark bark from young branches of *P. spinosa*. We looked for eggs during the period from January to March when *P. spinosa* was without leaves, marking the branches with loosely attached coloured ribbons, under the position of the egg. The eggs were examined regularly, when possible twice a week. We distinguished five categories: not rediscovered or missing; parasitized (a lateral hole); predated or damaged; closed; and open, with a characteristic, central hole on top. The timing of egg hatch was calculated solely by using the observed eggs which were first marked and later found open. During the same survey, the availability of food within 20 cm proximity of the egg was investigated. Presence of closed buds only indicated absence of food, whereas presence of one or more open buds indicated food was available. Each surrounding of an egg got only one score for buds: either all buds were closed or one or more buds were found open. Temperature data were extracted from the KNMI (Royal Netherlands Metereological Institute) database, using meteorological station De Bilt for the butterfly populations in Arnhem, Wageningen and Overlangbroek, Eelde for Zuidwolde and Steenwijk, Twente for Apeldoorn and Maastricht for the region of Zuid-Limburg (Fig. 2).

Temperature experiment

During 2006, we set up a laboratory experiment, with two temperature regimes, one at a standard temperature (T), using outside temperatures of 2004, and another at a higher temperature of $T + 5°C$. It started at day 59 (28th of February). Four *P. spinosa* plants and fourteen *T. betulae* eggs were kept at the standard temperature in a climate chamber, simulating minimum, intermediate, and maximum temperatures of the reference year 2004, for six, twelve, and 6 h respectively. Three times a week the temperatures were adjusted manually to the reference regime, using averages of two or 3 days. T_{max} of the reference year fluctuated in February 28–May 5th between +2.4 and +21.5°C, T_{min} between −5.9 and +12.7°C. In a second climate chamber, four plants and eleven eggs were kept in a similar, but warmer ($T + 5°C$) regime. The eggs were collected and installed during 2 days, day 59 and 65. The plants were delivered by Bronnen, an organisation specialising in indigenous plant material of shrubs and trees. Twice a week, the status of the buds and the hatch of the eggs were monitored using the same method as for field work. After the experiment, the hatched larvae were returned to the sites they had originated from.

Warmth sums

To measure the influence of temperature on date of egg hatch and budburst warmth sums were calculated using T_{max} of every day minus a development threshold of 5°C. From January 1st until day 100, the corrected T_{max} were summed, giving $T_{sum}100$ (in degree days). A second parameter for temperature was calculated using low temperatures. For this the numbers of days with frost for the first 100 days were counted for each year, giving Frost100.

Analyses

The dates of field and laboratory observations were transformed into day numbers, Day 1 being January 1st. Logistic frequency curves were fitted with the dates of egg hatch and budburst and used for further processing. The available data

Table 1 Investigated regions in different years

	2004		2005		2006		2007		2008	
	Eggs	Buds	Eggs	Buds	Eggs	Buds[a]	Eggs	Buds	Eggs	Buds
Apeldoorn-H	24	34	18	21						
Apeldoorn-J	26	40	24	29	24	9				
Apeldoorn-K	13	43	10	11	21	6				
Apeldoorn-Z	17	35	17	22	68	59				
Z-Limburg	62	85	32	87	49	25	46			
Overlangbroek	49	65	80	83	83	46	39		13	23
Steenwijk	89	153	31	50	109	25				
Zuidwolde							9			
Wageningen			20	27						
Arnhem							25			

Given are total numbers of hatched eggs and number of observations used for calculating presence of opened buds

[a] Less sample numbers, due to availability of averages per individual plant only

(day numbers) have all in common that the actual moments of opening of buds and eggs are not observed, but estimated from the dates the visits were done. The larger the number of eggs observed, the more accurate the estimates. The data of each location were used to estimate the point at which in a particular year 50% of the buds as well as 50% of the eggs were open. This point indicates the moment at which a population of eggs or buds shifts from mostly closed to mostly open. They were derived from interpolation between two points, and in two cases, from extrapolation (Overlangbroek 2004 and Apeldoorn 2006). Two other single dates of locations with 40–60% open eggs were available and added to the data set (57% for Zuid-Limburg 2007 and 44% for Arnhem in 2007).

Using a JMP Statistical Software package (release 5.0.1.2, SAS Institute Inc.) stepwise multiple regressions (Sokal and Rohlf 1981) were calculated with critical p-values smaller than 0.1. Two kinds of models were calculated, type A and B, with qualitative parameters and quantitative parameters, respectively. The explanatory variables Year (2004–2008) and Region (seven regions) were tested for the response variables egg hatch (50% point), budburst (50% point) and the timing difference between both, resulting in models type A. An additional Effect test gave summary results of Year and Region. Latitude, $T_{sum}100$ and Frost100 were tested for the same response variables resulting in models type B.

Results

Temperature sums and frost

The temperature ($T_{sum}100$) and frost (Frost100) varied considerable between years, at maximum 383.4 degree days and 44 days with frost respectively, but relative little between regions, at maximum 71.2 degree days and 8 days with frost respectively (Fig. 3a and b). The temperature sum was exceptionally low during the start of 2006, whereas a few months later this year became known for its extreme warm temperatures.

Egg hatch and budburst

Data were collected from 26 year-location combinations for egg hatch and 22 for budburst (Table 1). The egg hatch and budburst in Apeldoorn in 2005 is shown as an example in Fig. 4. Budburst started around day number 74 and egg hatch followed almost 3 weeks later. The results of the monitoring of egg hatch are presented in Fig. 5a. The day numbers at which 50% of the eggs were estimated to be opened, ranged from 105 to 132. A multiple regression showed that the day number of egg hatch varied both between Year and Region (Table 2). $T_{sum}100$ and Frost 100 both showed significant negative effects on day number of 50% egg hatch. Figure 5b presents the data of budburst of *P. spinosa* and shows that this takes generally place at day number 81–112. A multiple regression (Table 2) showed significant results for Year, but not for Region. $T_{sum}100$ showed a significant effect. The timing difference shows considerable variation within Year (Fig. 5c). In 2004, 2005, 2006, and 2008, the timing difference was on average 8, 20, 24, and 16 days, respectively. The timing difference showed in all cases that the point of 50% egg hatch was preceded by budburst, although in one case, Overlangbroek 2004, this is not certain, because it was close to zero and had to be estimated by extrapolation. For this response variable, a multiple regression (Table 2) showed a significant relation with the Year, but not with the Region. The multiple regression gave no significant results for timing difference.

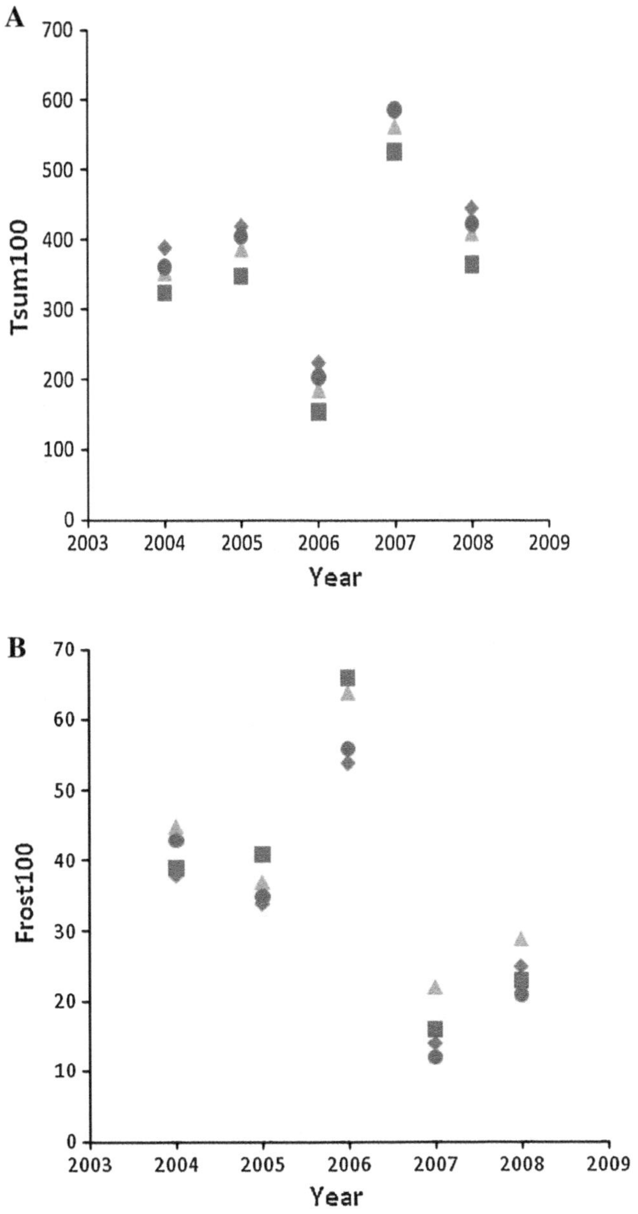

Fig. 3 Temperature sums (in degree days with 5°C threshold) from day 1–100 (**a**) and number of days with frost from day 1–100 (**b**) at four regions in years 2004–2008: De Bilt (*diamonds*), Eelde (*squares*), Twente (*triangles*), and Maastricht (*circles*)

Temperature experiment

Of the fourteen eggs at the normal temperature and the eleven eggs at the high temperature regime 11 and 8 hatched, respectively. We found that as a result of a 5°C rise both 50% points had on average an almost similar shift, being 21 and 19 days for egg hatch and budburst respectively (Fig. 6). Using the two developed models for egg hatch and budburst (Table 2), we predicted a difference between egg hatch and budburst for T standard and $T + 5$°C of 24 (124 minus 100) and 9 days (94 minus 85) respectively. During the temperature experiment we observed a difference of respectively 9 and 7 days.

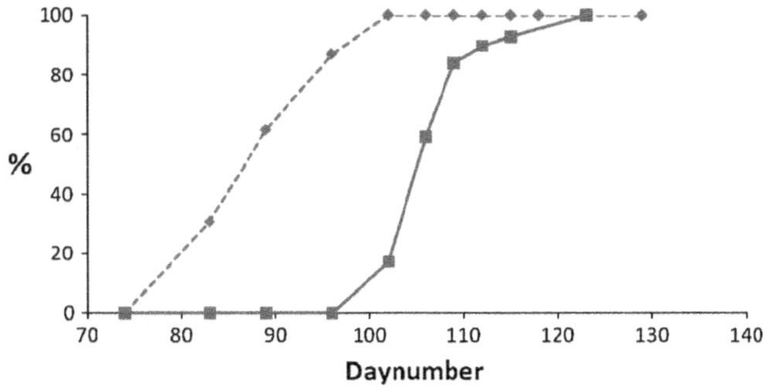

Fig. 4 Cumulative frequency distribution of monitoring results of egg hatch of *T. betulae* (*solid line*) and budburst of *P. spinosa* (*broken line*) in Apeldoorn (four sites summed) during 2005

The period at which 80% of the egg hatch took place was in the laboratory experiment under normal conditions 10 days ($n = 11$) and at the +5°C regime 7 days ($n = 8$; Table 3). The total period at which 80% budburst took place was under normal conditions 9 days ($n = 162$) and at the +5°C regime 20 days ($n = 202$; Table 3). This results in an egg hatch period being shortened by three days and a budburst period being prolonged by 11 days (Table 3; Fig. 6).

Discussion

At the start of this investigation, we hypothesised that during some years the egg hatch of some populations of *T. betulae* would be too early. This would perhaps result in days or maybe weeks of starvation for the young larvae, as seen in the Winter Moth (Visser and Holleman 2001; Van Asch 2007). We did not observe this in our study. In three out of 4 years, budburst of *P. spinosa* took place between day 100 and 110, apart from two local exceptions, and in most years, the point of 50% egg hatch had only minor deviations from day number 110. However, 2006 was an exception, with egg hatch taking place about 20 days later. For the budburst of *P. spinosa*, 2005 was an exceptional year, being much earlier, on between day 80 and 90 in seven out of eight locations. When we compare these results with the data on temperature (Fig. 3) it is clear that the first 100 days in 2006 was much colder compared to the other years, which explains quite well the delay in egg hatch in this particular year. The advancement of budburst during 2005 (day 80–90) cannot be explained by the temperature data. Perhaps calculations with other thresholds (e.g. mean 24-hourly basic temperatures: Wielgolaski 1999) should be used to explain this phenomenon.

The laboratory experiment indicated that 7–9 days difference in timing between egg hatch and budburst is

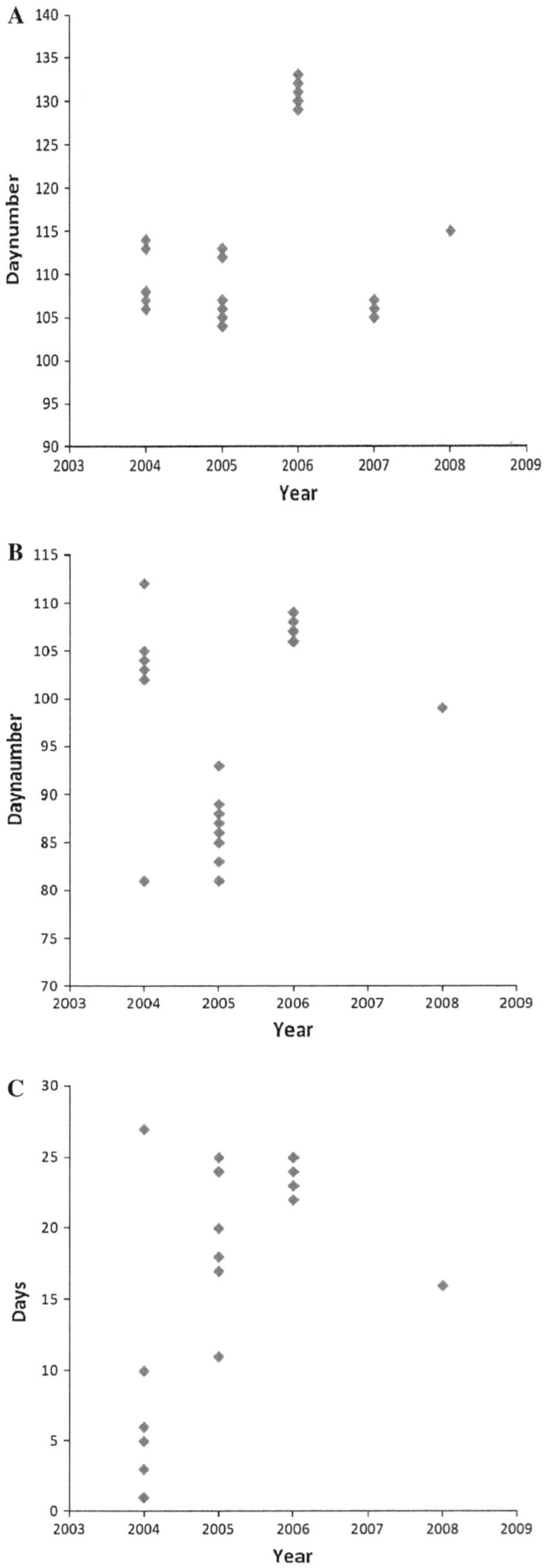

Fig. 5 The day of 50% egg hatch of *T. betulae* during 5 years (graph **a**, $n = 26$) and the day of 50% budburst of *P. spinosa* during 4 years (graph **b**, $n = 22$). *Points* indicate results from one location. Graph **c** Timing difference (*Y*-axis, hatch day minus day of budburst, using 50% points; $n = 22$) during investigated years (*X*-axis). The positive values indicate that the eggs hatched later than buds did burst

normal for this species, which is in accordance with Bink (1992). This result matches with the field observations in the reference year 2004, hence the experiment with standard temperatures mimics field conditions quite well. This difference is somewhat less than that observed during the other 3 years (Fig. 5c: 11–25 days timing difference), and much less than can be calculated with the two available ANOVA-models for the standard temperature (24 days timing difference). At $T + 5$°C the experiment and the ANOVA-models showed comparable results, which might indicate that the model predicts better with high temperature regimes. These predictions are partly based on negative relations between day number of egg hatch and budburst and warmth sum and, for egg hatch, a negative relation with days with frost. The sign of the latter is in accordance with the reduction of threshold warmth sums by frost days found in *O. brumata* (Visser and Holleman 2001). An accumulation of frost days reduces the warmth sums necessary for hatching, indicating that both species experience warmth sums partly as a relative condition. Under the selected laboratory conditions, a general temperature rise of 5°C showed no effect on the average time difference between egg hatch and budburst. However, the egg hatch period shortened and the budburst period prolonged as a result of the increased temperature. This has two consequences. There is a higher chance of ample food being available, but as a result of the absence of a sharp peak in the leaf flush, less chances of a perfect match between egg hatch and budburst. For the first instars of *T. betulae* this asynchrony could well lead to a significantly lower availability of a high quality food source of just opened buds and young leaves. Harrington et al. (1999) concluded that the interval between budburst and the start of larval feeding is an important factor in insect population success. Optimal growth should ensure a lower risk of mortality due to predators (Clancy and Prica 1987) and climate (Andrewartha and Birch 1954; Fordyce and Shapiro 2003) due to a shorter presence of vulnerable small stages. Our experiment indicates that the expected rise of temperature in the future (Loarie et al. 2009) could further reduce population success of *T. betulae*. To confirm this an experiment (e.g. Coley et al. 2006; Van Asch 2007) is needed to evaluate larval growth of *T. betulae* on leaves of different maturity.

Another factor involved in the feeding options of the larvae is their mobility (e.g. Forkner et al. 2008). In this

Table 2 Results from testing with ANOVA (calculated stepwise, using a standard least squares ANOVA with *P*-values < 0.1)

Response variable	Egg hatch				Budburst				Timing difference			
Models A with qualitative parameters												
R^2_{adj}	0.96				0.65				0.43			
	F-ratio		*df*	*P*	*F*-ratio		*df*	*P*	*F*-ratio		*df*	*P*
Regions	4.56		6	<0.01	0.83		4	0.53	0.26		4	0.90
Year	150.63		4	<0.0001	12.85		3	<0.0005	7.42		3	<0.005
Response variable	**Egg hatch**				**Budburst**				**Timing difference**			
Models B with quantitative parameters												
R^2_{adj}	0.72				0.40				0.09, NS			
	F-ratio	*B*	*df*	*P*	*F*-ratio	*B*	*df*	*P*	*F*-ratio	*B*	*df*	*P*
Latitude	3.67	−0.032	1	0.07	–	–	–	–	–	–	–	–
T_{sum} 100	12.92	−0.183	1	<0.005	13.13	−0.072	1	<0.005	–	–	–	–
Frost 100	4.99	−0.902	1	<0.05	–	–	–	–	–	–	–	–

Models A give the explanatory value of Years (4 groups, 5 for timing of eggs) and Regions (7 groups) for the three response variables Timing of eggs, Timing of buds, and Timing difference. Models B give for the same response variables the explanatory values of three quantitative variables: Latitude, Warmth sum T_{sum}100 (degrees above five, summed for each day until day 100), Frost 100 (number of days with frost until day 100). Dashes indicate an unapproved parameter during stepwise building of the model

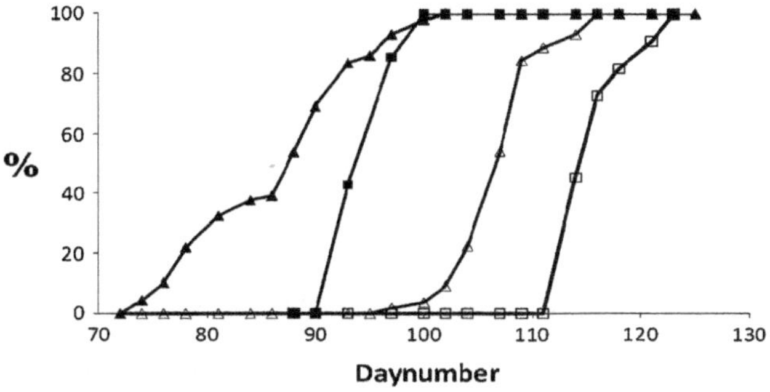

Fig. 6 Results from a laboratory experiment using a standard year (*open shapes*) and a standard year plus 5°C (*closed shapes*). A cumulative frequency distribution is presented of budburst (*triangles*) and egg hatch (*squares*)

study, we estimated the availability of food at 20 cm distance from the egg, but perhaps young instars are able to search further on the plant for food. According to Thomas (1974), the maximum distance larvae of *T. betulae* move is one meter, but the capability of the tiny newly hatched larvae is surely less. Clearly, a larger action radius could increase the availability and quality of food considerably, but this will depend on the spatial differentiation of budburst events. A wide phenological window caused by the continuous opening of different buds of an individual plant would give both early and late hatched larvae the opportunity to find food. In this way, temperature rise may even lead to unexpected adaptations in dispersal abilities.

The year 2006 was rather extreme in the sense that egg hatch took place on average between 20 and 25 days after budburst in all areas. This was remarkable because the population in the laboratory experiment showed only a little delay in egg hatch at standard temperatures (50% hatch was at day 115). It makes it highly likely that in 2006 in situ egg development will have been slowed down by the low outside temperature between the start of the experiment (day 59–65) and the normal time of egg hatch (around day 110). The observed minimum temperatures in the first half of March (day 60–75), several days of five degrees below zero or the high number of days with frost (twenty

Table 3 Estimated 50% opening date, minima, maxima and ranges of opening date of buds and eggs at a laboratory experiment with two temperatures, standard and +5°C

	Opening date (10th percentile)	Opening date (50th percentile)	Opening date (90th percentile)	Range (days)	Timing difference
T-standard-eggs	111	115	121	10	9
T-standard-buds	102	106	111	9	
T + 5°C-eggs	91	94	98	7	7
T + 5°C-buds	76	87	96	20	

The timing difference is the difference between the 50% point of eggs and buds

out of forty), could be responsible for this phenomenon. Budburst of Blackthorn was not delayed by this late cold period: outside and inside development under normal temperatures were almost the same.

Most of the observed variation in egg hatch, budburst and difference in timing could be explained by the climatic conditions of the year in question. The late hatch in 2006 and the average hatch in 2007 indicate that at the start of the season cold periods are a more effective factor in delaying the date of egg hatch in *T. betulae* than warm periods are in advancing it. It is possible that the population is already adapted to an early spring and has a setback when the season occasionally has a late start. A possible indication of the importance of these cold periods is that the late egg hatch of 2006 was followed by fewer eggs being counted during the national monitoring in 2007 (Fig. 1). Other decreases, as in the years 2000 and 2004, could also have been caused by mismatches or the late appearance of egg hatch. Continuing our observations to obtain a longer time series, could make it possible to get more data about the importance of cold periods for this butterfly. Apart from this aspect, possibly based on recent adaptation to warmer springs, the earlier mentioned lack of possibilities for good timing caused by enlarged phenological windows of budburst should be given attention. Moreover, as we have not seen any caterpillar which was too early, the effect of the maturing of leaves on caterpillar growth is expected to be important in both scenarios. Coley et al. (2006) have already provided good evidence of hampered growth in 85 tropical butterfly species on diets of mature leaves. This still has to be examined for *T. betulae*. Thus, so far we have shown with the laboratory experiment that the hatching date of this species is well adapted to phenological shifts in budburst during years with increased temperatures.

Acknowledgments We thank the volunteers of Dutch Butterfly Conservation, in particular those of the regions Steenwijk, Apeldoorn and Zuid-Limburg for their kind assistance, the State Forestry Department for their permission to collect eggs, Margriet van Asch for her help with the laboratory experiment, Claire Hengeveld for improving the text, and Michael Wallis de Vries for advice on the analysis. This paper has been kindly supported by a grant from the Erica Foundation in trust of the Prince Bernhard Cultural Foundation.

References

Andrewartha HG, Birch LC (1954) The distribution and abundance of animals. University of Chicago Press, Chicago

Bink F (1992) Ecologische Atlas van de Dagvlinders van Noordwest-Europa. Schuyt & Co, Haarlem

Bink F, Moenen R (2004) Life-history and habitat of the Brown hairstreak *Thecla betulae* (Lepidoptera: Lycaenidae). Entomol Ber (Amsterdam) 64(4):106–112

Bos F, Bosveld M, Groenendijk D, Van Swaay C, Wynhoff I (2006) De Dagvlinders van Nederland, verspreiding en bescherming (Lepidoptera: Hesperioidea, Papilionidae). Nationaal Natuurhistorisch Museum Naturalis, KNNV & EIS, Leiden

Both C, Van Asch M, Bijlsma RG, Van den Burg AB, Visser ME (2009) Climate change and unequal phenological changes across four trophic levels: constraints or adaptation? J Anim Ecol 78: 73–83

Buse ASJ, Good JEG (1996) Synchronization of larval emergence in winter moth (*Operophtera brumata L.*) and budburst in pedunculate oak (*Quercus robur L.*) under simulated climate change. Ecol Ent 21:335–343

Clancy KM, Prica PW (1987) Rapid herbivore growth enhances enemy attack: sublethal plant defenses remain a paradox. Ecology 68:733–737

Coley PD, Bateman ML, Kursar TA (2006) The effects of plant quality on caterpillar growth and defense against natural enemies. Oikos 115:219–228

Dewar RC, Watt AD (1992) Predicted changes in the synchrony of larval emergence and budburst under climatic warming. Oecologia 89:557–559

Dudler H, Kinkler H, Lechner R, Retzlaff H, Schmitz W, Schumacher H (1999) Rote Liste der gefährdeten Schmetterlinge (Lepidoptera) in Nordrhein-Westfalen. Landesanstalt für Ökologie. Bodenordnung und Forsten, Landesamt für Agrarordnung, Recklinghausen

Fartmann F, Timmermann K (2006) Where to find the eggs and how to manage the breeding sites of the Brown Hairstreak (*Thecla betulae* (Linnaeus, 1758)) in Central Europe? Nota Lepid 29(1/2): 117–126

Fordyce JA, Shapiro AM (2003) Another perspective on the slow-groth/high-mortality hypothesis: chilling effects on swallowtail larvae. Ecology 84:263–268

Forkner RE, Marquis RJ, Lill JT, Le Corff J (2008) Timing is everything? Phenological synchrony and population variability in leaf-chewing herbivores of Quercus. Ecol Ent 33:276–285

Harrington R, Woiwod I, Sparks T (1999) Climate change and trophic interactions. TREE 14:146–150

Klemola T, Ruohomäki K, Tanhuanpää M, Kaitaniemi P (2003) Performance of a spring-feeding moth in relation to time of oviposition and bud-burst phenology of different host species. Ecol Ent 28:319–327

Koschuh A, Savas V, Gepp J (2005) Winter-Eifunde von Zipfelfalterarten (Lepidoptera: Lycaenidae) in Graz und Umland (Steiermark, Österreich). Natursch Landschaftspl 3(2):46–53

Loarie SR, Duffy PB, Hamilton H, Asner GO, Field CB, Ackerly DD (2009) The velocity of climate change. Nature 462:1052–1105

Maes D, Van Dyck H (1996) Een gedocumenteerde Rode Lijst van dagvlinders van Vlaanderen. Instituut voor Natuurbehoud, Brussel

Merckx T, Berwaerts K (2010) What type of hedgerows do Brown hairstreak (*Thecla betulae L.*) butterflies prefer? Implications for European agricultural landscape conservation. Insect Conserv Diver 3:194–204

Myers JH (1992) Experimental manipulation of the phenology of egg hatch in cyclic populations of tent caterpillars. Can Ent 124: 737–742

Raupp MJ, Werren JH, Sodof CS (1988) Effects of short-term phenological changes in leaf suitability on the survivorship, growth, and development of gypsy moth. Environ Entomol 17: 316–319

Schroeder LA (1986) Changes in leaf quality and growth performance of Lepidopteran larvae. Ecology 67:1628–1636

Scriber JM, Slansky F Jr (1981) The nutritional ecology of immature insects. Annu Rev Entomol 26:183–211

Sokal RR, Rohlf FJ (1981) Biometry. Freeman and Company, New York

Stefanescu C (2000) New data on the ecology of *Thecla betulae* in the northeast of the Iberian Peninsula (Lycaenidae). Nota Lepid 23(1):64–70

Tauber MJ, Tauber CA (1976) Insect seasonality: diapause maintenance, termination, and postdiapause development. Annu Rev Entomol 21:81–107

Thackeray SJ, Sparks TH, Frederiksen M, Burthe S, Bacon PJ, Bell JR, Botham MS, Brereton TM, Bright PW, Carvalho L, Clutton-Brock T, Dawson A, Edwards M, Elliott JM, Harrington R, Johns D, Jones ID, Jones JT, Leech DI, Roy DB, Scott WA, Smith M, Smithers RJ, Winfield IJ, Wanless S (2010) Trophic level asynchrony in rates of phenological change for marine, freshwater and terrestrial environments. Glob Change Biol. doi: 10.1111/j.1365-2486.2010.02165.x

Thomas JA (1974) Factors influencing the numbers and distribution of the Brown Hairstreak, *Thecla betulae L.* (*Lepidoptera, Lycaenidae*) and the Black Hairstreak, *Strymonidia pruni L.* (*Lepidoptera, Lycaenidae*). Phd thesis, University Leicester

Van Asch M (2007) Seasonal synchronization between trophic levels under climate change. PhD thesis, Rijksuniversiteit Groningen

Van Asch M, Visser ME (2007) Phenology of forest caterpillars and their host trees: the importance of synchrony. Annu Rev Entomol 52:37–55

Van Asch M, Van Tienderen PH, Holleman LJM, Visser ME (2007) Predicting adaptation of phenology in response to climate change, an insect herbivore example. Glob Change Biol 13: 1596–1604

Van Dongen S, Backeljau T, Matthysen E, Dhondt AA (1997) Synchronization of hatching date with budburst of individual host trees (*Quercus robur*) in the winter moth (*Operophtera brumata L.*) and its fitness consequences. J Anim Ecol 66:113–121

Van Swaay CAM (2006) Basisrapport Rode lijst dagvlinders. Rapport VS2006.002 De Vlinderstichting, Wageningen

Van Swaay CAM, Plate CL, Van Strien A (2002) Monitoring butterflies in the Netherlands: how to get unbiased indices. Proc Exp Appl Entomol 13:21–27

Visser ME, Holleman JM (2001) Warmer springs disrupt the synchrony of oak and winter moth phenology. Proc R Soc Lond B 268:289–294

Wielgolaski FE (1999) Starting dates and basic temperatures in phenological observations of plants. Int J Biometeorol 42: 158–168

J Insect Conserv (2011) 15:241–258
DOI 10.1007/s10841-010-9341-z

ORIGINAL PAPER

Evidence based conservation of butterflies

J. A. Thomas · D. J. Simcox · T. Hovestadt

Received: 12 May 2010 / Accepted: 3 September 2010 / Published online: 25 September 2010

Abstract Few results of research aimed at solving questions arising from butterfly conservation are rigorously tested by manipulating populations and habitats in the field. Some factors common to successful conservation projects are analysed. In most non-migratory species, population density may vary by up to two orders of magnitude between sites or over time, and is primarily determined by the extent to which a subset of each species' foodplant (or ant host) exists in the optimum growth-form or micro-habitat preferred by its larvae. Successful conservation projects have identified the optimum subset of each species' larval resource before managing sites to increase its representation. In contrast, short-term fluctuations around a site's carrying capacity or equilibrium level are mainly attributable to variation in weather, and are generally two orders of magnitude smaller than that attributable to larval habitat quality. There is little evidence that changing the abundance of adult resources, apart from shelter, influences population size or trends. The main constraint of the adult stage is the inability of many species to track the generation of new habitat patches that arise across modern landscapes. Within-patch larval habitat quality is again critical at the meta-population scale, explaining slightly more examples of patch occupancy than site isolation. This is because the higher density populations supported by optimum habitat are less likely to go extinct, and immigrants to new high-quality patches have a higher probability of founding new populations. Such patches may also generate up to a hundred times more emigrants per hectare than low-quality source patches.

J. A. Thomas (✉) · T. Hovestadt
Department of Zoology, University of Oxford, South Parks Road, Oxford OX1 3PS, UK
e-mail: jeremy.thomas@zoo.ox.ac.uk

J. A. Thomas · D. J. Simcox
CEH Wallingford, Oxfordshire OX10 8BB, UK

T. Hovestadt
Muséum National d'Histoire Naturelle, CNRS UMR 7179, 1 Avenue du Petit Château, 91800 Brunoy, France

Keywords Habitat quality · Metapopulation · Butterfly population dynamics · Larval niche · Dispersal

Introduction

The first scientific efforts to assess the changing status of butterflies, and to attempt to conserve species through managing habitats in accordance with research results describing their ecology and behaviour, began in the UK in the late 1960s (Pyle 1976). Since then, there has been an extraordinary expansion of monitoring and conservation-orientated research across Europe, North America, Japan and elsewhere.

From the outset, scientists have been more competent at describing the changes that were occurring in species' distributions and abundance (e.g. in UK: Heath et al. 1984; Pollard and Yates 1993; Asher et al. 2001; Thomas 2005; Fox et al. 2006) than in understanding the mechanisms driving those changes; and despite some notable exceptions, the track record of reversing trends in declining species through targeted conservation is disappointing (see Thomas 1984a, 1991, 1995; Warren 1992; Pullin 1995; New et al. 1995; New 1997; van Swaay and Warren 1999; Kűhn et al. 2005; Settele et al. 2005, 2009; Thomas et al. 2009; Dennis 2010). Yet there is some cause for optimism. For example, by the late 1980 s the populations of six declining species had largely or entirely disappeared from

UK nature reserves, several of which had been established mainly to conserve them (Thomas 1991): today four of these species (*Maculinea arion*, *Polyommatus bellargus*, *Hesperia comma*, *Melitaea athalia*) have mostly or wholly returned (or been re-introduced) following enlightened conservation management (Table 1), and with the exception of the guild of butterflies that inhabits early seral stages on woodland floors, several other rare specialists have increased or maintained status disproportionately well in conservation areas and, increasingly, on land managed primarily for forestry or under agri-environmental schemes (e.g. Brereton et al. 2008; Thomas and Lewington 2010).

The approach taken by conservation scientists has shifted over 40 years. At the outset, intensive ecological studies were made of 10 European species, involving the compilation of sufficiently detailed life-tables—and the identification and quantification of every factor causing mortality or reduced natality—for a key factor analysis to be made (see Thomas 1984a, 1991; Thomas et al. 1998a for case histories and further references). Although, to our knowledge, these studies generated the only datasets from which mechanistic, as opposed to correlational, models of a butterfly's population dynamics can be constructed, each 3–7 year study was laborious and expensive. Moreover, in most cases the results were of limited usefulness for conservation, for they tended to describe the drivers of short-term fluctuations of a single, relatively stable, population, rather than the factors that determine the carrying capacities of different sites or which drive long-term population trends in species or assemblages across ranges. Most, however, provided the first insights into the key role played by the larval niche in butterfly population dynamics, owing to simple measurements made of the effect on larval survival of variation in two or more local growth-forms of the foodplant(s) or seral stage (e.g. Dempster 1969, 1983; Thomas 1974; Pollard 1979; Warren et al. 1986). Only in *M. arion* and *Maculinea rebeli* was variation in the quality of resource precisely quantified for every individual of the (1338, 1788 respectively) eggs whose survival was followed through each life-stage to adulthood (Thomas et al. 1998a, 2009). These latter studies enabled a series of mechanistic models to be constructed (reviewed by Thomas et al. 1998a; Clarke et al. 2005) in which habitat quality was a major variable, ranging from phenomenological or deterministic density-dependent interaction models (Hochberg et al. 1992; Mouquet et al. 2005a, b; Thomas et al. 2009) to others involving spatially explicit individual-based stochastic processes (e.g. Hochberg et al. 1994; Clarke et al. 1997).

With the realisation that variation in the quality of the larval resource was the known (or likely) driver of major population changes in studied species (Thomas 1984a), the focus of conservation research shifted to the less onerous precise description of species' realised larval niches, and to examining the extent to which variation in the quality or abundance of these and the adults' resources correlated with site occupancy or population size across a large number of habitat patches (for review of early studies see Thomas 1991). These studies also suggested that typically 10–40% of patches containing apparently suitable source habitat were not occupied by populations of colonial species (*sensu* Thomas 1984a) at any one time, and the constraint of low adult dispersal across landscapes was recognised early on (e.g. Ehrlich 1961; Thomas 1974, 1975, 1983a, b; Dempster et al. 1976; Dempster 1991; Warren 1987a; Thomas et al. 1992). Soon, a new science of meta-population ecology was developed by Hanksi and co-workers, largely using butterflies as model systems (for reviews see Hanksi 1999; Ehrlich and Hanski 2004). This also spawned numerous landscape-scale studies useful to practical butterfly conservation.

During the past decade, much research on butterflies has focused on adult resources, physiology, behaviour, and larval interactions with ants (myrmecophily) and parasitoids, providing an unparalleled knowledge of the multiple components in the environment, including the resource base exploited by most stages (Dennis and Van Dyck 2003, 2006; Turlure et al. 2010), that impact on a butterfly species' populations and with which it interacts (see Settele et al. 2009; Dennis 2010). An important development has

Table 1 Recent status of populations of rare butterfly species on UK nature reserves and conservation areas compared with the period 1960–1989 (from Thomas 1991; Asher et al. 2001; Fox et al. 2006; Thomas and Lewington 2010)

Species	% Loss 1960–1989	2000–2009 Status
Lycaena dispar	100	Still extinct
Melitaea cinxia	0	100% present
Maculinea arion	100	25% return + >30 new
Polyommatus bellargus	27	100% return + many new
Hesperia comma	24	100% return + many new
Thymelicus action	0	10–15% range and population decline
Satyrium pruni	0	100% present + several new
Carterocephalus palaemon	100 (England only)	Still extinct
Melitaea athalia	100	100% return + net increases

been to study and predict the impacts of climate change on the definition and availability of species' realised niches, ranges and abundance (e.g. Settele et al. 2008), and the extent to which species are regionally adapted to their local environment, including the rate at which phenotypes can adapt to habitat availability, climate or other changes following strong regional selection, for example for increased dispersal (e.g. Hill et al. 1999; CD Thomas et al. 2001; Hughes et al. 2003).

Here we attempt to analyse which factors have caused butterfly populations to decline to extinction or increase in density by the one to two orders of magnitude often observed in the wild. Except in the context of landscape-scale (meta-population) studies, we restrict analyses to the many semi-natural and other sites studied where a species' resources still apparently exist, rather than tabulate the fundamental destruction of biotopes (*sensu* Thomas 1991), such as the drainage and agricultural improvement of wetlands or the clearing of rain forests, that remains the main driver of butterfly extinctions and conservation priority in less developed regions of the world (e.g. Brown and Brown 1992; Stewart et al. 2007). In particular, we enquire whether just one or two key factors exist for study species that have such an over-riding impact on its population dynamics that all other interactions, while undeniably interesting, are of little relevance when it comes to manipulating habitats for practical conservation. In other words, we examine the evidence-based route successfully taken in entomological pest control, where the manipulation of one or two parameters may transform a site's carrying capacity and hence a species' abundance.

Intra-specific variation in population density

To understand the magnitude of variation that exists between the size of populations supported by a species' optimum and sub-optimal (usually occupied) source sites, we here extend a preliminary (Thomas 1984a) analysis of the UK Butterfly Monitoring Scheme (BMS) time-series of species' indexes of abundance, using the additional 24-years' data now available at http://www.ukbms.org (Fig. 1). For each species, a comparison was made of the mean adult counts per 100 m of 5 m wide transect over 10–33 years on the five sites that consistently supported the highest density of adults, and the five sites that supported the lowest densities. To avoid artefacts or sampling bias, considerable care was taken to compare like with like. Thus comparisons were restricted to sites containing single populations and similar habitat types, for example to more-or-less homogenous grassland containing the larval foodplant in the cases of *Melanargia galathea*, *Coenonympha pamphilus* and *Maniola jurtina*, thus excluding woodland

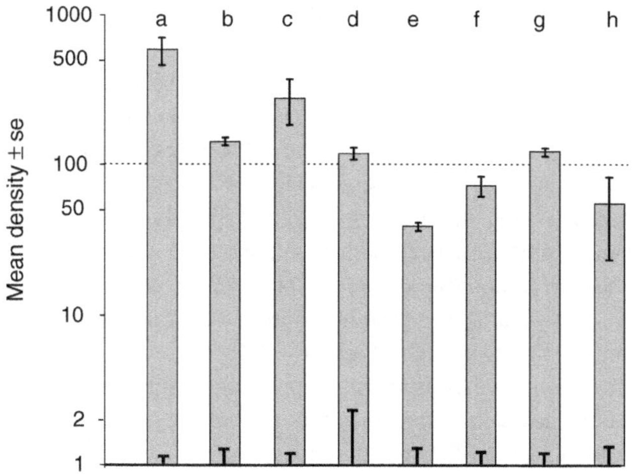

Fig. 1 Intraspecific variation in the mean densities of populations of eight butterfly species recorded on UK Butterfly Monitoring Scheme transects. Columns represent the mean population density (±SE) per 100 m of transect on the five highest-density sites and the five lowest-density sites, with the mean of the low density values = 1 (+SE). Methods follow Thomas (1991) and Thomas et al. (1994), with analyses restricted to sites occupied for 9 of 10 successive years by a species, and to those containing similar types of habitat, including larval foodplant, for each species in all sub-areas sampled by the transect. **a** *Melanargia galathea* **b** *Hipparchia semele* **c** *Argynnis paphia* **d** *Polyommatus coridon* **e** *Maniola jurtina* **f** *Coenonympha pamphilus* **g** *Boloria euphrosyne* **h** *Plebejus argus*

sites that contained occasional rides and glades; to calcareous grasslands for *M. galathea* and acid ones for *Hipparchia semele*; to sites where the transect was considered to sample the population of each species to an equal extent; and so on. In addition, for a site to qualify as supporting a low-density population, transect counts must record adults for at least nine out of ten consecutive years (see Thomas et al. 1994 for full criteria).

The results, presented here for eight species (Fig. 1), confirm the tentative earlier conclusion that the density of adults in the largest populations of a butterfly species is consistently about a hundred times higher than the density in its smallest ones. The same is true of *Maculinea nausithous*, *M. rebeli*, *Maculinea teleius*, *M. arion*, *Maculinea alcon*, *Melitaea cinxia*, *M. athalia*, *P. bellargus*, *Polyommatus coridon*, *Cupido minimus*, *Thecla betulae* and *H. comma* where population estimates have been based on egg or larval counts and contained low (usually 0–15%) standard errors (Thomas 1974, 1983a, b, 1984b; Elmes et al. 1996; Thomas et al. 1986, 2009; Thomas and Lewington 2010; Warren 1987b, 1991). Comparing the single highest density BMS site with the single lowest density one typically showed a 1,000-fold difference.

Hereafter we assume that the observed two-order-of-magnitude difference in butterfly density between the largest and smallest populations of a species in a landscape reflects the spectrum of environmental quality that exists between its optimum sites and those of the lowest source quality that are nevertheless capable of supporting a predominantly closed population for at least 9–32 generations. There is abundant evidence (Dennis 2010) that some differences are explained by fixed, typically abiotic, attributes, such as aspect, soil depth, shelter and local climate that are impossible or very expensive for conservationists to manipulate. For example, due to variation in soil attributes, the two highest quality UK restoration sites for *M. arion* have consistently supported 42 times more adult butterflies per square metre of *Thymus*/*Myrmica sabuleti* grassland than the two poorest ones, even when the latter were optimally managed for 10–21 years (our calculation from Thomas et al. 2009). On the other hand, there are many hundreds of examples of apparently suitable sites where the population size of a non-migratory species (including *M. arion*) dramatically decreased or (less often) increased by two orders of magnitude over a short number of generations before stabilising at a new density, whilst neighbouring populations remained relatively stable (e.g. Thomas 1983a, b; Thomas et al. 1986, 2009; Warren 1987b, 1991). We illustrate one of many examples from a wide range of species and populations to be found in the UK BMS dataset, involving a hitherto unreported comparison of changes on two UK sites supporting *P. coridon*. As is typical of many such shifts (Thomas 1991), the larval foodplant *Hippocrepis comosa* remained abundant and apparently unchanged in density on both sites during this period (Fig. 2).

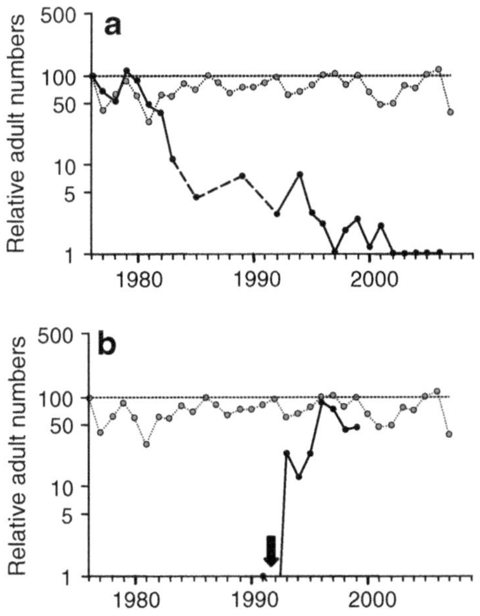

Fig. 2 Examples of change by two orders of magnitude in the population density of *Polyommatus coridon* on single UK sites following: **a** gradual habitat degradation and **b** conservation management. *Grey symbols*, *dotted line* national trend for *P. coridon*; *black lines* and *symbols* population at **a** Swanage and **b** Winchester following conservation management (*arrow*). *Dashed lines* gap of 2–3 years in recording

Our premise for the discussion below is that the differences shown in Figs. 1 and 2 define the order of magnitude that a population parameter must explain if it can be considered a key factor that drives long-term population change in butterfly species. For it is in the major drivers that can convert a potentially high-quality site into being a sub-optimal (or extinct) one, and especially vice versa, that the practical conservationist should be interested. We consider candidate factors.

Identifying factors determining abundance and drivers of population change

For most studied species of butterfly, analysis of the factors regulating population size or the drivers of population change is not straightforward, since typically only one or a few candidate factors were measured per study. However, it is possible to make a crude meta-analysis of 30 European species that have been studied more completely (Table 2), involving the compilation of life-tables, or comparatively complete comparisons being made both of site attributes and populations, or where population size was altered following manipulation of a site's environment through targeted conservation management or, in the case of *Pieris rapae* (Dempster 1969), pest control. For each species, we assigned a value of 0–3 to any of 11 non-exclusive variables that have historically been considered as a key population factor for butterflies, based on authors' assessments of the importance of each.

The results (Fig. 3) support previous reviews (e.g. Thomas 1991; New et al. 1995), as well as Hanski's (1999) statement that the realization that the constraints on species imposed by variation in the quality of the larval habitat and the ability of adults to colonise new patches across landscapes were "the two major reorientations in butterfly biology and conservation in 20 years". Below, we consider these two and certain other putative drivers at a detail commensurate with their ranking in Fig 3. Thus the potential impacts of butterfly collectors and insecticides, which both scored zero and have long been dismissed as negligible drivers of change (Thomas 1984a), are not discussed. They were, however, included in the meta-analysis because, for >100 years and a decade respectively prior to 1970, they were the main factors believed to threaten butterfly populations, and mitigating their impacts formed the rationale for many unsuccessful conservation projects.

Table 2 The principal source material for meta-analysis of Fig. 3

Species	Adult parameters						Larval parameters					Type of study	Tested in field?	Source
	Nectar honeydew	Mating	Natality weather	Dispersal	Collectors	Shelter	Food density	Subset of quality food	Parasitoids	Site area	Pesticide			
Carterocephalus palaemon	✔	✔	✔	✔	✔	✔	✔	✔	X	✔	X	C	X	14, 20, 30–33,
Thymelicus action	✔	✔	a	✔	✔	✔	✔	✔	✔	✔	a	C	✔$_{s}$	11, 12, 14, 20, 18, 58
Hesperia comma	✔	a	a	✔	✔	a	✔	✔	X	✔	a	C	✔$_{lsc}$	12, 14, 15, 25, 52, 53
Pyrgus malvae	✔	✔	a	✔	X	✔	✔	✔	X	✔	X	C	X	35
Papilio machaon	✔	✔	✔	✔	X	a	✔	✔	✔	✔	a	I_{lt}	✔$_{s}$	5, 6, 12, 14, 58
Leptidea sinapis	✔	✔	✔	✔	X	✔	✔	✔	✔	✔	a	I_{lt}	✔$_{lsc}$	12, 14, 40, 48, 58
Pieris rapae	a	a	✔	✔	X	✔	✔	✔	✔	✔	✔	Lt	X	1, 2, 3, 5, 12
Thecla betulae	✔	✔	✔	✔	✔	✔	✔	✔	✔	✔	a	I_{lt}	✔$_{s}$	8, 12, 14, 20, 59
Satyrium pruni	✔	✔	✔	✔	✔	✔	✔	✔	✔	✔	a	I_{lt}	✔$_{lsc}$	8, 9, 12, 14, 20, 58–59
Lycaena phlaeas	a	a	a	a	a	a	✔	✔	✔	✔	X	I_{lt}	X	4, 5, 14
L. dispar	✔	✔	✔	a	X	a	✔	✔	✔	✔	X	I_{lt}	✔$_{s}$	5, 7, 12, 14
Plebejus argus	a	a	a	✔	✔	✔	✔	✔	✔	✔	X	C	✔$_{s}$	20, 23, 24, 26, 57
Aricia agestis	✔	✔	a	✔	✔	a	✔	✔	X	✔	X	C	X	28
Polyommatus bellargus	✔	✔	✔	✔	✔	✔	✔	✔	✔	✔	a	C	✔$_{lsc}$	11, 12, 14, 18, 20, 37, 58
Maculinea arion	✔	✔	✔	✔	✔	✔	✔	✔	✔	✔	✔	I_{lt}	✔$_{lsc}$	5, 12, 14, 17, 19, 38, 58
M. rebeli	✔	✔	✔	✔	✔	a	✔	✔	✔	✔	X	I_{lt}	X	17, 29
M. nausithous	✔	a	a	✔	✔	✔	✔	✔	✔	✔	✔	I_{lt}	✔$_{s}$	13, 14, 17, 49, 50
M. teleius	✔	a	a	✔	✔	✔	✔	✔	✔	✔	✔	I_{lt}	✔$_{s}$	13, 14, 17, 49, 50
M. alcon	✔	a	a	✔	✔	a	✔	✔	✔	✔	X	I_{lt}	X	17, 39
Limenitis camilla	a	a	a	✔	X	a	✔	✔	✔	✔	a	I_{lt}	X	12, 14, 21
Boloria selene	✔	✔	a	✔	✔	✔	✔	✔	X	✔	a	C	✔$_{lsc}$	14, 16, 20, 27, 45, 58
B. euphrosyne	✔	✔	a	✔	✔	✔	✔	✔	X	✔	a	C	✔$_{lsc}$	14, 16, 20, 27, 45
Agrynnis adippe	✔	a	a	a	✔	✔	✔	✔	X	✔	a	C	✔$_{lsc}$	14, 16, 20, 27, 45, 47, 58
A. aglaja	✔	a	a	a	✔	✔	✔	✔	X	✔	X	C	X	14, 16, 20, 27
A. paphia	✔	✔	a	a	✔	✔	✔	✔	X	✔	a	C	✔$_{s}$	14, 16, 20, 45
Euphydryas aurinia	✔	✔	a	✔	✔	✔	✔	✔	✔	✔	X	Lt	✔$_{lsc}$	20, 22, 34, 51, 55, 56
Melitaea cinxia	✔	✔	a	✔	✔	✔	✔	✔	✔	✔	X	Lt	X	14, 18, 20, 46,
M. athalia	✔	✔	✔	✔	✔	✔	✔	✔	A	✔	X	Lt	✔$_{lsc}$	12, 14, 41–44, 58

Table 2 continued

Species	Adult parameters					Larval parameters						Type of study	Tested in field?	Source
	Nectar honeydew	Mating	Natality weather	Dispersal	Collectors	Shelter	Food density	Subset of quality food	Parasitoids	Site area	Pesticide			
Hipparchia semele	✓	✓	a	✓	✓	✓	✓	✓	X	✓	X	C	✓$_{lsc}$	20, 27, 58
Coenonympha tullia	✓	✓	a	✓	✓	✓	✓	✓	X	✓	X	C	X	36, 57

Columns 2–12, main variables assessed in population studies (✓ = precisely or crudely measured, a = anecdotal assessment, X = not measured), nectar = impact of available nectar on longevity or natality; *column 13* type of study (I_{lt} = individual-based life-tables, Lt = life tables, C = correlations); *column 14* key factor(s) validated or identified by testing in the field through conservation or other management at scale of landscape (✓$_{lsc}$) or single sites (✓$_s$); *column 15* main sources by number: (1–4) Dempster 1967, 1968, 1969, 1971; (5) Dempster and Pollard 1981; (6) Dempster et al. 1976; (7) Duffey 1968; (8–14) Thomas 1974, 1975, 1983a, b, 1984a, b, 1991; (15–19) Thomas et al. 1986, 1996, 1998a, 2009, JA Thomas et al. 2001; (20) JA Thomas and DJ Simcox unpublished; (21) Pollard 1979; (22) Porter 1981; (23, 24) Thomas 1985a, b; (25) Thomas and Jones 1993; (26) CD Thomas et al. 1999; (27) Elmes and Thomas 1992; (28) Bourn and Thomas 1992; (29) Hochberg et al. 1994; (30–33) Ravenscroft 1994a–c, 1995; (34) Lewis and Hurford 1997; (35) Brereton 1997; (36) Dennis and Eales 1999; (37) Roy and Thomas 2003; (38, 39) Mouquet et al. 2005a, b; (40–47) Warren 1985, 1987a–c, 1991, 1992, 1994, 1995; (48) Warren et al. 1986; (49, 50) Wynhoff 1998, 2001; (51) Wahlberg et al. 2002; (52) Wilson et al. 2009; (53) Davies et al. 2005; (54) Revels 2006; (55) Bulman et al. 2007; (56) Klapwijk 2008; (57) Dennis 2010 and references therein, (58) Thomas and Lewington 2010, (59) Bickmore and Thomas 2000)

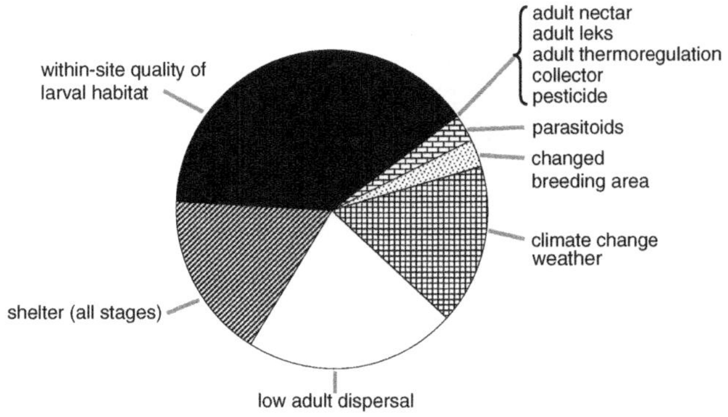

Fig. 3 The main drivers of long-term population trends in 30 European butterfly species where full population dynamic or similarly intensive studies were made and/or targeted conservation or pest control management has succeeded. Scores used in meta-analysis: *3* key driver, *2* secondary driver, *1* detectable effect, *0* no effect detected. See Table 2 for sources

Weather, climate change and population size

The observed and predicted impacts of climate change on different species are reviewed elsewhere by Dennis (1993), Parmesan et al. (1999), Roy et al. (2001), Warren et al. (2001), Thomas et al. (2006) and Settele et al. (2008), and are not repeated here. However, variation in the attributes and breadth of a species' larval niche under different climates is briefly considered below, and it is relevant to note that the short-term fluctuations (as opposed to long-term trends) of monitored species, which tend to be synchronised across regions (Pollard 1991) and are widely accepted as being driven mainly by variation in weather (Pollard and Yates 1993; Roy et al. 2001), are typically two orders of magnitude smaller than the consistent differences or long-term trends in population density between sites reported here (Figs. 1 and 2). Thus although short-term fluctuations are of significantly greater amplitude towards species' range margins (Thomas et al. 1994), the mean temporal Coefficient of Variation (CV) on the stable sites of the species represented in Fig. 1 ranges from 0.62 (*M. jurtina*) to 0.91 (*P. coridon*) (see Thomas et al. 1998a for CVs of all UK species that meet criteria for analysis).

Shelter

A lack of shelter scores as a depressor of population size in many studies. With the exception of Dover's (e.g. Dover 1996; Dover et al. 1997) studies in open farmland, this is the least precisely defined of the 11 variables considered, and with increased knowledge each example could probably be reclassified as equating to 'within-site larval quality' or 'low adult dispersal', or both. For variation in shelter can affect butterfly populations in either the immature or adult stages: in the former by altering the microclimate, especially temperature, of the larval habitat;

in the latter, by affecting the propensity of adults to fly across exposed or more sheltered examples of the matrix that lies between habitat patches, or within sites by providing shelter from inclement weather or refugia at night (Dover et al. 1997; Dennis 2010). Although often ill-defined (except by Dover 1996), shelter is a useful parameter to identify for the practical conservationist, for it can be manipulated on many sites, for example through scrub or woodland management (e.g. Thomas 1975; Dover et al. 1997).

Population-scale factors

Interactions with enemies or resources

Very few populations of recorded butterflies (Asher et al. 2001) exhibit the major cycles in abundance or patterns of short- to medium-term boom-bust changes that are characteristic of reciprocal interactions with resources or enemies in certain other insect taxa or in simpler communities breeding on intensively cultivated land. Moreover, most short-term fluctuations are attributable to weather effects (above). A striking exception, postulated by Pollard and Yates (1993) and confirmed empirically by Revels (2006), is the 4–6 year population cycling of *Celastrina argiolus* and its specific ichneumonid parasitoid *Listrodomus nycthemerus* (Fig. 4). Even here, it is possible that the equilibrium level around which both insects cycle is determined by the density or quality of the host's habitat. Similar dynamic interactions have been claimed for certain Melitaeninae or related fritillaries, but at present the evidence is equivocal (e.g. Porter 1981; Klapwijk 2008). In other butterfly species, cycles or declines due to over-exploitation of a resource, for example the weak cycling just detectable in *M. arion* populations around each site's equilibrium level (Thomas et al. 2009), are orders of magnitude below that needed to explain observed major population changes.

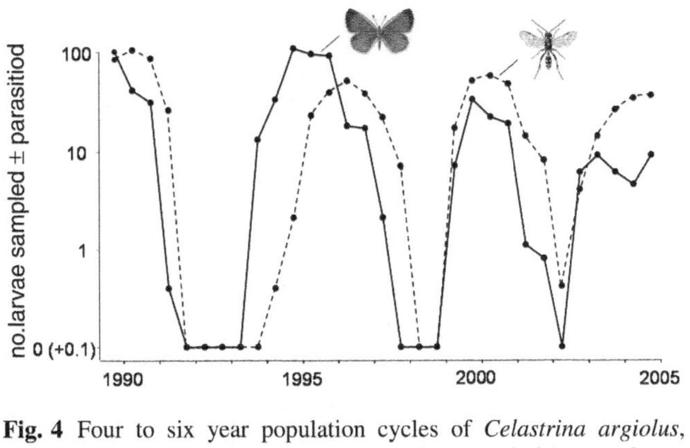

Fig. 4 Four to six year population cycles of *Celastrina argiolus*, tracked and apparently driven by its specific parasitoid *Listrodomus nycthemerus*. Data from Revels (2006), plotted here as 3-generation running averages to smooth seasonal fluctuations of spring and summer generations

Adult resources

Limited or declining adult resources are frequently evoked as the cause of small populations or long-term declines, but apart from shelter, the evidence for this is weak. Indeed, most putative factors (nectar, mating and thermoregulatory spots) scored zero in the meta-analysis of 30 species studied in the field (Fig. 3). Moreover, their impact, in the relatively few cases established under laboratory conditions, typically explains a <1 to 3-fold (under extreme unnatural conditions up to 7-fold) changes in the egg population (Boggs and Ross 1993; Fischer and Fiedler 2001; Bergstrom and Wiklund 2002; Mevi-Schütz and Erhardt 2003; Jervis and Boggs 2005) rather than the 50–100-fold or greater variations (Figs. 1, 2) which we seek to implicate here. Thus intra-specific variation in mean female natality seldom varied by more than 2- or 3-fold between years and sites in species for whom life-tables have been compiled (Table 2), and when it did, the effect was often diminished before the next adult emergence by density-dependent mortalities (or enhanced survival) in the intervening larval stage (e.g. Dempster 1983; Pollard 1991; Nowicki et al. 2009; Thomas et al. 2009). Moreover, observed variation in oviposition and population numbers is strongly correlated with the weather (e.g. Gossard and Jones 1977; Thomas 1974; Pollard 1979, 1991; Roy et al. 2001; Thomas et al. 2009) or with early adult death by enemies (Dempster 1983), both of which contribute to the mean 4–7 day survival rate of non-hibernating adults in temperate regions measured in the field, compared to the 20–30 days when protected in captivity. Nevertheless, two adult resources that are still influential in conservation planning are briefly considered.

Nectar. Prior to the 1970 s, a lack of nectar was hypothesised to be the third major factor (after collectors and insecticides) driving declines in the UK, to an extent that senior conservation advisors proposed the planting of (exotic) buddleja and other nectar sources as the main strategy to improve nature reserves where butterflies were declining. To our knowledge there is no scientific study, in which the mechanisms causing population dynamic change in a butterfly have been measured, that implicates variation in nectar as a driver. Certainly, several studies have reported a positive correlation between population size and the abundance of nectar sources on study sites (see Dennis 2010). On the other hand, no fewer than 12 of the 30 species included in the meta-analysis (Fig. 3 and references therein) show a significant inverse correlation between population size (or change) and the availability of their preferred adult nectar sources. This is perhaps simply explained by the fact that the optimum habitat for oviposition and larvae of these latter species involves growth-forms of the larval foodplant (or ant resource) existing either in very early (*H. comma*, *Plebejus argus*, *P. bellargus*, *M. arion*, *M. rebeli*, *Boloria euphrosyne*, *Argynnis adippe*, *M. cinxia*, *H. semele*) or late (*Thymelicus acteon*, *M. nausithous*, *Argynnis paphia*) seral stages of their biotopes, in which flowers generally occur at low densities: in contrast, nectar is usually most abundant in mid-successional swards (and other biotopes) (Erhardt 1985), coinciding with the optimum larval niches of *Pyrgus malvae*, *M. teleius*, *M. alcon*, *Boloria selene*, *Agrynnis aglaja*, *Euphydryas aurinia* and *Coenonympha tullia* of the species in Fig. 3. It should be noted that most of the above correlations were obtained from studies made in the UK or the northern half of Europe: several species that occupy flower-poor early seral stages in northern Europe occupy cooler mid-successional stages with abundant flowers under warmer climates further south (Thomas et al. 1998b; JA Thomas et al. 1999).

Mate-location. To our knowledge, there is no evidence of a direct impact on population size in European butterfly species from the availability of preferred mating sites. Moreover, although the females of certain species may be reluctant to lay eggs when not mated (Karlsson and Van Dyck 2009), the incidence of sterility is generally less than 2% of the egg population in even the smallest studied populations, and as described above, variation in natality, for which there is little evidence of any effect from the nutritional value of male nuptial gifts (Bergstrom and Wiklund 2002), or the reluctance of unfertilised females to oviposit, explains no more than small-scale fluctuations in well-studied species. For the same reason, the possibility of increased emigration of adults from patches with few mating sites (or other adult resources) does not explain variation in population size in species for which life-tables have been constructed. Nor would it necessarily be a bad thing, at the scale of meta-populations (below), if this were a major variable.

Larval habitat quality

In all but one (i.e. 97%) of the well-studied species in Fig. 3, and in the large majority of other species where the ecology of the young stages has been studied, the female restricts or concentrates oviposition on a subset of the larval foodplant that is growing in definable and predictable situations or growth-forms (see Thomas 1991, 2007; Dennis 2010; Thomas and Lewington 2010). The attributes of the preferred subset vary greatly between species. For example, each of the five European *Maculinea* species restricts oviposition to initial foodplants that possess flowerbuds of a narrow size range, development stage and prominence (Thomas and Elmes 2001); *Carterocephalus palaemon* oviposits on a minority of *Brachypodium* plants whose leaves are not only tall and rich in nitrogen, but also growing in warm sheltered spots (Ravenscroft 1994b); *Aglais urticae* (Pullin 1987) and *Polyommatus icarus* (Dennis 1985) use young, nitrogen rich growths of their foodplant, whereas *Aricia agestis* selects plants with exceptionally succulent leaves (Bourn and Thomas 1992); *T. acteon* selects only the largest tussocks of *Brachypodium pinnatum*, and of these only clumps growing in warm sheltered spots (Thomas 1983b); *B. selene* exploits medium-sized *Viola* plants growing in moist soil whereas *B. euphrosyne* selects younger, smaller-leaved *Viola* growing in warmer spots (Thomas et al. 1986); in woodland, the larval niche of many species whose young stages inhabit the ground or shrub layer is restricted to examples of foodplant growing within narrow limits of light or shade (e.g. Pollard 1979; Shreeve 1986; Warren 1987b,c, 1995; Thomas 1991; Thomas et al. 1986). And towards the northern edges of their ranges, a great many species oviposit only on examples of foodplant that are growing in exceptionally warm microclimates, typically restricting the larval stage to slopes with warm aspects and within them to spots of early successional vegetation, such as very short or sparse grassland swards or fresh clearings in woodland (Thomas 1991, 1993; Thomas and Lewington 2010).

Where measured, the optimum form of the larval resource is closely correlated with enhanced larval survival or fitness in the wild (e.g. Thomas et al. 1998a). So restrictive is this specialisation that, in at least 20 of the species included in the meta-analysis (Fig. 3), no correlation was found between the absolute abundance of larval foodplant and butterfly population size across sites or over time, and in the cases of *H. comma*, where oviposition was restricted to small *Festuca ovina* plants surrounded by bare ground, and in *M. nausithous* which oviposits in late successional wetlands where *Sanguisorba officinale* grows tall but typically sparsely, a significant inverse correlation was found between foodplant and butterfly densities (Thomas 1984b; Thomas et al. 1986). On the other hand, in all 20 species used in the meta-analysis where the availability of the preferred larval niche was adequately quantified,[1] a strong correlation existed between this, often minority, subset of the larval resource and butterfly population size. Importantly for conservation, the association in all these species explained a spectrum of populations sizes that spanned two or more orders of magnitude (Figs. 1 and 2),

[1] *Thymelicus acteon*, *Hesperia comma*, *Leptidea sinapis*, *Thecla betulae*, *Satyrium pruni*, *Aricia agestis*, *Polyommatus bellargus*, *Maculinea arion*, *M. nausithous*, *M. teleius*, *Boloria selene*, *B. euphrosyne*, *Agrynnis adippe*, *A. aglaja*, *A. paphia*, *Euphydryas aurinia*, *Melitaea cinxia*, *M. athalia*, *Hipparchia semele*, *Coenonympha tullia*. See Table 2 for references.

both between sites or within individual patches where management changed over time. In the few species where the preference of ovipositing females and larvae was precisely quantified across the full spectrum of its resource, ranging from optimum via sub-optimal sources to sink habitats, the correlation between larval habitat quality and butterfly density was close, with values of $R^2 = 51\%$ (*P. bellargus*), 54% (*M. cinxia*), 62% (*T. acteon*) and 99% (*M. arion*) (JA Thomas et al. 2001; Thomas et al. 2009).

The importance for conservation of identifying and manipulating this specialisation is illustrated in two case histories. In northern parts of its range, larvae of the bivoltine butterfly *P. bellargus* are restricted during the autumn/spring-feeding generation to a (usually small) subset of *H. comosa* plants growing in sheltered south-facing depressions where the sward is also very short (Fig 5a). These represent the warmest hotspots available within sites (Thomas 1983a), and the relative sizes of *P. bellargus* populations in the UK are closely correlated, across two orders of magnitude, with the number of foodplants growing in optimum or sub-optimal growth-forms (JA Thomas et al. 2001). In May–June, the resultant adults oviposit on plants growing in the same narrow range of absolute temperatures to those selected by autumn-flying adults, but because the weather is warmer, the eggs are distributed over a wider, partly different subset of foodplants, including many growing in relatively cool and unsheltered spots (Fig. 5b). On an average UK site, this permits the summer-feeding generation of larvae to exploit roughly twice as many *H. comosa* plants than in autumn-winter, and may explain why the mean adult population size of *P. bellargus* was about twice as numerous in the second generation on monitored sites in 1976–2002 (Roy and Thomas 2003).

Over much of northern Europe, *P. bellargus* populations declined and experienced frequent local extinctions during the first 80 years of the twentieth century, when many farmers abandoned unfertilised south-facing slopes and, after the mid 1950 s, rabbits disappeared as an effective grazing force due to myxomatosis. In the UK, the rate of population extinctions was so steep that by 1980 an extrapolation suggested the national extinction of this species early in the current century (Thomas 1983a). In practice, recognition that populations of *P. bellargus* survived only on sites where *H. comosa* grew in close-cropped hotspots, and that large populations persisted only where many plants occurred in that condition, caused wardens and sympathetic farmers to increase grazing on declining and former sites, applying regimes which conservationists had shunned because they reduced the flowering of most plant species (the more recent discovery that *P. bellargus* will inhabit taller turf during summer negated that conflict). This restored grazing, coupled with the gradual return of

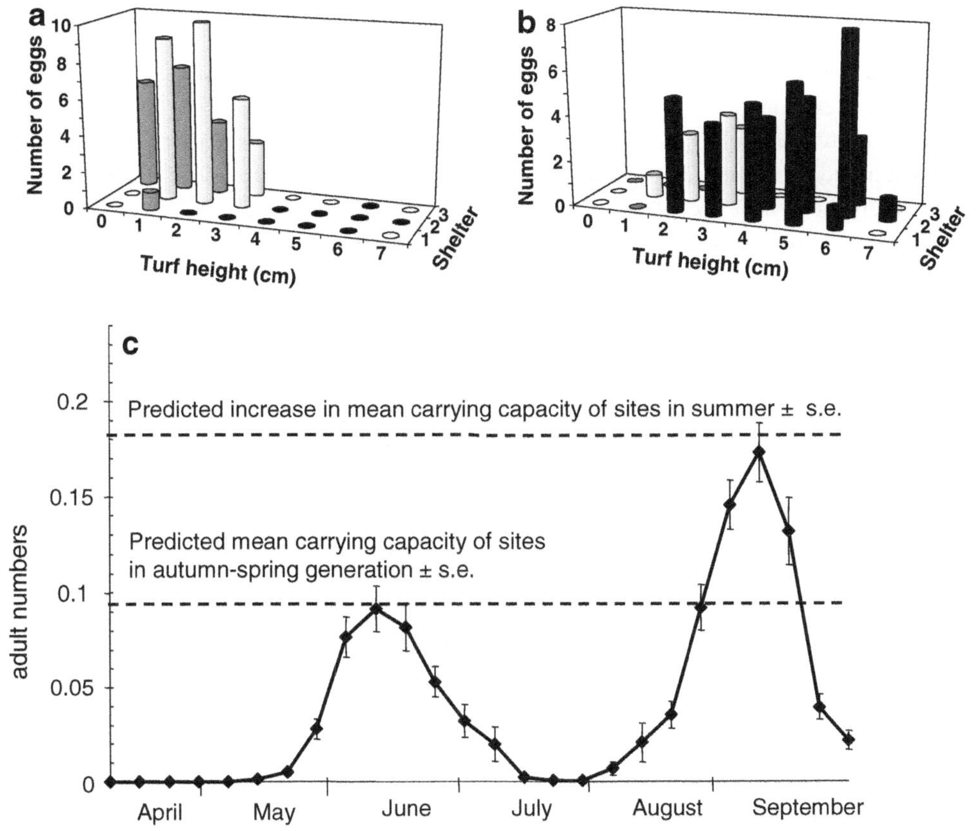

Fig. 5 Seasonal shifts in the larval niche of the bivoltine thermophilous butterfly *Polyommatus bellargus* in the UK, on a site where the distribution of its perennial foodplant *Hippocrepis comosa* was constant. **a** Autumn–spring-feeding larval generation, **b** summer-feeding larval generation; from Roy and Thomas (2003) where categories of shelter are defined (*3* most sheltered), *grey columns* growth-form of *Hippocrepis* exploited in autumn–spring only, *white* growth-forms used by both generations, *black* summer only. **c** Mean density (± SE) of *P. bellargus* in each adult generation on UK sites sampled by the UK butterfly Monitoring Scheme in 1976–2001; *dashed lines* predicted carrying capacity of sites in each generation calculated from the proportion of *Hippocrepis* plants available for exploitation by the preceding cohort of larvae

rabbits, transformed the sward structure of calcareous grasslands in favour of this butterfly's larval niche, as evidenced by the results of repeat surveys made of all potential sites across 80% of *P. bellargus*' occupied UK range in 1978 and 1999 (Thomas 1983a; JA Thomas et al. 2001). Coinciding with this shift (and incidental reduction in nectar availability), many of the ca. 75 surviving populations of 1980 increased greatly in numbers and spread to found about 75 new colonies by 1990 and a further 100 populations a decade later, mainly colonising former sites (Fox et al. 2006). A similar 4-fold recovery in site occupancy occurred over the same period with *H. comma* in the UK, for the same reasons (Davies et al. 2005), aided also, in both species, by a warming climate which broadened the larval niche and hence the number of exploitable foodplants (JA Thomas et al. 1999).

The evidence that the recovery of *P. bellargus* and *H. comma* in the UK resulted initially through targeted conservation management to restore their larval niches in grassland is persuasive, yet based on correlations. Greater certainty exists about the restoration of *M. arion* to the UK, following national extinction in 1979, for new grazing regimes were based on precise niche and population models resulting from an intensive 6-year study of the mechanisms affecting natality and survival in all stages of the life-cycle (see Introduction). Again, that study indicated that this most specialised of butterflies was an even greater specialist during its larval period than had been appreciated, and that apparently minor variation to the structure of its grassland habitat altered both the intrinsic growth rate of a population (λ) and the carrying capacity (K) of its sites from <1 to 5.9 and from 0 to 90,000 eggs per hectare respectively (Thomas et al. 2009), with both parameters being closely correlated ($R^2 = 85\%$) on the 13 sites where both have been measured (Fig. 6). Although confined to different soils—and requiring the additional resource of the ant *M. sabuleti* to coexist with the initial larval foodplants (*Thymus* or *Origanum vulgare*), occurring for optimal survival as high density, small colonies containing large-bodied workers and few or no queens—the vegetation structure required by *M. arion* larvae resembles that of *P. bellargus* and *H. comma* at northern latitudes and high altitudes, and requires similar grazing management (Thomas et al. 2009). Although low survival was recorded with other species of *Myrmica*, early models predicted that a threshold of 68% of the larval population must be adopted by *M. sabuleti* for the butterfly's intrinsic rate of increase to exceed one. *M. sabuleti* is a thermophilous ant that requires warm short open swards under UK climates (Fig. 7). For the same reasons that *P. bellargus* and *H. comma* declined across northern Europe, so did *M. arion*'s specific ant host, to the extent that it was undetectable on the large majority of former sites when surveyed in the

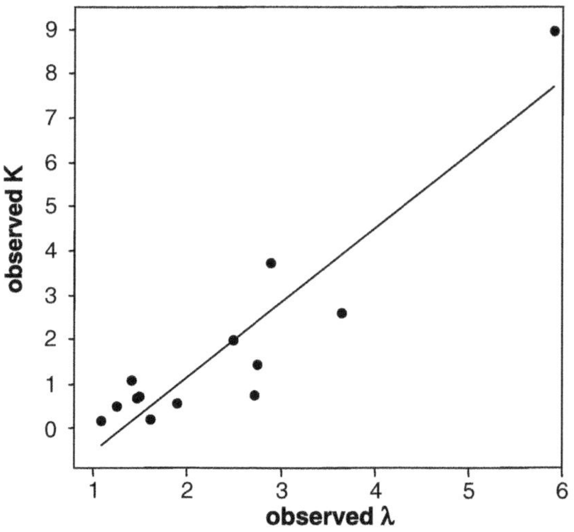

Fig. 6 Relationship between values of the observed intrinsic growth rate (λ) and the observed carrying capacity (or equilibrium level) (K) of *Maculinea arion* populations following colonisation of 13 unoccupied conservation sites in the UK. Values of K represent eggs per m^2, based on the mean of the three highest values of populations that stabilised and fluctuated round an equilibrium level (CV < 0.5, see Fig. 8). Plotted from data in Supporting Online Material (Thomas et al. 2009) and D.J. Simcox (unpublished): obs $K = -2.17 + 1.67$ (obs λ), $R^2 = 85\%$; $P = 0.000$ (omitting outlier, $R^2 = 59\%$; $P = 0.004$)

early 1970 s. Although research results came too late to save the last small UK *M. arion* population from extinction, attempts to restore its optimum larval habitat, through grazing sites to < 2 cm tall during the critical spring and autumn months for *M. sabuleti* larval development, began in the mid seventies and are currently being applied to about 100 former or neighbouring sites. The return of *M. sabuleti* over a 35 year period is illustrated for one restoration site (Fig. 7), and is typical of all others (Thomas et al. 2009): it is pertinent to note that the niche of *M. sabuleti* has broadened under the warmer climates since the early 1990 s (Fig. 7, dashed curve), and today the minimum density required to support *M. arion* encompasses sward heights of up to 4 cm tall, easing the delivery of successful conservation management.

By the early 1980s, a simple mechanistic model (Thomas et al. 2009), in which four of the five variables described different aspects of larval survival in ant nests (proportion of larvae adopted by *M. sabuleti*, other or no *Myrmica*; weight of ant brood; density-dependent mortalities through starvation in ant nests; impact of drought on ant carrying capacity) was tested by introducing Swedish *M. arion* to Dartmoor (Site X), where the original population study had been made and where the predicted carrying capacity (and value of λ) had been raised through the impact of altered management on *M. sabuleti* populations by nearly 10-fold. Figure 8 shows that for the 22 years over which this new population was monitored, the actual

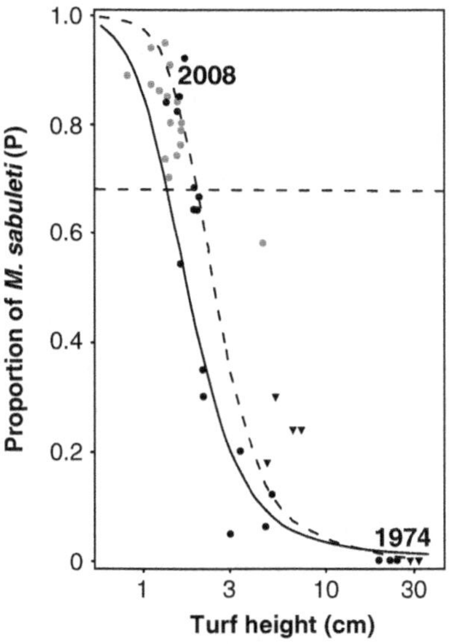

Fig. 7 Changes in the niche and abundance of the ant *Myrmica sabuleti* under past (*solid curve* 1974–1989) and current warmer (*dashed curve* 1990–2008) UK climates on one typical *Maculinea arion* restoration site. *Black dots* years when site did not support *M. arion*; *grey dots* years supporting *M. arion*; *triangles* ungrazed controls; *horizontal dashed line* minimum predicted density for *M. arion* to have a positive intrinsic growth rate ($\lambda > 1$). From Thomas et al. (2009)

number of *M. arion* closely followed the long-term theoretical predictions. Further validation was obtained from the second longest-running population, introduced to Green Down in 1992, where the predicted values of λ and K for *M. arion* were 0.3 and 0 prior to conservation management (representing sink habitat), but quickly rose to >3 and nearly 4,000 eggs per hectare, respectively, under the prescribed grazing. Again, the introduced population of *M. arion* closely followed long-term predictions derived from the quality of its larval habitat. The same was true of model predictions of 109 annual population changes on other sites in 1983–2009 ($N = 109$ population changes on 21 sites; GLM adjusted $R^2 = 77.7\%$, $P = 0.000$; from Thomas et al. 2009 plus 10 additional datapoints and two new sites for 2008–2009 changes). At the time of writing, *M. arion* has spread to >30 prescriptively managed sites in the UK, two of which have supported the highest known densities in the world of this globally threatened species for the past decade (Thomas et al. 2009). Together, these datasets establish that variation in the occurrence of the larval niche (and the occasional impacts of drought on larval resources) is the overriding variable that determines long-term trends and average population size in this Red Data Book species.

There is similar evidence of conservation success among other butterflies whose larval niche was precisely described and then deliberately created or increased through the management of individual sites. The longest-term or best documented examples of populations increasing by 10- to >100-fold in numbers after targeted management again come from the UK or the Netherlands. They include in woodland, scrub or scrub-grassland: *Satyrium pruni* (Thomas and Lewington 2010), *T. betulae* (Bickmore and Thomas 2000), *Hamearis lucina* (MS Warren pers comm), *Leptidea sinapis* (S Jeffcoate pers comm), *M. athalia*

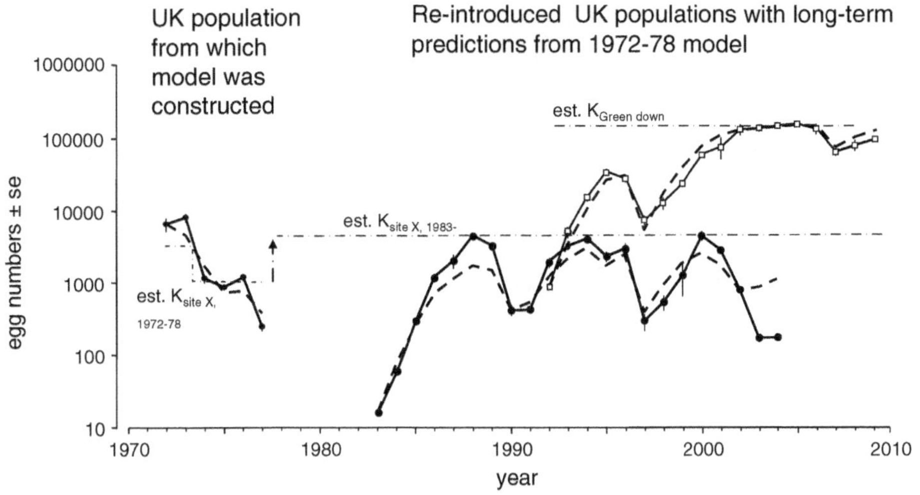

Fig. 8 Predicted (*dashed line*) and observed (*solid line*) population changes of *Maculinea arion* on two UK conservation sites. *Broken dotted lines* change in estimated carrying capacity of sites before and after conservation management ($K_{\text{Green Down}} = 0$ prior to 1990). *Solid symbols* Site X, Dartmoor: 1972–1978 = the last native UK population during the period of intensive population study; 1983– = introduced population from Sweden. *Open symbols* Green Down, Somerset. Predictions of population changes based on a mechanistic model describing annual variation in the quality of the resource, including impacts of drought in 5 years, with predictions made sequentially from dates of introductions (1983, 1992) without correction for observed butterfly numbers in subsequent years; data and model from Thomas et al. (2009) plus 2009 data for Green Down

(Warren 1991) and four species of *Viola*-feeding fritillary (Elmes and Thomas 1992; Asher et al. 2001; Thomas and Lewington 2010; DJ Simcox unpublished); in more open grasslands *P. coridon*, *H. semele* (Elmes and Thomas 1992) and *E. aurinia* (Bulman 2007); and in wetlands *M. nausithous* and *M. teleius* (Wynhoff 1998, 2001), *Lycaena dispar* (C Van Sway pers comm), and *Papilio machaon* (Fox et al. 2006). Only in *S. pruni*, *M. athalia*, *L. dispar* and perhaps *P. coridon* has this management been applied to enough sites to reverse or halt the national decline, but these examples provide evidence of the importance of maintaining this component of a species' habitat.

Although every species studied in this way possessed a unique, species-specific fundamental niche (albeit one that may vary under different climates), among the more hopeful results of recent conservation projects is that the targeted restoration of one declining species' habitat has often resulted in an improved, if not optimal, environment for other specialists, that had been declining for the same general reasons of altered land-use. In other words, where they belong to a similar guild, it has proved possible to manage for declining assemblages of butterflies (and other wildlife). Most accounts are anecdotal, but quantified examples include 1-2 order of magnitude increases by seven other Biodiversity Action Plan (BAP) butterfly species (*B. euphrosyne*, *B. selene*, *A. adippe*, *H. semele*, *P. malvae*, *Erynnis tages*, *T. betulae*) on UK sites managed for *M. arion*.

A more rigorous test of understanding a species' requirements is to attempt to create them from scratch on intensive arable farmland, or on the raw bedrock or ballast of major constructions, such as on the verges, embankments and cuttings of new roads or railways. It is beyond the scope of this paper to describe examples in detail, but most successful ones have involved inclusion into the design of the precise larval niches of an endangered butterfly species or, more often, of an assemblage of scarce or threatened species, including their preferred topographies, soil depths and nutrient levels (e.g. Morris et al. 1994). Successful UK examples include the restoration of species-rich calcareous grassland to the arable fields of Twyford Down and the adjoining creation on chalk ballast of >2 km grassland on the former A303 near Winchester, UK following the construction of the M3 in the 1990s, creating new sites that today support high densities of BAP species *P. coridon*, *C. minimus* and *E. tages* (Snazell et al. 2001; Thomas 2001; Thomas and Lewington 2010); the creation of scrub and clay grassland habitats on former arable fields isolated by the M40 in Oxfordshire, which today support high densities of *T. betulae* and the UK's largest population of *S. pruni* (Bickmore and Thomas 2000 and unpublished 2007–2009 monitoring); and the creation of two sites on new UK railway constructions that today support medium-sized populations of *M. arion* and other BAP species (D.J. Simcox, unpublished; Thomas and Lewington 2010).

Metapopulation-scale factors

Adult dispersal

A lack of occupancy of certain of a species' apparently suitable habitat patches was recognised early in empirical studies of butterflies, and attributed to low adult dispersal (e.g. Ehrlich 1961). Convincing empirical evidence for low vagility comes from the rates of spread of certain species after colonisation or introduction to unoccupied landscapes containing multiple islands of the species' larval habitat. Two early examples found that *S. pruni* and *M. athalia* spread at rates of about 1 and 1.6 km a decade (Fig. 9) in landscapes containing patches of variable habitat quality, the former being a similar rate to that recorded in *P. argus* after introduction to the Dulas valley, Wales, and the latter to the spreads of *P. bellargus*, *H. comma* and *M. arion* following the restoration of their habitats in patches across landscapes in England (Thomas 1991; Thomas et al. 1992; Davies et al. 2005; Thomas and Lewington 2010).

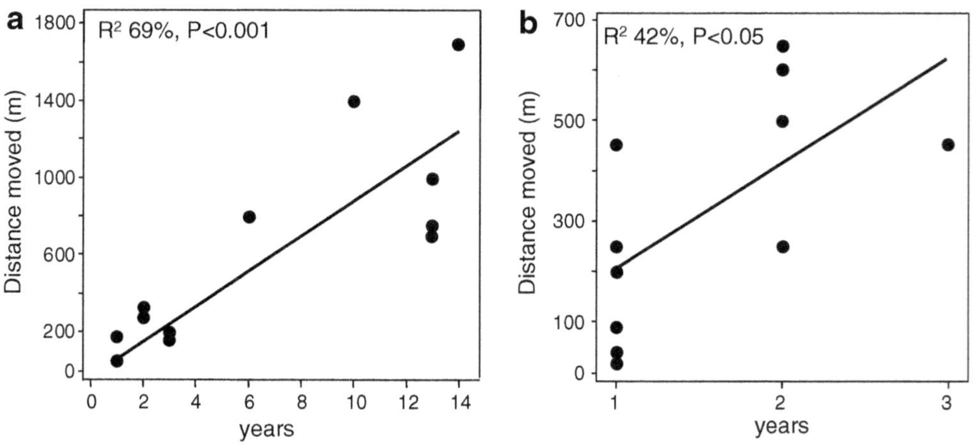

Fig. 9 Observed colonisation rates of habitat patches of **a** *Satyrium pruni*, **b** *Melitaea athalia* following conservation management of sites across landscapes or introduction to new landscapes. From Thomas et al. (1992)

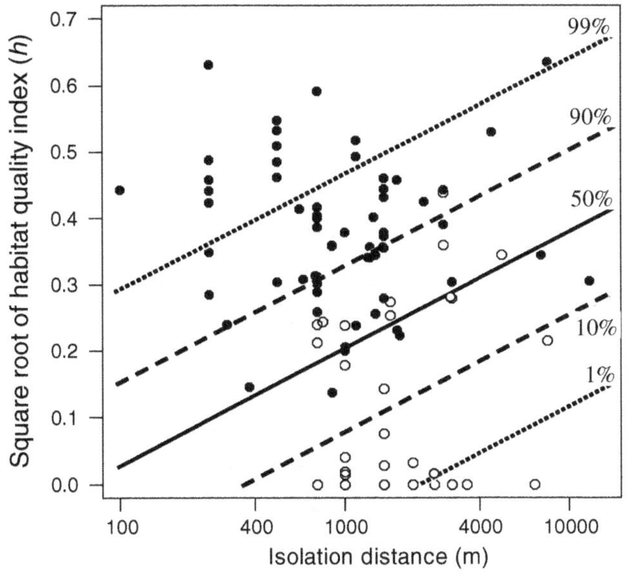

	area	isolation	habitat
GKG association	12%	51%	85%
Partial significance p	0.35	0.001	< 0.0001

Fig. 10 Occupancy pattern by *Polyommatus bellargus* of all (95) sites containing *Hippocrepis comosa* in swards of varying structure in a 1960 km^2 region of Dorset and Wiltshire, UK. The size, isolation and within-patch habitat quality for spring–autumn feeding larvae was measured on every site. From JA Thomas et al. (2001)

Similarly, only one (*Polygonia c-album*) of the UK butterflies whose climate envelope limits have shifted >100 km northwards with warming climates in recent decades has fully tracked the availability of new habitat; in contrast, 34 out of 46 study species failed to expand at all (Warren et al. 2001; Hill et al. 2009).

Field-based evidence supporting the predicted importance of a viable meta-population structure by stochastic patch occupancy (SPOM) and other models for species' persistence in more typical landscapes has been elusive, but applies to at least nine examples listed by Schtickzelle and Baguette (2009). Most involve Melitaeninae and related fritillaries (e.g. Hanski 1994; Bulman et al. 2007), with examples also from the Lycaenidae and Hesperiidae (CD Thomas et al. 2001; Davies et al. 2005)

The remaining evidence-base for metapopulation viability is too well established (e.g. Hanski 1999) to need repeating. Here we return to the less familiar impact of variation of larval habitat quality within patches in a landscape, in this case in the context of meta-population dynamics.

Patch area, isolation and habitat quality

Theoretical and empirical meta-population studies have typically considered the size, number, distribution or dynamics of a species' habitat patches in a landscape, and assumed that every source patch is of equal quality (Dennis 2010). We have seen, however, that the highest-quality sites consistently support a population density that may be 100 times greater than low quality ones of the same size (Figs. 1 and 2). On the few occasions during meta-population studies when within-patch variation in a species' larval habitat quality was simultaneously measured precisely (as opposed to the simple recording of nectar sources, larval foodplant density or other assumed correlates with population λ or K), it was generally the single variable that explained the highest proportion of patch occupancy, with site isolation typically a close second. We illustrate the example of *P. bellargus* in the UK (Fig. 10): similar patterns are reported for *C. tullia* (Dennis and Eales 1997, 1999), *Thymelicus acteon* and *M. cinxia* (JA Thomas et al. 2001), *Melitaea aurinia* (Betzholtz et al. 2006) and *M. alcon* (Habel et al. 2001). JA Thomas et al. (2001), Dennis (2010) and Turlure et al. (2010) point out that in many meta-population studies, site area is incorrectly used as a surrogate for habitat quality: in field studies where both parameters were independently measured, site area explained few patterns of site occupancy (e.g. Fig. 10).

The high explanatory value of larval habitat quality rather than patch size in metapopulations is attributable to the dramatic potential difference in a species' carrying capacity per unit area of site. For example, in the case of *M. galathea* (the species in Fig. 1 showing the greatest range of population densities), it would on average require a 586 ha patch of low quality source habitat to produce the same-sized population to that consistently supported by 1 ha of the species' optimum habitat. The equivalent patch sizes of poor habitat for the other species depicted are 278 ha (*H. semele*), 139 ha (*M. jurtina*), 124 ha (*P. coridon*), 119 ha (*C. pamphilus*), 73 ha (*Boloria euphrosyne*), 56 ha (*P. argus*), and 39 ha (*A. paphia*). There are three mechanisms why high quality patches are more likely to be occupied and to contribute to a species' persistence across landscapes, compared with low quality ones (JA Thomas et al. 2001):

(1) *Probability of extinction.* In theory and in practice, individual sites in a landscape that support small or low density populations have a higher probability of extinction (Hovestadt and Poethke 2006; Hovestadt et al. 2010). For example on BMS and other monitored UK sites, the 27 patches that initially supported a small (mean density 7.9 adults per km transect, max 17.2) population of *B. euphrosyne* experienced a combined total of 49 apparent extinctions (and 26 recolonisations) during the 10–28 years that they were monitored, with the average population occupying a site for just 5.9 (±SE 0.5) years. In contrast, the 13 sites that initially supported populations with *B. euphrosyne* indexes of >45 (max 61) adults per km transect experienced no

extinctions in the 14–38 (mean 22) years during which they were monitored (our calculation from http://www.ukbms.org and unpublished data).

(2) *Number of emigrants*. All else being equal, a habitat patch that supports a hundred times higher densities of a butterfly species than a same-sized low-quality patch is likely to send two orders of magnitude more emigrants into the landscape, with a correspondingly higher likelihood that one or more individuals will reach a new patch. In reality, individuals are more likely to emigrate from very low density populations (due to rejection of low quality habitat or failure to encounter mates) or from very high density ones (due to density effects); site area and edge effects may also modify emigration (Hovestadt and Poethke 2006; Gros et al. 2009; Hovestadt et al. 2010).

(3) *Probability of establishing a new population*. Due to the close correlation between the intrinsic rate of increase (λ) and the carrying capacity (K) of habitat patches of varying quality for a species (Fig. 6), it is likely that an immigrant reaching a new or vacant habitat patch in a landscape has a higher probability of founding a population in a high quality (high K) habitat than in a low-grade source patch. This is because growth over the first few generations is faster in the former situation: for example, the observed initial growth rate of *M. arion* was 55 times greater on the highest quality patch compared with the lowest quality example of source habitat created on UK nature reserves (Fig. 6). Thus, in optimum habitat, a colonising population is likely to move quickly through the vulnerable period of very low numbers, when stochastic and inverse density-dependent factors make extinction likely. We can find little empirical evidence to support or reject this deduction, but it is suggested in theory by model simulations of immigrants' success (Thomas and Hovestadt in prep), where habitat patch quality was parameterised to encompass the spectrum of values for λ measured on actual *M. arion* sites, and immigrant females were assumed to have mated, but thereafter experience the observed stochastic and inverse density-dependent mortality (e.g. in survival, natality, and unequal sex ratios) observed for butterflies at low densities (Thomas et al. 2009 and unpublished data). It is beyond the scope of this paper to provide full results, but a single typical simulation is presented in Fig. 11. In this example, the intrinsic rate of increase (λ) for low quality habitat (Fig. 11a) slightly exceeds the minimum observed for source habitat of *M. arion* in the field, yet it took 85 years to establish a population above a threshold size where persistence was likely, while after 100 years the population had reached only 20% of the carrying capacity for the site. In contrast, the same flow of immigrants invading mid-quality habitat (Fig. 11b, $\lambda = 2.5$) established a population within 22 years, and fluctuated around the site's carrying capacity in years 30–100, even though the size of this site was, by definition, under half that of Fig. 11a. Other simulations indicate that the probability of establishing a new population increases only marginally across the upper half of the spectrum in patch quality observed in *M. arion* (Fig. 6).

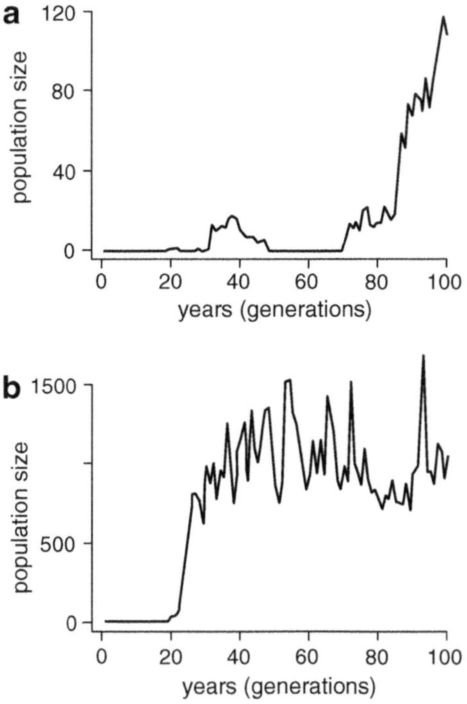

Fig. 11 Model simulations of the probability of founding a population in a patch of new habitat in a landscape. On average once every 10 years, an immigrant (fertilised, if ♀) reaches a patch with a carrying capacity (K) of 1,000 adults. **a** Low quality habitat ($\lambda = 1.1$), **b** medium–high quality habitat ($\lambda = 2.5$). From Thomas and Hovestadt (in prep)

Conclusion

In this analysis, variation in the intrinsic quality of habitat preferred by the larval stage of most of the better-studied species of temperate butterfly emerged as the most important factor determining the size and persistence of populations. Not only does the presence of optimum source habitat result in greatly enhanced intrinsic rates of population increase and carrying capacity within sites, but at the scale of a metapopulation, there is field evidence that populations which occupy high quality patches are more persistent and release more emigrants into the landscape than is the case with those breeding in low quality source habitat; and in theory, new patches of optimum habitat are also more likely to be successfully colonised by immigrants.

Although the density-dependent and other mechanisms responsible for increases in λ and K are understood in very few species, it is sufficient for the practical conservationist to identify the preferred immature habitat of a valued species (which may vary across species' ranges as well as under different regional climates) and, if possible, manipulate sites to encourage the generation or persistence of near-optimum conditions. Although in most species certain sites will be of intrinsically higher potential due to abiotic attributes that are expensive or impossible to change, the optimum habitat of most species studied correlates with a subset of their foodplants that is growing in a narrow seral stage in grassland, heathland, woodland or fenland successions. The latter properties can be manipulated by management, and there are several examples in most biotopes where targeted conservation management has resulted in the increase of a threatened species strongly against the national trend. Our studies of this variable at the metapopulation scale suggest that conservationists would do well to spend their limited resources in improving the habitat of low- but potentially high-quality sites in landscapes, but that there are diminishing returns from striving to convert good sites into great ones. In practice, it appears sufficient to generate habitat in the upper half of the spectrum spanning the poorest to the highest quality habitat for a species or assemblage: across the lower half, there is a high probability of extinctions on occupied sites and a low chance that new sites will be colonised; beyond mid-quality habitat the chances of persistence and establishment are high and do not greatly improve for optimum conditions, although the latter does generate additional emigrants. On balance, on current (incomplete) knowledge, it is more effective to spend limited resources to achieve adequate, upper-range quality habitat on several sites for a species or assemblage in a landscape than to spend the same resources achieving the ultimate on a single site.

Acknowledgments Research was conducted within the project CLIMIT (Settele and Kühn 2009), funded through the FP6 BiodivERsA Eranet by the German Federal Ministry of Education and Research (JAT), the French ANR (TH), and the UK NERC (DJS). We thank RHL Dennis for comments and suggestions.

References

Asher J, Warren MS, Fox R, Jeffcoate G, Jeffcoate S (2001) The millenium atlas of butterflies in Britain and Ireland. Oxford University Press, Oxford

Bergstrom J, Wiklund C (2002) Effects of size and nuptial gifts on butterfly reproduction. Behav Ecol Sociobiol 52:296–302

Betzholtz PE, Ehrig A, Lindeborg M, Dinnétz P (2006) Food plant density, patch isolation and vegetation height determine occurrence in a Swedish metapopulation of the marsh fritillary *Euphydryas aurinia* (Rottemburg, 1775) (Lepidoptera. Nymphalidae). J Insect Conserv 11:343–350

Bickmore CJ, Thomas JA (2000) The development of habitat for butterflies in former arable cultivation. Asp Appl Biol 58: 305–312

Boggs CL, Ross CL (1993) The effect of adult food limitation on life history traits in *Speyeria mormonia*. Ecology 74:433–441

Bourn N, Thomas JA (1992) The ecology and conservation of the Brown Argus butterfly in Britain. Biol Conserv 63:67–74

Brereton TM (1997) Ecology and conservation of the butterfly *Pyrgus malvae* (Grizzled Skipper) in south-east England. Dissertation, University of East London

Brereton TM, Warren MS, Roy DB, Stewart K (2008) The changing status of the Chalkhill Blue butterfly *Polyommatus coridon* in the UK: the impacts of conservation policies and environmental factors. J Insect Conserv 2:629–638

Brown KS, Brown GG (1992) In: Whitmore TC, Sayer JA (eds) Tropical forest deforestation and species extinction. Chapman & Hall, London, pp 119–142

Bulman CR, Wilson RJ, Holt AR, Bravo LG, Early RI, Warren MS, Thomas CD (2007) Minimum viable metapopulation size, extinction debt and the conservation of a declining species. Ecol Appl 7:1460–1473

Clarke RT, Thomas JA, Elmes GW, Hochberg ME (1997) The effects of spatial patterns in habitat quality on community dynamics within a site. Proc R Soc Lond B 264:247–354

Clarke RT, Mouquet N, Thomas JA, Hochberg ME, Elmes GW, Tesar D, Singer A, Hale J (2005) Modelling the local population dynamics of *Maculinea* and their spatial interactions with their larval foodplant and *Myrmica* ant species. In: Settele J, Kűhn E, Thomas JA (eds) Studies in the ecology & conservation of butterflies in Europe II. Pensoft, Sofia, pp 115–119

Davies ZG, Wilson RJ, Brereton TM, Thomas CD (2005) The re-expansion and improving status of the silver-spotted skipper butterfly (*Hesperia comma*) in Britain: a metapopulation success story. Biol Conserv 124:189–198

Dempster JP (1968) The control of *Pieris rapae* with DDT. I. The natural mortality of the young stages of *Pieris*. J Appl Ecol 4:485–500

Dempster JP (1967) The control of *Pieris rapae* with DDT. II. Survival of young stages of *Pieris* after spraying. J Appl Ecol 5:451–462

Dempster JP (1969) Some effects of weed control on the numbers of the small cabbage white (*Pieris rapae* L.) on Brussels sprouts. J Appl Ecol 6:339–345

Dempster JP (1971) Some observations on a population of small copper butterfly *Lycaena phlaeas* (Linnaeus). Ent Gaz 22:199–204

Dempster JP (1983) The natural control of populations of butterflies and moths. Biol Rev 58:461–481

Dempster JP (1991) Fragmentation, isolation and mobility of insect populations. In: Collins NM, Thomas JA (eds) The conservation of insects and their habitats. Academic Press, London, pp 143–154

Dempster JP, King ML, Lakhani KH (1976) The status of the swallowtail butterfly in Britain. Ecol Entomol 1:71–84

Dempster JP, Pollard E (1981) Fluctuations in resource availability and insect populations. Oecologia 50:412–416

Dennis RLH (1985) *Polyommatus icarus* on Brerteon Heath in Cheshire: voltinism and switches in resource exploitation. Ent Gaz 36:175–179

Dennis RLH (1993) Butterflies & climate change. Manchester University Press, Manchester

Dennis RLH (2010) A resource-based habitat view for conservation: butterflies in the British Landscape. Wiley-Blackwell, Chichester

Dennis RLH, Eales HT (1997) Patch occupancy in *Coenonympha tullia* (Lepidoptera: Satyridae): habitat quality matters as much as patch size and isolation. J Insect Conserv 1:167–176

Dennis RLH, Eales HT (1999) Probability of site occupancy in the large heath butterfly *Coenonympha tullia* determined from geographical and ecological data. Biol Conser 87:295–301

Dennis S, Van Dyck H (2003) Towards a resource-based concept for habitat: a butterfly biology viewpoint. Oikos 102:417–426

Dennis S, Van Dyck H (2006) Habitats and resources: the need for a resource-based definition to conserve butterflies. Biodivers Conserv 15:1943–1966

Dover JW (1996) Factors affecting the distribution of satyrid butterflies on arable farmland. J Appl Ecol 33:723–734

Dover JW, Sparks TH, Greatorex-Davies JN (1997) The importance of shelter for butterflies in open landscapes. J Insect Conserv 1:89–97

Duffey E (1968) Ecological studies on the large copper butterfly *Lycaena dispar* Haw. *batavus* Obth. at Woodwalton Fen National nature reserve, Huntingdonshire. J Appl Ecol 5:69–96

Ehrlich PR (1961) Intrinsic barriers to dispersal in checkerspot butterfly. Science 134:108–109

Ehrlich PR, Hanski I (2004) On the wings of checkerspots: A model system for population biology. OUP, New York

Elmes GW, Thomas JA (1992) The complexity of species conservation: interactions between *Maculinea* butterflies and their ant hosts. Biodivers Conserv 1:155–169

Elmes GW, Clarke RT, Thomas JA, Hochberg ME (1996) Empirical tests of specific predictions made from a spatial model of the population dynamics of *Maculinea rebeli*, a parasitic butterfly of red ant colonies. Acta Oecol 17:61–80

Erhardt A (1985) Diurnal Lepidpotera: sensitive indicators of cultivated and abandoned grassland. J Appl Ecol 22:849–861

Fischer K, Fiedler K (2001) Effects of adult feeding and temperature regime on fecundity and longevity in the butterfly *Lycaena hippothoe*. J Lep Soc 54:91–95

Fox R, Asher J, Brereton T, Roy D, Warren MS (2006) The state of butterflies in Britain and Ireland. Pisces, Newbury

Gossard TW, Jones RE (1977) The effects of age and weather on egglaying in *Pieris rapae* L. J Appl Ecol 14:65–71

Gros A, Poethke HJ, Hovestadt T (2009) Sex-specific spatio-temporal variability in reproductive success promotes the evolution of sex-biased dispersal. Theor Popul Biol 76:13–18

Habel JC, Schmitt T, Hardtle W, Lutkepohl M, Assmann T (2001) Dynamics in a butterfly–plant–ant system: influence of habitat characteristics on turnover rates of the endangered lycaenid *Maculinea alcon*. Ecol Entomol 32:536–543

Hanski I (1994) A practical model of metapopulation dynamics. J Anim Ecol 63:151–162

Hanski I (1999) Metapopulation ecology. Oxford University Press, Oxford

Heath J, Pollard E, Thomas JA (1984) Atlas of butterflies in Britain and Ireland. Viking, Harmondsworth

Hill JK, Thomas CD, Lewis OT (1999) Flight morphology in fragmented populations of a rare British butterfly, *Hesperia comma*. Biol Conserv 87:277–283

Hill JK, Ohlemüller R, Fox R, Thomas CD (2009) Climate warming and distribution changes in butterflies In: Ecology of butterflies in Europe. CUP, Cambridge, pp 315–321

Hochberg M, Thomas JA, Elmes GW (1992) The population dynamics of a Large Blue Butterfly, *Maculinea rebeli*, a parasite of red ant nests. J Anim Ecol 61:397–409

Hochberg ME, Clarke RT, Elmes GW, Thomas JA (1994) Population dynamic consequences of direct and indirect interactions involving a large blue butterfly and its plant and red ant hosts. J Anim Ecol 63:375–391

Hovestadt T, Kubisch A, Poethke HJ (2010) Information processing in models for density-dependent emigration: a comparison. Ecol Model 221:405–410

Hovestadt T, Poethke HJ (2006) The control of emigration and its consequences for the survival of populations. Ecol Model 190:443–453

Hughes CL, Hill JK, Dytham C (2003) Evolutionary trade-offs between reproduction and dispersal in populations at expanding range boundaries. Proc R Soc Lond B 270:147–150

Jervis MA, Boggs CL (2005) Linking nectar amino acids to fitness in female butterflies. TREE 20:585–587

Karlsson B, Van Dyck H (2009) Evolutionary ecology of butterfly fecundity. In: Settele J, Shreeve T, Konvicka M, Van Dyck H (eds) Ecology of butterflies in Europe. CUP, Cambridge, pp 189–197

Klapwijk MJ (2008) Host parasitoid interactions in the context of climate change and habitat fragmentation. Dissertation, University of Oxford

Kűhn E, Feldman R, Thomas JA, Settele J (eds) (2005) Studies on the ecology and conservation of butterflies in Europe. I. General concepts and case studies. Pensoft, Sofia

Lewis OT, Hurford C (1997) Assessing the status of the marsh fritillary butterfly (*Eurodryas aurinia*): an example from Glamorgan, UK. J Insect Conserv 1:159–166

Mevi-Schütz J, Erhardt A (2003) Effects of nectar amino acids on fecundity of the wall brown butterfly (*Lasiommata megera* L.). Basic Appl Ecol 4:413–421

Morris MG, Thomas JA, Ward L, Snazell RG, Pywell RF, Stevenson M, Webb NR (1994) Recreation of early successional stages for threatened butterflies—an ecological engineering example. J Environ Manage 42:119–135

Mouquet N, Thomas JA, Elmes GW, Clarke RT, Hochberg ME (2005a) Population dynamics and conservation of a highly specialized predator: a case study of *Maculinea arion*. Ecol Monogr 75:525–542

Mouquet N, Belrose V, Thomas JA, Elmes GW, Clarke RT, Hochberg ME (2005b) Conserving community modules: a case study of the endangered lycaenid butterfly *Maculinea alcon*. Ecology 86:3160–3173

New TR (1997) Butterfly conservation. OUP, Melbourne

New TR, Pyle M, Thomas JA, Thomas CD, Hammond PC (1995) Butterfly conservation management. Ann Rev Entomol 40:57–83

Nowicki P, Bonelli S, Barbero F, Balletto E (2009) Relative importance of density-dependent regulation and environmental stochasticity for butterfly population dynamics. Oecologia 161:227–239

Parmesan C et al (1999) Poleward shifts of species' ranges associated with regional warming. Nature 399:579–583

Pollard E (1979) Population ecology and change in range of the white admiral butterfly *Ladoga camilla* L. in England. Ecol Entomol 4:61–74

Pollard E (1991) Synchrony of population fluctuations: the dominant influence of widespread factors on local butterfly populations. Oikos 60:7–10

Pollard E, Yates TJ (1993) Monitoring butterflies for ecology and conservation. Chapman & Hall, London

Porter K (1981) The population dynamics of small colonies of the butterfly *Euphydryas aurinia*. Dissertation, Oxford Brookes

Pullin AS (1987) Changes in leaf quality following clipping and regrowth on *Urtica dioica*, and consequences for a specialist insect herbivore *Aglais urticae*. Oikos 49:39–45

Pullin AS (1995) Ecology and conservation of butterflies. Chapman & Hall, London

Pyle RM (1976) Conservation of Lepidoptera in the United States. Biol Conserv 9:55–75

Ravenscroft NOM (1994a) The ecology of the chequered skipper *Carterocephalus palaemon* in Scotland. I. Microhabitat. J Appl Ecol 31:613–622

Ravenscroft NOM (1994b) The ecology of the chequered skipper *Carterocephalus palaemon* in Scotland. II. Food plant quality and population range. J Appl Ecol 31:623–630

Ravenscroft NOM (1994c) The feeding behaviour of *Carterocephalus palaemon* (Lepidoptera: Hesperiidae) caterpillars: does it avoid host defences or maximize nutrient intake? Ecol Entomol 19:26–30

Ravenscroft NOM (1995) The conservation of *Carterocephalus palaemon* in Scotland. In: Pullin AS (ed) Ecology and conservation of Butterflies. Chapman & Hall, London, pp 165–179

Revels, R (2006) More on the rise and fall of the Holly blue. British Wildlife, 419–424

Roy DB, Rothery P, Pollard E, Moss D, Thomas JA (2001) Butterfly numbers and weather: the potential for predicting historical trends in abundance and the effects of climate change. J Anim Ecol 70:201–217

Roy DB, Thomas JA (2003) Seasonal variation in the niche, habitat availability and population fluctuations of a bivoltine thermophilous insect near its range margin. Oecologia 134:439–444

Schtickzelle N, Baguette M (2009) Metapopulation viability analysis: a chrystal ball for the conservation of endangered butterflies? In: Settele J, Shreeve T, Konvicka M, Van Dyck H (eds) Ecology of butterflies in Europe. CUP, Cambridge, pp 339–352

Settele J, Kűhn E, Thomas JA (eds) (2005) Studies in the ecology & conservation of butterflies in Europe II. Species ecology along a European gradient: *Maculinea* butterflies as a model. Pensoft, Sofia

Settele J, Kudrna O et al (2008) Climatic risk atlas of European butterflies. Pensoft, Sofia

Settele J, Kühn E (2009) Insect conservation. Science 325:41–42

Settele J, Shreeve T, Konvička M, Van Dyck H (eds) (2009) Ecology of butterflies in Europe. CUP, Cambridge

Shreeve TG (1986) Egglaying by the speckled wood butterfly (*Pararge aegeria*)—the role of female behaviour, host plant abundance and temperature. Ecol Entomol 11:229–236

Snazell RS, Rispin WE, Thomas JA, Elmes GW (2001) M3 Bar End to Compton: invertebrate monitoring: Arethusa Clump chalk grassland restoration. NERC, Huntingdon

Stewart AJA, New TR, Lewis OT (2007) Insect conservation biology. CABI, Wallingford

Thomas CD (1985a) Specializations and polyphagy of *Plebejus argus* (Lepidoptera: Lycaenidae) in North Wales. Ecol Entomol 33:29–51

Thomas CD (1985b) The status and conservation of the butterfly *Plebejus argus* L. (Lepidoptera: Lycaenidae) in North West Britain. Biol Conserv 33:29–51

Thomas CD, Jones CD (1993) Partial recovery of a skipper butterfly (*Hesperia comma*) from population refuges: lessons for conservation in a fragmented landscape. J Anim Ecol 62:472–481

Thomas CD, Thomas JA, Warren MS (1992) Distributions of occupied and vacant butterfly habitats in fragmented landscapes. Oecologia 92:563–567

Thomas CD, Glen SWT, Lewis OT, Hill JK, Blakely DS (1999) Population differentiation and conservation of endemic races: the butterfly. Plebejus argus Anim Conserv 2:15–210

Thomas CD, Badsworth EJ, Wilson RJ, Simons AD, Davies ZG, Musche M, Conradt L (2001) Ecological and evolutionary processes at expanding range margins. Nature 411:577–581

Thomas CD, Franco AMA, Hill JK (2006) Range retractions and extinction in the face of climate warming. TREE 21:415–416

Thomas JA (1974) Ecological studies of hairstreak butterflies. Dissertation, Leicester University

Thomas JA (1975) The black hairstreak butterfly. ITE, Huntingdon

Thomas JA (1983a) The ecology and conservation of *Lysandra bellargus* (Lepidoptera: Lycaenidae) in Britain. J Appl Ecol 20:59–83

Thomas JA (1983b) The ecology and status of *Thymelicus acteon* (Lepidoptera: Hesperidae) in Britain. Ecol Entomol 8:427–435

Thomas JA (1984a) The conservation of butterflies in temperate countries: past efforts and lessons for the future. In: Vane-Wright RI, Ackery P (eds) Biology of butterflies. Academic Press, London, pp 333–353

Thomas JA (1984b) The behaviour and habitat requirements of *Maculinea nausithous* (the dusky large blue butterfly) and *M. teleius* (the scarce large blue) in France. Biol Conserv 28:325–347

Thomas JA (1991) Rare species conservation: case studies of European butterflies. In: Spellerberg I, Goldsmith B, Morris MG (eds) The scientific management of temperate communities for conservation. Blackwells, Oxford, pp 149–197

Thomas JA (1993) Holocene climate change and warm man-made refugia may explain why a sixth of British butterflies inhabit unnatural early-successional habitats. Ecography 16:278–284

Thomas JA (1995) The conservation of declining butterfly populations in Britain and Europe: priorities, problems and successes. Biol J Linn Soc 56:55–72

Thomas JA (2001) Can ecologists recreate habitats and restore absent species? In: Smout TC (ed) Nature, landscape and people since the second world war: a celebration of the 1949 act. Royal Society of Edinburgh, Edinburgh, pp 150–160

Thomas JA (2005) Monitoring change in the abundance and distribution of insects using butterflies and other indicator groups. Phil Trans R Soc Lond B 360:339–357

Thomas JA (2007) Philip's guide to butterflies of Britain and Ireland. Philip's, London

Thomas JA, Elmes GW (2001) Foodplant niche selection rather than the presence of ant nests explains oviposition patterns in the myrmecophilous butterfly genus *Maculinea*. Proc R Soc Lond B 268:471–477

Thomas JA, Hovestadt T (in prep) Effects of habitat quality on the colonisation of unoccupied patches in landscapes

Thomas JA, Lewington R (2010) Butterflies of the British Isles. BWP, Gillingham

Thomas JA, Thomas CD, Simcox DJ, Clarke RT (1986) The ecology and declining status of the silver-spotted skipper butterfly (*Hesperia comma*) in Britain. J Appl Ecol 23:365–380

Thomas JA, Moss D, Pollard E (1994) Increased fluctuations of butterfly populations towards the northern edges of species' ranges. Ecography 17:215–220

Thomas JA, Snazell RG, Moy IL (1996) The conservation of violet-feeding fritillaries in southern Britain. ITE, Huntingdon

Thomas JA, Clarke RT, Elmes GW, Hochberg ME (1998a) Population dynamics in the genus *Maculinea* (Lepidoptera: Lycaenidae). In: Dempster JP, McLean IFG (eds) Insect population dynamics: in theory and practice. Chapman & Hall, London, pp 261–290

Thomas JA, Simcox DJ, Wardlaw JC, Elmes GW, Hochberg ME, Clarke RT (1998b) Effects of latitude, altitude and climate on the habitat and conservation of the endangered butterfly *Maculinea arion* and its *Myrmica* ant hosts. J Insect Conserv 2:39–46

Thomas JA, Rose RJ, Clarke RT, Thomas CD, Webb NR (1999) Intraspecific variation in habitat availability among ectothermic animals near their climatic limits and their centres of range. Funct Ecol 13:55–64

Thomas JA, Bourn NAD, Clarke RT, Whitfield KE, Simcox DJ, Pearman GS, Curtis R, Goodger B (2001) The quality and isolation of habitat patches both determine where butterflies persist in fragmented landscapes. Proc R Soc Lond B 268:1791–1796

Thomas JA, Simcox DJ, Clarke RT (2009) Successful conservation of a threatened *Maculinea* butterfly. Science 325:80–83

Turlure C, Choutte J, Van Dyck H, Baguette M, Schtickzelle N (2010) Functional habitat area as a reliable proxy for population

size: case study using two butterfly species of conservation concern. J Insect Conserv 14:379–388
Van Swaay CAM, Warren MS (1999) Red data book of European butterflies. Council of Europe, Strasbourg
Wahlberg N, Klemetti T, Hanski I (2002) Dynamic populations in a dynamic landscape: the metapopulation structure of the marsh fritillary butterfly. Ecography 25:224–232
Warren MS (1985) The influence of shade on butterfly numbers in woodland rides, with special reference to the wood white butterfly *Leptidea sinapis*. Biol Conserv 33:147–164
Warren MS (1987a) The ecology and conservation of the heath fritillary butterfly, *Mellicta athalia*. II. Adult population structure and mobility. J Appl Ecol 24:483–498
Warren MS (1987b) The ecology and conservation of the heath fritillary butterfly, *Mellicta athalia*. III. Population dynamics and the effect of habitat management. J Appl Ecol 24:499–513
Warren MS (1987c) The ecology and conservation of the heath fritillary butterfly, *Mellicta athalia.* I Host selection and phenology. J Appl Ecol 24:467–482
Warren MS (1991) The successful conservation of an endangered species, the heath fritillary butterfly, *Mellicta athalia*, in Britain. Biol Conserv 55:37–56
Warren MS (1992) The conservation of British butterflies. In: Dennis RLH (ed) The ecology of butterflies in Britain. OUP, Oxford, pp 246–274
Warren MS (1994) The UK status and suspected metapopulation structure of a threatened European butterfly, the marsh fritillary *Eurodryas aurinia*. Biol Conserv 67:329–349
Warren MS (1995) Managing local microclimates for the high brown fritillary, *Argynnis adippe*. In: Pullin AS (ed) Ecology and conservation of butterflies. Chapman & Hall, London, pp 198–210
Warren MS, Pollard E, Bibby TJ (1986) Annual and long-term changes in a population of the wood white butterfly *Leptidea sinapis*. J Anim Ecol 55:707–719
Warren MS, Hill JK, Thomas JA, Asher J, Fox R, Huntley B, Roy DB, Telfer MG, Jeffcoate S, Harding P, Jeffcoate G, Willis SG, Greatorex-Davies JN, Moss D, Thomas CD (2001) Rapid responses of British butterflies to opposing forces of climate and habitat change. Nature 414:65–69
Wilson RJ, Davies ZG, Thomas CD (2009) Modelling the effect of habitat fragmentation on range expansion of a butterfly. Proc R Soc B 276:1421–1427
Wynhoff I (1998) Lessons from the reintroduction of *Maculinea teleius* and *M. nausithous* in the Netherlands. J Insect Conserv 2:47–57
Wynhoff I (2001) At home on foreign meadows: the reintroduction of two Maculinea butterfly species. Dissertation, Wageningen Agricultural University, Wageningen

J Insect Conserv (2011) 15:233–240
DOI 10.1007/s10841-010-9340-0

ORIGINAL PAPER

Butterfly abundance in a warming climate: patterns in space and time are not congruent

Nick J. B. Isaac · Marco Girardello · Tom M. Brereton · David B. Roy

Received: 20 April 2010 / Accepted: 3 September 2010 / Published online: 1 October 2010

Abstract We present a model of butterfly abundance on transects in England. The model indicates a significant role for climate, but the direction of association is counter to expectation: butterfly population density is higher on sites with a cooler climate. However, the effect is highly heterogeneous, with one in five species displaying a net positive association. We use this model to project the population-level effects of climate warming for the year 2080, using a medium emissions scenario. The results suggest that most populations and species will decline markedly, but that the total number of butterflies will increase as communities become dominated by a few common species. In particular, *Maniola jurtina* is predicted to make up nearly half of all butterflies on UK Butterfly Monitoring Scheme (UKBMS) transects by 2080. These results contradict the accepted wisdom that most insect populations will grow as the climate becomes warmer. Indeed, our predictions contrast strongly with those derived from inter-annual variation in abundance, emphasizing that we lack a mechanistic understanding about the factors driving butterfly population dynamics over large spatial and temporal scales. Our study underscores the difficulty of predicting future population trends and reveals the naivety of simple space-for-time substitutions, which our projections share with species distribution modelling.

N. J. B. Isaac (✉) · M. Girardello · D. B. Roy
Centre for Ecology and Hydrology, Maclean Building, Benson Lane, Crowmarsh Gifford, Wallingford, Oxfordshire OX10 8BB, UK
e-mail: njbisaac@gmail.com

T. M. Brereton
Butterfly Conservation, Manor Yard, East Lulworth, Wareham, Dorset BH20 5QP, UK

Keywords Biotic homogenisation · Butterflies · Climate change · Climate envelope · Mixed models · Niche · Space-for-time substitution · UK butterfly monitoring scheme

Introduction

Climate change is now well-established as a major threat to biodiversity (Thomas et al. 2004) and species distributions are changing in a manner consistent with climate change (Hickling et al. 2006; Hitch and Leberg 2007). The most common way of predicting the effects of climate change is using species distribution modelling (also known as climate envelope or niche modelling, Guisan and Thuiller 2005; Araujo and Guisan 2006).

Distribution modelling works by mapping species patterns of spatial occurrence from a geographic to environmental space (the niche). The correlation between environmental gradients (especially temperature) is then used to predict the species distribution under some projected future combination of climate. The core assumption is that similar mechanisms operate across space and across time (the space-for-time substitution). Niche (bioclimate) models are increasingly the primary tool for identifying the risks of climate change and informing future conservation policy for biodiversity (Thomas et al. 2004), and specifically butterflies (Settele et al. 2008), in spite of criticism that they generally ignore population biology or biotic interactions (Araujo and Luoto 2007).

Projections based on population time series are much less common (Roy et al. 2001), reflecting the limited taxonomic coverage of high-quality monitoring data. Time series projections make predictions about changes in population abundance (as opposed to species distributions), and therefore provide a more ecologically meaningful measure of biodiversity change. However, parameterization of such models is intensive and may be impractical for large numbers of species.

In this paper, we use the spatial pattern of population abundance to make predictions about the population-level effects of climate change. We employ a space-for-time substitution to project changes in abundance among nearly 10,000 populations from sites in the UK Butterfly Monitoring Scheme (UKBMS). We compare our results with the time-based population model of Roy et al. (2001). The results allow us to test the robustness of the space-for-time substitution, and provide novel insights into the practice of using niche models to predict the effects of climate change.

Methods

Butterfly data

We measured butterfly population abundance as the geometric mean 'site index' for the years 2002–2006, excluding sites that were surveyed only once during the 5 year period and also site-species combinations in which butterflies were never observed in more than 1 week of the year. We divided these abundance indices by the transect length, in order to derive an estimate of population density. The resulting dataset contains 9,991 populations (site-by-species combinations) representing 46 species and 520 sites.

Model of butterfly abundance

Isaac et al. (in prep) infer a role for climate in butterfly abundance using multi-model inference (Burnham and Anderson 2002) and linear mixed effects models (Bolker et al. 2009). Their models describe the correlates of butterfly population density for these 9,991 populations, and include site traits (climate, topography, site management), species traits (measures of life history and niche breadth) and population traits (position in range) as predictor variables. Isaac et al. measured climatic conditions on UKBMS sites using two axes from a principal components analysis of 10 climate variables. The first principal component, which was not important in explaining butterfly abundance, was highly correlated with several rainfall measures. The second principal component of climatic variation was implicated as important in determining butterfly abundance: this axis was highly correlated with measures of site temperature (mean annual temperature, mean temperature of the coldest month, mean temperature of the warmest month, growing degree days).

We first fitted a model containing the five most important predictor variables in the candidate set of models considered by Isaac et al. (in prep). In addition to climate, these were butterfly species richness (of the site), larval growth rate and distributional extent (both species traits) and position in range (a population trait). We then refitted this model with site mean annual temperature instead of the principal component, which lead to a substantial improvement in model fit (∂AIC = 9). Mean annual temperature estimates were derived from gridded data on long term monthly temperature records (New 1999; Mitchell and Jones 2005) interpolated to 10 km resolution (Oliver et al. 2009). Removing mean annual temperature from the model lead to a substantially worse fit (∂AIC = 12.9, $\chi^2 = 14.9$, $df = 1$, $P = 0.0001$), thus confirming the correlation between temperature and abundance.

We further adapted this model to test whether the response to temperature is heterogeneous among species by fitting a model in which temperature appears in the random part of the model (a random slopes model). In this model, the abundance-temperature relationship is different for each species and is assumed to follow a normal distribution: the estimated parameters are the mean and variance of this distribution. Standard model comparison was used to test whether the random slopes model is a significant improvement over the simple model, in which the temperature response is fixed across all species. Following Enders and Tofighi (2007), we used temperature data centred on zero for each species, in order to avoid spurious correlations. We extracted the species-specific slopes (i.e. the abundance- temperature relationships) from this model (following Isaac and Carbone 2010) and used them for further analyses.

We conducted all analyses using the *lme4* package (Bates et al. 2008) in *R* (R Development Core Team 2008).

Future climate

We used UK Climate Projections (Defra 2010) for mean annual temperature (median estimate) for the year 2080 based on a moderate emissions scenario SRES A1 B (IPCC 2007), which envisions a balance between fossil fuels and renewable energy sources. The A1 storyline and scenario family describe a world of rapid economic growth, with a peak in global population around 2050 and the rapid introduction of efficient new technologies. Major underlying themes are convergence among regions, capacity building and increased cultural and social interactions, with a substantial reduction in regional differences in per capita income. This scenario predicts that mean annual temperatures in 2080 will be up to 3.6°C warmer than present in southern England, and around 2.8°C warmer in northern England.

The data for climate projections is gridded with a resolution of 25 km. We matched 477 UKBMS sites to the relevant grid cell in the climate projections dataset. We excluded 43 coastal sites that fell outside the grid extent for climate projections, including all five UKBMS sites of *Thymelicus acteon*. Our projected abundance dataset therefore covers 9,075 populations representing 45 species and 477 sites.

Predicted abundance under climate change

We derived predicted abundance for each of our 9,075 populations by using the species-specific temperature response, the site-specific temperature change and the current abundance (geometric mean for the years 2002–2006). For each species, we summed both current and predicted future abundance across all sites; for each site, we summed across species. These values estimate the total number of butterflies for each species and each site, for both 2002–2006 and 2080. We then expressed the difference between time periods as a percentage change in abundance.

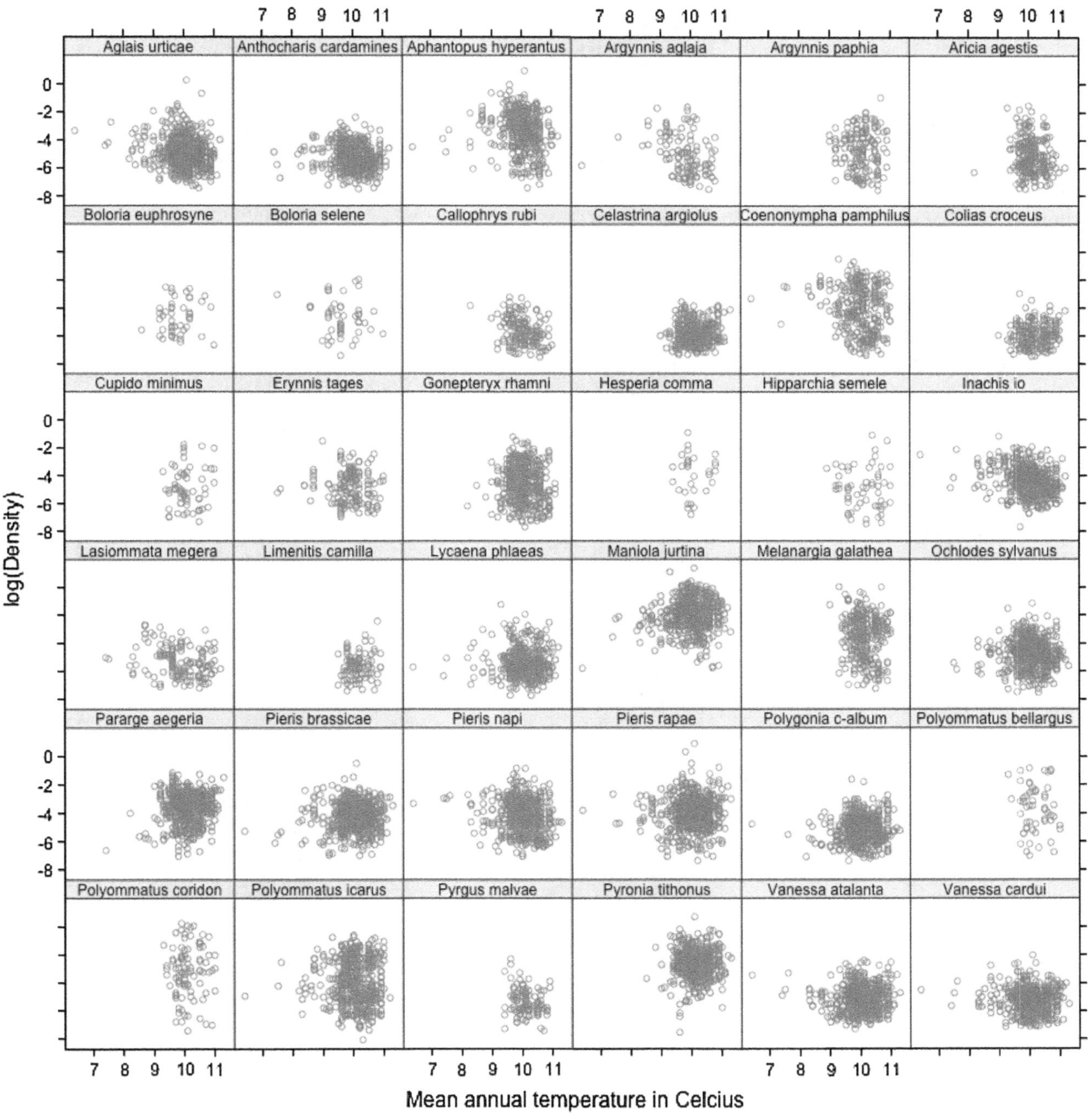

Fig. 1 Scatter plot of butterfly population density ($\log_e$ transformed) against mean annual temperature for 35 UK butterfly species. Population density is measured as the geometric mean 'site index' using 2002–2006, divided by transect length. Species found on fewer than 30 sites are not displayed but were included in the model

Results

The overall relationship between mean annual temperature and butterfly density is weak but generally negative (Fig. 1). Before accounting for variation among species, the parameter estimate is −0.178, which means that a rise of 1°C in mean annual temperature is accompanied by a drop in abundance of $(1 - e^{-0.178} =)$ 16.3%.

After accounting for variation among species (the random slopes model), the mean effect of temperature is estimated to be slightly steeper, at −0.213, corresponding to a drop in abundance of 19.2% for each degree rise in mean annual temperature. However, standard deviation in species' temperature responses (0.265) is large compared with the mean, indicating that a substantial proportion of species (20%) are more abundant at warmer sites (Table 1). This random slopes model with species-specific temperature responses represents a substantial improvement in fit over the simpler model containing a single slope (∂AIC = 102, $\chi^2 = 106$, $df = 2$, $P < 0.0001$), supporting the hypothesis that species respond differently to temperature.

Neither the mean temperature response nor the variance among species is substantially altered by removing the other non-species traits (richness and position in range) from the model. We also repeated our model comparison after excluding the five extreme sites with mean annual temperature below 8°C and found the results were qualitatively unchanged.

Combining the species-specific temperature responses with site-specific predictions for climate warming by 2080 yielded projected changes in abundance for 9,075 butterfly populations. These ranged from −93% to +71%, with a mean of −26% and median of −36%. However, the predicted change in total abundance (i.e. weighting each population by present abundance) is +21%. In other words, although most populations are predicted to decline, the total number of butterflies on UKBMS sites in England is predicted to rise.

Projected change in species abundance under climate change varies from −92% (*Argynnis aglaja*) to +68% (*Maniola jurtina*). However, species are distributed bimodally through this range (Fig. 2), with a median projected change of −50%. Sites cover a similar range, but a small majority are predicted to gain butterflies (Fig. 2): the median change is +0.5%. There is no obvious pattern to the spatial distribution of predicted change among sites: many sites predicted to suffer heavy losses are in close proximity to others that we predict will gain butterflies (Fig. 3). The variation in space reflects both turnover in community composition and local variation in the (current) abundance of individual species. The apparent contradiction between site and species projections is explained by

Table 1 Species responses to climate warming. Species are sorted in decreasing order of temperature response

Species	$Beta_T$	Current	Future	Change	%change
Maniola jurtina	0.149	111.76	187.72	75.96	68.0
Celastrina argiolus	0.114	1.27	1.89	0.62	49.1
Cupido minimus	0.112	1.07	1.59	0.52	48.6
Pyronia tithonus	0.079	45.78	60.44	14.66	32.0
Pararge aegeria	0.076	18.62	24.20	5.58	30.0
Pieris rapae	0.074	19.07	24.69	5.62	29.5
Polygonia c-album	0.067	3.92	4.94	1.02	26.1
Pieris brassicae	0.042	10.29	11.92	1.63	15.8
Vanessa atalanta	0.007	4.01	4.11	0.10	2.5
Euphydryas aurinia	−0.068	0.86	0.68	−0.18	−21.1
Lycaena phlaeas	−0.073	2.98	2.31	−0.67	−22.4
Colias croceus	−0.081	0.60	0.45	−0.15	−24.7
Polyommatus coridon	−0.097	11.43	8.11	−3.32	−29.0
Hesperia comma	−0.098	1.16	0.82	−0.34	−29.2
Polyommatus icarus	−0.111	20.54	13.94	−6.60	−32.1
Coenonympha tullia	−0.132	0.08	0.05	−0.03	−35.2
Ochlodes sylvanus	−0.127	6.34	4.08	−2.27	−35.7
Argynnis adippe	−0.157	0.67	0.41	−0.26	−39.0
Limenitis camilla	−0.157	0.52	0.30	−0.22	−42.3
Melitaea athalia	−0.167	0.30	0.17	−0.13	−43.2
Vanessa cardui	−0.188	2.96	1.56	−1.41	−47.5
Leptidea sinapis	−0.194	0.42	0.21	−0.21	−49.3
Plebejus argus	−0.196	0.63	0.31	−0.32	−50.2
Hamearis lucina	−0.207	0.27	0.13	−0.14	−51.5
Gonepteryx rhamni	−0.219	8.64	4.02	−4.62	−53.5
Aricia artaxerxes	−0.248	0.47	0.22	−0.25	−54.1
Anthocharis cardamines	−0.244	3.01	1.30	−1.71	−56.8
Argynnis paphia	−0.250	3.39	1.41	−1.98	−58.4
Hipparchia semele	−0.262	1.15	0.47	−0.69	−59.4
Polyommatus bellargus	−0.257	2.82	1.14	−1.68	−59.7
Pieris napi	−0.303	12.80	4.52	−8.28	−64.7
Boloria euphrosyne	−0.342	0.78	0.25	−0.53	−68.2
Erynnis tages	−0.335	2.09	0.66	−1.43	−68.3
Aglais urticae	−0.336	8.40	2.64	−5.76	−68.6
Melanargia galathea	−0.355	18.19	5.22	−12.96	−71.3
Aphantopus hyperantus	−0.383	32.13	8.52	−23.61	−73.5
Aricia agestis	−0.397	2.88	0.71	−2.17	−75.3
Boloria selene	−0.441	0.93	0.22	−0.71	−76.5
Pyrgus malvae	−0.444	0.82	0.17	−0.65	−78.8
Erebia aethiops	−0.496	0.52	0.11	−0.41	−79.0
Inachis io	−0.473	9.10	1.83	−7.26	−79.8
Coenonympha pamphilus	−0.495	13.15	2.36	−10.78	−82.0
Lasiommata megera	−0.596	1.01	0.15	−0.86	−85.4
Callophrys rubi	−0.566	0.82	0.12	−0.70	−85.8
Argynnis aglaja	−0.756	1.76	0.14	−1.62	−92.1

$Beta_T$ is the species-specific temperature response, measured as the change in log(density) for each degree of temperature increase. Current abundance is the sum of population density (per linear meter of transect) across 477 English UKBMS sites for 2002–2006; Future is the comparable value under climate warming in 2080. Change and %change are the simple differences between current and future abundance estimates, in absolute and percentage terms, respectively

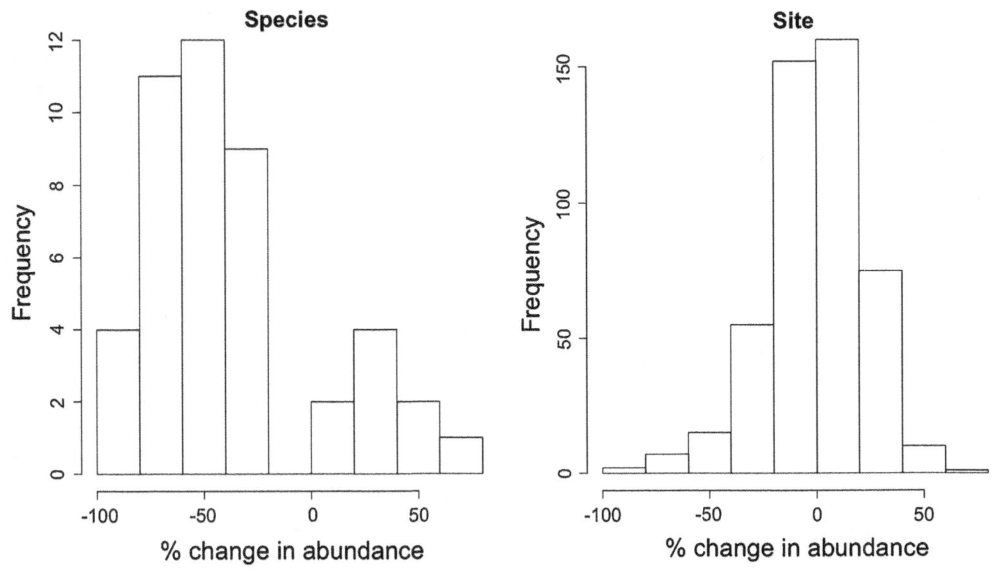

Fig. 2 Histogram of projected change in abundance due to climate change, expressed as a percentage. These data were generated from model predictions for 9,075 populations, summed across species ($n = 45$) and sites ($n = 477$)

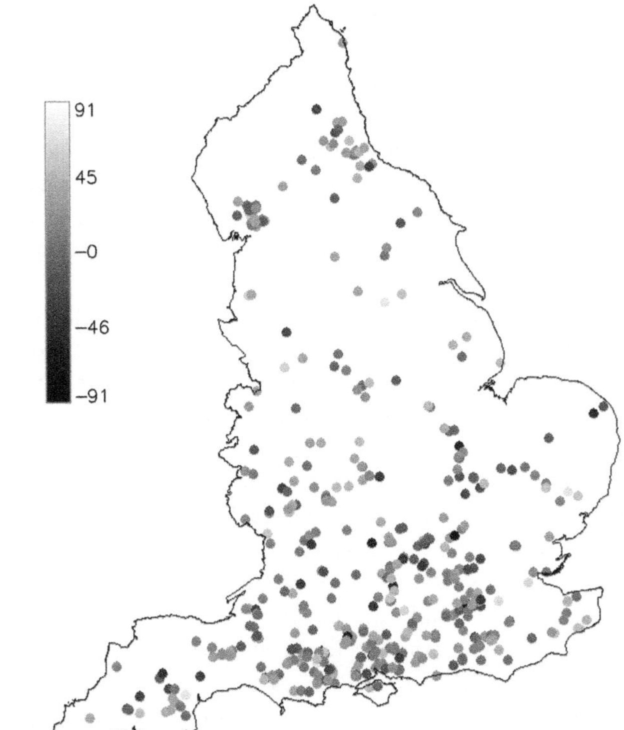

Fig. 3 Map of 477 UKBMS sites showing projected change in abundance due to warming (expressed as a percentage), summed across species ($n = 45$)

the response of just two species, Gatekeeper *Pyronia tithonus* and Meadow Brown *Maniola jurtina*. These are the two commonest species on UKBMS transects in England, and both are projected to increase in abundance by over 50% (Table 1). Our model suggests that *Maniola jurtina*, which currently makes up 29% of butterflies on UKBMS sites in England, will make up 48% of the total by 2080. Comparable figures for *Pyronia tithonus* are 12 and 15%. The 50% rarest species currently contribute 1 in 21 observations on the UKBMS: by 2080 this rate is projected to fall to 1 in 41 (data from Table 1).

Discussion

Our results suggest substantial future declines in the abundance of most UK butterfly species due to climate change. Our models predict that climate warming will lead to the increased domination of butterfly communities by a few generalist species, providing another example of biotic homogenisation (Keith et al. 2009; Pino et al. 2009; Ekroos et al. 2010). Taken at face value, these findings present a worrying picture for butterfly conservation and imply that conservation priorities should be reassessed urgently.

There are several reasons to suspect the predictions of our models are at best incomplete, or at worst misleading. First and foremost, our model describes patterns of population density in narrow strips on occupied sites, completely ignoring patterns of site occupancy and changes therein. Measuring density over larger spatial scales, incorporating unoccupied sites, might lead to a different set of conclusions (see Cowley et al. 2001a). Changes in occupancy over time are equally important: most UK butterflies are continuing to decline in their distributions (Warren et al. 2001; Fox et al. 2006), and future population extinctions would be in addition to the losses predicted here. However, 29% of species are increasing (Fox et al. 2006), and for these species any losses in abundance predicted here might be offset by colonization of previously unoccupied sites. A second important caveat is that our predictions are based solely on changes in temperature, ignoring other variables in the model that might conceivably change, notably position in range and species richness. Both of these traits would be affected by changes in species' distributions and both had much larger effects on butterfly abundance than did temperature. Moreover, seasonal shifts and changes in site management are probably more important than the crude index of mean annual temperature used here. We were also unable to model changes in microclimatic availability, which may show quite different patterns to mean annual temperature (Wallisdevries and Van Swaay 2006). Finally, we must question the correlation from which our predictions were generated. The negative relationship between abundance and temperature is highly counter-intuitive: climate warming is generally assumed to benefit ectotherms such as butterflies (Warren et al. 2001). UK butterflies are reported to be expanding their range margins northwards in response to climate change (Hickling et al. 2006) and most species are more abundant in warmer years (Roy et al. 2001). Why then, for most species, are there more butterflies in sites where it's colder?

Three broad classes of explanation exist for the generally negative relationship between abundance and temperature. One is that basal metabolic rates of ectothermic animals increase exponentially with temperature (Gillooly et al. 2001), so that more energy is required for each individual to remain alive. This means that energy supply (i.e. food availability) must rise with increasing temperature in order to sustain the same number of individuals (Brown et al. 2004). All other things being equal, metabolic theory therefore predicts the negative correlation between population density and temperature that we have observed (Storch 2003). For the range of mean annual temperatures across England, each degree rise in temperature raises individual metabolic rates by 9% (Gillooly et al. 2001), thus accounting for much of the 16–19% drop in abundance suggested by our models. However, the rise in temperature (in space or in time) is likely accompanied by longer growing seasons and greater primary productivity, thus offsetting the increase in individual metabolic requirements.

A second explanation is that abundance is mediated by natural enemies. It seems likely that most insects experience high rates of parasitism and disease (Roy et al. 2009), although few specific case studies for butterflies have been described (Revels 2006). The abundance and species richness of natural enemies probably declines with latitude (Rosenzweig 1995), such that butterflies on colder sites in northern England experience much lower exposure to natural enemies than those in the south (Menendez et al. 2008; Gaston 2009). If abundance in the absence of natural enemies (i.e. carrying capacity) is independent of temperature then a negative relationship would be observed with mean abundance, on account of including sites and years with high mortality. However, this would imply smaller inter-annual fluctuations in population size at colder sites, which is the opposite to that reported for the majority of species on the UKBMS (Thomas et al. 1994). However, the role of natural enemies on mean butterfly population size would be testable with comparative data on parasite load at multiple sites.

Finally, it is possible that the apparent correlation between abundance and temperature is an artefact of changes in occupancy and niche breadth with latitude. Most UK butterflies approach their northern range boundaries in the UK, and occupancy (measured as proportion of grid cells) is typically lower near the margins than in the centre of the species range (Gaston 2003, Fox et al. 2006). We also know that niche breadth in UK butterflies declines towards species' range margins (Oliver et al. 2009). These two observations suggest that butterfly populations near range margins are restricted to favoured biotopes: sites that would be suitable in warmer climates are unoccupied (see also Cowley et al. 2001b; Paivinen et al. 2005). If notional carrying capacity is biotope-specific but unrelated to climate then the relationship between mean density and temperature would be triangular, with both high- and low-density populations in the warmer south but only high-density populations in the colder north. Such heterogeneity is evident in the residual distribution for four species *Aglais urticae, Coenonympha pamphilus, Vanessa atalanta* and *V. cardui*, but does not appear to be a general phenomenon. Further investigation into the relationship between occupancy, temperature, niche breadth and density, at a range of spatial scales, is therefore to be encouraged.

Our predictions contrast strongly with those of Roy et al. (2001), who found that most butterfly species increased in abundance (measured as the national index) during warm years. These correlations had low power to predict subsequent population trends for most species, but for eight species the correlations were sufficiently strong to permit projections for abundance trends under climate warming up to the year 2080. Six species were predicted to increase, one decrease (*Pieris brassicae*) and one to remain approximately stable (*Lasiommata megera*). Of these species, we predict two species to increase (*Pieris brassicae* and *Pyronia tithonus*) and six to decline: i.e. seven out of eight predictions are in the opposite direction! The results of Roy et al. (2001) are quite intuitive: butterflies are more abundant in warm years because of lower winter mortality, rapid development and higher plant biomass. Moreover, Roy and Sparks (2000) reported rapid changes in butterfly life cycles in response to recent climate warming (e.g. advanced eclosion of adult butterflies, extended flight seasons, increased propensity for multivoltinism), all suggesting generally positive population responses to warming. The time-based population model of Roy et al. (2001) was based on fine-scale climatic data (e.g. specific months related to the butterfly life cycle), that would be difficult to generalise to large numbers of species. The approach described here, whilst lacking certain subtleties of butterfly ecology, has the advantage that it could be applied to any situation where there is spatial data on population density.

Although the explanation for our negative temperature-abundance relationship remains elusive, our results demonstrate clearly that different processes operate across space and over time. Put bluntly, we lack a mechanistic understanding about the drivers of butterfly population dynamics over large spatial and temporal scales. Without such knowledge, it is speculative to extrapolate from current patterns into the future. This message is profoundly important for interpreting the climate projections of niche models (Guisan and Thuiller 2005; Araujo and Guisan 2006), which share with our study the simple assumption that patterns in space are static, and can be used as a substitute for patterns in time. In this case we are fortunate that time-series information is available to verify our projections. Unfortunately, such reality checks are unavailable for the vast majority of modelling exercises.

Niche (bioclimate) models assume phenotypic plasticity of populations and constant responses across a species' entire range (Berry et al. 2002). However, evidence of widespread adaptation of butterfly populations to local climates is accumulating (Van Dyck and Wilding 2009) and gradients in butterfly biology along climatic gradients are well known (Nylin 2009). Many butterflies occurring towards their cooler range limit shift to inhabit warmer, narrower niches, whereas with core populations sample the landscape more widely (Thomas 1993; Thomas et al. 1999; Cowley et al. 2001b; Roy and Thomas 2003; Oliver et al. 2009). Other studies report gradients in size or melanism in butterflies across latitudes (Dennis and Shreeve 1989; Nylin and Svard 1991; Ayres and Scriber 1994). A fuller assessment of the extent of local adaptations within populations is therefore essential for more accurate predictions of the impacts of climate warming on biodiversity and the ecosystem services they support (Visser 2008), particularly for species with low dispersal ability.

Acknowledgments We are indebted to the hundreds of volunteers who collect data on the UKBMS. We are grateful to Stephen Freeman, Tom Oliver, Helen Roy and Jeremy Thomas for advice and discussion, and to two anonymous reviewers who provided insightful comments on previous versions of this manuscript. NJBI is supported by a NERC fellowship (NE/D009448/2). DR was partly funded by the Biodiversa project CLIMIT (Settele and Kuhn 2009; Thomas et al. 2009) within FP6 of the European Commission (EC). The UKBMS is funded by a multi-agency consortium led by Defra, and including CCW, JNCC, FC, NE, NERC, NIEA and SNH.

References

Araujo MB, Guisan A (2006) Five (or so) challenges for species distribution modelling. J Biogeogr 33:1677–1688

Araujo MB, Luoto M (2007) The importance of biotic interactions for modelling species distributions under climate change. Glob Ecol Biogeog 16:743–753

Ayres MP, Scriber JM (1994) Local adaptation to regional climates in Papilio canadensis (Lepidoptera, Papilionidae). Ecol Monogr 64:465–482

Bates DM, Maechler M, Dai B (2008) lme4: linear mixed-effects models using S4 classes

Berry PM, Dawson TP, Harrison PA, Pearson RG (2002) Modelling potential impacts of climate change on the bioclimatic envelope of species in Britain and Ireland. Glob Ecol Biogeog 11:453–462

Bolker BM, Brooks ME, Clark CJ, Geange SW, Poulsen JR et al (2009) Generalized linear mixed models: a practical guide for ecology and evolution. Trends Ecol Evol 24:127–135

Brown JH, Gillooly JF, Allen AP, Savage VM, West GB (2004) Toward a metabolic theory of ecology. Ecology 85:1771–1789

Burnham KP, Anderson DR (2002) Model selection and multi-model inference: a practical Information-theoretic approach. Springer, Berlin, p 496

Cowley MJR, Thomas CD, Roy DB, Wilson RJ, Leon-Cortes JL et al (2001a) Density-distribution relationships in British butterflies. I. The effect of mobility and spatial scale. J Anim Ecol 70:410–425

Cowley MJR, Thomas CD, Wilson RJ, Leon-Cortes JL, Gutierrez D et al (2001b) Density-distribution relationships in British butterflies. II. An assessment of mechanisms. J Anim Ecol 70: 426–441

Defra (2010) UK climate projections user interface. http://ukclimateprojections.defra.gov.uk

Dennis RLH, Shreeve TG (1989) Butterfly wing morphology variation in the British Isles—the influence of climate, behaviorial posture and the host plant habitat. Biol J Linn Soc 38:323–348

Ekroos J, Heliola J, Kuussaari M (2010) Homogenization of lepidopteran communities in intensively cultivated agricultural landscapes. J Appl Ecol 47:459–467

Enders CK, Tofighi D (2007) Centering predictor variables in cross-sectional multilevel models: a new look at an old issue. Psychol Meth 12:121–138

Fox R, Asher J, Brereton TM, Roy D, Warren M (2006) The state of butterflies in Britain and Ireland. Pisces Publications, Oxford

Gaston KJ (2003) The structure and dynamics of geographic ranges. In: Harvey PH, May RM (eds). Oxford: Oxford University Press

Gaston KJ (2009) Geographic range limits of species. Proc R Soc Lond B 276:1391–1393

Gillooly JF, Brown JH, West GB, Savage VM, Charnov EL (2001) Effects of size and temperature on metabolic rate. Science 293:2248–2251

Guisan A, Thuiller W (2005) Predicting species distribution: offering more than simple habitat models. Ecol Lett 8:993–1009

Hickling R, Roy DB, Hill JK, Fox R, Thomas CD (2006) The distributions of a wide range of taxonomic groups are expanding polewards. Glob Change Biol 12:450–455

Hitch AT, Leberg PL (2007) Breeding distributions of North American bird species moving north as a result of climate change. Conserv Biol 21:534–539

IPCC (2007) Climate change 2007: IPCC fourth assessment report. Cambridge University Press, Cambridge

Isaac NJB, Carbone C (2010) Why are metabolic scaling exponents so controversial? Quantifying variance and testing hypotheses. Ecol Lett 13:728–735

Keith SA, Newton AC, Morecroft MD, Bealey CE, Bullock JM (2009) Taxonomic homogenization of woodland plant communities over 70 years. Proc R Soc Lond B 276:3539–3544

Menendez R, Gonzalez-Megias A, Lewis OT, Shaw MR, Thomas CD (2008) Escape from natural enemies during climate-driven range expansion: a case study. Ecological Entomology 33:413–421

Mitchell TD, Jones PD (2005) An improved method of constructing a database of monthly climate observations and associated high-resolution grids. Int J Climatol 25:693–712

New M, Hulme M, Jones PD (1999) Representing twentieth century space-time climate variability. Part 1: development of a 1961–1990 mean monthly terrestrial climatology. J Climatol 12:829–856

Nylin S (2009) Gradients in butterfly biology. In: Settele J, Shreeve T, Konvicka M, Dyck Hv (eds) Ecology of butterflies in Europe. Cambridge University Press, Cambridge, pp 198–216

Nylin S, Svard L (1991) Latitudinal patterns in the size of European butterflies. Holarctic Ecol 14:192–202

Oliver T, Hill JK, Thomas CD, Brereton T, Roy DB (2009) Changes in habitat specificity of species at their climatic range boundaries. Ecol Lett 12:1091–1102

Paivinen J, Grapputo A, Kaitala V, Komonen A, Kotiaho JS et al (2005) Negative density-distribution relationship in butterflies. BMC Biology 3

Pino J, Font X, de Caceres M, Molowny-Horas R (2009) Floristic homogenization by native ruderal and alien plants in north-east Spain: the effect of environmental differences on a regional scale. Glob Ecol Biogeog 18:563–574

R Development Core Team (2008) R: a language and environment for statistical computing. R Foundation for Statistical Computing, Vienna

Revels R (2006) More on the rise and fall of the holly blue. British Wildlife 17:419–424

Rosenzweig ML (1995) Species diversity in space and time. Cambridge University Press, Cambridge

Roy DB, Sparks TH (2000) Phenology of British butterflies and climate change. Glob Change Biol 6:407–416

Roy DB, Thomas JA (2003) Seasonal variation in the niche, habitat availability and population fluctuations of a bivoltine thermophilous insect near its range margin. Oecologia 134:439–444

Roy DB, Rothery P, Moss D, Pollard E, Thomas JA (2001) Butterfly numbers and weather: predicting historical trends in abundance and the future effects of climate change. J Anim Ecol 70: 201–217

Roy HE, Hails RS, Hesketh H, Roy DB, Pell JK (2009) Beyond biological control: non-pest insects and their pathogens in a changing world. Insect Conservation and Diversity 2:65–72

Settele J, Kuhn E (2009) Insect Conservation. Science 325:41–42

Settele J, Kudrna O, Harpke A, Kühn I, van Swaay C et al (2008) Climatic risk atlas of European butterflies. 1–710 p

Storch D (2003) Comment on "Global biodiversity, biochemical kinetics, and the energetic-equivalence rule". Science 299:346b

Thomas JA (1993) Holocene climate changes and warm man-made refugia may explain why a 6th of British butterflies possess unnatural early-successional habitats. Ecography 16:278–284

Thomas JA, Moss D, Pollard E (1994) Increased fluctuations of butterfly populations towards the northern edges of species ranges. Ecography 17:215–220

Thomas JA, Rose RJ, Clarke RT, Thomas CD, Webb NR (1999) Intraspecific variation in habitat availability among ectothermic animals near their climatic limits and their centres of range. Funct Ecol 13:55–64

Thomas CD, Cameron A, Green RE, Bakkenes M, Beaumont LJ et al (2004) Extinction risk from climate change. Nature 427:145–148

Thomas JA, Simcox DJ, Clarke RT (2009) Successful conservation of a threatened Maculinea butterfly. Science 325:80–83

Van Dyck H, Wilding JJ (2009) Adaptation and plasticity in butterflies: the interplay of genes and environment. In: Settele J, Shreeve T, Konvicka M, Dyck Hv (eds) Ecology of butterflies in Europe. Cambridge University Press, Cambridge, pp 159–170

Visser ME (2008) Keeping up with a warming world; assessing the rate of adaptation to climate change. Proc R Soc Lond B 275:649–659

Wallisdevries MF, Van Swaay CAM (2006) Global warming and excess nitrogen may induce butterfly decline by microclimatic cooling. Glob Change Biol 12:1620–1626

Warren MS, Hill JK, Thomas JA, Asher J, Fox R et al (2001) Rapid responses of British butterflies to opposing forces of climate and habitat change. Nature 414:65–69

J Insect Conserv (2011) 15:327–339
DOI 10.1007/s10841-010-9323-1

SHORT COMMUNICATION

Declines of prairie butterflies in the midwestern USA

Scott R. Swengel · Dennis Schlicht · Frank Olsen · Ann B. Swengel

Received: 4 March 2010 / Accepted: 3 August 2010 / Published online: 13 August 2010

Abstract Tallgrass prairie butterfly surveys in recent decades in four states in the USA indicate numerous declines of prairie-specialist butterflies including *Speyeria idalia*, *Oarisma poweshiek*, *Atrytone arogos*, *Hesperia dacotae*, and *H. ottoe* in fire-managed preserves, including large high-quality ones. These results replicate previous findings, indicating that upon initiation of conservation action, both cessation of prior management and inception of new management affect specialists negatively and that butterfly declines can be as great on reserves as non-reserves. Results at Wisconsin sites with species-specific management protocols, including permanent non-fire refugia, were more favorable for the specialists (*S. idalia*, *Lycaeides melissa samuelis*) the protocols were specifically designed to benefit. Butterfly declines after preservation will likely continue unless the conservation approach changes to include consideration of individual species' required resources and management tolerances. The ecosystem approach assumes that habitat specialists are co-evolved with processes such as fires assumed to maintain those ecosystems. Data presented here indicate that tallgrass prairie specialist butterflies are not co-evolved with current fire regimes. An alternate perspective views ecological processes as resetting vegetation to current climate and landscape conditions. Over geologic time, relict vegetation associations persist as outliers until an event resets them. In modern times, human disturbances (especially soil-exposing ones) can reset sites to favour the more generalist species (plants and butterflies) found in the prevailing, human-degraded landscape.

Keywords *Atrytone arogos* · *Hesperia dacotae* · *Hesperia ottoe* · *Lycaeides melissa samuelis* · *Oarisma poweshiek* · *Speyeria idalia* · Butterfly declines · Prairie management · Burning

S. R. Swengel (✉) · A. B. Swengel
909 Birch Street, Baraboo, WI 53913, USA
e-mail: swengel@naba.org

D. Schlicht
Iowa Lepidoptera Project, 1108 First Avenue, Center Point, IA 52213, USA

F. Olsen
1513 Parkwood Lane NE, Cedar Rapids, IA 52402, USA

Introduction

Since European contact about 99% of North American tallgrass prairie (predominately herbaceous flora) has been destroyed primarily by conversion to agriculture (Curtis 1959; Howe 1994; Samson and Knopf 1994). Unmanaged patches of prairie today often become overgrown by woody species ("succession") and accumulate plant litter. As a result, periodic processes are widely considered necessary for prairies to persist today (Curtis 1959; Vogl 1974; Anderson 1982). Because of habitat destruction, these processes are disrupted and difficult to reconstruct.

Many infer that fire is the dominant process maintaining the open condition of prairie (Sauer 1950; Vogl 1974), although the primary cause (lightning or humans), season, and frequency of these fires remain in dispute (Higgins 1984, 1986; Howe 1994; Umbanhowar 1996; Russell 1997). Because lightning-caused ignitions are relatively rare in prairies (Higgins 1984), the fire-maintenance hypothesis relies on human-caused fires (Hulbert 1973) to explain prairie persistence. There is little evidence that American Indians set frequent landscape-scale fires (Higgins 1986; Umbanhowar 1996), and fire frequency actually increased when Europeans settled the prairie (Umbanhowar 1996; Russell 1997). Because of the short period (<0.1% of prairie history) that humans have lived in North America (Pielou 1991), they cannot have caused the origin of prairie and were unlikely to be the main force maintaining it. Alternative theories assert that this openness was maintained by climate (Transeau 1935; Weaver 1954; King 1981), herbivory by grazers and browsers (Larson 1940; Moore 1988), seed consumption and mechanical destruction of forests by superabundant but now extinct passenger pigeons (*Ectopistes migratorius*) (Ellsworth and McComb 2003), or a combination of these factors with soil and/or topography (Anderson 1982; Howe 1994).

The effects of ecosystem management with fire on insects are subjects both of research and controversy. Prairie-specialist butterflies are often assumed to be fire-adapted, with data on particular populations viewed as corroboration (Dana 1991; Shuey 1997; Panzer 2002). Others who found higher abundances of specialized butterflies in places with alternative managements to fire concluded that frequent burning is risky or harmful to prairie-specialist butterflies, and recommend greater reliance on less lethal managements like mowing and light grazing (Orwig 1992; Schlicht and Orwig 1998; Swengel 1996, 1998; Schlicht 2001; Swengel and Swengel 2001a).

Prairie-specialist butterflies have undergone disproportionately large declines during the last 1–2 centuries compared to other groups of butterflies (Orwig 1992). Although this is consistent with overall patterns of specialists declining more than generalists (e.g. Pollard and Eversham 1995; Fox et al. 2006; Kuussaari et al. 2007) prairie specialists are in even greater peril than butterflies specialized to other midwestern USA biomes (Schlicht and Orwig 1998).

Long-term monitoring is necessary to assess butterfly species' status and trend because of their great fluctuations in abundance due to climate and other factors (Dennis 1993; Pollard and Yates 1993; Thomas et al. 2002). To obtain long-term data in midwestern USA prairies, survey results from different teams must be combined. An underlying premise of most status/trend assessments is that data from different or informal (variable) methodologies can be pooled (Saarinen et al. 2003; Shuey 2005; van Swaay and van Strien 2005; van Swaay et al. 2006; Kuussaari et al. 2007). Schlicht et al. (2009) reported that during 1993–1996, two teams—Schlicht (one surveyor, fixed-width transect) and Swengels (two surveyor, unlimited width transect)—surveyed the same Minnesota prairies in the same seasonal timing in the same years, but without any coordination of sites, transect routes, survey methods, dates, and results between teams. Since strong covariance occurred in the butterfly abundance indices between the teams, thus validating the pooling of transect data for abundance analysis, a calibration of indices from the two-surveyor team to one-surveyor teams (the usual situation) was calculated.

Here, we expand the long-term trend analysis in Schlicht et al. (2009) by adding more datasets, states, and species. As in that study, here we also calibrate abundance indices from two-surveyor unlimited-width transect surveys to one-surveyor fixed-width transect surveys, and compare results for prairie-specialist butterflies to those of an outgroup of non-specialist species recorded in the same surveys. We also extend long-term surveys from Swengel and Swengel (2007), which evaluated species-specific prairie management guidelines including a permanent non-fire refugium (a unit kept unburned through cycles of rotational fire management elsewhere in the site). Results for a federally endangered butterfly, the Karner blue *Lycaeides melissa samuelis*, provides two opportunities for contrast with prairie: (1) This butterfly is a specialist of pine barrens (herbaceous flora mixed with trees and brush), a vegetation which has declined, but less than prairie (Curtis 1959; Borgerding et al. 1995). (2) Federal regulation allows two tiers of protection (described in "Methods") designed specifically for this butterfly, in contrast to an ecosystem approach in prairie. These results should provide useful information about the ecology and conservation of these species and the prairie ecosystem generally.

Methods

We amassed our own transect data in four states and located other datasets from the same study region (see Table 1), which encompassed a total east–west by north–south span of 600 × 600 km (Fig. 1). Specific methods of most surveyors are detailed in Schlicht et al. (2009). Where survey methods in Minnesota were not transects, we attempted to correct them to approximate transect observation rates (Schlicht et al. 2009).

The prairie survey sites were managed primarily with an ecosystem approach; i.e., based on implementing processes thought in some hypotheses to naturally maintain the vegetation (Iowa State Preserves Board 1981; TNC Minnesota 1994, 2010; TNC Wisconsin 1997; Wendt 1984; Illinois DNR 2005; TNC Iowa 2010). This ecosystem approach primarily used fire, usually on average about 20–35% burned per year (Swengel 1996; Schlicht et al. 2009). However, the proportion burned in a given year, or series of years, varied greatly, ranging from 0 to 100% of the prairie patch burned per year. That is, most or all of a site might be burned in consecutive years, followed by no fire for several years. Alternatively, varying proportions of a site might be burned in most or all years. It was unusual for a portion of a site to be unburned for >8–10 years. A reason for rotational burning (as opposed to 100% burns)

Table 1 Summary information on butterfly survey datasets analyzed in Schlicht et al. (2009) and this study, by team and state: survey years, dates, N sites (all followed by N analyzed), vegetation type, N observers on a survey, survey method, and species surveyed

Team	State[a]	Year	Dates	Sites	Type[b]	N Obs	Method[c]	Species[d]
Britten/Glasford	MN[e]	1998	702–721	9/1	P	1	PA	Hd
Dana	MN[e]	1979–1981	626–731	1/1	P	1	MRR	Hd, Ho, Op, Aa
Mason	MN[e]	1998–1999	719–822	13/2	P	1	FW	Si
Olsen	IA	2004–2009	420–1016	87/30	P	1	FW	All
Saunders	IA[f]	1993–1994	611–725	54/30	P	1	FWa	All
Schlicht	IA	2004	628–709	4/3	P	1	FW	All
	MN[e]	1993–1997, 2000	624–823	80/6	P	1	FW	All
Selby	MN[e]	1988–1990	524–905	1/1	P	1	FW	Many
	MN[e]	2003–2005	623–814	16/7	P	1	FW	All
Skadsen	MN[e]	2001	627–709	1/1	P	1	FWa	Si, Op, Hd
Swengels[g]	IL	1991–1997	627–901	6/2	P	2	UW	All
	IA	1989,1991–1997	628–821	9/8	P	2	UW	All
	MN[e]	1988–1997	618–820	30/7	P	2	UW	All
	WIs	1988–2009	426–913	39/12	P, G	2	UW	All
	WIc	1988–2009	413–910	150[i]/14	B	2	UW	All
	WInw	1991–2009	426–817	50[i]/11	B	2	UW	All
Wilder	WIc[h]	1996–2009	522–820	13/10	B	1	FW	Lms

[a] State abbreviations: *IA* Iowa, *IL* Illinois, *MN* Minnesota, *W* Wisconsin (*s* southern, *c* central, *nw* northwestern)

[b] Vegetation types: *P* prairie, *G* grassland (old field), *B* barrens

[c] Survey method: *FW* fixed width transect, *FWa* transect strip width not stated but assumed fixed (the prevailing method for most teams), *MRR* mark-release-recapture, with estimates of observation rates excluding handling time, *PA* presence/absence derived from collection, *UW* unlimited width transect

[d] Species surveyed indicated by initials of scientific name (e.g. Hd *Hesperia dacotae*); many = Si, Op, Aa, Hd, and Ho in 1988; those and many others in 1989–1990

[e] Schlicht et al. (2009)

[f] Saunders (1995)

[g] Swengel (1998), Swengel and Swengel (2005, 2007)

[h] more description in Swengel and Swengel (2005)

[i] Approximations due to challenges of defining a site in less fragmented vegetation

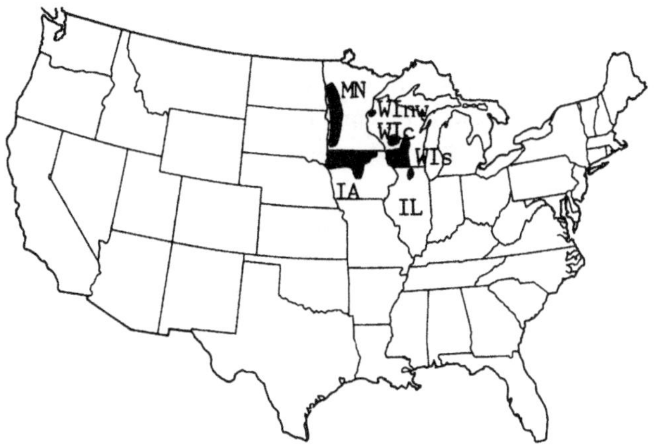

Fig. 1 Location of sites analyzed in Schlicht et al. (2009) and this study. For state abbreviations, see Table 1

was to allow for survival of prairie-specialist insects which might be isolated to the managed prairie patch (Panzer 2002). But it did not appear that recent survey data on immature or adult concentrations or population size affected burn location and extent. Other managements also occurred in many of these prairies, including mowing (leaving clippings), haying (removing clippings), brush-cutting (leaving clippings, removing clippings, or burning clippings in a brush pile in situ), and herbiciding (stump or wick treatment), usually in areas also managed with fire before and/or after other treatment(s). Livestock grazing only occasionally occurred at very few sites, more likely in private easements and wildlife hunting areas, very rarely in preserves, parks, and natural areas.

In some Wisconsin prairies, grasslands, and barrens, species-specific management for a state- or federally listed butterfly modified the ecosystem approach to reserve management (primarily burning). A cap ≤20% is placed on amount burned per year for regal fritillary (*Speyeria idalia*) sites, with encouragement to designate a permanent non-fire refugium in a core area for the butterfly (Wisconsin DNR 2000a). Less lethal managements (cutting, light grazing, rotational mowing/haying, spot herbiciding) are encouraged over more lethal ones (burning, heavy grazing). Butterfly monitoring is encouraged, and required for more lethal regimes. It encourages mapping of caterpillar food plant and butterfly distribution (usually adult but can be immatures), so as to avoid concentrating the butterfly into one or a few management units and, in the *L. melissa samuelis* protocol, to ensure proximity of an occupied patch for recolonization into a lethally managed unit. At Buena Vista Grassland, wildlife management happened to fit the species-specific protocol, with lethal management far below caps and primarily less lethal managements used. For *L. melissa samuelis* (federally listed), two levels of species-specific protocols exist but both levels allow more cutting than burning and cap burning at <33% of the habitat patch per year (Wisconsin DNR 2000b). Management in reserves (R) is intended to secure and restore ("recover") the butterfly's population, in conjunction with management for nature value of the ecosystem. Fort McCoy management, while not technically recovery, includes many conservation measures and is more analogous to R than SM or PH. A lesser standard applies to "shifting mosaic" (SM) and "permanency of habitat" (PH) sites. These are "working landscape" sites in timber management, rights-of-way, and other economic uses, as well as public lands (such as wildlife hunting areas) not included in recovery. This protocol aims not to jeopardize recovery of the butterfly and must be "with consideration for *L. melissa samuelis*" (as described further in Swengel and Swengel 2005).

We calculated observation rates for each butterfly species as the number of individuals/km if all surveys in the analysis had set routes and route length was available, or individuals/h if route length was unavailable for any surveys. Swengels' indices (from two-surveyor unlimited-width transects) were divided by the calibration constant (2.4) calculated in Schlicht et al. (2009) for fairer comparisons with indices from one-surveyor fixed-width teams. For each species, the peak index found by any team per year was identified to represent the butterfly's abundance at a site, if any surveys took place during the local main flight period that year. We were conservative about designating surveys as being during main flight and did not use data if unsure. One survey during main flight period has been adequate for producing representative indices for comparisons of relative abundance within and among sites (Thomas 1983; Swengel and Swengel 2005; Schlicht et al. 2009).

In Iowa, the target species were all analyzable prairie butterflies of conservation concern listed in Schlicht and Orwig (1998): Poweshiek skipperling *Oarisma poweshiek*, arogos skipper *Atrytone arogos*, Ottoe skipper *Hesperia ottoe*, *S. idalia*, and Aphrodite fritillary *S. aphrodite*. The outgroup comprised four widely occurring species in prairies selected for their general conservation interest (monarch *Danaus plexippus*), biogeographical interest (eyed brown *Satyrodes eurydice*) or predominance in grassland (common wood-nymph *Cercyonis pegala*, long dash *Polites mystic*); none were identified as at risk in Schlicht and Orwig (1998). The Wisconsin targets are the two analyzable prairie-specialists (*S. idalia*, *H. ottoe*) and the outgroups are *Lycaeides melissa samuelis* (pine barrens specialist) and *S. aphrodite* (widely occurring in Wisconsin prairies and other vegetations). The Minnesota study (Schlicht et al. 2009), summarized here as additional context, analyzed five prairie specialists (Dakota skipper *H. dacotae*, *H. ottoe*, *A. arogos*, *O.poweshiek*, *S. idalia*) versus an outgroup of five "common" (most frequently recorded non-specialist) species (*S. aphrodite*, meadow fritillary *Boloria bellona*, *C. pegala*, *D. plexippus*, *P. mystic*). The outgroups serve to test for systematic bias between earlier and later datasets.

Two opportunities for validation were available. First, while not matching sites as in the Schlicht and Swengel validation in Schlicht et al. (2009), the Swengel (central Wisconsin) and Fort McCoy datasets come from contiguous counties. Population indices matched by brood were correlated between the two datasets since abundance fluctuations should be similar between the two datasets. Swengel sites were divided into three categories: reserve (R) ($N = 4$ sites), "shifting mosaic" (SM) ($N = 5$ sites), and "permanency of habitat" (PH) ($N = 5$ sites).

Second, in five cases in 1993–1994 Saunders and Swengels surveyed the same Iowa prairies (five different sites) in the same year on dates averaging 12 days (range 0–16 days) apart, Swengels always surveying later than Saunders in the four instances when there was a difference (mean date 27 July and 14 July, respectively). Unlike the more robust validation in Schlicht et al. (2009), this comparison has too few iterates and too great an interval in dates within most iterates for statistically conclusive analysis that overcomes sampling errors due to variation in route, weather, date, and so on. However, for descriptive purposes, we present this comparison. Since this difference in dates was too great to calculate a calibration constant between the two teams, the same calibration constant as in Schlicht et al. (2009) was used for Swengels' results here. We examined how this calibration worked. First, we predicted which team should have higher observation rates for each analyzed species based on survey date. We then averaged the observation rates of each analyzed species at the five sites by team to test the prediction. Second, at one site (Cayler on 4 July 1994), both teams surveyed on the same day but did not see each other and did not follow the same routes. The observation rates of all analyzed species were averaged by team, to determine how similar the results were between teams.

All statistics were calculated using ABstat 7.20 (1994 Anderson-Bell, Parker, Colorado, USA). All tests were two-tailed, with statistical significance set at $P < 0.05$. Since significant results occurred much more frequently than expected due to Type I statistical errors, we did not lower the critical P value further, as far more Type II errors (biologically meaningful patterns lacking statistical significance) would then be created than Type I errors eliminated. The Spearman rank correlation was used for all correlations, Mann–Whitney U test for all tests between two categories, Chi Square Goodness of Fit test for all tests of current (observed) butterfly distribution within a site compared to past (expected) distribution, and binomial probability test for all tests for a preponderance of negative or positive correlation coefficients in a set of correlations (random distribution = 50% positive, 50% negative). In Wisconsin, continuous time series were available for each site. In Iowa, data from more sites were available, but usually without enough years to assemble a time series. Instead, we averaged all peak indices for each species from all sites ever recording the species during the study having at least one index from both the earlier (1989–1996) and later periods (2004–2007), separately for each period. All years with a survey during the flight period were included for each site.

Results

At five Iowa prairies surveyed by both Saunders and Swengels in the same year, four of seven species predicted to have higher observation rates in Saunders' surveys did, one species was the same (0 in both: *P. mystic*), and two were higher in Swengels' calibrated rates, against expectation (*S. idalia* and *S. aphrodite*, the largest of these seven species and most readily identified at the greatest distance, as covered in the unlimited-width transects of Swengels). The one species expected to have higher rates in Swengels' surveys did (*D. plexippus*). Thus, results for 2/8 species went against expectation, one was neutral (counting as 0.5 against and 0.5 with expectation), and results for 5/8 species were consistent with expectation, resulting in 5.5/8 (69%) in favor of the *a priori* prediction. This is distinctly greater than random (50%) despite the inadequate sample. It does not establish a statistical correlation but provides no basis to question this correlation. For the site surveyed by both teams on the same day, the average rate of all analyzed species was 13.2 individuals/h for Saunders and 9.9 (after applying the calibration) for Swengels. The calibration constant for Swengel results was possibly higher than needed, or this may represent expectable variation in a single example with two teams not following the same routes. But if the calibration lowers Swengel indices too much, then it would bias against a negative population trend versus time in this analysis since Swengel data weight earlier in the trend analyses.

In prairies with ecosystem management, specialists strongly declined in contrast to outgroups. In all states, all *H. ottoe* trends were negative (binomial $P = 0.001$), 7/11 significantly so (Table 2). In Iowa (Fig. 2), target species all declined >45–95% between the earlier and later periods, for two species significantly (*O. poweshiek*, *S. aphrodite*). The outgroup species either decreased non-significantly <50% or increased significantly >50%. In Minnesota in Schlicht et al. (2009), 22/27 trends of four target species (excluding *H. ottoe*, covered above) were negative (binomial $P = 0.001$), while the five outgroup "common" species had an even (random) distribution of positive and negative trends.

In Wisconsin, species-specific management produced some more favorable results. *S. idalia* had 5/8 negative trends (a non-significant distribution), none significant, while one positive trend was significant (Hogback) (Fig. 3). *S. aphrodite* had no significant trends, with 3/8 negative. In central Wisconsin (Fig. 4), *L. melissa samuelis* indices covaried strongly between Fort McCoy and R ($r = +0.731$, $P < 0.001$, $N = 23$ spring and summer broods), providing cross-validation of the two datasets. *L. melissa samuelis* indices at both Fort McCoy and R had slightly negative trends ($r = -0.147$, $N = 23$ broods and $r = -0.242$, $N = 27$, respectively) that were far from significant. By contrast, PH and SM declined drastically ($r = -0.767$ and -0.657, respectively, $P < 0.001$ and $N = 27$ for both). In northwestern Wisconsin, in the earlier period (1991–1997), *L. melissa samuelis* occurred in numbers proportional to survey effort in the unit that would become the non-fire refugium (but number of years since last fire was still similar to comparison fire-managed units), with 28/253 individuals (11%) occurring on the 11% of total survey length that occurred in that unit. Later (1998–2005), when the refugium was longer unburned than typical for fire-managed units there, significantly more *L. melissa samuelis* (15%) were recorded in the refugium than expected based on the earlier period (as reported in Swengel and Swengel 2007). This pattern became even stronger during 2006–2009: 29 individuals (59%) in the refugium versus 20 (41%) in the comparison units (Chi square $P = 0.0000$), or 13 times as many as expected in the refugium.

We surveyed other prairies that historically supported target species but did not find the target species in many recent years during their flights (Table 3). *O. poweshiek* at Puchyan (Table 3) demonstrates the persistence of a specialist in a very isolated site with no apparent recent management or vegetation change but strongly declining in other ecosystem-managed prairies.

Table 2 Spearman rank correlation coefficients (r) of year versus relative abundance of *H. ottoe*/h on peak survey per site per year, highest index and year recorded, last index, and consecutive years in this study's time series that last index was recorded

State/site	N (years)	r	P	Highest index	Highest index	Last index	Consecutive years last index recorded[a]
Illinois							
Harlem Hills	5	−0.707	>0.10	0.6	1993	0.0	1994–1997
Iowa							
Freda Haffner	5	−0.707	>0.10	1.0	1989	0.0	1993–1994, 2004–2005
Minnesota[b]							
Hole-in-the-mountain (new)	7	−0.802	**<0.05**	1.5	1990	0.0	1992, 1994–1996, 2005
Hole-in-the-mountain (old)	10	−0.688	**<0.05**	8.3	1979	0.0	1996
Prairie Coteau	6	−0.131	>0.10	0.3	1995	0.0	1996, 2000, 2005
Wisconsin							
Dewey Heights	19	−0.529	**<0.05**	30.5	1997	0.0	2008–2009
Hogback	18	−0.592	**<0.01**	0.8	1994	0.0	2005–2009
Muralt Bluff	22	−0.865	**<0.01**	9.5	1990	0.0	1998–2009
Oliver	21	−0.540	**<0.05**	18.4	1990	0.0	1993–2009
Rush Creek	9	−0.317	>0.10	12.0	1991	0.5	2009
Spring Green	12	−0.694	**<0.05**	9.0	1990	0.0	1997–1998, 2005–2006, 2008

Binomial probability of all correlations being negative is $P = 0.001$

Statistically significant declines ($P < 0.05$) are boldfaced

[a] The time series constructed from these datasets have years of missing data, so that a consecutive string of zero indices may not cover a consecutive-year period. E.g., for Freda Haffner, 5 years are in the time series, the highest index occurred in the first year (1989), and the index was zero in the four remaining years (1993–1994, 2004–2005) in the time series

[b] Minnesota data from surveys analyzed in Schlicht et al. (2009)

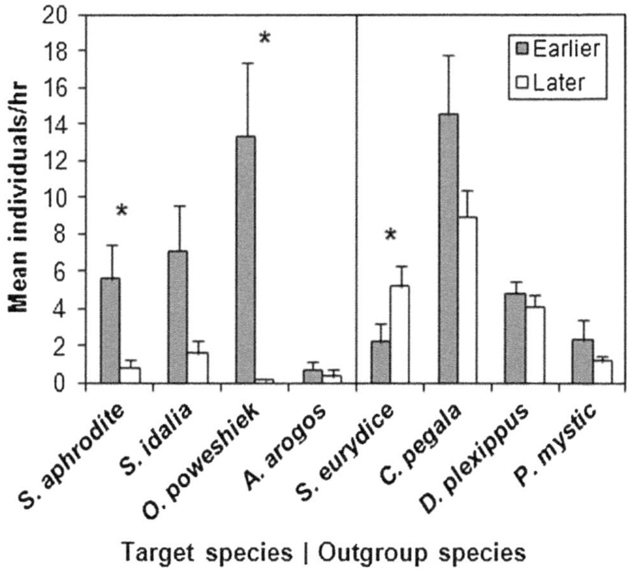

Fig. 2 Mean relative abundance (individuals/h) with SE *bars* in Iowa prairies in the earlier period (1989–1996) and later (2004–2007), for target species (prairie species of conservation concern) and outgroup species (not of conservation concern). Included sites have at least one survey result in each period, using whichever survey team produced the peak in as many years as survey data were available. * = statistically significant difference between the two periods ($P < 0.001$ for all). $N = 17, 5, 21, 3, 17, 26, 29$, and 19 sites, respectively

Discussion

Summary of butterfly results

In prairies with the ecosystem approach to management (primarily burning), species of conservation concern had a non-random pattern of declining trends while outgroup species did not (Tables 2 and 3; Fig. 2; Schlicht et al. 2009). Recent surveys by other researchers within our Minnesota study region confirm even larger declines in *O. poweshiek*, *H. ottoe*, and *A. arogos* after our study period ended and at sites that did not qualify for analysis here (Selby 2006; Minnesota DNR 2007; Dana 2008a, b). This is consistent with other studies where specialized butterflies were more sensitive to management and/or had declined more sharply than widespread species (Thomas 1984; Pollard and Yates 1993; Swengel 1998; Fox et al. 2006; van Swaay et al. 2006; Kuussaari et al. 2007). *S. idalia* declines on Iowa preserves had already become obvious by the mid-1990s (the earlier period of this study), with fire a likely cause at some sites (Debinski and Kelly 1998; Kelly and Debinski 1998). In Wisconsin prairies, *H. ottoe* (with no species-specific management) had 5/5 negative trends, while *S. idalia* (state-listed, with species-

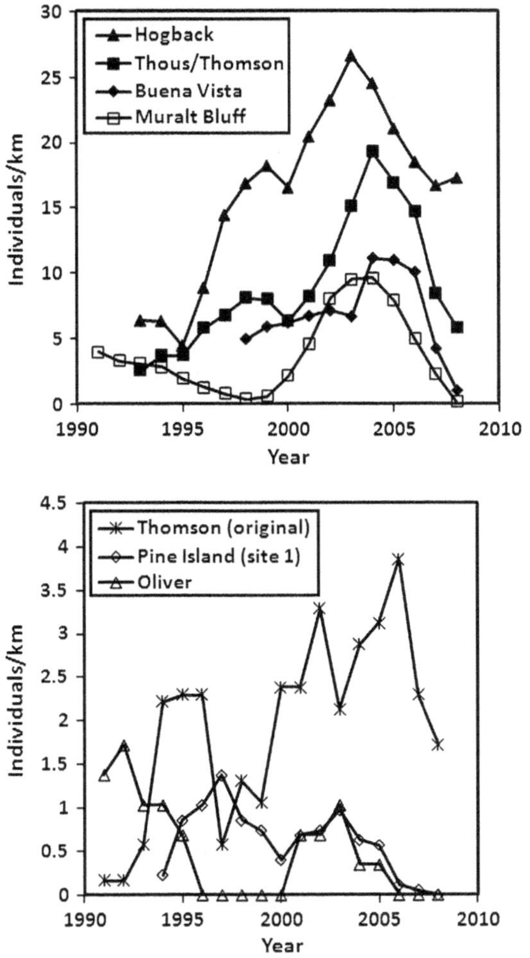

Fig. 3 *S. idalia* relative abundance (individuals/km) in Wisconsin prairies on the peak survey per site per year, smoothed as three-year running average (plotted in middle year) since there are no missing years. An eighth site, Pine Island 2, was not graphed because during 1993–2009, it had positive values only in 1995–1996

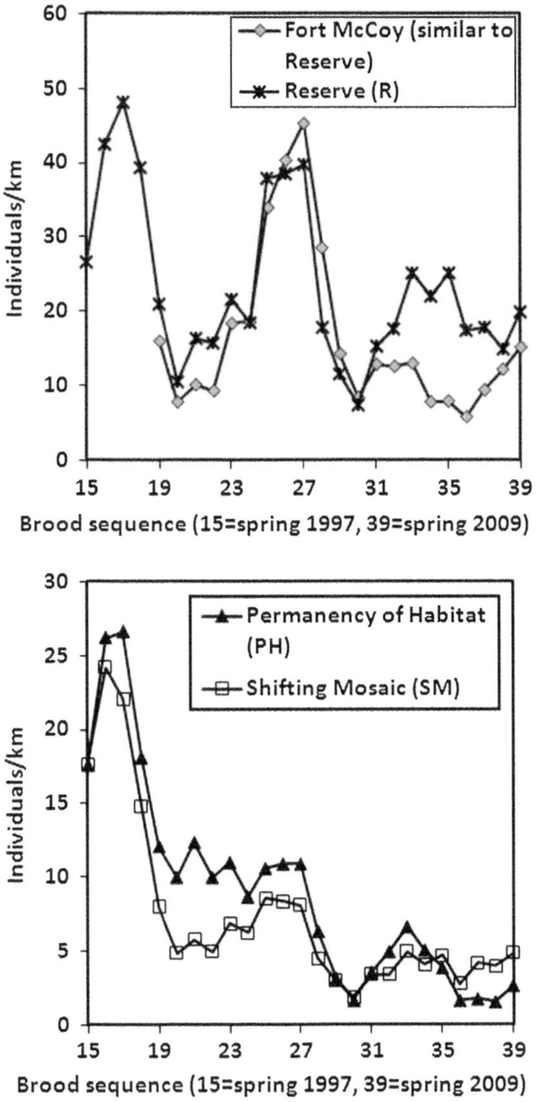

Fig. 4 *L. melissa samuelis* relative abundance (individuals/km) in four groups of sites in central Wisconsin pine barrens on the peak survey per site per brood (two broods per year)

specific management including non-fire refugia) had 5/8 negative trends, with the only significant trend positive. Closely related and covarying in occurrence (Swengel and Swengel 2001c), but more widely occurring in Wisconsin, *S. aphrodite* had similar results (3/8 negative trends, none significant).

Wisconsin *L. melissa samuelis* illustrates the benefits of increasing levels of species-specific conservation measures: Fort McCoy and R had better trends than SM and PH (Fig. 4) as did the non-fire refugium tailored to the butterfly compared to fire-managed units where burning was also modified to accommodate the butterfly (see "Results"). SM and PH, with a lower standard of conservation, had as negative a trend as specialist butterflies in the more fragmented, ecosystem-managed prairies (cf. Fig. 4 with Tables 2 and 3; Fig. 2, and Schlicht et al. 2009). At SM sites it's not clear whether the shifting mosaic concept (colonizations in new forestry cuts offsetting local extirpations from afforestation) is not working or whether the survey design of fixed sites is not happening to detect increases in recently treated sites. However, PH sites are intended to maintain populations long-term in one place. Declines there may be attributable at least in part to activities outside the protocol (as described in Swengel and Swengel 2005), rather than the protocol itself being inadequate.

H. ottoe declines in preserves (Table 2) confirm the insight of McCabe (1981) that preservation in practice involves two management impacts. First, light agricultural management (if any), such as grazing and haying, prior to preservation is discontinued. If long-standing, this prior management is implicated as favorable for specialists occurring in the site in good numbers at preservation. Second, new management (burning) is started. In this

Table 3 Historical sites, with value of last population index in surveys analyzed in this study. For currently undetected populations, consecutive years in this study's time series that an index of zero was recorded is provided; for extant population, the mean index for survey years in this decade is provided. Table does not include *H. ottoe* and *S. idalia* populations covered in Table 1; Fig. 2

Species		Last index	Consecutive years in time series index was recorded[c]
State[a]	Site[b]		
Undetected in last index			
Ottoe skipper	*Hesperia ottoe*		
Iowa	Cayler	0.0	1989, 1993–1996, 2004–2005
Dakota skipper	*Hesperia dacotae*		
Iowa	Cayler	0.0	1989, 1993–1996, 2004–2005
Minnesota	Chippewa	0.0	1996–1998
Arogos skipper	*Atrytone arogos*		
Iowa	Cayler	0.0	1996, 2004–2005
Minnesota	Bicentennial	0.0	1992–1993, 1995–1996, 2005
	Glacial Lakes	0.0	2003–2005
Melissa blue	*Lycaeides melissa melissa*		
Iowa	Doolittle	0.0	2004
	Freda Haffner	0.0	1989, 1993–1994, 2004–2005
	Gitchie Manitou	0.0	1993, 2004
Regal fritillary	*Speyeria idalia*		
Wisconsin	Spring Green	0.0	1990–1998, 2005–2006, 2008
Illinois	Byron	0.0	1993–1994
	Harlem Hills	0.0	1993–1997
Common ringlet	*Coenonympha tullia*		
Iowa	Cayler	0.0	1993–1994, 2004
	Doolittle	0.0	2004
Extant population			
Poweshiek skipperling	*Oarisma poweshiek*		
Wisconsin	Puchyan	Mean 6.6	2001–2002, 2005–2009

[a] Minnesota data as assembled and analyzed in Schlicht et al. (2009)

[b] Historical records from 1990s and earlier are documented by surveys in this study, Ferge 1997, sources summarized in Swengel and Swengel (2001b, c) and Schlicht et al. (2007), J. Nekola pers. comm., and J. Wiker pers. comm. These sources, Coffin and Pfannmuller (1988), and Swengel and Swengel (2001b, c) also provide documentation of consistent occurrence of these species previously at sites where undetected now

[c] See footnote [a] in Table 2

study, at most sites, surveys usually began after preservation. But they began within a year of preservation at Hole-in-the-Mountain both new and old tracts, and by far the highest *H. ottoe* rates occurred in those first years after preservation than later. This implies that management before preservation was more favorable than after. After preservation, most but not all sites were fire-managed with either no non-fire refugium or a refugium not in core *H. ottoe* habitat (Swengel and Swengel 2007; Schlicht et al. 2009). However, at Hogback, where *H. ottoe* was present in low numbers in the 1990s, light grazing was discontinued after preservation in the late 1990s. While the *H. ottoe* core remained never burned through summer 2009, this skipper still declined to non-detection from 2005 onward. Dana (1991) found that *H. ottoe* prefers shorter grass, with burning leading to taller grass, which is unfavorable as a vegetative structure and also increases fire mortality due to higher fuel loads. Removal of grazing also leads to unfavorable taller grass. These results echo Warren's (1993) equal rates of butterfly decline in reserves and non-reserves, even though different kinds of unfavorable management led to the declines. Although this study does not directly document losses on non-reserves, *H. ottoe* losses would appear considerable on unconserved land, based on the fragile population at Hogback pre-preservation.

Species-specific management protocols started in Wisconsin during the 1990s (Wisconsin DNR 2000a, b) showed more benefit but may not be sufficient for long-term viability of specialist populations. The protocols may not be sympathetic enough to the butterfly and are often not adequately implemented in the field, especially regarding the maximum amount burned per year. On the one hand, in some instances more than the target species benefited. Although not of conservation concern in Wisconsin (but it is in some other states), *S. aphrodite* had at least as favorable trends as *S. idalia*; unfortunately, *H. ottoe* did not benefit (Table 2). Many specialists benefited from the permanent non-fire refugium at Crex Meadows (Swengel and Swengel 2007). On the other hand, results were uneven as to whether decline was being averted. For *S. idalia*, the most favorable outcome was at Hogback, which has the largest refugium, plus it is never-burned rather than formerly burned. By contrast, at Muralt Bluff, both the formerly burned refugium (exceeding typical number of years since last fire only in the late 1990s) and the entire site are relatively small. Perilously low *S. idalia* numbers have persistently occurred there and at Oliver, Pine Island, and Thomson (original). For *L. melissa samuelis*, while declines could be expected in SM sites due to afforestation, declines were as great in PH sites, where consistently suitable vegetation is expected (Fig. 4). Concepts of how to maintain reserves need to include adequate maintenance of specific resources and conditions required by the butterflies in ways the butterflies can tolerate.

Ecological interpretations

Is this recent prairie butterfly decline primarily attributable to habitat loss and fragmentation or to other factors? Most prairie destruction occurred rapidly in the nineteenth century. For example, in Iowa, settlement by people of European extraction began in earnest in 1833, and in ca. 70 years (by about 1900), native prairie was essentially destroyed, with just scattered fragments remaining (Smith 1981). However, loss and deterioration of those fragments continues through today (cf. Smith 1981 and Roosa 1981). At the time of preservation sometime during the last 60 years, many populations of specialist butterflies occurred in these sites of varying sizes. As populations, including large high-density ones, were declining and disappearing at some sites preserved longer ago, substantial populations still existed in not-yet-preserved and more recently preserved sites imbedded in the same highly fragmented landscape. Examples include *S. idalia* declining in the late 1980s and early 1990s at Spring Green and Harlem Hills (Table 3) while still relatively abundant at more recently preserved Thomson and Hogback (Fig. 3); *O. poweshiek* and *A. arogos* declining at Bicentennial and Blazingstar in the late 1980s and early 1990s while still more abundant at Hole-in-the-Mountain and Prairie Coteau (Schlicht et al. 2009); and *H. ottoe* declining in the 1980s and 1990s at Cayler, Freda Haffner, Hole-in-the-Mountain, Prairie Coteau, and Spring Green (Table 2) while still more abundant in the Loess Hills (Vogel et al. 2010). These declines and losses are not limited to small sites. This suggests that management is an important contributing factor, with pre-preservation management happening to be more favorable in these specific sites than post-preservation. At a minimum, these results indicate that current ecosystem approaches to prairie conservation are ineffective at stopping declines and losses of prairie-specialist butterflies. More favorable management is documented and feasible for these butterflies (see "Conservation conclusion").

While burning is frequently stated to favor native flora and reduce and exclude brush and weeds (Curtis 1959; Hoffman and Kearns 1997; Packard and Mutel 1997), numerous studies indicate minor brush control or actual increases even with very frequent burning. Canopy cover in repeatedly burned prairies remained stable (Becker 1989) or increased as much as in unmanaged sites (Robertson et al. 1995), even after decades of annual burning (Towne and Owensby 1984). Burning as the primary management reduced native prairie/savanna plant species richness while other managements (e.g., cutting, haying) were more effective at canopy control (Collins et al. 1998; Neilsen et al. 2003). Overall non-native grass cover was similar before and after burning (Choi and Pavlovic 1994) and bare soil after burning allowed adventive plants to increase, including invasive weeds adapted to harsh conditions (e.g. Curtis and Partch 1948; Smith 1993; Diboll 1997). Some ecosystem management guides recommend burning 33–50% to even 100% per year, year after year, to initiate weed control, with cutting or other treatments in addition to burning for greater weed control (Smith 1993; Hoffman and Kearns 1997; Packard and Mutel 1997). These recommendations imply relative ineffectiveness of an individual burn and are opposite of the Wisconsin DNR (2000a) regulations that emphasize more cutting and reduced fire for sensitive insects like *S. idalia*. Minor or no control of brush and weeds for major investment in burning indicates the need to develop an alternative understanding of how fire functions in prairie.

The ecosystem approach to conservation assumes that specialized species are co-evolved with processes (e.g., fire, grazing) thought to maintain those ecosystems. However, prairie-specialist butterflies have significantly more negative responses than non-specialists to fire, the dominant process used to manage tallgrass prairie preserves (Swengel 1996, 2001; Panzer 2002; Swengel and Swengel 2007; Schlicht et al. 2009). Either the assumption of co-evolution is wrong, or these species are co-evolved with some other process like grazing (Williams 1997), not fire. Life history traits such as number of generations per year, degree of specialization, location during fire, and response of key plants to fire (Swengel 1996) statistically explain insect responses to fire, rather than the insects' ecosystem affiliations (reviewed by Swengel 2001). Because specialists have all or nearly all of their population within these fragments that become burned, specialized butterflies can only rebuild their populations if occupied unburned refugia exist nearby, burned vegetation is suitable for their recolonization, and enough time is allowed to rebuild numbers (Swengel 1996, 2001; Harper et al. 2000; Panzer 2002). Since most above-ground insects in the affected fuel appear to die in fires (Swengel 2001), prairie fire management incinerates generalist as well as specialist butterflies. However, generalists are more widely distributed both within and outside preserves (e.g. in roadsides, fields, flood plains, and so on) and are more flexible about vegetative composition and structure. Thus, for generalists a lower proportion of a population would be affected by prairie fire management, burned vegetation would more likely be suitable habitat, and more sources of recolonization would exist in the vicinity, making generalist populations better able than specialists to tolerate these fires. Since invertebrate specialists as a group are most averse to fire management (Swengel 1996, 1998, 2001; Nekola 2002), this disproves the hypothesis of species co-evolution with prairie fire.

An alternate perspective posits that some ecological processes may instead reset vegetation to current climate and landscape conditions. Over geologic time, relict vegetative associations persist as outliers until a dramatic event (fire, tree-felling windstorm) resets them to a different vegetation appropriate for the current climate and landscape composition (Pielou 1991). Today, dramatic events (especially soil-exposing ones) in native prairie could favor native and non-native ruderals (weeds and brush) of the prevailing human-degraded and fragmented landscape, and more generalist species (including butterflies). Repeated cutting near ground level in a Kentucky powerline right-of-way to inhibit tree growth resulted instead in strong similarities in plant species composition among communities in the corridor and adjacent forest interior (Luken et al. 1991). This management didn't change vegetation type, but reset growth habit and abundance of vegetation already prevailing in the landscape. Approximately similar vegetative responses occur from similarly timed mowing and burning (Daubenmire 1968), suggesting analogous impacts from fire. Stand-replacing (canopy-killing) fire may result in an increase in herb cover (Nuzzo et al. 1996; Neilsen et al. 2003) but not necessarily as a change of vegetation type but as a reset of the canopy structure of the same vegetation type.

A "reset" concept as described in Pielou (1991) may apply to modern anthropogenic landscapes, where dramatic managements "reset" to the current landscape of invasive and generalist weeds and brush. The prairie biome was not a relict in the paleontological sense used in Pielou (1991). Although conceived by some now as an early successional stage that would be forest but for frequent anthropogenic fires (Curtis 1959; Vogl 1974), the prairie biome has been a consistently occurring vegetative configuration for millions of years before humans arrived in North America (Weaver 1954). This occurrence approximately corresponded to certain climatic parameters, in the absence of human inhabitants but with an abundance of grazing and browsing animals (Kurtén 1971). This suggests that over most of prairie's existence, climate and possibly herbivory explain the primarily herbaceous vegetative composition of prairie. In peripheral areas, inertia could contribute so long as an event does not reset the vegetation (Pielou 1991). However, prairie is now a kind of vegetative relict occurring in relatively small isolated patches in an otherwise anthropogenic landscape consisting primarily of intensive agriculture. It stands to reason that unintensive land uses are more likely to forestall unfavorable resets in conserved vegetations due to anthropogenic climate change (Dennis 1993; Forister et al. 2010) than dramatic ones like fire.

Conservation conclusion

On a different continent, Clarke (2008) and New et al. (2010) address a similar combination of (1) orders of magnitude differences in estimates of prehistoric fire frequency and extent, and (2) relative paucity of data on long-term population trends, habitat requirements, and management tolerances of rare/specialist flora and fauna. As a result they advise a scientific approach with controls and outgroups to hedge against both what is and isn't known, so as to try to optimize outcomes now for biodiversity. New et al.'s (2010) recommendations are highly appropriate for North American prairie: (1) Sites that are small or isolated or have listed invertebrates should never be burned without carefully assessing specialist zoologist advice. (2) Micro-mosaic burns no more than a few ha each and staggered over years should be the norm instead of larger burns. (3) At least 20% of a site should be permanently protected from deliberate burns. (4) Small sites (<5 ha) should only be burned under exceptional circumstances, and then only with surveying and monitoring to investigate risks.

Enough monitoring and management data on prairie-specialist butterflies exist to indicate promising approaches to manage more favorably for them. The species-specific protocols developed in Wisconsin for *S. idalia* and *L. melissa samuelis* include particularly beneficial approaches. They recognize that livestock grazing in any form is not categorically undesirable (a persistent attitude: e.g., Curtis 1959; Henderson 1998) but can be done beneficially for conservation, not in consistent heavy loads as in agriculture (Williams 1997). They encourage less burning and more mowing/haying, but all of these on a rotation, and also encourage localized brush-cutting and spot-herbiciding. Burns need to be near known occupied patches not recently burned. Burns need to avoid affecting in a given year more than about 20% (*S. idalia*) to 33% (*L. melissa samuelis*) of both host patches and areas recently occupied by the butterflies (better if less in most or all years). This in turn requires ongoing monitoring of the butterfly (Wisconsin DNR 2000a, b). When monitoring results indicate low population size (regardless of possible causes, including adverse weather), future management activities need to be less lethal and even more favorable to the butterfly. Especially necessary are permanent non-fire refugia located in core areas for the butterfly population and managed with alternative less lethal managements (Swengel and Swengel 2007) that have been documented to maintain both the required vegetation and higher abundances of the butterflies: light grazing for *S. idalia*, rotational haying for *S. idalia*, *A. arogos*, *H. dacotae*, and *O. poweshiek* (a longer rotation), and long-term idling with localized brush treatment as needed for *O. Poweshiek* (McCabe 1981; Swengel 1996, 1998, 2008; Swengel and Swengel 1999, 2001c; Powell et al. 2007).

In addition, we propose a change in conservation philosophy. Currently, prairie-specialist butterflies are used as indicators of sites worthy of conservation. But after conservation, management is usually changed from recent land uses (although possibly no longer occurring) to an attempt to restore prehistoric processes, however, imperfectly known and replicable. Instead, we advise a focus on retaining taxa special to a site, rather than on processes thought to occur more generally. To do this, we recommend the concept of conserving not only sites but also their historic management history (pre-conservation), such as an unintensive haying or light grazing regime, that was critical to maintaining the flora and fauna in these sites more successfully than elsewhere in the landscape. This will foster the beneficial combination of management consistency within a patch but differences among patches of the same vegetative classification (Kirby 1992). This is desirable given the narrow vegetative and management tolerances of specialist butterflies but these vary among specialists co-occurring in vegetation of like classification (Swengel 1998; Swengel and Swengel 1997, 1999, 2001a; Schlicht et al. 2007). Species-specific protocols implemented in Wisconsin show some evidence of benefit to other specialist species, much more so in barrens (Swengel and Swengel 2007), which had more intact, diverse specialist faunas. This was less so in prairie: *H. ottoe* did not show benefit from the *S. idalia* efforts (Swengel and Swengel 2007). This shows the urgent need for more individual prairie specialists to receive more species-specific management, if what is left of the specialist butterfly fauna is to be retained in tallgrass prairie.

Acknowledgments The Olsen surveys in Iowa were funded by the Iowa Department of Natural Resources, Iowa Natural Heritage Foundation, and Seed Savers Exchange. The Swengel surveys in Wisconsin were funded in part by the Lois Almon Small Grants Research Program, Wisconsin Department of Natural Resources (WDNR), U.S. Fish and Wildlife Service, Jed Bromfield and Henya Rachmiel, Sandra McKibben, and especially Drs William and Elsa Boyce. We gratefully acknowledge funding for the Minnesota surveys as itemized in Schlicht et al. (2009). We appreciate that staff from WDNR provided data on management history for Buena Vista, Crex Meadows, and Muralt Bluff. We thank the Zinkle family for allowing access to Hogback when they owned it. We are very grateful to Timothy T. Wilder, Endangered Species Biologist, for providing us with unpublished monitoring survey data from the Department of the Army, Fort McCoy, Wisconsin. We greatly appreciate the many lepidopterists who have surveyed prairie and barrens butterflies and shared their observations with us personally and the Iowa Lepidoptera Project, especially Mike Saunders, Jeff Nekola, and James Wiker, and to the agencies that have supported surveys. We greatly appreciate many helpful comments from two anonymous reviewers and the Editor, Tim Shreeve.

References

Anderson RC (1982) An evolutionary model summarizing the roles of fire, climate, and grazing animals in the origin and maintenance of grasslands: an end paper. In: Estes JR, Tyrl RJ, Brunken JN (eds) Grasses and grasslands: systematics and ecology. University of Oklahoma Press, Norman, pp 297–308

Becker D (1989) Five years of annual prairie burns. In: Bragg TG, Stubbendieck J (eds) Proceedings of the eleventh North American prairie conference. University of Nebraska, Lincoln, pp 163–168

Borgerding EA, Bartelt GA, McCown WM (eds) (1995) The future of pine barrens in northwest Wisconsin: a workshop summary. Wisconsin Dept Nat Res PUBL-RS-913-94

Choi YD, Pavlovic ND (1994) Comparison of fire, herbicide, and sod removal to control exotic vegetation. Nat Areas J 14:217–218

Clarke MF (2008) Catering for the needs of fauna in fire management: science or just wishful thinking? Wildl Res 35:385–394

Coffin B, Pfannmuller L (eds) (1988) Minnesota's endangered flora and fauna. University of Minnesota Press, St. Paul

Collins SL, Knapp AK, Briggs JM et al (1998) Modulation of diversity by grazing and mowing in native tallgrass prairie. Science 280:745–747

Curtis JT (1959) The vegetation of Wisconsin: an ordination of plant communities. University of Wisconsin Press, Madison

Curtis JT, Partch ML (1948) Effect of fire on the competition between blue grass and certain prairie plants. Am Midl Nat 39:437–443

Dana RP (1991) Conservation management of the prairie skippers *Hesperia dacotae* and *Hesperia ottoe*: basic biology and threat of mortality during prescribed burning in spring. Minnesota Agric Exp Stn Bull 594–1991:1–63

Dana RP (2008a) Southwest Minnesota prairie Lepidoptera survey, 2006. Unpublished report to Minnesota Department of Nat Res, Minnesota County Biol Surv. Cited with permission

Dana RP (2008b) Southwest Minnesota prairie Lepidoptera survey, 2007. Unpublished report to Minnesota Dept of Nat Res, Minnesota County Biol Surv. Cited with permission

Daubenmire R (1968) Ecology of fire in grasslands. Adv Ecol Res 5:209–266

Debinski DM, Kelly L (1998) Decline of Iowa populations of the regal fritillary (*Speyeria idalia*) Drury. J Iowa Acad Sci 104:16–22

Dennis RLH (1993) Butterflies and climate change. Manchester University Press, Manchester

Diboll N (1997) Weed control. Missouri Prairie J 18(3):14–15

Ellsworth JW, McComb BC (2003) Potential effects of passenger pigeon flocks on the structure and composition of presettlement forests of eastern North America. Conserv Biol 17:1548–1558

Ferge L (1997) 1996 Wisconsin Lepidoptera season summary. Newsl Wisc Entomol Soc 24(1):4–7

Forister ML, McCall AC, Sanders NJ et al (2010) Compounded effects of climate change and habitat shift patterns of butterfly diversity. Proc Nat Acad Sci (Philadelphia) 107:2088–2092

Fox R, Asher J, Brereton T, Roy D, Warren M (2006) The state of butterflies in Britain and Ireland. Butterfly Conservation and the Centre for Ecology and Hydrology, Newbury

Harper MG, Dietrich CH, Larimore RL, Tessene PA (2000) Effects of prescribed fire on prairie arthropods: an exclosure study. Nat Areas J 20:325–335

Henderson RA (1998) To graze or not? Restor Manag Notes 16:6–7

Higgins KF (1984) Lightning fires in North Dakota grasslands and in pine-savanna lands of South Dakota and Montana. J Range Manag 37:100–103

Higgins KF (1986) Interpretation and compendium of historical fire accounts in the northern Great Plains. US Fish and Wildl Serv Resource Publ 161:1–39

Hoffman R, Kearns K (eds) (1997) Wisconsin manual of control recommendations for ecologically invasive plants. Bureau of Endangered Res Dept of Nat Res, Madison

Howe HF (1994) Managing species diversity in tallgrass prairie: assumptions and implications. Conserv Biol 8:691–704

Hulbert LC (1973) Management of Konza Prairie to approximate pre-whiteman fire influences. In: Hulbert LC (ed) Third midwest prairie conference proceedings. Kansas State University, Manhattan, pp 14–17 http://images.library.wisc.edu/EcoNatRes/EFacs/NAPC/NAPC03/reference/econatres.napc03.lhulbert2.pdf

Illinois DNR (2005) Illinois comprehensive wildlife conservation plan & strategy. Illinois Dept of Nat Res, Springfield. http://dnr.state.il.us/orc/wildliferesources/theplan/final/Illinois_final_report.pdf. Accessed 21 July 2010

Iowa State Preserves Board (1981) Directory of state preserves. Iowa Dept of Nat Res, Des Moines

Kelly L, Debinski DM (1998) Relationship of host plant density to size and abundance of the regal fritillary *Speyeria idalia* Drury (Nymphalidae). J Lepid Soc 52:262–276

King JE (1981) Late quaternary vegetational history of Illinois. Ecol Monogr 51:43–62

Kirby P (1992) Habitat management for invertebrates: a practical handbook. RSPB, Sandy

Kurtén B (1971) The age of mammals. Columbia University Press, New York

Kuussaari J, Heliölä J, Pöyry J, Saarinen K (2007) Contrasting trends of butterfly species preferring semi-natural grasslands, field margins and forest edges in northern Europe. J Insect Conserv 11:351–366

Larson F (1940) The role of the bison in maintaining the short grass plains. Ecology 21:113–121

Luken JO, Hinton AC, Baker DG (1991) Assessment of frequent cutting as a plant-community management technique in power-line corridors. Environ Manag 15:381–388

McCabe TL (1981) The Dakota skipper *Hesperia dacotae* (Skinner): range and biology, with special reference to North Dakota. J Lepid Soc 38:179–193

Minnesota Department of Natural Resources (2007) Draft amendments to Minnesota's list of endangered, threatened, and special concern species-January 2, 2007. http://files.dnr.state.mn.us/input/rules/ets/all.pdf. Minnesota Dept of Nat Res, St. Paul. Accessed 4 Jan 2008

Moore CT (1988) Mid-nineteenth century short grass expansion in the central and southern plains. In: Davis A, Stanford G (eds) The prairie: roots of our culture, foundation of our economy. Proceedings of the 10th North American Prairie conference. Native Prairies Assoc of Texas, Dallas, section 01.04

Neilsen S, Kirschbaum C, Haney A (2003) Restoration of midwest oak barrens: structural manipulation or process-only. Conserv Ecol 7(2):article 10 (14 pp). www.consecol.org/vol7/iss2/art10 viewed 15 Dec 2009

Nekola JC (2002) Effects of fire management on the richness and abundance of central North American grassland snail faunas. Anim Biodiver Conserv 25(2):53–66

New TR, Yen AL, Sands DPA et al. (2010) Planned fires and invertebrate conservation in south east Australia. J Insect Conserv (in press)

Nuzzo VA, McClain WE, Strole T (1996) Fire impact on groundlayer flora in a sand forest 1990–1994. Am Midl Nat 136:207–221

Orwig TT (1992) Loess hills prairies as butterfly survivia: opportunities and challenges. In: Smith DD, Jacobs CA (eds) Proceedings of the twelfth North American Prairie conference 1990. University of Northern Iowa, Cedar falls, pp 131–135

Packard S, Mutel CF (eds) (1997) The tallgrass restoration handbook: for prairies, savannas, and woodlands. Island Press, Washington

Panzer R (2002) Compatibility of prescribed burning with the conservation of insects in small, isolated prairie reserves. Conserv Biol 16:1296–1307

Pielou EC (1991) After the ice age: the return of life to glaciated North America. University of Chicago Press, Chicago

Pollard E, Eversham BC (1995) Butterfly monitoring 2—interpreting the changes. In: Pullin AS (ed) Ecology and conservation of butterflies. Chapman & Hall, London, pp 23–36

Pollard E, Yates TJ (1993) Monitoring butterflies for ecology and conservation. Chapman & Hall, London

Powell AFLA, Busby WH, Kindscher K (2007) Status of the regal fritillary (*Speyeria idalia*) and effects of fire management on its abundance in northeastern Kansas, USA. J Insect Conserv 11:281–285

Robertson KR, Schwartz MW, Olson JW et al. (1995) 50 years of change in Illinois hill prairies. Erigenia 14: 41–52, as posted on www.inhs.illinois.edu/~kenr/hillprairie.html. Accessed 17 Dec 2009

Roosa DM (1981) Iowa natural heritage preservation: history, present status, and future challenges. Proc Iowa Acad Sci 88:43–47

Russell EWB (1997) People and the land through time: linking ecology and history. Yale University Press, New Haven

Saarinen K, Lahti T, Marttila O (2003) Population trends of Finnish butterflies (Lepidoptera: Hesperioidea, Papilionoidea) in 1991–2000. Biodiver Conserv 12:2147–2159

Samson F, Knopf F (1994) Prairie conservation in North America. Bioscience 44:418–421

Sauer C (1950) Grassland climax, fire and management. J Range Manag 3:16–20

Saunders M (1995) Species status report for the Dakota skipper (*Hesperia dacotae*) and the Poweshiek skipperling (*Oarisma poweshiek*) in Iowa. Report to the Iowa Nongame Wildlife Program, Iowa Dept of Nat Res, Cedar Falls

Schlicht DW (2001) The decline of the Arogos skipper (*Atrytone arogos*) at Prairie Coteau in Pipestone County, Minnesota. In: Bernstein N, Ostrander L (eds) Proceedings of the Seventeenth North American Prairie Conference: seeds for the future, roots of the past. North Iowa Area Community College, Mason City, pp 197–200 http://digicoll.library.wisc.edu/cgi-bin/EcoNatRes/EcoNatRes-idx?type=div&did=ECONATRES.NAPC17.DSCHLICHT&isize=text. Accessed 5 Feb 2008

Schlicht DW, Orwig TT (1998) The status of Iowa's Lepidoptera. J Iowa Acad Sci 105(2):82–88

Schlicht DW, Downey JC, Nekola JC (2007) The butterflies of Iowa. University of Iowa Press, Iowa City

Schlicht D, Swengel A, Swengel S (2009) Meta-analysis of survey data to assess trends of prairie butterflies in Minnesota, USA during 1979–2005. J Insect Conserv 13:429–447

Selby G (2006) Effects of grazing on the Dakota skipper butterfly; prairie butterfly status surveys 2003-2005. Report to the Minnesota Dept of Nat Res, St. Paul. http://files.dnr.state.mn.us/eco/nongame/projects/consgrant_reports/2006/2006_selby.pdf. Accessed 25 June 2010

Shuey JA (1997) Dancing with fire: ecosystem dynamics, management, and the Karner blue (*Lycaeides melissa samuelis* Nabokov) (Lycaenidae). J Lepid Soc 51:263–269

Shuey JA (2005) Assessing the conservation value of a complementary system of habitat reserves relative to butterfly species at risk and divergent populations. Am Midl Nat 153:110–120

Smith DD (1981) Iowa Prairie-an endangered ecosystem. Proc Iowa Acad Sci 88:7–10

Smith TE (ed) (1993) Missouri vegetation management manual. Missouri Dept of Conserv, Jefferson City

Swengel AB (1996) Effects of fire and hay management on abundance of prairie butterflies. Biol Conserv 76:73–85

Swengel AB (1998) Effects of management on butterfly abundance in tallgrass prairie and pine barrens. Biol Conserv 83:77–89

Swengel AB (2001) A literature review of insect responses to fire, compared to other conservation managements of open habitats. Biodiver Conserv 10:1141–1169

Swengel AB (2008) Poweshiek paradise lost. Am Butterflies 16(4):16–32

Swengel AB, Swengel SR (1997) Co-occurrence of prairie and barrens butterflies: applications to ecosystem conservation. J Insect Conserv 1:131–144

Swengel AB, Swengel SR (1999) Observations on prairie skippers (*Oarisma poweshiek*, *Hesperia dacotae*, *H. Ottoe*, *H. leonardus pawnee*, and *Atrytone arogos iowa*) (Lepidoptera: Hesperiidae) in Iowa, Minnesota, and North Dakota during 1988–1997. Great Lakes Entomol 32:267–292

Swengel AB, Swengel SR (2001a) Effects of prairie and barrens management on butterfly faunal composition. Biodiver Conserv 10:1757–1785

Swengel AB, Swengel SR (2001b) A ten-year study of the status and trend of the regal fritillary (*Speyeria idalia*) (Lepidoptera: Nymphalidae) in Wisconsin, USA. Great Lakes Entomol 34(1):111–127

Swengel AB, Swengel SR (2001c) A ten-year study to monitor populations of the regal fritillary, *Speyeria idalia*, (Lepidoptera: Nymphalidae) in Wisconsin, USA. Great Lakes Entomol 34(2):97–115

Swengel AB, Swengel SR (2005) Long-term population monitoring of the Karner Blue (Lepidoptera: Lycaenidae) in Wisconsin, 1990–2004. Great Lakes Entomol 38:107–134

Swengel AB, Swengel SR (2007) Benefit of permanent non-fire refugia for Lepidoptera conservation in fire managed sites. J Insect Conserv 11:263–279

Thomas JA (1983) A quick method for estimating butterfly numbers during surveys. Biol Conserv 27:195–211

Thomas JA (1984) Conservation of butterflies in temperate countries: past efforts and lessons for the future. In: Vane-Wright RI, Ackery PR (eds) The biology of butterflies. Princeton University Press, Princeton, pp 333–353

Thomas CD, Wilson RJ, Lewis OT (2002) Short-term studies underestimate 30-generation changes in a butterfly metapopulation. Proc Royal Soc London B 269:563–569

TNC (The Nature Conservancy) Minnesota Chapter (1994) Minnesota chapter preserve guide. The Nature Conservancy, Minnesota Chapter, Minneapolis

TNC (The Nature Conservancy) Wisconsin Chapter (1997) The places we save. The Nature Conservancy Wisconsin Chapter, Madison

TNC (The Nature Conservancy) Iowa Chapter (2010) The Loess Hills. http://www.nature.org/wherewework/northamerica/states/iowa/preserves/art19446.html. Accessed 21 July 2010

TNC (The Nature Conservancy) Minnesota Chapter (2010) The Nature Conservancy Minnesota Chapter. http://www.nature.org/wherewework/northamerica/states/minnesota/index.html. Accessed 21 July 2010

Towne G, Owensby C (1984) Long-term effects of annual burning at different dates in ungrazed Kansas tallgrass prairie. J Range Manag 37:392–397

Transeau EN (1935) The prairie peninsula. Ecology 16:423–437

Umbanhowar CE Jr (1996) Recent fire history of the northern Great Plains. Am Midl Nat 135:115–121

van Swaay C, van Strien A (2005) Using butterfly monitoring data to develop a European grassland butterfly indicator. In: Kuehn E, Thomas JA, Feldmann R, Settele J (eds) Studies in the ecology and conservation of butterflies in Europe. Proceedings of the conference held in UFZ Leipzig, 5–9th December, 2005, vol. 1: general concepts and case studies. Pensoft Publishers, Sofia, pp 106–108

van Swaay C, Warren M, Loïs G (2006) Biotope use and trends of European butterflies. J Insect Conserv 10:189–209

Vogel JA, Koford RR, Debinski DM (2010) Direct and indirect responses of tallgrass prairie butterflies to prescribed burning. J Insect Conserv (in press)

Vogl RJ (1974) Effect of fire on grasslands. In: Kozlowski TT, Ahlgren CE (eds) Fire and ecosystems. Academic Press, New York, pp 139–194

Warren MS (1993) A review of butterfly conservation in central southern Britain: I. Protection, evaluation, and extinction on prime sites. Biol Conserv 64:25–35

Weaver JE (1954) North American prairie. Johnsen Publishing Co, Chicago

Wendt KM (1984) A guide to Minnesota Prairies. Minnesota Dept of Nat Res Nat Heritage Program, St. Paul

Williams AH (1997) In praise of grazing. Restor Manag Notes 15:116–118

Wisconsin Department of Natural Resources (2000a) DNR endangered species consultation for taking authorization for grassland or savanna management on public or private lands. Wisconsin Dept Nat Res, Madison. http://dnr.wi.gov/org/land/er/take/Grassland_Savanna_Protocol.htm. Accessed 2 Feb 2010

Wisconsin Department of Natural Resources (2000b) Wisconsin statewide Karner blue butterfly habitat conservation plan and environmental impact statement. PUBL-SS-947-00, Wisconsin Dept Nat Res, Madison

J Insect Conserv (2011) 15:103–110
DOI 10.1007/s10841-010-9327-x

ORIGINAL PAPER

Habitat preferences of *Maculinea arion* and its *Myrmica* host ants: implications for habitat management in Italian Alps

Luca Pietro Casacci · Magdalena Witek · Francesca Barbero · Dario Patricelli · Gaetano Solazzo · Emilio Balletto · Simona Bonelli

Received: 24 March 2010 / Accepted: 3 August 2010 / Published online: 17 August 2010

Abstract Projections of climate-change scenarios indicate that many *Maculinea arion* populations will disappear from Europe over the next 50 years. Extinctions will be particularly concentrated around the southern limits of the species' range, such as the Italian peninsula, except mountain populations, mainly in the Alps. *M. arion* is a social parasite of *Myrmica* ants and its survival depends on the presence and abundance of two resources, i.e. (1) a specific larval foodplant and (2) a suitable *Myrmica* host ant. We collected data on *Thymus* plants distribution, *Myrmica* ants abundance, turf height around *Myrmica* nests, distance between nests and the closest thyme plant (*Thymus pulegioides*), and vegetation structure at 14 patches of a *M. arion* fragmented population occurring in the Western Italian Alps (Val Ferret: Aosta). We found that patches with the highest abundance of *M. arion* had significantly higher foodplant abundance. Among 407 nests belonging to nine *Myrmica* species, we found *M. arion* larvae in one *Myrmica lonae* and in two *Myrmica sulcinodis* nests. The distribution of *M. sulcinodis* nests showed the best spatial overlap with *Thymus* plants. *M. sulcinodis* was also significantly more abundant on pastures showing high *M. arion* density, and therefore offered higher adoption probabilities to develop butterfly larvae. On *M. arion* habitat patches, the mean turf height around *Myrmica* nests varied from 10.5 to 22.3 cm and best matched the habitat requirements of *T. pulegioides*. Thus, we suggest that extensive grazing is needed to maintain the two most important resources for *M. arion*. Moreover, patches where *M. arion* was abundant included more numerous plant species characteristic of these mountain pastures, as well as higher plant diversity (Shannon index). As a consequence, *M. arion* represents an umbrella species and a good indicator for the conservation status of mountain grasslands.

Keywords Host ant specificity · *Myrmica* · Butterfly conservation · Habitat quality · Foodplant distribution · Turf height

L. P. Casacci (✉) · M. Witek · F. Barbero · D. Patricelli · G. Solazzo · E. Balletto · S. Bonelli
Department of Animal and Human Biology, University of Turin, Via Accademia Albertina 13, 10123 Turin, Italy
e-mail: luca.casacci@unito.it

Introduction

Butterflies are important for nature conservation due to their vulnerability to habitat change (Van Swaay and Warren 1999; Thomas et al. 2004; Settele et al. 2008) and show continuous decline throughout Europe (Conrad et al. 2006; Settele et al. 2008). Some of the most endangered butterfly species belong to the genus *Maculinea* (Munguira and Martin 1999; Thomas 1995) and are listed, accordingly, in the annexes of the EU-Habitat Directive (van Helsdingen et al. 1996), as well as in the European Red List (Van Swaay et al. 2010). However, despite special management, *Maculinea* populations are decreasing in many European countries, with such a fast rate that during the past 10 years the European status of *Maculinea arion* has changed from Near Threatened (Van Swaay and Waren 1999) to Endangered (Van Swaay et al. 2010).

M. arion populations depend on the presence and abundance of two main resources, i.e. a specific larval foodplant and a suitable *Myrmica* host ant. They occur on dry grasslands, where females lay eggs on the flower heads of *Thymus* spp. or *Origanum vulgare*. The first three larval instars feed on the flowers, but after three weeks they drop to the ground and wait for *Myrmica* ant workers to carry them into their nest. *M. arion* is an obligate predator on the *Myrmica* ants' brood and, inside the ant colonies, experiences strong scramble competition and high mortality rate (Thomas and Wardlaw 1992). *Maculinea* larvae are carried to nests by the workers of any *Myrmica* species that happen to encounter them, so that this event is not species-specific (Elmes et al. 1991, 2002; Akino et al. 1999; Thomas 2002). Integration into the nest, in contrast, is strictly species-specific and it is crucial for the survival and development of *Maculinea* larvae (Thomas et al. 1989; Elmes et al. 2004; Schönrogge et al. 2004). Earlier studies suggest that each of the five European *Maculinea* species had a different *Myrmica* host-ant (Thomas et al. 1989) and that the primary host of *M. arion* was *Myrmica sabuleti* (Thomas 1980; Thomas et al. 1989). The latter information is mostly based on English populations, which finally became extinct because the cessation of grazing caused changes in the vegetation structure and soil temperature and led, in turn, to changes in *Myrmica* ants composition. Nests of *M. sabuleti*, in fact, were replaced by those of the less xero-thermophilous, but unsuitable, *Myrmica scabrinodis*.

Projections of climate-change scenarios predict that many *M. arion* populations will disappear from all Europe over the next 50 years and that those closer to southern limits of the species' range will be the most threatened. While the vast majority of the populations currently occurring in the Italian peninsula are predicted to become extinct, some will survive in the more mountainous areas, mainly in the Alps, whose conservational importance is bound to increase (Settele et al. 2008).

We investigated the habitat preferences of *M. arion* in a fragmented population occurring in the Western Italian Alps. More in particular, we investigated whether habitat patches having higher *M. arion* abundance differ from those where butterflies are less abundant and related relative abundances to factors such as larval foodplant—*Thymus pulegioides*, (a foodplant species of the investigated *M. arion* population) abundance and *Myrmica* ants community composition. We analyzed host ant specificity and turf height around the nests in relation to the preferences of particular *Myrmica* species. Finally, we compared vegetation structures at sites showing different abundances of *M. arion* adults, to test if those showing high butterfly abundance have specific habitat characteristics and/or higher plants diversity.

Materials and methods

Study area

Our study was conducted in Val Ferret, a relatively broad E -W valley on the S slopes of the Alps, in the Aosta area (NW Italy: 45°50′N, 7°00′E, 1,619 m). The site is part of a prospective Special Protection Area (SPA IT1204030) within the NATURA-2000 European network (Fig. 1). In

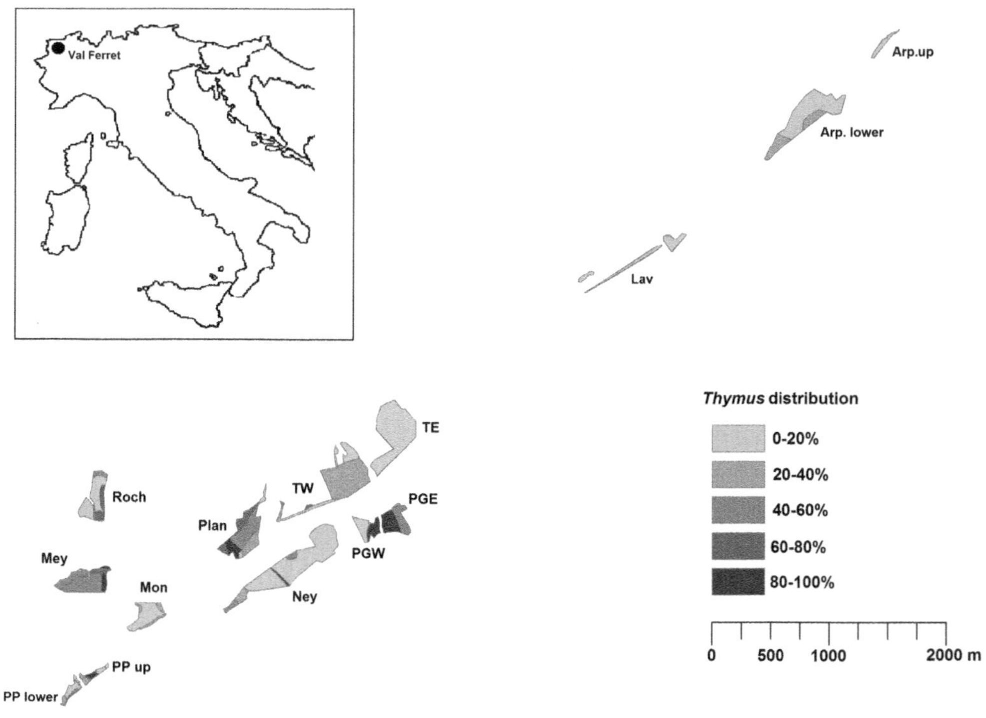

Fig. 1 Map of *M. arion* habitat patches and *T. pulegioides* distribution in Val Ferret (*PP lower*—Lower Pont Pailler, *PP up*—Upper Pont Pailler, *Mey*—Meyen, *Roch*—Rochefort, *Mon*—Montitaz, *Plan*—Planplincieux, *Ney*—Neyron, *TW*—West Tronchey, *TE*—East Tronchey, *PGW*—West Pontgolf, *PGE*—East Pontgolf, *Lav*—Lavachey, *Arp. lower*—Lower Arpnova, *Arp. up*—Upper Arpnova)

this area, *M. arion* inhabits a complex of mountainside pastures forming a fragmented population system.

Butterfly sampling

In 2007 and 2008 we carried out a pilot survey of *M. arion* abundances in several parts of the valley. The main criterion for the initial selection of sites was the abundance of *T. pulegioides*. In 2009, we carried out a more detailed study on 14 selected patches during the *M. arion* flight period, between June 29 and July 20. Patches were monitored every sunny day, and the time we spent at each patch varied from 30 min to 2 h, depending on patch size and butterfly abundance. We used the Mark–Release–Recapture (MRR) method to estimate population parameters (results of this study will be published elsewhere). The total number of adults observed at each patch is shown in Table 1. Based on these data we classified sites into two classes: (1) patches with a high number of *M. arion* adults, (2) patches with no or a few *M. arion* adults. This division was mostly based on data coming from 2009 but also from previous observations made in 2007 and 2008. For all patches belonging to group I, the high number of *M. arion* was observed during all three years of our study. Information on patches subdivision and relative *M. arion* abundances is shown in Table 1.

Thymus surveys

In 2007 and 2008, we evaluated the percentage area covered by *T. pulegioides* immediately before *M. arion* adults started to fly. Every patch was subdivided into a 10 × 10 m grid. Two operators walked across the meadows and recorded the GPS geographical coordinates of every square. *Thymus* presence/absence and areas covered by this plant in each square were visually assessed by Braun-Blanquet's method (Braun-Blanquet 1932) and recorded on a scale ranging from 1 to 5 (1: 0–20%, 2: 20–40%, 3: 40–60%, 4: 60–80%, 5: 80–100%). According to these data, a map of *T. pulegioides* distribution was drawn for all patches (Fig. 1).

Myrmica ant composition and host ant specificity

Field sampling was conducted in June, just before the flight period of *M. arion*, from 2007 to 2009. In order to detect host ant specificity pattern, *Myrmica* nests were searched for within a distance of up 2 m from *T. pulegioides* plants. Since this is the average foraging range of *Myrmica* ants (Elmes et al. 1998), this procedure ensured that all nests examined were potentially available for parasitism by *M. arion* larvae. In the years 2007 and 2008 we measured the distance of *Myrmica* nests found from the nearest *T. pulegioides* plant (the ant nests were searched from 0 to 5 m from *Thymus)* on 254 ant colonies, as well as the turf height surrounding these nests. Those *Myrmica* colonies which were found at a distance higher than 2 m from *Thymus* plants were not taken into account for establishing host ant specificity pattern. In 2009, all 223 nests found were within 2 m from *M. arion* foodplant. The total number of nests investigated was 477 and 407 of them were used to assess *M. arion* host ant specificity in the study area. From each ant nest found, we collected 10–20 workers, which we preserved in alcohol and identified at species level by the keys by Czechowski et al. (2002) and Seifert (2000, 2005).

Table 1 Description of study patches of *M. arion* metapopulation in Val Ferret

Name of patch	Area (ha)	Altitude (m a.s.l.)	Number of captured *M. arion* adults	*M. arion* abundance classes	Weighted means for thyme abundance
East Pontgolf	0.55	1,636	73	1	4.5
West Pontgolf	0.17	1,639	35	1	3
Lower Pont Pallier	0.11	1,445	53	1	1.7
Upper Pont Pallier	0.38	1,493	28	1	2.5
Planplincieux	0.49	1,613	32	1	3.2
Neyron	0.43	1,628	19	2	1.4
East Tronchey	0.78	1,647	19	2	1
West Tronchey	0.43	1,653	11	2	1.9
Meyen	0.49	1,619	8	2	4.2
Rochefort	0.72	1,801	6	2	1.9
Montitaz	0.19	1,543	4	2	1.1
Lavachey	0.06	1,653	0	2	1
Upper Arpnova	0.37	1,760	0	2	1
Lower Arpnova	0.22	1,697	0	2	1.3

Vegetation sampling

Within each patch, we randomly chose two 3 × 3 m areas and fixed their position by GPS. The survey was carried out on a total of 28 squares during the *M. arion* flight period, in 2008 and 2009. For each plant species a Braun-Blanquet visual cover value was recorded within every sampling square, using the same scale as for assessing thyme abundance. All plant species were identified using Pignatti's (1982) identification keys. The total number of plant species found in the investigated area was 101. Additionally, five measurements of the vegetation height were taken at random in each patch.

Statistical analysis

To asses *T. pulegioides* abundance, a weighted mean of abundance indexes was calculated for each patch. We used Mann–Whitney test to compare abundances between groups of patches having high or low/absence number of *M. arion* adults, respectively. Additionally, Spearman correlation between the weighted mean of *Thymus* abundance indexes and the number of *M. arion* adults for all patches was performed.

One-way ANOVA was used to asses differences among *Myrmica* species in nest distance from *T. pulegioides* plant and turf height around nests (square root transformed data for both measurements). These data were collected for nine *Myrmica* species but since only four nests of *Myrmica ruginodis* were found, the sample from these nests was considered too small to be included in the analysis.

We performed a cluster analysis (average linkage) on a matrix of Sørenson similarities, based on the presence/absence of plant species, to examine patterns of species assemblages across patches.

Shannon–Wiener diversity indexes were computed for all plant communities present in our vegetational surveys (squares), as

$$H' = \sum_{i=1}^{i=S} p_i * \ln p_i$$

where p_i is the fraction of number of individuals of the *i*th species to the total number of individuals in the community, and *S* is the species richness (total number of species occurring in the community).

A SIMPER analysis was also performed, to define which species contribute more to similarity/dissimilarity within and between samples. One-way ANOVA was used to asses the differences in vegetation height between the patches with high and low number of *M. arion*. Additionally, Spearman correlation between the vegetation height and the number of *M. arion* adults for all patches was performed.

All statistics were carried out by SPSS18 and PRIMER 6β.

Results

T. pulegioides abundance

The weighted means for *Thymus* abundance, calculated for each patch, are shown in Table 1. *T. pulegioides* abundance is significantly higher on patches having a high abundance of *M. arion* adults (Mann–Whitney test, $N_1 = 5$, $N_2 = 9$, $U = 6.0$, $P = 0.027$) and where *Thymus* coverage is on average 50%, whereas on patches with a lower number of *M. arion Thymus* coverage is on the average 20% only (Fig. 1). Also the correlation test between the abundance of *Thymus* and the number of *M. arion* found in all investigated patches was significant ($r = +0.64$, $P = 0.014$).

Myrmica ant composition and host-ant specificity

Within our study area, we found a total of nine *Myrmica* species. The percentages of nests of particular *Myrmica* species in patches with high and low number of *M. arion* are presented in Fig. 2. Out of 407 *Myrmica* nests found, only three were infested by *M. arion* and each of them contained one butterfly pupa. All infested *Myrmica* colonies were found in one patch (East Pontgolf). Two of the nests belonged to *Myrmica sulcinodis* while one was of *Myrmica lonae*. The rate of infection was very low and less than 1% of the nests were parasitized by pre-emerging *M. arion* throughout the study area.

We observed a significant difference in average distances measured between the *Myrmica* nests and *T. pulegioides* plants ($F_{7,243} = 2.28$, $P = 0.03$, square root transformed data Fig. 3). Moreover, the fraction of *M. sulcinodis* nests observed within the nearest distance from a *Thymus* plant is significantly higher on patches where *M. arion* adults fly in large numbers ($\chi^2 = 8.75$, $P = 0.003$). There is also a significant difference in the turf height surrounding nests of different *Myrmica* species ($F_{7,244} = 3.27$, $P = 0.002$, square root transformed data, Fig. 3). The nests of *Myrmica lobulicornis* occur in the shortest turf heights of about 10 cm, whereas *Myrmica rubra* built their nests in the highest sward heights of about 22 cm (Fig. 3).

Vegetation structure

Cluster analysis based on the presence/absence of individual plant species allows to classify patches into two groups (Fig. 4). The first cluster group consists of all patches containing a high number of *M. arion* adults, plus two other sites (i.e. Meyen and Rochefort) where *M. arion* was less abundant, whereas the second cluster group consists of all the other patches, which contained few or no *M. arion*

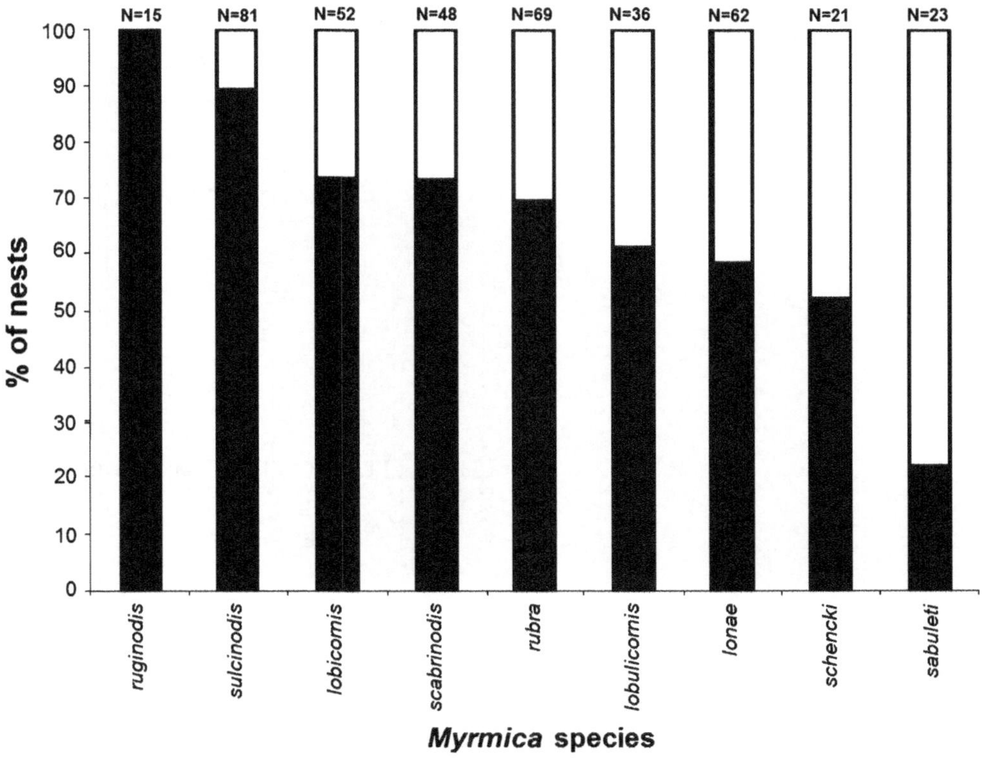

Fig. 2 Percentage of nests of particular *Myrmica* species on patches with high (*black bars*) and low (*white bars*) *M. arion* abundance in Val Ferret (*N* is the total number of nests found in all study area)

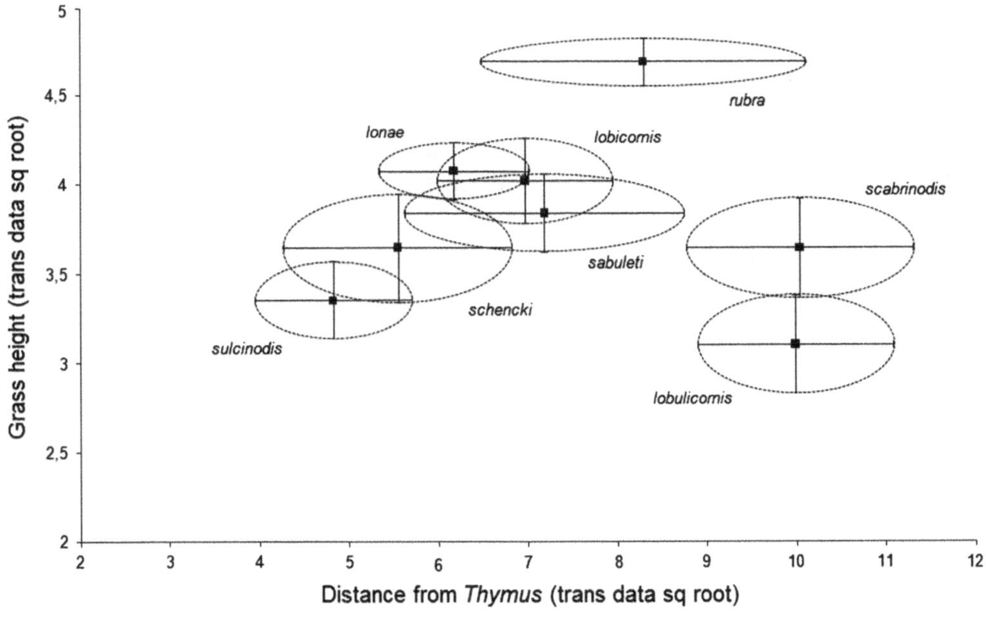

Fig. 3 The mean distance of nests of eight *Myrmica* species from *Thymus* plants plotted against the grass height around the nests (both data in the graph are square root transformed). The *ellipses* were built using the SE of each measures

adults. Shannon diversity indexes obtained for sites belonging to the first cluster were significantly higher than those from the second cluster (Mann–Whitney tests, $N_1 = 14$, $N_2 = 14$, $U = 41.00$, $P = 0.009$). SIMPER analysis shows that the former group hosts more plant species characteristic of mountain grasslands, such as: *Chaerophyllum hirsutum*, *Agrostis alpina*, *Geranium sylvaticum*, *Carex sempervirens*, *Leontodon montanus*. The other group of patches, in contrast, is characterised by plants frequently found in the intensively managed habitats, such as *Alchemilla vulgaris*, *Trifolium repens*, *Trifolium pratense*, *Anthriscus sylvestris* (a strongly nitrophilous species), or *Ranunculus acris.* Vegetation is significantly higher on patches belonging to the first cluster (one way ANOVA, $F_{1,69} = 5.61$, $P = 0.021$) than in those belonging to second one. Similarly, the correlation between the vegetation height and the number of *M. arion* for all patches is significant ($r = +0.612$, $P = 0.035$).

Fig. 4 Dendrogram of patches similarities based on presence/absence of plant species (*white squares* patches with high number of *M. arion*; *black squares* patches with low/absent number of *M. arion*)

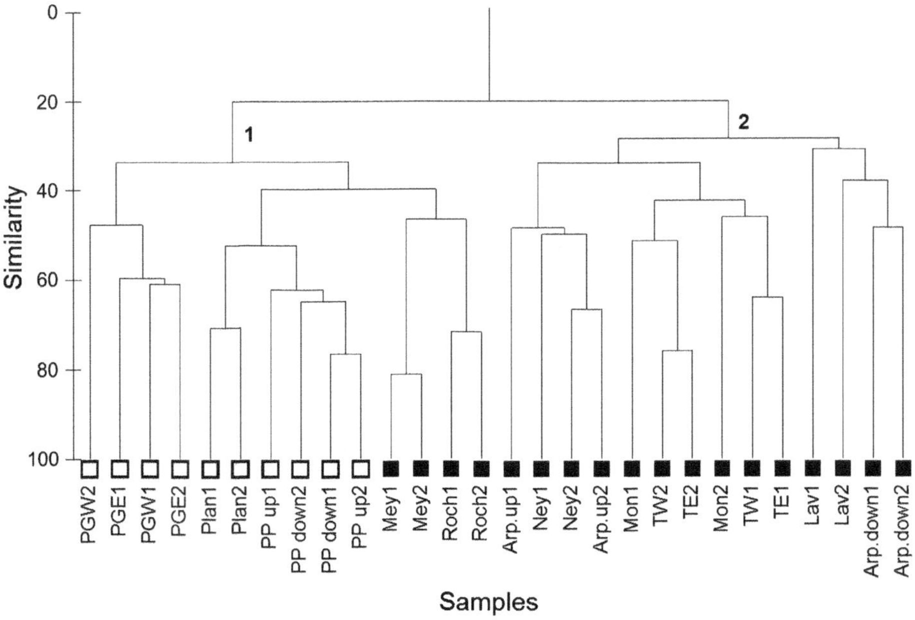

Discussion

In the Western Italian Alps, *M. arion* is more densely present in patches characterised by abundant *Thymus* foodplants, and where this plant's spatial distribution is more overlapped with the one of *Myrmica* host ant. Food-plant vicinity to host-ant nests is apparently the most important factor for population survival. This finding is in agreement with previous observations by Thomas et al. (2009), who have shown that in English populations there is no significant relationship between population survival and patch size or isolation rate, while the most significant factor is a sufficient overlap between the larval foodplants (*Thymus*) and host (*M. sabuleti*) nests. The relatively simple model emerging from life-table studies of *M. arion* in England shows that the intrinsic population growth rate is higher than 1.0 where a *Thymus* coverage larger than 68% occurs within the foraging range of *M. sabuleti* ants (Thomas 2009). Results from our study show that colonies of different *Myrmica* species occur at different distances from the *T. pulegioides* plants, probably as a consequence of their different microhabitat requirements. *M. sulcinodis* colonies occur in the nearest range with respect to *M. arion* foodplants, which can suggests that they have similar habitat demands as those of *T. pulegioides*. Therefore, larger overlapping and smaller distances can results in higher probability of *M. arion* larvae adoption since a higher number of foraging workers can be found in near surrounding of *Thymus* plants. *M. sulcinodis* is, moreover, also the most abundant *Myrmica* species on pastures with a high number of *M. arion* adults, whereas it is a scarce species on patches with lower butterfly abundance. It seems, therefore, that *M. arion* mostly inhabits patches where these two resources fully overlap each other.

We have found *M. arion* pupae in three nests belonging to *M. sulcinodis* and *M. lonae*. Data obtained from various *M. arion* populations show that *M. sabuleti* is the main European host ant species (Thomas et al. 1989, 1998). Other studies carried out in Poland, however, show that *M. arion* can use there as hosts five *Myrmica* species, i.e. *M. schencki*, *M. lobicornis*, *M. rugulosa*, *M. hellenica* and *M. lonae* (Sielezniew and Stankiewicz 2008; Sielezniew et al. 2009, 2010a, b). Our data suggest that the probability that *M. sabuleti* is host of *M. arion* in Val Ferret is small, since the proportion of this *Myrmica* species is on average very low (3%), especially on meadows showing a high number of adult *M. arion*, and no pupae were found in its nests. Moreover, none of *M. sabuleti* nest was found on patch (East Pont Golf) with the highest number of *M. arion* and where pupae were found in nests of other *Myrmica* species. *M. sulcinodis* and *M. lonae* are the most abundant *Myrmica* species in Val Ferret and their nests spatially overlaps well *T. pulegioides* plants. This makes the probability of *M. arion* adoption higher with respect to other *Myrmica* species. Unfortunately, the very small number of infested nests observed in our study area is insufficient to demonstrate whether or not these two *Myrmica* species can be consider as primary hosts (sensu Thomas et al. 1989) of *M. arion* in Val Ferret or multiple host use occurs there.

Thomas et al. (1998) point out that before drawing conservation plans, it is essential to have precise information on the turf structure and niche preferences of the local host-ant species. In England, a strong connection between turf height and the *Myrmica* community has been demonstrated (Thomas et al. 1998, 2009). In particular, *M. sabuleti* has very narrow niche, with optimal turf height of 0–3 cm, whereas the ant's density decreases with increasing height and the species totally disappears when the sward is higher than 7 cm (Thomas et al. 2009). In Val Ferret, the host-ant specificity of *M. arion* was apparently not limited to one *Myrmica* species. Moreover, it seems that most of the *Myrmica* species occurring there have relatively broad niches, in relation to turf height. For example, *M. sabuleti* nests were found in grass as short as 5–10 cm, but there were also many of them in much higher turf, from 20 to more than 30 cm high (Fig. 3). In the Italian Alps, the broader tolerance of *Myrmica* species towards variable turf heights may be a result of the much higher monthly and daily temperature excursion. Data obtained in Val Ferret in July 2009 show that the mean daily maximum temperature was 27°C and the minimum was 11°C (C. Cerrato, personal information). Data from SE England show that the mean daily maximum temperature is 19.5°C and the minimum is 13.5°C. Thus, in July, daily temperature changes are much larger in Val Ferret, where ant species are apparently more tolerant to temperature changes and better adapted to its oscillations. It is relevant to observe, that Elmes et al. (1998) point out that niches of all *Myrmica* species become narrower toward the edge of the European *Maculinea* range, which can explain why in England *M. sabuleti*'s niche is so narrow in comparison to Val Ferret, where the optimal turf height for this species is between 5 and 20 cm and between 10 and 20 cm for most *Myrmica* species. The habitat demands of *T. pulegioides* (*M. arion*'s local foodplant) are similar in both geographical regions and in the Alps this plant usually reaches 20–30 cm in height (Pignatti 1982).

We suggest, therefore, that a suitable habitat management to maintain both the *Myrmica* host-ants and the *M. arion* larval foodplant, is to keep turf height between 10 and 20 cm, which in practice means to maintain a not very intensive grazing regime. Moreover, it is known that grassland management intensity, including grazing, influences biodiversity. Marini et al. (2009) have demonstrated that intensive management decreases plant diversity, as well as orthopteran and butterfly species richness in the Italian Alps (see also Balletto et al. 1982). Similar data by Kruess and Tscharntke (2002) show that reducing grazing intensity promotes insect diversity in some German grasslands. Our vegetational observations have shown that where many plants characteristic of intensive grazing are found, abundance of *M. arion* adults is low. In contrast, patches with high numbers of *M. arion* adults are characterised by higher number of grass and herb species typical of the warm, mountain grasslands of the Western Alps. These pastures are grazed only once a year and for a short period of time, or even once every 2 years.

These findings may have important conservation value, because it may be possible, on their basis, to predict which habitats may be potentially occupied by *M. arion* in Alpine conditions. It is worth to notice, in this connection, that according to their vegetation, two patches, i.e. Rochefort and Meyen, belong to the same "vegetational cluster" (Fig. 4) as patches having dense populations of this butterfly species. *T. pulegioides* plants are also quite abundant on these two sites, which may suggest that they are potentially suitable habitat for *M. arion*. One of the host ant, *M. sulcinodis*, however, is absent from both patches and low numbers of *M. arion* were observed.

Our findings show that, in the Italian Alps, the highest abundance of *M. arion* occurs in patches characterised by high plant diversity, a feature that other studies have demonstrated to occur together with rich insect (and butterfly) communities (Suominen and Olofsson 2000; Välimäki 2005). In the Czech Carpathians, *M. arion* only occurs on grasslands supporting high plant- and butterfly diversity (Spitzer et al. 2009), while Randle et al. (2005) show that *M. arion* is an indicator of rare niches in some UK grasslands. In our study area, accordingly, *Parnassius apollo*, and the very rare myrmecophilous syrphid fly *Microdon analis* were observed on patches having high abundance of adult *M. arion* (data not shown). Our results confirm that *M. arion* is an umbrella species, which means that protecting sites occupied by this butterfly will also allow to protect other rare plants and invertebrate communities.

Acknowledgments We wish to thank Elena Giacometti and dr Piotr Nowicki for their help during field work in Val Ferret. The Italian Ministry for the Environment and Regione Autonoma Valle d'Aosta issued the relevant permissions for our studies. This study was partly funded by the BiodivERsA project CLIMIT (Settele and Kühn 2009; Thomas et al. 2009). The research was conducted within the cooperation project Interreg IIIA Alcotra "Cogeva Vahsa" and partly funded by the Valle d'Aosta Autonomous Region—Department of Agriculture and Natural Resources".

References

Akino T, Knapp JJ, Thomas JA, Elmes GW (1999) Chemical mimicry and host specificity in the butterfly *Maculine rebeli*, a social parasite of *Myrmica* ant colonies. Proc R Soc Lond B 266:1419–1426. doi:10.1098/rspb.1999.0796

Balletto E, Barberis G, Toso GG (1982) Aspetti dell'ecologia dei lepidotteri ropaloceri nei consorzi erbacei delle Alpi italiane. Quaderni sulla "Struttura delle Zoocenosi Terrestri" CNR, Roma, 2(II.2). Pubbl. AQ/1/193, pp 11–95

Braun-Blanquet J (1932) Plant sociology: the study of plant communities. McGraw Hill, New York

Conrad KF, Warren MS, Fox R, Parsons MS, Woiwod IP (2006) Rapid declines of common, widespread Britishmoths provide evidence of an insect biodiversity crisis. Biol Conserv 132:279–291. doi:10.1016/j.biocon.2006.04.020

Czechowski W, Radczenko A, Czechowska W (2002) The ants (Hymenoptera, Formicidae) of Poland. MIZ PAN, Warszawa

Elmes GW, Thomas JA, Wardlaw JC (1991) Larvae of *Maculinea rebeli*, a large blue butterfly, and their *Myrmica* host ants: wild adoption and behaviour in ant-nests. J Zoo 223:447–460. doi:10.1111/j.1469-7998.1991.tb04789.x

Elmes GW, Thomas JA, Wardlaw JC, Hochberg ME, Clarke RT, Simcox DJ (1998) The ecology of *Myrmica* ants in relation to the conservation of *Maculinea* butterflies. J Insect Conserv 2:67–78. doi:10.1023/A:1009696823965

Elmes GW, Akino T, Thomas JA, Clarke RT, Knapp JJ (2002) Interspecific differences in cuticular hydrocarbon profiles of *Myrmica* ants are sufficiently consistent to explain host specificity by *Maculinea* (large blue) butterflies. Oecologia 130:525–535. doi:10.1007/s00442-001-0857-5

Elmes GW, Wardlaw JC, Schönrogge K, Thomas JA, Clarke RT (2004) Food stress causes differential survival of socially parasitic caterpillars of *Maculinea rebeli* integrated in colonies of host and non-host *Myrmica* ant species. Entomol Exp Appl 110:53–63. doi:10.1111/j.0013-8703.2004.00121.x

Kruess A, Tscharntke (2002) Grazing intensity and the diversity of grasshoppers, butterflies, and trap-nesting bees and wasps. Conserv Biol 16(6):1570–1580

Marini L, Fontana P, Klimek S, Battisti A, Gaston KJ (2009) Impact of farm size and topography on plants and insect diversity of managed grasslands in the Alps. Biolo Conserv 142:394–403. doi:10.1016/j.biocon.2008.10.034

Munguira ML, Martin J (1999) Action plan for the *Maculinea* butterflies in Europe. Nature and Environment, No. 97. Council of Europe Publishing, Strasbourg

Pignatti S (1982) Flora d'Italia, vols I, II, III. Edagricole, Bologna

Randle Z, Simcox DJ, Schönrogge K, Wardlaw JC, Thomas JA (2005) *Myrmica* ants as keystone species and *Maculinea arion* as an indicator of rare niches in UK grasslands. In: Settele J, Kühn E, Thomas JA (eds) Studies on the ecology and conservation of butterflies in Europe. Species ecology along a European gradient: *Maculinea* butterflies as a model, vol 2. Pensoft, Sofia, Moscow, pp 26–27

Schönrogge K, Wardlaw JC, Peters AJ, Everett S, Thomas JA, Elmes GW (2004) Changes in chemical signature and host specificity from larval retrieval to full social integration in the myrmecophilous butterfly *Maculinea rebeli*. J Chem Ecol 30:91–107. doi:10.1023/B:JOEC.0000013184.18176.a9

Seifert B (2000) Myrmica lonae Finzi, 1926—a species separate from Myrmica sabuleti Meinert, 1861 (Hymenoptera: Formicidae). Abh Ber Naturkundemus Görlitz 72:195–205

Seifert B (2005) Rank elevation in two European ant species: *Myrmica lobulicornis* Nylander, 1857, stat. n. and *Myrmica spinosior* Santschi, 1931, stat. n. (Hymenoptera: Formicidae). Myrmecol Nachr 7:1–7

Settele J, Kühn E (2009) Insect conservation. Science 325:41–42

Settele J, Kudrna O, Harpke A, Kühn I, Van Swaay C, Verovnik R, Warren M, Wiemers M, Hanspach J, Hickler T, Kühn E, Van Halder I, Veling K, Vliegenthart A, Wynhoff I, Schweiger O (2008) Climatic risk atlas of European butterflies. Pensoft, Sofia-Moscow

Sielezniew M, Stankiewicz A (2008) Myrmica sabuleti (Hymenoptera: Formicidae) not necessary for the survival of the population of Phengaris (Maculinea) arion (Lepidoptera: Lycaenidae) in eastern Poland: lower host-ant specificity or evidence for geographical variation of an endangered social parasite? Eur J Entomol 105:637–641

Sielezniew M, Włostowski M, Dziekańska I (2009) *Myrmica schencki* (Hymenoptera: Formicidae) as the primary host of *Phengaris* (*Maculinea*) *arion* (Lepidoptera: Lycaenidae) at Heathlands in eastern Poland. Sociobiology 55(IB):95–106

Sielezniew M, Dziekańska I, Stankiewicz-Fiedurek AM (2010a) Multiple host-ant use by the predatory social parasite *Phengaris* (*=Maculinea*) *arion* (Lepidoptera, Lycaenidae). J Insect Conserv 14:141–149. doi:10.1007/s10841-009-9235-0

Sielezniew M, Patricelli D, Dziekańska I, Barbero F, Bonelli S, Casacci LP, Witek M, Balletto E (2010b) The first record of *Myrmica lonae* (Hymenoptera: Formicidae) as a host of the socially parasitic large blue butterfly *Phengaris* (*Maculinea*) *arion* (Lepidoptera: Lycaenidae). Sociobiology (in press)

Spitzer L, Benes J, Dandova J, Jaskova V, Konvicka M (2009) The large blue butterfly, *Phengaris* [*Maculinea*] *arion*, as a conservation umbrella on a landscape scale: The case of the Czech Carpathians. Ecol Indic 9(6):1056–1063. doi:10.1016/j.ecolind.2008.12.006

Suominen OL, Olofsson J (2000) Impacts of semi-domesticated reindeer on structure of tundra and forest communities in Fennoscandia: a review. Ann Zool Fenn 37:233–249

Thomas JA (1980) Why did the large blue become extinct in Britain? Oryx 15:243–247

Thomas JA (1995) The ecology and conservation of *Maculinea arion* and other European species of the large blue butterfly. In: Pullin AS (ed) Ecology and conservation of butterflies. Chapman & Hall, London

Thomas JA (2002) Larval niche selection and evening exposure enhance adoption of a predacious social parasite, *Maculinea arion* (large blue butterfly) by *Myrmica* ants. Oecologia 132:531–537. doi:10.1007/s00442-002-1002-9

Thomas JA, Wardlaw JC (1992) The capacity of a *Myrmica* ant nest to support a predacious species of *Maculinea rebeli* inhabitants in ant nests. Oecologia 91:101–109. doi:10.1007/BF00317247

Thomas JA, Elmes GW, Wardlaw JC, Woyciechowski M (1989) Host specificity among *Maculinea* butterflies in *Myrmica* ant nests. Oecologia 79:452–457. doi:10.1007/BF00378660

Thomas JA, Simcox DJ, Wardlaw JC, Elmes GW, Hochberg ME, Clarke RT (1998) Effects of latitude, altitude and climate on the habitat and conservation of the endangered butterfly *Maculinea arion* and its *Myrmica* ant hosts. J Insect Conserv 2:39–46

Thomas JA, Telfer MG, Roy DB, Preston CD, Greenwood JJD, Asher J, Fox R, Clarke RT, Lawton JH (2004) Comparative losses of British butterflies, birds, and plants and the global extinction crisis. Science 303:1879–1881. doi:10.1126/science.1095046

Thomas JA, Simcox DJ, Clarke RT (2009) Successful conservation of a threatened *Maculinea* butterfly. Science 325:80–83

Välimäki P (2005) The effects of the reindeer grazing on Lepidopteran communities in two fells in northern Fennoscandia. In: Jokinen M (ed) The effects of reindeer herding and protection in the Malla Nature Reserve. Metsäntutkimuslaitoksen tiedonantoja 941:179–227

Van Helsdingen PJ, Willemse L, Speight MCD (1996) Background Information on invertebrates of the Habitats Directive and the Bern Convention. Part I—Crustacea, Coleoptera and Lepidoptera. Nature and Environment 79. Council of Europe Publishing, Strasbourg

Van Swaay C, Warren MS (1999) Red Data book of European butterflies (Rhopalocera). Nature and the Environment, No. 99. Council of Europe Publishing, Strasbourg

Van Swaay C, Cuttelod A, Collins S, Maes D, Lopez Munguira M, Šašić M, Settele J, Verovnik R, Verstrael T, Warren M, Wiemers M, Wynhof I (2010) European red list of butterflies. Publications Office of the European Union, Luxembourg

J Insect Conserv (2011) 15:189–206
DOI 10.1007/s10841-010-9337-8

ORIGINAL PAPER

Not only the butterflies: managing ants on road verges to benefit *Phengaris* (*Maculinea*) butterflies

Irma Wynhoff · René van Gestel · Chris van Swaay · Frank van Langevelde

Received: 12 April 2010 / Accepted: 19 August 2010 / Published online: 19 September 2010

Abstract Obligate myrmecophilic butterfly species, such as *Phengaris (Maculinea) teleius* and *P. nausithous,* have narrow habitat requirements. Living as a caterpillar in the nests of the ant species *Myrmica scabrinodis* and *M. rubra*, respectively, they can only survive on sites with both host ants and the host plant Great Burnet *Sanguisorba officinalis*. After having been reintroduced into a nature reserve in the Netherlands in 1990, both butterfly species expanded their distribution to linear landscape elements such as road verges and ditch edges outside this reserve. As additional habitat of both butterfly species, vegetation management of these landscape elements became important. Our results show that a management beneficial for *Phengaris* butterflies should aim to increase the nest density of *Myrmica* species, at the same time reducing the density of nests of the competitor *Lasius niger* or at least keeping them at a low density. Unfavourable grassland management under which *L. niger* thrives, includes not mowing or flail-cutting the grass, or depositing dredgings along the side of the ditch. Management favourable for the two *Myrmica* species differs, demanding some flexibility if both species are to benefit. *M. scabrinodis* is best supported with early mowing of the road verge vegetation or late mowing in the nature reserve, both of which result in an open vegetation and warm microclimate. In contrast, the nest sites of *M. rubra* should be left undisturbed during the summer, and mown in late autumn. Mowing of butterfly habitat should be avoided between mid-June and mid-September as this would remove the flowerheads of the *Sanguisorba* plants, on which the butterflies lay their eggs.

Keywords *Myrmica* · Ant community · Conservation management · Road verge management · Reintroduction

Introduction

Sometimes actions in nature conservation have surprising results. The reintroduction of two Large Blue butterfly species, *Phengaris (Maculinea) teleius* and *P. nausithous,* into the Netherlands is a case in point. Released in 1990 into the nature reserve Moerputten in the southern province Noord-Brabant (Wynhoff 1998), they expanded their distribution outside the protection of this reserve, occupying the verges of roads running between agricultural fields. Until the 1970s, both species had occurred on meadows and pastures in stream valleys in the middle and south of the Netherlands; the last populations were eradicated by agricultural intensification. Since both species have an obligate relationship with *Myrmica* ants (Thomas and Settele 2004), the success of the reintroduction depended not only on the host plant *Sanguisorba officinalis* being abundant, but also large numbers of the host ant species were needed to provide suitable habitat. The new populations developed well and showed a considerable increase in numbers (Van Langevelde and Wynhoff 2009). *P. teleius* only colonized the meadow where it was released and hardly dispersed to other locations. Recently, however,

I. Wynhoff (✉) · R. van Gestel · C. van Swaay
Dutch Butterfly Conservation, P.O. Box 506, 6700 AM Wageningen, The Netherlands
e-mail: Irma.Wynhoff@vlinderstichting.nl

I. Wynhoff · C. van Swaay
Butterfly Conservation Europe, P.O. Box 506, 6700 AM Wageningen, The Netherlands

R. van Gestel · F. van Langevelde
Resource Ecology Group, Wageningen University, Droevendaalsesteeg 3a, 6708 PB Wageningen, The Netherlands

various subpopulations developed outside Moerputten at distances ranging from 1.5 to 5 kilometres, mostly on road verges and in another nearby nature reserve. The Dusky Large Blue, *P. nausithous* also settled first in Moerputten nature reserve, establishing a subpopulation on road verges south of the reserve after only 3 years. Later, two new subpopulations of *P. nausithous* were found at distances of 2 and 5 kilometres from the reintroduction site, one on a road verge. It became obvious that the habitat quality of the road verges, ditch edges and embankments was very important for the long-term persistence of both species (Van Langevelde and Wynhoff 2009). The road verges could fulfil a role not only as a dispersal corridor but also as temporary or permanent habitat (Haddad 2000). The question was how should such linear landscape elements as road verges be managed to facilitate the spread of these two butterfly species?

The caterpillars of *P. teleius* and *P. nausithous* spend about 10 months in the nests of *Myrmica scabrinodis* and *M. rubra* respectively feeding on ant larvae. The eggs of both butterfly species are deposited on the only host plant *Sanguisorba officinalis* (Thomas 1984). Hence, the suitability of a certain habitat patch is determined by the abundance of both the host plant and the specific host ants (Van Dyck et al. 2000 (for *P. alcon*); Wynhoff et al. 2008; Anton et al. 2008; Dierks and Fischer 2009; Batáry et al. 2009). In and around the nature reserve Moerputten, the density of ant nests is much lower than that of the host plants (Wynhoff et al. 2008; Van Langevelde and Wynhoff 2009). On road verges, too, and in some other nature reserves, the host plant density is also high, whereas it is missing in the pastures and fields under agricultural use. As both these plants and the ants are sensitive to management (Dahms et al. 2005; Grill et al. 2008), the province of Noord-Brabant designated the "*Maculinea* Action Plan Area" to stimulate an increase in the density of *Myrmica* ant nests and further dispersal of the *Phengaris* populations (Provincie Noord-Brabant 1997).

Within Europe, ants of the genus *Myrmica* and the competitive species of the genus *Lasius* are common and widespread (Seifert 1994, 1996; Van Loon and Mabelis 1996). Although each species occupies an ecologically distinct habitat, there is overlap in the distribution of species. On many sites, two or three *Myrmica* or *Lasius* species may occur in competition with each other (Elmes and Wardlaw 1982; Elmes et al. 1998). *Myrmica rubra* prefers warm and moist habitats consisting of rough grassy vegetation along edges of bogs, rivers and woods. *Myrmica scabrinodis* occurs on the short and open vegetation of slightly more thermophilous grasslands and on bogs and moors. The workers of *Lasius flavus* build very large nest mounds and are, like *Myrmica* species, sensitive to disturbance. *Lasius niger* is a rapid colonizer of disturbed ground. Able to stand high levels of disturbance, it is a serious competitor of *Myrmica* ants (Peeters et al. 2004).

Besides geographical differences, ant communities are structured by vegetation characteristics in relation to prey abundance and intraspecific and interspecific competition (Dauber and Wolters 2000). The vegetation may be changed by various factors such as disturbances or management, leading to direct and indirect effects on the ant communities. Direct effects include destruction of nests and reduction of ant biomass (Heller and Rohe 2000). However, the indirect effects are considered to be much more important (Elmes and Wardlaw 1982; Curry 1994; Elmes et al. 1998; Kruess and Tscharntke 2002; Armbrecht et al. 2004). The changes in vegetation structure (height and density of plants, plant species richness) and soil conditions brought about by disturbance, leads to changes in microclimate, food supply and habitat structure. The indirect effects of various types of management on ant communities have been studied on the scale of agricultural fields under various types of agricultural use (Dahms et al. 2005; Dauber et al. 2006) and in an experiment on meadows (Grill et al. 2008). But we know very little about the effect of various types of management on the presence and abundance of *Myrmica* ants on road verges and stream borders, where the vegetation is similar to that in hay meadows and extensively grazed pastures, and which is subject to frequent management.

In the years after reintroducing the butterflies, we monitored the distribution of the adults intensively in order to study factors influencing the selection of oviposition sites (Wynhoff et al. 2008) and the limitations for their spread (Van Langevelde and Wynhoff 2009). We also monitored the presence and abundance of ant nests in and around the nature reserve Moerputten over several years. In this paper, we seek to explain what determines the presence and abundance of *Myrmica* ants over time by testing various factors. We study whether there are management measures that can be taken to influence the presence and abundance of ants positively. Our results will help government and NGOs to design refined management strategies to promote the highly specialized *Phengaris* butterflies.

Materials and methods

Our study area covers the nature reserve Moerputten in the Dutch province of Noord-Brabant (51°41′N, 5°15′E, altitude 2 m above sea level, for further description see Wynhoff 1998) and surroundings, extending to the river Maas in the north to the Drongelens Kanaal in the south (Fig. 1a). In the Action Plan Area (Fig. 1b), we randomly chose 179 plots with host plants to monitor the ant

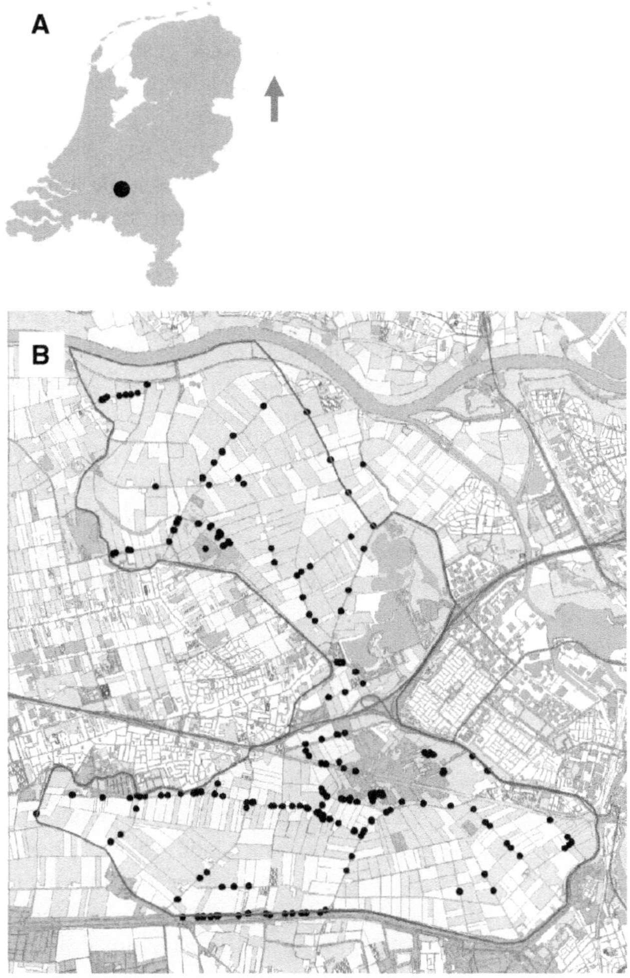

Fig. 1 **a** Location of the study area, Moerputten nature reserve and surroundings in the Netherlands, and **b** the distribution of ant plots in the study area, i.e. in the nature reserves, on the road verges, on the railway embankment and along streams

Table 1 Overview of results of ant monitoring in the nature reserve Moerputten and surrounding road verges, ditch sides and other nature reserves per year of investigation

	1996	2000	2003	2006	2009
Number of plots	179	179	193	193	208
Number of plots with ants	139	112	85	117	157
% Occupied with ants	77	63	44	61	75
Number of plots with *Myrmica*	65	67	40	68	85
% Occupied with *Myrmica*	36	37	21	35	41
Nr. of species found	10	9	10	9	9
Mean nr of species per plot	2,32	1,90	1,28	1,79	2,39
±SE	1,54	1,63	1,55	1,61	1,69
Max. nr of species per plot	6	5	6	6	7

community (up to 208 plots in 2009; Table 1). Using GPS, we recorded the coordinates of each plot, in addition to field maps and site descriptions.

Ant monitoring was carried out in June, July or August in the years 1996, 2000, 2003, 2006, and 2009 on dry days warmer than 17°C. Each plot was searched during exactly 15 min for ants in an area of about ten square metres around one or more *Sanguisorba* plants. Since it is impossible to determine the exact density of ant nests on plots of equal size and achieve high sample sizes at the same time, we chose to keep the search effort of 15 min equal to be able to compare the results. All plots in nature reserves were distinguished from the road verge plots. On plots along road sides we searched the whole gradient from dry strips close to the asphalt to the humid bottom of the road side ditches. Worker ants appear from their nest in the soil by disturbing the soil with a knife and by plucking at vegetation. We recorded both the presence of an ant species as well as the (relative) density of their nests (here called abundance). Some of the worker ants were collected for later identification in the lab, using keys of Van Boven and Mabelis (1986) and Seifert (1994, 1996). By actively searching for ants, the results are independent from the behaviour of the worker ants and largely independent from weather conditions.

Apart from the ant community, we also recorded the direction of the slope of the plot and its distance to the nearest tree or shrub as proxy of microclimatic conditions possibly affecting the ant community (Armbrecht et al. 2004). Furthermore, during 2006 and 2009 we consistently assessed the management by the responsible municipality. Since during the flight period from the end of July to mid-August, host plants with buds must be available for oviposition, the time of mowing is an important aspect of management. It means that cutting must not happen later than early June, to avoid that regrowth would be too young to produce new flowers. Furthermore, the host plants should not be cut before mid-September after the caterpillars have left the flowerheads and stay safe in the ant nests. We distinguished seven management types: (1) early mowing before mid-June, (2) late mowing after mid-September, (3) partly mown, (4) mown and removed between mid-June and mid-September, (5) not mown, (6) flail cutting, and (7) covered with dredged ditch sediment. Usually, grass is cut and the hay is removed immediately or within a week except for flail cutting, which tears the grass into pieces and leaves it behind on the spot. Since 1997, management was done according to the *Maculinea* Action Plan which implies that mowing is avoided between mid-June and mid-September in order to increase the amount of habitat of the butterflies. Also, flail cutting should be avoided.

Statistical analysis

We tested the influence of the variables mentioned above on the presence-absence and the abundance of all ant species combined, all *Myrmica* species together, and the four main ant species separately (*Myrmica rubra*, *M. scabrinodis*, *Lasius flavus* and *L. niger*). Because the mobility of ant colonies is limited, it is likely that an ant species is still present during the next three or 4 years after it has been found during a certain year. To correct for this dependence between years, we used Generalized Estimation Equations (GEEs, Zeger et al. 1988). When running tests on presence-absence patterns, we used the binomial distribution with the logit-link function. The abundances were studied assuming a normal distribution using the identity-link function. We used the categorical variables year, municipality, slope direction and distance to trees (because we approximated this distance, we used distance classes) as predictor variables. We distinguished several expositions of the slope: north, northeast, east, southeast, south & southwest (added together), west, northwest and flat. As the ubiquitous ant species *L. niger* can be an important competitor of *Myrmica* ants that can easily colonize disturbed sites, we used its abundance as an additional covariate when working on the presence-absence and abundance of all *Myrmica* or *Lasius* species together or separately (*M. rubra*, *M. scabrinodis* and *L. flavus*). For the significant categorical variables, we tested the pairwise comparisons using the Least Significant Differences adjustment (LSD) to find significant differences between the groups. By adding and removing predictor variables, we selected the best model using the Corrected Quasi Likelihood under Independence Model Criterion (QICC) that combines the highest number of factors and covariates explaining the variation in the tested dependent variable (low QICC values indicate more parsimonious models).

To test the effect of the various types of road verge management on the presence-absence and abundance of all ant species combined, all *Myrmica* ants, and the four main ant species (*M. rubra*, *M. scabrinodis*, *L. flavus* and *L. niger*) for the years 2006 and 2009 together, we used Generalized Linear Models (GLM). For presence-absence, we again used the binomial distribution with the logit-link function, and we applied a normal distribution with identity-link function for the abundances. We used the categorical variable management type as a predictor variable. We applied LSD pairwise comparisons to find significant differences between the given types of management. Plots in nature reserves were excluded from the analysis of management effects because the ant community was found to be too deviating from the communities on the road verges. We could not prove that this was independent from factors other than management.

Results

We found that most of the plots were occupied with ants in 1996 and 2009, with the lowest level of occupation in 2003 (Table 1). The percentage of plots with *Myrmica* ants was also lowest in 2003, while in the other years occupation was similar. The number of ant species found per year was remarkably stable: 9 or 10 ant species per year, with maximally 5 to 7 species in one plot on 1 day. Mean ant species richness between 1.3 and 2.4 seems to be low, but one has to consider the small plot size, when comparing the data with that in the literature. The most widespread and abundant ant species in all years were *M. scabrinodis*, *M. rubra*, *L. flavus* and *L. niger*. Their changes in occupation are given in Fig. 2. Furthermore, in all years, we also found *M. ruginodis*, *M. sabuleti*, *M. rugulosa*, *Formica fusca* and *F. rufibarbis*, but only on a small number of plots with a few colonies. The rare *M. rugulosa* is the only ant species that is found in the study area only in the nature reserves. Some species, namely *Lasius umbratus*, *Tetramorium caespitum*, *Leptothorax acervorum*, *Leptothorax nylanderi* and *Ponera coarctata* were even rarer, being found in only one or two of the years we monitored.

Striking differences appeared, when comparing the percentage of occupied plots of the four most common species between the municipalities and the nature reserves (Fig. 3). *L. flavus* is the only ant species found in almost equal percentages in all three municipalities and in the nature reserves. *M. scabrinodis* occurs much more in nature reserves than on road verges and ditch edges, whereas *L. niger* reaches its highest occupation outside nature reserves. *M. rubra* was not encountered in the municipality of Den Bosch, but only in those of Heusden and Vught, as well as the nature reserves, with the highest percentage of occupation in Vught.

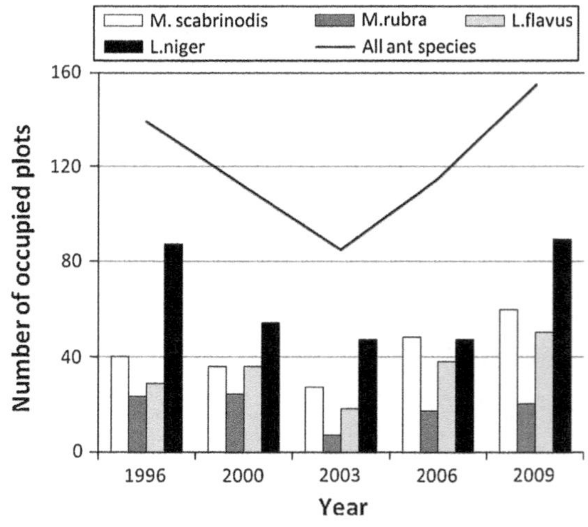

Fig. 2 Number of occupied monitoring plots in each of the 5 years investigated for the four main ant species and the number of all plots occupied by any ant species. The data were collected on 179 plots in Moerputten nature reserve and its surroundings in the Netherlands

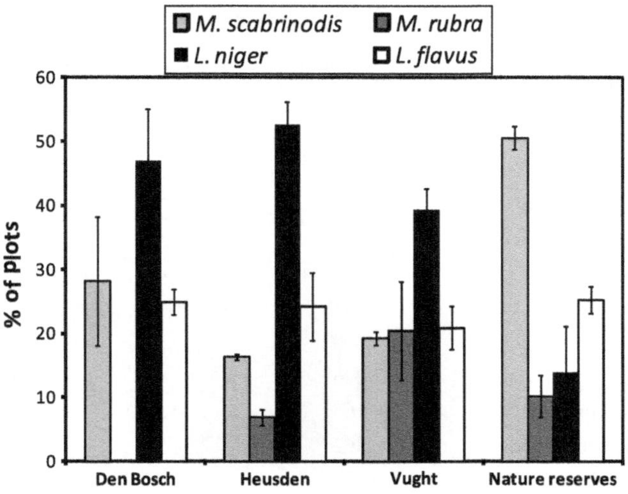

Fig. 3 Comparison of the percentage (and standard deviation) of monitoring plots occupied by *Myrmica scabrinodis*, *M. rubra*, *Lasius niger* or *L. flavus* between the municipalities and the nature reserves in the study area Moerputten and its surroundings in the Netherlands

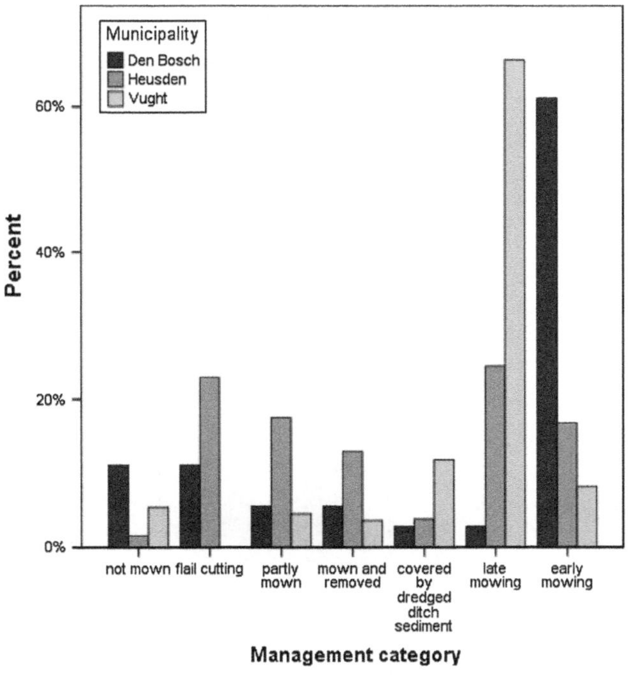

Fig. 4 Percentage of management category on road verge plots in the municipalities Den Bosch, Heusden and Vught (Province of Noord-Brabant, The Netherlands)

Management practices of the municipalities

Management of the road verges differed between the various municipalities (Fig. 4). Within the borders of Den Bosch, most plots were mown before mid-June. A number of plots were not mown or were subject to flail cutting. In contrast, the municipality of Vught has most plots mown late in the year after mid-September. Here, on the one hand, flail cutting is not used within the borders of the *Maculinea* Action Plan Area, but on the other hand, the percentage of plots covered with dredged ditch material is highest. In the municipality of Heusden, the management techniques applied were most diverse, with no type being applied to more than 25% of the plots.

Changes in the ant community

The results of the GEE's are presented in Appendix A (presence-absence) and Appendix B (abundance). Mostly, predictors that are significant for the presence-absence patterns are also significant predictors of abundances, although in case of abundances, relationships are more pronounced. The presence-absence pattern of the whole ant community is explained by a combination of all variables investigated (Table 2, Appendix A1). In 2009 and 1996, the probability of finding an ant nest of any species on a plot was highest, followed by the years 2000 and 2006, whereas the probability was lowest in 2003. Nature reserves were more populated by ants than road verges and ditch edges, with the verges and ditch edges in the municipality of Vught having the lowest probability. Slopes with a southern or western exposition had a higher probability of being occupied than those with a northern one. Also, the distance to the nearest tree was of influence, although without showing a clear trend.

With the exception of the distance to the next tree, the abundance of ant species was explained by the same variables (Table 2, Appendix B1). Most colonies were found in nature reserves. The best year by far was 2009 and the highest density was found on slopes with a southern exposition. The management of the road verges also appeared to have an impact on the presence-absence pattern (GLM, Wald $\chi^2 = 14.486$, df = 6, $P = 0.025$) and the abundance of the whole ant community (GLM, Wald $\chi^2 = 424.341$, df = 6, $P < 0.001$). Most ant nests were found on partly mown plots, or on those plots mown late in the year (Fig. 5). The probability of finding an ant nest was also highest in these management types. However, in all types it was higher than zero.

Changes in the *Myrmica* community

The probability of finding a *Myrmica* nest on a plot was equal for all years, except for 2003, when it was significantly lower (Appendix A2). Furthermore, there is a much higher probability of finding a nest in nature reserves compared to road verges and ditch edges. For the *Myrmica* community as a whole, the exposition of a plot could not be added to the model, but the distance to a tree appears to be important: The closer the tree, the greater

Table 2 Results of the statistical analysis of the impact of habitat characteristics on presence/absence patterns of ant species and abundance of ant nests by using generalized estimation equations (GEE's)

Dependent variable	P/A	*n*	QICC	Predictor	Wald chi-square	df	*P*
All ant species	P	955	1119,097	Year	62,640	4	<0.0001
				Municipality	15,195	3	0.002
				Slope direction	26,590	7	<0.0001
				Distance to tree	14,079	5	0.015
All ant species	A	606	7655,069	Year	43,662	4	<0.0001
				Municipality	20,663	3	<0.0001
				Slope direction	43,390	7	<0.0001
All *Myrmica* species	P	955	1020,500	Year	27,276	4	<0.0001
				Municipality	68,167	3	<0.0001
				Distance to tree	15,362	5	0.009
All *Myrmica* species	A	606	3257,416	Municipality	60,156	3	<0.0001
				Abundance *L. niger*	6,023	1	0.014
Myrmica scabrinodis	P	955	874,669	Municipality	33,583	3	<0.0001
				Slope direction	21,144	7	0.004
				Distance to tree	12,975	5	0.024
Myrmica scabrinodis	A	606	2001,333	Municipality	16,012	3	0.001
				Slope direction	26,501	7	<0.0001
				Abundance *L. niger*	16,687	1	<0.0001
Myrmica rubra	P	955	569,867	Year	17,552	4	0.002
				Municipality	11,753	3	0.008
Myrmica rubra	A	606	1433,194	Year	16,499	4	0.002
				Municipality	12,229	3	0.007
				Slope direction	22,654	7	0.002
				Abundance *L. niger*	13,465	1	<0.0001
Lasius niger	P	955	1057,108	Year	47,933	4	<0.0001
				Municipality	52,685	3	<0.0001
				Slope direction	24,329	7	0.001
Lasius niger	A	606	2523,641	Year	55,538	4	<0.0001
				Municipality	86,675	3	<0.0001
				Slope direction	16,795	7	0.019
Lasius flavus	P	955	727,982	Year	21,039	4	<0.0001
				Municipality	23,339	3	<0.0001
				Slope direction	53,003	7	<0.0001
				Abundance *L. niger*	26,132	1	<0.0001
Lasius flavus	A	606	3082,072	Year	14,225	4	0.007
				Slope direction	45,867	7	<0.0001
				Distance to tree	13,612	5	0.018

For the analysis of the abundance, all plots without ants were excluded. The data were collected during 5 years on 179 plots in The Netherlands. Explanation of field names: *P/A* presence-absence (P) or abundance (A) analysis, *n* sample size, *QICC* corrected quasi likelihood under independence model criterion, *df* degrees of freedom, *P* significance

the chance of finding *Myrmica* ants. Similarly, a high probability of occurrence was found at distances greater than 20 metres. The density of nests on a plot could only be explained by the variable municipality and by the abundance of *L. niger*. Densities are highest in nature reserves, and also when there are low densities of the competitor ant species (Table 2, Appendix B2). While no significant effect of management type on the occupation pattern of the *Myrmica* ants could be found (GLM, Wald $\chi^2 = 8.467$, df = 5, $P = 0.132$), there was a clear effect on their abundance (GLM, Wald $\chi^2 = 17.215$, df = 6, $P = 0.009$). Even though the *Myrmica* ants do not make up the major part of the ant community, most of them were found in the same management types, that is, on partly mown plots, or those plots that were mown late in the year.

M. scabrinodis and M. rubra

M. scabrinodis is found mainly in the nature reserves, even though it is the second-most abundant species on road verges (Appendix A3). On the meadows in the nature reserves it also reaches its highest densities. In addition, level plots are preferred to those on a slope. The distance to trees has a significant impact on the occupation probability but it does not show a clear pattern. A great abundance of *L. niger* clearly affects *M. scabrinodis* negatively (Table 2, Appendix B3).

In 2003, the probability of finding *M. rubra* and its density of colonies were significantly lower relative to all other years (Table 2, Appendixes A4, B4). We found the highest occupation probabilities and abundances in the nature reserves and along roads in the municipality of Vught. In addition, nest densities decreased when there were many colonies of *L. niger* nearby. Management of the road verges had no effect on the probability of finding a nest of *M. scabrinodis* (GLM, occupation Wald $\chi^2 = 1.860$, df = 66, $P = 0.868$; abundance Wald $\chi^2 = 5.186$, df = 6, $P = 0.52$). In contrast, plot occupation by *M. rubra* was linked to management (GLM, Wald $\chi^2 = 7.730$, df = 2, $P = 0.021$). The highest probability was found on plots mown late in the year. Mowing early or partly had a significantly lower probability, while in plots with all other management types this species was absent. This result is even more pronounced when analysing the abundance (GLM, Wald $\chi^2 = 15.606$, df = 6, $P = 0.016$, Fig. 5).

L. flavus and L. niger

L. flavus and *L. niger* show contrasting results with regard to changes in occupation and abundance over the years, even though the factors of importance are the same (Table 2, Appendixes A5, A6, B5, B6). *L. flavus* is a species mostly found in nature reserves, while *L. niger* has both highest probability of occupation and abundance on road verges. *L. niger* is also the only ant species with a high abundance in the year 2003. On the road verges, management had no effect on the presence-absence patterns of either species (GLM *L. flavus*, Wald $\chi^2 = 6.042$, df = 5, $P = 0.302$, GLM *L. niger* Wald $\chi^2 = 10.313$, df = 6, $P = 0.112$), but it was important in explaining the abundance. Most nests of *L. flavus* were found on the partly mown plots, whereas the species seems to avoid the unmown plots and the plots covered with ditch sediment (GLM, Wald $\chi^2 = 64.692$, df = 6, $P < 0.0001$, Fig. 5). *L. niger* found its highest abundance in the partly mown plots or where flail cutting was applied (GLM, Wald $\chi^2 = 87.435$, df = 6, $P < 0.0001$, Fig. 5).

Discussion

Our study shows that not only nature reserves, but also road verges and ditch edges, provide a habitat for the host ant species of the butterflies *P. nausithous* and *P. teleius*. However, their occurrences and densities are lower in the linear landscape elements. Although a comparison is difficult due to differences in geographical location and site characteristics, the meadows in our nature reserves and the road verges seem to be of moderate ant species richness. In Germany, 3–12 ant species of open habitats were found in traditionally managed grassland (Dauber and Wolters 2000; Seifert 1986; Dahms et al. 2005). In Poland, Woyciechowski and Miszta (1976) and Petal (1974, 1980) found 4–14 ant species on the moist meadows where the source populations of the reintroduced butterflies occur. On such meadows, *M. scabrinodis* usually dominates the ant community (Grill et al. 2008; Witek et al. 2008), which agrees with our findings. On road verges, however, *L. niger* is by far the commonest species. This species can withstand the disturbance associated with roads, such as frequent management, whereas *Myrmica* ants prefer less dynamic sites with mowing only once or twice a year, or even less, as in nature reserves or agricultural land under traditional management (Peeters et al. 2004). In such stable conditions, they can outcompete *L. niger* (Dauber and Wolters 2004). On road verges, *L. niger* is able to reduce the abundance of *Myrmica* ants in general, and *M. scabinodis* in particular, confirming its role in structuring the ant community.

In the summer of 2001, the road verges with the largest subpopulation of *P. nausithous* and the highest presence and abundance of *M. rubra,* was covered unintentionally by a layer of sand because of road works. Many monitoring plots were affected. The occurrence of ants was reduced from 78% in 2001 to 32% in 2002 and nest densities also decreased with more than 50%. While the *Myrmica* species suffered most, *L. niger* had already increased its nest density in 2002, later followed by increases in occupation (Wynhoff, unpublished data). The low values found for presence-absence and abundance of *M. rubra* in 2003 are probably also due to this accidental habitat destruction in 2001.

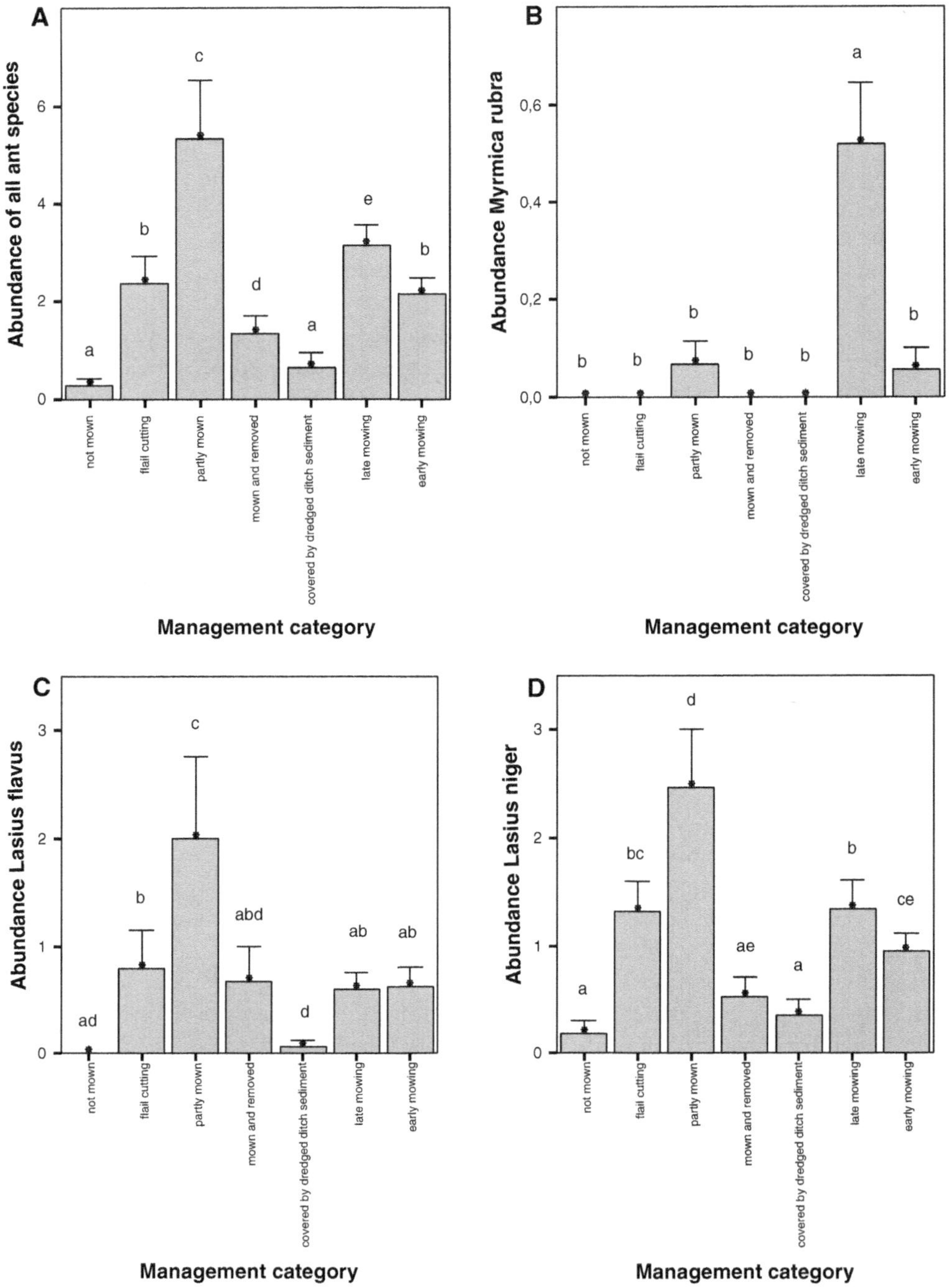

Fig. 5 Mean abundance of ant species on monitoring plots on the road verges in the municipalities of Den Bosch, Heusden and Vught in relation to plot management (Province of Noord-Brabant, The Netherlands). **a** all ant species, **b** *Myrmica rubra*, **c** *Lasius flavus*, **d** *Lasius niger*

Although the host plant *S. officinalis* is widespread on the road verges in the *Maculinea* Action Plan Area, the ants are much rarer. Therefore the impression that the majority of the road verges with the host plant are potential habitat for the expanding *Phengaris* butterflies is not supported by our data. Only between 20 and 40% of the plots were occupied by *Myrmica* ants. No ants could be found in the vicinity of 25–50% of the studied host plants. Caterpillars on these plants thus have hardly any chance of being found and adopted by a *Myrmica* ant. Earlier studies showed that the neighbouring agricultural land is virtually ant-free (Wynhoff et al. 2008). Due to modern agricultural techniques, such as frequent ploughing, manure injection and heavy grazing by cattle, invading ant colonies have only little chance of survival. When ants are found in arable land, both species-richness and nest densities are low (Dauber and Wolters 2004; Dahms et al. 2005; Dauber et al. 2006).

Site characteristics independent from management were found to have an impact on the local ant community. These are mainly related to the local microclimate. Since it was not possible to measure soil temperature or insolation at so many locations, we used the slope of the plot and its distance to the nearest tree as a proxy for microclimate. When the exposition of the slope was found to be significant, the highest probability of occupation or the highest predicted abundance was found on slopes with southern or western exposition, which receive sunlight for the longest time of the day and are therefore warmest. This result is to be expected for thermophilous insects like ants. The distance to trees affected presence-absence patterns, with a higher probability of occupation at closer distance but only up to 10 m as a tree close to an ant colony gives shade in the hottest hours of the day. Further away, the benefit seems to be rather coincidental.

The management types investigated have various effects on the ant species. Not only is this revealed for a certain management, but also illustrated by the differences between the municipalities, each of which is responsible for its own road verge management. In the nature reserve, management usually remains the same for years and is beneficial for nature. Hay meadows are cut once a year and the hay is removed after about a week of drying. Since 1990, cutting between mid-June and mid-September has been avoided in all meadows occupied by *Phengaris* butterflies. With one mowing a year, *M. scabrinodis* can dominate the community and can reach high densities with locally more than one colony per square metre, thus providing a high quality habitat for *P. teleius*. This supports the results found by Grill et al. (2008) in *Phengaris* habitats in Bavaria. On the road verges, factors such as traffic safety, water flow through the ditches and efficient use of mowing machines, like flail cutters, also affect management regimes, resulting in a much more dynamic environment for the ants, and consequently, in different ant communities. We found that highest abundance of ants is achieved by mowing only parts of the vegetation. In the unmown parts, ants are not disturbed and their nests are not destroyed, making it easy to recolonize the short cut part. The fast colonizer *L. niger* mainly profits from this management, but also *L. flavus* is able to sustain with a high nest density, probably because the high nest mounds of *L. flavus* are situated in the abandoned part of the vegetation and are therefore not disturbed. However, our data do not allow to prove this because we searched for ants along the whole gradient from the asphalt layer over the flat grassy verge down to the bottom of the neighbouring ditch. In addition, variation in vegetation structure and the amount of edge habitat are increased by mowing only the strip along the road, which is also beneficial for the ant community (Dauber and Wolters 2004; Golden and Crist 2000). The host ant of *P. teleius*, *M. scabrinodis,* seems to profit from mowing early in June, even though this effect was not significant in our analysis. On the meadows in the nature reserves, vegetation is usually mown in autumn, and nest densities of this ant species are even higher. These meadows are on poor soils without fertilization, that allows for late mowing. The vegetation cover stays open and does not prevent the sun from warming the soil. However, since late mowing does not prevent bush encroachment and colonization of these wet meadows by willows, every fifth year meadows are cut early, without having a negative effect on *M. scabrinodis* so far. If cut sufficiently early, the *Sanguisorba* plants are able to produce new buds in time for the females of *P. teleius* to deposit their eggs. On the road verges the soil is nutrient richer and has a denser vegetation. Here, *M. scabrinodis* is best supported by early mowing. Statistical analysis did not reach significance, but the probability of occupation on road verges was highest in the municipality of Den Bosch, where most plots are mown early in the year.

In contrast, *M. rubra,* the host ant of *P. nausithous,* profits from mowing late in autumn. This ant species is less thermophilous than *M. scabrinodis*. High vegetation provides sheltered nest locations where the ants are less subject to the heat during summer. In the nature reserve, the species thrives only on edges along high vegetation, such as beds of tall reed and willow forest edges. Probably due to climatic requirements, at the study site, *M. rubra* co-occurs only rarely with *M. scabrinodis*, as it usually does within the range of the butterflies in Europe (Elmes et al. 1998). Mowing only parts of the vegetation could be expected to be beneficial for *M. rubra*, because the colonies can survive in the unmown parts. However, there are more beneficial effects for the *Lasius* species, which are able to reduce the nest density of *Myrmica* species. A disadvantage of regular late mowing is that nutrients are not removed from the plots. This in turn allows shrubs, tall herbs and grasses to overgrow the *Sanguisorba* plants and prevent them from flowering. Therefore, once every 3–6 years, depending on the nutrient richness of the soil and the productivity of the vegetation, road verges should be mown early.

L. niger profits most from flail cutting, a method, which is not beneficial to the other ant species but most cost-efficient for the municipalities. It leaves a layer of dead organic material on the vegetation preventing the sun to warm up the soil and the ant nests. Furthermore, *L. niger* has a significant negative effect on the abundance of all *Myrmica*-species. As a strong competitor, it not only shapes the ant community on the road verges and ditch edges, but also the habitat quality of these linear elements for *Phengaris* butterflies.

Recommendations for conservation

Our results suggest that management benefitting *Phengaris* butterflies should aim to increase nest densities of *Myrmica* species (see also Anton et al. 2008; Dierks and Fischer 2009) whereas *L. niger* should decrease or at least be kept at a low level of occupation with low nest densities. Furthermore, it is important not to mow the habitat of the butterflies between mid-June and mid-September, because this would destroy temporarily locations with both host ants and flowering host plants, and females would not being able to deposit their eggs on these sites. Although our results show that certain types of management should not be applied, they are less clear about what to do best. Unfavourable actions include not mowing or using road verges to deposit ditch detritus. Here, microclimate will change into shady and cool conditions, where ant nests cannot persist anymore. In the habitat and the surroundings of the Dutch *Phengaris* populations, the best option to support *Myrmica* species is mowing once per year. However, this still leaves three possibilities open: cutting only parts of the vegetation, mowing before mid-June or mowing after mid-September. Our results suggest that *M. scabrinodis* is best supported by early mowing of the road verge vegetation, or by late mowing in the nature reserves. The best management for the host ant of *P. nausithous*, *M. rubra*, appears to be mowing late in September or October, as the municipality of Vught does on most of its road verges. With different mowing regimes required for the two *Myrmica* species, some flexibility is needed to promote both butterfly species. The best mosaic pattern of early cut road verge vegetation with parts left until autumn would be realized by mowing one side of the road early, and the other one late. The butterflies can cross rural roads, and when they leave an ant nest on the mown side after pupation, they can find mates and host plants at the other side. Since *P. nausithous* is much more dependent on road verges as a permanent habitat (Van Langevelde and Wynhoff 2009), a scheme of late mowing with early mowing periodically, should be realised here, at least for the sites within the dispersal range of the species. In the Moerputten, *P. teleius* is much more dependent on meadows for reproduction, and it uses road verges mainly for dispersal (Van Langevelde and Wynhoff 2009).

Restoring wet meadows and marshes for the butterflies requires not only suitable habitat conditions, but also good colonization probabilities. Plant species which are not able to colonize by themselves can be brought to the sites by mowing the meadows in the nature reserves and transporting the seed by spreading the hay in the restored sites (Klimkowska et al. 2007; Rasran et al. 2007). The ants have to colonize these sites mainly from the linear elements by walking to them after having split off from the colony. Therefore, to facilitate the restoration, careful management of vegetation in these linear elements is needed. Up till now, translocation of ant nests has proven to be difficult (Pontin 1969; Elmes 1971). Until new methods have been developed, it will be crucial to build up densities of *Myrmica* nests in restored sites that are needed to make sustainable populations of *Phengaris* butterflies. Even when road verge populations will be at close distance, *L. niger* will be the first species to colonize the new sites. Our results suggest that only when the vegetation of the meadow is dense, and a constant management of one or two cuts a year has been established, *Myrmica* species will be able to outcompete *L. niger*.

Acknowledgments We want to thank Sicco Ens, Sergej Jansen, Saskia Janssen, Gerard Oostermeijer, Nicoliene Peet, Maarten van Steenis and Agnieszka Bucholc for field work assistance and Nicoliene Peet for her help with the ant identification. Claire and Rob Hengeveld kindly corrected the English. National State Forestry and Natuurmonumenten gave us permission to enter their nature reserves. IW was funded by the Province of Noord-Brabant.

Appendix A

Probability of occupation of ant species as significant predictors on dependent variables, analyzed by Generalized Estimation Equations (GEE's, see text for details). The data were collected on 179 monitoring plots during 5 years of investigation in the Moerputten nature reserve and surrounding road verges and ditch edges in The Netherlands. The variables are the year of investigation, the municipality responsible for the management, the slope direction of the plot, the distance to the nearest tree or shrub in distance classes and the abundance of *Lasius niger*. Only significant results are presented.

A1: all ant species combined, A2: all *Myrmica* species combined, A3: *Myrmica scabrinodis*, A4: *Myrmica rubra*, A5: *Lasius niger*, A6: *Lasius flavus*

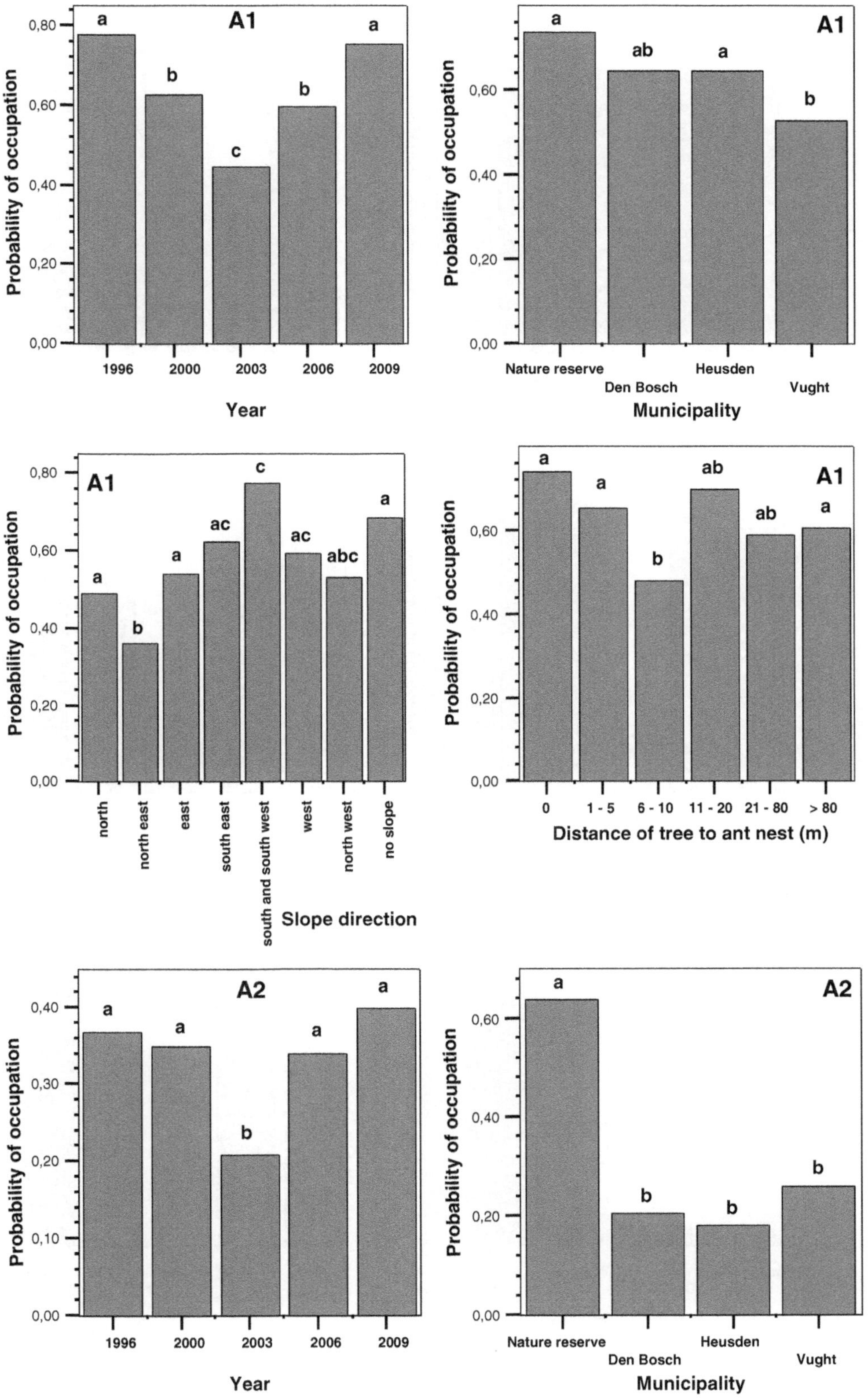
A1
Probability of occupation
0,00
0,20
0,40
0,60
0,80
a
b
c
b
a
1996
2000
2003
2006
2009
Year
A1
Probability of occupation
a
ab
a
b
Nature reserve
Den Bosch
Heusden
Vught
Municipality
A1
Probability of occupation
a
b
a
ac
c
ac
abc
a
north
north east
east
south east
south and south west
west
north west
no slope
Slope direction
A1
Probability of occupation
a
a
b
ab
ab
a
0
1 - 5
6 - 10
11 - 20
21 - 80
> 80
Distance of tree to ant nest (m)
A2
Probability of occupation
0,10
0,30
a
a
b
a
a
1996
2000
2003
2006
2009
Year
A2
Probability of occupation
a
b
b
b
Nature reserve
Den Bosch
Heusden
Vught
Municipality

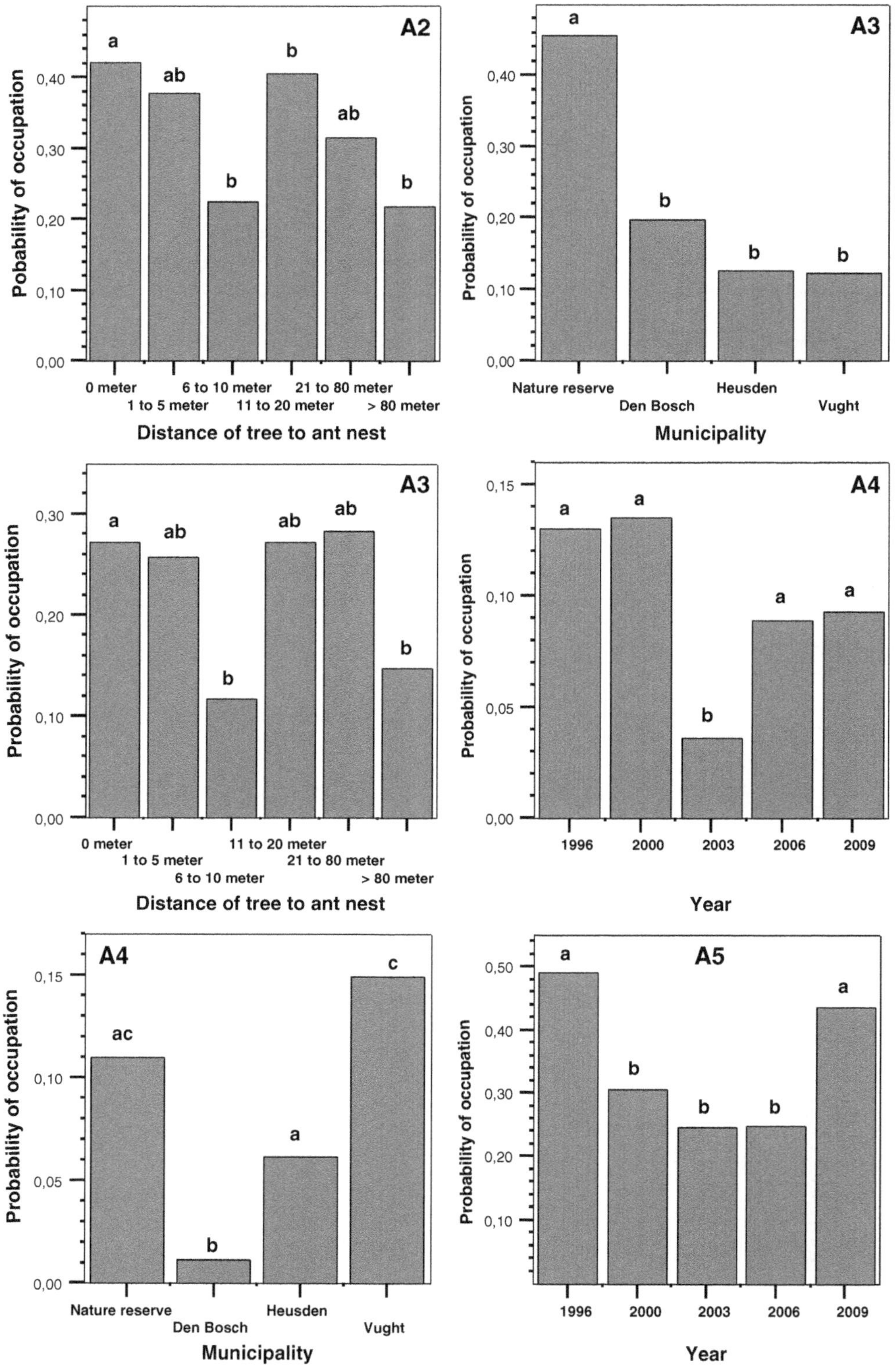
A2
Pobability of occupation
0,00
0,10
0,20
0,30
0,40
a
ab
b
b
ab
b
0 meter
1 to 5 meter
6 to 10 meter
11 to 20 meter
21 to 80 meter
> 80 meter
Distance of tree to ant nest
A3
Probability of occupation
0,00
0,10
0,20
0,30
0,40
a
b
b
b
Nature reserve
Den Bosch
Heusden
Vught
Municipality
A3
Probability of occupation
0,00
0,10
0,20
0,30
a
ab
b
ab
ab
b
0 meter
1 to 5 meter
6 to 10 meter
11 to 20 meter
21 to 80 meter
> 80 meter
Distance of tree to ant nest
A4
Probability of occupation
0,00
0,05
0,10
0,15
a
a
b
a
a
1996
2000
2003
2006
2009
Year
A4
Probability of occupation
0,00
0,05
0,10
0,15
ac
b
a
c
Nature reserve
Den Bosch
Heusden
Vught
Municipality
A5
Probability of occupation
0,10
0,20
0,30
0,40
0,50
a
b
b
b
a
1996
2000
2003
2006
2009
Year

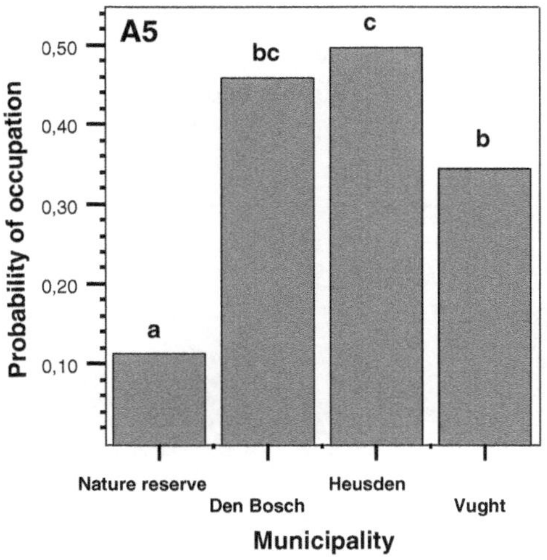
A5
Probability of occupation
0,50
0,40
0,30
0,20
0,10
a
bc
c
b
Nature reserve
Den Bosch
Heusden
Vught
Municipality

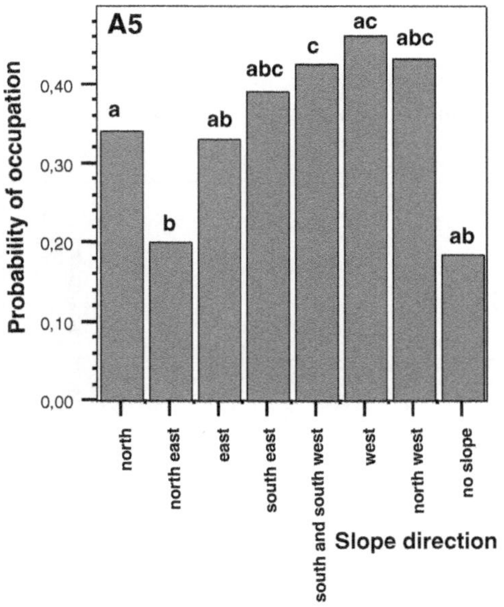
A5
Probability of occupation
0,40
0,30
0,20
0,10
0,00
a
b
ab
abc
c
ac
abc
ab
north
north east
east
south east
south and south west
west
north west
no slope
Slope direction

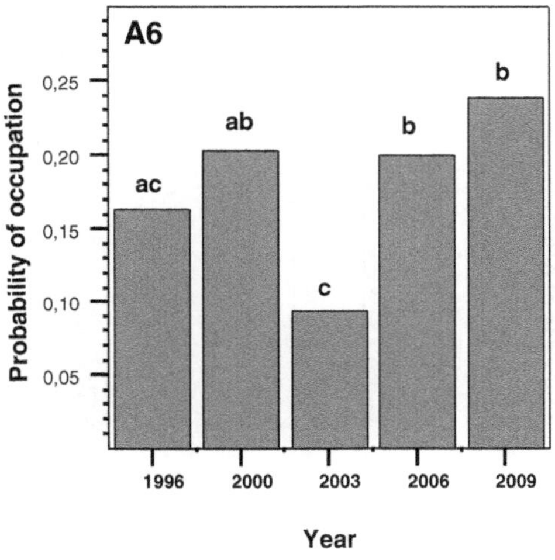
A6
Probability of occupation
0,25
0,20
0,15
0,10
0,05
ac
ab
c
b
b
1996
2000
2003
2006
2009
Year

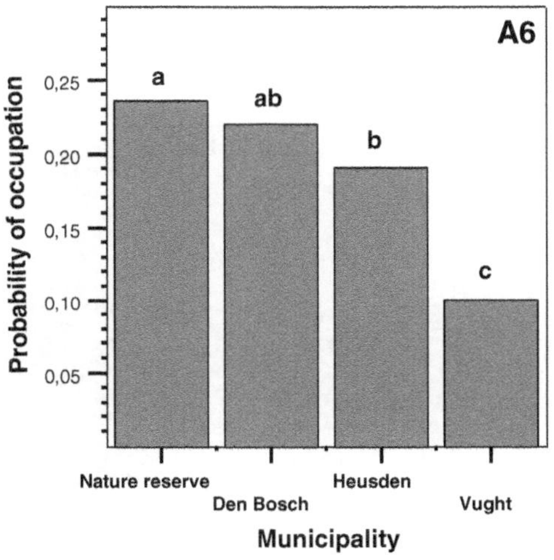
A6
Probability of occupation
0,25
0,20
0,15
0,10
0,05
a
ab
b
c
Nature reserve
Den Bosch
Heusden
Vught
Municipality

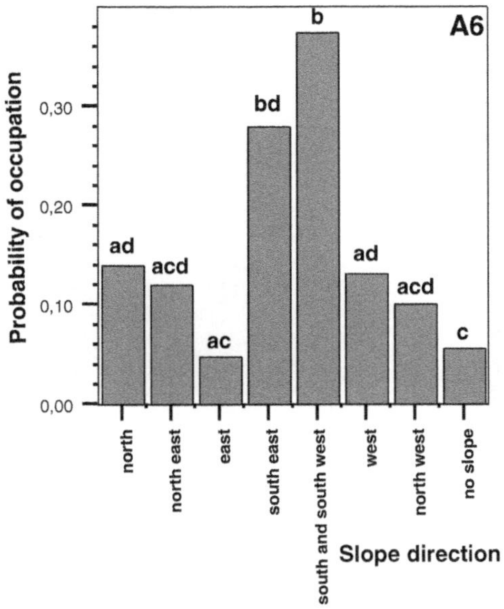
A6
Probability of occupation
0,30
0,20
0,10
0,00
ad
acd
ac
bd
b
ad
acd
c
north
north east
east
south east
south and south west
west
north west
no slope
Slope direction

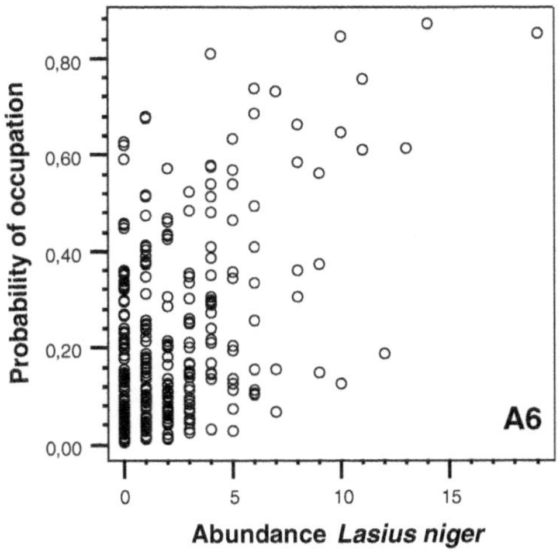
A6
Probability of occupation
0,80
0,60
0,40
0,20
0,00
0
5
10
15
Abundance Lasius niger

Appendix B

Predicted abundance of ant nests as significant predictors in relation to dependent variables, analyzed by Generalized Estimation Equations (GEE's, see text for details). The data were collected on 179 monitoring plots during 5 years of investigation in the Moerputten nature reserve and surrounding road verges and ditch edges in The Netherlands.

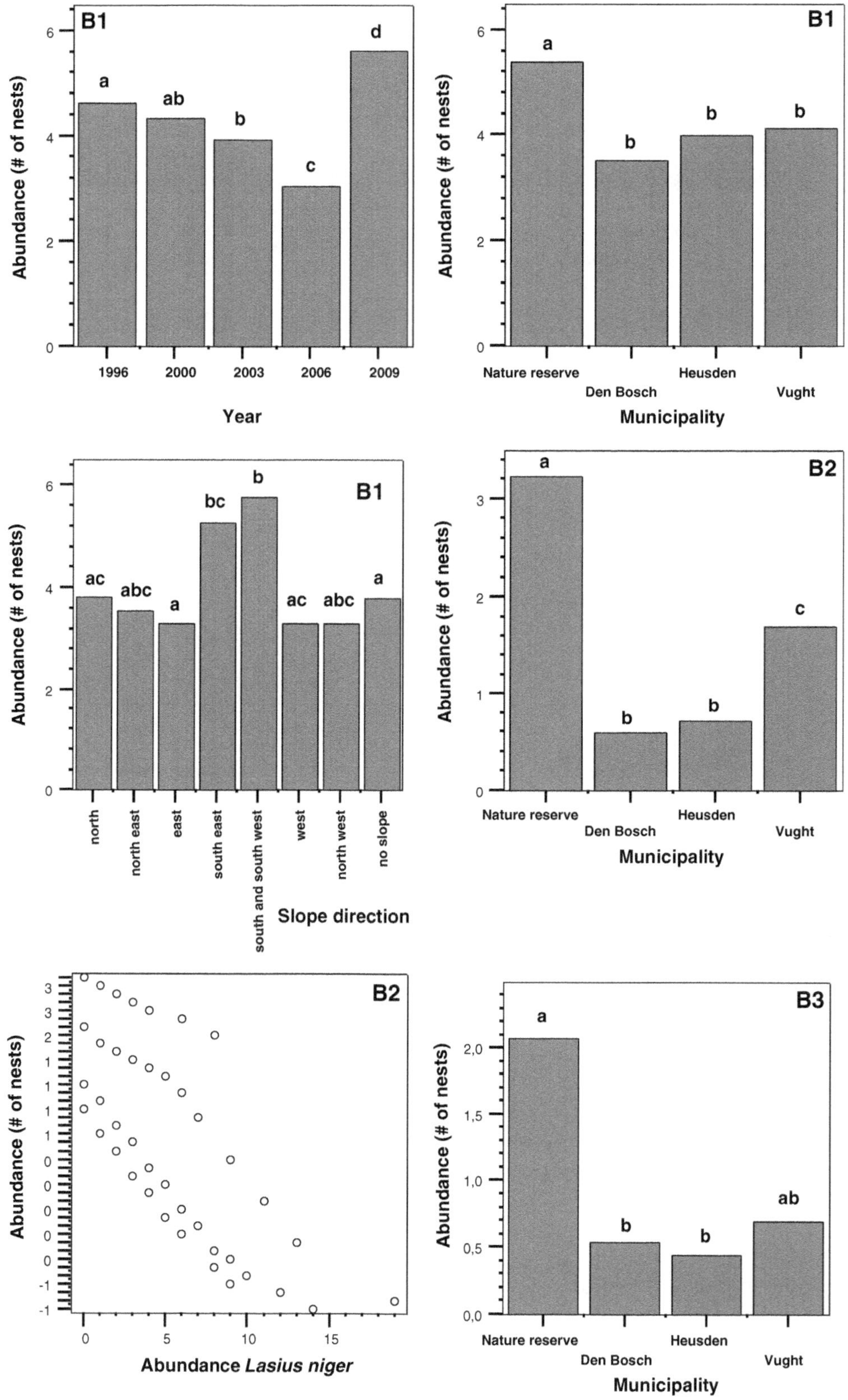

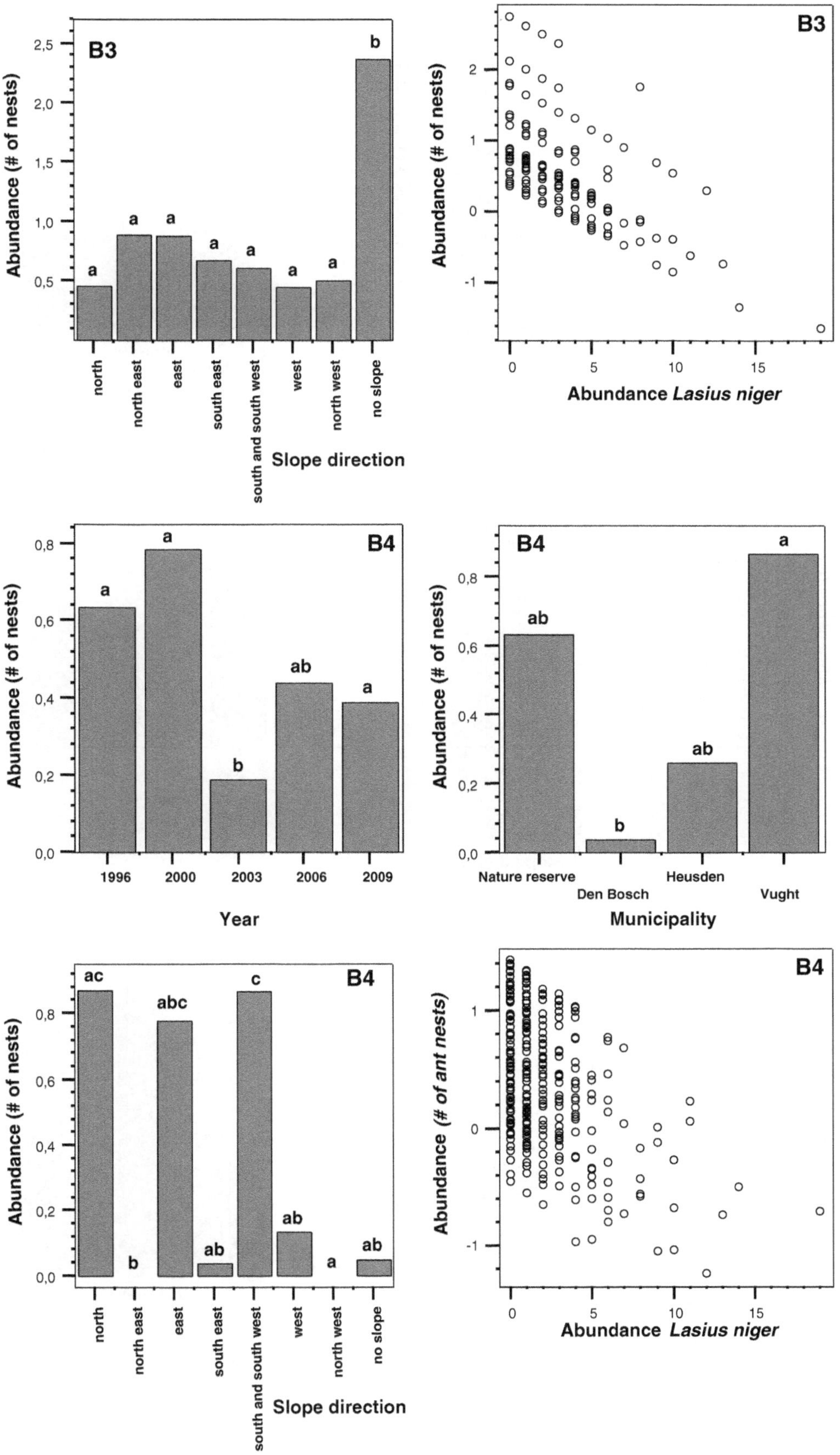
B3
Abundance (# of nests)
north
north east
east
south east
south and south west
west
north west
no slope
Slope direction
B3
Abundance (# of nests)
Abundance *Lasius niger*
B4
Abundance (# of nests)
1996
2000
2003
2006
2009
Year
B4
Abundance (# of nests)
Nature reserve
Den Bosch
Heusden
Vught
Municipality
B4
Abundance (# of nests)
north
north east
east
south east
south and south west
west
north west
no slope
Slope direction
B4
Abundance (*# of ant nests*)
Abundance *Lasius niger*

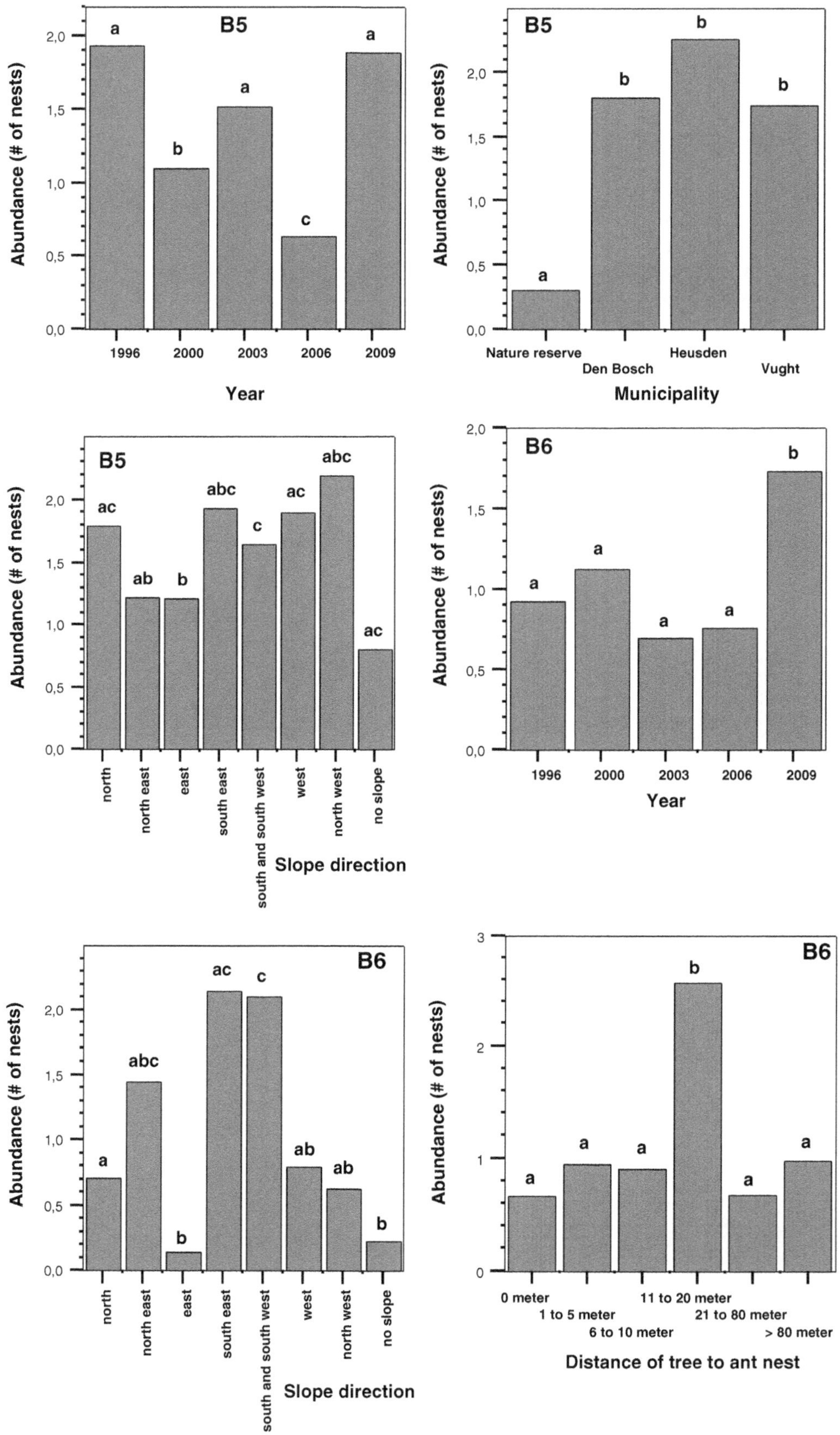
B5
Abundance (# of nests)
Year
1996
2000
2003
2006
2009
a
b
a
c
a
B5
Abundance (# of nests)
Municipality
Nature reserve
Den Bosch
Heusden
Vught
a
b
b
b
B5
Abundance (# of nests)
Slope direction
north
north east
east
south east
south and south west
west
north west
no slope
ac
ab
b
abc
c
ac
abc
ac
B6
Abundance (# of nests)
Year
1996
2000
2003
2006
2009
a
a
a
a
b
B6
Abundance (# of nests)
Slope direction
north
north east
east
south east
south and south west
west
north west
no slope
a
abc
b
ac
c
ab
ab
b
B6
Abundance (# of nests)
Distance of tree to ant nest
0 meter
1 to 5 meter
6 to 10 meter
11 to 20 meter
21 to 80 meter
> 80 meter
a
a
a
b
a
a

The variables are the year of investigation, the municipality responsible for the management, the slope direction of the plot, the distance to the nearest tree or shrub in distance classes and the abundance of *Lasius niger*. Only significant results are presented.

B1: all ant species combined, B2: all *Myrmica* species combined, B3: *Myrmica scabrinodis*, B4: *Myrmica rubra,* B5: *Lasius niger*, B6: *Lasius flavus*

References

Anton C, Musche M, Hula V, Settele J (2008) Myrmica host-ants limit the density of the ant-predatory large blue *Maculinea nausithous*. J Insect Conserv 12:511–517

Armbrecht I, Perfecto I, Vandermeer J (2004) Enigmatic biodiversity correlations: ant diversity responds to diverse resources. Science 304:284–286

Batáry P, Kőrösi Á, Őrvőssy N, Kővér S, Peregovits L (2009) Species-specific distribution of two sympatric *Maculinea* butterflies across different meadow edges. J Insect Conservation 13(2):223–230

Curry JP (1994) Grassland invertebrates. Chapman & Hall, Cambridge

Dahms H, Wellstein C, Wolters V, Dauber J (2005) Effects of management practices on ant species richness and community composition in grasslands (Hymenoptera: Formicidae). Myrmecologische Nachrichten 7:9–16

Dauber J, Bengtsson J, Lenoir L (2006) Evaluating habitat loss and land-use continuity on ant species richness in seminatural grassland remnants. Conserv Biol 20(4):1150–1160

Dauber J, Wolters V (2000) Diversität der Ameisenfauna im Landnutzungsmosaik einer peripheren Region. Mitteilungen der Deutschen Gesellschaft für allgemeine und angewandte Entomologie 12:281–284

Dauber J, Wolters V (2004) Edge effects on ant communty structure and species richness in an agricultural landscape. Biodivers Conserv 13:901–915

Dierks A, Fischer K (2009) Habitat requirements and niche selection of *Maculinea nausithous* and *M. teleius* (Lepidoptera: Lycaenidae) within a large sympatric metapopulation. Biodivers Conserv 18:3663–3676

Elmes GW (1971) An experimental study on the distribution of heathland ants. J Anim Ecol 40(2):495–499

Elmes GW, Thomas JA, Wardlaw JC, Hochberg ME, Clarke RT, Simcox DJ (1998) The ecology of *Myrmica* ants in relation to the conservation of *Maculinea* butterflies. J Insect Conserv 2:67–79

Elmes GW, Wardlaw JC (1982) A population study of the ants *Myrmica sabuleti* and *Myrmica scabrinodis,* living at two sites in the south of England. I. A comparison of colony populations. J Anim Ecol 51:651–664

Golden DM, Crist TO (2000) Experimental effects of habitat fragmentation on rove beetles and ants: patch area or edge? Oikos 90:525–538

Grill A, Cleary DFG, Stettmer C, Bräu M, Settele J (2008) A mowing experiment to evaluate the influence of management on the activity of host ants of *Maculinea* butterflies. J Insect Conserv 12(6):617–627

Haddad N (2000) Corridor length and patch colonization by a butterfly, *Junonia coenia*. Conserv Biol 14:738–745

Heller G, Rohe W (2000) Vergleichende Untersuchung zur Ameisenfauna von Grünlandstandorten in Rheinland-Pfalz. Mainzer Naturwissenschaftliches Archiv 38:123–173

Klimkowska A, Van Diggelen R, Bakker JP, Grootjans AP (2007) Wet meadow restoration in Western Europe: a quantitative assessment of the effectiveness of several techniques. Biol Conserv 140:318–328

Kruess A, Tscharntke T (2002) Grazing intensity and the diversity of grasshoppers, butterflies, and trap-nesting bees and wasps. Conserv Biol 16:1570–1580

Peeters TMJ, van Achterberg C, Heitmans WRB, Klein WF, Lefeber V, van Loon AJ, Mabelis AA, Nieuwenhuijsen H, Reemer M, de Rond J, Smits J, Velthuis HHW (2004) De wespen en mieren van Nederland (Hymenoptera: Aculeata)–Nederlandse Fauna 6. Nationaal Natuurhistorisch Museum Naturalis, Leiden, KNNV Uitgeverij, Utrecht & European Invertebrate Survey Nederland, Leiden

Petal J (1974) Analysis of a sheep pasture ecosystem in the Pieniny Mountains (the Carpathians) XV. The effect of pasture management on ant population. Ekologia Polska 22:679–692

Petal J (1980) Ant populations, their regulation and effect on soil in meadows. Ekologia Polska 28:297–326

Pontin AJ (1969) Experimental Transplantation of nest mounds of the ant *Lasius flavus* (F.) in a habitat containing also *L. niger* (L.) and *Myrmica scabrinodis* Nyl. J Anim Ecol 38(3):747–754

Provincie Noord-Brabant (1997) Beheersconvenant in het kader van het Beschermingsplan Pimpernelblauwtjes bij Den Bosch. Provincie Noord-Brabant, Den Bosch

Rasran L, Vogt K, Jensen K (2007) Effects of topsoil removal, seed transfer with plant material and moderate grazing on restoration of riparian fen grasslands. Appl Veg Sci 10:451–460

Seifert B (1986) Vergleichende Untersuchungen zur Habitatwahl von Ameisen (Hymenoptera: Formicidae) im mittleren und südlichen Teil der DDR. Abh Ber NaturkundemusGörlitz 59: 1–124

Seifert B (1994) Die freilebenden Ameisen Deutschlands (*Hymenoptera: Formicidae*) und Angaben zu deren Taxonomie und Verbreitung. Abh Ber Naturkundemus Görlitz 67:1–44

Seifert B (1996) Ameisen beobachten, bestimmen. Naturbuch Verlag, Augsburg

Thomas JA (1984) The behaviour and habitat requirements of *Maculinea nausithous* (the Dusky Large Blue Butterfly) and *M. teleius* (the Scarce Large Blue) in France. Biol Conserv 28:325–347

Thomas JA, Settele J (2004) Butterfly mimics of ants. Nature 432:283–284

Van Boven JKA, Mabelis AA (1986) De mierenfauna van de Benelux (*Hymenoptera: Formicidae*). Wetenschappelijke Mededelingen KNNV 173:1–64

Van Dyck H, Talloen W, Oostermeijer JGB, Feenstra V, Van der Hidde A, Wynhoff I (2000) Does the presence of ant nests matter for oviposition to a specialized myrmecophilous *Maculinea* butterfly? Proc R Soc Lond B Biol Sci 267:861–866

Van Langevelde F, Wynhoff I (2009) What limits the spread of two congeneric butterfly species after their reintroduction: quality or spatial arrangement of habitat? Anim Conserv 12:540–548

Van Loon AJ, Mabelis AA (1996) Flora en Fauna 2030. Fase III: Cryptobiota. Mieren (*Formicidae*). Mededelingen EIS-Nederland 83:1–34

Witek M, Sliwinska E, Skórka P, Nowicki P, Wantuch M, Vrabec V, Settele J, Woyciechowski M (2008) Host ant specificity of large blue butterflies *Phengaris* (*Maculinea*) (Lepidoptera: Lycaenidae) inhabiting humid grasslands in East-central Europe. Eur J Entomol 105:871–877

Woyciechowski M, Miszta A (1976) Spatial and seasonal structure of ant communities in a mountain meadow. Ekologia Polska 24:577–592

Wynhoff I (1998) Lessons from the reintroduction of *Maculinea teleius* and *M. nausithous* in the Netherlands. J Insect Conserv 2:47–57

Wynhoff I, Grutters M, Van Langevelde F (2008) Looking for the ants: selection of oviposition sites by two myrmecophilous butterfly species. Anim Biol 58:371–388

Zeger SL, Liang K-Y, Albert PS (1988) Models for longitudinal data: a generalized estimating equation approach. Biometrics 44:1049–1060

J Insect Conserv (2011) 15:153–163
DOI 10.1007/s10841-010-9334-y

ORIGINAL PAPER

Butterflies on the brink: habitat requirements for declining populations of the marsh fritillary (*Euphydryas aurinia*) in SW England

Melanie Smee · Wesley Smyth · Mark Tunmore · Richard ffrench-Constant · Dave Hodgson

Received: 25 March 2010 / Accepted: 24 August 2010 / Published online: 6 September 2010

Abstract 1. The marsh fritillary *Euphydryas aurinia* is one of our most endangered butterflies, and the only to be protected under European legislation as well as British. It persists in fragile subpopulations threatened by habitat fragmentation and degradation. 2. A combination of swaling and cattle grazing are accepted to be best practice for managing wet, unimproved grasslands—the favoured habitat for *E. aurinia* in Cornwall. These two well-endorsed methods of management were used to increase and improve the quality of habitat for *E. aurinia* over a 5 years period, 2004–2008, at a stronghold network of habitat patches in mid Cornwall, south-west England. 3. Analyses of adult and larval densities over 5 years in fifty-four transects across nine sites found *E. aurinia* to favour habitat patches with higher densities of the larval food plant (Devil's-bit scabious *Succisa pratensis*), higher sward height in autumn, and intermediate optimum levels of stock grazing. 4. Main findings indicated most sites experienced significant declines in numbers. Unfavourable weather in the last 2 years of monitoring was likely to have had a significant impact on the response of individual subpopulations to habitat management though poor recovery rates may also reflect a time-lag in colonisation events after habitat improvement has occurred. 5. Habitat management produced an improvement, albeit an inconsistent improvement in habitat variables across patches—*S. pratensis* shows a clear recovery at some sites. Autumn sward height increased significantly at one site, and a quadratic relationship between stock grazing and important habitat variables has been found which will aid further improvement over all sites for the long term persistence of *E. aurinia*.

Keywords Metapopulation dynamics · Adaptive management · Conservation · Monitoring · Habitat fragmentation · Stock grazing

M. Smee · R. ffrench-Constant · D. Hodgson (✉)
Centre for Ecology and Conservation, Biosciences,
University of Exeter, Cornwall Campus Tremough,
Penryn, Cornwall TR10 9EZ, UK
e-mail: d.j.hodgson@exeter.ac.uk

W. Smyth
Natural England, Pydar House, Pydar Street, Truro,
Cornwall TR1 1XU, UK

M. Tunmore
Trewhella Cottage, Cury Cross Lanes, Helston,
Cornwall TR12 7AZ, UK

Introduction

The marsh fritillary, *Euphydryas aurinia*, is vulnerable to local extinction from much of its current distribution across the UK and Europe. It is the only British butterfly species to be listed under the EC Habitats and Species Directive as well as the Bern Convention, fully protected under the Wildlife and Countryside Act 1981 and a priority species in the UK Biodiversity Action Plan (BAP). Agricultural improvement or abandonment, and changes in management practices have meant a loss of 92% of south-west England's damp pasture, the main habitat used by *E. aurinia* (Hobson et al. 2001). Having suffered serious declines across most of its range in Europe, *E. aurinia* has undergone a 55% loss over a period of 30 years (Asher et al. 2001) in the UK—currently considered a stronghold for the species, supporting 5–15% of the European distribution (van Swaay and Warren 1999). This dramatic decline is mirrored in Cornwall, with 62% of known populations going extinct between 1990 and 2000 (Hobson and Budd 2001). These trends have caused the butterfly to contract to a handful of core geographical areas (Asher et al. 2001), one of these being a set of subpopulations (temporary populations as part of a metapopulation) found across the mid-Cornwall Moors.

Metapopulation theory states that persistence in a fragmented landscape requires a balance between stochastic extinctions and re-colonisations (Hanski 1994, 1998, 1999; Hanski and Gaggiotti 2004; Levins 1969). Processes of disturbance and succession cause natural patch networks to be dynamic, with regional extinction being prevented through the process of habitat tracking, whereby a species is able to follow the shifting mosaic of suitable environmental conditions (Thomas 1994). In an anthropogenically modified landscape, restoration or creation of new habitat can allow such habitat tracking to occur. As a species widely accepted to persist as metapopulations in fragmented landscapes (Hula et al. 2004; Wahlberg et al. 2002; Warren 1994) *E. aurinia* thereby requires restoration of habitat through appropriate management at a landscape-scale, to reduce the probability of subpopulation extinction, and to promote connectivity among suitable habitat patches.

E. aurinia occupies two main biotypes in the UK: damp, neutral or acidophilous grasslands, and dry, calcicolous grasslands (Barnett and Warren 1995). Consequently, 'appropriate' management is not the same for both and has to be assessed on a 'per site' basis. Traditional farming methods benefitted the species by creating heterogeneous landscapes, providing open, low vegetation for basking larvae (Porter 1982), as well as tall nectar-rich flowering herbs for adults on the wing—a combination some rural areas can still provide (Liu et al. 2006). However, the advent of modern farming techniques and intensive grazing of cattle in western Europe have caused habitat to become too homogeneous and unsuitable for *E. aurinia* (van Swaay and Warren 1999). Two common methods, which have now become adaptive tools used to create suitable, but usually temporary, habitats (Wahlberg et al. 2002) are extensive cattle grazing and swaling (controlled burning). However, responses to grazing are equivocal (Pöyry et al. 2005), and burning can cause heavy mortality to *E. aurinia* larvae at certain times of the year (Warren 1994) emphasizing the need for more detailed study of the effect of these techniques on important habitat variables for *E. aurinia*. Studies on the habitat requirements of *E. aurinia* in Germany and the Czech Republic have showed that females prefer to lay egg clutches on large, prominent *S. pratensis* plants surrounded by low, cushion-forming vegetation (Anthes et al. 2003; Konvicka et al. 2003).

The aim of this study was to determine the most important habitat requirements of *E. aurinia* on the mid-Cornwall Moors, and whether management can achieve this and consequentially increase population numbers to secure a positive future for this species. Our a priori expectations include a significant positive effect of the density of the larval host plant *S. pratensis* and an effect of sward height on occurrence of *E. aurinia,* as these have been shown to be important in previous studies (Betzholtz et al. 2007; Bulman 2001; Hula et al. 2004; Konvicka et al. 2003). A medium, optimum level of sward height is likely to be important for *E. aurinia*. Too short and *S. pratensis* may become too small or isolated and cause higher mortality of larval groups through starvation or predation, but too tall and rank and *S. pratensis* may become too difficult for adult butterflies to find or be outcompeted and lost altogether (Konvicka et al. 2003). Either extreme of sward height could also create unfavourable microclimatic conditions for both larval and adult stage. We also hypothesised that an important quadratic relationship may exist between intensity of management and grass sward height, as over-grazing can result in very short, unsuitable host plants and areas of bare ground, whereas under-grazing and abandonment can lead to rank or scrubby, unsuitable habitat. The management undertaken aimed not only to improve the quality of the habitat, but also sought to increase the connectivity between suitable habitat patches. Here we report on the successes and failures of these management techniques, and the consequent trajectory of the mid-Cornwall Moors metapopulation. We predicted that habitat management would improve key predictors of *E. aurinia* density with the aim of a subsequent network-wide increase in *E. aurinia* densities. However, this prediction should be considered against the cyclic nature of *E. aurinia* subpopulations (Ford and Ford 1930) whereby a peak in numbers was likely attained in 2006, followed by lower numbers in more recent summers of 2007 and 2008, alongside substantial declines in many other UK butterfly species in the same time period (Botham et al. 2008).

Materials and methods

Study species

E. aurinia (Lepidoptera; Nymphalidae) is a univoltine species with its flight period from late May until early July in the study area. Although some individuals have been known to move up to 15–20 km (Warren 1994) the adult butterfly is generally sedentary (Junker and Schmitt 2010; Porter 1981). Females usually lay an initial large batch of c. 300 eggs soon after emergence in the natal patch, and smaller subsequent batches on plants further away (Porter 1981). Eggs are laid on the underside of leaves of the sole host plant used in this area, Devil's-bit scabious *S. pratensis* Moench (Dipsacaceae). The larvae live gregariously in silken webs from late July, through winter diapause, until spring the following year (Porter 1982). Larvae then become solitary in the penultimate instar to feed and eventually final 6th instar larvae pupate.

Study area

The study area was located in south-west England, encompassing an area known as the mid-Cornwall Moors (Fig. 1). Previous surveys showed an extant set of sub-populations of *E. aurinia* to be breeding across areas of Goss Moor National Nature Reserve (NNR) (SW 9560) and Breney Common Cornwall Wildlife Trust Reserve (SX 0561). These areas are a wet mosaic of unimproved purple moor grass (*Molinia caerulea*) pasture, heath and scrub, and are a Designated Special Area of Conservation (SAC), partly for the *E. aurinia* subpopulations they support. The project was a partnership with several organisations (Butterfly Conservation, Natural England, Cornwall Wildlife Trust, Highways Agency and Environment Agency) supported by the EU LIFE Natura fund. Prior to the establishment of the LIFE Project a survey was carried out to find other suitable habitat in the surrounding areas. Transect areas defined as 'suitable' for the study were open, unimproved grassland with the host plant present (Wahlberg et al. 2002), or areas that showed the potential to become suitable with management. In addition to the sites at Goss Moor and Breney Common a further seven sites were selected for inclusion in the LIFE Project area, making a total of nine sites encompassing fifty-four individual transect areas over a total area of 1,050 ha, for inclusion in a five-year monitoring programme.

Field techniques

Two field visits were made each year; the first to survey adult butterflies in spring (late May through June); and the second to survey pre-hibernating larval webs in autumn (late August through September).

Adult surveys

Adult monitoring surveys were carried out using the standard Butterfly Conservation methodology (Tunmore et al. 2008). Across each transect area a zig-zag path was walked, covering both suitable and unsuitable areas of habitat, recording numbers of adult *E. aurinia* and the time taken to complete the survey (minutes). Two field visits were made to each transect area ensuring the peak flight period was covered. Using the flight area of each transect (Ha), relative adult density was calculated for each visit to each transect area, and an average of the two relative densities was taken and used in the analyses. Because of poor weather, some transects were not repeated in the final year of surveys, so just the single value was used where this occurred. Sites were visited between 10:30 h and 16:00 h on days of suitable weather (minimum of 13°C with at least 60% sun or a minimum of 17°C with at least 40% sun, and little or no wind) (Tunmore et al. 2008).

Larval web surveys

Most larval web counts were carried out using the standard Butterfly Conservation transect methodology; which involves walking a zig-zag transect across the patch of habitat recording numbers of occupied webs seen 1 metre either side of the transect line (Tunmore et al. 2008). The length (metres) of the transect was also measured using a pedometer and the time taken to walk the transect (minutes) was recorded. As part of another continuing monitoring programme, more larval web counts were undertaken by a second field surveyor using area searches of fixed plots. During 2 years of the current study, a number of comparative surveys (both surveyors conducting

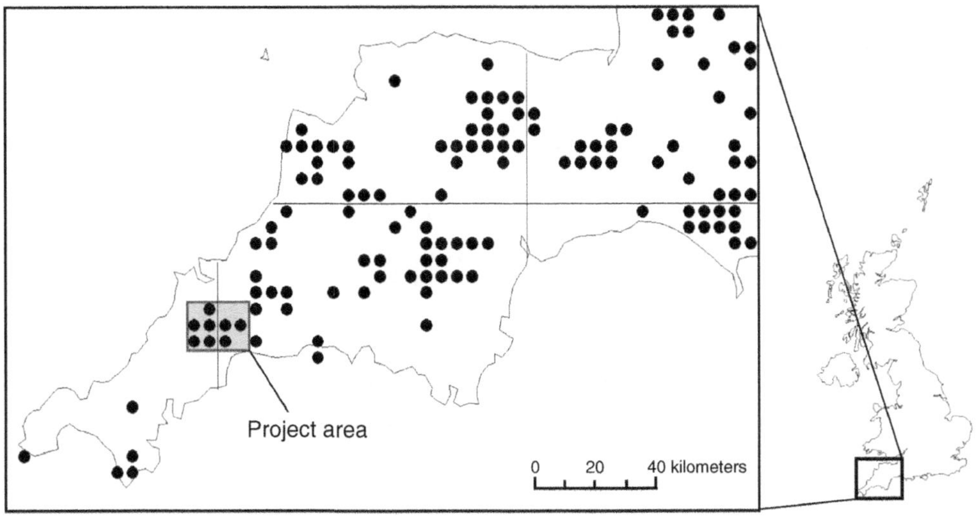

Fig. 1 South-west England, showing the project area across the Mid-Cornwall Moors. It is the most westerly stronghold for the species, as they have already declined from more coastal areas in Cornwall and only a couple of diminishing small colonies remain otherwise. Distribution map courtesy of Butterfly Conservation

larval web counts of the same transect area but using different methodologies) were carried out in suitable areas (where larval webs had been found) in order to calculate a 'conversion factor' to convert the number of webs found during area searches of fixed plots to a transect equivalent. Regression analysis gave the best conversion factor (Transect based search = 3.4236*Area search of fixed plot) to allow statistical analysis using both sets of data.

Habitat condition monitoring

Several habitat variables were measured on larval web transects to assess the habitat quality in which larval webs were found. Sward height (cm) was measured using a drop disk at thirty random points along each transect (Stewart et al. 2001). We also recorded the presence/absence of *S. pratensis*, *M. caerulea* tussocks, invading scrub and evidence of stock grazing at each of these thirty points (Brereton et al. 2005). The percentage of points where each habitat variable was present, out of the thirty measured, gave a proxy of density for *S. pratensis*, *M. caerulea* tussocks and invading scrub and a measure of intensity of stock grazing over each transect area. Three measures of habitat management were used in analyses: the years in which grazing was commenced on each site, the number of swaling events over the 5 years management periods, and the number of years since the most recent swaling event.

Data analysis

To investigate the relationship between adult or larval densities and habitat variables, we used a restricted dataset of only those transect areas with at least one occurrence of either adults or webs over the 5 years period (thirty-three out of fifty-four transects) as we were interested in effects on density not presence/absence. The response variables in our analyses were adult densities and larval web densities. The explanatory variables were intensity of stock grazing (% of transect points that showed evidence of grazing), density of host plant (% of transect points where *S. pratensis* was observed), scrub density (% of points), *M. caerulea* density (% of points), and spring and autumn sward height (in cm). We included quadratic terms for all fixed effect explanatory variables due to our a priori expectations of some humped relationships. Modelling the impacts of habitat variables on *E. aurinia* densities was performed in two phases. First, we used model simplification of multiple regressions, to account for covariance among explanatory variables and to clarify the habitat variables that best predicted density. Second, we analysed each habitat variable in separate regressions, to highlight individual links between habitat variables and *E. aurinia*.

In analyses of the effects of management on habitat variables we used the entire dataset of all fifty-four transect areas, as we were interested in whether management could improve the habitat variables of importance across all sites, to allow potential re-colonisation events. To determine if there had been a positive response to the management, temporal trends of habitat variables were analysed as response variables, with year as a fixed effect explanatory variable, at an individual site level, as often this is more informative for individual landowners, and an overall average can be misleading. Temporal trends of adult and web densities were also analysed at both individual sites and across all nine sites to give an overall trend.

All data analysis was carried out using R version 2.9.0 (R development core team (2010) r: A language and environment for statistical computing. R foundation for statistical computing, vienna, austria. Isbn 3-900051-07-0, url http://www.R-project.Org. Accessed 21 May 2010) (R Development Core Team). Linear mixed effects models were used in all analyses and we accounted for spatial structuring and repeated measures by modelling survey area, nested within site, as random effects on intercepts. Year was modelled as a random slope. Response variables (adult or web density) were log-transformed to reduce skew and improve homogeneity of variance in residuals, and models were checked for homoscedasticity of residuals and Gaussian errors. We used likelihood ratio tests and simplified maximum likelihood versions of the mixed effects models to test for significance of fixed effect explanatory variables at the $P < 0.05$ level (Crawley 2007).

Results

Of the nine sites selected for inclusion in the study throughout the 5 years period, only five had extant populations of *E. aurinia*, and only thirty-three of the fifty-four designated transects had at least one occurrence of either adults or webs over the period.

Can habitat variables explain the presence of *E. aurinia*?

As expected, a higher density of the host plant *S. pratensis* significantly increased the number of webs found, as did higher sward height in autumn (Table 1). Single term regressions (Table 2) revealed a humped relationship between web density and level of stock grazing: above an optimum level, grazing intensity became detrimental (Fig. 2). This relationship became non-significant when considered in the same model as host plant density and sward height, implying that grazing effects are indirect: it seems likely that optimal grazing can maximise sward

Table 1 Results of a linear mixed effects model using multiple regression to relate habitat variables with larval web densities and adult butterfly densities. Models used data only from occupied transect areas

	Coefficient	X_1^2	P
Webs			
Significant terms			
Autumn sward height (cm)	0.041	10.82	<0.001
S. pratensis density (%)	0.015	4.58	0.032
Non-significant terms			
Quadratic autumn sward height (cm)		1.18	0.28
Spring sward height (cm)		0.19	0.66
Quadratic spring sward height (cm)		0.21	0.65
Quadratic *S. pratensis* density (%)		0.16	0.69
M. caerulea density (%)		3.10	0.08
Quadratic *M. caerulea* density (%)		1.18	0.28
Scrub density (%)		0.30	0.58
Quadratic scrub density (%)		1.84	0.18
Intensity of stock grazing (%)		1.60	0.21
Quadratic intensity of stock grazing (%)		2.98	0.08
Adults			
Autumn sward height (cm)		0.06	0.80
Quadratic autumn sward height (cm)		0.06	0.81
Spring sward height (cm)		2.45	0.12
Quadratic spring sward height (cm)		0.31	0.58
S. pratensis density (%)		1.79	0.18
Quadratic *S. pratensis* density (%)		0.01	0.91
M. caerulea density (%)		2.62	0.11
Quadratic *M. caerulea* density (%)		2.56	0.11
Scrub density (%)		0.61	0.43
Quadratic scrub density (%)		0.75	0.39
Intensity of stock grazing (%)		1.99	0.16
Quadratic intensity of stock grazing (%)		0.11	0.74

Table 2 Results of linear mixed effects models using single regressions to determine the gross influence of each habitat variable on density of larval webs and adult butterflies. Models used data only from occupied transect areas

	Coefficient	X_1^2	P
Webs			
Significant terms			
Autumn sward height (cm)	0.0380	8.97	0.003
Intensity of stock grazing (%)	0.0290	3.91	0.048
Quadratic intensity of stock grazing (%)	−0.0004	6.02	0.014
Non-significant terms			
Quadratic autumn sward height (cm)		3.49	0.06
Spring sward height (cm)		2.42	0.12
Quadratic spring sward height (cm)		1.73	0.19
S. pratensis density (%)		2.74	0.10
Quadratic *S. pratensis* density (%)		0.65	0.42
M. caerulea density (%)		0.14	0.71
Quadratic *M. caerulea* density (%)		0.67	0.41
Scrub density (%)		0.01	0.94
Quadratic scrub density (%)		2.90	0.09
Adults			
Autumn sward height (cm)		0.59	0.44
Quadratic autumn sward height (cm)		0.00	0.99
Spring sward height (cm)		2.65	0.10
Quadratic spring sward height (cm)		0.42	0.52
S. pratensis density (%)		0.52	0.47
Quadratic *S. pratensis* density (%)		0.11	0.75
M. caerulea density (%)		1.80	0.18
Quadratic *M. caerulea* density (%)		1.60	0.21
Scrub density (%)		0.01	0.91
Quadratic scrub density (%)		0.89	0.35
Intensity of stock grazing (%)		2.30	0.13
Quadratic intensity of stock grazing (%)		0.14	0.71

height and host plant density, rather than affecting webs directly. In single term regressions, larval web densities were still positively influenced by the height of autumn sward. However, without accounting for the effect of the other habitat variables, the influence of *S. pratensis* was lost ($\chi_1^2 = 2.74$, $P = 0.098$). There were no significant habitat variables associated with adult butterfly densities.

Although none of the measured habitat variables had a significant effect on the densities of adult butterflies seen, the number of larval webs the previous year was found to be a strong predictor of the densities of adult butterflies seen in the current year (linear coefficient 0.01, $\chi_1^2 = 12.59$, $P < 0.001$). Adult butterfly densities in the current year also predict numbers of larval webs in the current year (linear coefficient 0.033, $\chi_1^2 = 23.08$, $P < 0.001$), so that management for one life stage also benefits the other.

Has management achieved optimum levels of habitat variables favoured by *E. aurinia*?

Three measures of management were analysed: the year grazing started; the number of times a transect area was burnt during the 5 years project; and how many years since the most recent burning of a transect area. The evidence of stock grazing, measured over each transect area, was also included in analyses as an indicator of the intensity of grazing over each area.

Over all 5 years of the project and over all nine sites whether occupied by *E. aurinia* or not, the year in which grazing started was the only management measure to affect the density of *S. pratensis* ($\chi_1^2 = 15.29$, $P = 0.033$) with less host plant available the more recently grazing had been started and if the area hadn't been grazed at all. In areas where grazing had occurred for longer (continuous grazing

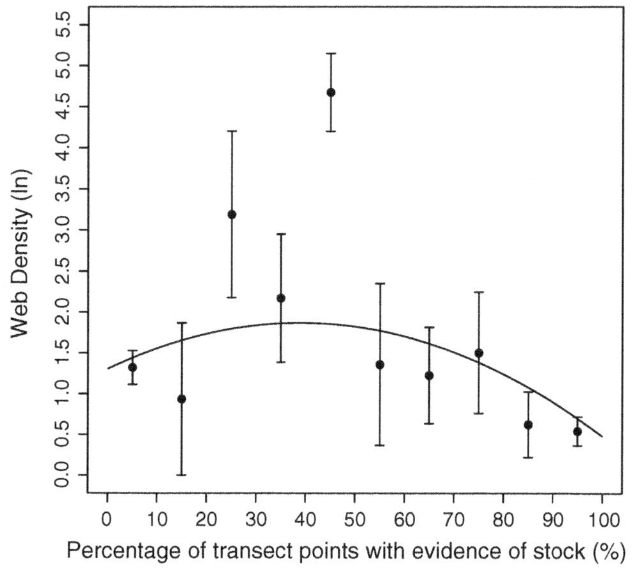

Fig. 2 The significant quadratic effect of stock grazing on density of larval webs. Larval web density (ln) is averaged across all transects, ± standard error, for each ten percent 'bin' of stock grazing intensity

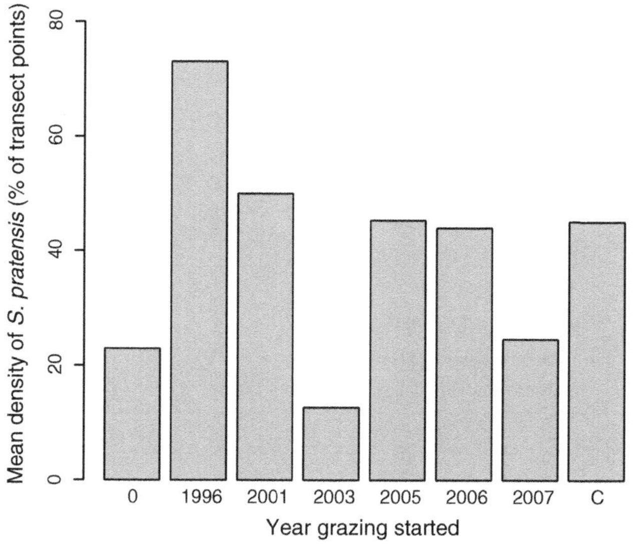

Fig. 3 The year when grazing was started has a significant effect on the occurrence of *S. pratensis* over the transects ($\chi_1^2 = 15.29$, $P = 0.033$) with less host plant being available for *E. aurinia* the more recently grazing had started or if it was not started at all. *0* never grazed, *C* continuous grazing

and grazing since 1996 and 2001) the density of the host plant was highest (Fig. 3).

Unfortunately the host plant *S. pratensis* did not increase significantly over the 5 years period at all nine sites despite management ($\chi_1^2 = 2.83$, $P = 0.09$). However, when considering sites individually, which is often more informative, there are opposing significant results. As Fig. 4 illustrates, at three of the sites there was a considerable increase in *S. pratensis* over the 5 years (Goss Moor: linear coefficient 2.51, $\chi_1^2 = 7.71$, $P = 0.006$; Retire Common: linear coefficient 1.395, $\chi_1^2 = 4.65$, $P = 0.031$; and Tregonetha: linear coefficient 1.681, $\chi_1^2 = 3.86$, $P = 0.049$). At one site, Criggan Moor, there was a vast decline in the density of *S. pratensis* found (linear coefficient −8.263, $\chi_1^2 = 5.21$, $P = 0.022$) possibly due to under grazing as the areas are very wet and difficult for cattle to access.

There was no evidence for either increasing or decreasing trends in the intensity of stock grazing across all nine sites over all 5 years ($\chi_1^2 = 1.55$, $P = 0.21$). Eight out of the nine sites appeared to be diverging from the optimal level of stock grazing (determined from quadratic relationship = stock grazing at 42% of stop points along the transect). The one site that had moved towards the optimum was Goss Moor (linear coefficient = 7.99 $\chi_1^2 = 8.41$, $P = 0.004$), with 50.8% of stop points having evidence of stock grazing where originally in 2004 it had 14.3%. Other sites significantly decreased (Breney Common: linear coefficient = −21.03 $\chi_1^2 = 26.61$, $P < 0.001$; Redmoor: linear coefficient = −28.42 $\chi_1^2 = 17.41$, $P < 0.001$) towards 0% and one (Quoit Farm: linear coefficient = 27.14 $\chi_1^2 = 14.74$, $P < 0.001$) increased, almost reaching 100%.

In an analysis of habitat management variables (when grazing started, years since the last burn and evidence of stock grazing), two of the management variables had a significant impact on the height of autumn sward over the fifty-four transect areas. The more recently grazing had been started, the higher the average sward height at sites ($\chi_1^2 = 14.95$, $P = 0.037$, Fig. 5a), although if started in 1996 the average was slightly higher and similar to at sites where grazing had never occurred. Highest autumn sward height is found at sites with continuous grazing occurring, which could be a result of low-level extensive grazing rather than intensive grazing in other areas. This agrees with the relationship between the evidence of stock found along transect walks and autumn sward height (linear coefficient = 0.006, $\chi_1^2 = 8.43$, $P = 0.004$; quadratic coefficient = −0.0001, $\chi_1^2 = 17.64$, $P < 0.001$, Fig. 5b) whereby higher levels of stock grazing decrease the average autumn sward height, but at a medium level, an optimum sward height was achieved.

As described in both multiple regression models (Table 1) and single regression models (Table 2), numbers of webs increase with higher sward height in autumn. Over all nine sites there was a significant decrease in autumn sward height (linear coefficient −1.49, $\chi_1^2 = 19.20$, $P < 0.001$), although at one site, Redmoor, there was a positive increase (linear coefficient 1.87, $\chi_1^2 = 7.50$, $P = 0.006$).

Has *E. aurinia* responded to management across the mid-cornwall moors?

Figure 6a clearly shows the dramatic decline in numbers of larval webs found (linear coefficient −0.32, $\chi_1^2 = 15.45$,

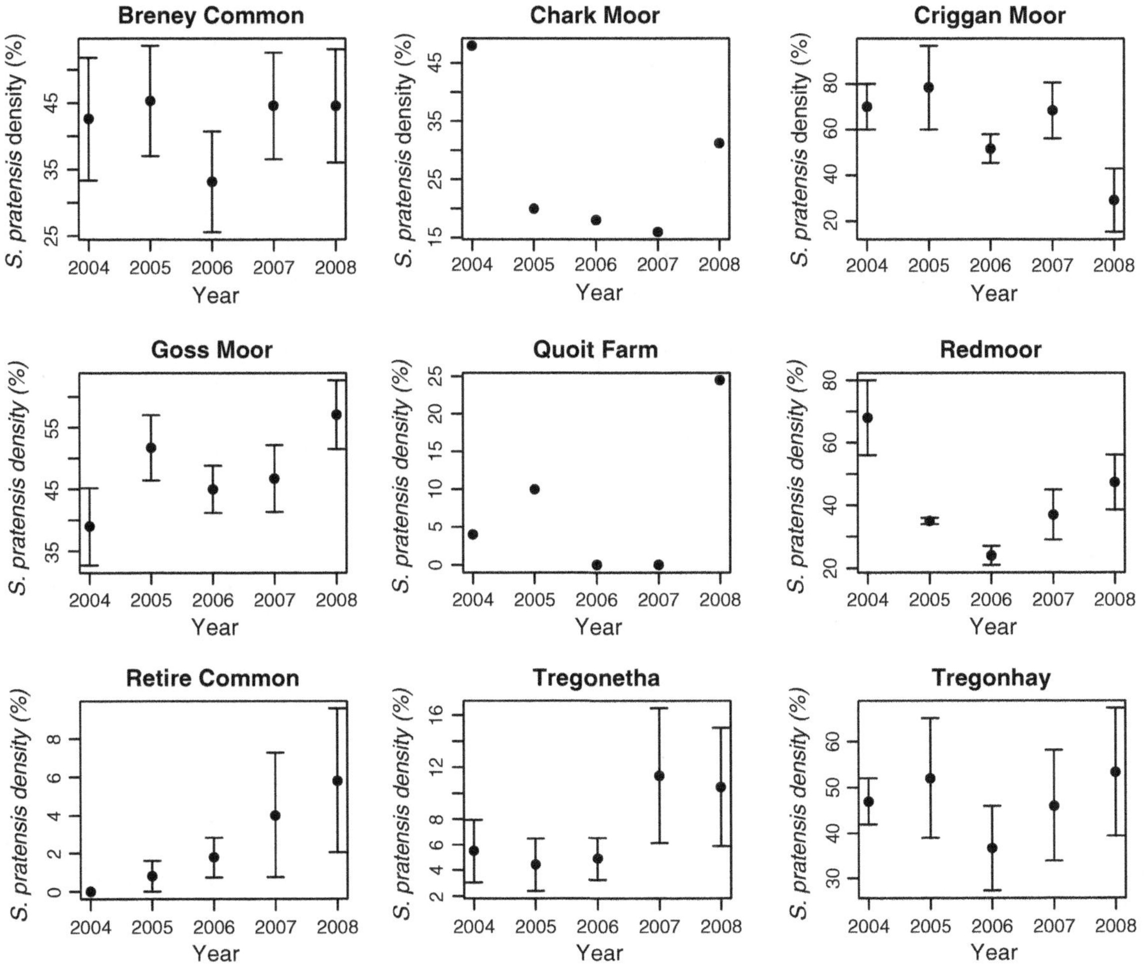

Fig. 4 Temporal comparison of *S. pratensis* density for each site (% of transect points where *S. pratensis* was present) over the 5 years of the project. ± standard errors of mean *S. pratensis* density calculated across transect areas within sites. Analysis across all nine sites showed no significant change in *S. pratensis* density ($\chi_1^2 = 2.83$, $P = 0.09$)

$P < 0.001$) and adult butterflies seen (linear coefficient −0.19, $\chi_1^2 = 7.83$, $P = 0.005$) over the entire study site despite management, at those transects occupied at least once during the study. However, during the project there was a colonisation event at Redmoor (Fig. 6e) and consequently there was an increase in web density (linear coefficient 0.321, $\chi_1^2 = 1.94$, $P = 0.164$) and adult density (linear coefficient 0.776, $\chi_1^2 = 6.49$, $P = 0.011$), but only the latter significant.

Discussion

Across the thirty-three transects occupied at least once during the study period, larval webs preferred areas of habitat with higher density of their host plant *S. pratensis*, in combination with a suitably tall sward height in autumn and optimum levels of stock grazing. However, across the entire nine sites there could be a possible conflict of interest between achieving the desired levels of *S. pratensis* and autumn sward, as more recent grazing seems to increase autumn sward height but lower the density of *S. pratensis*. Being an inferior competitor (Buhler and Schmid 2001), *S. pratensis* may be getting outcompeted by taller vegetation such as *M. caerulea* by reducing access to light and nutrients, and reducing opportunities for germination. Caution should be taken when interpreting results of sward height, as it is intuitively unlikely that an ever-increasing height of sward would continue to have positive effects on numbers of larval webs. Previous studies have also shown an optimum level of sward over which the likelihood of *E. aurinia* presence decreases (Konvicka et al. 2003). A lack of data, or the short period over which the study was done, may mean that a quadratic relationship between larval webs and autumn sward height was not found. It should be noted that 'continuous' levels of grazing produce

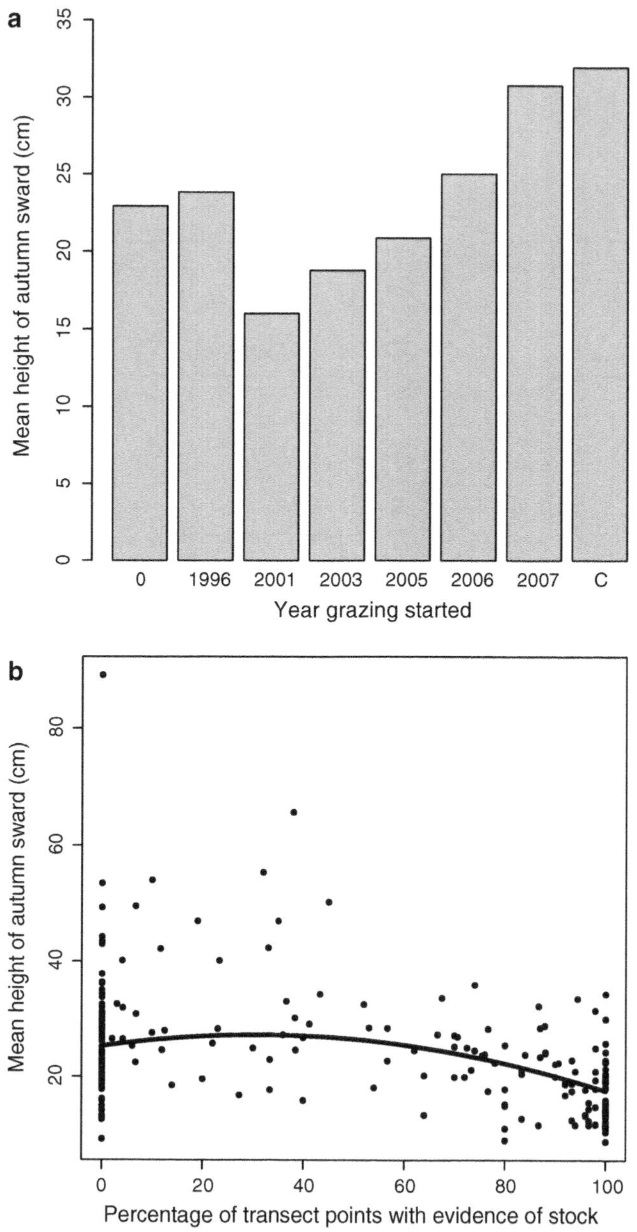

Fig. 5 In an analysis using data from all fifty-four transect areas, autumn sward height is **a** significantly affected by the year in which grazing started (*C* continuous grazing; *0* never grazed), and **b** the intensity of grazing it is subjected to

both desired outcomes, and without knowing the particular stocking densities could be inferred as low-level extensive grazing over prolonged periods. This compliments the quadratic relationship found between evidence of stock grazing and both numbers of larval webs and autumn sward height—both responding favourably to optimum medium levels of stock grazing. Although densities of adult *E. aurinia* were not affected by the habitat variables measured, they were a direct result of the density of webs the previous year, and so by managing for one the other will also benefit. There will undoubtedly be other habitat variables that were not measured in the current study, but which will be of huge importance for the presence of adult *E. aurinia*, such as availability or density of nectar sources.

Using management tools to create suitable habitat

Due to differences between the needs of adults and larvae, it is often difficult to determine the 'optimal' level of habitat variables needed for the species as a whole. Previously, much work to decipher what constitutes as 'suitable habitat' and 'favourable condition' has been done on the basis of the adult butterfly's preferences (Betzholtz et al. 2007), but this is often not the most useful measure. The density of the host plant is irrelevant for adult butterflies' survival as they feed opportunistically upon the nectar of other flowering plants (Anthes et al. 2003), as *S. pratensis* does not flower during the flight period. Here we presented both larval web counts and adult surveys, as it is still crucial to maintain a favourable habitat for the adult stage. However, it appears the adults rely on none of the habitat variables measured, but densities are directly related to the number of webs the previous year. Consequently, if the habitat suits the requirements of the larvae, it will also have a positive, if indirect, effect on adult butterfly numbers.

Intermediate levels of grazing were seen to maximise autumn sward height, perhaps to an optimum level, and density of *S. pratensis*—both important for larval webs, but neither showing significant improvements over all sites. Although Lewis and Hurford (1997) found no significant effect of management on population size, Peet et al. (1999) found that plant species richness was higher in managed plots, which were also more structurally heterogeneous. Peet et al. therefore, suggested that patches could be left unmanaged on a 2-year rotation without significantly affecting the composition of the plant community. Cutting, or mowing, is also used as a management tool on some *E. aurinia* sites and is reported to be similarly suitable (Dolek and Geyer 1997) yet intrinsically difficult on damp grasslands due to uneven terrain. Swaling can open up vegetation that was previously too dense, but eventually leads to a reduction in *S. pratensis* abundance by encouraging growth of *M. caerulea*. With the lack of swaling influencing any habitat variables on the mid-Cornwall Moor sites, it seems the key for management of these wet grasslands is grazing at an intermediate level so as not to allow over-grazing but not leave sites abandoned. Achieving this optimum is difficult as it is not easy to arrange suitable grazing regimes on damp grassland sites of low agricultural value.

Response of the butterfly—are we doing enough?

Bulman et al. (2007) demonstrated that to achieve population persistence in the long term more than 70 ha of

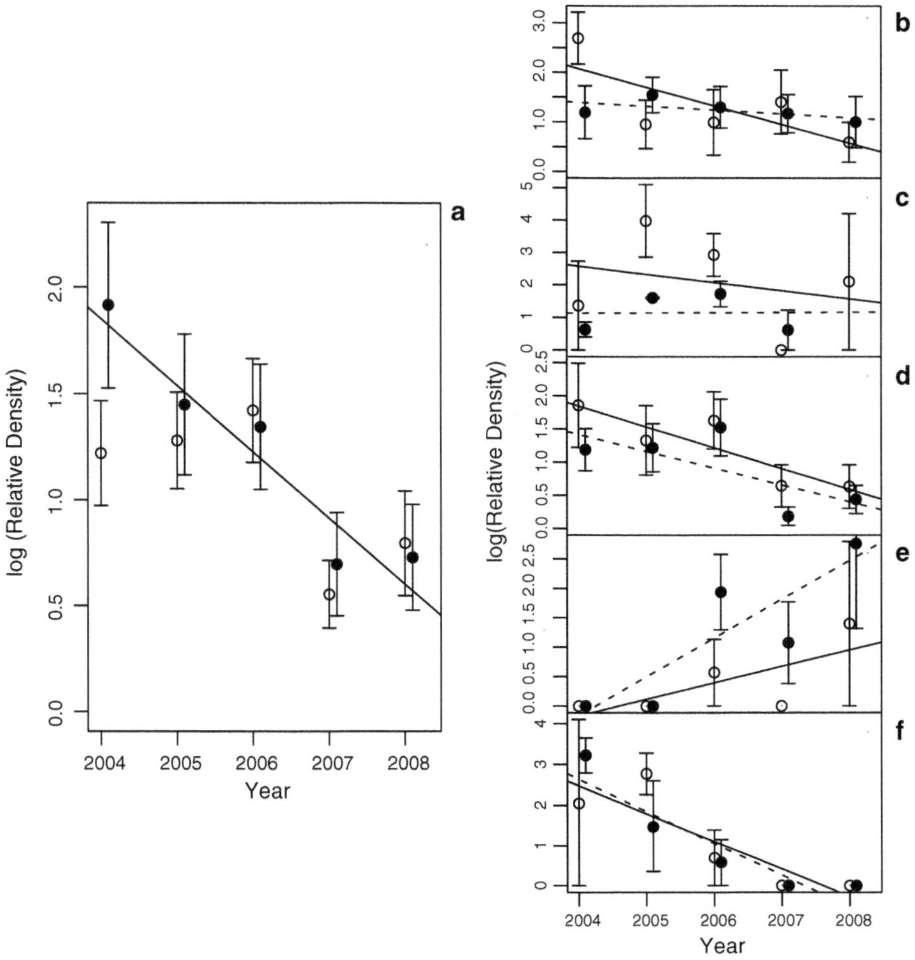

Fig. 6 **a** Temporal comparison of log transformed larval web densities (*solid circles* and *solid line*) and adult densities (*open circles* and *dashed line*) across the 5 years of monitoring at all sites. **(b-f)** Individual temporal trends of web densities at only those sites where *E. aurinia* has an extant colony. **b** Breney common. **c** Criggan moor. **d** Goss moor. **e** Redmoor. **f** Tregonhay

suitable breeding habitat is needed to be in favourable condition. 1,050 ha of land was encompassed by the nine mid-Cornwall Moors project sites, but not all was in the right condition. In 2004 it was estimated from digitised National Vegetation Classification (NVC) maps that there was 315.8 ha of potential habitat across the nine sites. At the start of the project 15 ha of this was thought to be in favourable condition. By the end of the project, this figure had increased by 115–130 ha in 2008. Although there seems to be enough suitable habitat to support the continued functioning of the metapopulation, numbers of *E. aurinia* cast doubt on the ability of the metapopulations to survive over the longer term.

It is possible that, regardless of the management and potentially improved habitat, *E. aurinia* simply cannot respond because of other limiting factors. Ford and Ford (1930) reported huge fluctuations in numbers of *E. aurinia* across the years, from swarms covering hedgerows to dwindling numbers close to extinction. Such large fluctuations may be attributed to unmanageable climatic fluctuations, or to top-down effects of natural enemies, especially parasitoids (Porter 1983).

It is recommended to use larval web counts as the most reliable indicator of population size (Lewis and Hurford 1997), but our results show a steep significant decline in both adult and larval densities across all nine sites of the mid-Cornwall Moors, except at Redmoor where there was an apparent colonisation event during the lifetime of the project. The increase in webs and adults at this site was not found to be significant, but if the population continues to grow over the next generation or two, it will be a huge success for the project.

It is possible that the overall trend presented here is not wholly correct, as the last 2 years of monitoring experienced unusually high levels of rainfall and poor weather during the flight period. Recent densities may have been low for this reason alone, and good weather for the next couple of generations could see a massive improvement due to better condition of the habitat.

The significant decline could also be a result of severe inbreeding (Saccheri et al. 1998). While there are ad hoc records of wandering adults outside of the known breeding subpopulations it is not known whether the mid-Cornwall Moors populations intersperse with each other, and hence whether the metapopulation is still functioning and fulfilling the four necessary conditions for persistence (Hanski 1999; Hanski et al. 1995). If it is not, there is likely to be a significant increase in homozygosity, or loss of genetic variation. Hanski et al. (1994) estimated that approximately 80% of *Melitaea cinxia* butterflies spend their entire lifetime in the natal patch—a pattern that is likely to be mirrored in *E. aurinia* and will undoubtedly increase the probability of inbreeding depression. Recent studies (Joyce and Pullin 2003; Sigaard et al. 2008) have indicated restrictions of gene flow within *E. aurinia* metapopulations and emphasize the need for networks of suitable habitat. Increased levels of emigration as a mechanism of the 'Allee effect' (Kuussaari et al. 1998) may also be exacerbating the problems of small populations by reducing numbers further, or we may be witnessing the payback of the mid-Cornwall Moors' 'extinction debt' (Bulman et al. 2007; Tilman et al. 1994).

Implications for the conservation of *E. aurinia* across the mid-cornwall moors, and beyond

Much is now known about the ecology and conservation of *E. aurinia* (e.g. Bulman 2001; Porter 1981; Warren 1994) and the microhabitat preferences of both adults and larvae (e.g. Anthes et al. 2003; Betzholtz et al. 2007; Konvicka et al. 2003) as well as the viability of some populations across the species' range (e.g. Bulman 2001; Schtickzelle et al. 2005; Wahlberg et al. 2002) using the incidence function model (IFM) (Hanski 1994) to predict occurrences of *E. aurinia* over a network of habitat patches. A next step for the mid-Cornwall Moors metapopulation would be to conduct a population viability analysis and determine what possible future *E. aurinia* has there. Combined with the usage of habitat distribution models (Early et al. 2008) and studying the gene flow within and between populations, a solid management plan for the area could be envisaged.

It is sensible to conclude that short term studies are not enough for the responses of species to be measured on, or decisions based on for correct management. We may either be underestimating the re-colonisation of restoratively grazed patches (Pöyry et al. 2005) or alternatively we could be underestimating levels of population turnover and consequently, underestimating extinction threats (Thomas et al. 2002). Most assessments of metapopulation dynamics have based inferences on one or two year abundances or dynamics (e.g. Anthes et al. 2003; Hanski 1994; Hanski and Thomas 1994; Wahlberg et al. 1996). We show that even after 5 years of monitoring, more time is needed to allow restoration of habitats and responses of populations. However, our analyses of the mid-Cornwall population time series demonstrate that key, manageable habitat variables can be used to predict population densities, and manipulated to increase the suitability of habitats for population growth and establishment (whether by natural dispersal processes or via reintroduction schemes).

Acknowledgments Many thanks go to a multitude of helpers in the field; R. Hobson, M. Davey, M. Easter, H. Silver, P. Brown and C. Berry. Also to Natural England, Cornwall Wildlife Trust and all landowners allowing access to land. The habitat management and monitoring schemes were part funded by the EU LIFE Nature Programme. M.S., R.f–C. and D.J.H are funded by BBSRC.

References

Anthes N, Fartmann T, Hermann G, Kaule G (2003) Combining larval habitat quality and metapopulation structure: the key for successful management of pre-alpine *Euphydryas aurinia* colonies. J Insect Conserv 7:175–185

Asher J, Warren M, Fox R, Harding P, Jeffcoate G, Jeffcoate S (2001) The millenium atlas of butterflies in Britain and Ireland. Oxford University Press, Oxford

Barnett LK, Warren MS (1995) Species action plan: marsh fritillary. *Eurodryas aurinia.* Butterfly Conservation

Betzholtz PE, Ehrig A, Lindeborg M, Dinnetz P (2007) Food plant density, patch isolation and vegetation height determine occurrence in a Swedish metapopulation of the marsh fritillary *Euphydryas aurinia* (Rottemburg, 1775) (Lepidoptera, Nymphalidae). J Insect Conserv 11:343–350. doi:10.1007/s10841-006-9048-3

Botham MS, Brereton TM, Middlebrook I, Cruickshanks KL, Harrower C, Beckmann B, Roy DB (2008) United Kingdom butterfly monitoring scheme report for 2008. CEH Wallingford

Brereton TM, Brook S, Hobson R (2005) Habitat condition monitoring for butterflies: 2004 pilot study. Butterfly conservation report No. S05-06, Wareham

Buhler C, Schmid B (2001) The influence of management regime and altitude on the population structure of *Succisa pratensis*: implications for vegetation monitoring. J Appl Ecol 38:689–698

Bulman CR (2001) Conservation biology of the marsh fritillary butterfly *Euphydryas aurinia.* Ph.D. Thesis, University of Leeds

Bulman CR, Wilson RJ, Holt AR, Bravo LG, Early RI, Warren MS, Thomas CD (2007) Minimum viable metapopulation size, extinction debt, and the conservation of a declining species. Ecol Appl 17:1460–1473

Crawley MJ (2007) The R book. Wiley, Chichester

Dolek M, Geyer A (1997) Influence of management on butterflies of rare grassland ecosystems in Germany. J Insect Conserv 1:125–130

Early R, Anderson B, Thomas CD (2008) Using habitat distribution models to evaluate large-scale landscape priorities for spatially dynamic species. J Appl Ecol 45:228–238. doi:10.1111/j.1365-2664.2007.01424.x

Ford HD, Ford EB (1930) Fluctuation in numbers, and its influence on variation, in *Melitaea aurinia*, Rott (Lepidoptera). Trans R Entomol Soc Lond 78:345–352

Hanski I (1994) A practical model of metapopulation dynamics. J Anim Ecol 63:151–162

Hanski I (1998) Metapopulation dynamics. Nature 396:41–49

Hanski I (1999) Metapopulation ecology. Oxford University Press, Oxford

Hanski I, Gaggiotti OE (2004) Ecology genetics and evolution of metapopulations. Academic Press, San Diego

Hanski I, Thomas CD (1994) Metapopulation dynamics and conservation—a spatially explicit model applied to butterflies. Biol Conserv 68:167–180

Hanski I, Kuussaari M, Nieminen M (1994) Metapopulation structure and migration in the butterfly *Melitaea-cinxia*. Ecology 75:747–762

Hanski I, Pakkala T, Kuussaari M, Lei GC (1995) Metapopulation persistence of an endangered butterfly in a fragmented landscape. Oikos 72:21–28

Hobson R, Budd P (2001) The marsh fritillary in cornwall: site dossier. Incorporating the bodmin moor larval survey 2000. Report No. S01-32, Butterfly Conservation, Wareham

Hobson R, Bourn NAD, Warren MS, Brereton TM (2001) The marsh fritillary in England: a review of status and habitat condition. Report no. S01-31, Butterfly Conservation, Wareham

Hula V, Konvicka M, Pavlicko A, Fric Z (2004) Marsh fritillary (Euphydiyas aurinia) in the Czech Republic: monitoring, metapopulation structure, and conservation of an endangered butterfly. Entomol Fenn 15:231–241

Joyce DA, Pullin AS (2003) Conservation implications of the distribution of genetic diversity at different scales: a case study using the marsh fritillary butterfly (*Euphydryas aurinia*). Biol Conserv 114:453–461. doi:10.1016/s0006-3207(03)00087-9

Junker M, Schmitt T (2010) Demography, dispersal and movement pattern of *Euphydryas aurinia* (Lepidoptera: Nymphalidae) at the Iberian Peninsula: an alarming example in an increasingly fragmented landscape? J Insect Conserv 14:237–246. doi: 10.1007/s10841-009-9250-1

Konvicka M, Hula V, Fric Z (2003) Habitat of pre-hibernating larvae of the endangered butterfly *Euphydryas aurinia* (Lepidoptera : Nymphalidae): what can be learned from vegetation composition and architecture? Eur J Entomol 100:313–322

Kuussaari M, Saccheri I, Camara M, Hanski I (1998) Allee effect and population dynamics in the glanville fritillary butterfly. Oikos 82:384–392

Levins R (1969) Some demographic and genetic consequences of environmental heterogeneity for biological control. Bull Entomol Soc Am 15:237–240

Lewis OT, Hurford C (1997) Assessing the status of the marsh fritillary butterfly (Eurodryas aurinia): an example from Glamorgan, UK. J Insect Conserv 1:159–166

Liu WH, Wang YF, Xu RM (2006) Habitat utilization by ovipositing females and larvae of the marsh fritillary (*Euphydryas aurinia*) in a mosaic of meadows and croplands. J Insect Conserv 10:351–360. doi:10.1007/s10841-006-9009-x

Peet NB, Watkinson AR, Bell DJ, Sharma UR (1999) The conservation management of imperata cylindrica grassland in Nepal with fire and cutting: an experimental approach. J Appl Ecol 36:374–387

Porter K (1981) The population dynamics of small colonies of the butterfly *Euphydryas aurinia*. Ph.D. Thesis, Oxford University

Porter K (1982) Basking behaviour in larvae of the butterfly *Euphydryas-aurinia*. Oikos 38:308–312

Porter K (1983) Multivoltinism in *Apanteles-bignellii* and the influence of weather on synchronization with its host *Euphydryas-aurinia*. Entomol Exp Appl 34:155–162

Pöyry J, Lindgren S, Salminen J, Kuussaari M (2005) Responses of butterfly and moth species to restored cattle grazing in semi-natural grasslands. Biol Conserv 122:465–478. doi:10.1016/j.biocon.2004.09.007

R Development Core Team (2010) R: a language and environment for statistical computing. R foundation for statistical computing, Vienna, Austria. ISBN 3-900051-07-0, URL http://www.R-project.org. Accessed 21 May 2010

Saccheri I, Kuussaari M, Kankare M, Vikman P, Fortelius W, Hanski I (1998) Inbreeding and extinction in a butterfly metapopulation. Nature 392:491–494

Schtickzelle N, Choutt J, Goffart P, Fichefet V, Baguette M (2005) Metapopulation dynamics and conservation of the marsh fritillary butterfly: population viability analysis and management options for a critically endangered species in Western Europe. Biol Conserv 126:569–581. doi:10.1016/j.biocon.2005.06.030

Sigaard P, Pertoldi C, Madsen AB, Sogaard B, Loeschcke V (2008) Patterns of genetic variation in isolated danish populations of the endangered butterfly *Euphydryas aurinia*. Biol J Linn Soc 95:677–687. doi:10.1111/j.1095-8312.2008.01078.x

Stewart KEJ, Bourn NAD, Thomas JA (2001) An evaluation of three quick methods commonly used to assess sward height in ecology. J Appl Ecol 38:1148–1154

Thomas CD (1994) Extinction, colonization, and metapopulations—environmental tracking by rare species. Conserv Biol 8:373–378

Thomas CD, Wilson RJ, Lewis OT (2002) Short-term studies underestimate 30-generation changes in a butterfly metapopulation. Proc R Soc Lond B Biol Sci 269:563–569. doi:10.1098/rspb.2001.1939

Tilman D, May RM, Lehman CL, Nowak MA (1994) Habitat destruction and the extinction debt. Nature 371:65–66

Tunmore M, Robinson F, Hodgson D (2008) Marsh fritillary monitoring on nine sites in mid-cornwall: flight and larval web surveys. A report to mid-cornwall moors LIFE project. Cornwall Environmental Consultants Ltd, Truro

Van Swaay CAM, Warren MS (1999) Red data book of European butterflies (*Rhopalocera*), 99th edn. Council of Europe Publishing, Strasbourg

Wahlberg N, Moilanen A, Hanski I (1996) Predicting the occurrence of endangered species in fragmented landscapes. Science 273:1536–1538

Wahlberg N, Klemetti T, Hanski I (2002) Dynamic populations in a dynamic landscape: the metapopulation structure of the marsh fritillary butterfly. Ecography 25:224–232

Warren MS (1994) The UK status and suspected metapopulation structure of a threatened European butterfly, the marsh fritillary *Eurodryas-aurinia*. Biol Conserv 67:239–249

J Insect Conserv (2011) 15:269–277
DOI 10.1007/s10841-010-9344-9

ORIGINAL PAPER

The effects of habitat fragmentation on niche requirements of the marsh fritillary, *Euphydryas aurinia*, (Rottemburg, 1775) on calcareous grasslands in southern UK

M. S. Botham · D. Ash · N. Aspey · N. A. D. Bourn · C. R. Bulman · D. B. Roy · J. Swain · A. Zannese · R. F. Pywell

Received: 19 April 2010 / Accepted: 3 September 2010 / Published online: 19 September 2010

Abstract The marsh fritillary, *Euphydryas aurinia*, has declined greatly in distribution across its range within Europe, resulting in its designation as a protected species under Annex II of the 1979 Bern Convention and the EC Habitats and Species Directive. The decline has been linked to a marked reduction in the extent of suitable calcareous and wet grassland habitats, habitats which have been lost through conversion of land to agriculture or urban areas, or reduced in quality due to inappropriate management. The UK is now one of the major strongholds for this butterfly in Europe, although much of the remaining habitat is small, isolated and highly fragmented. *E. aurinia* populations fluctuate greatly due to the combined effects of biotic (e.g. parasitoids) and abiotic (e.g. climate change) factors. We quantified the habitat associations of larval webs of *E. aurinia* on fragmented versus extensive (unfragmented) calcareous grassland habitat in southern England to test the hypothesis that habitat requirements of *E. aurinia* are more constrained within fragmented landscapes. Within both fragmented and unfragmented landscapes the quality and quantity of its main host plant in the UK, *Succisa pratensis*, was positively related to numbers of *E. aurinia* larval webs found. The sward height was also important at predicting the distribution of larval webs in both landscapes, although the heights were greater within sites in the unfragmented ($\approx$20 cm) compared to fragmented ($\approx$15 cm) landscape. We also found significant effects of elevation and the cover of bare ground on numbers of larval webs. Elevation was strongly correlated with the availability of host plant, whilst bare ground was only significant on sites within the fragmented landscape, showing a negative relationship with number of larval webs. Our results further emphasise the importance of not only maintaining the habitat quality of extant calcareous grassland sites for *E. aurinia* in the UK, but also increasing the size and connectivity of these sites to increase the chances and rate of (re)colonisation of unoccupied but suitable habitat. In addition, we show that the habitat requirements of *E. aurinia* on sites in a large unfragmented landscape may be less specific and thus require less extensive management than that required to create optimal conditions necessary at smaller, more isolated sites in fragmented landscapes.

Keywords *Euphydryas aurinia* · Marsh fritillary · Calcareous grassland · Butterfly · Metapopulation · Habitat fragmentation · Salisbury plain

M. S. Botham (✉) · N. Aspey · D. B. Roy · A. Zannese · R. F. Pywell
Centre For Ecology & Hydrology Wallingford, Maclean Building, Benson Lane, Crowmarsh Gifford, Wallingford, Oxon OX10 8BB, UK
e-mail: math2@ceh.ac.uk

D. Ash · J. Swain
Defence Estates, Westdown Camp, Tilshead, Salisbury, Wilts SP3 4RS, UK

N. A. D. Bourn · C. R. Bulman
Butterfly Conservation, Manor Yard, East Lulworth, Wareham, Dorset BH20 5QP, UK

Introduction

The marsh fritillary, *Euphydryas aurinia*, is protected under the 1979 Bern Convention (Annexe II) and the EC Habitats and Species Directive (Annexe II). Whilst it is widely distributed throughout Europe, Morocco and Algeria through temperate areas of Asia to Korea, it is threatened in much of its range, especially in northern

Europe where it has declined to between 20 and 50% of its former area of occupancy within a 25-year period (van Swaay and Warren 1999). Within Europe, the UK population is considered one of the largest and most important; thought to constitute 5–15% of the European distribution in the 1990s (van Swaay and Warren 1999). However, despite recent trends of increase in abundance on monitored sites within the UK (Botham et al. 2008) the distribution of *E. aurinia* continues to decline; by approximately 46% between 1970–1982 and 1995–2004 (Fox et al. 2006, 2010).

Loss of habitat caused by agricultural intensification and development is thought to be the major cause of this decline, with recent studies showing a significant association between a gradient in decline across *E. aurinia's* European range and a corresponding gradient in agricultural intensity (Thomas et al. 2008). *E. aurinia* is relatively sedentary, rarely dispersing beyond 750 m, although colonisation may rarely take place over 5–20 km (Warren 1994). It shows a metapopulation structure which is sensitive to the negative effects of habitat fragmentation caused by loss of habitat (Bulman et al. 2007; Bulman 2001).

Habitat fragmentation has a range of negative impacts on butterfly populations depending on the species (Wilson and Roy 2009). Beyond simply reducing the amount of habitat available, for poorly dispersive species, it can greatly restrict the ability for empty but suitable habitat patches to be re-colonised (Warren 1994). Isolated populations in small fragments of habitat are also considered as more vulnerable to extinction events and may be subject to the negative impacts of inbreeding as population sizes crash in years when climatic and/or pathogenic factors are prevalent (Gibbs and Van Dyck 2009; Bulman et al. 2007; Sigaard et al. 2008). Inbreeding and resultant loss of genetic diversity within small isolated populations is also likely to have an effect on the habitat requirements of less mobile species, potentially resulting in greater levels of specialism and thus greater sensitivity to changes in the quality of the remaining habitat (Gibbs and Van Dyck 2009). Thus fragmented populations of *E. aurinia* may differ in their habitat specificity to those populations found on less fragmented habitats. This may have important implications for the management of remaining sites, which are already known to vary greatly in the precise habitat requirements of *E. aurinia* (e.g. Konvicka et al. 2003).

In the UK *E. aurinia* occurs on two distinct habitat types: wet, *Molinia caerulea* dominated grasslands and dry calcareous grasslands (often dominated by *Bromus erectus*). On both habitats its main foodplant is devil's bit scabious, *Succisa pratensis*, although larvae have occasionally been found on field scabious, *Knautea arvensis* and small scabious, *Scabiosa columbaria* (Porter 1981). Most of the available habitat, of both types, is considerably fragmented throughout the UK. Calcareous grassland sites in the south west of England (Dorset) are a classic example of a fragmented landscape within which *E. aurinia* persists in a metapopulation, with local extinction and colonisation events common on some of the smaller sites. Further north in England, but still within 25 km, there is what is considered to be a large and stable population of *E. aurinia* inhabiting the expansive calcareous grassland of Salisbury Plain, Wiltshire; approximately 40,000 ha (Toynton and Ash 2002). This site, owned by the Ministry of Defence and used for military training purposes, may be regarded as an unfragmented habitat.

The aim of the current study was to quantify whether there are any differences in the habitat requirements of *E. aurinia* within fragmented (Dorset) versus unfragmented (Salisbury Plain) landscapes, where the geographical location within the UK is similar and the type of habitat (calcareous grassland dominated by *B. erectus*) and main host plant (*S. pratensis*) are the same. To do this we surveyed 40 × 40 m quadrats at 45 sites in Dorset which had an average area of 6.5 hectares and a total of eighty-one 40 × 40 m quadrats across Salisbury Plain, recording the number of larval webs found, the height of the vegetation, and the number, size and coverage of host plants amongst other habitat variables.

Methods

Larval web surveys were conducted between 14th and 28th August 2009. At this time of year *E. aurinia* larvae are gregarious and form dense webs around suitable host plants making them easy to locate. Because the host plant is distributed in patches across the habitats, particularly on Salisbury Plain, we used a nested quadrat sampling method (see below for details).

Site choice

For our fragmented landscape, 45 calcareous grassland sites in Dorset were chosen from a list of sites on which *E. aurinia* has been found within the last 10 years, and on which the main host plant, *S. pratensis*, is present (Bulman 2001). For our unfragmented landscape we chose a total of 81 plots across the Defence Training Estate Salisbury Plain(DTE SP) based on previous records of *E. aurinia* provided by Defence Estates and on the presence of host plants derived from vegetation surveys carried out by the Centre for Ecology & Hydrology (Pywell et al. 2006). The DTE SP is approximately 40,000 ha in size, of which there is 14,000 ha of good quality calcareous grassland in amongst a mosaic of species-rich mesotrophic grassland, extensively managed farmland, tracks and plantations (Toynton and Ash 2002). It can be sub-divided into three distinct sections (West, Central Impact Area and East) which differ in the intensity and type of military activity and management, which has resulted in localised differences in the composition and structure of the vegetation communities and density of *S. pratensis*. (Toynton and Ash 2002). In this study we surveyed a large (approximately 70%) proportion of the DTE SP searching for areas of suitable habitat with the host plant and of the 81 plots we achieved a good coverage across the whole area (West: 21, Central Impact Area: 39, and East: 21 plots respectively). Hereafter, sites in the fragmented landscape are also referred to as plots.

Sampling

Nine 1 × 1 m quadrats were nested in a regular pattern within each 40 × 40 m plot (Fig. 1).

Within each 1 m^2 quadrat we measured the following variables: sward height, host plant quantity, host plant quality, adult nectar plant availability, species richness and the percentage of bareground cover. Sward height was measured by two methods: using a drop disk (300 mm of diameter, 200 g) and by estimating the height of non flowering leaves using a meter ruler (Stewart et al. 2001). The final measurement used in the analysis was an average of both measures for each quadrat. We defined host plant quantity by the percentage cover of *S. pratensis*. Host plant quality was defined by both the number of *S. pratensis* plants in each quadrat divided into four categories (1–5 plants; 6–10 plants; 11–20 plants and >20 plants) and the predominant size class of the plants in the quadrat divided into 3 categories (small if the basal rosette was less than one hand span in diameter, medium if the basal rosette was approximately one hand span in diameter and large if it was greater than one hand span). A *Succisa* quality index (SQI) for each quadrat was then calculated as the product of the number and size of plants within that quadrat. Adult nectar plant availability was determined by recording the species of flowering plants on which *E. aurinia* adults were nectaring during a survey carried out in May/June 2009 on DTE SP site. 77% were found on *Leontodon* spp (*L. hispidus* and *L. taraxicoides*) with the remaining 23% found on one of the following plants: *Hippocrepis comosa*, *Lotus corniculatus*, *Pilosella officinarum*, *Ranunculus* spp and *Helianthemum nummularium*. The percentage cover of these plants was therefore recorded in each quadrat. Species richness was measured as the number of forb species present in each quadrat.

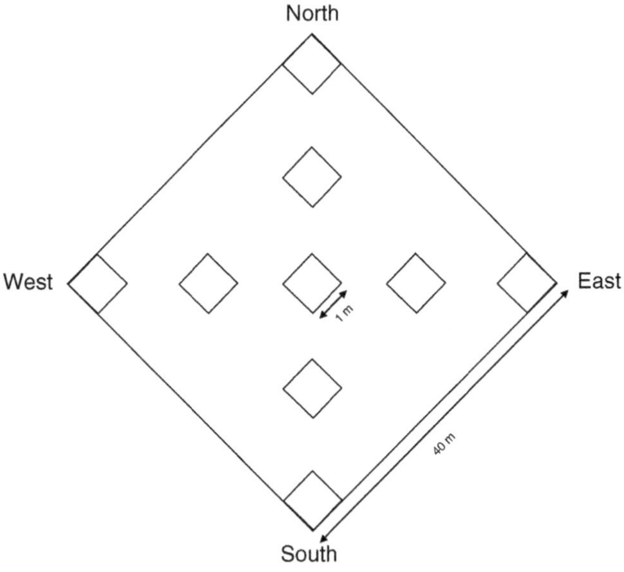

Fig. 1 The nested quadrat design used to survey larval webs of *E. aurinia* and associated habitat variables

These measurements, recorded in each of the nine 1 m^2 quadrats were averaged to give single measurements for each 40 × 40 m plot. The number of larval webs was recorded by walking a 1 m wide zigzag transect throughout the plot. Finally, we also recorded aspect and elevation at the plot level.

Statistical analysis

We investigated the relationship between the number of larval webs in plots and the environmental variables by fitting a generalised linear model with a log link function and quasi Poisson error distribution to account for overdispersion. Host plant quality and quantity (percentage cover) were positively correlated with one another. We therefore decided to use a composite variable made by the product of the two (*Succisa* composite = SQI*percentage cover of *S. pratensis*) in our analysis.

The model was fitted with sward height and its quadratic term, *Succisa* composite, percentage of bare ground, adult plant nectar availability, species richness and its quadratic term as covariables. *Succisa* composite and bare ground were log transformed as there were some extreme values of these variables. Because we were mainly interested in exploring whether habitat requirements differed between the fragmented landscape (Dorset) and the non-fragmented landscape (DTE SP), we included 2-way interactions between the previously mentioned variables and location (categorical variable with 2 levels). We had no a priori reason to expect the number of larval webs to vary differently as a function of aspect or elevation between the two locations. To avoid over-fitting the model we simply added elevation and aspect as single effects. To turn aspect into a covariable we transformed each aspect into radians and replaced each value by cos (aspect in radians) + sin (aspect in radians).

Quadrats with missing values for more than one environmental variable were excluded and the analysis was performed with a total of 114 plots (73 for DTE SP and 41 for Dorset).

We checked for spatial auto-correlation of larval webs and residuals from the model among the quadrats within each location using correlograms. We did not find any significant pattern, therefore did not include a correlation structure within our model.

We performed backwards, stepwise regression using F tests to remove any non significant term and to define the minimum adequate model. We used $P < 0.05$ as a selection criteria.

There was one outlier in the data set (number of larval webs equal to 115 in DTE SP). The model was performed again, excluding this data point. We found similar significant effects and model predictions without the data point and therefore decided to present results from the model including all the data. The heteroscedasticity and normality of the residuals were checked graphically. The residuals were not normally distributed. We attributed this result to the high proportion of quadrats that were recorded without any larval webs (almost 70% of zeros). All statistical analyses were carried out using R version 2.10.1 (R development core team 2009).

Results

Larval webs were found in 18% of plots in the fragmented (Dorset), and 46% of plots in the unfragmented (DTE SP) landscape (total of 34 and 541 webs, respectively). Table 1 displays the coefficients, standard errors and P-values of the model investigating the effects of environmental variables and location on the number of larval webs. Below, we describe only those variables which had significant effects on the number of larval webs found.

Table 1 Coefficients of a minimum adequate generalised linear model investigating the effects of environmental variables and location on the number of larval webs with associated standard errors (SE) and P values

Effects	Coefficient	SE	P value
Intercept	−36.9497	17.5635	0.038
Average sward height	2.8527	1.8716	0.130
Average sward height^2	−0.0972	0.0662	0.145
Succisa composite	1.7446	0.2795	0.000
Bare ground	−3.509	3.3879	0.303
Location DTE SP	9.1112	13.6395	0.506
Elevation	0.249	0.146	0.091
Elevation^2	−0.0009	0.0005	0.082
Average sward height × location DTE SP	−1.7746	1.92	0.357
Average sward height^2 × location DTE SP	0.0689	0.0673	0.308
Bare ground × location DTE SP	5.3816	3.5221	0.130

There was a significant difference in the response of larval web numbers to average sward height between plots in fragmented and unfragmented landscapes (average sward height^2 × location $F_{1,103} = 4.1$, $P = 0.04$, average sward height × location $F_{1,103} = 2.9$, $P = 0.09$). The relationship was quadratic indicating that larval webs were most often found in intermediate heights of sward. In unfragmented plots, the optimal sward height was higher than in fragmented plots (approximately 20 cm for DTE SP vs. 15 cm for Dorset) and their distribution occurred at a broader range of heights than in fragmented plots (Fig. 2; Table 1).

The effect of bare ground on the variation in larval web numbers differed between fragmented and unfragmented plots ($F_{1,103} = 6.5$, $P = 0.01$). In fragmented plots, there was a decrease in the number of larval webs with increasing levels of bare ground whereas the proportion of bare ground did not seem to affect the number of larval webs in unfragmented plots (Fig. 3; Table 1).

There was a quadratic relationship between elevation and the number of larval webs, indicating that the larval webs occurred at an intermediate altitude (elevation^2: $F_{1,103} = 12.0$, $P < 0.001$, elevation: $F_{1,103} = 11.0$, $P < 0.001$ (Fig. 4; Table 1). However, this relationship was associated with higher proportions of *Succisa* composite at those intermediate altitudes (Fig. 5).

Finally, there was a significant increase in the number of larval webs with an increase in *Succisa* composite (quantity and quality) that was similar in both fragmented and unfragmented plots [$F_{1,103} = 83.3$, $P < 0.001$ (Fig. 6; Table 1)].

Discussion

We found that a greater percentage of the unfragmented plots were occupied compared to fragmented plots. Caution in interpretation must be made because we collected data from only 1 year and *E. aurinia* populations are known to vary greatly from year to year and from site to site (Bulman 2001; Porter 1981; Warren 1994). However, it is commonly appreciated that fragmented sites are vulnerable to local extinctions and take time to be recolonised depending on isolation and patch size (Bulman et al. 2007; Hanski 1998, 1999). At any point in time therefore, it is highly unlikely that all sites in the fragmented landscape would be occupied, even in years that *E. aurinia* produces large abundances, and especially after poor years such as 2007

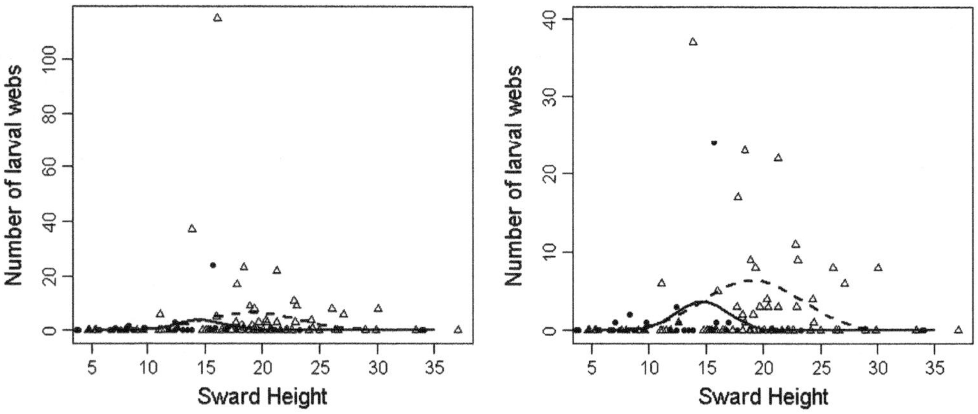

Fig. 2 Number of larval webs as a function of average sward height in Dorset (*closed circles*) and DTE SP (*open triangles*). *Lines* represent model predictions (*bold line* for Dorset and *dashed line* for DTE SP). *Left panel* shows all data, *right panel* focus on data excluding the plot with 115 larval webs

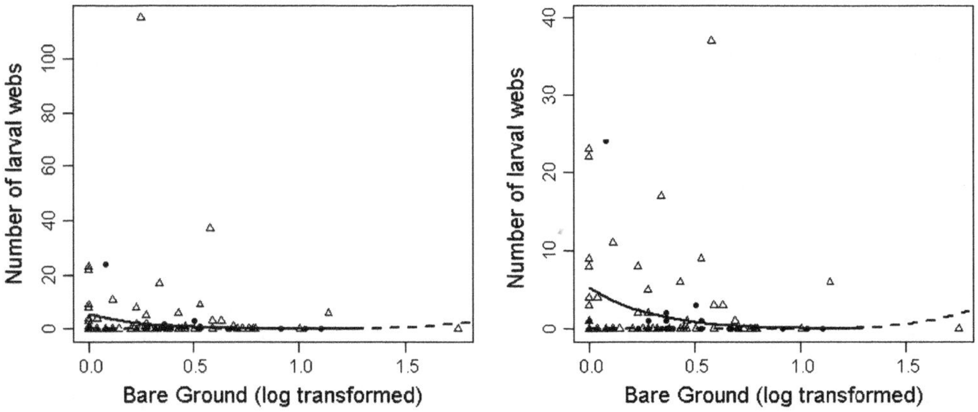

Fig. 3 Number of larval webs as a function of bare ground (log transformed) in Dorset (*closed circles*) and DTE SP (*open triangles*). *Lines* represent model predictions (*bold line* for Dorset and *dashed line* for DTE SP). *Left panel* shows all data, *right panel* focus on data excluding the plot with 115 larval webs

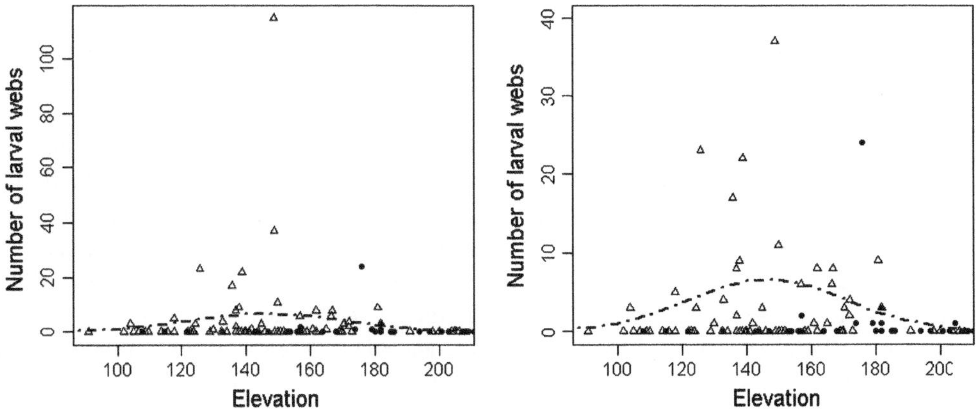

Fig. 4 Number of larval webs as a function of elevation in Dorset (*closed circles*) and DTE SP (*open triangles*). *Line* represents model predictions. *Left panel* shows all data, *right panel* focus on data excluding the plot with 115 larval webs

(Botham et al. 2007) in which many small, isolated populations are likely to have become locally extinct (Hula et al. 2004). In contrast, plots throughout the unfragmented landscape of DTE SP are likely to be recolonised more quickly in such years since suitable or semi-suitable habitat is close by, allowing the core populations where habitat is most suitable, to disperse into unoccupied habitat.

As with previous studies on populations of *E. aurinia* throughout its range (e.g. Bulman 2001; Konvicka et al. 2003; Liu et al. 2006; Anthes et al. 2003; Betzholtz et al. 2007; Stefanescu et al. 2006), we found that both the quantity and quality of host plant are important. Patches with better quality and/or a higher density of host plant were associated with greater densities of larval webs within both fragmented and unfragmented calcareous grassland landscapes. *E. aurinia* lays its eggs in batches on suitable host plants and the larvae are gregarious for much of their lives (Porter 1981). In addition, ovipositing females are often attracted to host plants which already have eggs (e.g. Stefanescu et al. 2006). Thus large quantities of host plant are essential if the larvae are to avoid starvation. This can be achieved by laying eggs on large plants (Porter 1981) or on small plants where plant density is high (typical of grazed calcareous grassland habitats).

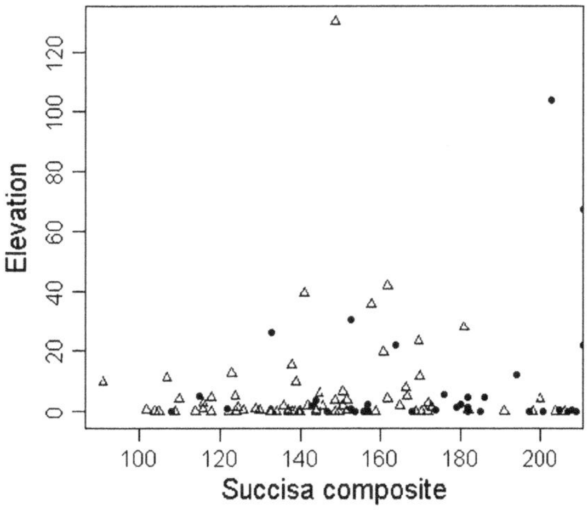

Fig. 5 The relationship between *Succisa* composite and elevation in Dorset (*closed circles*) and DTE SP (*open triangles*)

Sward height, a measure of the vegetation structure, was also a significant factor at both sites with an optimal intermediate height being associated with greater numbers of larval webs. Previous studies have shown that density of host plant alone is not enough to predict the presence of *E. aurinia* (Anthes et al. 2003; Konvicka et al. 2003). Vegetation structure has been shown to be important as the height of the surrounding vegetation is likely to be important in creating and maintaining the optimal microclimatic conditions necessary for larval survival (Porter 1981; Konvicka et al. 2003; Fowles and Smith 2006). A significant interaction between location and sward height suggests that the optimal sward height differed between plots in the fragmented and unfragmented landscapes, with a shorter sward in fragmented plots (approximately 15 cm) than at unfragmented plots (approximately 20 cm). The range of sward heights at which larval webs were found was also greater at unfragmented plots than at fragmented plots. Whilst this difference could represent a difference in microclimates between the fragmented and unfragmented landscapes (Betzholtz et al. 2007; Ravenscroft and Gaywood 1996), it is most likely to reflect the differences in management. Most of the calcareous grassland sites in Dorset are under grazing management, usually in a Higher Level Scheme with prescriptions specific for *E. aurinia*—low density cattle grazing during May to September, with the aim of producing a varied sward height between 5 and 15 cm (C. Bulman personal communication). In contrast the DTE SP is grazed infrequently on a rotational cycle, using a penning system which results in a greater range of and longer sward heights (Toynton and Ash 2002). *E. aurinia* can use calcareous grassland sites with vegetation ranging from 5 to 25 cm in sward height (Bulman 2001). It is possible that in Dorset *E. aurinia* has contracted to its 'core' optimal sites, whereas on larger, less fragmented sites such as the DTE SP, where there is greater and more heterogeneous resources available, it may use a wider range of 'suitability'.

We also found that both elevation and the percentage cover of bare ground had a significant effect on where larval webs were found. We suggest that the elevational effect can be attributed to the strong association between elevation and host plant distribution. We found that the same elevational band at which the greatest number of larval webs was found was also associated with where the greatest values of *Succisa* composite were recorded. The elevational range of our sites is well within that at which *E. aurinia* is found throughout its European range (Tolman and Lewington 2008) and thus it is unlikely that elevation *per se* has an effect on the distribution at the sites we surveyed. The valley bottoms and hill tops on our sites are generally less floristically rich and dominated by more competitive species which creates unsuitable conditions for the host plant (Grime et al. 1986). Thus, the suitable structure and availability of host plants for *E. aurinia* are confined to certain elevations throughout our sites.

A significant interaction between location and bare ground cover shows that at fragmented plots in Dorset larval web numbers were greatest where there is little or no bare ground and fewest where there is the greatest cover of bare ground, whilst there was no relationship between the

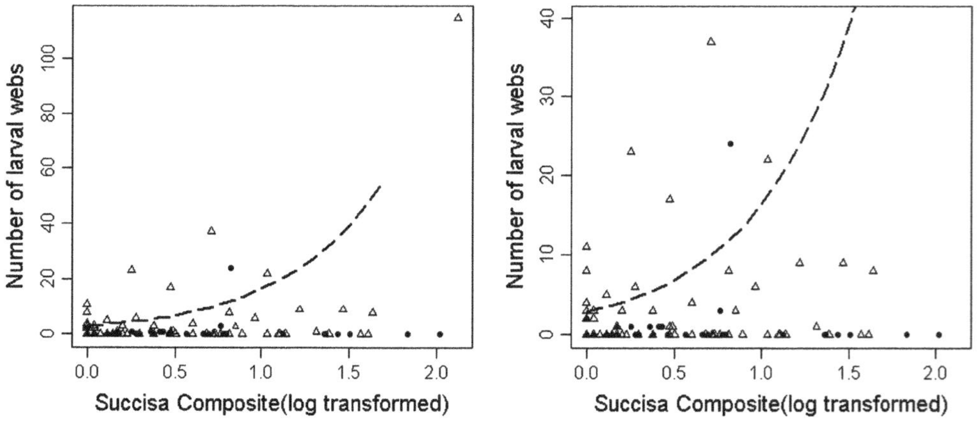

Fig. 6 Number of larval webs as a function of *Succisa* composite (log transformed) in Dorset (*closed circles*) and DTE SP (*open triangles*). *Line* represents model predictions. *Left panel* shows all data, *right panel* focus on data excluding the plot with 115 larval webs

number of larval webs and bare ground on unfragmented plots in DTE SP. A greater cover of bare ground on Dorset sites is likely to be associated with high levels of grazing which creates a less suitable turf structure. Why we did not see the same pattern on DTE SP plots is most likely because of the high levels of disturbance caused by military vehicles and explosions/fires from military training exercises (Warren et al. 2007). These events not only facilitate the germination of host plants (Grime et al. 1986) creating locally suitable patches, but also occur relatively randomly in both space and time. Military training grounds are often associated with high biodiversity because of the heterogeneous habitats created by military activities (Warren et al. 2007). On the DTE SP some of greatest 'core' densities of larval webs are located in the highly disturbed areas in the Central Impact Area where military activity is greatest (Toynton and Ash 2002). Fires from shells and flares prevent the sward from becoming too rank, and craters formed by shells also create patches of bare ground which are colonised by *S. pratensis* and subsequently used by *E. aurinia*.

Beyond ascertaining the habitat requirements of *E. aurinia* on calcareous grassland, an aim of this study was to investigate whether fragmentation affects this by using the DTE SP as an unfragmented landscape and numerous small sites in Dorset as the fragmented landscape. All sites are calcareous grassland sites and are in the south west of the UK. Thus, we expected, if fragmentation has no effect that the habitat requirements would be very similar at all sites. We found differences in the effects of sward height and bare ground on the distribution of larval webs between the two landscapes. However, there are a number of caveats which we must consider. Firstly, differences *per se* are not indicative of an effect of fragmentation. Studies on different populations of *E. aurinia* across its distributional range have shown that different populations using the same host plant on very similar habitats often show different habitat requirements, with sward height often being the variable that differs (Betzholtz et al. 2007) In addition, there are differences in the available habitat and the management of the habitat between the two landscapes which are more likely to account for such differences in vegetation structure as discussed previously. Patch size and isolation have also been shown to be important in other studies (Bulman 2001; Bulman et al. 2007; Betzholtz et al. 2007). We could not compare patch size and isolation between Dorset and DTE SP plots because whilst we have good information on these parameters for the fragmented Dorset landscape we do not have the equivalent for the DTE SP. We should also consider whether the DTE SP truly represents a contiguous unfragmented habitat for *E. aurinia*? Clearly, it is highly unlikely that the entire habitat on the DTE SP is suitable, both in terms of vegetation structure and the distribution of the host plant. However, the matrix between suitable patches contains suitable adult nectaring sources and offers little in terms of physical barriers to dispersal. Distances in between patches of host plant are unlikely to exceed that of even the lower ranges at which *E. aurinia* commonly disperses (Warren 1994). In addition, whilst *S. pratensis* is known to be the major host plant for *E. aurinia* on calcareous grassland in the UK, it also uses *Knautia arvensis* and *Scabiosa columbaria*, though there has been limited success when rearing larvae on these alternative host plants in captivity (Porter 1981). We found a substantial number of larval webs on both of these host plants across the DTE SP in 2009 both in areas where *S. pratensis* was absent and also where it was present. We suggest therefore, that it is feasible to consider the DTE SP as unfragmented in the context of this study particularly since it represents the closest example, certainly in the UK, of a large unfragmented *E. aurinia* habitat. Finally, we have only sampled 1 year. Populations of *E. aurinia* are known to fluctuate greatly from year to year and thus differences in occupancy and habitat requirements may represent populations at different points in this cycle (Bulman 2001). However, previous data collected on larval web abundance on the DTE SP (Botham unpublished data) and the Dorset sites (N. Bourn personal communication) suggests 2009 was a year in which abundances were generally high at all sites compared to the previous year. In addition, data from the UK Butterfly Monitoring Scheme suggests that *E. aurinia* generally increased at monitored sites throughout the UK in 2009 compared to 2008 following low abundances in 2007 (Botham et al. unpublished data). A major cause of population fluctuations in *E. aurinia* is attributed to mortality caused by parasitoids (*Cotesia* spp) (Klapwijk 2008). The population dynamics of host-parasitoid systems and their interaction with climatic factors are extremely important and can have pronounced effects on the habitat associations of butterflies (Klapwijk 2008; Klapwijk et al. 2010; Menendez et al. 2008). Unfortunately, we were not able to investigate parasitoids within the restrictions of the current study.

Differences in the variability of habitat requirements between Dorset and DTE SP plots are more likely to indicate effects of fragmentation. One hypothesised effect of fragmentation on butterfly populations is a reduction in genetic diversity within the population due to isolation and subsequent inbreeding (Joyce and Pullin 2003). This could result in narrower, more specialised habitat requirements in small, isolated habitat fragments (Joyce and Pullin 2003). To date there has been little genetic evidence of such differences in *E. aurinia* populations across the UK (Joyce and Pullin 2003). However, small, isolated patches may also be more prone to environmental effects (Gibbs and Van Dyck 2009) and thus eggs laid in sub-optimal habitats will have an even more reduced chance of survival adding further to the specialisation of habitat choice in future generations. The range of sward heights within which larval webs were found was far greater in the unfragmented than the fragmented landscape. We also regularly found larval webs on alternative host plants. In addition, whereas there was a significant negative relationship with bare ground in the fragmented landscape, this relationship was not evident in the unfragmented landscape. This overall greater variability in habitat requirements may be driven by underlying differences in genetic diversity or may be more directly a result of differences in survivorship on sub-optimal habitat between the fragmented and unfragmented landscapes driven by differences in vulnerability to local abiotic factors. These factors would require further investigation including genetic analyses. A more likely explanation is that small, isolated fragmented patches of habitat are vulnerable to local extinction and slow rates of recolonisation, a point discussed above in regards to the differences in occupancy of plots in fragmented versus unfragmented landscapes. *E. aurinia* abundances have increased since a poor year in 2007 (Botham et al. unpublished data). In years where abundances are high competition for resources will increase and thus more females are likely to lay eggs on sub-optimal plants (including alternative host plants) increasing the variability of the measured habitat requirements and dispersal is likely to be greater resulting in a greater number of occupied sites and the potential for an increase in the overall heterogeneity of occupied sites. Our data, collected in 2009, likely represents the expansion from core populations of the DTE SP metapopulation into previously unoccupied and sub-optimal sites resulting in greater variability in observed habitat associations at that point in time compared to fragmented sites in Dorset where recolonisation of unoccupied sites from the core populations is slower because of barriers to dispersal between small, isolated fragments. In addition, the habitat on the DTE SP is much more heterogeneous than that on the Dorset sites.

Our study further emphasises the need to consider increasing the connectivity and size of existing *E. aurinia* habitats (Bulman et al. 2007; Wahlberg et al. 2002). In the UK, where most suitable habitat patches for *E. aurinia* are small, fragmented and often isolated, we suggest that addressing connectivity and isolation rather than habitat quality may be a more important concern in determining the persistence of *E. aurinia* populations on calcareous grassland habitats in the UK. When abundances increase suboptimal habitat is used if it can be reached by ovipositing females and/or where suboptimal habitat is used lower chances of survival may be buffered on larger habitat patches by lower vulnerability to abiotic factors. Thus unoccupied habitats on large unfragmented sites are more likely to be quickly filled unlike on small, isolated fragments of habitat. Metapopulations in fragmented landscapes are likely to find it difficult to recover from population declines and local extinctions caused by sequences of years with unsuitable climatic conditions and/or high densities of parasitoids and thus may be lost altogether in the long-term. Increasing the size and connectivity of existing patches of habitat need not necessarily require management to create optimal habitat. Suboptimal habitat in the vicinity of core *E. aurinia* populations is likely to be colonised when butterfly abundances are high. Where it is not possible to expand habitats, small isolated patches of suitable habitat should continue to be managed to create the optimal habitat requirements (e.g. Barnett and Warren 1995).

Acknowledgments The work was funded by Defence Estates. We would like to thank Bill Meek and Pete Nuttall for their assistance in the field and Keith Porter for his useful comments throughout the project. We would also like to thank the two anonymous referees for their helpful comments on the paper.

References

Anthes N, Fartmann T, Hermann G, Kaule G (2003) Combining larval habitat quality and metapopulation structure: the key for successful management of pre-alpine *Euphydryas aurinia* colonies. J Insect Conserv 7:175–185

Barnett LK, Warren MS (1995) Species action plan: marsh fritillary, *Eurodryas aurinia*. Butterfly Conserv

Betzholtz PE, Ehrig A, Lindeborg M, Dinnetz P (2007) Food plant density, patch isolation and vegetation height determine occurrence in a Swedish metapopulation of the marsh fritillary *Euphydryas aurinia* (Rottemburg, 1775) (Lepidoptera, Nymphalidae). J Insect Conserv 11:343–350

Botham MS, Brereton T, Middlebrook I, Cruickshanks KL, Roy D (2007) United Kingdom butterfly monitoring scheme annual report 2007. Centre for Ecology and Hydrology Wallingford

Botham MS, Brereton T, Middlebrook I, Cruickshanks KL, Harrower C, Beckmann B, Roy D (2008) United Kingdom butterfly monitoring scheme annual report 2008. Centre for Ecology and Hydrology Wallingford

Bulman CR (2001) Conservation biology of the marsh fritillary butterfly, *Euphydryas aurinia*. PhD, Leeds, Leeds

Bulman CR, Wilson RJ, Holt AR, Bravo LG, Early RI, Warren MS, Thomas CD (2007) Minimum viable metapopulation size, extinction debt, and the conservation of a declining species. Ecol Appl 17:1460–1473

Fowles AP, Smith RG (2006) Mapping the habitat quality of patch networks for the marsh fritillary *Euphydryas aurinia* (Rottemburg, 1775) (Lepidoptera, Nymphalidae) in Wales. J Insect Conserv 10:161–177

Fox R, Asher J, Brereton T, Roy D, Warren M (2006) The state of butterflies in Britain and Ireland. Information Press, Oxford

Fox R, Warren M, Brereton T (2010) The butterfly red list for Great Britain. *Species Status*. Joint Nature Conservation Committee, Peterborough, pp 1–32

Gibbs M, Van Dyck H (2009) Reproductive plasticity, oviposition site selection, and maternal effects in fragmented landscapes. Behav Ecol Sociobiol 64:1–11

Grime JP, Hodgson JG, Hunt R (1986) Comparative plant ecology: a functional approach to common British species. Unwin Hyman, London
Hanski I (1998) Metapopulation dynamics. Nature 396:41–49
Hanski I (1999) Habitat connectivity, habitat continuity, and metapopulations in dynamic landscapes. Oikos 87:209–219
Hula V, Konvicka M, Pavlicko A, Fric Z (2004) Marsh Fritillary (*Euphydiyas aurinia*) in the Czech Republic: monitoring, metapopulation structure, and conservation of an endangered butterfly. Entomologica Fennica 15:231–241
Joyce DA, Pullin AS (2003) Conservation implications of the distribution of genetic diversity at different scales: a case study using the marsh fritillary butterfly (*Euphydryas aurinia*). Biol Conserv 114:453–461
Klapwijk MJ (2008) Host-parasitoid interactions in the context of climate change and habitat fragmentation. PhD, Oxford, Oxford
Klapwijk MJ, Grobler BC, Ward K, Wheeler D, Lewis OT (2010) Influence of experimental warming and shading on host-parasitoid synchrony. Global Change Biol 16:102–112
Konvicka M, Hula V, Fric Z (2003) Habitat of pre-hibernating larvae of the endangered butterfly *Euphydryas aurinia* (Lepidoptera : Nymphalidae): what can be learned from vegetation composition and architecture? Eur J Entomol 100:313–322
Liu WH, Wang YF, Xu RM (2006) Habitat utilization by ovipositing females and larvae of the Marsh fritillary (*Euphydryas aurinia*) in a mosaic of meadows and croplands. J Insect Conserv 10:351–360
Menendez R, Gonzalez-Megias A, Lewis OT, Shaw MR, Thomas CD (2008) Escape from natural enemies during climate-driven range expansion: a case study. Ecological Entomol 33:413–421
Porter K (1981) The population dynamics of small colonies of the butterfly *Euphydryas aurinia*. PhD, Oxford, Oxford, UK
Pywell R, Swetnam R, Walker K, Roy D (2006) Assessing floristic change on Salisbury Plain between 1996 and 2004. Centre For Ecology and Hydrology
Ravenscroft NOM, Gaywood MJ (1996) The status and habitat of the marsh fritillary in western Scotland. Scott Nat Herit
Sigaard P, Pertoldi C, Madsen AB, Sogaard B, Loeschcke V (2008) Patterns of genetic variation in isolated Danish populations of the endangered butterfly *Euphydryas aurinia*. Biol J Linn Soc 95:677–687
Stefanescu C, Penuelas J, Sardans J, Filella I (2006) Females of the specialist butterfly *Euphydryas aurinia* (Lepidoptera : Nymphalinae: Melitaeini) select the greenest leaves of *Lonicera implexa* (Caprifoliaceae) for oviposition. Eur J Entomol 103:569–574
R Development Core Team (2009) R: a language and environment for statistical computing. R Foundation for Statistical Computing, Vienna, Austria. ISBN 3-900051-07-0, URL http://www.R-project.org
Thomas CD, Bulman CR, Wilson RJ (2008) Where within a geographical range do species survive best? A matter of scale. Insect Conserv Divers 1:2–8
Tolman T, Lewington R (2008) Collins butterfly guide: the most complete field guide to the butterflies of Britain and Europe Collins, London
Toynton P, Ash D (2002) Salisbury plain training area—the British steppes? Br Wildl 13:335–343
van Swaay CAM, Warren MS (1999) Red data book of European butterflies (Rhopalocera). Nature and environment. Council of Europe, Strasbourg
Wahlberg N, Klemetti T, Hanski I (2002) Dynamic populations in a dynamic landscape: the metapopulation structure of the marsh fritillary butterfly. Ecography 25:224–232
Warren MS (1994) The UK status and suspected metapopulation structure of a threatened European butterfly, the marsh fritillary *Eurodryas aurinia*. Biol Conserv 67:239–249
Warren SD, Holbrook SW, Dale DA, Whelan NL, Elyn M, Grimm W, Jentsch A (2007) Biodiversity and the heterogeneous disturbance regime on military training lands. Restor Ecol 15:606–612
Wilson RJ, Roy DB (2009) Butterfly population structure and dynamics. In: Settele J, Shreeve TG, Konvicka M, Van Dyck H (eds) Ecology of butterflies in Europe. Cambridge University Press, Cambridge, pp 81–97

J Insect Conserv (2011) 15:23–35
DOI 10.1007/s10841-010-9300-8

ORIGINAL PAPER

Leptidea sinapis (Wood White butterfly) egg-laying habitat and adult dispersal studies in Herefordshire

S. A. Clarke · D. G. Green · J. Joy · K. Wollen · I. Butler

Received: 24 March 2010 / Accepted: 31 May 2010 / Published online: 24 June 2010

Abstract The Wood White butterfly *Leptidea sinapis*, a UK BAP Priority species, is present in a number of Forestry Commission woodlands in the West Midlands and these constitute an important part of the national resource of this butterfly. A joint SITA Trust funded project between the Forestry Commission and Butterfly Conservation (linking with the National Wood White Conservation Project) is researching the use of habitat by *Leptidea sinapis* within these sites with regard to targeting conservation management. Two aspects of this research are presented: the results of a mark-recapture study and an analysis of egg-laying habitat. Mark-recapture results show considerable movement between areas of concentrations of adults and between egg-laying areas, with males moving further and faster than females. Movements do occur across potential habitat barriers. Egg locations were found by following females and by searching for eggs. The habitat used for egg-laying is extremely variable even within a site. A number of foodplants species are used and habitat ranges from low height foodplants within bare ground to a foodplant height of over 1 m within thick vegetation and scrub. These results are being integrated into an ongoing project to restore *Leptidea sinapis* habitat within managed Forestry Commission woodlands.

Keywords Conservation management · Mark-recapture · Forestry · Habitat associations

Introduction

In England, *Leptidea sinapis* Wood White butterfly uses a range of sheltered habitat including woodland rides, scrub edges and is often found on sites such as disused railways and coastal undercliffs. The larval foodplants are legumes such as *Lathyrus pratensis* Meadow Vetchling, *Lotus pedunculatus* Greater Bird's-foot-trefoil, *Lotus corniculatus* Bird's-foot-trefoil and *Lathyrus linifolius* Bitter-vetch (McCracken and Warren 2007). The larvae move off the foodplant to pupate in surrounding vegetation and the species over-winters as a pupa.

Leptidea sinapis is identified as a UK Biodiversity Action Plan Priority Species in Biodiversity Reporting & Information Group (2007). It has very localised distribution in England and Wales and has declined rapidly in Britain over the past few decades, with a long-term population decline (1977–2004) of 64% (Fox et al. 2006). Despite this overall decline, in some regions, particularly Shropshire and Herefordshire, the species appears to be currently expanding its range at present (Joy and Williams 2008) and a series of surveys has highlighted the extent and importance of *L. sinapis* within West Midland Forestry Commission woodlands in this area (Joy 2000, 2006). Many of these, including those with the strongest populations, are plantation woodlands. The early successional habitats that this species requires are largely

Guest Editor: Dr. M. S. Warren.

S. A. Clarke (✉) · D. G. Green
Wessex Environmental Associates, Hampshire, UK
e-mail: SClarke@wessexenv.mail1.co.uk

J. Joy
West Midlands Butterfly Conservation, Wareham, UK

K. Wollen
Forestry Commission, Forest of Dean District, Coleford, UK

I. Butler
Ian Butler Photography, Malvern, UK

restricted to ride and forest track edges at these sites and are consequently heavily reliant upon sympathetic management. As a result, in 2007 a joint SITA Trust funded project between the Forestry Commission and Butterfly Conservation initiated management within selected Herefordshire woodlands to both maintain existing *L. sinapis* habitat and to create additional areas. Partly because of uncertainty concerning the exact requirements of this species, the project also included a series of investigations to examine both the response of the butterfly to management and also how the species utilises available woodland resources. Two aspects of this research are presented here: the results of a mark-recapture study and an investigation into egg-laying habitat. In addition some incidental observations on mating behaviour were made and these are also presented.

Research into *L. sinapis* in Britain was carried out by Martin Warren in the 1980s at sites in Northamptonshire (Warren 1981, 1984, 1985, 1986) and this work informed subsequent conservation management. More recent studies of habitat use were undertaken by Stephen Jeffcoate in Chiddingfold Forest, Surrey (Jeffcoate 2005, 2006) and this suggested that the butterfly may utilise a wider range of foodplants and egg-laying conditions than previously considered. It also suggested that a strong second brood may now occur more regularly and widely than was seen during Warren's research in the 1980s. At that time *L. sinapis* was mainly univoltine in northern parts of England, with a partial late summer second brood in southern sites, but only in some years (Warren 1984). Work on the butterfly at Forestry Commission woodland in Shropshire during 2007 (Clarke and Green 2008) confirmed that a second brood now occurs even in the most northerly part of the species range and also showed that the butterfly was using a greater range of habitats and foodplants than was suggested in much of the literature. It is possible that the butterfly's phenology and habitat requirements may be changing or that this species habitat associations are more site-specific than was previously realised.

The aim of the work presented here was to help define some aspects of *L. sinapis* requirements within the Herefordshire woodland strongholds and to integrate this information into the on-going conservation management at these sites. The studies of egg-laying habitat are intended to inform the precise types of habitat management that will provide maximum benefit for the species. The information from the mark-recapture work will help determine the positioning and scale of the required management. The results may also help to provide some answers as to why the species is expanding in this region and how this success might be applied to other regions.

Methods

The study sites

Eight woodland areas in Herefordshire were included in this project. Two (Haugh Wood SO592366 and Wigmore Rolls SO395695), both woodlands of mostly mixed plantation, were used as study sites for the work presented here. These two woodlands contain the largest and probably the most persistent populations of *L. sinapis* and it was considered that examining habitat use within strong populations would more accurately reflect the species optimum requirements, as failing colonies may be forced to use marginal habitat.

Both sites were used for the studies of egg-laying habitat, but Haugh Wood was chosen for the mark-recapture work. This was for two main reasons: firstly, in order to obtain a high percentage of recaptures, it is important to be able to adequately cover the study site with the time and effort available. At Haugh Wood a single recorder was able to sample all the main *L. sinapis* areas and some of the less used areas in a single day. Secondly, at Haugh Wood adults were concentrated in one area in the north and one area (approximately 850 m away at the nearest point) in the very south of the wood, with only scattered individuals seen between these two areas. Much of the habitat between these areas appeared to be only marginally suitable for *L. sinapis* and a well shaded road separated the north and south of the wood. One aim of the study was to investigate whether *L. sinapis* in these two areas were effectively isolated or whether adults did move between the two areas and through the potential habitat barriers.

Egg-laying studies

An investigation into egg-laying habitat was carried out at Haugh Wood and Wigmore Rolls in both 2008 and 2009, aiming to locate egg-laying areas and to record details of the food-plant, egg location and surrounding vegetation/habitat. At Haugh Wood the entire site was included in the search for eggs and egg-laying activity, but at the larger site of Wigmore Rolls the search was within part of the site only.

The preferred method for locating eggs was to follow females at a slight distance to avoid the possibility of interference and to place a marker alongside any eggs they laid. This method was highly successful at Bury Ditches (Shropshire) in 2007 (Clarke and Green 2008) and was also used by Jeffcoate (2006) in Surrey and Friberg et al. (2008a) in Sweden. It removes much of the potential human search bias from the study. In many other studies of Lepidoptera, eggs are found by searching known foodplants in likely habitat. (Random searching of whole sites is usually too time-consuming to consider.) For many species this is the only possible method, but if searches are made where eggs are predicted by the recorder this will of course bias the results towards the expected habitat characteristics and known food-plants.

While the preferred method was to follow females and observe egg-laying, during conditions when adults were inactive some searches for previously laid eggs were carried out. For every record it was noted if the egg had been found or if it has been seen being laid.

For each egg, its exact location (10 figure grid reference using hand-held GPS), distance from any ride edge, height above ground, height of the food-plant, food-plant species, shading of the food-plant and flowering status were recorded. It was also noted if any other *L. sinapis* eggs were found on the same food-plant. Then within a 1 m quadrat around each egg, measurements were made of vegetation (species, percentage cover and mean height), the percentage of bare ground and estimates of direct shade. Direct shade was defined as the area directly under the canopy of scrub or tree. It did not include shade from shadows of nearby structures or vegetation. Scrub was defined as woody growth of more than 1.5 m high.

In addition, in 2009, the overall abundance of potential larval food-plants within four areas used by *L. sinapis* at Haugh Wood and three occupied areas at Wigmore Rolls, was estimated using the DAFOR scale of abundance (D—dominant, A—abundant, F—frequent, O—occasional, R—rare). While this is an inaccurate measure of actual abundance, it is a useful means of showing relative abundance particularly within large areas. The Haugh Wood areas were sampled during both the first and second brood flight periods, the Wigmore Rolls areas were only sampled during the first brood flight period.

Mark-recapture studies

In 2009, a mark-recapture study was undertaken at Haugh Wood on both first and second brood adults. There were 9 survey events over a 44 day period for the first brood (between 12 May and 24 June) and 3 survey events over a 14 day period for the second brood (between 6 August and 19 August).

On each survey event as many of the main rides and glade areas together with a sample of minor rides were walked during weather conditions suitable for adult flight. As well as areas that appeared suitable for *L. sinapis*, occasional visits were made to areas of Haugh Wood that appeared unsuitable in order to check for the presence of adults that might be moving outside of the usual habitats. An attempt was made to capture every individual seen and because *L. sinapis* is not difficult to net most sightings resulted in capture.

Each butterfly was marked on the underside using the 1,2,4,7 system (Watt et al. 1977) which consists of a series of small dots which provide a unique number for each individual. Red, green, blue or black dots were made with a fine tip (0.6 mm), fast-drying, permanent marker pen. Although the survival of any first brood adults into the second brood seemed unlikely, to ensure that any overlap would be apparent, the second brood was numbered consecutively to the first brood. All butterflies were released immediately after marking by placing them gently on the nearest vegetation to the capture point.

For each capture or recapture, the location (10 figure grid reference using hand-held GPS), date, time, individual unique number and sex were recorded together with brief notes giving the behaviour observed directly before capture (flight, resting, nectaring etc.).

Results

Egg-laying habitat

Distribution of eggs

At Haugh Wood, concentrations of eggs were found largely within 6 main areas although small numbers of were also recorded outside of these patches (Fig. 1). There was no evidence of any change in egg-laying areas between the broods, although in 2009 one area (south Bear's Wood) was not available to the second brood because ride widening had temporary removed the vegetation.

Egg distribution at Wigmore Rolls was not analysed because the sampling was too patchy across the larger area of suitable habitat.

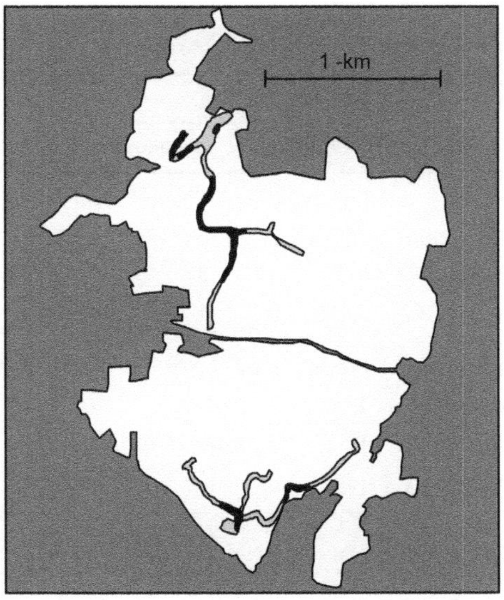

Fig. 1 The main egg-laying areas and main adult flight areas at Haugh Wood. Egg-laying areas in *black*. Adult flight areas in *light grey*

Larval food-plants

Table 1 shows the choice of plant for egg-laying. To investigate the possibility of any bias, the results were divided according to the method of location. The "found" eggs were those located by searching and the "laid" eggs were those seen laid after following females (see Sect. "Method").

Initially, using the results for all eggs, *Lathyrus pratensis* appeared to be the most frequently used food-plant for the first brood at both sites and *Lotus pedunculatus* for the second brood at Haugh Wood. A more careful look at the results suggests this might be a result of sample bias. The "laid" eggs results showed a different pattern, with *Lotus pedunculatus* more frequently used than *Lathyrus pratensis* in both broods. The sample sizes are too low to statistically test this difference between "found" and "laid" eggs, but 30 out of a total of 35 "found" eggs on *Lathyrus pratensis* at Haugh Wood (first brood) were all recorded by the same team of surveyors, on a single day and along one ride verge and this alone would strongly influence the result.

The result suggested that eggs are more easily found by humans when they are on *Lathyrus pratensis* than on other potential food-plants. In addition, *Lathyrus pratensis* has long been stated in the literature as a particularly favoured food-plant and this might have an influence on search behaviour. Surveys of *L. sinapis* food-plant use should avoid depending on egg searching if possible.

It seems that both *Lathyrus pratensis* and *Lotus pedunculatus* are the main larval food-plant at both sites and that there was no evidence for any changes in food-plant choices between the first and second broods.

Table 1 Use of larval food-plants, at Haugh Wood and Wigmore Rolls, 2008 & 2009

Plant	Number of eggs					
	First brood		Second brood		Total	
Haugh Wood	Found	Laid	Found	Laid	Found	Laid
Lotus pedunculatus	10	8	6	10	16	18
Lathyrus pratensis	31*	4	4	0	35	4
Lotus corniculatus	3	0	1	1	4	1
Wigmore Rolls	Found	Laid			Found	Laid
Lotus pedunculatus	15	21			15	21
Lathyrus pratensis	17	9			17	9
Lotus corniculatus	13	0			13	0
Others**	0	3			0	3
		Total both sites			156	

Results are divided according to the method of location

* 30 were recorded on same day, on same ride, by three recorders

** Others—*Potentilla reptans, Vicia cracca* and grass sp.

It is likely that food-plant choice is greatly dependant on the abundance of the different food-plant species at each site. To test this, the relative abundance of potential food-plants was estimated within seven areas occupied by *L. sinapis* within the study sites (see Sect. "Method"). The results as presented in Table 2 suggest that relative abundance probably does account for the observed use of food-plants. *Lotus pedunculatus* is the most common species, closely followed by *Lathyrus pratensis* and other potential food-plants are relatively uncommon.

The single eggs laid on *Potentilla reptans* and on a blade of grass appeared to be mistakes on the part of the butterfly. On both occasions the butterfly was seen to initially land on *Lotus pedunculatus* and then lay on the adjacent and overlapping vegetation.

Egg and food-plant height

Figures 2 and 3 show the position of eggs in relation to the height of the food-plant, height of surrounding vegetation and height above ground. The relative difference between egg height and food-plant height remained similar between the sites and between broods and strongly suggests that the height of the egg above ground is solely a function of the height of the food-plant.

Eggs were laid towards the top of all food-plants species, but there was a slight variation in the actual distance from the tip between the plant species. Eggs laid on *Lotus corniculatus* or *Lotus pedunculatus* were consistently 2–3 cm below the tip of the food-plant stem (*Lotus pedunculatus* mean 2.4 cm ± 0.4 SE, *L. corniculatus* mean 2.8 cm ± 0.8 SE). Eggs on *Lathyrus pratensis* tend to be laid lower on the stem (mean of 9.0 cm ± 1.0 SE, below food-plant tip).

As would be expected, the height of the food-plants increases between the first and second brood periods (Fig. 2) and so therefore does the height of eggs from the ground. Similarly, food-plant heights (and therefore egg heights) differ between Haugh Wood and Wigmore Rolls.

Warren (1984) noted that eggs tended to be laid on tall food-plants protruding above the surrounding vegetation. Figures 2 and 3 show that at Haugh Wood and Wigmore Rolls this was true for *Lathyrus pratensis* but it varied for the other plant species. *Lotus pedunculatus* was used as frequently, if not more often, than *Lathyrus pratensis* and its mean height was actually considerably lower than the surrounding vegetation at Haugh Wood during the second brood and also at Wigmore Rolls during the first brood.

Surrounding vegetation height, bare ground and shade

The mean height of surrounding vegetation, percentage cover of bare ground and percentage of direct shade were

Table 2 Estimate of food-plant abundance at Haugh Wood and Wigmore Rolls, 2009

Estimated abundance (DAFOR scale)	Potential food-plants			
	Lotus pedunculatus	*Lathyrus pratensis*	*Lotus corniculatus*	Other species *Vicia cracca, V. sativa & V. sepium*
Haugh Wood (4 sample areas)	O (LF)	O	R	R
Wigmore Rolls (3 sample areas)	F	O (LF)	R	R

D dominant, *A* abundant, *F* frequent, *LF* locally frequent, *O* occasional, *R* rare

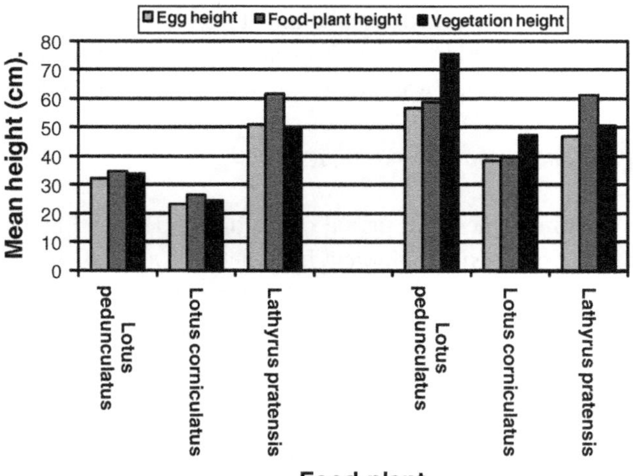

Fig. 2 Egg, food-plant and surrounding vegetation heights. Haugh Wood first and second brood, 2008 and 2009. Sample sizes are given in Table 1

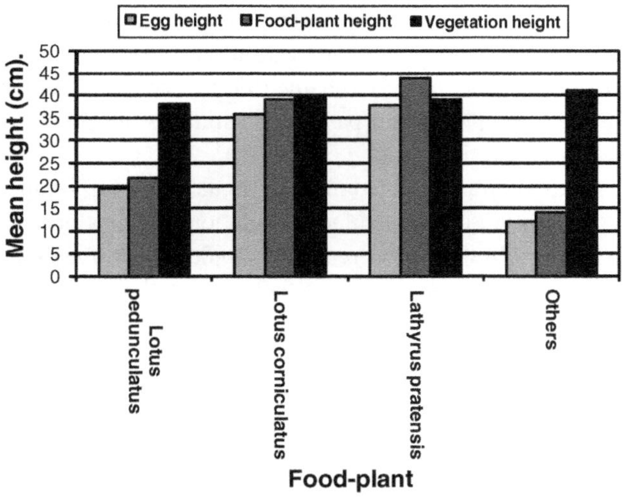

Fig. 3 Egg, food-plant and surrounding vegetation heights. Wigmore Rolls first brood, 2008 and 2009. Sample sizes are given in Table 1

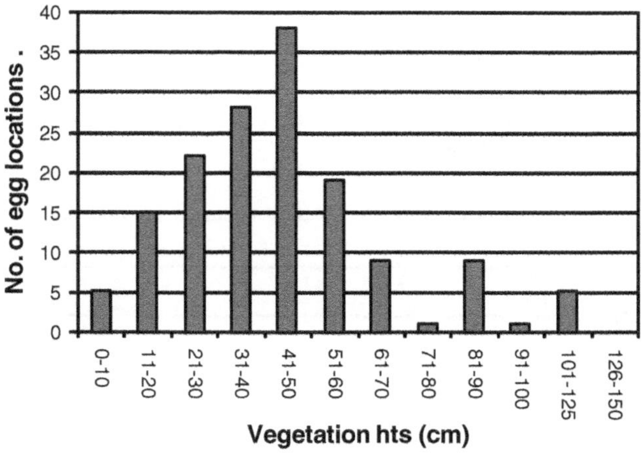

Fig. 4 Vegetation heights at egg locations. Both sites and both broods, 2008 and 2009

recorded within a 1 m quadrat around each egg. A *t*-test was used to test for differences in these variables between broods and between study sites. No significant difference was found in: surrounding vegetation height between broods ($t = 4.3E-05$, $P > 0.05$) or between sites ($t = 0.08$, $P > 0.05$), percentage cover of bare ground between broods ($t = 0.0004$, $P > 0.05$) or between sites ($t = 0.4$, $P > 0.05$), direct shade between broods ($t = 0.004$, $P > 0.05$) or between sites ($t = 0.2$, $P > 0.05$). Therefore the results for both broods and sites can be combined and are shown in Figs. 4, 5, and 6.

Figure 4 shows that a wide range of vegetation heights were used as egg-laying locations, with a peak at heights between 40 and 50 cm. The lowest recorded vegetation height surrounding an egg was 6 cm and the highest was 120 cm.

Figure 5 shows that although most eggs were recorded amongst a vegetation structure that contained no bare ground, all levels of bare ground were used up to a maximum recorded of 96%.

Figure 6 shows the percentage of shade at egg locations. Females did tend to pick unshaded locations, with over 60% of eggs recorded within quadrats which had no direct shade, but a small numbers of eggs did occur over a wide range of shade conditions. The maximum shade recorded was 100% for a single egg that was seen being laid at the base of a 1.5 m high young conifer. The conifer branches conferred 100% direct shade over the 1 m quadrat around the egg, but due to the ride verge slope and the time of day the location was in full sun at the time the egg was laid. Additionally, the surrounding conifer scrub sheltered the area so despite the measured 100% direct shade the

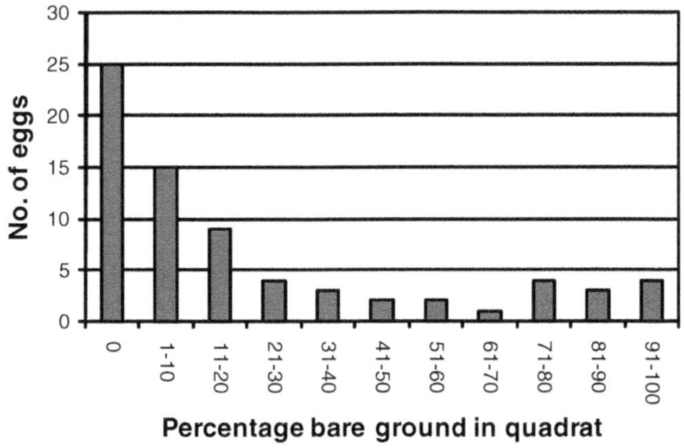

Fig. 5 Percentage of bare ground at egg locations. Both sites and both broods, 2008 and 2009

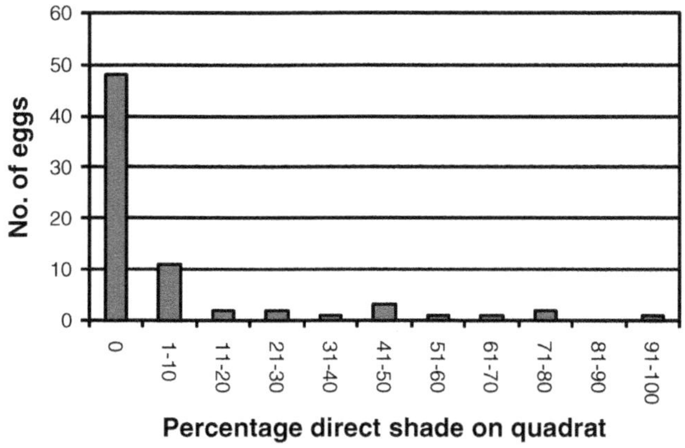

Fig. 6 Percentage of direct shade at egg locations. Both sites and both broods, 2008 and 2009

food-plant was actually within a particularly a warm spot at that time.

Mark-recapture studies

Handling effect observations

In some studies an "escape" reaction or "handling effect" has been recorded for butterflies after capture, with the individual exhibiting abnormally rapid, direct and/or high flight (Singer and Wedlake 1981; Gall 1984; Mallet et al. 1987). If this reaction occurs during mark-recapture then this will affect the distances moved and may cause individuals to leave the study area when they otherwise would have remained (Morton 1982).

An attempt was made to limit such a reaction by placing the butterfly gently on vegetation on release. This was generally successful with three commonly observed reactions. 1. Grasping the vegetation and remaining stationary for a few seconds before resuming what appeared to be "normal" flight. 2. "Playing dead" with the butterfly curling its legs and falling motionless to one side, occasionally remaining in this position for nearly a minute before resuming activity. 3. Immediate flight on release but the flight appeared "normal".

A very small minority (estimated at less than 5 individuals per capture day) did show what would be described as "escape" reaction and engage in rapid flight, occasionally flying high into the surrounding woodland. However, the quick return of most individuals to "normal" activity was well illustrated by red 67 (male) which was captured and marked on 14 June at 16.01 h and then was seen in copulation just 16 min later, 133 m further along the same ride, and was still in copulation just over 1 h after.

First and second brood flight periods

To estimate the flight period of the first and second broods at Haugh Wood (insufficient information was available for Wigmore Rolls), the dates of all capture events from the mark-recapture project were combined with the results of a Pollard Butterfly Transect (carried out by the West Midland branch of Butterfly Conservation, data provided courtesy of Butterfly Conservation). This gave a minimum timing for first brood of between 1st May and 10th July (71 days minimum) and second brood of between 4th August and 19th August 2009 (15 days minimum).

Population size

Estimates of population size were calculated using POPAN (Schwarz and Arnason 1996) within the software package MARK (White and Burnham 1999). The full method, model assumptions and parameter estimates are presented and discussed in Clarke et al. (2010).

The data was collected in order to maximise information on mobility, not to calculate population and was not sufficiently robust to give realistic population estimates for each sample occasion. Nevertheless, for the first brood only, the best fit model gives an estimate of super population size (the total number of individuals in existence over the time period of the mark-recapture) of 224 ± 42 individuals of females and 281 ± 23 for males. There was not sufficient data for the second brood to attempt a similar estimation.

Distribution of adults at Haugh Wood

The positions of all adults captured were mapped and the results showed there were two areas containing the main concentrations of adults (Fig. 1). One area was along open aspect rides, scallops and glades in the far south of Haugh Wood (centred around SO588355) and the other was in the north-west (centred around SO586371), along a wide ride with scallops and several small tracks with adjacent scrubby glades. The two areas were approximately 850 m apart at the closest point and much of the intervening habitat consisted of mature closed-canopy plantations with shady rides which supported only limited potential food-plants. A well shaded public road was a further potential habitat barrier, separating the north and south of the wood. There were scattered captures of both males and females recorded outside of these areas. The furthest of these captures were just over 500 m from the nearest main area and included two males and two females around the roadside gateway to south Haugh Wood. One of the males (black 44) was later recaptured in the main habitat area in the north.

No evidence was found to suggest any separation in distribution between the first and second brood or between male and female.

Adult range and mobility

Table 3 shows the number of *L. sinapis* captured and recaptured at Haugh Wood in 2009 together with the maximum recorded time between first and last capture (known residence time). There was a fairly high recapture rate with a minimum of 21% for first brood females and a maximum of 63% for second brood males.

The maximum time between first and last capture are the "known residence times" and provide a rough estimate of the maximum adult life-span. They are likely to be underestimates as it is possible the individuals emerged some time before the first capture and may have survived after the last capture.

For each individual, the distance between all pairs of capture points was calculated. All the distances given in the study are direct distances between points. The actual movement will be greater than the direct distance since sequences of flights between capture times are unlikely to be directional. The butterflies might sometimes fly directly over the top of mature woodland (they are occasionally seen to fly towards the canopy before being lost from sight) but more usually they probably tend to follow rides, tracks and glades moving along indirect routes.

Figure 7a, b show the distances between all capture points against the length of time taken to move between them. This maximises the number of movement observations but does weight the data distribution strongly towards individuals with multiple recaptures. It could therefore be misleading if such individuals happen to be atypical. Figure 8 is a subset of the data in Fig. 7 where each individual is represented once only by its maximum distance moved between any two captures plotted against the corresponding time interval. The maximum distance is described here as the range over which the individual was recorded moving. While this will almost certainly be less than the individual's true range, it will be bounded by the true range and therefore statistically dependant on it. The highest possible range that could be recorded within Haugh Wood was approximately 2,700 m (the maximum distance across the study site).

Both sexes show a similar pattern of distance and range against time, but females generally move slightly shorter distances. The median in Fig. 7 is mostly less than 200 m for females but often greater than 200 m for males. There is also less scatter within both the range data and the data for all capture pairs. For females there is a clear overall increase in distance between same day recaptures and next day recaptures (Fig. 7b). For males a similar pattern is seen in Fig. 7a but is less clear because of the wider distribution for same day recaptures. It is not evident at all for males in the range data (Fig. 8), where there are fewer observations for short time intervals. For both sexes there is little or no correlation between distance and time interval at longer time scales. If movements were directional, that is dispersing away from the point of emergence then distance, and therefore range would increase with time, but this is not seen for the population as a whole.

The relatively large distances moved by many individuals of both sexes within 2 days show that these individuals at least are capable of dispersing widely if directional movement were to continue. However, most seem to remain within a limited area. 75% of all individuals have observed ranges less than 400 m. Nevertheless, some individuals (mainly males) do undertake long directional movements, sometimes over short periods of time. The movements of these individuals are detailed in Table 5.

The maximum and mean observed ranges are summarised in Table 4. Because of the tendency for distances to

Table 3 Summary of captures and known residence times

	First brood			Second brood		
	Male	Female	Unknown sex	Male	Female	Unknown sex
Initial captures only	172	95	1	35	40	0
Recapture occasions (not individuals)	169	30	0	27	16	0
No. of individuals recaptured (% of initial captures)	84 (49%)	20 (21%)	0	22 (63%)	14 (35%)	0
Maximum time between first and last capture (days)	26	27	–	6	12	–

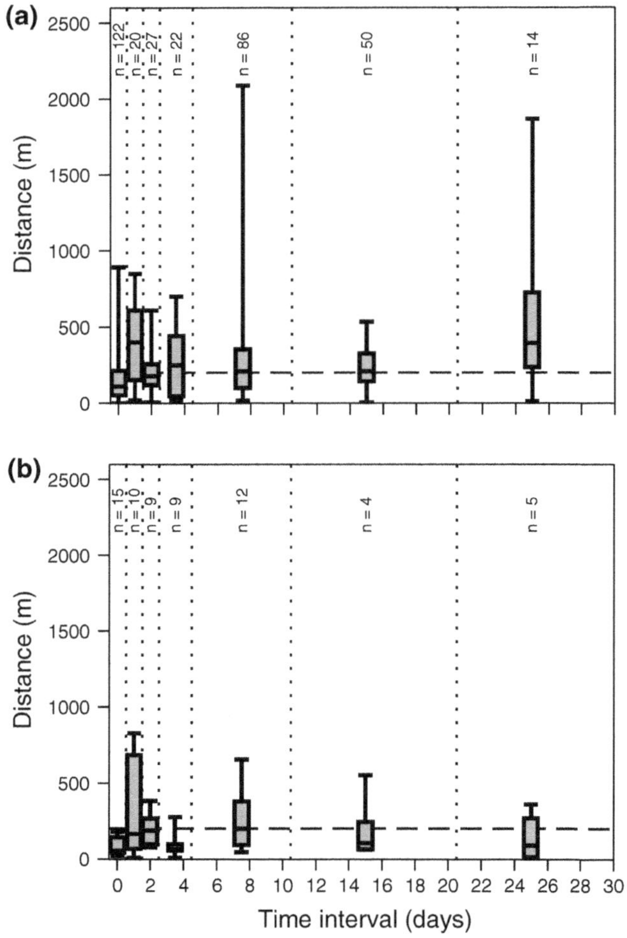

Fig. 7 Variation of distance between capture points with time interval between captures for males (**a**) and females (**b**). Time intervals are rounded to integer days and data are divided into variable width time interval bins, delimited by *vertical lines* (dotted). Bins are: same day, 1, 2, 3–4, 5–10, 11–20 and 21–30 days. The bin widths are chosen according to data availability; *n* gives the number of movements in each bin. Data distributions show median, upper and low quartiles, minima and maxima. A *reference line* (dashed) indicates 200 m separation between capture points

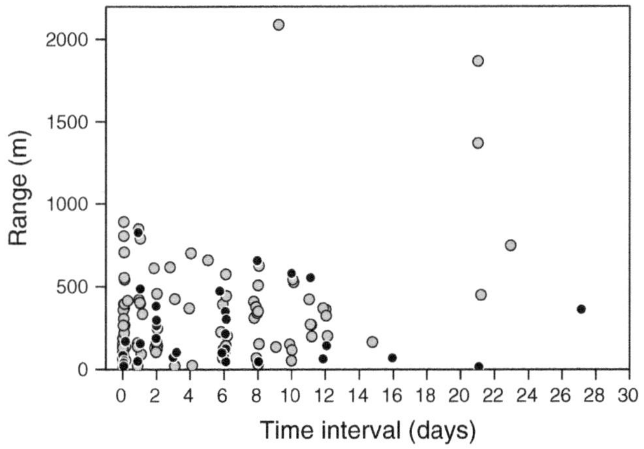

Fig. 8 Variation of individuals' maximum distance between any two capture points (i.e. their recorded range) with time interval between captures. Males are shown in *grey* and females in *black*

be lower for same day recaptures, ranges based on these are likely to be unrepresentative and are excluded to avoid sampling bias. The results show that males in both broods had a slightly larger mean range than females. For both sexes, range was slightly greater for the first brood than the second brood but the sample size and also the length of capture period are considerably smaller in the second brood which could account for the difference. While the maximum ranges of some individuals were considerable, over 2,000 m for one-first brood male and over 800 m for one-first brood female, the mean range of the population was less than 400 m for both broods and both sexes.

As well as calculating the range and distances moved, the tracks of all recaptured individuals were also examined. Table 5 shows details of just the ten males and ten females with the maximum recorded ranges and Fig. 9a, b give a visual representation of the movements seen for these 20 individuals. The results showed that all recorded females remained within or close to the areas of adult concentrations and often showed routes that turned back on themselves. Many males also remained within or close to the areas of adult concentrations showing routes that turned back on themselves, but a proportion of males did move out of these areas or even between the north and south adult areas.

No females were recorded moving between north and south Haugh Wood. However, the female with the longest recorded range moved over 800 m across the main southern adult area, so they appear to be capable of moving substantial distances.

Four males were recorded moving between south and north Haugh Wood. Two moved from the main area of adult concentration in the south to that in the north. One moved from the southern wood entrance into the main adult area in the north. One moved from the main area in the north to that in the south and then moved back again to the north (over 17 days). The other males showed in Table 5 and Fig. 9a moved in a similar pattern to the females, but the distances travelled between captures were slightly greater.

Observations of mating behaviour

Mating behaviour was not specifically studied during this project, but there were two noteworthy observations, both made at Haugh Wood in 2009 during the second brood.

The first observation was of a marked female who was recorded mating twice, and with different males. On 7th August (at 15.16 h) red 236 female was found in copulation with red 237, then on 13th August she was found just

Table 4 Summary of observed range for recaptured individuals

	Maximum range (metres)		Mean range ± SE (metres)	
	Male	Female	Male	Female
First brood	2,090 in 9 days	828 in 1 day	396.5 ± 51.3	284.6 ± 63.6
Second brood	574 in 6 days	488 in 1 day	232.6 ± 59.9	225.0 ± 47.3

Table 5 Summary of capture details for the ten males and ten females with the maximum observed ranges

Code	Sex	Range (metres)	Sum of distances between all captures (metres)	Maximum single distance (metres)	Days between first & last capture	No. of captures	Comments
Red 9	M	701.5	707.2	621.4	4.0	4	
Red 130	M	709.1	709.1	709.1	0.07	2	Remained in south, slightly outside of main adult area
Black 44	M	746.7	887.9	735.9	23.0	4	Moved from southern gate into the north
Blue 6	M	790.3	1,551.3	761.5	12	6	
Red 66	M	807.5	807.5	807.5	0.05	2	
Black 34	M	847.8	847.8	847.8	0.9	2	
Red 42	M	891.6	1,745.2	891.6	10.1	4	
Black 36	M	1,371.3	1,371.3	1,371.3	21.0	2	Moved from south to north
Black 39	M	1,868.2	2,112.7	1,868.2	23.0	3	Moved from south to north
Red 1	M	2,089.6	4,015.5	2,051.0	17.1	5	Moved from north to south and then back to north
Red 236	F	304.1	304.1	304.1	6.1	2	
Red 203	F	350.8	527.0	350.8	7.1	3	
Red 12	F	361.4	385.4	278.0	27.2	4	
Red 75	F	383.3	383.3	383.3	2.0	2	
Red 243	F	474.3	474.3	474.3	5.7	2	
Red 202	F	488.0	488.0	488.0	1.0	2	
Green 12	F	553.7	553.7	553.7	11.1	2	
Black 60	F	581.3	581.3	581.3	10.0	2	
Red 107	F	656.1	656.1	656.1	8.0	2	
Black 21	F	828.3	1,019.5	828.3	0.9	4	

Code numbers of under 200 are first brood and of over 200 are second brood individuals

over 300 m along the same ride, now in copulation with red 238 (at 18.16 h).

The second observation was an act of copulation. The complex behaviour of male *L. sinapis* in the presence of perched females has been well documented (Wiklund 1977; Friberg 2009). This behaviour has been typically described as courtship and such displays were often seen during this project. However, none were seen that resulted in mating, although copulating pairs were regularly found. One act of copulation only was seen and that did not appear to follow from a complex courtship display. On 7th August (15.00 h) a male flew down to a resting female. The male appeared to partially flick his wings as he landed but no further display was visible and the two were in copulation within less than 2 s.

The record of a marked female copulating with two different males confirms the findings of Wiklund and Forsberg (1991) that more than one mating may occur. The observation of a mating that was not preceded by a courtship display extends the findings of Friberg et al. (2008b) who noted for both *L. sinapis* and *L. reali* that "Many females of both species accepted mating after only a few seconds of courtship, before the courting male had preformed any wing beats."

Discussion

Egg-laying habitat

Egg distribution

Jeffcoate (2006) found a spatial separation in egg-laying habitat between the broods at Chiddingfold Forest (Surrey)

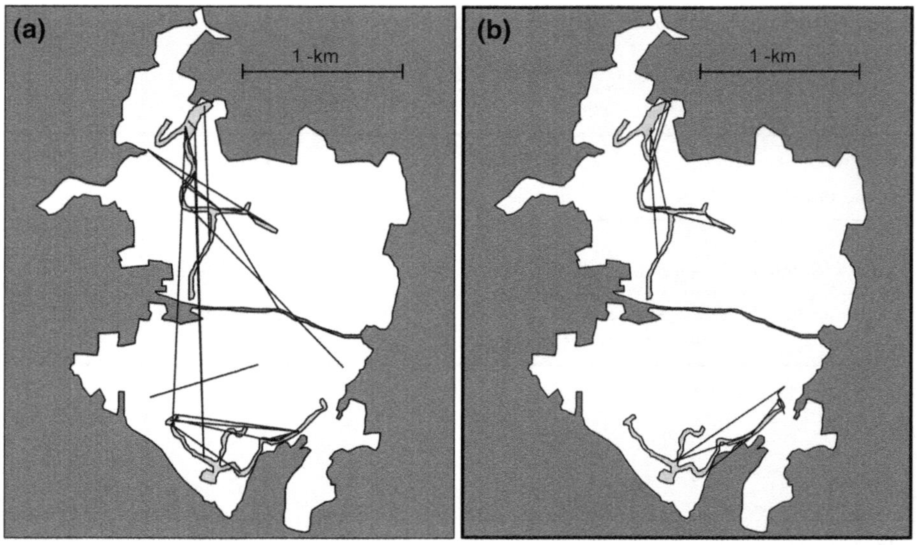

Fig. 9 Haugh Wood. Movements over between captures for the individuals with the maximum observed ranges, males (**a**) and females (**b**). Adult flight areas in *light grey*, movements shown as *black lines*

where recently cleared habitat was preferred for egg-laying by the first brood but was seldom used by the second brood. At Haugh Wood (not examined at Wigmore Rolls), there was no evidence that different egg-laying areas were used between the broods. There were recently created clearings at Haugh Wood similar to that used on the Surrey site, but they had a very low abundance of larval food-plants so a similar brood choice was not available. It is possible that this situation might change as the new glades develop.

Wiklund (1977) had found a spatial separation between adult nectaring and egg-laying areas but Warren (1984) did not find this at the Northamptonshire sites and this was also not the case at Haugh Wood. At Haugh Wood all egg-laying sites were within the main adult areas (Fig. 1).

Larval food-plants

At both Haugh Wood and Wigmore Rolls, *Lathyrus pratensis* and *Lotus pedunculatus* were the most commonly used food-plants for egg-laying and no evidence was found that the first brood differed from the second brood in their food-plant choices. The choice of food-plant seemed to correlate well with the relative abundance of these species, as was also found by Jeffcoate (2006) at Chiddingfold Forest, although he additionally noted that in one part of his study site *Lotus* spp seemed to be ignored in favour of *Lathyrus pratensis*.

L. sinapis is known to use a number of larval food-plants and at Haugh Wood and Wigmore Rolls, egg-laying also occurred on *Lotus corniculatus*, though this plant was not common on either site and was only rarely used. At Wigmore Rolls, three other egg-laying plants were recorded, *Potentilla reptans* (one egg), grass species (one egg) and a *Vicia cracca* (one egg). *Vicia cracca* was rare at Wigmore Rolls so its limited use as a larval food-plant is expected.

Egg and food-plant height

It has been suggested that the height the egg is laid above ground changes between the first and second broods, and that this is correlated to seasonal temperature (Jeffcoate 2006). Our study suggests that the egg height is determined primarily by the height of the available food-plants and there is no evidence that lower growing plants are selected as egg-laying sites by the first brood. Eggs were laid towards the top of the food-plants, regardless of the species used and this has been documented by previous studies (Warren 1984; Joy 2000; Clarke and Green 2008). This position near to the top of the foodplant could allow the newly-hatched caterpillar easy access to young leaves, and/or might be a warmer position likely to speed egg development.

Food-plant height does increase between the period of the first and the second brood and so therefore does the height of eggs from the ground, as was noted by Jeffcoate (2006). Additionally, the height of the food-plants will also be somewhat site-specific, as was seen in the comparison between Haugh Wood and Wigmore Rolls (Figs. 2, 3). This has also been noted by other surveys (Clarke and Green 2008) and is not surprising as many aspects of a site, including soil conditions and managements such as mowing will affect the growth form of food-plants.

There was a slight difference between food-plants in egg position relative to the plant tip. Eggs laid on *Lathyrus pratensis* tended to be positioned lower on the plant stem than eggs on *Lotus corniculatus* or *Lotus pedunculatus*. From observation of egg-laying it is suggested that this could be a function of the plant structure. The terminal bud, leaves and tendril of *Lathyrus pratensis* seemed to make it difficult for a female to hold on so they grasp the stem at the base of the terminal leaves and lay on the leaflets below. In contrast, on both *Lotus* species females usually seem to be able to grasp the very top leaflets and so the eggs are positioned closer to the plant tip.

Surrounding vegetation, bare ground and shade conditions

Warren (1984) noted that eggs tended to be found on tall food-plants that protrude from the surrounding vegetation. At Haugh Wood and Wigmore Rolls this was the case when *Lathyrus pratensis* was used, but for *Lotus pedunculatus* eggs were often laid on plants that were considerably lower than the surrounding vegetation. *Lathyrus pratensis* was the most commonly used food-plant on the Northamptonshire sites used by Warren so that may account for the difference in results. Warren (1986) found that clearing the vegetation away from the food-plant did make it more likely to be used for egg-laying. Observations at Haugh Wood and Wigmore Rolls suggest that much does depend on the structure of the surrounding vegetation and correlated with this will be the temperature of the food-plant. In loosely-structured and open vegetation whatever the height, females are regularly seen flying below the top of the vegetation. In a tight dense structure, the females are unable to fly amongst the vegetation so they will only find the plants that protrude above.

At both Haugh Wood and Wigmore Rolls a wide range of habitat conditions were utilised for egg-laying and no significant difference was found in vegetation height, amount of bare ground or amount of shade at egg locations between either broods or sites. Eggs were laid within a wide range of vegetation heights of between 6 and 120 cm, with a peak of between 40 and 50 cm. Most eggs were within well vegetated habitat but all amounts of bare ground were seen to be used, up to the maximum recorded of 96%.

Most eggs were recorded in locations with no direct shade, but despite this strong preference, small numbers of eggs were found in other shade conditions, including 100% direct shade. Warren (1986) stated "It seems that the females will search for egg-laying sites in a wide range of shade conditions, but the eggs laid depends on the frequency of encounters with food-plants." This also describes what was seen in this study. If the butterfly was able to encounter a food-plant of any species during its flight and if that plant was a sunny, warm spot *at that moment*, then egg-laying was likely to occur. At Haugh Wood, there were some grassy scallops along a north/south orientated ride that were usually well shaded and sheltered by mature trees and scrub, but for a short time each day these scallops became sunny "hot-spots". Egg-laying was regularly seen here while the sun was on the patch, although for the rest of the day adults moved through the spot without stopping. The females may well chose whatever is a "hot-spot" at the moment of egg-laying and most likely cannot make a judgement as to the daily warmth offered at that location.

Conservation implications of food-plant choice and egg-laying location

This study suggests that *L. sinapis* can use a wide and comparatively non-specific range of conditions for egg-laying, but with a preference for unshaded locations. Management should aim at creating a high density of different potential food-plants growing within open but sheltered conditions within a variety of early successional vegetation structures from bare ground to young scrubby growth.

If site management can create and maintain a complex of early succession habitats where potential food-plants grow in sheltered, warm locations then it seems *L. sinapis* will be able to find plants suitable for egg-laying. The overall impression from this study and from the variation in findings of other studies is that choice and variation may be a key feature in the long-term survival of this species at individual sites. Weather conditions in particular are likely to have a large effect in determining whether locations are suitable or even which food-plant species survive best. In some years plants in dense vegetation may be too cool over the spring and summer, in other years, plants in bare ground conditions may become too dry.

Adult distribution and movement at Haugh Wood

Adult distribution

In both broods, two discrete areas about 850 m apart, contained the main concentrations of adults, one in the north-west and one in the far south of the wood. Adults were not completely limited to these two areas as small numbers of both sexes were distributed over much of the woodland, often some distance away from the two main concentrations. As expected, these two areas did also appear to contain the main concentrations of suitable habitat and much of the intervening woodland was judged to be of limited suitability for *L. sinapis* with a potentially major barrier to movement in the form of a well shaded public road separating the north and south of the wood.

It has been suggested that in species where male harassment of females is high, the females sometimes will utilise separate areas within a site or disperse away from areas of high male density (Parker 1978; Baguette et al. 1998). Male harassment of females in *L. sinapis* does seem to be costly, especially as females appear to mate only once or twice, so most mating attempts will be unwanted (Wiklund and Forsberg 1991; Wiklund et al. 2001). Males were often observed interrupting egg-laying and the failed courtship displays are time consuming for both sexes. Friberg (2009) found some evidence on a site in Sweden that *L. sinapis* females might be using less optimal areas so as to remove themselves from the attentions of patrolling males. This did not seem to occur at Haugh Wood, possibly the density of males was not so high as to force the females to move, or they might have been moving into scrubby areas where they were hidden from the attentions of both the *L. sinapis* males and the field workers.

Adult range and mobility

The mark-recapture study at Haugh Wood was successful, with 343 individuals marked over two broods of which 41% were recaptured at least once. An estimated population size for the first brood suggests approximately 500 individuals.

Adults were easy to net, easy to mark and most did not exhibit any rapid flight "escape" reaction on release, suggesting that marking did not substantially interfere with their subsequent behaviour.

Mark-recapture results showed that adults tended to remain within one of the two main habitat areas and within these they appeared to move back and forward along the linear rides and tracks. This impression is likely to be only part of the actual pattern of movement. The scrubby areas and mature woodland alongside tracks were almost impossible to enter to make captures, but adults were frequently seen to fly both over and into these areas.

Males showed a much greater tendency than the females to move into, over or through habitat that seemed marginal or unsuitable. Females showed a strong tendency to remain within an area that had suitable adult resources, but would move freely between egg-laying patches. Scattered records of females additionally occurred across the woodland in what seemed to be marginal or unsuitable habitat suggesting that they will move out of a habitat patch, so both sexes will explore habitat outside of the main resource areas to at least some extent.

Males in both broods had a slightly larger mean range than females. Male *L. sinapis* have been recorded as spending more than twice the amount of time in flight than females (Warren 1984) as they patrol the site looking for mating opportunities, so a larger potential range might be expected. The maximum range recorded for males was 2,087 m (one-first brood male) which is fairly close to the highest possible range of 2,700 m that could have been measured (the maximum distance across the study site). For females the maximum range was smaller at 828 m but this distance was covered in less than 1 day. The distance between the two main areas used by adults was 850 m so movement of both males and females between these patches is clearly possible.

Four individuals, all males, were actually recorded moving between south and north Haugh Wood, with one moving from the main area in the north to that in the south and then moving back again to the north. These butterflies must have either flown along shaded rides (which contained limited resources for *L. sinapis*) or over the top of mature woodland and all must have crossed the intervening public road. Two females and two males were recorded in the southern gateway on the road that divides north and south Haugh Wood so it is not unlikely that some females as well as males did move across the road and perhaps between the two habitat areas. Although no females were actually recorded moving between north and south Haugh Wood, they do appear to be capable of moving substantial distances.

Long movements for male *L. sinapis* have previously been noted but usually this has been inferred from finding individuals some distance from a known colony. However, Warren (1984) did mark adult *L. sinapis* at Yardley Chase and his marked individuals were occasionally observed to have moved a direct distance of up to 3 km, which is comparable to the 2,078 m direct distance recorded at Haugh Wood.

Conservation implications of the observed *L. sinapis* movements

A key objective for this study was to assess the ability of *L. sinapis* to reach patches of new habitat. The findings at Haugh Wood suggest that although there is a strong tendency for adults to remain within a habitat patch, a proportion of both sexes do disperse. Potential habitat barriers such as shaded rides without nectar or food-plants, mature woodland or roads can be crossed, although the potential for crossing areas of totally unsuitable landscape such as arable land was not investigated.

Males seem to be able to move between areas that are 2 km or more distant, but females may have a more limited range. It seems realistic to suggest that any newly available habitat within 400 or 500 m of present habitat will probably be found quickly by both males and females, but areas more than 1 km from suitable habitat would have a lower probability of rapid colonisation. Almost certainly much would depend on the intervening habitat, but this study suggests that if, even sub-optimal habitat, can be created to make "stepping stones" then *L. sinapis* is very capable of colonising areas more than 1 km distance.

Conclusions

The results of this study confirm several previously known aspects of the ecology of *L. sinapis* in the UK but also give some support to Jeffcoate's recent suggestions that some aspects of the butterfly's ecology differ from that previously believed. The widely observed change to partial bi-voltinism, in even the most northerly parts of the species' UK range, is particularly striking. In addition, at the two study sites at least, *L. sinapis* appears to be a using a wider range of vegetation heights and structures, foodplant species and shade conditions. It appears that either, some changes have taken place since Warren's studies in the 1980s, or that the species requirements may be more geographically specific or even site specific than previously thought. All are possible. If change is taking place, and particularly if these changes are climatically driven, then this is likely to be a continuing process and we must remain alert to the species shifting requirements. Additional examination of the use of habitats at further sites and in other geographical areas is highly desirable in order to investigate this further. It must also be borne in mind that the study concentrated entirely on adult butterflies and their eggs. The requirements of the other stages of the life-cycle will of course be critical factors and these remain imperfectly understood at present.

At Haugh Wood and Wigmore Rolls, we can now offer more confident guidance on conservation management for *L. sinapis*. We can also suggest appropriate locations within these sites for future habitat creation in the knowledge that these will be found by the butterfly. We need to discover if our findings are equally applicable elsewhere in the UK.

Acknowledgments This work was commissioned by the Forestry Commission under the SITA Trust funded project "Herefordshire's Wood White Butterflies" with advice from Butterfly Conservation. The Forestry Commission nominated officer was Kate Wollen and the nominated officer for Butterfly Conservation was Jenny Joy (West Midland Senior Regional Officer, Butterfly Conservation).

The authors would like to thank Robin Hemming for his help and advice on Haugh Wood and Stephen Jeffcoate for his most useful comments and discussions on our results. Many thanks are due to John Hemmings for help with the mark-recapture data analyses.

References

Baguette M, Vansteenwegen C, Convi I, Nève G (1998) Sex-biased density-dependent migration in a metapopulation of the butterfly *Proclossiana eunomia*. Acta Oecol 19:17–24

Biodiversity Reporting & Information Group (2007) Report on the Species and Habitat Review. Report to the UK Biodiversity Partnership

Clarke SA, Green DG (2008) A preliminary investigation into Wood White *Leptidea sinapis* (Linnaeus, 1758) habitat on Shropshire sites 2007. Butterfly Conservation Report No. S08-15

Clarke SA, Green DG, Butler I (2010) Wood White *Leptidea sinapis* (Linnaeus, 1758) habitat in Herefordshire Forestry Commission woodlands 2008 and 2009. Forestry Commission report

Fox R, Asher J, Brereton T, Roy D, Warren MS (2006) The state of butterflies in Britain and Ireland. Book. Pisces Publications, Newbury

Friberg M (2009) The evolutionary ecology of niche separation. Studies on the sympatric butterflies *Leptidea sinapis* and *Leptidea reali*. PhD thesis, Stockholm University

Friberg M, Olofsson M, Berger D, Karlsson B, Wiklund C (2008a) Habitat choice precedes host plant choice—niche separation in a species pair of a generalist and a specialist butterfly. Oikos 117:1337–1344

Friberg M, Vongvanich N, Borg-Karlson A-K, Kemp DJ, Merilaita S, Wiklund C (2008b) Female mate choice determines reproductive isolation between sympatric butterflies. Behav Ecol Sociobiol 62:873–886

Gall LF (1984) The effects of capture and marking on subsequent activity in *Boloria acrocnema* (Lepidoptera: Nymphalidae), with a comparison of different numerical models that estimate population size. Biol Conserv 28:139–154

Jeffcoate S (2005) The Wood White butterfly in Surrey in 2004. A report to Forest Enterprise

Jeffcoate S (2006) Seasonal variation in the use of vegetation resources by *Leptidea sinapis* (Linnaeus, 1758) (Lepidoptera: Pieridae), a multivoltine species in southern Britain: implications for its conservation at the edge of its range and in the context of climate change. Ent Gaz 57:69–82

Joy J (2000) The status of the Wood White butterfly (*Leptidea sinapis*) in South Shropshire during 2000. Butterfly Conservation report

Joy J (2006) Shropshire biodiversity action plan. Wood White (*Leptidea sinapis*). Update. Shropshire Biodiversity Steering Group. Report to Shropshire County Council

Joy J, Williams M (2008) Butterfly conservation regional action plan for the West Midlands. Butterfly Conservation report S08-19

Mallet J, Longino JT, Murawski D, Murawski A, Simpson de Gamboa A (1987) Handling effects in *Heliconius*: where do all the butterflies go? J Anim Ecol 56:377–386

McCracken M, Warren M (2007) Wood White *Leptidea sinapis* factsheet. Butterfly Conservation, Wareham

Morton AC (1982) The effect of marking and capture on recapture frequencies of butterflies. Oecologia 53:105–110

Parker GA (1978) Evolution of competitive mate searching. Annu Rev Entomol 23:173–196

Schwarz CJ, Arnason AN (1996) A general methodology for the analysis of capture-recapture experiments in open populations. Biometrics 52:860–873

Singer MC, Wedlake P (1981) Capture does affect probability of recapture in a butterfly species. Ecol Entomol 6:215–216

Warren MS (1981) The ecology of the Wood White butterfly *Leptidea sinapis* (L.) (Lepidoptera: Pieridae). PhD thesis, University of Cambridge

Warren MS (1984) The biology and status of the Wood White butterfly *Leptidea sinapis* (L.) (Lepidoptera: Pieridae), in the British Isles. Ent Gaz 35:207–223

Warren MS (1985) The influence of shade on butterfly numbers in woodland rides, with special reference to the Wood White *Leptidea sinapis*. Biol Conserv 33:147–164

Warren MS (1986) Annual and long-term changes in a population of the Wood White butterfly *Leptidea sinapis*. J Anim Ecol 55:707–719

Watt WB, Chew FS, Snyder LRG, Watt AG, Rothchild DE (1977) Population structure of Pierid butterflies. I Numbers and movements of some Montane *Colias* species. Oecologia 27:1–22

White GC, Burnham KP (1999) Program MARK: survival estimation from populations of marked animals. Bird Study 46 Supplement:120–138

Wiklund C (1977) Courtship behaviour in relation to female monogamy in *Leptidea sinapis* (Lepidoptera). Oikos 29:275–283

Wiklund C, Forsberg J (1991) Sexual size dimorphism in relation to female polygamy and protandry in butterflies: a comparative study of Swedish Pieridae and Satyridae. Oikos 60:373–381

Wiklund C, Karlsson B, Leimar O (2001) Sexual conflict and cooperation in butterfly reproduction: a comparative study of polyandry and female fitness. Proc R Soc Lond B 268:1661–1667

J Insect Conserv (2011) 15:221–232
DOI 10.1007/s10841-010-9339-6

ORIGINAL PAPER

New tools to boost butterfly habitat quality in existing grass buffer strips

Robin J. Blake · Ben A. Woodcock · Duncan B. Westbury · Peter Sutton · Simon G. Potts

Received: 25 March 2010 / Accepted: 3 September 2010 / Published online: 17 September 2010

Abstract There are approximately 29,000 ha of grass buffer strips in the UK under Agri-Environment Schemes; however, typically they are floristically poor and as such are of limited biodiversity value. Introducing a sown wildflower component has the potential to increase dramatically the value of these buffer strips for a suite of native species, including butterflies. This study investigates management practices aiming to promote the establishment and maintenance of wildflowers in existing buffer strips. The effectiveness of two methods used to increase the establishment of wildflowers for the benefit of native butterfly species were tested, both individually and in combination. The management practices were: (1) the application of a selective graminicide (fluazifop-P-butyl) which reduces the dominance of competitive grasses; and (2) scarification of the soil which creates germination niches for sown wildflower seeds. A wildflower seed mix consisting of nine species was sown in conjunction with the scarification treatment. Responses of wildflowers and butterflies were monitored for two years after establishment. Results indicate that the combined scarification and graminicide treatment produced the greatest cover and species richness of sown wildflowers. Butterfly abundance, species richness and diversity were positively correlated with sown wildflower species richness, with the highest values in the combined scarification and graminicide treatment. These findings have confirmed the importance of both scarification as a means of introducing wildflower seed into existing buffer strips, and subsequent management using graminicides, for the benefit of butterflies. Application of this approach could provide tools to help butterfly conservation on farmland in the future.

Keywords Buffer strips · Wildflowers · Butterflies · Management · Agri-environment scheme · Graminicide

Introduction

Butterflies (Lepidoptera) can provide several important ecosystem services in terrestrial farmland habitats (Hammond and Miller 1998). As key members of food webs, particularly as larvae, they represent an essential dietary component for many farmland bird species, providing protein for chick growth and development (Potts 1986; Wilson et al. 1999). Some adult butterflies can play a role in crop and wildflower pollination (Corbet 1987), although their economic importance in temperate regions is likely to be low compared to bees (Losey and Vaughan 2006). Butterflies are also sensitive to environmental change, and thus can be indicative of the ecological health and quality of a habitat (van Swaay et al. 2006).

Despite their importance, widespread declines have been documented in the abundance and species diversity of farmland butterflies (Fox et al. 2007). UK Butterfly Monitoring Scheme data revealed a 23% decline in farmland species such as *Thymelicus sylvestris* Poda (small skipper) and *Anthocharis cardamines* Linneaus (orange tip)

R. J. Blake (✉) · D. B. Westbury · S. G. Potts
Centre for Agri-Environmental Research, School of Agriculture, Policy & Development, University of Reading, Earley Gate, Whiteknights Road, Reading, Berks RG6 6AR, UK
e-mail: r.blake@reading.ac.uk

B. A. Woodcock
NERC Centre for Ecology & Hydrology, Maclean Building, Crowmarsh Gifford, Wallingford, Oxon OX10 8BB, UK

P. Sutton
Syngenta, Jealott's Hill International Research Centre, Bracknell, Berks RG42 6EY, UK

between 1990 and 2007 (Butterfly Conservation et al. 2009). A similar trend has been documented in Europe with an 11% decline in species distribution over the last 25 years (van Swaay et al. 2006). Declines in other farmland taxa have been observed. For example, the number of UK bumblebee species has declined by approximately 50% since the 1940s, and three species have gone extinct (Williams 1982; Edwards and Jenner 2005). Arable plants have also suffered marked declines and are now the rarest group of Britain's vascular plant species (Cheffings and Farrell 2005). Plants are an integral part of arable ecosystems (Marshall et al. 2003), and many plant species are important hosts for butterfly larvae, with their flowers providing nectar for adult butterflies (Asher et al. 2001).

The decline of farmland butterflies and other taxa has been partially attributed to agricultural intensification (O'Connor and Shrubb 1986; Robinson and Sutherland 2002). Loss and fragmentation of foraging and nesting habitats, including species-rich meadows and hedgerows, have been implicated in the decline of butterflies and bees (Feber et al. 1996; Pywell et al. 2005). In addition, increased use of herbicides and fertilisers has caused detrimental effects on butterfly host plants (Wilson 1992), thus reducing available larval resources. The reduction in leguminous crops, e.g. *Trifolium* spp. (clover), following changes in rotation practices, has also reduced available floral resources (Goulson et al. 2008).

The development of Agri-Environment Schemes were one of the policy instruments developed in response to the loss of biodiversity in farmed landscapes and work on the basis where farmers are given financial incentives in return for environmentally sensitive farming (Ovenden et al. 1998). The most recent scheme in the UK is Environmental Stewardship which was introduced in 2005, and includes Entry Level Stewardship (ELS) and Higher Level Stewardship (HLS) (Defra 2008). A popular option has been the establishment of perennial grass buffer strips, with approximately 29,000 ha currently under UK agri-environment agreements (Natural England 2009). These strips are usually two to six metres wide and typically established by sowing a mixture of fine and tussocky grass species such as *Festuca rubra* L. (red fescue) and *Dactylis glomerata* L. (cocksfoot) (Critchley et al. 2006). The strips have several important functions. They help reduce pesticide and fertiliser drift into more sensitive features of the field boundary, such as hedgerows and water courses (Marshall and Moonen 2002). They have also been shown to be important habitats for a variety of generalist predatory invertebrates, in particular ground beetles and spiders (Meek et al. 2002; Thomas et al. 2002). However, botanically they tend to be species poor, typically lacking a wildflower component (Marshall 2003; Field et al. 2006). Although agri-environment options do exist for establishing sown grass and wildflower mixtures, for example the English HLS option HE10, uptake has been low with only 1,300 ha currently under agreement (Natural England 2009). Introducing wildflowers into the existing grass buffer strips has the potential to increase their biodiversity value, through the provision of foraging resources for insect pollinators (Pywell et al. 2004; Potts et al. 2007). However, due to the competitive dominance of the existing grass component, sowing wildflower seed mixtures is likely to have limited success, unless germination niches are created. Management has included the use of scarification of the buffer strip surface to create germination niches (Woodcock et al. 2008). Many grass species will rapidly colonise scarified ground, limiting the effectiveness of this approach to create germination niches. However, the use of graminicides, such as fluazifop-P-butyl (Fusilade Max, Syngenta Crop Protection Ltd.), may have the potential to suppress, rather than eliminate, susceptible tussocky grasses, such as *D. glomerata* L., in existing buffer strips, whilst leaving the more resilient fine-leaved grass species such as *F. rubra* L. intact (Syngenta Crop Protection Ltd. 2003). Fluazifop-P-butyl has no effect on wildflower species (Westbury and Dunnett 2008), and so in suppressing the grasses, this in turn can promote the development of sown wildflowers benefitting a greater range of invertebrate species (Marshall and Nowakowski 1994, 1996; Potts et al. 2007).

This study tested the following hypotheses:

(1) Scarification can create germination niches into which sown wildflower seeds can establish.
(2) Graminicide (fluazifop-P-butyl) can reduce grass dominance in buffer strips.
(3) There is an interaction between scarification and graminicide.
(4) The introduction of sown wildflowers will increase the abundance, diversity and species richness of butterflies.

Materials and methods

Study sites

Experiments were performed on two arable farms in Berkshire, UK. These were at Aborfield (N51°23′39″, W0°55′08″) and North Sydmonton (N51°21′31″, W1°17′21″). The Aborfield soil type is gravel over London clay, with freely-draining lime-rich loam at North Sydmonton. The grass buffer strips were established in 2004 at both sites, with a basic grass mixture including *D. glomerata*, *Phleum pratense* L. (timothy), and *F. rubra*, and sown at a rate of 20 kg ha^{-1}. In 2005, both farms entered into ELS agreements. The buffer strips were situated adjacent to fields which have been cropped in rotation with winter wheat, winter barley, oilseed rape, rye, winter beans and grass (Aborfield), and winter wheat, winter barley, winter beans and winter oats (North Symonton), since 2004. Prior to the start of this study, the buffer strips at both sites were managed by cutting once per year in the autumn.

Experimental design

The study was initiated in spring 2008 at both study sites, when four buffer strip management treatments were established on the outer 4 m (i.e. adjacent to the crop) of the existing 6 m grass buffer strips. This relates to current English ELS management prescriptions for 6 m buffer strips whereby the outer 3–4 m is cut annually, and the inner 2–3 m is cut less often to provide different habitats (Defra 2008). The inner 2 m was not used in the study. Each of the four management treatments was randomly assigned to one of four plots. The plots were arranged in three separate replicate blocks with each plot represented once in each block, thus 12 plots at each study site. Plots within a block shared the same aspect. Each plot measured 25 m × 4 m, with a 5 m untreated area between plots. All blocks were situated with the long axis running parallel and adjacent to the hedgerow, which formed the field boundary. Three of the four plots received a different treatment applied individually, and in combination with, two management techniques: (a) scarification to create germination niches into which sown wildflower seeds can establish; (b) selective graminicide (fluazifop-P-butyl) to reduce the competitive dominance of grasses. The fourth plot was the control treatment and did not receive scarification or graminicide. Thus, the four treatments were: (1) control (no scarification or graminicide); (2) scarification only; (3) graminicide only; and (4) scarification and graminicide in combination. All management treatments were applied in spring 2008. The wildflower seed mixture was only applied to plots which had been scarified (i.e. treatments two and four). Scarification was applied once in March 2008 with a power harrow to a depth of approximately 5 cm to break up the existing grass sward and create approximately 60% bare ground. Following scarification, a perennial wildflower seed mixture containing nine species known to be important pollinator forage plants was sown at a total rate of 2.4 kg ha^{-1} (Table 1), using a battery-operated WE-B Universal Hand Power Spreader (Wolf-Garten, Betzdorf, Germany). The graminicide fluazifop-P-butyl (Fusilade Max 125 g l^{-1} EC) was applied at a rate of 93.75 g a.i. ha^{-1} (scarification and graminicide treatment), and 187.5 g a.i. ha^{-1} (graminicide only treatment) in April 2008. Fluazifop-P-butyl is a selective, non-residual post-emergence phenoxy herbicide used to control grasses in broad-leaved crops and other situations including non-cropped buffer strips (Syngenta Crop Protection Ltd. 2003). It is systemic, being rapidly absorbed into the leaves of target grasses and translocated to the site of active growth where it disrupts lipid synthesis through the inhibition of acetyl coenzyme A carboxylase (ACCase) (Herbert et al. 1997). Both applications were performed with a 2 m plot sprayer, operating at a pressure of 2.4 bars, and a water volume of 200 l ha^{-1}. The application rate of 93.75 g a.i. ha^{-1} was chosen as it represented the half label rate permitted for use in non-cropped buffer strips and the aim of this application was to suppress, not eliminate, the susceptible grass species. In addition, the action of the scarification had already suppressed the grass structure to some extent, therefore a lower rate was considered appropriate. The higher rate of 187.5 g a.i. ha^{-1} was considered to be suitable for use in the graminicide only treatment to suppress the high biomass of grasses present, open up the sward, and provide maximum opportunities for the species in the seed bank. Treatments were allocated randomly within blocks.

During the autumn of 2008 and 2009, all plots were cut with a tractor-mounted flail mower to a height of approximately 15 cm, and the cuttings left in situ. Grass cuttings are typically left on plots by arable farmers as they tend not to have appropriate machinery for their removal. This cut had the aim of reducing standing biomass and opening up the sward, and followed current ELS management recommendations (Defra 2008).

Botanical composition

The vegetation was assessed once in June 2008 and June 2009 using 0.5 × 0.5 m (0.25 m^2) quadrats. Ten replicate quadrats were randomly positioned within each plot, leaving a buffer of approximately one metre to account for edge effects. All species were identified and assigned a percentage cover value based on an eight point scale (1 = <1%, 2 = 1–5%, 3 = 6–10%, 4 = 11–20%, 5 = 21–40%, 6 = 41–60%, 7 = 61–80%, 8 = 81–100%). Plant nomenclature followed Stace (1997).

Butterfly sampling

Butterfly abundance, diversity and species richness was determined on the wing during standard transect walks on each treatment plot. The methods followed represented a modified version of those described by Pollard et al. (1975), Pollard and Yates (1993), and Westphal et al. (2008). All individual butterflies occurring either in-flight, or on a flower, were recorded to species, up to 2 m either side of a 25 m × 4 m transect route running along the centre line of each plot. This ensured that the entire 4 m buffer strip width was sampled. Each plot was sampled

Table 1 Details of wildflower seed mixture and common farmland butterfly species which utilise these plants as larval hosts and adult nectar sources

Seed species	Plant family	Larval host plant preferences[a]	Adult nectar sources[b]	Sowing rate (kg ha^{-1})
Achillea millefolium L.	Compositae	*Vanessa cardui* L. *Polyommatus icarus* Rottemburg	*Coenonympha pamphilus* L. *Aglais io* L. *Lasiommata megera* L. *Lycaena phlaeas* L. *Maniola jurtina* L. *Melanargia galathea* L. *Ochlodes sylvanus* Esper	0.04
Centaurea nigra L.	Compositae	–	*Aglais urticae* L. *Colias croceus* Fourcroy *V. cardui* L. *Gonepteryx rhamni* L. *M. jurtina* L. *M. galathea* L. *O. sylvanus* Esper *Pieris brassicae* L. *Polygonia c-album* L. *P. icarus* Rottemburg *Pyronia tithonus* L. *Thymelicus sylvestris* Poda	0.50
Leucanthemum vulgare Lam.	Compositae	–	*V. cardui* L. *G. rhamni* L. *M. jurtina* L. *Pieris rapae* L. *P. tithonus* L. *T. sylvestris* Poda	0.10
Plantago lanceolata L.	Plantaginaceae	*V. cardui* L.	–	0.33
Lotus corniculatus L.	Fabaceae	*C. croceus* Fourcroy *P. icarus* Rottemburg	*V. cardui* L. *O. sylvanus* Esper *P. icarus* Rottemburg *P. rapae* L. *T. sylvestris* Poda	0.40
Rumex acetosa L.	Polygonaceae	*L. phlaeas* L.	–	0.10
Galium verum L.	Rubiaceae	–	–	0.10
Silene dioica L.	Caryophyllaceae	–	*G. rhamni* L. *P. napi* L. *Anthocharis cardamines* L. *P. rapae* L.	0.25
Trifolium pratense L.	Fabaceae	*C. croceus* Fourcroy *P. icarus* Rottemburg	*V. cardui* L *L. phlaeas* L. *M. galathea* L. *P. icarus* Rottemburg *P. rapae* L *P. tithonus* L. *Thymelicus lineola* Ochsenheimer *T. sylvestris* Poda	0.57

[a] From Dennis (1992)

[b] From Porter et al. (1992) and Feber et al. (1996)

four times per year between the months of May and September. To increase the number of butterfly observations, the transect was walked twice on each sampling day, once in the morning, and once in the afternoon, at a rate of 5 min per plot, giving eight observations per plot per year. Although the infrequency of sampling may have underestimated the butterfly responses, it was timed to coincide with peak levels of butterfly activity from May to September. Transect walks were conducted between 10:00 and 17:00 h BST. The temperature was above 13°C with at least 60% clear sky, or 17°C in any sky conditions, and it was not raining. In addition the wind speed was not in excess of five on the Beaufort scale. Air temperature, percentage cloud cover and wind speed were recorded at the start and end of each transect walk.

Data analysis

The adult butterfly abundance counts from the two transect walks conducted on each sampling day were averaged to provide a mean abundance value for each sampling month and site. This resulted in four mean values, which were then summed to provide total abundance counts per site and per year. Total abundance, diversity (Shannon-Wiener) and species richness for each plot and year were then determined. The abundance and species richness data were $\log_e N + 1$ transformed to normalise the data sets. Values of species diversity were performed using Multi Variate Statistical Package Version 3.1 (MVSP 1999). Summed percentage cover values were calculated separately for sown and unsown wildflowers, and total grass species, in each experimental plot to provide information on the available foraging resources. These were based on average percentage cover values derived from the ten quadrats recorded from each experimental plot. Summed percentage cover values for wildflowers and grasses could be greater than 100% to take into account the three dimensional structure of the sward, and overlap of species in the quadrat. Species richness ($\log_e$) of sown and unsown wildflowers, and unsown annual and perennial species, was also determined. The treatment responses of summed percentage covers, wildflower species richness, and butterfly abundance, diversity and species richness were analysed using general linear mixed models in SAS 9.2 (SAS 2008). Repeated-measures was used to account for sampling across years (2008 and 2009). Fixed effects in all models were year, scarification, and graminicide, and the interactions between them. Random effects were site and block nested within site. Year was included as a repeated measure. Using this same model structure the effect on butterfly abundance, diversity and species richness in response to the sown wildflower species richness was also assessed. Interactions between year, percentage cover of grasses and percentage cover of wildflowers were treated as fixed effects. All models were simplified by deletion of non significant factors (except where part of a significant interaction) from a type III model, and degrees of freedom were calculated by the iterative Satterthwaite method (Schabenberger and Pierce 2002). Between-treatment differences in response variables were tested for significance using Tukey's multiple comparison test ($P = 0.05$).

Results

Over the two experimental years, 389 adult butterflies of 18 species, were observed (Aborfield 2008: 77 individuals, 9 species; 2009: 141 individuals, 13 species; North Sydmonton 2008: 95 individuals, 11 species; 2009: 76 individuals, 10 species) (Table 2). About 90% of the overall abundance was composed of the following species: *Maniola jurtina* L. (meadow brown, 104 individuals), *Aphantopus hyperantus* L. (ringlet, 56), *Pyronia tithonus* L. (gatekeeper, 40), *Pieris brassicae* L. (large white, 40), *Pieris rapae* L. (small white, 32), *Thymelicus sylvestris* Poda (small skipper, 28), *Vanessa cardui* L. (painted lady, 23), *Polyommatus icarus* Rottemburg (common blue, 21)

Table 2 Abundance of butterfly species observed per treatment across both sites in 2008 and 2009

Observed butterfly species	2008				2009			
	T1	T2	T3	T4	T1	T2	T3	T4
Thymelicus lineola Oschsenheimer						2	1	2
Ochlodes sylvanus Esper					5	4	1	8
Thymelicus sylvestris Poda					5	3	2	18
Pieris napi L.	1			1				
Pieris brassicae L.	4		4	6	5	6	5	10
Pieris rapae L.	3	2	8	10	4	2	2	1
Polyommatus icarus Rottemburg	1	1		1	3	2	1	12
Lycaena phlaeas L.							2	
Polygonia c-album L.			1					
Vanessa cardui L.						4		19
Aglais io L.				3		1		
Vanessa atalanta L.								1
Aglais urticae L.								1
Pyronia tithonus L.	3	5	2	13	6	3	2	6
Melanargia galathea L.	1			3	3			2
Maniola jurtina L.	17	13	15	14	14	6	4	21
Aphantopus hyperantus L.	12	11	5	10	8	1	2	7
Pararge aegeria L.		1		1				

T1, Existing buffer strip (control); T2, Existing buffer strip + scarification + wildflower seed; T3, Existing buffer strip + graminicide; T4, Existing buffer strip + scarification + wildflower seed + graminicide

and *Ochlodes sylvanus* Esper (large skipper, 18). All other species comprised less than 10 individuals. Eight of the nine sown wildflower species were observed during the botanical assessments, the exception being *Rumex acetosa* L. (common sorrel). The unsown wildflower species observed comprised both annual and perennial species (Aborfield 2008: 23 annuals, 15 perennials; 2009: 10 annuals, 11 perennials; North Sydmonton 2008: 12 annuals, 15 perennials; 2009: 1 annual, 9 perennials). The number of sown and unsown grasses was 20.

Responses of vegetation to management treatments

Significant effects of sown wildflower cover were observed for the scarification × graminicide, scarification × year, and graminicide × year interactions (Table 3). Thus treatment 4, which received scarification (and wildflowers) and graminicide, produced the highest cover of sown wildflowers in both years, with a significant increase from 2008 to 2009 ($P = 0.05$) (Fig. 1a). In addition, this treatment also produced the highest sown species richness (Fig. 2a) across both years, although no significant year effects were observed (Table 4). Although sown wildflowers were observed in treatment 2 which received scarification (and wildflowers), but no graminicide, both the sown cover and species richness was significantly lower than in treatment 4 ($P = 0.05$). No sown wildflowers were observed in the existing grass buffer strip (treatment 1) in both years, or in the treatment which received only graminicide (treatment 3) in 2008. This is to be expected as these treatments did not receive sown wildflowers.

For the unsown cover, a significant effect was observed for the graminicide treatments (Table 3; Fig. 1b). The unsown species richness model revealed a scarification × year interaction (Table 4), with significantly higher values in the establishment year (2008) when scarification was applied, compared to 2009 ($P = 0.05$) (Fig. 2b). Treatments 1 (control) and 3 (graminicide only), which received no scarification, resulted in significantly lower species richness values in both years (Fig. 2b). Significant scarification × year interactions were observed for both the unsown annual and perennial species richness (Table 4), with lower values in 2009 compared to 2008 ($P = 0.05$) (Fig. 3a, b).

Table 3 Summary of wildflower and grass percentage cover model outputs

Model	Sown wildflower cover ($\log_e n + 1$)	Unsown wildflower cover ($\log_e n + 1$)	Grass cover ($\log_e n + 1$)
1	Scar: $F_{1,20} = 44.53$***	Scar: ns	Scar: ns
	Gram: $F_{1,20} = 12.23$**	Gram: $F_{1,22} = 14.17$**	Gram: $F_{1,21} = 29.29$***
	Year: $F_{1,21} = 23.57$***	Year: ns	Year: $F_{1,22} = 15.12$***
	Scar × Gram: $F_{1,20} = 8.65$**	Scar × Gram: ns	Scar x Gram: ns
	Scar × Year: $F_{1,21} = 17.40$***	Scar × Year: ns	Scar × Year: ns
	Gram × Year: $F_{1,21} = 5.61$*	Gram × Year: ns	Gram × Year: $F_{1,22} = 15.30$***
	Scar × Gram × Year: ns	Scar × Gram × Year: ns	Scar × Gram × Year: ns
	AIC: 82.9	AIC: 133.6	AIC: 131.2

Scar Scarification, *Gram* Graminicide. Non-significant terms removed from models by stepwise deletion

* $P < 0.05$, ** $P < 0.01$, *** $P < 0.001$, ns $P > 0.05$

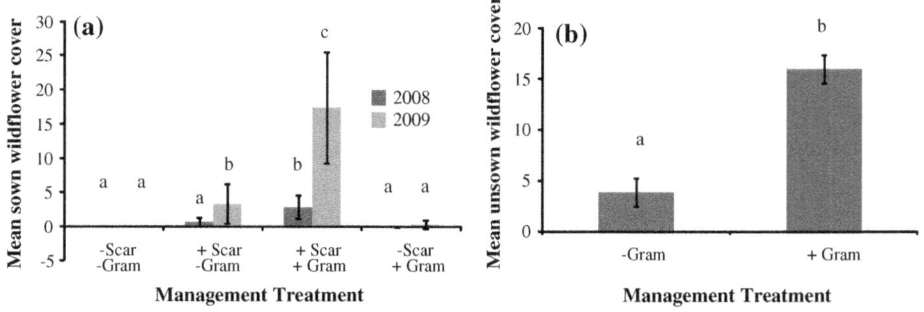

Fig. 1 The response of mean wildflower cover to management treatments in 2008 and 2009 (**a**) Sown wildflower cover (±SE) showing scarification (*Scar*) × graminicide (*Gram*) interaction responses across both sites and years (**b**) Unsown wildflower cover (±SE) showing graminicide responses across both sites and years. Wildflower mixture only sown in scarified plots (+*Scar*). Based on post hoc Tukey's tests. Treatment means which do not differ significantly ($P > 0.05$) share the same letter (*a*, *b* or *c*). Graphs show untransformed data

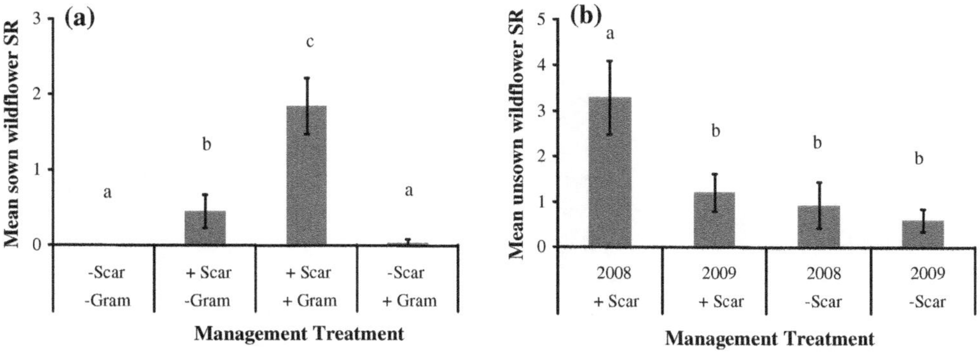

Fig. 2 The response of mean wildflower species richness (SR) to management treatments in 2008 and 2009 (**a**) Sown wildflower species richness (±SE) showing scarification (*Scar*) × graminicide (*Gram*) interaction responses across both sites and years (**b**) Unsown wildflower species richness (±SE) showing scarification (*Scar*) × year interaction responses across both sites. Wildflower mixture only sown in scarified plots (+*Scar*). Based on post hoc Tukey's tests. Treatment means which do not differ significantly ($P > 0.05$) share the same letter (*a* or *b*). Graphs show untransformed data

Table 4 Summary of wildflower species richness model outputs

Model	Sown wildflower species richness ($\log_e$)	Unsown wildflower species richness ($\log_e$)	Unsown annual wildflower species richness ($\log_e$)	Unsown perennial wildflower species richness ($\log_e$)
2	Scar: $F_{1,15} = 82.15$***	Scar: $F_{1,20} = 28.38$**	Scar: $F_{1,20} = 27.93$***	Scar: $F_{1,16} = 10.91$**
	Gram: $F_{1,15} = 25.06$***	Gram: $F_{1,20} = 25.27$***	Gram: $F_{1,20} = 18.89$***	Gram: $F_{1,16} = 13.91$**
	Year: ns	Year: $F_{1,22} = 23.30$***	Year: $F_{1,21} = 41.79$***	Year: $F_{1,22} = 10.56$**
	Scar × Gram: $F_{1,15} = 20.07$***	Scar × Gram: ns	Scar × Gram: ns	Scar × Gram: ns
	Scar × Year: ns	Scar × Year: $F_{1,22} = 10.01$**	Scar × Year: $F_{1,21} = 10.55$**	Scar × Year: $F_{1,22} = 4.59$*
	Gram × Year: ns	Gram × Year: ns	Gram × Year: $F_{1,21} = 11.46$**	Gram × Year: ns
	Scar × Gram × Year: ns	Scar × Gram × Year: ns	Scar × Gram × Year: ns	Scar × Gram × Year: ns
	AIC: −0.9	AIC: 42.8	AIC: 13.2	AIC: 21.5

Scar Scarification, *Gram* Graminicide. Non-significant terms removed from models by stepwise deletion

* $P < 0.05$, ** $P < 0.01$, *** $P < 0.001$, ns $P > 0.05$

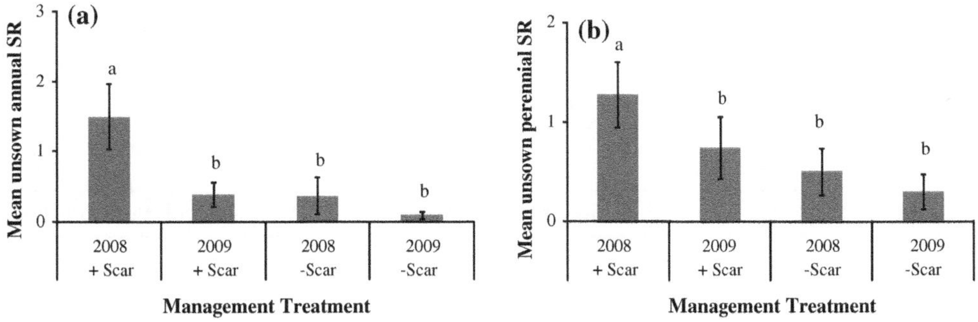

Fig. 3 The response of mean unsown wildflower species richness (SR) to management treatments in 2008 and 2009 (**a**) Unsown annual wildflower species richness (±SE) showing scarification (*Scar*) × year interaction (**b**) Unsown perennial wildflower species richness (±SE) showing scarification (*Scar*) × year interaction. Based on post hoc Tukey's tests. Treatment means which do not differ significantly ($P > 0.05$) share the same letter (*a* or *b*). Graphs show untransformed data

Significant effects of grass cover were observed for the graminicide × year interaction (Table 3). Tukey's test revealed a significantly lower mean cover of grasses in the treatments where graminicide was applied ($P = 0.05$), with 2008 producing significantly lower values than 2009 ($P = 0.05$) (Fig. 4).

Responses of butterflies to management treatments

Significant interactions were observed for the scarification × graminicide treatment on butterfly abundance, species richness and diversity (Table 5). Tukey's test revealed that where scarification was combined with a

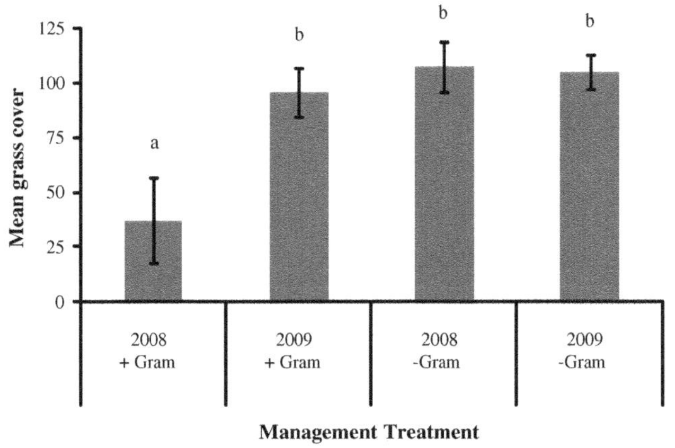

Fig. 4 The response of mean total grass cover to management treatments in 2008 and 2009 (±SE). Graph shows overall treatment responses across both sites. Graminicide (*Gram*) was only applied in 2008. Based on post hoc Tukey's tests. Treatment means which do not differ significantly ($P > 0.05$) share the same letter (*a* or *b*). Graph shows untransformed data

graminicide application (treatment 4), both the abundance and diversity of butterflies was significantly higher ($P = 0.05$), compared to the treatments where graminicide was applied in the absence of scarification (treatment 3), and where scarification was applied in the absence of graminicide (treatment 2) (Fig. 5a, b). Tukey's test also revealed a significantly higher butterfly species richness ($P = 0.05$), compared to the other three treatment combinations (Fig. 5c), implying that a response to species richness is conditional upon the presence of both scarification and graminicide. Significant effects were observed for sown wildflower species richness on butterfly abundance, species richness and diversity (Table 5), highlighting the importance of the presence of scarification, and therefore the sown wildflowers, for butterflies.

Discussion

This study assessed the effectiveness of two management treatments (scarification and graminicide), and their interaction, on butterfly abundance, species richness and diversity in existing grass buffer strips on two arable farms.

For overall treatment effects, the combination of scarification and graminicide produced the greatest establishment in terms of cover, and species richness, of sown wildflowers. Scarification is frequently used in grassland restoration as a method for creating gaps in existing swards, into which wildflower seeds can be sown (Hopkins et al. 1999). While studies have shown a greater emergence of sown species, and increased plant species richness, following scarification (Hofmann and Isselstein 2004; Edwards et al. 2007), Pywell et al. (2007) suggested that the successful establishment of desirable species is difficult without additional treatments to reduce the levels of competitive species such as grasses. In this study, the combined scarification and graminicide treatment produced a greater cover of sown species than was seen for the simply the scarification only treatment, supporting Pywell's conclusions, and suggesting that the action of scarification alone was insufficient to allow the sown species to effectively colonise the existing grass swards. Despite this, there was still a significant increase in sown cover in this treatment between year one and two, although not to the extent of the combined treatment, as potential germination niches disappeared following the recovery of the grasses after scarification (Grubb 1977). These results also suggest that the practice of leaving the cuttings in situ following the autumn cut in year one had not adversely affected the development of the sown species. Further assessments in years three and four may help determine if the sown wildflower species can persist. As expected there was a

Table 5 Summary of butterfly model outputs

Model	Butterfly abundance ($\log_e n + 1$)	Butterfly species diversity ($\log_e n + 1$)	Butterfly species richness ($\log_e$)
3	Scar: $F_{1,15} = 5.72$*	Scar: ns	Scar: $F_{1,15} = 4.81$*
	Gram: ns	Gram: ns	Gram: ns
	Scar × Gram: $F_{1,15} = 30.00$***	Scar × Gram: $F_{1,15} = 20.59$***	Scar × Gram: $F_{1,15} = 28.20$***
	AIC: 76.0	AIC: 83.7	AIC: 56.9
4	Sown wildflower species richness: $F_{1,23.1} = 9.52$**	Sown wildflower species richness: $F_{1,21.9} = 8.04$**	Sown wildflower species richness: $F_{1,21.9} = 9.58$**
	Year: ns	Year: ns	Year: ns
	Wildflower species richness*year: $F_{1,23} = 6.72$*	Wildflower species richness*year: ns	Wildflower species richness*year: ns
	AIC: 80.3	AIC: 91.7	AIC: 66.4

Scar Scarification, *Gram* Graminicide. Non-significant terms removed from models by stepwise deletion

* $P < 0.05$, ** $P < 0.01$, *** $P < 0.001$, ns $P > 0.05$

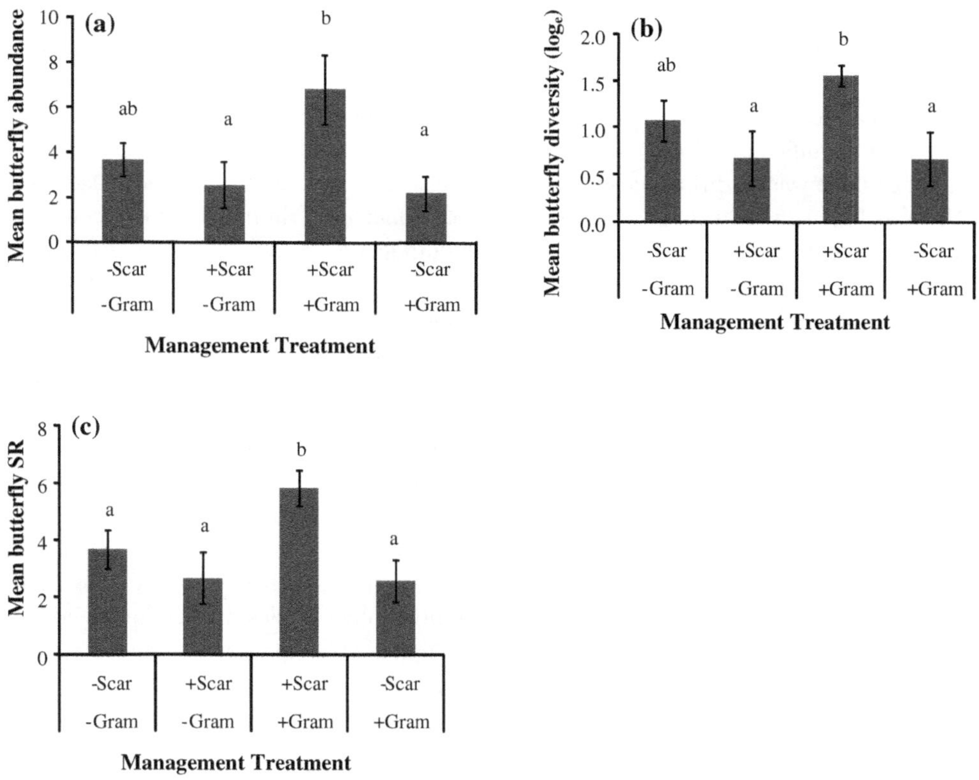

Fig. 5 The response of butterflies to management treatments in 2008 and 2009 (**a**) Mean butterfly abundance (±SE) (**b**) Mean species diversity (±SE) (**c**) Mean species richness (SR) (±SE). *Scar* Scarification, *Gram* Graminicide. Graphs show overall treatment responses across both sites and years. Wildflower mixture only sown in scarified plots (*+Scar*). Based on post hoc Tukey's tests. Treatment means which do not differ significantly ($P > 0.05$) share the same letter (*a* or *b*). Graphs show untransformed data unless stated

lack of desirable sown species in both the control treatment, and the graminicide only treatment, as no wildflowers were sown. The few sown wildflowers observed in the graminicide only treatment in 2009 were likely due to invasion from an adjacent sown plot.

The importance of applying graminicide after scarification was demonstrated through a dramatic and significant increase in sown wildflower cover, and a significant reduction in total grass cover, compared to plots not treated with graminicide. This is consistent with previous work which has shown that graminicide applications can reduce levels of competitive grasses in buffer strips (Marshall and Nowakowski 1994, 1996). Mixed results have been shown for sown and unsown plant species diversity following a single graminicide application in year one. Marshall and Nowakowski (1991) showed beneficial effects on sown communities, whilst Boatman (1992) found that higher species diversity persisted into the second year in graminicide-treated plots. However, Westbury and Dunnett (2008) found that a single initial application was not sufficient to enhance species diversity of sown wildflowers during their two year study period, mainly due to the promotion of unsown competitive species such as *Cirsium* spp. (thistles). It is clear that scarification, and graminicide application, can be successful in promoting and maintaining sown plant species diversity, as our study shows, but may not be suitable on sites with intrinsic weed problems as this may promote undesirable competitive species (Peel and Christian 2009).

Graminicide applications were only conducted during the first year of the study, and by the second year the grass cover had recovered to levels similar to the treatments which had received no graminicide. This further emphasises the competitive dominance of the grasses within the sward, and is supported by Westbury and Dunnett (2008) who observed a greater grass biomass in year two in the graminicide-treated plots compared to the control following graminicide application in year one. However, it is important to reiterate that the aim of this study was to initially suppress the susceptible grasses, thereby reducing competition and creating gaps in the sward for colonisation by the sown and unsown species. The grasses do provide biodiversity benefits, although typically for predatory invertebrate taxa e.g. spiders and ground beetles (Meek et al. 2002; Thomas et al. 2002). Therefore it is important that they are not eliminated completely, and instead allowed to grow back when the sown species have successfully established themselves.

In terms of unsown species, wildflower cover was not influenced by the scarification, but was by the graminicide application, with higher values in graminicide-treated plots. As with the sown species, the graminicide was crucial in suppressing the competitive dominance of the susceptible grass species, thereby shifting the competitive advantage in favour of the wildflowers, and promoting seedling establishment (Hopkins et al. 1999; Hofmann and Isselstein 2004). The unsown species richness however was influenced by the scarification. Treatments which were not scarified produced significantly lower unsown species richness, although there was no difference between year one and two. This was likely due to the lack of soil surface disturbance in these treatments, which meant that the specific requirements to trigger seed dormancy and germination were not met (Edwards and Crawley 1999). Furthermore, there were higher values of both annual and perennial species in scarified plots in year one compared to year two. Annual species require regular soil disturbance for successful germination (Grubb 1977), and the absence of further scarification after year one, likely selected against these species. As buffer strips are perennial habitats, and the aim of this study is to promote sown perennial wildflower species within the sward, the maintenance of annual unsown wildflowers may be better suited to other management prescriptions, e.g. uncropped, cultivated margins (Defra 2008). The reduction in unsown perennial species may be due to their inability to compete with the grasses (Grubb 1977).

The butterflies responded to scarification and graminicide with the highest abundance, species richness and diversity found within this treatment combination. In addition, a positive correlation was observed between the wildflower species richness and the butterfly responses. Perhaps unexpectedly, a high abundance of butterflies was also observed in the grass dominated control plots, indicating the importance of grass buffer strips for certain butterfly species (Asher et al. 2001). Many grasses found within existing buffer strips are important larval host plants including the meadow grasses such as *Poa pratensis* L. (smooth meadow grass) and the fescues such as *F. rubra* (Dennis 1992). The three most abundant species within the control plots were the Satyridae species *M. jurtina* (31 individuals), *A. hyperantus* (20 individuals), and *P. tithonus* (9 individuals), all of which utilise these grass species as larval host plants (Dennis 1992; Asher et al. 2001). However, the butterfly diversity and species richness was significantly lower suggesting that species that do not utilise grasses will preferentially avoid these areas when wildflowers are present elsewhere. For example, members of the family Nymphalidae, such as *V. cardui* and *Aglais io* L. (peacock), were absent as their larvae favour broad-leaved species including *Urtica* spp. (nettles), *Cirsium* spp., and *Plantago lanceolata* L. (ribwort plantain) (Dennis 1992). The lower species richness and diversity may also be due to the limited nectar foraging resources available for adult butterflies, as existing buffer strips tend to lack a sown wildflower component (Critchley et al. 2006).

Butterfly species richness and diversity was significantly higher in the treatment where scarification and graminicide were both present. Five Nymphalidae species were observed in this treatment, whereas in the other three treatments a maximum of only two species were observed. Other species that utilise wildflowers as larval hosts were also present, including *P. icarus* (Dennis 1992). Interestingly, five Satyridae species (77 individuals) were present in this treatment compared to only four (64 individuals) in the control, suggesting that the action of the scarification and graminicide did not have a negative impact on these species, even though the combination of both treatments initially reduced the level of grasses. The higher number of individuals in the combined scarification and graminicide treatment supports a study by Meek et al. (2002) who showed that *M. jurtina* and *A. hyperantus* abundance was higher when both wildflowers and grasses were present. In addition, other studies found that butterfly species richness was highest in sown wildflower strips (Feber et al. 1996; Haaland and Gyllin 2010). It is likely that the higher species richness in the combined treatment was also due to the presence of nectar foraging plants for the adult butterflies, as confirmed by the positive correlation between butterfly species richness and wildflower cover. Wildflowers such as *Leucanthemum vulgare* Lamarck and *Centaurea nigra* L. are known to be important nectar sources for the Satyridae and other butterflies (Porter et al. 1992) and both were present, as part of the wildflower seed mixture in the scarified plots, at the two sites and across both years. As the wildflower mixture was sown in March during year one, it is unlikely that the sown perennial species would have flowered during the first year, and so the mixture would have contributed little, if any, nectar resources in that year. Previous studies have highlighted the importance of providing nectar resources for butterflies. For example, Feber et al. (1996) showed that a lack of nectar resources can lead to detrimental effects on adult butterfly fecundity, whilst Dover (1996) found that sub-optimal adult resources are likely to be a more important limiting factor for butterfly population size than a lack of host plants. Thus, the provision of nectar resources for adults is crucial to maintain and increase current populations of butterflies within the agricultural landscape (Pywell et al. 2004).

Conclusions

Despite the introduction of agri-environment schemes, farmland biodiversity continues to decline. There is a need to find widespread and commonly applied tools to meet the revised EU target of halting and reversing biodiversity loss in Europe by 2020. This study has highlighted how two such potential tools, scarification with wildflowers, and graminicide, could help achieve this. While cost and the reluctance to use more agrochemicals might be a barrier to adoption of these tools, the low application rate used in this study (93.75 g a.i. ha^{-1}) represents a relatively small cost for a large biodiversity benefit. It is also considerably less than the quantity of graminicide applied in-field where the maximum recommended rate is 375 g a.i. ha^{-1}. Further long-term studies, at a national level, together with a thorough cost benefit analysis, may provide an opportunity to incorporate these management tools into existing ELS and HLS options. In addition, widespread education and engagement of farmers to support uptake of these management tools is imperative. If this is achieved then this could deliver widespread benefits to butterflies and other important components of biodiversity.

Acknowledgments This project was funded by the Biotechnology and Biological Sciences Research Council and Syngenta. We would like to thank the farmers Mark Robins (Farley Estates, Reading) and Peter Clarke (Adbury Park Farm, Newbury) for allowing access to their land for data collection.

References

Asher J, Warren M, Fox R, Harding P, Jeffcoate G, Jeffcoate S (2001) The millenium atlas of butterflies in Britain and Ireland. Oxford University Press, Oxford

Boatman ND (1992) Improvement of field margin habitat by selective control of annual weeds. Asp Appl Biol 29:431–436

Butterfly Conservation, CEH, DEFRA, Joint Nature Conservation Committee (2009) Populations of butterflies on farmland in England. http://www.defra.gov.uk/environment/biodiversity/documents/indicator/200904a1b.pdf. Accessed 26 May 2010

Cheffings CM, Farrell L (2005) Species status report no. 7: the vascular plant red data list for Great Britain. Joint Nature Conservation Committee, Peterborough

Corbet S (1987) More bees make better crops. New Sci 115:40–43

Critchley CNR, Fowbert JA, Sherwood AJ, Pywell RF (2006) Vegetation development of sown grass margins in arable fields under a countrywide agri-environment scheme. Biol Conserv 132: 1–11

Defra (2008) Entry level stewardship handbook, 2nd edn. Natural England, UK

Dennis RLH (1992) Appendix 1. A check list of British butterflies and their hostplants. In: Dennis RLH (ed) The ecology of butterflies in Britain. Oxford University Press, Oxford

Dover JW (1996) Factors affecting the distribution of satyrid butterflies on arable farmland. J Appl Ecol 33:723–734

Edwards GR, Crawley MJ (1999) Herbivores, seed banks and seedling recruitment in mesic grassland. J Ecol 87:423–435

Edwards M, Jenner M (2005) Field guide to the bumblebees of Great Britain and Ireland. Ocelli, Eastbourne

Edwards AR, Mortimer SR, Lawson CS, Westbury DB, Harris SJ, Woodcock BA, Brown VK (2007) Hay strewing, brush harvesting of seed and soil disturbance as tools for the enhancement of botanical diversity in grasslands. Biol Conserv 134:372–382

Feber RE, Smith H, Macdonald DW (1996) The effects on butterfly abundance of the management of uncropped edges of arable fields. J Appl Ecol 33:1191–1205

Field RG, Gardiner T, Mason CF, Hill J (2006) Countryside stewardship scheme and butterflies: a study of plant and butterfly species richness. Biodivers Conserv 15:443–452

Fox R, Warren MS, Asher J, Brereton TM, Roy DB (2007) The state of Britain's butterflies 2007. Butterfly Conservation and the Centre for Ecology and Hydrology, Wareham

Goulson D, Lye GC, Darvill B (2008) Decline and conservation of bumble bees. Annu Rev Ent 53:191–208

Grubb PJ (1977) Maintenance of species-richness in plant communities—importance of regeneration niche. Biol Rev Camb Phil Soc 52:107–145

Haaland C, Gyllin M (2010) Butterflies and bumblebees in greenways and sown wildflower strips in southern Sweden. J Insect Cons 14:125–132

Hammond PC, Miller JC (1998) Comparison of the biodiversity of Lepidoptera within three forested ecosystems. Ann Ent Soc Am 91:323–328

Herbert D, Walker KA, Price LJ, Cole DJ, Pallett KE, Ridley SM, Harwood JL (1997) Acetyl-CoA carboxylase—a graminicide target site. Pestic Sci 50:67–71

Hofmann M, Isselstein J (2004) Seedling recruitment on agriculturally improved mesic grassland: the influence of disturbance and management schemes. Appl Veg Sci 7:193–200

Hopkins A, Pywell RF, Peel S, Johnson RH, Bowling PJ (1999) Enhancement of botanical diversity of permanent grassland and impact on hay production in environmentally sensitive areas in the UK. Grass For Sci 54:163–173

Losey JE, Vaughan M (2006) The economic value of ecological services provided by insects. Bioscience 56:311–323

Marshall EJP (2003) Methods to diversify field margin plant communities, BD1607. Marshall Agroecology Ltd., UK. http://www.agroecol.co.uk/marshallagro1BD1607.htm. Accessed 25 January 2009

Marshall EJR, Moonen AC (2002) Field margins in northern Europe: their functions and interactions with agriculture. Agric Ecosyst Environ 89:5–21

Marshall EJP, Nowakowski M (1991) The use of herbicides in the creation of a herb-rich field margin. Brighton Crop Protection Conference—Weeds

Marshall EJP, Nowakowski M (1994) The effects of Fluazifop-P-Butyl and cutting treatments on the establishment of sown field margin strips. Field margins: integrating agriculture and conservation, British Crop Protection Council, Surrey, UK, Brighton 58, pp 307–312

Marshall EJP, Nowakowski M (1996) Interactions between cutting and graminicide on a newly-sown grass and wild flower field margin strip. Asp Appl Biol 44:307–312

Marshall EJP, Brown VK, Boatman ND, Lutman PJW, Squire GR, Ward LK (2003) The role of weeds in supporting biological diversity within crop fields. Weed Res 43:77–89

Meek B, Loxton D, Sparks T, Pywell R, Pickett H, Nowakowski M (2002) The effect of arable field margin composition on invertebrate biodiversity. Biol Conserv 106:259–271

MVSP (1999) Multivariate statistical package version 3.1. Kovach Computing Services, Anglesey

Natural England (2009) Agri-environment schemes in England 2009: a review of results and effectiveness

O'Connor RJ, Shrubb M (1986) Farming and birds. Cambridge University Press, Cambridge

Ovenden GN, Swash ARH, Smallshire D (1998) Agri-environment schemes and their contribution to the conservation of biodiversity in England. J Appl Ecol 35:955–960

Peel S, Christian M (2009) Sward enhancement: selection of suitable sites. Natural England Technical Information Note TIN061

Pollard E, Yates TJ (1993) Monitoring butterflies for ecology and conservation. Chapman and Hall, London

Pollard E, Elias DO, Skelton MJ, Thomas JA (1975) A method of assessing the abundance of butterflies in Monks Wood National Nature Reserve in 1973. Ent Gaz 26:77–88

Porter K, Steel CA, Thomas JA (1992) Butterflies and communities. In: Dennis RLH (ed) The ecology of butterflies in Britain. Oxford University Press, Oxford

Potts GR (1986) The partridge: pesticides, predation and conservation. Collins, London

Potts SG, Westbury DB, Woodcock BA, Ramsay AJ, Harris SJ, Springate S, Pywell R, Meek B, Carvell C, Hulmes L, Warman L, Sparks T, Cook SK, Henderson IG (2007) Experiment 2—management of the non-cropped margin structure to maximise biodiversity

Pywell RF, Warman EA, Sparks TH, Greatorex-Davies JN, Walker KJ, Meek WR, Carvell C, Petit S, Firbank LG (2004) Assessing habitat quality for butterflies on intensively managed arable farmland. Biol Conserv 118:313–325

Pywell RF, Warman EA, Carvell C, Sparks TH, Dicks LV, Bennett D, Wright A, Critchley CNR, Sherwood A (2005) Providing foraging resources for bumblebees in intensively farmed landscapes. Biol Conserv 121:479–494

Pywell RF, Bullock JM, Tallowin JB, Walker KJ, Warman EA, Masters G (2007) Enhancing diversity of species-poor grasslands: an experimental assessment of multiple constraints. J Appl Ecol 44:81–94

Robinson RA, Sutherland WJ (2002) Post-war changes in arable farming and biodiversity in Great Britain. J Appl Ecol 39:157–176

SAS (2008) Statistical analysis system: version 9.2. SAS Institute, Cary

Schabenberger O, Pierce FJ (eds) (2002) Contemporary statistical models for the plant and soil sciences. CRC Press, London

Stace C (1997) New flora of the British Isles. Cambridge University Press, Cambridge

Syngenta Crop Protection Ltd (2003) Fusilade Max Product Label. http://www.syngenta-crop.co.uk/pdfs/products/FusiladeMax_uk_product_label.pdf. Accessed 13 February 2009

Thomas SR, Noordhuis R, Holland JM, Goulson D (2002) Botanical diversity of beetle banks effects of age and comparison with conventional arable field margins in southern UK. Agric Ecosyst Environ 93:403–412

van Swaay C, Warren M, Lois G (2006) Biotope use and trends of European butterflies. J Insect Cons 10:189–209

Westbury DB, Dunnett NP (2008) The promotion of grassland forb abundance: a chemical or biological solution? Basic Appl Ecol 9:653–662

Westphal C, Bommarco RGC, Lamborn E, Morison N, Petanidou T, Potts SG, Roberts SPM, Szentgyörgyi H, Tscheulin T, Vaissière BE, Woyciechowski M, Biesmeijer JC, Kunin WE, Settele J, Steffan-Dewenter I (2008) Measuring bee diversity in different European habitats and biogeographical regions. Ecol Monogr 78:653–671

Williams PH (1982) The distribution and decline of British bumblebees. J Apic Res 21:236–245

Wilson PJ (1992) Britain's Arable weeds. Br Wildl 3:149–161

Wilson JD, Morris AJ, Arroyo BE, Clark SC, Bradbury RB (1999) A review of the abundance and diversity of invertebrate and plant foods of granivorous birds in northern Europe in relation to agricultural change. Agric Ecosyst Environ 75:13–30

Woodcock BA, Westbury DB, Tscheulin T, Harrison-Cripps J, Harris SJ, Ramsey AJ, Brown VK, Potts SG (2008) Effects of seed mixture and management on beetle assemblages of arable field margins. Agric Ecosyst Environ 125:246–254

J Insect Conserv (2011) 15:301–309
DOI 10.1007/s10841-010-9353-8

ORIGINAL PAPER

What can sown wildflower strips contribute to butterfly conservation?: an example from a Swiss lowland agricultural landscape

Christine Haaland · Louis-Félix Bersier

Received: 7 April 2010 / Accepted: 8 September 2010 / Published online: 23 September 2010

Abstract The objective of this study was to compare butterfly abundances and diversity between wildflower strips and extensively used meadows to identify which butterfly species can be supported by establishing wildflower strips. Butterflies were recorded along transects during one season in twenty-five sown wildflower strips and eleven extensively used meadows in a Swiss lowland agricultural landscape (600 ha). In total 1,669 butterflies of 25 species were observed (25 in the strips, 18 in meadows). This can be related to 38 species recorded in the region (lowland part of Kanton Fribourg) within the Swiss Biodiversity Monitoring Programme. In wildflower strips the number of butterflies per transect meter was significantly higher than in the meadows, but there was no significant difference in species richness. Butterfly communities, though, were quite different between the two habitat types. Habitat type, abundances of flowering plants and presence of forest within 50 m were identified as factors influencing butterfly species richness. Butterfly abundances were affected by habitat type and abundance of flowering plants. In wildflower strips, 65% of all flower visits by butterflies were observed on *Origanum*. It can be concluded that sown wildflower strips can support a substantial part of a regions species pool. This is mostly true for common species, but can apply to rare species when, for example, larval food plant requirements are met.

C. Haaland (✉)
Department of Landscape Management, Design and Construction, Swedish University of Agricultural Sciences, Box 66, 230 53 Alnarp, Sweden
e-mail: christine.haaland@ltj.slu.se

L.-F. Bersier
Department of Biology, Unit Ecology and Evolution, University of Fribourg, Chemin du Musée 10, 1700 Fribourg, Switzerland

Keywords Agri-environmental schemes · Flower visits · Grassland management · Larval food plant · Larval habitat · Regional species pool · Switzerland

Introduction

The intensification of agriculture has caused severe species losses and declines due to habitat loss, habitat fragmentation and habitat degradation. The decline of butterflies in agricultural landscapes has been particularly well documented (e.g. Van Dyck et al. 2009; Van Swaay et al. 2006, 2009). To counteract the negative effects of intensive agricultural use and to maintain extensive management forms, agri-environmental schemes have been introduced within the EU and other European countries several decades ago. Sown wildflower strips are an agri-environmental measure intended to benefit in particular insects, not least because of their importance as pollinators, in pest control and their crucial role in food webs (Marshall et al. 2006; Pfiffner and Wyss 2004; Pywell et al. 2006; Woodcock et al. 2005). The diminishing of nectar resources, for example, has been acknowledged as a cause for decline of bumblebees (Carvell et al. 2006). Wildflower strips have been, or are planned to be, established in several European countries (e.g. Carvell et al. 2007; Haaland et al. 2010; Pywell et al. 2007). In Switzerland they were introduced as agri-environmental scheme in 1993. Schemes for wildflower strips vary between regions and countries regarding their seed mixture, length of funding and management. Seed mixtures can comprise few or up to 30 plant species, some include grasses, some only flowering herbs. The objectives can range from providing nectar resources for certain pollinators to enhancing overall biodiversity in intensively agricultural areas by providing larval, adult and overwintering habitats for a wider range of species (Nentwig 2000).

The benefit of wildflower strips on insect diversity has been studied especially in countries where wildflower strips have been established for a longer period of time, as in the UK and Switzerland (e.g. Aviron et al. 2007a; Carvell et al. 2007; Pfiffner and Wyss 2004; Pywell et al. 2005; Woodcock et al. 2008). The results of these studies have so far shown that sown wildflower strips have often higher insect diversity or abundances than other types of subsidised field margins, as natural regeneration or grass margins (e.g. Carvell et al. 2004; Meek et al. 2002; Pywell et al. 2005). Factors identified as influencing insect diversity or abundances are, for example, abundance of flowering plants (Carvell et al. 2004; Feber et al. 1996; Heard et al. 2007; Kohler et al. 2008), management (e.g. Feber et al. 1996; Woodcock et al. 2008), age of the strips (e.g. Carvell et al. 2007; Frank and Künzle 2006) and landscape factors such as patch size, connectivity, and landscape context (e.g. Aviron et al. 2007a, 2010; Heard et al. 2007). Butterflies have been shown to occur in higher abundances in strips containing wildflowers than other strips with grasses only, or with natural regeneration (Feber et al. 1996; Haaland and Gyllin 2010; Meek et al. 2002; Pywell et al. 2007). Aviron et al. (2007a) found in a study area in Switzerland higher diversity and abundance of butterflies in wildflower strips than in conventional grasslands or wheat fields. Jacot et al. (2007) observed in another Swiss study butterfly species and individuals numbers in wildflower strips that are comparable to biodiversity hotspots. The authors also noted that wildflower strips with grasses had higher butterfly species richness and abundances than wildflower strips without grasses.

The objective of this study was firstly to investigate butterfly species richness and abundances in Swiss sown wildflower strips in comparison to extensively used meadows, a habitat that was assumed to be one of the species richest habitats (regarding butterflies) in lowland agricultural landscapes. It was thought that the meadows could act as a source for individuals in wildflower strips, which are regularly ploughed up after a maximum of 7 years and newly established at the same place or somewhere else. Thus, a hypothesis was that wildflower strips would contain a subset of butterfly species presented in extensively used meadows. Number of flowering plant species, flower abundance, habitat area and the presence/absence of forest in 50 m buffers were analysed as possible factors influencing species richness and abundances. Secondly, the observed species richness was to be compared with the regional species pool to evaluate the proportion of butterfly species found in the wildflower strips in the study area. Finally, it was aimed to identify factors influencing butterfly diversity and abundances in the strips and in the meadows.

Methods

Study area

The study area was situated in the Western part of Switzerland in the municipality of Bösingen, Kanton Fribourg (Fig. 1). It comprised a size of about 600 ha and the highest point lies 650 m above sea level. The area belongs to the *Schweizer Mittelland* which covers the Western lowlands of Switzerland between the Jura and the Alps. The landscape was characterized by rather small-scale agriculture with small fields, some woodlands and spread settlements. The agricultural land was dominated by arable with cereal, corn and potato production, but there was also considerable amount of grassland (ca 20%, estimated from orthophotographs), most of it was intensively used. Areas with agri-environmental schemes (grasslands, orchards, sown wildflower strips, grass strips along woodlands and watercourses and hedges) covered about 15% of the study area (administrative map on agri-environmental payments of the municipality Bösingen). Areas with this type of agri-environmental scheme are called ecological compensation areas in Switzerland. All together, the study area represented a rather less intensively used part of the Mittelland with a large amount of ecological compensation area.

The sown wildflower strips were created using with a standard seed mixture often applied in Switzerland containing 24 different plant species (species list in Appendix A). None of the wildflower strips showed signs of mowing during the study period. The age of the strips varied between one and 7 years. Within the study area all sown wildflower strips were investigated. Among meadows, a selection was made. Only meadows that had the least intensive management, according to the agricultural schemes, were selected. These meadows were cut at least twice a year, but due to regulation not before the middle of June. During the study 25 sown wildflower strips and 11 meadows were investigated. The mean size of the wildflower strips was 0.41 ha (min 0.15, max 1.16) and of the meadows 0.76 ha (min 0.21, max 1.64); the difference in means is significantly different (t-test: t-value: -3.08, df 34, $P = 0.004$). Of the 11 meadows, 9 had a polygon shape and 2 were strips. The wildflower strips had in 15 cases a strip shape and in 10 cases a polygon shape.

Butterfly recording

Butterflies were recorded along transects five times between May and September 2008, once each month. Transects were placed in the middle of the strips or patches and were walked along the whole strip or patch. Butterflies were recorded three meters each side of the transect line. The width of total six meters for butterflies was chosen

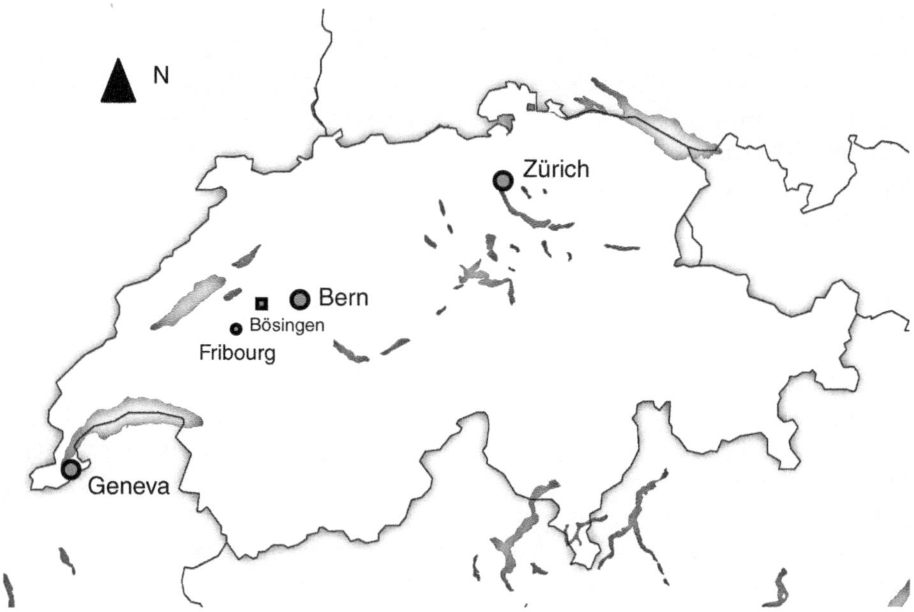

Fig. 1 Map with location of the study area in the municipality of Bösingen, Kanton Fribourg, Switzerland

since it corresponded to the width of several wildflower strips. Usually butterflies were identified to species level, but *Colias hyale* and *Colias alfacariensis* were not distinguished and also not the two *Thymelicus* species *T. lineola* and *T. sylvestris*. Recordings were carried out between 10 am and 5 pm and only in good weather conditions (Pollard and Yates 1993). When butterflies were observed using flowers, the visited species were recorded. To account for the differences in transect lengths, we expressed species richness and abundance as density per sampled meter (note that the relationship between transect length and species richness was best described by a linear model).

Regional butterfly species pool

The butterfly species lists obtained from the recordings for both habitat types were compared with monitoring results for the lowland part of Kanton Fribourg of the Swiss Biodiversity Monitoring Programme for butterflies. The comparison was carried out to obtain information on how many butterfly species could be potentially expected in the study area and for which butterfly species larval food plants were available or not in the two investigated habitats in the study area. Further information on the methodology on recording within the monitoring programme can be gained from the Koordinationsstelle Biodiversitäts-Monitoring Schweiz (2008). Larval food plants for butterflies were obtained from Schweizerischer Bund für Naturschutz (1994).

Plant recording including larval food plants

In each transect (6 m width) all flowering plants were recorded during the five sampling sessions and classified into five groups according to their flower abundance. Classes ranged from 1 to 5, whereby

1 = very rare (one flowering individual)
2 = rare (few flowering individuals)
3 = spread (flowering individuals frequently spread)
4 = abundant (many flowering individuals, but not dominating)
5 = very abundant (very many flowering individuals, dominating).

Not all plants could be identified to species level; in several cases plants were identified only to genus. The total number of flowering species per site and a flower abundance index was calculated. The flower index is the sum of all flower abundance classes per investigated habitat patch (wildflower strip/meadow). At the first visit, the abundance of *Rumex* spp., *Urtica dioica*, *Viola* spp. and grasses (*Poaceae*) were also recorded independently of their flowering status. These taxa are larval food plants and might otherwise be missed while recording flowering plants.

Spatial variables

The investigated wildflower strips and meadows were digitized in the geographical information system (GIS) ArcGIS (Esri 2006) using an ortophotograph from 2006. The spatial variables were transect length, area of wildflower strips and of meadows, and presence/absence of forest in a 50 m buffer around the transect.

Analysis

Statistical analysis were carried out in R (R Development Core Team 2009) and Statistica (StatSoft 1997).

A stepwise multiple linear regression was applied to analyse factors influencing butterfly abundance and species richness in sown wildflower strips and meadows. Additionally, a correspondence analysis (CA) was carried out with the R package "vegan" (Oksanen et al. 2010) to analyse species composition of the study sites and factors influencing species composition.

Results

Butterfly abundance and diversity in wildflower strips and meadows

In total the transect length was 6.7 km of which 4.9 km (73%) were in wildflower strips and 1.8 km (27%) in meadows. The mean transect length in wildflower strips was 200 m (min 70 m, max 450 m) and did not differ significantly from that in meadows (mean 165 m, min 85 m, max 310 m; *t*-test : *t*-value: 0.94, *df* 34, $P = 0.355$). Altogether 1,669 butterflies of 25 species were observed (Appendix B). By far the most butterflies (88%, $n = 1{,}464$) were recorded in wildflower strips. The number of butterflies per transect meter was significantly higher in wildflower strips than in the meadows (*t*-test: *t*-value 3.47, *df* 34, $P = 0.0014$, Fig. 2a). All 25 species were recorded in the wildflower strips and 18 species in the meadows. The difference in species richness per transect meters was marginally significant (*t*-test: *t*-value: 1.97, *df* 34, $P = 0.057$, Fig. 2b). Figure 3 shows the distribution of species within both types of habitats. The first ordination axis explains 16.3% of the variation in butterfly abundances within the 36 sites, the second 13.1%. It is apparent that the wildflower strip and meadow sites naturally segregate according to species composition. Species that were more often observed in the meadows than expected (taking into account the differences in transect length in the two studied habitats) were the mazarine blue (*Cyaniris semiargus*), common blue *(Polyommatus icarus*), pale clouded yellow (*Colias hyale/alfacariensis*) marbled white (*Melanargia galathea*) and the small heath (*Coenonympha pamphilus*). A species that exclusively occurred in the wildflower strips was the mallow skipper (*Carcharodus alceae*). The mallow skipper was listed as "critically endangered" in the Swiss red data book (Gonseth 1994), but is now one of the most abundant butterfly species in wildflower strips in the study area. This species was absent in the investigated meadows, where its food plants (*Alcea rosea, Malva alcea, M. moschata, M. neglecta, M. sylvestris*, Ebert 1993) rarely occurred.

Flowering plant species

Eighty-five species and 11 genera of flowering plants were identified. The mean number of flowering plant taxa was 27.7 (±7.9 s.d.) on transects in wildflower strips, a figure significantly larger than the 17.9 (±5.5 s.d.; *t*-test: *t*-value: 3.72, *df* 34, $P < 0.001$) in meadows. Between 6 and 21 of the 24 sown species were recorded in flower in the wildflower strips. The species sown in the wildflower strips did not usually occur, or occurred in much lower frequencies, in the meadows, the only exception was *Leucanthemum vulgare*. In the meadows, several species were much more abundant than in the wildflower strips, these were *Trifolium* spp. (mostly *T. pratense* and *T. repens*, but also *T. campestre*), *Ranunculus* spp., *Plantago lanceolata*, *Knautia arvensis* and *Leontodon autumnalis*. Even though grasses were not sown in the wildflower strips, grasses were very abundant in most strips due to succession.

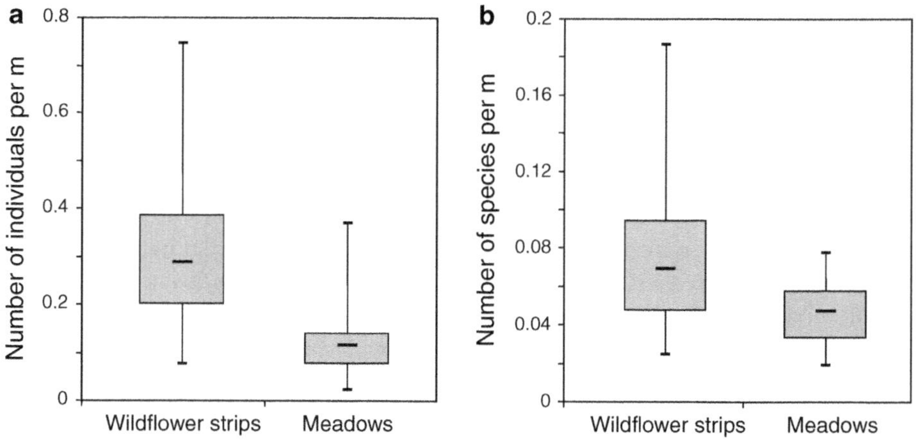

Fig. 2 **a** *Box* plot (minimum, 25th percentile, median, 75th percentile, maximum) of number of butterflies per meter in wildflower strips and meadows (*t*-test: *t*-value: 3.47, *df* 34, $P = 0.0014$). Total number of butterflies recorded in wildflower strips was 1,464 and 205 in meadows, **b** *Box* plot of butterfly species richness per meter in wildflower strips and meadows (*t*-test: *t*-value 1.97, *df* 34, $P = 0.057$). Total number of butterfly species recorded in wildflower strips was 25 and 18 in meadows

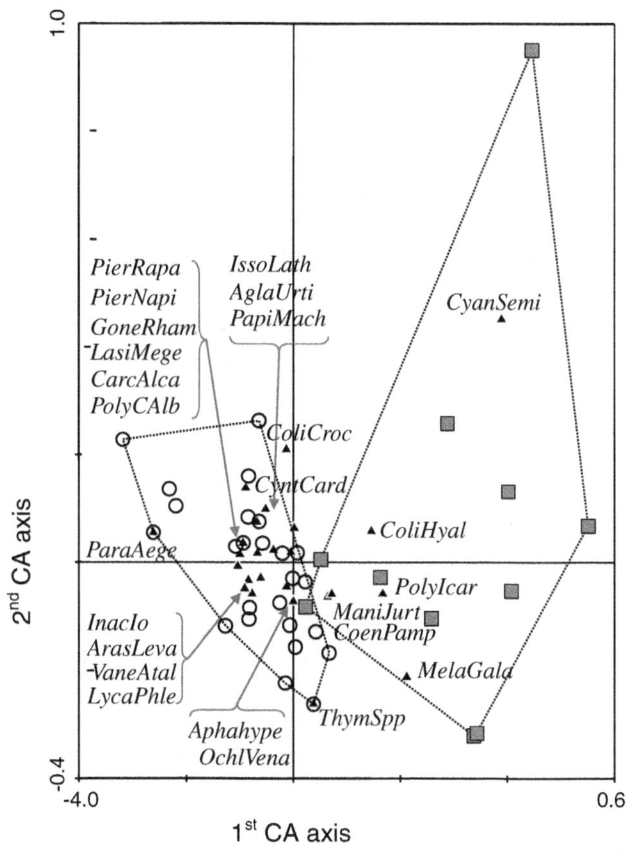

Fig. 3 Results of correspondence analysis. *Black triangles*: butterfly species (four-first letters of genus and of species; see Appendix B for complete names); *circles*: wildflower strips, *grey squares*: extensively used meadows; *dotted lines* enclose the corresponding sites

Factors influencing butterfly species richness and abundance

The results of the multiple linear regressions (stepwise forward) showed that the explanatory variables can explain between 75 and 66% of the variation in butterfly abundance and species richness respectively (Table 1). Significant factors related with butterfly abundances were habitat type (meadows having less species than wildflower strips) and flowering plant species richness. There was also a positive effect of transect length on abundances. Butterfly species richness was influenced significantly by the factor habitat type (same trend as above) and flower abundance (expressed as flower index). Additionally, the presence of forest in the 50 m buffer positively affected butterfly species richness. There is a linear relationship between the factors flowering plant species and flower index ($R^2 = 0.78$).

Flower visits

In wildflower strips, 42% ($n = 609$) of all butterfly recordings concerned individuals perched on a flower, whereas in meadows only 19% ($n = 39$) of all butterfly observations were flower visits. In wildflower strips 65% of all flower visits occurred on *Origanum vulgare*, 11% on *Centaurea jacea* and 9% on *Epilobium ciliatum* (Fig. 4).

Comparison with regional species pool and availability of larval food plants

The Swiss Biodiversity Monitoring Programme of the lowland part of Kanton Fribourg, which includes the study region, comprised 4101 butterfly observations of 38 species between 2003 and 2007. Of these 38 species two-third ($n = 25$) were recorded in the wild flower strips of the study area. Amongst the species not observed in the study area in Bösingen were the five rarest species (with only 1 individual observed during the monitoring: *Apatura ilia*, Red list (RL) category according to Gonseth (1994) 2, endangered; *Cupido minimus*, RL 3, vulnerable; *Leptidea*

Table 1 Results of the multiple linear regression (stepwise forward); explanatory variables entered in the analysis were: habitat_type (M = meadow), transect length, area, number of flowering species (flower_species), flower_index (sum of all abundance classes of all flowering species over the whole season), presence/absence of forest within a 50 m buffer (forest_50 m)

Dependent variable	Independent variable	Beta	Multiple *R*-square	*F*—to entr	*P*-level
Butterfly abundance (log)	Flower_species	0.461	0.60	50.5	<0.001***
	Transect length	0.275	0.65	5.3	0.027*
	Habitat_type (M)	−0.387	0.71	5.8	0.022*
	forest_50 m	0.152	0.74	3.8	0.060
	Area	0.140	0.87	0.75	0.222
$R^2 = 0.75$, Adjusted $R^2 = 0.71$; $F(5,30) = 18.17$, $P < 0.001$, Std. Error of estimate: 0.198					
Butterfly species richness (log)	Flower_index	0.435	0.47	29.8	<0.001***
	Habitat type	−0.495	0.55	6.1	0.018*
	Forest_50m	0.237	0.63	6.7	0.014*
	Area	0.211	0.66	2.9	0.097
$R^2 = 0.66$, Adjusted $R^2 = 0.62$; $F(4,31) = 15.08$ $P < 0.001$ Std. Error of estimate: 0.092					

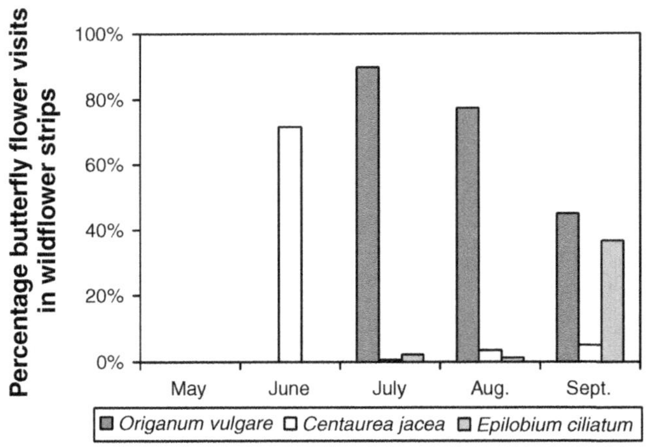

Fig. 4 Change of flower visits in wildflower strips during the season with regard to the three most visited flowering plant species (number of butterfly visits = 609)

sinapis, *Melitaea athalia* RL 3; *Nymphalis antiopa,* RL 3), further two species with woody plant species as larval food plant (*Celastrina argiolus*, *Limenitis camilla*), three fritillary species (*Argynnis paphia*; *Brenthis daphne,* RL 2; *Clossiana dia,* RL 2) and *Lycaena tityrus*. It can be noted that in particular fritillaries were not well represented within the wildflower strips and in the meadows.

A comparison of presence and absence of butterflies in the study area (in relation to the regional species pool) with the larval food plant availability in the studied sites (based only on absence/presence data for all transects combined) is shown in Table 2. For 67% of the 33 species recorded within the monitoring programme (excluding species whose larvae feed on woody plants) the adult butterfly and the larval food plant were observed along transects in wildflower strips of the study area. For 21% the larval food plant was present in the wildflower strips, but the butterfly was not observed. These were species, which either are regionally rare, have different habitat requirements than open agriculture landscapes and/or a larval food plant that occurred in very low frequencies in the wildflower strips. Both the larval food plant and the butterfly were not found for 6% of the 33 species (mostly species where larvae feed on woody species) and in a further 6% the butterfly species was observed, but not the larval food plant.

Table 2 Presence and absence of larval food plant and butterfly species in the two studied habitats in the study area as percentage of total number of species recorded in the lowland part of Kanton of Fribourg (Swiss Biodiversity Monitoring Programme, 2003–2007)

	Percentage species of total number of species recorded in Kanton Fribourg (total number of species = 33[a])	
	Sown wildflower strips (*n* = 25) (%)	Extensively managed meadows (*n* = 11) (%)
LFP present B present	67	45
LFP present B absent	21	30
LFP absent B present	6	9
LFP absent B absent	6	15

LFP larval food plant; *B* adult butterfly. Butterfly species with woody species as larval food plant (the 5 species *Apatura ilia*, *Celastrina argiolus*, *Gonepteryx rhamni*, *Limenitis camilla*, *Nymphalis antiopa*) have been excluded from the analysis since young trees and trees seedlings were not recorded as potential larval food plant

[a] Four species are pooled into two groups (*Colias hyale* and *C. alfacariensis*; *Thymelicus lineola* and *T. sylvestris* which were not separately recorded)

Discussion

Three major conclusions can be drawn. Firstly, a considerable proportion (66%) of butterflies present in the regional species pool were found in the wildflower strips. This has to be evaluated from the background that the number of butterfly species recorded within the Swiss monitoring programme in Kanton Fribourg is small compared to the total butterfly diversity in Switzerland (nearly 200 species). It has been noted that wildflower strips largely support common species (e.g. for bumblebees Pywell et al. 2005, for butterflies Aviron et al. 2010). It has also been pointed out that agri-envrinmental schemes in general often tend to support butterfly species of the wider-countryside (Settele et al. 2009). This applies also to a certain degree in this study. It is mostly widespread and common species that were found in the wildflower strips, but this is also true for the extensively managed meadows. Of seven red list species recorded within the monitoring scheme in Kanton Fribourg, only one species, the mallow skipper, was recorded within wildflower strips in the study area. However, even widespread and common butterfly species are in decline in landscapes with intensive use (Van Dyck et al. 2009) and wildflower strips are likely to be a measure to stabilise or even enhance the abundances of these species. Fritillaries are in decline in many European countries; they do not profit from the wildflower strips, at least not in the study area, probably because specific habitat requirements are not met.

Secondly, the butterfly communities in wildflower strips and the extensively used meadows were quite different, with higher abundances and species richness in the wildflower strips. It is important to remember that sampling intensity was different between the two habitat types, with almost three-fourths of the transect length being situated in the wildflower strips. Difference in total species richness are therefore not directly comparable. However, species richness per meter showed a close to significant differences

($P = 0.057$) between extensively used meadows and wildflower strips. With higher sampling efforts in the meadows differences might have become more visible. Also, the number of butterflies (per transect meter) was 2.5 times higher in the wildflower strips than in the meadows.

Species communities in the extensively used meadows are characterized by the presence of butterflies with grasses as larval food plant (*Melanargia galathea, Coenonympha pamphilus, Maniola jurtina*) and the two blues *Polyommatus icarus* and *Cyaniris semiargus*. The number of certain blues could probably be enhanced in wildflower strips by including certain *Fabaceae* (other than *Onobrychis*) in the seed mixture. From the result of the correspondence analysis (Fig. 3), it is clear that the hypothesis that wildflower strips contain a subset of species occurring in the meadows was not confirmed. In such a case, the wildflower strip sites would have been included inside the envelope (dotted line) of the meadow sites.

Thirdly, sown wildflower strips were used as larval habitat for several species. Butterfly larvae were not specifically investigated during this study, but several species, including the mallow skipper and the swallowtail, were observed using the larval hostplants present. The case of the mallow skipper shows that the provision of the larval food plant (in addition to adult food resources) can increase abundances considerably (see Appendix B, Wermeille and Carron 2005). The mallow skipper would probably not have been judged as a species with high dispersal ability, but the current density of wildflower strips seems to be sufficient to allow increases in abundances. Larvae of the swallowtail are very visible and were observed in wildflower strips on parsnip (*Pastinaca sativa*) and wild carrot (*Daucus carota*). It is thus very likely that the wildflower strips are beneficial for swallowtails in the study area.

The use of wildflower strips as larval habitat has implications for the design of wildflower strips regarding composition of seed mixture and duration of agri-environmental scheme (time span between sowing and eventual removal). When choosing seed mixtures, inclusion of larval food plants should be carefully considered as they can have a major positive impact of butterfly species. It would thus not be sensible to design the seed mixtures based only on needs as adult food resources (Feber et al. 1996). Non-adult life stages seem less often considered when designing agri-environmental schemes. This has, for example, been pointed out by Lye et al. (2009) regarding nesting sites for bumblebees compared to the provision of adult food resources. It is of great importance to acknowledge that in cases where wildflower strips are used as larval habitat, a short time span between sowing and ploughing (often the same time span as the duration of the agri-environmental scheme) can be detrimental since eggs might be laid in high numbers in habitats that are ploughed in late autumn or early spring. Short schemes for wildflower strips—for one or 2 years—are not unusual in some countries. Nevertheless, 5 years should be a minimum and removal should be rotational, or staggered.

The density of wildflower strips was comparatively high in the study area and they were quite evenly distributed, both factors that result in relatively short distances between strips. The success of wildflower strips for butterfly conservation may thus be dependent to a certain extent on their spatial distribution. Aviron et al. (2010) demonstrated a linear increase in butterfly diversity with percent cover of wildflower strips in Swiss agricultural landscapes. The authors therefore emphasised the importance of density and connectivity of wildflower strips in a landscape for butterfly diversity. In our study area, this was reached with the help of the administrators of the agri-environmental schemes, who explicitly choose a network approach. This meant informing and advising all farmers in the area for establishment of ecological compensation areas (an approach probably comparable with the high level stewardship schemes in the UK). This approach has previously shown success regarding moth conservation (Merckx et al. 2009).

The study also showed that the extensively used meadows in the study area were less favourable for butterfly conservation. We believe that they were still too intensively used and/or that the land use prior to entering the agri-environmental scheme was too intensive to support large numbers of butterflies (see also Aviron et al. 2007b). The current regulation allows several sward cuts starting in the middle of June as well as limited fertilizer application. Such practices are not beneficial for many butterfly species (Aviron et al. 2007c). 30% of butterfly species recorded within the monitoring scheme have their larval food plant present in the meadows, but were not recorded there (Table 2); this may be an indication of a suboptimal management. Similar problems for butterfly species conservation caused by early mowing of grassland over large areas allowed by regulations in agri-environmental schemes have been pointed out before (Konvicka et al. 2008; Settele et al. 2009; Walter et al. 2007).

Acknowledgments We want to thank Jacques Studer and Christian Imesch from the Ökobüro in Fribourg for providing all information on ecological compensation areas in the study area including maps. We are also very thankful to Matthias Plattner, Hintermann & Weber AG, Basel, for the provision of the butterfly data from the Swiss Biodiversity Monitoring Programme (BDM). All farmers are thanked for allowing access to their land. Two anonymous reviewers have contributed to improve the manuscript. CH was financed by the Swedish Research Council FORMAS and LFB by the Swiss National Science Foundation (3100A0-113843).

Appendix A

Plant species sown in wildflower strips in the study area

Achillea millefolium
Agrostemma githago
Anthemis tinctoria
Centaurea cyanus
Centaurea jacea
Cichorium intybus
Daucus carota
Dipsacus fullonum
Echium vulgare
Fagopyrum esculentum
Hypericum perforatum
Legousia speculum-veneris
Leucanthemum vulgare
Malva moschata
Malva sylvestris
Melilotus alba
Onobrychis viciifolia
Origanum vulgare
Papaver rhoeas
Pastinaca sativa
Silene pratensis
Tanacetum vulgare
Verbascum lychnitis
Verbascum thapsus

Appendix B

Butterflies in sown wildflower strips and meadows

	Sown wildflower strips	Meadows
Aglais io	51	2
Aglais urticae	111	6
Aphantopus hyperantus	357	12
Araschnia levana	23	1
Carcharodus alceae[a]	149	
Coenonympha pamphilus	47	24
Colias croceus	8	1
Colias hyale/alfacariensis	23	28
Cyaniris semiargus	1	10
Gonepteryx rhamni	1	
Issoria lathonia	8	1
Lasiommata megera	26	4
Lycaena phlaeas	7	
Maniola jurtina	118	44
Melanargia galathea	5	5
Ochlodes sylvanus	36	3
Papilio machaon	28	4
Pararge aegeria	1	
Pieris napi	252	17
Pieris rapae	172	20
Polygonia c-album	4	
Polyommatus icarus	16	22
Thymelicus lineola/sylvestris	3	
Vanessa atalanta	8	
Vanessa cardui	9	1
Sum	1,464	205

[a] *Critically endangered* according to the Swiss red data book (Gonseth 1994)

References

Aviron S, Herzog F, Klaus I, Luka H, Pfiffner L, Schupbach B, Jeanneret P (2007a) Effects of Swiss agri-environmental measures on arthropod biodiversity in arable landscapes. Aspects Appl Biol 81:101–109

Aviron S, Jeanneret P, Schüpbach B, Herzog F (2007b) Effects of agri-environmental measures, site and landscape conditions on butterfly diversity of Swiss grassland. Agr Ecosyst Environ 122:295–304

Aviron S, Kindlmann P, Burel F (2007c) Conservation of butterfly populations in dynamic landscapes: The role of farming practices and landscape mosaic. Ecol Model 205:135–145

Aviron S, Herzog F, Klaus I, Schüpbach B, Jeanneret P (2010) Effects of wildflower strip quality, quantity, and connectivity on butterfly diversity in a Swiss arable landscape. Restor Ecol. doi: 10.1111/j.1526-100X.2010.00649.x

Carvell C, Meek WR, Pywell RF, Nowakowski M (2004) The response of foraging bumblebees to successional change in newly created arable field margins. Biol Conserv 118:327–339

Carvell C, Roy DB, Smart SM, Pywell RF, Preston CD, Goulson D (2006) Declines in forage availability for bumblebees at a national scale. Biol Conserv 132:481–489

Carvell C, Meek WR, Pywell RF, Goulson D, Nowakowski M (2007) Comparing the efficacy of agri-environment schemes to enhance bumble bee abundance and diversity on arable field margins. J Appl Ecol 44:29–40

Ebert G (1993) Die Schmetterlinge Baden-Würtembergs. Band 2. Verlag Eugen Ulmer, Stuttgart

ESRI (2006) ArcGIS

Feber RE, Smith H, Macdonald DW (1996) The effects on butterfly abundance of the management of uncropped edges of arable fields. J Appl Ecol 33:1191–1205

Frank T, Künzle I (2006) Effect of early succession in wildflower areas on bug assemblages (Insecta : Heteroptera). Eur J Entomol 103:61–70

Gonseth Y (1994) Rote Liste der gefährdeten Tagfalter der Schweiz. In: Duelli P (ed) Rote Liste der gefährdeten Tierarten der Schweiz. Bundesamt für Umwelt, Wald und Landschaft. Bern, Switzerland, pp 48–51

Haaland C, Gyllin M (2010) Butterflies and bumblebees in greenways and sown wildflower strips in southern Sweden. J Insect Conserv 10:125–132

Haaland C, Naisbit RE, Bersier, L-F (2010) Sown wildflower strips for insect conservation—a review. Insect Conserv Divers (in press). doi:10.1111/j.1752-4598.2010.00098.x

Heard MS, Carvell C, Carreck NL, Rothery P, Osborne J, Bourke AFG (2007) Landscape context not patch size determines bumble-bee density on flower mixtures sown for agri-environment schemes. Biol Lett 3:638–641

Jacot K, Eggenschwiler L, Junge X, Luka H, Bosshard A (2007) Improved field margins for a higher biodiversity in agricultural landscapes. Aspects Appl Biol 81:277–283

Kohler F, Verhulst J, van Klink R, Kleijn D (2008) At what spatial scale do high-quality habitats enhance the diversity of forbs and pollinators in intensively farmed landscapes? J Appl Ecol 45:753–762

Konvicka M, Benes J, Cizek O, Kopecek F, Konvicka O, Vitaz L (2008) How too much care kills species: Grassland reserves, agri-environmental schemes and extinction of Colias myrmidone (Lepidoptera : Pieridae) from its former stronghold. J Insect Conserv 12:519–525

Koordinationsstelle Biodiversitäts-Monitoring Schweiz (2008) Anleitung für die Feldarbeit zum Indikator « Z7-Tagfalter » . Bern, Bundesamt für Umwelt. http://www.biodiversitymonitoring.ch/pdfs/downloads/875%20Anleitung%20Z7-Tagf%20v14.pdf

Lye G, Park K, Osborne J, Holland J, Goulson D (2009) Assessing the value of Rural Stewardship schemes for providing foraging resources and nesting habitat for bumblebee queens (Hymenoptera: Apidae). Biol Conserv 142:2023–2032

Marshall EJP, West TM, Kleijn D (2006) Impacts of an agri-environment field margin prescription on the flora and fauna of arable farmland in different landscapes. Agr Ecosyst Environ 113:36–44

Meek WM, Loxton D, Sparks TH, Pywell R, Pickett H, Nowakowski M (2002) The effect of arable filed margin composition on invertebrate biodiversity. Biol Conserv 106:259–271

Merckx T, Feber RE, Riordan P, Townsend MC, Bourn NAD, Parsons MS, Macdonald DW (2009) Optimizing the biodiversity gain from agri-environment schemes. Agr Ecosyst Environ 130:177–182

Nentwig W (ed) (2000) Streifenförmige ökologische Ausgleichsflächen in der Kulturlandschaft: Ackerkrautstreifen, Buntbrachen, Feldränder. Verlag Agrarökologie, Bern Hannover, p 293

Oksanen J, Blanchet FG, Kindt R, Legendre P, O'Hara RG, Simpson GL, Solymos P, Stevens MHH, Wagner H (2010) vegan: Community Ecology Package. R package version 1:17

Pfiffner L, Wyss E (2004) Use of sown wildflower strips to enhance natural enemies of agricultural pests. In: Gurr GM, Wratten SD, Altieri MA (eds) Ecological engineering for pest management: advances in habitat manipulation for arthropods. Collingwood Victoria, CSIRO Publishing, pp 165–186

Pollard E, Yates TJ (1993) Monitoring butterflies for ecology and conservation. Chapman & Hall, London, UK

Pywell RF, Warman EA, Carvell C, Sparks TH, Dicks LV, Bennett D, Wright A, Critchley CNR, Sherwood A (2005) Providing foraging resources for bumblebees in intensively farmed landscapes. Biol Conserv 121:479–494

Pywell RF, Warman EA, Hulmes L, Hulmes S, Nuttall P, Sparks TH, Critchley CNR, Sherwood A (2006) Effectiveness of new agri-environment schemes in providing foraging resources for bumblebees in intensively farmed landscapes. Biol Conserv 129:192–206

Pywell RF, Meek WM, Carvell C, Hulmes L, Nowakowski M (2007) The Buzz project: biodiversity enhancement on arable land under the new agri-environment schemes. Aspects Appl Biol 81:61–68

R Development Core Team (2009) R: A language and environment for statistical computing. R Foundation for Statistical Computing, Vienna, Austria. ISBN 3-900051-07-0, URL http://www.R-project.org

Schweizerischer Bund für Naturschutz (ed) (1994) Tagfalter und ihre Lebensräume. Arten—Gefährdung—Schutz. 4. Auflage. Holliger, Egg/ZH, Switzerland

Settele J, Dover J, Dolek M, Konvicka M (2009) Butterflies of European ecosystems: impact of land use and options for conservation management. In: Settele J, Shreeve T, Konvicka M, Van Dyck H (eds) Ecology of butterflies in Europe. Cambridge University Press, Cambridge, pp 353–370

StatSoft (1997) Statistica 5.1 StatSoft, Tusla, USA

Van Dyck H, Van Strien AJ, Maes D, Van Swaay CAM (2009) Declines in common, widespread butterflies in a landscape under intense human use. Conserv Biol 23:957–965

Van Swaay CAM, Warren MS, Lois G (2006) Biotope use and trends of European butterflies. J Insect Conserv 10:189–209

Van Swaay CAM, Maes D, Warren MS (2009) Conservation status of European butterflies. In: Settele J, Shreeve T, Konvicka M, Van Dyck H (eds) Ecology of butterflies in Europe. Cambridge University Press, Cambridge, pp 322–338

Walter T, Schneider K, Gonseth Y (2007) Schnittzeitpunkt in Ökowiesen: Einfluss auf die Fauna. Agrarforschung 14:114–119

Wermeille E, Carron G (2005) Value of fallows for the Mallow Skipper (*Carcharodus alceae*) and some other butterfly species. Revue Suisse d'Agricul 37:175–182

Woodcock BA, Westbury DB, Potts SG, Harris SJ, Brown VK (2005) Establishing field margins to promote beetle conservation in arable farms. Agr Ecosyst Environ 107:255–266

Woodcock BA, Westbury DB, Tscheulin T, Harrison-Cripps J, Harris SJ, Ramsey AJ, Brown VK, Potts SG (2008) Effects of seed mixture and management on beetle assemblages of arable field margins. Agr Ecosyst Environ 125:246–254

J Insect Conserv (2011) 15:207–220
DOI 10.1007/s10841-010-9338-7

ORIGINAL PAPER

Land-use, environment, and their impact on butterfly populations in a mountainous pastoral landscape: individual species distribution and abundance

J. W. Dover · A. Rescia · S. Fungariño · J. Fairburn · P. Carey · P. Lunt · C. Arnot · R. L. H. Dennis · C. J. Dover

Received: 24 March 2010 / Accepted: 3 September 2010 / Published online: 24 September 2010

Abstract Butterflies were studied, at the species level, in 47 mountain meadows in a 1.5 × 1.6 km study area in the Picos de Europa National Park, Spain. Butterfly transects were carried out on nine occasions in June and July 2004 and the summed data used in binary logistic and stepwise multiple regression analyses using 28 biotic and abiotic parameters. Models were created for 37 species in total: 24 using logistic regression and 24 with multiple regression; models from both approaches were obtained for 11 species. Abiotic factors dominated many analyses with factors such as proximity to water, aspect and altitude being prominent. Abiotic factors may reflect acceptable minimum conditions for presence of a species and interact with biotic factors to determine habitat quality. Classification of the meadows as either under hay or summer grazing management, or 'winter grazing or abandoned' was not particularly revealing probably due to inherent variability in management intensity within meadows and degree of abandonment. Features that reflected management influences, lack of management, disturbance, and sward condition featured in many analyses. Whilst many meadows are still actively managed, features that can be related to abandonment are evident for many species. The early stages of relaxation of management intensity can be positive for butterflies, but if management is not restored losses are likely as succession proceeds. The implications of this are briefly discussed.

Keywords Hay · Grazing · Management · Abandoned · Grassland · Prime butterfly area · High nature value farmland · Land-use

J. W. Dover (✉) · J. Fairburn · P. Lunt · R. L. H. Dennis · C. J. Dover
Institute for Environment, Sustainability and Regeneration, Staffordshire University, College Road, Stoke-on-Trent ST4 2DE, UK
e-mail: j.w.dover@staffs.ac.uk

A. Rescia · S. Fungariño
Departamento de Ecología, Complutense University, Madrid, Spain

P. Carey
CEH Monks Wood, Abbots Ripton, Huntingdon PE28 2LS, UK

P. Carey
Bodsey Ecology Ltd, Ramsey PE26 2XH, UK

P. Lunt
University of Plymouth, Drake Circus, Plymouth PL4 8AA, UK

C. Arnot
RPS Group PLC, Mallams Court, 18 Milton Park, Abingdon OX14 4RP, UK

Introduction

Underlying species' distribution and abundance are the classic determining factors of the abiotic and the biotic. The latter is strongly influenced by the former, but also by human intervention. Human cultural history (Brook and Bowman 2006) can be dominant in current land-use or relict, and this continuum is clearly evident in Europe (Plieninger et al. 2006). In places, for example in much of lowland UK, cultural (traditional) management appears to have been virtually obliterated, whereas in others it is still dominant, such as in the northern mountains of Spain (Rescia et al. 2008). The term 'High Nature Value' (HNV) farmland has been coined to describe farming systems where traditional, extensive, agricultural management maintains high levels of biodiversity (Cooper et al. 2007; Dover et al. 2010a). As traditional agricultural management is replaced by modern intensive techniques, structural heterogeneity is lost and landscapes become homogeneous (Jongman 2002; Renetzeder et al. 2010). The same process is evident when agricultural management is abandoned: in both cases with negative consequences for biodiversity (Concepcion et al. 2008; Macdonald et al. 2000; Schneider and Fry 2001, 2005).

From the standpoint of the early twenty-first century, the shift in agricultural intensification since the early 1950s (Krausmann et al. 2003; Mottet et al. 2006) has led to large areas of western European farmland becoming relatively homogeneous with only tiny features, such as narrow field boundaries, surviving as reminders of a more heterogeneous past, and supporting fewer species with specialised habitat requirements. With the accession of more eastern-European states into the European Union (Gorton et al. 2009) landscape simplification is likely to accelerate in those states as well, and will probably be completed faster than in the western states (early adopters have to wait for each new technological advance, whilst late adopters acquire mature technology). Studies of organic farming, a philosophically-driven attempt to de-intensify agriculture, have demonstrated the importance of heterogeneity within landscapes for biodiversity (Rundlof and Smith 2006; Rundlof et al. 2008; Weibull et al. 2000). Agri-environment schemes are essentially attempts to de-intensify and, whilst laudable, unless carefully targeted are unlikely to conserve species with exacting habitat requirements (Hodge and Reader 2010; Pywell et al. 2004; Warren et al. 2005).

Attempts to re-create landscape features that support a rich biodiversity in a vacuum, i.e. without understanding the ecological characteristics and constraints from exemplars is, if not doomed to failure, exceptionally difficult. Identifying such characteristics is becoming less easy as economic pressures force farmers to either intensify agricultural practices or abandon marginal land. The Picos de Europa, northern Spain, currently retain, within a forest matrix, an impressive suite of unimproved grasslands managed as traditional hay meadows (Farino 1999). As in many mountainous regions these are likely to decline in number and quality because of socioeconomic factors (Rescia et al. 2008), but currently retain a substantial butterfly fauna (Farino 1999). Dover et al. (2010b) found a minimum of 75 species of butterfly in their small study area in the Picos de Europa and described the factors affecting species richness and butterfly abundance in the study meadows at different levels of aggregation from 'total butterflies' down to 'Family-level' and a host-plant grouping. Only one species-level analysis was carried out: that of the Duke of Burgundy *Hamearis lucina*, and only because this was the single representative present from the Riodinidae. This study examines the factors that affect the distribution and abundance of butterflies in the open grasslands of the Picos de Europa at the species level in order to examine the impacts of abiotic factors and management practices.

Methods

Study site

The study site (Fig. 1) was adjacent to Fuente Dé at the head of the Deva Valley in the Picos de Europa, northern Spain (43°05′N-04°35 W to 43°18′N-05°05′W). It was approximately 1.5 × 1.6 km in extent and the 47 study meadows ranged in altitude from 1,000 to 1,400 m. Meadow land-use was defined as being: management for hay (24 meadows), summer grazing (7), or 'winter grazing or abandoned' (16). The latter term identifies meadows which were neither summer grazed or cut for hay over the summers of 2003–2005; many were clearly abandoned with much scrub whilst others may have been in earlier stages of abandonment-related succession. These meadows may also have had some grazing pressure from stock or wild animals in autumn/early winter or spring. Full details of the study

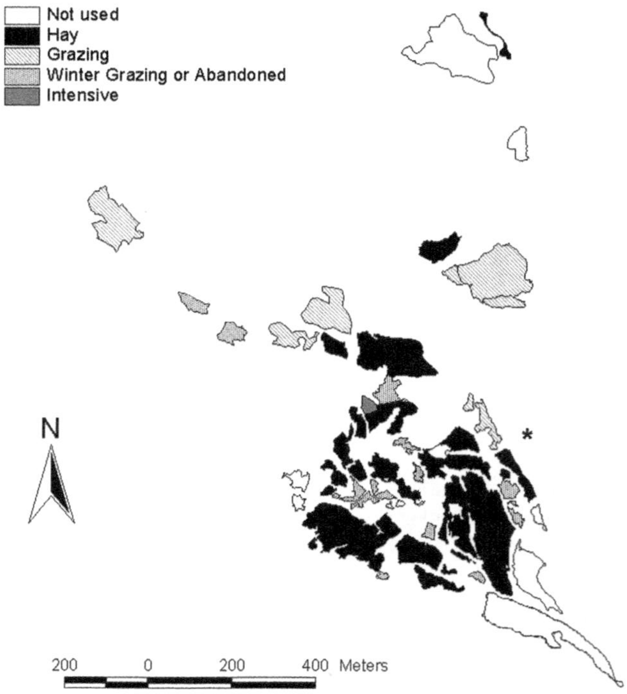

Fig. 1 Map of the study site showing hay, summer grazing, and 'winter grazing or abandoned' meadows used in the study. Other meadows included on the map were not studied and are given for context. *Note*: *asterisk* indicates other, undigitised, meadows were present to the east (and *down-slope*) of the block of meadows in the *lower right-hand* side of the figure. The meadows were surrounded by forest dominated by beech (*F. sylvatica*)

site and meadow definitions can be found in Dover et al. (2010b, c).

Butterfly data

Butterfly data were derived from transects (Pollard et al. 1975) around the perimeter of the study meadows conducted on nine occasions from the beginning of June to late July 2004. Transects were restricted to the perimeters to avoid crop damage. The meadows yielded 15,046 butterflies in total from a minimum of 75 species (see Dover et al. 2010c and the appendix of Dover et al. 2010b). Data from the different meadows were considered independent: a Spearman correlation analysis on the difference in total butterfly abundance (Dover et al. 2010b) between pairs of meadows with Euclidean distance between meadow pairs gave a correlation coefficient of 0.045 and a *P*-value of 0.141 (n = 1,081).

Exploratory analyses

As the current study includes a large number of potential explanatory variables for regression analyses, initially, the relationships were explored between variables using principal components analysis (PCA), factor analysis (FA) using a varimax rotation to focus on variable loadings, and non metric scaling (NMS). These techniques were conducted on 26 abiotic and biotic predictors (landform, management and plant variables); the butterfly species were entered as supplementary to analysis. Relationships proved to be very complex (see results) and for this reason it proved necessary to use the original variables in regression models rather than components or factors derived from PCA and FA. Owing to the limited amount of variance accounted by the first two PCA axes, relationships are illustrated using NMS for Euclidean distances on standardised variables; as a technique exploiting full computer power for iterative optimisation, NMS has greater capacity for reducing 'stress' in plots and of identifying structures in relationships (McCune and Grace 2002). Analyses were carried out in Statistica 7.1 (Statsoft, Inc).

Regression analyses

For the purposes of this paper two data sets were prepared, one for binary logistic regression, to explore variables affecting colonisation (presence/absence), and one for stepwise multiple regression, to explore variables related to relative population size. In both cases analyses were only performed where a minimum of seven meadows (15% of the total) were occupied by any given species (see Dover et al. 2010b); for logistic regression an upper limit of a maximum of 40 occupied meadows (85% of the total) was also set.

For multiple regression, the same set of 26 biotic and abiotic variables were used as in the analyses of Dover et al. (2010b) (Table 1), and butterfly data were standardised to the number of individuals per 100 m of perimeter. For logistic regression the variable: 'area of meadow' was dropped in favour of the 'length of the meadow perimeter'. Some initial analyses revealed multicollinearity between the two variables and perimeter was preferred over area as it was directly measured (by pedometer) rather than being estimated from a 2-D GIS map of the, topographically variable, study area (derived from aerial photography and a differential GPS survey, Dover et al. 2010b). Length of perimeter may also be more directly related to the likelihood of colonisation than area, although area may be more important in the retention of an individual of a given species (Dover and Settele 2009).

Analyses were performed using SPSS v16 (SPSS Inc., Chicago). For logistic regression, models with the lowest −2 log likelihood were checked for goodness-of-fit using the Hosmer-Leamshow test. Multicollinearity was checked for using Tolerance/Variance Inflation Factor (VIF) statistics followed by inspection of the loadings of the variance proportions on the model eigenvalues (Field 2009). For multiple regression, residuals were checked for normality using the Kolmogorov–Smirnov test in the 'Explore' function of SPSS 16, and the dependents transformed as required.

Results

Ordination

Principal components analysis extracted ten components with eigenvalues >1 and seven which account for >5% of the variance. The first two components accounted for only 27% of the variance and ten components 77% of the variance. Only three of the predictor variables have high loadings on components and a further 17 variables have modest loadings >0.5. Varimax rotated factor analysis resolved some closely related pairs. For instance, three pairs of variables are significantly correlated; three highly so ($r > 0.70$) (e.g. Watln and WatNr, Owners and Area, South and East, the latter negatively associated). But other pairs are only modestly associated ($r < 0.32$–0.48) (e.g. Hialt and NearM, ScrubS and ScrubB, SumG and M100 m (negative), Soffic and Cent) or even not significantly correlated (Rumex and Planc, ScrubS and Paqui) despite jointly high loadings. Variable definition on components was considered too poor to substitute components or factors for original variables. As accounted variance (three

Table 1 Variables used in regression analyses of butterfly data

Variable	Metric	Variable contraction
Abiotic		
Highest altitude	Metres	Hialt
Water in meadow	Presence/absence	WatIn
Water in or near to meadow	Presence/absence	WatNr
Principal aspect	S, E	South, East
Multiple aspects	Number/meadow	MultA
Slope	Degrees	Slope
Area[a]	Hectares	Area
Perimeter length[b]	Metres	Perim
Distance to nearest meadow	Metres	NearM
Number of meadows within 100 m	Number	M100 m
Management		
Ownership parcels	Number/meadow	Parcl
Hay meadow	Yes/no	HayM
Summer grazing	Yes/no	SumG
Winter grazing or abandoned	Yes/no	GorAb
Management-related		
Asphodelus albus[c]	0,1,2	Aalba
Pteridium aquilinum around edges[c]	0,1,2	Paqui
Rhinanthus spp.[c]	0,1,2	Rmino
Scrub around meadow edge	Yes/no	ScrS
Scrub in (body of) meadow	Yes/no	ScruB
Sward components		
Centaurea spp.	Plants/0.0625 m^2	Cent
Plantago lanceolata	Plants/0.0625 m^2	Planc
Primula spp.	Plants/0.0625 m^2	Prims
Rumex spp.	Plants/0.0625 m^2	Rumex
Sanguisorba minor	Plants/0.0625 m^2	Smin
Sanguisorba officinalis	Plants/0.0625 m^2	Soffi
Viola spp.	Plants/0.0625 m^2	Viola

[a] Not logistic regression

[b] Not multiple regression

[c] Three-point ordinal scale: 0 = absent, 1 = present but low density, 2 = present but high density

axes, 36%) and communalities of variables on the first three axes are low in a PCA, the relationship among variables are illustrated in a non metric three dimensional plot based on Euclidean distances between standardised variables (Fig. 2); the Euclidean distances correlate highly with the resemblance coefficient (1-Pearson r = 0.92). There are no distinct clusters of variables or species in the axes.

Binary logistic regression

Using the maximum and minimum occupancy criteria, analyses were carried out on 36 species; however, models were only derived for 24 species (Table 3). Across the dataset, nine abiotic variables accounted for the majority of significant relationships (26 out of 46). Most associations identified were positive, but topographical variation (multiple aspects) was identified as a negative variable for *Lycaena tityrus* and *Hamearis lucina*, slope for *Lycaena coridon* (but not for *Vanessa atalanta* or *Coenonympha arcania*) whilst 'the number of meadows within 100 m', a fragmentation variable, was negative for *Aglais urticae and Gonepteryx rhamni* but not for *Pyronia tithonus* (Table 2).

There were only four instances of direct links with meadow management: summer grazing was negatively correlated with *Boloria selene*; winter grazing/abandonment was positively correlated with *B. selene* but negatively with *Pyrgus malvae* and *Vanessa cardui*. Several variables reflect management or the lack of it. The number of ownership parcels, a complex variable combining factors including area and management issues (Dover et al. 2010b), was positively associated with *Melitaea phoebe*. *Rhinanthus*, generally associated with active hay management, was positively related to *Leptidea sinapis*, whereas variables indicative of abandonment: scrub in the body of the meadow and *P. aquilinum* were positively related to

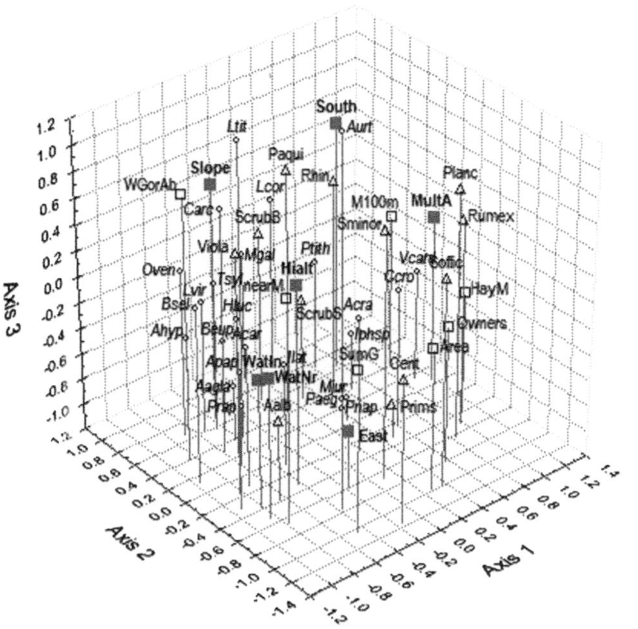

Fig. 2 Non-metric three dimensional scaling plot of associations among variables and 25 butterfly species within the study area. Associations are Euclidean distances on standardised variables; Kruskal stress 0.22. *Key*: *Grey squares* abiotic variables; *white squares* management variables; *white triangles* plant variables; *small circles* butterfly species

L. coridon, *V. cardui*, *G. rhamni*, *Pieris rapae* and *Boloria dia*. *A. albus*, a plant considered to be associated with relaxed management pressure (see Dover et al. 2010b) was negatively associated with *V. cardui* but positively with *P. tithonus*. Four sward variables accounted for a further seven associations: with the nectar variable *Centaurea* positively associated with *Lasiommata maera;* the disturbance variable *P. lanceolata* was positively associated with *Celastrina argiolus* but negatively with *Lycaena virgaureae* and *L. sinapis; S. minor* with *Cupido minimus* and *B. selene* and *Primula* with *B. selene*. None of the species which can be considered to have more open population structures (Warren 1992) (Table 2) have models that include sward components (*Centaurea*, *P. lanceolata*, *S. minor* or *Primula spp.*) and most of the satyrids (generally regarded as closed population species) do not have models with sward components—the exception being *L. maera*.

Multiple regression

Multiple regression analysis was carried out on 44 species, with models successfully created for 24 species (Table 3), including *H. lucina* which was first reported in Dover et al. (2010b).

Across the dataset 10 abiotic variables accounted for half of the identified relationships (31 out of 61) and for all but six species (*L. coridon*, *V. cardui*, *Iphiclides spp.*, *P. napi*, *P. rapae* and *Melanargia galathea*) most of the explained variation was attributable to abiotic factors. In the majority of the cases the direction of the relationship varied depending on the species in question, but topographical variation (multiple aspects) was identified as always a negative variable (for *Ochlodes sylvanus*, *Anthocharis cardamines* and *H. lucina*) and highest altitude as always positive (for *L. virgaureae*, *Argynnis paphia*, *Boloria euphrosyne*, *Issoria lathonia*, *A. cardamines*, *Aphantopus hyperantus* and *Coenonympha arcania*). The latter was also the most consistently identified variable, although the combined associations with water ('water in the meadow' and 'water in/near meadow') were also present in six models (Table 3). 'Area' was only identified twice, and in both instances as a negative relationship with *Thymelicus sylvestris* and *Maniola jurtina*. Smaller meadows may be less economic to manage and so more likely to be abandoned, or less actively managed, which may make them attractive to these grass-feeding species. However, the legume crop pest *Colias crocea* (Thomas 1986) responded positively to the number of ownership parcels in a meadow, a variable strongly correlated with area (see above).

Hay management appeared to favour only two species (*Aporia crataegi* and *M. jurtina*), whilst summer grazing negatively affected three (*O. sylvanus*, *P. rapae and A. hyperantus*). Variables associated with lack of management were identified for 8 species. Scrub in the body of the meadow appeared to favour *O. sylvanus*, *T. sylvestris* and *L. virgaureae;* whilst scrub on the perimeter negatively affected *C. arcania*. *A. albus* was positively related to *B. euphrosyne* and *I. lathonia*, but negatively to *V. cardui*. *P. aquilinum* positively influenced the *Iphiclides spp.*

Sward variables were more in evidence in the multiple regression models compared with the logistic regression models. The most prominent variable was *P. lanceolata* negatively related to *A. paphia*, *A. hyperantus*, *M. galathea* and *P. tithonus* but positively related to *A. urticae*. *Centaurea* was negatively related to *O. sylvanus* and *M. galathea* but positively to *H. lucina*. The latter species is not shown to be related to its *Primula* hostplants (though *Iphiclides* and *P. napi* are), but it is positively associated with *Viola* as is *I. lathonia*, whilst *Iphiclides spp.* are negatively associated with *Viola*. *Sanguisorba minor* had a negative association with *Iphiclides spp.*; *Sanguisorba officinalis* has a positive association, but only with *P. napi*. *Rumex* was only associated (positively) with *M. jurtina*.

Distribution vs. distribution plus abundance

Eleven species produced both logistic regression and multiple regression models. A comparison of these models (Tables 2 and 3) show a substantial degree of overlap in the abiotic variables with seven out of ten species models having

Table 2 Binary logistic regression of 24 butterfly species found in 47 study meadows at Fuente De, Picos de Europa 2004

Factor type	Predictor	HE	Lycaenidae						Nymphalidae							
		*P. malvae**	*C. argiolus*	*C. minimus*	*L. bellargus*	*L. coridon*	*L. tityrus*	*L. virgaureae*	*A. urticae*	*B. dia**	*B. selene*	*I. lathonia**	*M. phoebe*	*P. c-album*	*V. atalanta*	*V. cardui*
Abiotic	Constant	0.873	−2.114	−1.654	−3.027	1.376	3.212	−34.875	1.210	−4.473	−15.140	−20.885	−1.860	−2.015	−3.579	3.004
	Aspect S															
	Aspect E															
	Multiple aspects						**−1.457a**									
	Water in meadow		2.761b											1.861a		
	Water in/near meadow									3.469c	3.142a					
	Highest altitude							0.029a				0.017a				
	Perimeter length				0.005a			0.016b	0.005a		0.025a				0.006b	
	Slope					**−0.075a**									0.052a	
	No. meadows/100 m								**−0.264a**							
Man	Ownership parcels												0.371a			
	Summer grazing										**−8.176a**					
	WinterG/abandoned	**−2.534b**									7.040a					**−2.359a**
Man-rel	*Rhinanthus*															
	Scrub in meadow					2.460b										2.202a
	P.aquilinum									1.956a						
	A. albus															**−1.772a**
Sward	*Centaurea spp.*															
	P. lanceolata		0.741a					**−1.108a**								
	S. minor			0.468a							1.150a					
	Primula spp.										5.058a					
	N meadows occupied	20	23	16	10	32	34	27	28	19	18	21	14	10	18	36
	Model X^2	24.172	17.46	4.635	13.47	14.01	5.82	30.03	15.29	23.94	36.38	14.25	7.39	6.08	13.83	15.07
	df	4	2	1	1	2	1	3	2	4	4	2	1	1	2	3
	Model P-value	0.000	0.000	0.031	0.003	0.001	0.016	0.000	0.000	0.000	0.000	0.001	0.007	0.014	0.001	0.002
	Cox and Snell R^2 (%)	40.2	31	9.4	16.9	25.8	11.6	47.2	27.8	39.5	53.9	26.2	14.7	12.1	25.5	27.4
	Nagelkerke R^2 (%)	54	41.4	13	26.3	36.1	16.8	63.4	37.5	53.9	73.2	35	20.7	18.8	34.6	41.4
	Population structure	C	C	C	C	C	C	C	O	C	C	O	C	O	O	O

Table 2 continued

Factor type	Predictor	Pieridae				RI	Satyridae			
		G. rhamni	*L. sinapsis*	*P. brassicae*	*P. rapae*	*H. lucina*	*A. hyperanuts*	*C. arcania*	*P. tithonus**	*L. maera**
Abiotic	Constant	−0.078	−4.917	−0.480	0.539	−1.830	−37.230	−0.548	−3.489	−3.406
	Aspect S									2.472a
	Aspect E					2.713a				
	Multiple aspects					**−2.402a**				
	Water in meadow			1.684a						
	Water in/near meadow						2.396a			
	Highest altitude						0.32b			
	Perimeter length		0.010b			0.009b				
	Slope							0.095a		
	No. meadows/100 m	**−0.194a**							0.243a	
Man	Ownership parcels									
	Summer grazing									
	WinterG/abandoned									
Man-rel	*Rhinanthus*		2.344b							
	Scrub in meadow	2.084b			2.026a					
	P.aquilinum									
	A. albus								1.729a	
Sward	*Centaurea spp.*									4.883a
	P. lanceolata		**−1.385a**							
	S. minor									
	Primula spp.									
	N meadows occupied	20	16	23	38	21	29	37	32	20
	Model X^2	13	27.17	5.86	6.49	28.43	19.52	8.46	15.51	32.46
	df	2	3	1	1	3	2	1	3	6
	Model P-value	0.002	0.000	0.016	0.011	0.000	0.000	0.004	0.001	0.000
	Cox and Snell R^2 (%)	24.2	43.9	11.7	12.9	45.4	34.0	16.5	28.1	49.9
	Nagelkerke R^2 (%)	32.5	60.7	15.6	20.7	60.7	46.2	25.5	39.3	67
	Population structure	O	C	O	O	C	C	C	C	C

HE Hesperiidae; *R* Riodininae

a = $P < 0.05$, b = $P < 0.01$, c = $P < 0.001$; − = negatively associated variable (also given in bold)

Population structure: *C* closed; *O* open

Man management variable; *Man-rel* management-related variable

L. bellargus, *Lycaena bellargus*; *P c-album*, *Polygonia c-album*; *P. brassicae*, *Pieris brassicae* (other scientific names in full in the text)

* The full model X^2 values are given, but only significant variables are included in the table

Table 3 Multiple regression of 24 butterfly species in 47 study meadows at Fuente De, Picos de Europa 2004

Factor type	Predictor	HE		Lycaenidae			Nymphalidae					
		O. sylvanus	*T. sylvestris*	*L. coridon*†	*L. tityrus*	*L. virgaureae*^{BC}	*A. aglaja*	*A. paphia*	*A. uticae*†	*B. euphrosyne*†	*I. lathonia*†	*V. cardui*
Abiotic	Constant	3.232	0.930	0.316	0.665	4.230	6.231	−25.083	0.197	−1.296	−2.156	1.670
	Aspect E								**−0.156 (14b)**			
	Aspect S	1.440 (10a)					**−2.941 (13a)**					
	Multiple aspect	**−0.952 (8a)**										
	Water in meadow	1.784 (14c)										
	Water in/near meadow		2.399 (11c)								**−0.140 (6a)**	
	Area		**−1.385 (10a)**									
	Slope			**−0.007 (9a)**	0.030 (10a)					0.003 (12a)		
	Highest altitude					0.003 (26c)		0.026 (19c)		0.001 (14c)	0.002 (17c)	
	No. meadows/100 m											
Man	Ownership parcels											
	Hay meadow											
	Summer grazng		**−1.830 (6a)**									
Man-related	Scrub in meadow		2.216 (20c)	0.275 (13b)		0.202 (8a)						
	Scrub on perimeter											
	A. albus									0.063 (8a)	0.19 (13c)	**−0.573(18b)**
	P. aquilinum											
Sward	*Centaurea spp.*	**−1.273 (6a)**										
	P. lanceolata							**−0.885 (13b)**	0.039 (10a)			
	Rumex spp.											
	S. minor											
	S. officinalis											
	Primula spp.											
	Viola spp.										0.150 (8b)	
	Total R^2 (%)*	38	47	22	10	34	13	32	23	34	39	18
	N meadows occupied	44	38	32	34	27	47	43	28	26	21	36
	Model F-value	6.31	9.21	6.25	4.96	11.32	6.82	10.25	6.74	7.29	8.20	9.94
	df (regression, residual)	4,42	4,42	2,44	1,45	2,44	1,45	2,44	2,44	3,43	4,42	1,45
	Model *P*-value	0.000	0.000	0.004	0.031	0.000	0.012	0.000	0.003	0.000	0.000	0.003

Table 3 continued

Factor type	Predictor	PA	Pieridae					RI	Satyridae					
		Iphiclides spp.	*A. cardamines*†	*A. crataegi*†	*C. crocea*	*P. napi*	*P. rapae*†	*H. lucina*	*A. hyperantus*[BC]	*C. arcania*	*M. galathea*	*M. jurtina*	*P. aegeria*	*P. tithonus*†
Abiotic	Constant	0.026	−1.426	0.386	1.934	0.897	0.265	0.356	3.125	−6.918	27.629	22.026	−1.490	0.160
	Aspect E							0.365 (10b)				17.915 (12c)		
	Aspect S		0.211 (11c)								11.667 (18b)			
	Multiple aspect		**−0.110 (9b)**					**−0.359 (22c)**						
	Water in meadow		0.206 (27c)											
	Water in/near meadow						0.219 (10b)		0.292 (10b)					
	Area											**−9.133 (6a)**		
	Slope									0.033 (12b)				
	Highest altitude		0.002 (13c)						0.002 (14b)	0.007 (10b)				
	No. meadows/100 m												**−0.097 (16b)**	0.020 (15b)
Man	Ownership parcels				0.326 (25c)									
	Hay meadow			0.127 (8a)								18.103 (16c)		
	Summer grazng						**−0.246 (13b)**		**−0.305 (8a)**					
Man-related	Scrub in meadow													
	Scrub on perimeter									**−1.213 (5a)**				
	A. albus													
	P. aquilinum	0.131 (8b)												
Sward	*Centaurea spp.*							0.333 (6a)			**−11.484 (10b)**			
	P. lanceolata								**−0.070 (7a)**		**−3.675 (11b)**			**−0.051 (9a)**
	Rumex spp.											**−7.137 (11b)**		
	S. minor	**−0.057 (8b)**												
	S. officinalis					1.368 (10a)								
	Primula spp.	0.330 (17c)				1.235 (14b)								
	Viola spp.	**−0.108 (7a)**						0.351 (15c)						
	Total R^2 (%)*	39	59	8	25	24	23	53	39	30	39	46	40	24
	N meadows occupied	10	38	44	44	44	38	21	29	37	47	47	16	32
	Model F-value	6.81	15.035	4.13	15.23	6.85	6.64	11.88	6.66	6.15	9.15	8.83	8.42	7.00

Table 3 continued

Factor type	Predictor	PA	Pieridae					RI	Satyridae					
		Iphiclides spp.	*A. cardamines*†	*A. crataegi*†	*C. crocea*	*P. napi*	*P. rapae*†	*H. lucina*	*A. hyperantus*[BC]	*C. arcania*	*M. galathea*	*M. jurtina*	*P. aegeria*	*P. tithonus*†
	df (regression, residual)	4,42	4,42	1,45	1,45	2,44	2,44	4,42	4,42	3,43	3,43	4,42	1,45	2,44
	Model *P*-value	0.000	0.000	0.048	0.000	0.003	0.003	0.000	0.000	0.001	0.000	0.000	0.006	0.002

Data includes regression coefficients followed by R^2 values and significance levels in parentheses

Negative variables in bold

Only reported for species where standardised residuals were normal (following transformation if necessary)

a = $P < 0.05$, b = $P < 0.01$, c = $P < 0.001$; † = Log10 transformation

* Rounded; *HE* Hesperiidae; *RI* Riodininae; *PA* Papilionidae; *Man* management

one or two variables identified in common, one species had no abiotic factor identified as a significant influence. The 'perimeter' variable is excluded from comparison as the dependents were expressed as abundance per unit length of perimeter. In some cases variables were not consistently identified between analyses (e.g. *A. urticae* had number of meadows/100 m as a negative binary variable, but aspect East as a negative multiple regression (MR) variable). *I. lathonia, P. rapae* and *C. arcania* all gained additional abiotic variables from MR analyses.

Of the directly management-related variables no 'common' associations were identified between the analyses. The association with 'winter grazing or abandoned' evident in the binary analyses for *V. cardui* did not persist in the MR analysis, but *V. cardui* and *A. hyperantus* both gained summer grazing (−ve).

The abandonment-related variables retained two associations in common between analyses (scrub in the body of the meadow for *L. coridon* and *A. albus* for *V. cardui*). The (+ve) binary association with scrub did not persist in the MR analysis for *V. cardui* and *P. rapae*, nor did the association with *A. albus* for *P. tithonus*. The MR analysis identified an association (+ve) with *A. albus* for *I. lathonia,* and scrub on the perimeter (−ve) for *C. arcania*. The sward variables had no associations in common between the analyses.

Generally the sward variables showed a greater incidence in the MR analyses than the binary, although the MR analyses identified no sward variables for the three lycaenids.

Discussion

Although some relationships among predictor variables are expected to be, and have been found to be, clearly defined (Watln and Watlr, Owners and Area, South and East), most others are not. The overwhelming impression is one of complex associations between sets of predictors, plant, management and abiotic variables. Indications are that relationships are subtle and change with the shift in focus from incidence of species to their abundances. The logistic regression approach reflects whether a meadow is likely to contain a species, irrespective of its abundance, whilst the multiple regression approach, in addition to presence and absence, uses the abundance of the species in the meadows. Generally, the species models presented here using the logistic regression approach are less likely to include 'sward-level' and 'management' variables than models derived from multiple regression, strongly suggesting that whilst abiotic factors are dominant in whether a meadow is likely to be colonised or not, the population size is determined by management, and other factors, which drive the condition/structure of the sward.

Abiotic factors

Of the 37 species for which models were constructed only eleven species yielded models using both binary logistic regression and stepwise multiple regression. For these models, as far as abiotic factors are concerned, over 50% of the variables identified were the same irrespective of the method. Nevertheless, the large number of species models reported here demonstrate how variable members of a community are in their responses to local conditions. Whilst the validity of the division of the variables into crude groupings of abiotic, management, management-related, and sward can be debated, within those groupings the majority of variation explained, for the majority of species, appears to reside with the abiotic (which in this analysis includes landscape structure/fragmentation). General trends do indicate the dominance of some factors: water was identified as an important influence for eleven species and, with one exception, always as a positive influence. This may reflect the need to prevent desiccation of host plants, and the majority of species with water in the models use grasses, violets or other easily desiccated hosts. Water availability cuts across management regimes: many of the summer grazing meadows had water as did some hay meadows and 'winter grazed/abandoned' meadows. Aspect was important for nine species, but the direction of the relationship and the specific aspect depended on the individual species; however, the relationship of a species with multiple aspects, reflecting topographical variability within a meadow, was always negative. Highest altitude was also a frequently identified variable and this may be related to less active management and lower-yielding swards at higher elevations. The logistic regressions can be seen as identifying, on the one hand, factors that influence the presence of a species in a meadow; but, on the other hand, must also reflect the ability of a meadow to retain that species (indicating some basal level of habitat quality).

Spatial issues

The prominence of the length of the perimeter in the logistic regressions (seven species) as a positive influence is interesting as the longer the perimeter of a meadow, the more likely a given species is likely to 'encounter' and (perhaps) enter it. Also, depending on the shape of the meadow, the longer the perimeter the more likely it is that a meadow has a large area: and larger meadows are more likely to retain an organism than smaller meadows. This is partly because of the likelihood of finding appropriate resources in large meadows, but also because of the reduced likelihood of re-encountering the perimeter and leaving the meadow (see refs in Dover and Settele 2009). Few species appeared to be explicitly affected by fragmentation, but two wide-ranging species *G. rhamni* and *A. urticae* and the woodland specialist *Pararge aegeria* were negatively associated with the 'number of meadows within 100 m', implying that these species were more likely to be found in the more isolated meadows, whilst *P. tithonus* was more associated with clusters of meadows. The latter species is closely associated with shrubs/hedgerows (Dover and Sparks 2000; Dennis 2004), and the narrower belts of vegetation separating clustered meadows may favour this species. For *P. aegeria*, isolated meadows are more likely to have woodland than scrub separating them from other open spaces, and may therefore provide more appropriate edge habitat. The association with more isolated meadows for the other species is less obvious, especially as *G. rhamni* uses rather shrubby species as hosts (Thomas 1986), but may reflect the concentrating effect of small pockets of a resource within a wooded matrix; or, as both species overwinter as adults, be an association with more suitable hibernation sites.

Meadow management

Several of the species associations with management regimes showed negative relationships and particularly with summer grazing (*T. sylvestris*, *B. selene*, *P. rapae*, *A. hyperantus*). In all these cases it seems likely that the loss of juvenile stages can be invoked as an explanation. Hay management was a positive influence on *A. crataegi* and *M. jurtina*; the former species probably for the nectar resources in the meadows (as its host plants are shrubs), and the latter because hay meadows maintain both host and nectar plant resources, although hay harvesting may have some negative impacts (Dover et al. 2010c). Winter grazing/abandonment was only explicitly identified as negative for two species: *P. malvae* (a species requiring more open swards than would be found in abandoned meadows (Thomas 1986)) and *V. cardui*, and few negative associations were identified with abandonment-related variables (*A. albus* for *V. cardui*, and scrub on the perimeter of meadows for *C. arcania*). Whilst *B. selene* was the only species explicitly identified as prospering under the winter grazing/abandoned regime, probably because of a low likelihood of juvenile species destruction and a more lush sward, several species (*T. sylvestris, L. coridon, L. virgaureae, B. dia, B. euphrosyne, I. lathonia, Iphiclides spp.* and *P. tithonus*) were positively influenced by variables related to abandonment/relaxed management. This response is well documented for grassland species whose habitat is in the early stages of abandonment (e.g. Erhardt 1985), and it has been suggested that conservation management should seek to incorporate reduced-intensity management within managed landscapes specifically to capture its beneficial aspects (Balmer and Erhardt 2000;

Cremene et al. 2005; Kruess and Tscharntke 2002; Marini et al. 2009; Schwarzwälder et al. 1997). Rescia et al. (2008) and Dover et al. (2010b) demonstrated that meadow extent had decreased since the 1950s at the study site. As a result, the identification of abandonment-related variables affecting the butterfly community is unsurprising, and currently appears to be affecting more species positively than negatively. However, if unchecked, abandonment will ultimately be detrimental to the butterfly community.

Disturbance

Disturbance of grassland swards can modify plant communities positively or negatively. Some disturbance, by stock, is considered valuable in hay meadows following grass harvesting to ensure there are gaps in the sward for germination (Jones and Hayes 1999); however, too much disturbance can result in the colonisation of undesirable species (Augustine and McNaughton 1998). *P. lanceolata* is an excellent coloniser, and hence indicator, of disturbed grassland (Grime et al. 2007; Kuiper and Bos 1992) and is generally negatively associated with species (*L. virgaureae*, *A. aglaja*, *L. sinapis*, *H. lucina*, *C. arcania* and *P. tithonus*) although some species: *A. urticae* and *C. argiolus*, showed positive relationships. If *P. lanceolata* is acting as an indicator of high levels of disturbance in managed swards at the study site (hay and summer grazing) this may not necessarily be a reflection of overgrazing by stock as wild mammals, such as wild boar (*Sus scrofa*), water vole (*Arvicola terrestris*) or deer, can cause extensive damage (Dover et al. 2010b). In abandoned meadows, disturbance by wild mammals may delay the process of succession (invasion by coarse grasses, shrubs, then trees (Erhardt 1985)) and maintain sward quality by creating germination niches.

Is categorisation by management useful?

The relatively small number of species models which directly incorporate management type (hay management, summer grazing, winter grazing or abandoned), as opposed to variables describing the consequences of management or the lack of it, at first seems rather odd. However, meadow management is more complex and dynamic than the three categories used would suggest. For example, a 'summer grazing' meadow need not have been wholly managed as summer grazing—or even used as summer grazing every year (see Dover et al. 2010b). Likewise, meadows that were actively managed for hay could include areas of scrub. The categorisation of winter grazing/abandoned indicates a level of uncertainty about the extent of abandonment, and the possibility that such meadows may be grazed on an opportunistic basis (MacDonald et al. 2000). In addition, Dover et al. (2010c) provided evidence to suggest that many butterflies found in the adjacent tracks originated in the meadows, implying either that they readily dispersed from natal meadows, or that they were engaged in complementation or supplementation of resources (Dunning et al. 1992; Ouin et al. 2004). Given that many of the meadows were spatially close to one another, short distance movements between them for dispersal and/or complementation/supplementation seems likely, blurring distinctions (especially with presence/absence data). Also, transects were walked around the perimeter of meadows rather than randomly within them, which, it could be argued, is exactly where such short distance dispersers/resource seekers would most likely to be detected. The complexity of the abiotic: biotic: management: ownership interactions in these extensively managed mountain meadows, the needs of the farmers to adjust the timing of farming operations to local environmental conditions (Dover et al. 2010b), and the subtlety of butterfly responses, make the simple categorisations used here less useful in predicting the distribution/abundance of a species than might be expected.

Implications

The results of this work are, in some respects, predictable: species' reactions to landscape and management are complex and individual. Nevertheless, worrying trends are evident in the data that strongly suggest that the traditional pastoral management in the meadows of the Picos de Europa is under threat, as evidenced by the classic responses of butterflies to elements of vegetation change associated with abandonment (including meadow shrinkage and loss (Dover et al. 2010b; Rescia et al. 2008)). Previous socioeconomic work (Rescia et al. 2008) suggests that unless intervention in agricultural support systems, or the monetisation of landscape management (for example through taxation of tourism, Dover et al. 2010b), occurs this abandonment trajectory will continue with ultimately negative consequences for biodiversity. This scenario is, unfortunately, being played-out in many countries in Europe (MacDonald et al. 2000) and without decisive action it is likely that future editions of the compendium volume '*Prime Butterfly Areas in Europe*' (van Swaay and Warren 2003) will be much slimmer in the future.

Acknowledgments We are grateful to the Earthwatch Institute for funding; and to the large number of Earthwatch volunteers, and MSc students from Staffordshire University who helped us with fieldwork; to our employers for allowing us to participate in the research; to Profs Bob Bunce and Francisco Pineda for helping us frame the project; to Graham Smith for GIS processing; and to the Picos de Europa National Park and the Government of Cantabria for giving us permission to work in the Park. We thank Dr Tim Sparks for statistical advice. We are indebted to National Park staff for their help and support.

References

Augustine DJ, McNaughton SJ (1998) Ungulate effects on the functional species composition of plant communities: herbivore selectivity and plant tolerance. J Wildlife Manage 62:1165–1183

Balmer O, Erhardt A (2000) Consequences of succession on extensively grazed grasslands for central European butterfly communities: rethinking conservation practices. Conserv Biol 14:746–757

Brook BW, Bowman DMJS (2006) Postcards from the past: charting the landscape-scale conversion of tropical Australian savanna to closed forest during the 20th century. Landsc Ecol 21:1253–1266

Concepcion ED, Diaz M, Baquero RA (2008) Effects of landscape complexity on the ecological effectiveness of agri-environment schemes. Landsc Ecol 23:135–148

Cooper T, Arblaster K, Baldock D, Farmer M, Beaufoy G, Jones G, Poux X, McCracken D, Bignal E, Elbersen B, Wascher D, Angelstam P, Roberge J-M, Pointereau P, Seffer J, Galvanek D (2007) Final report for the study on HNV indicators for evaluation. Institute for European Environment Policy (Report to DG Agriculture 2006-G4-04), London, p 187

Cremene C, Groza G, Rákosy L, Schileyko AA, Baur A, Erhardt A, Baur B (2005) Alterations of steppe-like grasslands in Eastern Europe: a threat to regional biodiversity hotspots. Conserv Biol 19:1606–1618

Dennis RLH (2004) Butterfly habitats, broad scale biotope affiliations and structural exploitation of vegetation at finer scales: the matrix revisited. Ecol Entomol 29:744–752

Dover J, Settele J (2009) The influences of landscape structure on butterfly distribution and movement: a review. J Insect Conserv 13:3–27

Dover J, Sparks T (2000) A review of the ecology of butterflies in British hedgerows. J Environ Manage 60:51–63

Dover JW, Spencer S, Collins S, Hadjigeorgiou I, Rescia A (2010a) Grassland butterflies and low intensity farming in Europe. J Insect Conserv. doi:10.1007/s10841-010-9332-0

Dover JW, Rescia A, Fungariño S, Fairburn J, Carey P, Lunt P, Arnot C, Dennis RLH, Dover CJ (2010b) Land-use, environment, and their impact on butterfly populations in a mountainous pastoral landscape: species richness and Family-level abundance. J Insect Conserv. doi:10.1007/s10841-010-9331-1

Dover JW, Rescia A, Fungariño S. Fairburn J, Carey P, Lunt P, Dennis RLH, Dover CJ (2010c) Can hay harvesting detrimentally affect adult butterfly abundance? J Insect Conserv. doi: 10.1007/s10841-010-9267-5

Dunning J, Danielson B, Pulliam H (1992) Ecological processes that affect populations in complex habitats. Oikos 65:169–175

Erhardt A (1985) Diurnal Lepidoptera: sensitive indicators of cultivated and abandoned grassland. J Appl Ecol 22:849–861

Farino T (1999) Flora and Fauna of the Picos de Europa. Teresa Farino, Potes

Field A (2009) Discovering statistics using SPSS. Sage Publications, London

Gorton M, Hubbard C, Hubbard L (2009) The folly of European Union policy transfer: why the common agricultural policy (CAP) does not fit central and Eastern Europe. Reg Stud 43:1305–1317

Grime JP, Hodgson JG, Hunt R (2007) Comparative Plant Ecology: a functional approach to common British species, 2nd edn. Castlepoint Press, Dalbeattie

Hodge I, Reader M (2010) The introduction of entry level stewardship in England: extension or dilution in agri-environment policy? Land Use Pol 27:270–282

Jones AT, Hayes MJ (1999) Increasing floristic diversity in grassland: the effects of management regime and provenance on species introduction. Biol Conserv 87:381–390

Jongman RHG (2002) Homogenisation and fragmentation of the European landscape: ecological consequences and solutions. Landsc Urban Plann 58:211–221

Krausmann F, Haberl H, Schulz NB, Erb KH, Darge E, Gaube V (2003) Land-use change and socio-economic metabolism in Austria—part I: driving forces of land-use change: 1950–1995. Land Use Pol 20:1–20

Kruess A, Tscharntke T (2002) Grazing intensity and the diversity of grasshoppers, butterflies, and trap-nesting bees and wasps. Conserv Biol 16:1570–1580

Kuiper PJC, Bos M (1992) Plantago: a multidisciplinary study. Springer, Berlin

MacDonald D, Crabtree JR, Wiesinger G, Dax T, Stamou N, Fleury P, Lazpita JG, Gibon A (2000) Agricultural abandonment in mountain areas of Europe: environmental consequences and policy response. J Environ Manage 59:47–69

Marini L, Fontana P, Battisti A, Gaston KJ (2009) Agricultural management, vegetation traits and landscape drive orthopteran and butterfly diversity in a grassland-forest mosaic: a multi-scale approach. Insect Conserv Divers 2:213–220

McCune B, Grace JB (2002) Analysis of ecological communities. MjM Software Design, Oregon

Mottet A, Ladet S, Coque N, Gibon A (2006) Agricultural land-use change and its drivers in mountain landscapes: a case study in the Pyrenees. Agric Ecosyst Environ 114:296–310

Ouin A, Aviron S, Dover J, Burel F (2004) Complementation/supplementation of resources for butterflies in agricultural landscapes. Agric Ecosyst Environ 103:473–479

Plieninger T, Höchtl F, Spek T (2006) Traditional land-use and nature conservation in European rural landscapes. Environ Sci Policy 9:317–321

Pollard E, Elias DO, Skelton MJ, Thomas JA (1975) A method of assessing the abundance of butterflies in monks wood national nature reserve in 1973. Entomol Gaz 26:79–88

Pywell RF, Warman EA, Sparks TH, Greatorex-Davies JN, Walker KJ, Meek WR, Carvell C, Petit S, Firbank LG (2004) Assessing habitat quality for butterflies on intensively managed arable farmland. Biol Conserv 118:313–325

Renetzeder C, Schindler S, Peterseil J, Prinz MA, Mucher S, Wrbka T (2010) Can we measure ecological sustainability? Landscape pattern as an indicator for naturalness and land use intensity at regional, national and European level. Ecol Indic 10:39–48

Rescia AJ, Pons A, Lomba I, Esteban C, Dover JW (2008) Reformulating the social-ecological system in a cultural rural mountain landscape in Picos de Europa region (northern Spain). Landsc Urban Plan 88:23–33

Rundlof M, Smith HG (2006) The effect of organic farming on butterfly diversity depends on landscape context. J Appl Ecol 43:1121–1127

Rundlof M, Bengtsson J, Smith HG (2008) Local and landscape effects of organic farming on butterfly species richness and abundance. J Appl Ecol 45:813–820

Schneider C, Fry GLA (2001) The influence of landscape grain size on butterfly diversity in grasslands. J Insect Conserv 5:163–171

Schneider C, Fry G (2005) Estimating the consequences of land-use changes on butterfly diversity in a marginal landscape in Sweden. J Nat Conserv 13:247–256

Schwarzwälder B, Lörtscher M, Erhardt A, Zettel J (1997) Habitat utilization by the heath fritillary butterfly, *Mellicta athalia ssp. celadussa* (Rott) (Lepidoptera: Nymphalidae) in montane grasslands of different management. Biol Conserv 82:157–165

Thomas JA (1986) RSNC Guide to butterflies of the British Isles country life, Twickenham

van Swaay CAM, Warren MS (2003) Prime butterfly areas of Europe: priority sites for conservation. National reference centre for

agriculture, nature and fisheries. Ministry of Agriculture, Nature Management and Fisheries, The Netherlands
Warren MS (1992) Butterfly populations. In: Dennis RLH (ed) The ecology of butterflies in Britain. Oxford University Press, Oxford, pp 73–92
Warren M, Brereton T, Wigglesworth T (2005) Do agri-environment schemes help butterflies? experience from the UK. In: Kühn E, Feldmann R, Thomas JA, Settele J (eds) Studies on the ecology and conservation of butterflies in Europe, vol 1, General Concepts and Case Studies, pp 121–123
Weibull A-C, Bengtsson J, Nohlgren E (2000) Diversity of butterflies in the agricultural landscape: the role of farming system and landscape heterogeneity. Ecography 23:743–750

J Insect Conserv (2011) 15:179–188
DOI 10.1007/s10841-010-9336-9

ORIGINAL PAPER

Landscape scale conservation: resources, behaviour, the matrix and opportunities

T. G. Shreeve · R. L. H. Dennis

Received: 24 March 2010 / Accepted: 31 August 2010 / Published online: 24 September 2010

Abstract Landscape scale conservation efforts are becoming more commonplace in conservation, with a move from single species to multi-species initiatives. These initiatives are reliant on modelling processes, largely underpinned by metapopulation models. We argue that generic models developed for individual species in particular landscapes over selected time periods may only be applicable to alternative landscapes and time periods in restricted circumstances. Variability in species responses to landscapes and environmental conditions is dependent on a range of species-specific intrinsic characteristics, dependent on their responses to resources, (including weather) and also individual states. We propose that the behavioural component of how species respond to resources needs to be taken into account in modelling species responses to landscape, and therefore how limited resources for conservation are deployed. Species behaviours are inherently complex. We argue that because of this complexity the conservation of the majority of species, especially of the least rare, may be best served if conservation effort is additionally focused on increasing landscape heterogeneity and disturbance. This may also facilitate persistence in the face of climate change. We suggest that heterogeneity should be promoted through agri-environment schemes.

T. G. Shreeve (✉) · R. L. H. Dennis
School of Life Sciences, Oxford Brookes University, Headington, Oxford OX3 0BP, UK
e-mail: tgshreeve@brookes.ac.uk

R. L. H. Dennis
Institute for Environment, Sustainability and Regeneration, Staffordshire University, Mellor Building, College Road, Stoke-on-Trent ST4 2DE, UK

Keywords Butterfly conservation · Landscape-scale conservation · Behaviour and conservation · Resources · Mobility · Space-time heterogeneity

Introduction

The persistence of populations at the landscape scale is intricately connected with landscape configuration, specifically the abundance and distribution of exploitable resources (e.g. Ricketts 2001; Fahrig and Nuttle 2005; Nabe-Nielsen et al. 2010; Swift and Hannon 2010). Declines of the majority of butterfly species, especially in the Palaeractic, are well monitored (van Swaay and Warren 1999; Asher et al. 2001; Van Swaay and Van Strien 2005, 2008). Habitat loss and fragmentation are identified as the primary causes of these declines (e.g. Thomas et al. 2001; Stefanescu et al. 2004). In the Palaearctic this is a result of degradation of sites and landscape simplification through changes to the agricultural landscape and increasing development. Within the Palaearctic this has led to a loss of traditional management within woodlands, loss of heterogeneity within the lowland cultivated landscape, and a reduction of species rich upland and montane grassland (e.g. van Swaay and Warren 2003; Stefanescu et al. 2005). Erosion of concentrations of resources (e.g. exploitable host plants), reduces the spatial occurrences of populations, reduces population sizes and increases the chance of local extinction through a number of stochastic and population processes. Fragmentation effectively increases inter-patch distance and therefore decreases the chances of recolonization and rescue effects (Hames et al. 2001).

Whilst many species are in decline, the realised patterns of species' declines are not simple. Attention is often focused on already scarce species, but there is evidence that what are generally regarded as widespread species are declining as rapidly or even more rapidly than the already vulnerable species (León-Cortés et al. 1999, 2000; Asher et al. 2001; Van Dyck et al. 2009) and it is suggested that these declines of common species may be related more to changes to the landscape other than to fragmentation effects per se (e.g. Aviron et al. 2007; Lütolf et al. 2009). There are also instances of some species increasing in spatial extent, and some classically 'sedentary' species having recolonized locations within their ranges (e.g. *Hesperia comma*, Davies et al. 2005), while others have increased in range (e.g. *Pararge aegeria*, Hill et al. 1999; *Arashnia levana*, Mittka et al. 2009) despite simplification of landscapes and loss of resources in the areas that have been colonized. Together, declines and increases point to complex responses to the spatial distributions of resources within landscapes, and to a temporal component to distribution changes (e.g. Mittka et al. 2009). The maintenance of multi-species communities in modern landscapes is therefore problematic (Settele et al. 2009), as individual species have unique responses to specific site changes, landscape changes, climate and weather.

Historically, conservation effort has focused on species of conservation concern, with emphasis on maintenance of species within existing sites. More recently the landscape context of conservation has come to the fore. This is fostered by the understanding that populations have finite extinction probabilities as part of natural and anthropogenic processes, and the persistence and enhancement of species is best served by landscape scale efforts, focusing on habitat networks and issues related to connectivity (Dennis 2010). Such an approach could facilitate recovery of individual species and biodiversity in response to past and current landscape change, especially if combined with appropriately targeted agri-environment schemes (Donald and Evans 2006). However the variability in species' responses to specific landscape configurations, including specific resource patch configurations, linear features and barriers, makes this a daunting challenge as specific landscape features may act positively for some species and negatively for others and their influence on individual species may vary temporally.

Landscape scale approaches to conservation

The landscape approach to conservation has been enhanced by the development and application of metapopulation modelling (e.g., Ehrlich and Hanski 2004), representing a significant shift in conservation thinking. However, the focus has been largely on species of conservation concern and of declining biotopes. Whilst such an approach might address the attainment of biodiversity targets embodied in legislation and targeted directives, there are two potential pitfalls. The first is that a focus on target species and biotopes may not promote overall biodiversity sufficiently well. Common species, are generally assumed to be of less concern than priority species, and it is tacitly assumed that catering for target species may act as regenerators for overall biodiversity, though this remains to be adequately tested (Lindenmeyer et al. 2002; but see Lambeck 1997; Roberge and Angelstam 2004). Assumptions are often made on the basis of observed associations of species with particular biotopes rather than comprehensive analyses that associate species in relation to precise resource requirements (utilities and consumables) and shared traits (see Shreeve et al. 2001, 2004; Dennis et al. 2003, 2006, 2007). The second potential problem is that our understanding of how species respond to landscape configuration, including spatial and temporal variation within regions has a limited evidence base and specific model parameters may be taken from different species and landscapes (see Schtickzelle and Baguette 2009).

There are good metapopulation models for some species (e.g. *Melitaea cinxia*, *Proclossiana eunomia*, Hanski 1994; Schtikzelle et al. 2006); the best models are for species with prescribed and limited patches within landscapes where the matrix is generally devoid of resources for the selected species (Dennis et al. 2003). Even then, resource-based models of patch suitability (i.e. those that include a component of quality) may be better descriptors of patch carrying capacity than area measures (Dennis and Eales 1997, 1999; Turlure et al. 2010). Whilst these models are applicable to particular species with highly specific requirements for resources that are scarce within landscapes, the components that go into these models may not be representative for the majority of species especially those with wide resource ranges. Without empirical evidence on species' responses to landscapes there is a risk that landscape model approaches, however sophisticated the models, may not provide the best solution to overall biodiversity loss.

Within the field of landscape planning models are sometimes proposed that are based on assumptions that may be divorced from reality, even though the approaches seem to offer a shortcut to conservation. The approaches fall into three main categories. First, there are those that use an umbrella species, based on the assumption that the response of the umbrella species is representative of other species with which it co-occurs (but see Mac Nally et al. 2002) The second approach uses focal species in which a limited suite of species is assumed to represent the majority. In both these approaches there are issues in selecting either single or suites of representative taxa, if only because all species have unique resource requirements and responses to environmental variables (Bifolchi and

Lodge 2005; Seddon and Leech 2008; but see Spitzer et al. 2009). Further difficulties may arise if the taxa are selected as being representative on the basis of biotope or vegetation associations in common with other species since these associations may not represent similarity of resource requirements; the latter define habitats (Dennis et al. 2003, 2006, 2007) The third approach uses ecological profiles, sometimes generic profiles, in which responses to landscape configurations are derived on the basis of how organisms with generalised (often averaged) traits respond to specific landscape features. Again, this approach needs to be treated with caution especially if the traits are poorly defined. The tools to construct landscape models are widely available, but the data on which to base these models are scarce. Rightly, models are approximations, but a fundamental issue is whether they should, and can, be made to be more realistic using parameters that can be applied across landscapes and time scales.

A corollary issue is whether there are alternative, more cost-effective, methods for promoting biodiversity, ones that may secure eco-system services in the face of limited information on how and why species respond to landscape features. Historic changes of land-use have undoubtedly contributed to species' declines. Because of socio-economic pressures, wholesale reversion of the landscape is not possible. In lowland areas conservation and biodiversity promotion is therefore faced with the challenge of maintaining biodiversity in marginal and linear areas within a mostly intensively exploited agricultural landscape. In upland areas the problems are different, where the challenge is not necessarily a lack of land area that could be used to promote biodiversity, but a lack of resources due to a lack of management and a paucity of land-users to combat the changes of open areas to scrub and woodland (van Swaay and Warren 2003).

Current approaches to modelling species reactions to landscape

The simplest models for most species regard the landscape as a patch-matrix configuration. Where resources occur in discrete patches and the intervening matrix is devoid of resources the persistence of populations at the landscape scale can be modelled with a Stochastic Patch Occupancy Model, or Incidence Function Model (e.g. Hanski 1994; Wahlberg et al. 2002). Simple models may be extended to include spatial realism by incorporating differing connectivity of patches and population dynamics within patches (Spatially Realistic Metapopulation Theory) (Hanski 2004; Hanski and Ovaskainen 2003). More complex models, such as Individual Based Models have also been developed (e.g. Keeling 2002; Ovaskainen and Hanski 2004), in which movement rules, patch shapes and landscape arrangement are integrated. These different models have contributed greatly to our understanding of how species react to patch arrangements in landscapes and have helped move the focus of conservation from the patch to the landscape. They have been tested against empirical evidence, and can explain the persistence of some species in specific landscapes. However, they may do little to help us understand how changes of landscape configuration, and temporal changes of environmental conditions, which may change mobility patterns, actually influence patch connectivity and whether specific landscape features decrease or promote connectivity. In addition they may have limitations in explaining how species with diffuse resources react to landscape configuration.

Patch connectivity can be modelled in different ways, and is often one of the most elusive elements in modelling species in fragmented landscapes. Most models incorporate a distance decay function in which the probability of individuals successfully moving into a patch may then be approximated by the following general expression:

$$P_i = \sum_{j=1}^{n} M_j * \exp(-\lambda * d_{ij})$$

where P_i = probability of immigration into a patch, M_j = number leaving a patch, λ, = migration mortality and d_{ij} = distance between two patches. M_j is usually some function (often density dependent) of the number of individuals in a patch (N_t). N_t has been modelled in a number of ways, usually incorporating information about N_{t-1}, intrinsic growth rate r and carrying capacity K (which may be area or resource based). Patch population growth rate may be modelled as a density dependent function of population size, or may assume that populations persist at a maximum, dependent on weather. Modelling of both N_t and P_i may also incorporate random variables.

In such a system the number of individuals successfully moving between two landscape points declines exponentially with distance and the only landscape effect on patch connectivity is distance itself. However, the parameters in such a model may be derived from real landscape studies, but caution is required in applying the parameters to either a different landscape or to the same landscape at a different time because a different landscape configuration and/or set of overall environmental conditions (e.g. temperature) may influence the ease with which individuals can traverse a landscape and how quickly they do so. This may be a direct landscape effect itself or may be because of within-patch changes of traits that influence dispersal capacity (see Schtickzelle and Baguette 2009).

In Individual Based Models, mobility is usually modelled as a form of a random walk in which step lengths follow a Poisson distribution and turn angles follow a random distribution. Whilst this begins to introduce a more complex movement pattern, it makes the assumption that step lengths and turn angles are largely independent of the landscape. If resources, landscape configuration and weather influence the speed of movement though a landscape (e.g. Dover and Fry 2001), distances moved, and movements around topographic features, then models based on these functions may oversimplify how real organisms interact with the landscape.

In modelling populations within landscapes the components involved are not just distances, connectivity and movement patterns, but also the propensity for individuals to move from existing locations. Most models make the assumption that such out-movements are dependent on density; often a density-dependent function is incorporated, but the strength of density-dependence may vary (e.g. Baguette et al. 1998; Stevens et al. 2010). Although population size in relation to area is often invoked as a critical factor for movements between patches, critical resources, rather than area may be involved (Turlure et al. 2010). In addition movements of individuals from patches may not be a simple density-dependent process. For example Shreeve et al. (1995) determined that the residence times of *Maniola jurtina* in resource patches, was low at both low and high population density. Two factors were invoked to explain this: for males a lack of contact with conspecifics at low density leads to movement away from patches, as does a shortage of resources and harassment at high density.

Whilst the dynamics of some populations over time have been successfully modelled, well parameterized models are best applied to species that use scarce, patchily distributed resources, and for which the matrix between the patches is essentially empty. Problems arise when the matrix is structurally heterogeneous (H. Van Dyck et al., in preparation), irrespective of whether the matrix contains consumable resources for the species, since movements of individuals will no longer be a random process, structural features imparting directionality and generating changes in velocity through the matrix. Exploitable resources further complicate connectivity if they influence behaviour of individuals within the matrix. For many butterfly species which currently occur within landscapes in discrete patches, explicit modelling of connectivity is thus difficult because the landscape is a complex of spatially and temporally diverse resources and utilities (Dennis 2010).

Irrespective of connectivity issues, the propensity of an individual to leave a patch may also be dependent on the nature of the boundary between the patch and the matrix: For example, individuals of *Maniola jurtina* and *Pyronia tithonus* alter their flight behaviour at patch boundaries (Conradt and Roper 2006) and related *Maculinea* species have different responses to specific edge types (Batary et al. 2009). For some species, and in some landscapes, corridors are being promoted as features that enhance connectivity of fragmented populations, and facilitate species reaching and finding new, but unoccupied, patches. In a few cases corridors and their key features for particular species have been specifically identified, using resource selection functions, least-cost models (e.g. Adriaensen et al. 2003; Chetkiewicz and Boyce 2009), and individual based models (e.g. Tracey 2006). However, the role of corridors and their physical makeup remains equivocal (e.g. Davies and Pullin 2007; H. Van Dyck et al., in preparation) although physical landscape features undoubtedly influence movements of butterflies (Dover 1997; Dover and Fry 2001). By analogy, if landscapes have features that positively influence permeability, there will also be some features that act negatively; barriers, through the influence of any particular feature, may vary temporally and with encounter frequency (H. Van Dyck et al., in preparation).

Clearly modelling how populations persist in landscapes is complex, but for some species in some landscapes models have been successfully parameterized and applied (see Schtickzelle and Baguette 2009). However, transferring models to different landscapes or even to the same landscape at a different time is potentially problematic if movements are affected by landscape features or weather. Thus, a landscape configuration suitable at one time and one place may not be adequate for other times and places.

Understanding the consequences of landscape configuration for patch connectivity for most animal species, especially butterflies whose dispersive movements are hard to track, is a relatively poorly researched area (but see Cant et al. 2005; Hovestadt and Nowicki 2008; Ovaskainen et al. 2008). Two approaches can be taken; the first is examining the effect of landscape configuration (including within patch factors) on inter-patch movements (e.g. Roland and Matter 2007; Fred and Brommer 2009). The second is to adopt a behavioural approach (e.g. Cant et al. 2005; Baguette and Van Dyck 2007) which recognises that movement is a behavioural response to individual requirements and resource configurations. For many taxa, especially insects, our knowledge of the behavioural responses to landscape is limited, though the resultant processes might be approximated (Dennis and Hardy 2007).

We therefore suggest that a behavioural approach to patch connectivity and movement away from patches may identify those factors that we need to understand in order to determine how and why species will respond to landscape configurations and why these responses may differ spatially and temporally (Dennis 2004). We advocate this approach because behaviourally detailed individual-based models should be sensitive to changes of landscape configuration and population parameters. This may be particularly important as species are not only faced with current landscape changes but with climate change. The latter may mean that some species are unlikely to persist in current locations but we currently have insufficient information on species/landscape interactions to fully understand what we need to do to facilitate movement through current landscapes.

A behavioural model of landscape use

A model of connectivity in a patch-matrix landscape, should involve three processes, (1) why individuals leave resource patches that support populations, (2) how they respond to the matrix and (3) how they detect and move into potential patches. Such an approach adopts a habitat patch/hostile matrix configuration of the landscape. For some species the landscape is not entirely hostile and resources may be present in the matrix (Dennis 2004; Dennis and Hardy 2007). These matrix resources may provide opportunities for some essential functions and may theoretically increase connectivity if they increase either individual longevity or reduce travel time. Alternatively they may decrease connectivity if they impede the movement of individuals between high quality locations (Dennis 2010). Consumables and utilities in the matrix can be considered as a multidimensional continuum (each resource = 1 dimension); from a complete absence, to one in which the distribution, abundance and quality of resources is equal to that of prime resource patches. If consumables and utilities (e.g. temperature, solar radiation) are spatially and temporally variable then modelling of connectivity and landscape use needs to take this into account. Because patch and matrix resource quality and quantity can vary then even in a metapopulation context sites may temporally shift from sources to sinks and vice versa (Thomas and Kunin 1999), making landscape models difficult to construct and apply over different time scales. At an extreme, a patch may become matrix and matrix may become patch, especially with rotation in landuse. More fundamentally, the response of individuals to the matrix could be modelled in the same way as the response of individuals to factors that influence residence and emigration within conventionally conceived patches. The reason for this is that individuals' behaviours emanate as responses to resources, immediate requirements and interactions with other individuals, factors that are continuously variable throughout the landscape.

Behavioural responses to resource configurations, and thus to landscapes, are dependent on a range of factors, from genetics, developmental history, physiological state, past experience, future reproductive potential, perceptual ability and interactions with conspecifics and other interacting species. To fully understand how individuals react both temporally and spatially to particular landscape configurations therefore requires an understanding of these factors that impinge on the immediate behaviour of individuals. Critical understanding is needed of why individuals stay in a location or move, and, if the latter, how movement behaviours are influenced by resources and conditions in the landscape. Because organisms' states can change over time, and resources and utilities can also change with time, there is no expectation of a uniform response to a particular landscape amongst individuals of a particular species, nor is there any expectation of the same range of responses within individuals in a particular landscape at different times.

A complicating factor in modelling movement in landscapes is that movements may fall into different categories, dispersive and non-dispersive movements. The former may be categorised by linear flight direction, with relatively few turns and changes in orientation to landscape features and resources (Cant et al. 2005). Individuals adopting these different flight modes may also react differently to identical features in the landscape. Thus, structural features that may act as barriers for 'non-dispersive' individuals may not act in the same way for dispersive individuals, especially if their flight patterns and flight heights are different. Butterflies from different landscapes may also have different flight behaviours and abilities to orientate to landscape features and also have different propensities to cross landscape and patch features (Ross et al. 2005; Van Dyck and Baguette 2005; Merckx and Van Dyck 2007; Schtickzelle et al. 2007). The tendency to make dispersive flights, and the distances moved and directions taken, may also be linked to local weather conditions (Mitikka et al. 2008). In addition, rapid changes in the propensity of individuals from populations in different landscapes to disperse represent evolutionary changes to landscape structure (e.g. Hill et al. 1996). Although differences of flight behaviours are recognised and differences may exist between populations of the same species in different landscapes, the factors that determine which flight mode individuals adopt are obscure.

As species can make dispersive flights, the critical issues are what determines when individuals make dispersive flights, what they do once they make this decision and how landscape features influence movement through the landscape or residence in the landscape. We outline the main factors involved in Fig. 1. There are many interacting factors within the process-response model, with the number of relationships indicating complexity. However, the major relationships are between landscape structure, resources, the thermal environment, genotype and morphology, flight, perception and behaviour within resource patches and a patch leaving threshold. Each of these major variables is further subdivided into component parts. If behaviourally detailed individual-based models can be parameterised then there is the chance of generating powerful tools to

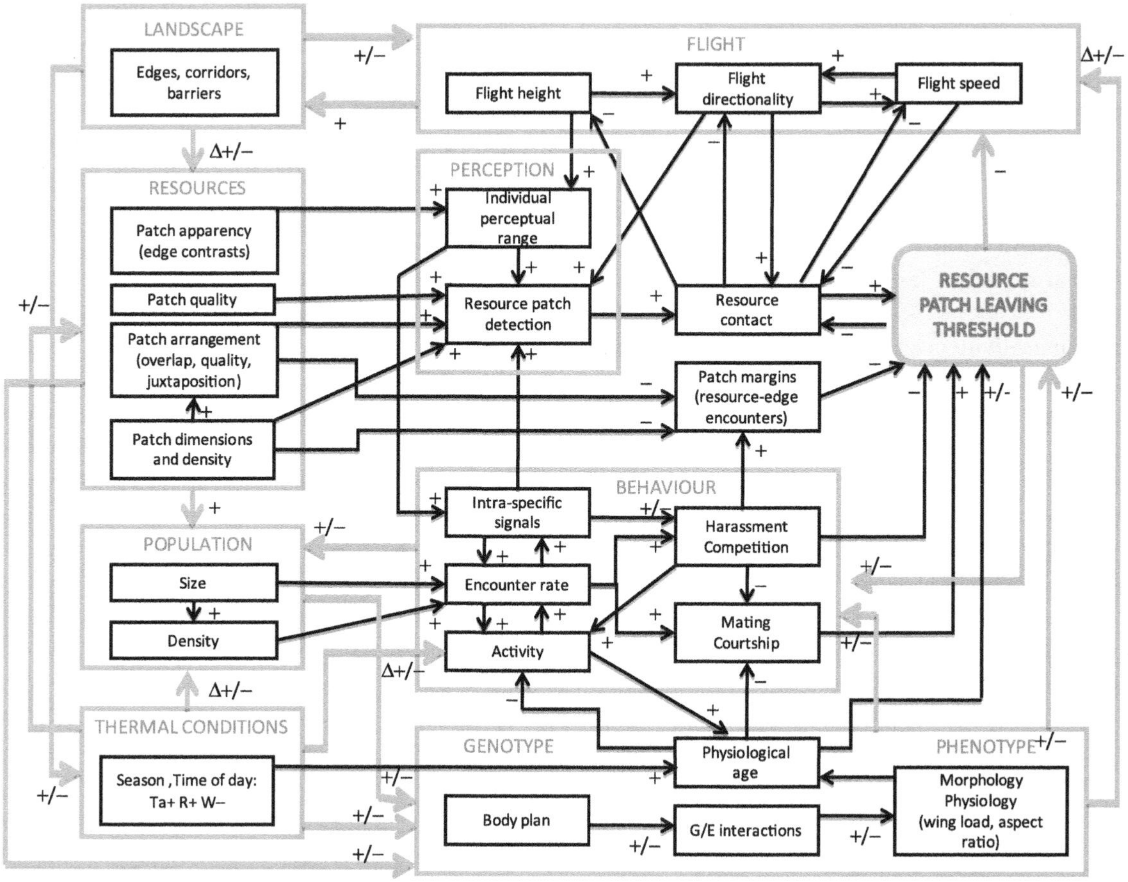

Fig. 1 A process-response model of the behavioural components of adult butterfly responses to landscape characteristics. Major variable subsystems and behavioural responses are indicated by thick box margins and the component parts of each by thin box margins. Arrows indicate connections between processes and responses and a + or − indicates either a positive or a negative relationship between factors. A ± indicates the relationship could be either. Central to the model is the decision butterflies make to stay or move from a resource (resource patch leaving threshold) which can lead to increased or reduced residence times at a resource and a cascade of consequences for use of patches, and how individuals interact with landscape features and other individuals within population units or elsewhere

predict species responses to specific landscape configurations and conditions. However a challenge for such behavioural models is that link-strengths and variables will differ between species. The spatial configuration of consumables in the landscape and the state of utilities (e.g. temperature, solar radiation intensity) will determine the likelihood of individual reproductive success and longevity in the landscape. Thus a landscape configuration that benefits one species may be of different value to another as behavioural responses to resource distributions will differ. This is a complex problem and, as with more conventional models, the issues for conservation are immediate and cannot wait for multispecies models to be developed.

Landscape configuration: maximizing the potential

So, do we design landscapes for species of concern, or do we try and configure landscapes for the benefit of multiple species? Although agri-environment schemes may offer some hope via targeted restoration the land available for conservation is limited in intensive agricultural areas or the resources for management are severely restricted in less intensively farmed areas (e.g. uplands). One approach to maximising benefit is to identify species with similar ecological attributes. Such an approach (Shreeve et al. 2001) has identified ecological attributes of species and identified species groups that may react in a similar way to specific landscape changes. For the resident butterflies in the UK four groups, each with overlapping resource requirements and traits are identified, those with resources usually associated with tall grassland and its structural component, those associated with resources which are restricted to forest structures, those with resources that commonly occur in xeric open areas, and a group of species associated with ruderal resources that occur in disturbed locations. Identification of precise resource requirements also determines the likelihood of similarity of response to landscape change, but biotope association does not (Shreeve et al. 2001; Dennis 2010). However, how do we maximise benefits for species groups with conflicting requirements when land availability and resources for conservation effort are limited? A solution may come from the work on individual species' abundances and species' richness in relation to spatial heterogeneity (Steffan-Dewenter et al. 2002; Cozzi et al. 2008; Kumar et al. 2009), where spatial heterogeneity equates to the heterogeneity of resources and structures. Enhancing butterfly richness in the landscape should therefore be a matter of increasing spatial heterogeneity in resources within the landscape, but the emerging issue is at what scale? Studies of social Hymenoptera (Steffan-Dewenter et al. 2002), butterflies in south west France (Rossi and Van Halder 2010), fritillary species in the Swiss Alps (Cozzi et al. 2008) and shrubland butterflies in Colorado (Kumar et al. 2009) indicate a relationship between species richness and landscape heterogeneity and that the spatial scale of heterogeneity individual species relate to is dependent on their mobility. If heterogeneity is consistent within a landscape then the issue becomes one of what is the minimum scale of heterogeneity required to maximise species richness within landscapes? The studies (see above) of responses of individual species or groups of species indicate that the minimum scale of land-use and management diversity to which species respond may be between 250 and 3,000 m. Because responses of species to resources in the landscape are dependent on individuals detecting sources for them, or major landscape features, the question of scale is intricately linked with perceptual ability, which is a behavioural characteristic. Those few studies of butterfly perceptual ability indicate that species may be able to detect landscape features from a distance of 100–200 m (Cant et al. 2005; Merckx and Van Dyck 2007). Thus, small scale heterogeneity, with patch dimensions in the order of 100–200 m may be sufficient to promote species in heavily modified environments. Within-patch structural and botanical diversity may also promote species richness (e.g. Spitzer et al. 2009; Dennis 2010).

Increasing heterogeneity does not imply that existing sites should be heavily modified, but is an issue about how to make changes to those land-parcels and marginal areas where effort may offer the chance to promote overall biodiversity in the landscape and facilitate persistence of existing population units. Although there are some higher tier agri-environment schemes within the EU which aim to reduce inputs to agricultural systems and restore more traditional land-use patterns, such schemes are relatively rare and also dependent on acceptance of socio-economic constraint by land-users (but see Aviron et al. 2009). The issue is therefore how to make the best of the intervening units with limited resources and space. Because of the complexity of species' responses to landscapes in both space and time predictive modelling on the basis of limited information may not be the best approach. Increasing heterogeneity should maximise resource types and configurations for more species than efforts targeted to single species or specific vegetation types, although individual species may not use all apparently appropriate resource configurations at any one time. Some of the resource requirements for butterflies in the British Isles are now collated (see Shreeve et al. 2001; Dennis et al. 2008; Dennis 2010; Hardy 2010) and these data should be used to help decide what resources could be re-introduced into the landscape, or promoted, to maximise heterogeneity.

We do not suggest that the creation of large patches of 'new habitat' is the only solution to increasing species occurrences but advocate the additional strategy of increasing heterogeneity by the creation of more small exploitable resource patches within the landscape. For similarly sized areas the approach of more small patches rather than a fewer larger patches may increase richness (e.g. Tscharntke et al. 2002) and connectivity by reducing individual search times (Doak et al. 1992; Tischendorf and Fahrig 2000) and decreasing mortality probability when individuals are in resource free areas of the landscape. Increasing spatial heterogeneity therefore becomes the logical extension of increasing the number of small resource patches when precise resource requirements for a range of species cannot be adequately defined.

Species with specialised resource requirements and with scarce resources are generally of conservation concern and there is a legitimate reason for trying to focus effort at the landscape scale to maintain and enhance their abundance and spatial occurrence. Formerly common species have also undergone large scale reductions in the general landscape. Preservation of the flagship species and enhancement of the common need not conflict. Most specialised species are confined to specific sites, some of which are extremely isolated. However, identification of resources and structural features that may enhance the connectivity of the existing patches of priority species may provide opportunities for enhancement of the more common species. General adoption of less intensive (but variable) management of marginal, unproductive land, such as verges (e.g. Ries et al. 2001; Valtonen et al. 2007; Noordijk et al. 2009) field margins (e.g. Dover 1997; Field et al. 2005, 2006, 2007; Merckx et al. 2009) and linear features (Dover and Sparks 2000; Dover et al. 2000; Ouin and Burel 2002; Croxton et al. 2005) and incorporation of disturbance to encourage natural resource changes should promote resource heterogeneity. This may also address issues around marginality of core and peripheral populations (Shreeve et al. 1996). Ultimately we need greater understanding of individual species responses to particular landscape configurations to predict the viability of suites of species to particular configurations via behavioural responses to resources.

We argue that incorporating behavioural responses to landscape, population and resource conditions could provide a powerful tool for predicting species responses to both landscape and climate change. However, the processes and relationships governing these behavioural responses, outlined in our process-response model, are complex and many await quantification. We argue that conventional models, although applicable in specific circumstances, may not be transferrable across space and time because the behavioural elements of species responses to landscape and resource configuration are not explicitly incorporated or fully parameterised. However the need for intervention is urgent if we are to halt, or even reverse, species' declines and we therefore advocate increasing spatial resource heterogeneity to enhance landscapes as the most appropriate conservation tool when knowledge of the requirements, behaviours and tolerances of the majority of species are poorly known. We also suggest that this can be achieved if vegetational structural and compositional diversity is incorporated within cross-compliance agri-environment schemes. Such structural and compositional diversity will promote resources, both consumables and utilities (see Dennis et al. 2003), thereby facilitating opportunities for species richness of butterflies and other taxa.

References

Adriaensen F, Chardon JP, De Blust G, Swinnen E, Villalba S, Gulinck H, Matthysen E (2003) The application of 'least-cost' modelling as a functional landscape model. Landsc Urban Plan 64:233–247

Asher J, Warren MS, Fox R, Harding P, Jeffcoat G, Jeffcoat S (2001) The millennium atlas of butterflies in Britain and Ireland. Oxford University Press, Oxford

Aviron S, Kindlmann P, Burel F (2007) Conservation of butterfly populations in dynamic landscapes: the role of farming practices and landscape mosaic. Ecol Model 205:135–145

Aviron S, Nitsch H, Jeanneret P, Buholzer S, Luka H, Pfiffner L, Pozzi S, Schuepbach B, Walter T, Herzog F (2009) Ecological cross compliance promotes farmland biodiversity in Switzerland. Front Ecol Environ 7:247–252

Baguette M, Van Dyck H (2007) Landscape connectivity and animal behavior: functional grain as a key determinant for dispersal. Landscape Ecol 22:1117–1129

Baguette M, Vansteenwegen C, Convi I, Neve G (1998) Sex-biased density-dependent migration in a metapopulation of the butterfly *Proclossiana eunomia*. Acta Oecol 19:17–24

Batary P, Koroesi A, Oervoessy N, Koever S, Peregovits L (2009) Species-specific distribution of two sympatric Maculinea butterflies across different meadow edges. J Insect Conserv 13:223–230

Bifolchi A, Lodge T (2005) Efficiency of conservation shortcuts: an investigation with otters as umbrella species. Biol Conserv 126:523–527

Cant E, Smith AD, Reynolds DR, Osborne JL (2005) Tracking butterfly flight paths across the landscape with harmonic radar. Proc R Soc B Biol Sci 272:785–790

Chetkiewicz CLB, Boyce MS (2009) Use of resource selection functions to identify conservation corridors. J Appl Ecol 46:1036–1047

Conradt L, Roper TJ (2006) Non-random movement behavior at habitat boundaries in two butterfly species: implications for dispersal. Ecology 87:125–132

Cozzi G, Mueller CB, Krauss J (2008) How do local habitat management and landscape structure at different spatial scales affect fritillary butterfly distribution on fragmented wetlands? Landsc Ecol 23:269–283

Croxton PJ, Hann JP, Greatorex-Davies JN, Sparks T (2005) Linear hotspots? The floral and butterfly diversity of green lanes. Biol Conserv 121:579–584

Davies ZG, Pullin AS (2007) Are hedgerows effective corridors between fragments of woodland habitat? An evidence-based approach. Landsc Ecol 22:333–351

Davies ZG, Wilson RJ, Brereton TM, Thomas CD (2005) The re-expansion and improving status of the silver-spotted skipper butterfly (*Hesperia comma*) in Britain: a metapopulation success story. Biol Conserv 124:189–198

Dennis RLH (2004) Butterfly habitats, broad scale biotope affiliations and structural exploitation of vegetation at finer scales: the matrix revisited. Ecol Entomol 29:744–752

Dennis RLH (2010) A resource-based habitat view for conservation. Wiley-Blackwell, Chichester

Dennis RLH, Eales HT (1997) Patch occupancy in *Coenonympha tullia* (Lepidoptera: Satyridae): habitat quality matters as much as patch size and isolation. J Insect Conserv 1:167–176

Dennis RLH, Eales HT (1999) Probability of patch site occupancy in *Coenonympha tullia* (Müller) (Lepidoptera: Satryinae) determined from geographical and ecological data. Biol Conserv 87:295–301

Dennis RLH, Hardy PB (2007) Support for mending the matrix: resource seeking by butterflies in apparent non-resource zones. J Insect Conserv 11:157–168

Dennis RLH, Shreeve TG, Van Dyck H (2003) Towards a functional resource-based concept for habitat: a butterfly biology viewpoint. Oikos 102:417–426

Dennis RLH, Shreeve TG, Van Dyck H (2006) Habitats and resources: the need for a resource-based definition to conserve butterflies. Biodivers Conserv 15:1943–1966

Dennis RLH, Shreeve TG, Sheppard A (2007) Species conservation and landscape management. In: Stewart AJA, New TR, Lewis OT (eds) Insect conservation biology: proceedings of the royal entomological society's 23rd symposium. CABI Biosciences, Wallingford, pp 92–126

Dennis RLH, Hardy PB, Shreeve TG (2008) The importance of resource databanks for conserving insects: a butterfly biology perspective. J Insect Conserv 12:711–719

Doak DF, Marion PC, Kareiva PM (1992) Spatial scale mediates the influence of habitat fragmentation on dispersal success—implications for conservation. Theor Popul Biol 41:315–336

Donald PF, Evans AD (2006) Habitat connectivity and matrix restoration: the wider implications of agri-environment schemes. J Appl Ecol 43:209–218

Dover JW (1997) Conservation headlands: effects on butterfly distribution and behaviour. Agric Ecosyst Environ 63:31–49

Dover JW, Fry GLA (2001) Experimental simulation of some visual and physical components of a hedge and the effects on butterfly behaviour in an agricultural landscape. Entomol Exp Appl 100:221–233

Dover J, Sparks T (2000) A review of the ecology of butterflies in British hedgerows. J Environ Manage 60:51–63

Dover J, Sparks T, Clarke S, Gobbett K, Glossop S (2000) Linear features and butterflies: the importance of green lanes. Agric Ecosyst Environ 80:227–242

Ehrlich PR, Hanski I (2004) On the wings of Chekerspots: a model system for population biology. Oxford University Press, Oxford

Fahrig L, Nuttle WK (2005) Population ecology in spatially heterogeneous environments. In: Lovett GM, Jones CG, Turner MG, Weathers KC (eds) Ecosystem function in heterogeneous landscapes. Springer, New York, pp 95–118

Field RG, Gardiner T, Mason CF, Hill J (2005) Agri-environment schemes and butterflies: the utilisation of 6 m grass margins. Biodivers Conserv 14:1969–1976

Field RG, Gardiner T, Mason CF, Hill J (2006) Countryside stewardship scheme and butterflies: a study of plant and butterfly species richness. Biodivers Conserv 15:443–452

Field RG, Gardiner T, Mason CF, Hill J (2007) Agri-environment schemes and butterflies: the utilisation of two metre arable field margins. Biodivers Conserv 16:465–474

Fred MS, Brommer JE (2009) Encroaching forests decouple alpine butterfly population dynamics. Proc Natl Acad Sci USA 104:13702–13704

Hames RS, Rosenberg KV, Lowe JD, Dhondt AA (2001) Site reoccupation in fragmented landscapes: testing predictions of metapopulation theory. J Anim Ecol 70:182–190

Hanski I (1994) A practical model of metapopulation dynamics. J Anim Ecol 63:151–162

Hanski I (2004) Metapopulation theory, itause and misuse. Basic Appl Ecol 5:225–229

Hanski I, Ovaskainen O (2003) Metapopulation theory for heterogeneous lanscapes. Theor Popul Biol 64:119–127

Hardy PB (2010) Butterfly nectar and hostplant database. http://www.staffs.ac.uk/schools/sciences/geography/links/IESR/staff_honfellows_rd_butterflydatabase

Hill JK, Thomas CD, Lewis OT (1996) Effects of habitat patch size and isolation on dispersal by *Hesperia comma* butterflies: implications for metapopulation structure. J Anim Ecol 65:725–735

Hovestadt T, Nowicki P (2008) Resources influence dispersal and population structure in an endangered butterfly. Isr J Ecol Evol 54:137–154

Keeling MJ (2002) Using individual-based simulations to test the Levins metapopulation paradigm. J Anim Ecol 71:270–279

Kumar S, Sionson SE, Stohlgren TJ (2009) Effects of spatial heterogeneity on butterfly species richness in Rocky Mountain National Park, CO, USA. Biodivers Conserv 18:739–763

Lambeck RJ (1997) Focal species: a multi-species umbrella for nature conservation. Conserv Biol 11:849–856

León-Cortés JL, Cowley MJR, Thomas CD (1999) Detecting decline in a formerly widespread species: how common is the common blue butterfly *Polyommatus icarus*? Ecography 22:643–650

León-Cortés JL, Cowley MJR, Thomas CD (2000) The distribution and decline of a widespread butterfly *Lycaena phlaeas* in a pastoral landscape. Ecol Entomol 25:285–294

Lindenmeyer DB, Manning Ad, Smith PL, Possingham HP, Fischer J, Oliver I, McCarthy MA (2002) The focal-species approach and landscape restoration: a critique. Conserv Biol 16:338–345

Lütolf M, Guisan A, Kienast F (2009) History matters: relating land-use change to butterfly species occurrence. Environ Manage 43:436–446

Mac Nally R, Bennett AF, Brown GW, Lumsden LF, Yen A, Hinkley S, Lillywhite P, Ward DA (2002) How well do ecosystem-based planning units represent different components of biodiversity. Ecol Appl 12:900–912

Merckx T, Van Dyck H (2007) Habitat fragmentation affects habitat-finding ability of the speckled wood butterfly, *Pararge aegeria* L. Anim Behav 74:1029–1037

Merckx T, Feber RE, Riordan P, Townsend MC, Bourn NAD, Parsons MS, Macdonald DW (2009) Optimizing the biodiversity gain from agri-environment schemes. Agric Ecosyst Environ 130:177–182

Mitikka V, Heikkinen RK, Lutot M, Araujo MB, Saarinen K, Poyry J, Fronzek S (2008) Predicting range expansion of the map butterfly in Northern Europe using bioclimatic models. Biodivers Conserv 17:623–641

Nabe-Nielsen J, Sibly RM, Forchhammer MC, Forchhammer MC, Forbes VE, Topping CJ (2010) The effects of landscape modifications on the long-term persistence of animal populations. Plos One 5:e8932

Noordijk J, Delille K, Schaffers AP et al (2009) Optimizing grassland management for flower-visiting insects in roadside verges. Biol Conserv 142:2097–2103

Ouin A, Burel F (2002) Influence of herbaceous elements on butterfly diversity in hedgerow agricultural landscapes. Agric Ecosyst Environ 93:45–53

Ovaskainen O, Hanski I (2004) From individual behavior to metapopulation dynamics: unifying the patchy population and classic metapopulation models. Am Nat 164:364–377

Ovaskainen O, Smith AD, Osborne JL, Reynolds DR, Carreck NL, Martin AP, Niitpold K, Hanski I (2008) Tracking butterfly movements with harmonic radar reveals an effect of population age on movement distance. Proc Natl Acad Sci USA 105:19090–19095

Ricketts TH (2001) The matrix matters: effective isolation in fragmented landscapes. Am Nat 158:87–99

Ries L, Debinski DM, Wieland ML (2001) Conservation value of roadside prairie restoration to butterfly communities. Conserv Biol 15:401–411

Roberge JM, Angelstam P (2004) Usefulness of the umbrella species concept as a conservation tool. Conserv Biol 18:76–85

Roland J, Matter SF (2007) Encroaching forests decouple alpine butterfly population dynamics. Proc Natl Acad Sci USA 104:13702–13704

Ross JA, Matter SF, Roland J (2005) Edge avoidance and movement of the butterfly *Parnassius smintheus* in matrix and non-matrix habitat. Landscape Ecol 20:127–135

Rossi JP, Van Halder I (2010) Towards indicators of butterfly biodiversity based on a multiscale landscape description. Ecol Indic 10:452–458

Rushton SP, Ormerod SJ, Kerby G (2004) New paradigms for modelling species distributions? J Appl Ecol 41:193–200

Schtickzelle N, Baguette M (2009) (Meta)population viability analysis: a crystal ball for the conservation of endangered butterflies? In: Settele J, Shreeve T, Konvička A, Van Dyck H (eds) Ecology of butterflies in Europe. Cambridge University Press, Cambridge, pp 339–352

Schtickzelle N, Mennechez G, Baguette M (2006) Dispersal depression with habitat fragmentation in the bog fritillary butterfly. Ecology 87:1057–1065

Schtickzelle N, Joiris A, Van Dyck H, Baguette M (2007) Quantitative analysis of changes in movement behaviour within and outside habitat in a specialist butterfly. BMC Evol Biol 7:4

Seddon PJ, Leech T (2008) Conservation short cut, or long and winding road? A critique of umbrella species criteria. Oryx 42:240–245

Settele J, Dover J, Dolek M, Konvicka M (2009) Butterflies of European ecosystems: impact of land use and options for conservation management. In: Settele J, Shreeve T, Konvička A, Van Dyck H (eds) Ecology of butterflies in Europe. Cambridge University Press, Cambridge, pp 353–370

Shreeve TG, Dennis RLH, Williams WR (1995) Uniformity of wing spotting of *Maniola jurtina* (L.) (Lepidoptera: Satyrinae) in relation to environmental heterogeneity. Nota Lep 18:77–92

Shreeve TG, Dennis RLH, Pullin AS (1996) Marginality: scale determined processes and the conservation of the British butterfly fauna. Biodivers Conserv 5:1131–1141

Shreeve TG, Dennis RLH, Roy DB, Moss D (2001) An ecological classification of British butterflies: ecological attributes and biotope occupancy. J Insect Conserv 5:145–161

Shreeve TG, Dennis RLH, Van Dyck H (2004) Resources, habitats and metapopulations—whither reality? Oikos 106:404–408

Spitzer L, Benes J, Dandova J, Jaskova V, Konvička M (2009) The large blue butterfly, *Phengaris [Maculinea] arion*, as a conservation umbrella on a landscape scale: the case of the Czech Carpathians. Ecol Indic 9:1053–1063

Stefanescu C, Herrando S, Paramo F (2004) Butterfly species richness in the north-west Mediterranean Basin: the role of natural and human-induced factors. J Biogeogr 31:905–915

Stefanescu C, Penuelas J, Filella I (2005) Butterflies highlight the conservation value of hay meadows highly threatened by land-use changes in a protected Mediterranean area. Biol Conserv 126:234–246

Steffan-Dewenter I, Munzenberg U, Burger C, Thies C, Tscharnke T (2002) Scale-dependent effects of landscape context on three pollinator guilds. Ecology 83:1421–1432

Stevens VM, Pavoine S, Baguette M (2010) Variation within and between closely related species uncovers high intra-specific variability in dispersal. Plos One 5:e11123

Swift TL, Hannon SJ (2010) Critical thresholds associated with habitat loss: a review of the concepts, evidence, and applications. Biol Rev 85:35–53

Thomas CD, Kunin WE (1999) The spatial structure of populations. J Anim Ecol 68:647–657

Thomas JA, Bourn NAD, Clarke RT, Stewart KE, Simcox DJ, Pearman GS, Curtis R, Goodger B (2001) The quality and isolation of habitat patches both determine where butterflies persist in fragmented landscapes. Proc R Soc Lond B Biol Sci 268:1791–1796

Tischendorf L, Fahrig L (2000) On the usage and measurement of landscape connectivity. Oikos 90:7–19

Tracey JA (2006) Individual-based modeling as a tool for conserving connectivity. In: Crooks KR, Sanjavan MA (eds) Connectivity conservation: maintaining connections for nature. Cambridge University Press, Cambridge, pp 343–368

Tscharntke T, Steffan-Dewenter I, Kruess A, Thies C (2002) Contribution of small habitat fragments to conservation of insect communities of grassland-cropland landscapes. Ecol Appl 12:354–363

Turlure C, Choutt J, Van Dyck H, Baguette M, Schtickzelle (2010) Functional habitat area as a reliable proxy for population size: case study using two butterfly species of conservation concern. J Insect Conserv 14:379–388

Valtonen A, Saarinen M, Jantunen J (2007) Intersection reservations as habitats for meadow butterflies and diurnal moths: guidelines for planning and management. Landsc Urban Plan 79:201–209

Van Dyck H, Baguette M (2005) Dispersal behaviour in fragmented landscapes: routine or special movements? Basic Appl Ecol 6:535–545

Van Dyck H, Van Strien AJ, Maes D, Van Swaay CAM (2009) Declines in common, widespread butterflies in a landscape under intense human use. Conserv Biol 23:957–965

Van Swaay CAM, Van Strien AJ (2005) Using butterfly monitoring data to develop a European grassland butterfly indicator. In: Kuhne E, Feldmann R, Thomas JA, Settele J (eds) Studies on the ecology and conservation of butterflies in Europe. Vol 1, general concepts and case studies. Pensoft, Moscow, pp 106–108

Van Swaay CAM, Van Strien AJ (2008) The European butterfly indicator for grassland species: 1990–2007 report VJ2008.22 De Vlinderstichtung. Wageningen, The Netherlands

Van Swaay CAM, Warren MS (1999) Red data book of European butterflies (Rhopalocera). Nature and environment no 99. Council of Europe, Strasbourg

Van Swaay CAM, Warren MS (2003) Prime butterfly areas of Europe: an initial selection of priority sites for conservation. J Insect Conserv 10:1572–1753

Wahlberg N, Klemetti T, Hanski I (2002) Dynamic populations in a dynamic landscape: the metapopulation structure of the marsh fritillary butterfly. Ecography 25:224–232

J Insect Conserv (2011) 15:37–54
DOI 10.1007/s10841-010-9307-1

ORIGINAL PAPER

Habitat resources, remnant vegetation condition and area determine distribution patterns and abundance of butterflies and day-flying moths in a fragmented urban landscape, south-west Western Australia

Matthew R. Williams

Received: 19 March 2010 / Accepted: 30 June 2010 / Published online: 27 July 2010

Abstract The creation of cities, towns and farms following European settlement of Australia has fragmented the original vegetation. Many native species that were previously widespread are now found only within isolated remnants of their original habitat. These relictual populations are at increased risk of decline and local extinction, so identifying the factors that determine their persistence is important for ongoing management and conservation. I compared the effects of site area, connectivity, vegetation condition and habitat resources on the presence, abundance and total number of species of butterflies and day-flying moths within 46 urban fragments of remnant vegetation in south-west Western Australia. Site area and vegetation condition were the dominant determinants of species presence: large sites with more undisturbed vegetation favoured 16 of 20 native species and only one (*Geitoneura minyas*) benefited from disturbance. Another nine species that were not sufficiently widespread or abundant to enable individual analysis were collectively more prevalent in larger sites. Resource quality and quantity dominated the patterns of site occupancy, consistent with habitat resources, not metapopulation effects, determining current distribution patterns. The total number of species at each site reflected the collective responses of the individual species: increasing with area and declining with vegetation disturbance. The effects of area and vegetation condition were not simply additive: disturbance had a far greater impact on small remnants. Restoration or maintenance of high vegetation condition will be essential to maintain regional species diversity and to prevent local extinctions of butterflies and day-flying moths, especially in small remnants.

Keywords Connectivity · Disturbance · Metapopulation · Site occupancy · Vegetation condition

M. R. Williams
Centre for Ecosystem Diversity and Dynamics,
Department of Environmental Biology, Curtin University,
PO Box U1987, Perth, WA 6845, Australia

M. R. Williams (✉)
Department of Environment and Conservation,
Locked Bag 104, Bentley, WA 6983, Australia
e-mail: Matthew.Williams@dec.wa.gov.au

Introduction

Patches of remnant vegetation are now common features in urban and rural landscapes throughout the world. Populations of fauna in these fragments are frequently isolated and the resources necessary for breeding and reproduction may have been altered, potentially causing decline and local extinction of some species. A variety of threats result from human pressures on remnant vegetation—recreation, trampling of vegetation, altered fire regimes and the introduction of weeds and fungal pathogens—which cause changes in the structure and composition of plant communities and consequent loss of native fauna (New 1984; Samways 2007). The faunal assemblages that persist in fragments are the result of the operation of several factors, including the characteristics of the fragments (area, isolation, level of disturbance, and the presence of resources for individual taxa) and characteristics of the species (requirements for specific levels of resources, ability to adapt to altered conditions, vagility, colonization ability, and occurrence in nearby source areas). The intervening matrix typically contains few resources for many species and may present a substantial barrier to dispersal, particularly in the southern hemisphere where urbanisation has been more rapid and intensive (McKinney 2008; New and Sands 2002a).

The majority of studies of habitat fragmentation have examined the responses of vertebrates and higher plants, with less emphasis on invertebrates despite their numerical predominance and importance in ecosystem functioning (Akite 2008; McKinney 2008; New 1991). Butterflies are one of the best-known invertebrate groups and where baseline data are available show marked declines in distribution that often exceed declines in other taxa: for example, half the resident butterfly species of Britain are threatened or extinct and most have reduced geographic ranges (Ewers and Didham 2006; Warren et al. 2001). Many species of butterflies now depend on remnant vegetation for survival, especially in urban areas (Blair 1999; Connor et al. 2002; Hardy and Dennis 1999; Newland 2003; Ruszczyk and De Araujo 1992). The threats to butterflies are generally the same as for other fauna: habitat destruction or alteration; changes to management practices; isolation of remnant habitat; pollution and use of chemical insecticides and herbicides; climate change; and in some cases overcollecting or trade in specimens (Beaumont and Hughes 2002; Brown and Freitas 2004; New 1991; Pollard and Yates 1993). More than 400 species of butterflies are recorded for Australia and few more are likely to be discovered, making them the best-known insect group (Braby 2000). Several studies have documented the conservation status of individual taxa (Braby 2000; Kitching et al. 1999; New 1993) and a major review of the Australian fauna has been conducted (Sands and New 2002). These cite the lack of systematic surveys in conservation reserves and the lack of a national database as the major impediments to establishing the conservation status of many species. Information about the requirements of butterflies within fragments is needed to conserve these remnant populations (Dennis et al. 2003, 2006b).

Two dominant paradigms have been proposed to account for insect distribution patterns (Gripenberg and Roslin 2005; Grundel and Pavlovic 2007; Thomas et al. 2001): in the first, site occupancy depends on sufficient quantity and quality of resources required by the larvae and adults (habitat effects); in the second, site occupancy varies over time with successive extinction and recolonization events, and depends on the isolation or connectivity of habitat patches (metapopulation effects). Although not necessarily mutually exclusive (Dennis 2010; Dennis and Eales 1997, 1999), the two models lead to different conservation priorities: the first emphasizes the need to maintain optimal habitat within fragments, whereas the second gives priority to preserving numerous remnants that are within each species' dispersal range.

The conservation value of remnants is often determined on the basis of total species number, which often reflects species-area relationships. Approaches that examine species-area relationships to determine biodiversity hotspots and priority areas for conservation overlook the fact that area per se may not be the only, or indeed the most important, predictor of total species number (Dapporto and Dennis 2008; Ewers and Didham 2006; Fattorini 2006; Ulrich 2003; Ulrich and Buszko 2005; Veech 2000). Insects are often more sensitive to habitat quality than site area, and many decline rapidly after disturbance (Dennis 2004b; Dennis and Eales 1997, 1999; Gibb and Hochuli 2002; Maes and Van Dyck 2001; McKinney 2008; Thomas et al. 2001). Studies of insect faunas have shown that habitat area, quality, isolation and connectivity may all be important (Samways 2007); that the level of urbanization, which diminishes from the urban core to suburban areas, almost always reduces species numbers (Hardy and Dennis 1999; McKinney 2008); but that responses may also be species-specific (Dapporto and Dennis 2008; Ewers and Didham 2006; Garden et al. 2006; Grundel and Pavlovic 2007; McKinney 2008). Particular site atttributes, including both structural elements (such as tree cover) and species-specific resources (such as host plants) are also important determinants of persistence for many butterflies and day-flying moths (Baz and Garcia-Boyero 1995; Dennis et al. 2004; Dover et al. 1997; Gripenberg and Roslin 2005; Kadlec et al. 2008; Krauss et al. 2004). To manage and conserve fauna within fragments therefore requires knowledge of individual species responses to site characteristics and disturbance (Dennis 2004a; Dennis and Hardy 2007; Dover et al. 2008; Dover and Rowlingson 2005; Garden et al. 2006; Grundel and Pavlovic 2007). Lack of information about the resources required to sustain these isolated populations has been identified as a limiting factor in conserving fauna in urban fragments (Garden et al. 2006). Determining those aspects of the size, quality and the spatial arrangement of remnants that account for species persistence is important in guiding the management and conservation of remaining faunal populations.

This study examines the effects of site area, connectivity, vegetation condition and habitat resources on the presence, abundance and number of butterfly and day-flying moth species in urban remnants, with the aims of (i) testing whether quality and quantity of vegetation and resources, or metapopulation effects dominate, and (ii) identifying the important factors for conservation and management of remaining populations.

Methods

Study area

The urban area of the Perth metropolitan region in south-west Western Australia is an example of a disturbed landscape. Prior to European settlement the Swan Coastal

Plain portion of the region consisted of continuous tracts of low open woodland and open forest dominated by *Banksia* and *Eucalyptus* (Beard 1990). Land clearing for agriculture and urbanization has reduced the native vegetation to 18% of its original extent, which now consists of fragments varying in size and level of disturbance (Government of Western Australia 2000). The intervening matrix consists mainly of low-density residential suburbs dominated by non-native plant species. Historically, the urban area has expanded along a north–south axis, bounded in the west by the Indian Ocean and in the east by the Darling Range, so that linear distance from the city centre is a good indicator of the time of fragmentation (Kennewell and Shaw 2008). The region has a Mediterranean-type climate with a pronounced summer drought and is prone to periodic fires (McCaw and Hanstrum 2003).

The study focused on the 36 species of butterflies and day-flying moths regularly recorded from the study area, referred to as 'butterflies' herein. There are few day-flying moths in the study area, the most notable of which are the sun-moths (*Synemon* spp., Castniidae), active and conspicuous moths that are often confused with butterflies (Edwards 1997a, b). One, *Synemon gratiosa*, is listed as endangered (Burbidge 2004) but relatively little is known about the others.

Forty-six patches of remnant vegetation were selected for surveys, from approximately one hundred located within the study region. Four were included because they were known to have populations of the endangered *Synemon gratiosa*, one because it is the closest remnant to the urban centre, and the remainder were selected at random from a major review of the conservation values of remnant vegetation within the study area (Government of Western Australia 2000) (37 sites) or added opportunistically (4 sites). All sites were within 35 km of Perth (population 1.5 million), the major urban centre in Western Australia, and ranged in area from 0.7 to 362 ha (Appendix 1; Fig. 1). Several of the fragments were closely located, but were separated by cleared areas or roads, were joined by narrow corridors of highly disturbed remnant vegetation, or had parts subjected to different management practices, and were treated as separate sites.

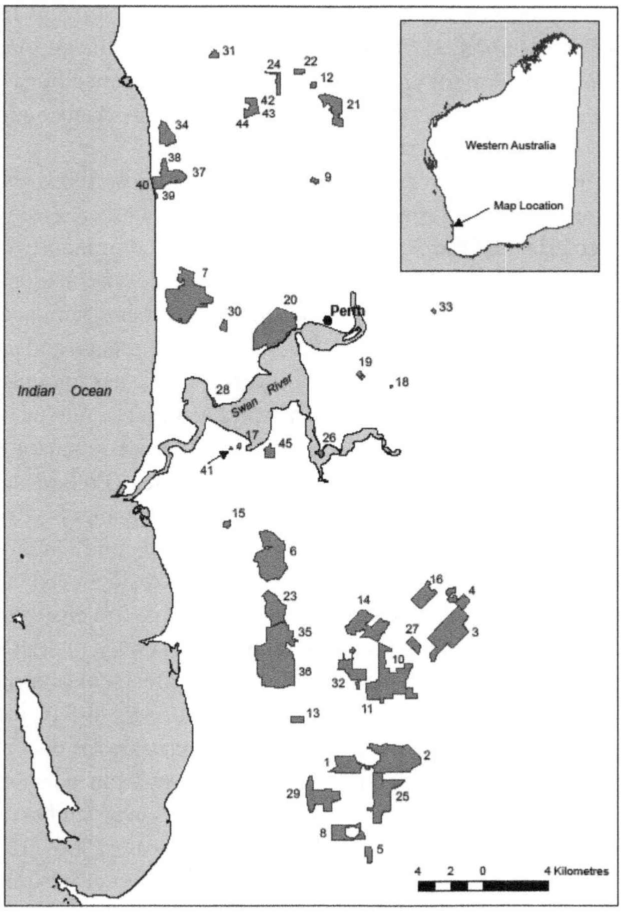

Fig. 1 Location of sites surveyed for butterflies and day-flying moths in the Swan Coastal Plain, Perth metropolitan region, south-west Western Australia. Each site is identified by the site number in Appendix 1. The total extent of each site is depicted, not the remaining native vegetation; the area of remnant vegetation is given in Appendix 1

Field surveys

At each site one to three transects were established, positioned to sample as many vegetation types and seral stages as possible including any recently or long un-burnt areas. Although transects were predominantly restricted to existing walk tracks and fire breaks to protect the native vegetation from trampling and reduce the spread of weeds or fungal pathogens, parts of some transects traversed the vegetation. Transect length varied with site area, ranging from 260 to 5,100 m, and smaller sites typically had shorter transects but a higher sampling fraction (the fraction of the site area sampled by the transect).

To sample the fauna, surveys were conducted at fortnightly intervals during the two main butterfly flight periods in the Austral spring (late September—mid December, usually six surveys) and in autumn (late February—early April, two surveys), between September 2003 and April 2005. All surveys were conducted by the author, with the occasional assistance of 1 or 2 volunteers, in fine weather with forecast maximum temperatures above 21°C. This sampling regime detects over 85% of the resident fauna (Williams 2008). The butterfly fauna of the region is well documented and phenological data for each species are given by Braby (2000) and Williams (2009). Each transect was walked at a steady pace and all butterflies and day-flying moths observed within a rectangle 5 m to either side and ahead of the observer(s) were identified and tallied. Two species (*Motasingha dirphia* and *M. trimaculata*) that are difficult to distinguish in the field were recorded as the more common *M. trimaculata*; this is the conventional approach (Pollard and Yates 1993). The abundance of each species was measured by the number of individuals counted per 100 m of transect.

The species were divided into two groups: those that breed entirely or predominantly on native plants ('resident species'); and those that breed entirely or predominantly on introduced plants ('non-resident species'). The resident species are almost always restricted to bushland remnants, as their host plants rarely occur outside of remnants and are not widely cultivated. The non-resident species included those that are regular migrants and vagrants, and occur throughout the urban matrix. This division is equivalent to the concepts of 'island' and 'matrix' species (Pollard and Yates 1993), 'urban avoiders' and 'urban adapters' (Koh and Sodhi 2004), and 'matrix sensitive' and 'matrix occupying' species (Garden et al. 2006). Resident (island, urban avoider, matrix sensitive) species have low dispersal rates, are local or uncommon, and are generally restricted to remaining suitable habitat patches. Non-resident (matrix, urban adapter, matrix occupying) species are highly dispersive, common, occur in both remnant vegetation and intervening areas, and are of little conservation concern. This division follows the approach of previous studies of Australian butterflies (Braby and Edwards 2006; Hill 1987). Taxonomy follows Braby (2000) for butterflies, and Common (1990) and Edwards (1997b) for moths.

Site attributes

For each site several physical, vegetation and habitat resource attributes that potentially affect species presence, abundance and the total species number were measured. Site physical attributes were area (A), two measures of site connectivity (distances to the nearest (N) and nearest larger (NL) patch of remnant vegetation, whether surveyed or not), and a measure of the level of urbanization and time since isolation (the distance of the site from the urban centre (D)). Vegetation and habitat resource attributes were: the density of shade trees (T, the number of stems of *Eucalyptus, Casuarina* and *Allocasuarina* spp. > 10 cm dbh, per 100 m of transect); two measures of vegetation condition, as described below; and the density of three host plants (*Daviesia divaricata* (Dd), *Jacksonia sternbergiana* (Js) and *Patersonia occidentalis* (Po)) that are used by four butterfly species (*Mesodina cyanophracta* (Po), *Hypochrysops halyaetus* (Dd and Js), *Neolucia agricola* (Dd and Js), and *Lampides boeticus* (Js)) (Braby 2000; Dover et al. 2008; Dover and Rowlingson 2005). Host plants of other species were not recorded because they are either cryptic, difficult to identify, incompletely known, or ephemerals that were not evident during the sampling period. Tree and host plant densities were recorded on the same transects used to survey fauna.

The vegetation condition at the sites had been evaluated < 5 years prior to the surveys as part of a comprehensive study of the conservation significance of all remnant vegetation in the Perth metropolitan region (Government of Western Australia 2000). Vegetation condition at each site was scored using an objective rating scale (Keighery 1994) and the fraction of the vegetation in each of five condition classes was determined: (i) no obvious signs of disturbance (rated 9); (ii) vegetation structure intact, some disturbance affecting individual plant species, and weeds only non-aggressive species (7); (iii) vegetation structure altered, obvious signs of disturbance such as evidence of logging, grazing, repeated fires, presence of aggressive weeds, or dieback caused by fungal pathogens (5); (iv) vegetation structure significantly altered by very obvious signs of multiple disturbance, such as very frequent fires, high densities of some very aggressive weeds, partial clearing, dieback or grazing (3); and (v) as for the previous category, but vegetation structure severely impacted by disturbance and unlikely to regenerate without intensive management (1). Intermediate ratings were also used, so a typical condition assessment would be '> 75% rated 5–7, < 25% 1–3'. Two measures of vegetation condition were calculated: the average vegetation condition (AVC), and the fraction of high condition vegetation (HCV, the fraction rated > 5). An advantage of using the existing vegetation condition ratings was that they could then be used to make predictions of species occupancy and total species number for the total 287 vegetation remnants within the region that had been assessed.

Data analysis

The relationship between the site attributes, the incidence and abundance of each species and the number of species present was examined using logistic and linear regression models. For individual species presence and abundance, two complementary models were fitted (Fletcher et al. 2005): first, the relationship between species presence and the site attributes was examined using a logistic regression model; second, within the sites it occupied, the log(abundance) of each species was regressed on the same factors using linear regression. Including only occupied sites eliminated the bias that arises from a high proportion of sites having zero abundances and the need to use the arbitrary log $(x + 1)$ transformation. For the species-incidence models the complementary log–log function was used in place of the usual logit link, as this has been shown to be a more appropriate null model, although this choice has a negligible effect on results (Williams 1995). The number of resident and non-resident species at each site was modelled using the cumulative extreme-value function (EVF) model, the multiple species equivalent of the complementary log–log link function, although results using the power function model gave identical results. To improve the linearity of the relationship between the covariates and the dependent variables, positively-skewed variables were transformed for analysis: abundance, A, D, N, and NL were log-transformed; and T, Dd, Js and Po were square root transformed. Host plant densities were only included as predictors for the relevant species.

All possible subsets of the predictor variables were examined, although the maximum number of attributes included in each model was limited by the amount of data available for each species. To obtain reliable results using logistic regression approximately 10 occurrences or absences, and in multiple regression approximately 10 cases, are needed for each fitted parameter (Barry and Elith 2006; Burnham and Anderson 1998, 2002; Hosmer and Lemeshow 2000; Stevens 1992). Therefore, for the individual species models only those taxa occurring in more than five sites were analysed and each model was restricted to an appropriate number of explanatory variables (one, if $n < 14$, or two variables for the logistic regression models, and similarly up to four variables for the linear regressions). Nine resident species that were not sufficiently widespread or abundant to be analysed individually were analysed collectively as 'rare resident species'. A potential problem in multiple regression models is high correlation (collinearity) between predictor variables (Vittinghoff et al. 2005). Two measures were used to assess if high correlation necessitated removal of any of the site attributes: condition index and tolerance. A condition index > 10 or tolerance < 0.2 are indicative of problematic collinearity (Fry 1993; Stevens 1992).

The best fitting main effects models were determined using information-theoretic (IT) analysis and the relative importance of each predictor was assessed by multi-model averaging (Burnham and Anderson 1998, 2002). IT analysis and multi-model averaging is superior to other methods of identifying important predictor variables and outperforms traditional methods such as stepwise regression (Whittingham et al. 2006). The IT method compares a set of alternative models by calculating a likelihood weight for each (the Akaike weight, w), equal to the probability that it would be the best model were the data collected again. Values of w vary between 0 and 1, corresponding to the weight of evidence in support of a candidate model within the set; a value > 0.9 indicates a clearly superior model (Burnham and Anderson 1998). These weights were used to compute the weighted average (multi-model averaged) set of parameter estimates for each model and to identify the best supported variables. The relative importance of the individual site attributes was assessed by the Wald statistic, the ratio of parameter estimate to its standard error, with greater absolute values indicating greater importance and values > 2 corresponding to approximate statistical significance (Hosmer and Lemeshow 2000). For the total species models, the interactions between variables determined to be important main effects were also examined.

Model adequacy was assessed using the chi-squared lack-of-fit test (Hosmer and Lemeshow 2000). To assess the goodness-of-fit of the species incidence models the area under the receiver-operator characteristic curve statistic (*AUC*) was calculated. *AUC* takes values between 0.5 and 1, and an approximate guide for *AUC* values in terms of oodness-of-fit is: values > 0.9 indicate excellent fit; > 0.8 good; > 0.7 fair; > 0.6 poor; and < 0.6 no or negligible fit (Fawcett 2006; Swets 1988). To assess the goodness-of-fit of the best fitting species abundance and total species models the usual R^2 statistic was calculated. Because of the small number of sites relative to the number of predictor variables, the small-sample adjusted form of Akaike's Information Criterion (AIC_c) statistic was used. It is important to note that the models and attributes selected as the "best" predictors were only the best within the set examined, and may not include other important but unmeasured factors.

Results

Sites more distant from the city centre were typically larger, had vegetation in better condition, lower tree densities, were less isolated from other remnants and had more resident species (Table 1). Although many of the predictors were significantly correlated they were sufficiently distinct that all could be included in the regression models (condition number = 4.0, tolerances > 0.25). For the total 702 km covered in the surveys 6,540 individuals of 33 species (28 butterflies and 5 day-flying moths) were recorded (Appendices 1, 2). Only three butterfly species regularly recorded from the study area were not detected: *Ogyris idmo*, *Cephrenes trichopepla* and *C. augiades* (the first, a myrmecophage, occurs in very low numbers in only one site, whereas the latter two are introduced taxa usually restricted to suburban gardens). Most species were restricted to half or less of the fragments, with only three species (*Vanessa kershawi*, *Pieris rapae*, and *Geitoneura klugii*) occurring in more than 75% of sites. Total site species number (S_{tot}) varied from 1 to 18 species, and resident species (S_{res}) from 0 to 13. Twenty of the butterfly and all of the day-flying moth species were resident taxa and there were eight non-resident butterfly species recorded. Sufficient presence-absence data were available for analysis of 20, and abundance data for 22, of the species.

Individual species presence and abundance

Site area and vegetation condition were the dominant predictors of species presence, being identified as important for eight and seven taxa respectively, and together being identified as important for 14 of the 20 species analysed (Table 2; full results for the models are given in Appendix 3). Larger sites were positively associated with presence in all of these cases and only one species (*Geitoneura minyas*) was favoured by degraded vegetation. Collectively, the nine rare resident species were also more frequent in larger sites. All four butterfly species for which host plant density was assessed were more prevalent at sites with higher host plant densities. The density of trees was an important predictor for five species with higher tree density positively associated with the presence of four. Distance from the urban centre was positively related to the presence of only one species and increased site connectivity did not increase the likelihood of any species being present. Goodness-of-fit ranged from poor to excellent, with good or better fit ($AUC > 0.8$) for six species. Models that included host plant density as a predictor produced the four best-fitting models (AUC 0.83–0.93).

Site area was also the dominant predictor of abundance with four species being more abundant at larger sites and three non-resident species being more abundant in smaller sites (Table 1; full results for the models are given in Appendix 4). Where it was recorded, host plant density was the also the best predictor of increased abundance. Four species were more abundant in more isolated sites and three in sites with lower tree densities. Only three species were more abundant in sites with higher vegetation condition. No species were found to be more abundant in sites with higher connectivity. Overall, 14 of the 16 resident species showed a positive response in their site occupancy or abundance to increased site area or vegetation condition.

Table 1 Spearman rank correlations (above) and associated *P*-values (below) between site attributes, and with the number of species present

	A	D	N	NL	AVC	HCV	T	Dd	Js	Po
A	1	**0.49**	−0.14	0.10	**0.29**	0.21	−0.19	−0.25	0.02	0.10
		<0.01	0.34	0.51	0.05	0.15	0.21	0.09	0.91	0.49
D		1	**−0.34**	**−0.29**	**0.44**	**0.39**	**−0.38**	**−0.43**	−0.08	0.28
			0.02	0.05	<0.01	0.01	0.01	<0.01	0.60	0.06
N			1	**0.70**	−0.11	−0.24	**0.31**	**0.32**	0.15	0.13
				<0.01	0.45	0.11	0.04	0.03	0.32	0.40
NL				1	−0.06	−0.16	**0.30**	**0.32**	0.16	0.07
					0.70	0.29	0.05	0.03	0.29	0.65
AVC					1	**0.89**	**−0.30**	−0.14	0.03	**0.54**
						<0.01	0.04	0.35	0.86	<0.01
HCV						1	**−0.34**	−0.20	−0.03	**0.57**
							0.02	0.18	0.86	<0.01
T							1	**0.59**	**0.37**	−0.27
								<0.01	0.01	0.07
Dd								1	0.23	**0.15**
									0.12	0.33
Js									1	−0.09
										0.57
S_{res}	**0.43**	**0.38**	−0.13	0.01	**0.48**	**0.53**	−0.03	0.19	0.18	**0.51**
	<0.01	<0.01	0.38	0.93	<0.01	<0.01	0.85	0.22	0.24	<0.01
S_{nres}	0.14	−0.19	0.16	**0.38**	−0.07	0.01	**0.42**	**0.34**	**0.46**	−0.01
	0.35	0.20	0.30	0.01	0.66	0.93	<0.01	0.02	<0.01	0.97
S_{tot}	**0.40**	0.26	−0.09	0.11	**0.39**	**0.46**	−0.07	**0.35**	−0.23	0.17
	0.01	0.08	0.57	0.48	0.01	<0.01	0.66	0.02	0.13	0.27
Tolerance	0.65	0.63	0.41	0.42	0.27	0.28	0.56	0.55	0.65	0.58

Statistically significant correlations are in bold type. Site attributes are area (A, ha), distance from the urban centre (D, km), distances to the nearest and nearest larger remnants (N, NL, km), average vegetation condition (AVC), fraction of high condition vegetation (HCV), density of *Eucalyptus*, *Casuarina* and *Allocasuarina* spp. trees (T, number of stems > 10 cm dbh per 100 m of transect), density (number per 100 m of transect) of three host plants, *Daviesia divaricata* (Dd), *Jacksonia sternbergiana* (Js), and *Patersonia occidentalis* (Po), and number of resident (S_{res}), non-resident (S_{nres}) and total (S_{tot}) species. Tolerances indicate the degree of collinearity between the site attributes

Table 2 Factors related to site occupancy and abundance of butterflies and day-flying moths in 46 fragments of remnant vegetation, Swan Coastal Plain, south-west Western Australia

Species	Family or subfamily	Sites occupied	Total count	Occupancy	*AUC*	Abundance	R^2
Resident species							
Trapezites sciron	Hesperiidae	7	36	+HCV	0.78	−T	0.45
Anisynta sphenosema	Hesperiidae	9	45	+A	0.72	−D	0.52
Motasingha trimaculata	Hesperiidae	7	13	+NL	0.77	+AVC	0.34
Mesodina cyanophracta	Hesperiidae	15	79	+Po, +D	0.93	+Po, +N	0.70
Taractrocera papyria	Hesperiidae	27	96	+A, +T	0.70	-	
Geitoneura klugii	Satyrinae	37	1,289	+A	0.81	+A	0.15
Geitoneura minyas	Satyrinae	21	1,807	+A, −AVC	0.81	+NL, −N	0.56
Heteronympha merope	Satyrinae	21	120	+A, +T	0.78	+A	0.31
Hypochrysops halyaetus	Lycaenidae	7	104	+Dd	0.89	+(Dd + Js)	0.30
Candalides acastus	Lycaenidae	9	34	−T	0.78	+AVC	0.52
Nacaduba biocellata	Lycaenidae	17	41	+HCV	0.68	−D, −T	0.61
Neolucia agricola	Lycaenidae	23	326	+A, +(Dd + Js)	0.83	+(Dd + Js), +A	0.42
Zizina labrudus	Lycaenidae	21	124	+A	0.68	−T	0.16
Synemon sp. (Perth)	Castniidae	17	206	+AVC, +T	0.75	-	
Synemon gratiosa	Castniidae	6	19	+HCV	0.63	+NL	0.28
Pollanisus cupreus	Zygaenidae	26	291	+AVC	0.77	+N, +D, +AVC	0.46
Rare resident species	9 taxa	14	75	+A	0.66	−D	0.40
Non-resident species							
Pieris rapae[a]	Pieridae	43	695	N/a		−A	0.18
Junonia villida	Nymphalinae	7	16	+HCV	0.59	−A	0.59
Vanessa kershawi	Nymphalinae	45	988	N/a		−D, +A	0.19
Danaus chrysippus	Danainae	6	7	–	-	−A	0.68
Danaus plexippus[a]	Danainae	11	15	+A	0.68	+D	0.23
Lampides boeticus	Lycaenidae	24	93	+Js, +T	0.92	+Js	0.22

For each species, the factor(s) from the best-fitting one- or two-variable models are given, which are also the best-supported variables from multi-model averaging. For models with two or more variables the best-supported variable is stated first. Attribute codes as in Table 1. Goodness-of-fit of each occupancy and abundance model is indicated by *AUC* (area under the receiver-operator characteristic curve) and R^2 (coefficient of determination), respectively. Rare resident species (not individually listed) that were recorded in five or fewer sites were *Trapezites argenteoornatus*, *Hesperilla donnysa*, *H. chrysotricha*, *Croitana croites* (all Hesperiidae), *Ogyris amaryllis*, *Theclinesthes miskini*, *T. serpentata* (all Lycaenidae), *Hecatesia thyridion*, and *Periscepta polysticta* (both Noctuidae)

[a] Introduced species

Total species numbers

The total number of both resident and non-resident species was higher in larger sites, but for non-resident species the relationship with area was negligible (Fig. 2). The slopes of the power function species-area curves for both resident and total species number (0.19 and 0.15) were typical of the values for mainland areas ($z = 0.12$–0.18), not true islands ($z = 0.25$–0.35) (Rosenzweig 1995).

The number of resident species increased with site area, the fraction of high condition vegetation and the density of trees (Table 3). The best fitting model for non-resident species included both area and the density of trees, although the relationship with area was weak and model fit was poor ($R^2 = 0.2$, Table 3). The chi-squared test of fit indicated that the models for both resident and non-resident species numbers were adequate ($P > 0.05$). Inclusion of interactions between these main effects resulted in a best-fitting model for resident species that included site area, HCV and their interaction; tree density was not included in this final model (Table 4). The interaction indicated that sites with low vegetation condition had steeper species-area curves with lower intercepts so that the impact of vegetation degradation on the number of resident species was proportionally greater in small fragments (Fig. 3).

Discussion

This study has shown that the distribution of butterflies and day-flying moths in urban remnants varies in relation to site characteristics. For resident species that rely on remnant

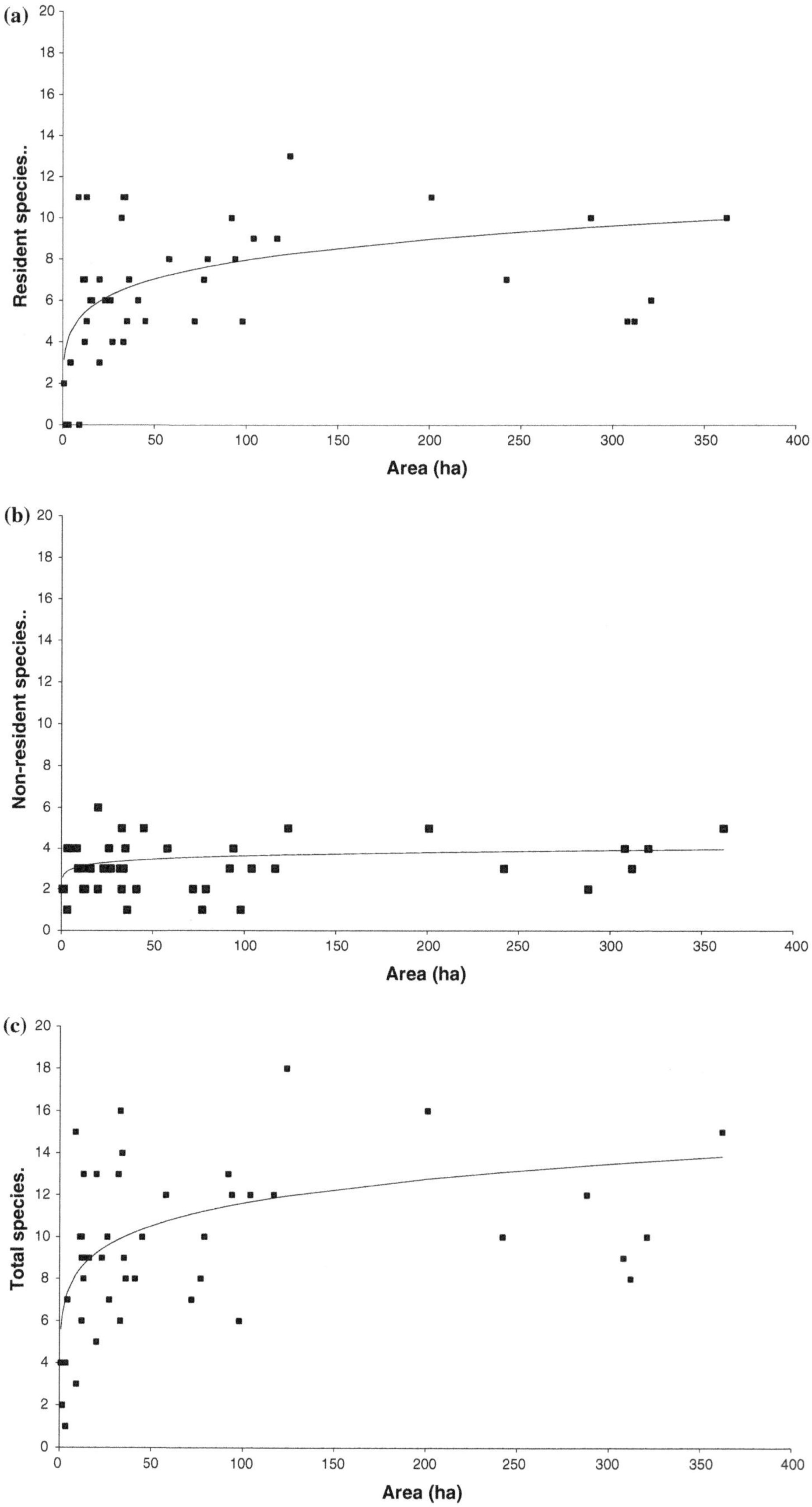

Fig. 2 Species-area relationships of **a** resident, **b** non-resident and **c** all species of butterflies and day-flying moths in urban vegetation fragments, south-west Western Australia. Fitted lines are for the EVF model, $S = P[1 - \exp(-\exp(-b_0 + b_1\ln(A)))]$, where P = the number of species in the regional source pool, but the power function model $S = cA^z$ gave practically identical results. **a** $S_{res} = 32 \times [1 - \exp(-\exp(-2.19 + 0.20 \times \ln(A)))]$ or $S_{res} = 3.39 \times A^{0.19}$, $R^2 = 0.27$; **b** $S_{nres} = 13 \times [1 - \exp(-\exp(-1.49 + 0.08 \times \ln(A)))]$ or $S_{nres} = 2.641 \times A^{0.07}$, $R^2 = 0.04$; and **c** $S_{tot} = 45 \times [1 - \exp(-\exp(-1.96 + 0.16 \times \ln(A)))]$ or $S_{tot} = 5.942 \times A^{0.15}$, $R^2 = 0.27$

Table 3 Multi-model averaged parameter estimates and standard errors (*SE*) of the site attributes predicting the number of butterfly and day-flying moth species in 46 fragments of remnant vegetation, Swan Coastal Plain, south-west Western Australia

	Resident species ($AUC = 0.61$, $R^2 = 0.38$)			Non-resident species ($AUC = 0.58$, $R^2 = 0.20$)		
	Estimate	*SE*	Wald statistic	Estimate	*SE*	Wald statistic
Intercept	**−2.93**	0.53	−6.65	−1.65	0.47	−3.53
A	**0.18**	0.05	3.83	**0.06**	0.04	1.34
D	0.07	0.06	1.07	−0.02	0.04	−0.43
N	0.00	0.01	0.02	0.01	0.02	0.46
NL	0.00	0.01	0.28	0.04	0.03	1.19
AVC	−0.03	0.03	−0.87	−0.02	0.04	−0.63
HCV	**0.95**	0.29	3.28	0.20	0.19	1.05
T	**1.18**	0.55	2.16	**1.60**	0.72	2.22

The site attributes (as in Table 1) in the best fitting model are bolded

Table 4 Parameter estimates, standard errors (*SE*) and Wald statistics for the model predicting the number of butterfly and day-flying moth species using site area, fraction of high condition vegetation and their interaction

	Resident species		
	Estimate	*SE*	Wald statistic
Intercept	−3.09	0.34	9.20
A	0.35	0.09	3.97
HCV	1.95	0.53	3.68
A*HCV	−0.35	0.14	2.50

$AUC = 0.61$, $R^2 = 0.37$

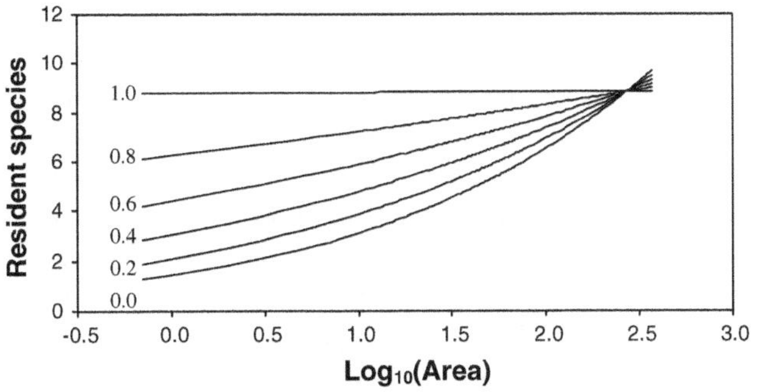

Fig. 3 Predicted species-area curves of fragments across the range of fragment areas in this study (0.7–360 ha), with varying fractions of high condition vegetation between 1.0 (100%) and 0.0 (0%)

vegetation, fragment size and the level of vegetation disturbance most often determined the patterns of incidence, abundance and the total number of species present at each site. Although the habitat relations of the rare species could not be determined individually, these were also more likely to be present at larger sites. For those species for which it was assessed, the density of host plants was also an important predictor of both site occupancy and abundance. Every resident species analysed showed a positive response in either occupancy or abundance to one or more of these factors. Tree density was important for some species but responses were idiosyncratic. Increased distance from the urban centre favoured the presence of only one species and increased site connectivity was not important for any. Site resources were thus the dominant determinants of both site occupancy and abundance of the resident species. In contrast, increased site connectivity did not increase the occupancy or abundance of any species so there is no evidence to support a metapopulation structure for any of these species. Removal of species that occur within the matrix is expected to increase the likelihood of detecting connectivity effects (Ewers and Didham 2006) but the lack of any such effects for the resident species indicates that they consist of predominantly relictual populations. However, the least widespread and abundant species could not be analysed individually and some species known from the region were not detected, and these may prove to be sensitive to site connectivity effects. The patterns of site occupancy in this region contrast with butterfly faunas on true islands, which are often dominated by the effects of contemporary and historical geography on dispersal so that island isolation is a major factor affecting both island occupancy and total species number (Dapporto and Dennis 2008; Dennis et al. 2000, 2008). It also contrasts with several studies of mainland areas that found site connectivity to be a major factor controlling species occupancy or diversity in habitat fragments (Brown and Freitas 2004; Fartmann 2006; Krauss et al. 2003, 2005; Thomas et al. 2001). It is, however, consistent with the increasingly recognized importance of habitat quality in explaining site occupancy in habitat fragments even for species that do exist as metapopulations (Dennis and Eales 1997; Dover et al. 2008; Dover and Rowlingson 2005; Fred and Brommer 2003; Krauss et al. 2004, 2005; Thomas et al. 2001; Wahlberg et al. 2002).

Total species number integrated the responses of the individual species: area, vegetation condition and their interaction were the dominant predictors. The interaction between site area and vegetation condition suggests that there are more species in larger sites not in response to area per se, but because there is more suitable habitat within them. Small sites (< 10 ha) with degraded vegetation had the fewest species and this is consistent with declining habitat quality. The results suggest that declining area has little impact on sites with pristine vegetation, but there was some confounding between area and vegetation condition as there were few sites < 10 ha that also had high condition vegetation and all sites < 5 ha were degraded (HCV $\leq$ 0.2, Appendix 1). This reinforces the principle that the key to survival of any species is the maintenance of a minimum area of suitable habitat (Atmar and Patterson 1993; Lamont et al. 1993).

Changes in habitat quality caused by urbanization may benefit some butterfly species (Connor et al. 2002; Hardy and Dennis 1999; Shapiro 2002) and in some urban areas most species that remain are those that have adapted to introduced host plants (New and Sands 2002a; Shapiro 2002). In this study there was only one such native species (*Geitoneura minyas*), and declining vegetation condition caused by disturbance disadvantaged many more taxa than it benefited. Although some of the butterflies in this study use alien host plants (various introduced weeds) and were more prevalent in small or disturbed sites, these were predominantly introduced or 'urban-adapter' species. Overall, this conforms with previous studies that have demonstrated the negative impacts of disturbance and associated habitat degradation on butterfly species richness in urban areas with relatively few species favoured (Brown and Freitas 2004; Dennis and Hardy 2004; Hardy and Dennis 1999; Hogsden and Hutchinson 2004; McKinney 2008; Nelson and Nelson 2001; New and Sands 2002b).

The non-resident species showed no species-area effect and some were more prevalent in smaller sites. Because these taxa are adapted to plants within the matrix, they may well avoid all but the edge of fragments. Transects within small sites inevitably sample more areas close to edges and these species may be recorded more frequently as a result. This supports treating the resident and non-resident species separately because they had different responses to fragmentation.

The resources required by the immature stages are critical habitat components for butterflies and other insects in temperate areas (Dennis 2010; Thomas et al. 2001). For the four species where it could be assessed, host plant density was an important factor both for site occupancy and abundance. The importance of host plant density has often been demonstrated (Dennis 2004a; Dennis and Hardy 2004; Grundel and Pavlovic 2007; Krauss et al. 2004; Wahlberg et al. 2002) and it is likely that better predictive models for many of the species in this study could be obtained if the density of their respective host plants were determined.

For two main reasons, the models for some species should be viewed cautiously. First, a potential shortcoming in species incidence modelling is that many species are imperfectly detected, leading to negative bias in estimates of their site occupancy (Dennis et al. 2006a; MacKenzie et al. 2003). Detectabilities (the probability that the species will be detected at a site, given that it is present) for the species analysed in this study have been quantified (Williams 2009). They were typically lowest and hence most problematic for the least abundant species and those occurring at few sites, notably *Anisynta sphenosema* (0.12), *Danaus chrysippus* (0.28), *D. plexippus* (0.28), *Junonia villida* (0.53), and *Synemon gratiosa* (0.62). A greater number of surveys and a larger sample of occupied sites would be needed to construct models that incorporate parameters to estimate the seasonal variation in site occupancy and detectability for these species. For the remaining species detectability was high (> 0.86). Second, a common source of error in habitat models is the absence of important covariates (Barry and Elith 2006). Host plant density was only measured for four species but was consistently the best predictor of presence. Similarly, other important covariates may better explain site occupancy and abundance for individual species—factors such as host plant quality or the presence and abundance of ant mutualists may be needed to explain the distributional patterns of some taxa (Dover et al. 2008; Dover and Rowlingson 2005). These additional covariates need to be assessed before a full understanding of each species' patterns of occupancy and abundance can be achieved.

There are several implications for conservation of butterflies within this region. Maintenance or restoration of high condition vegetation will be essential to conserve these remnant populations and to maintain local and regional species richness. Larger remnants harboured more species, including more of the regionally rare taxa, so maintaining or improving vegetation condition is especially important for these sites. Although many species still occur in small (< 10 ha) remnants it is likely that stochastic events will take a proportionally greater toll on smaller fragments (Ewers and Didham 2006; New and Sands 2002a). Similarly, this study found that such sites are more prone to species losses following disturbance and consequent decreases in vegetation condition. Small patches also have a greater proportion of edge to interior, making them generally more exposed to exogenous disturbances and more difficult to maintain (Samways 2007). Several of the fragments in this study were separated by cleared areas such as roads but even such relatively narrow features may be significant barriers to dispersal, especially if the contrast between the fragment and the matrix is high (Ewers and Didham 2006). Although establishing and maintaining connectivity between these fragments may increase the total effective population size within each subfragment, and provide greater resilience for the combined population, maintaining and improving vegetation condition is of greater importance.

The ongoing SLOSS (Single Large or Several Small) debate, which contrasts alternative strategies for maximising species diversity within reserve networks, parallels the habitat resource/metapopulation paradigms of species persistence in fragments. The results of this study support the preservation of a few large, rather than many small remnants. Several large reserves are generally considered to be superior to small reserves for the persistence of area- and extinction-prone species, despite some studies that show greater species numbers in small, widespread reserves (Ewers and Didham 2006; Gibb and Hochuli 2002; Tscharntke et al. 2002). This study found that the impact of disturbance was proportionally much greater in small remnants. However, the responses shown by butterflies and day-flying moths need to be considered together with the distribution patterns of other fauna in these remnants.

If improvement or restoration of habitat can be achieved within the degraded sites, deliberate translocations of selected taxa may be possible to re-establish locally extinct populations. The individual models for each species, summarized in this paper, will assist in determining those species at greatest risk of local extinction and in identifying the sites where they are notably absent. However, there are also species-specific factors that need to be considered to conserve particular taxa (Garden et al. 2006). Host plant density was a crucial factor, and a better understanding of this relationship will help to guide future conservation. Further studies that examine the interaction between site area and vegetation condition in small (< 10 ha) fragments would clarify the roles of these factors on species presence and persistence. The responses of fauna to habitat fragmentation may take many years or decades to manifest (Garden et al. 2006; Kadlec et al. 2008), and there is no evidence that the fauna in these fragments is at equilibrium. Species presence within a fragment may not equate to long-term viability. Currently, there are no species known to be locally extinct within the region but the extent of vegetation clearing (82% removed) suggests that there is potentially an unresolved extinction debt. Periodic monitoring of these sites, and comparison of any changes in fauna with management practices and disturbance (fencing, weed control and changes in vegetation condition) or stochastic events (fires), offer an opportunity to identify causes of species loss or recruitment.

Acknowledgments Byron Lamont, Ian Abbott, Colin Yates and two anonymous reviewers provided helpful comments on the manuscript.

Appendix 1

See Table 5.

Table 5 Attributes of 46 urban habitat fragments sampled for butterflies and day-flying moths, south-west Western Australia

Site	*n*	L	A	D	N	NL	AVC	HCV	T	Dd	Js	Po	S_{res}	S_{nres}	S_{tot}
1	8	3,098	288	26.9	0.17	0.49	5.2	0.75	0.6	0.0	0.0	5.3	10	2	12
2	8	2,699	92	27.4	0.02	0.17	5.2	0.75	2.8	0.1	12.9	3.0	10	3	13
3	8	2,115	312	20.1	0.03	0.21	7.0	0.75	0.0	0.0	10.9	0.8	5	3	8
4	8	1,861	77	19.7	0.03	0.03	5.2	0.80	1.6	0.0	0.1	4.1	7	1	8
5	8	2,640	32	33.1	0.49	0.49	6.8	0.90	4.0	0.0	15.8	0.9	10	3	13
6	8	2,320	27	14.9	0.03	0.03	1.6	0.00	6.2	3.1	0.0	0.9	4	3	7
7	7	5,095	362	7.4	1.47	16.3	4.4	0.60	7.6	0.5	0.0	0.0	10	5	15
8	8	2,380	117	31.5	0.49	0.75	7.6	0.90	6.1	0.0	0.8	5.9	9	3	12
9	8	1,424	11	7.8	1.20	1.20	5.0	0.75	6.0	1.6	13.6	25.4	7	3	10
10	8	1,613	41	20.8	0.13	0.13	4.4	0.60	1.2	0.0	0.0	1.3	6	2	8
11	8	1,301	79	22.2	0.02	0.25	4.4	0.60	0.1	0.0	0.0	8.1	8	2	10
12	8	1,070	8.5	13.7	0.62	0.62	5.4	0.85	1.6	5.9	6.9	16.0	11	4	15
13	8	1,480	26	25.2	0.46	0.46	4.8	0.70	3.5	0.0	4.3	2.3	6	4	10
14	8	1,687	72	19.3	0.13	0.69	6.0	0.75	0.8	0.0	0.0	1.8	5	2	7
15	8	769	13	14.0	2.24	2.24	4.4	0.00	13.7	6.8	10.0	0.0	5	3	8
16	8	1,748	98	18.3	0.03	0.36	5.0	0.60	2.8	0.0	0.1	0.7	5	1	6
17	8	957	4.2	9.4	0.28	1.47	5.0	0.75	10.4	0.0	11.2	0.0	3	4	7
18	8	346	0.7	6.5	1.65	1.65	3.0	0.00	0.0	0.0	0.0	0.9	2	2	4
19	8	1,035	9.1	4.6	1.65	4.91	5.0	0.75	7.6	4.4	2.4	35.8	0	3	3
20	8	4,810	321	1.3	1.62	3.96	4.4	0.15	16.8	3.6	8.9	0.0	6	4	10
21	8	4,044	124	11.5	1.50	1.50	5.0	0.75	5.3	7.6	32.3	75.3	13	5	18

Table 5 continued

Site	*n*	L	A	D	N	NL	AVC	HCV	T	Dd	Js	Po	S_{res}	S_{nres}	S_{tot}
22	8	1,732	16	14.6	0.92	0.92	6.8	0.90	4.3	3.6	0.8	4.6	6	3	9
23	8	1,655	4.3	17.4	0.02	0.02	2.8	0.20	2.4	0.0	3.6	0.1	3	4	7
24	8	2,281	33	13.5	0.92	1.30	4.5	0.00	9.5	2.4	0.6	3.4	11	5	16
25	8	695	242	29.9	0.49	4.18	7.6	0.90	0.6	0.0	0.7	8.6	7	3	10
26	8	2,658	12	8.3	0.01	0.01	4.4	0.60	4.3	0.0	1.1	0.0	7	3	10
27	8	1,403	36	21.0	0.69	0.69	6.6	0.80	5.1	0.0	0.0	4.8	7	1	8
28	7	900	3.3	8.4	0.97	0.97	1.8	0.00	2.0	0.0	0.0	0.0	0	1	1
29	8	3,962	201	28.8	0.08	0.08	5.2	0.80	3.3	0.0	1.8	3.0	11	5	16
30	9	1,450	20	5.5	0.15	0.15	4.0	0.50	7.3	0.1	0.0	0.6	3	2	5
31	8	1,184	15	16.7	0.85	0.85	3.2	0.00	17.9	5.5	2.1	0.0	6	3	9
32	8	2,296	104	21.6	0.25	0.25	6.0	0.80	1.2	0.0	0.0	1.7	9	3	12
33	8	849	3.3	7.1	1.77	1.77	2.5	0.00	5.4	1.9	0.0	1.5	0	4	4
34	6	3,194	94	13.5	0.78	3.28	3.0	0.00	4.5	0.4	15.9	0.0	8	4	12
35	8	2,190	58	19.5	0.01	2.61	4.3	0.30	7.9	0.3	4.0	0.0	8	4	12
36	8	2,052	308	20.6	0.01	0.03	4.3	0.30	5.3	0.0	2.4	0.0	5	4	9
37	7	2,336	33	11.5	0.02	0.02	5.0	0.75	0.0	0.0	0.0	0.0	4	2	6
38	7	1,210	20	12.6	0.14	0.22	5.0	0.75	26.7	0.0	13.5	0.0	7	6	13
39	7	1,641	45	12.0	0.02	1.78	5.0	0.75	2.6	0.0	0.0	0.0	5	5	10
40	7	1,379	12	12.5	0.03	0.03	5.0	0.75	2.7	0.0	0.0	0.0	4	2	6
41	8	257	1.5	9.9	0.28	0.28	5.0	0.00	13.6	7.4	6.6	0.0	0	2	2
42	8	1,363	13	13.3	0.01	0.02	5.4	0.85	6.1	7.3	0.0	11.7	11	2	13
43	8	3,221	34	12.6	0.02	3.82	5.4	0.85	3.3	3.8	0.3	15.2	11	3	14
44	8	1,242	12	12.6	0.01	0.02	5.4	0.85	4.3	4.8	0.0	1.6	7	2	9
45	8	2,144	35	8.7	1.47	4.56	3.1	0.00	10.7	5.2	0.5	0.2	5	4	9
46	8	2,164	23	17.9	0.02	0.14	2.8	0.20	5.5	0.3	16.3	0.0	6	3	9

Site number (Site), number of surveys (*n*), transect length (L), site area (A, ha), distance from the urban centre (D, km), distances to the nearest and nearest larger remnants (N, NL, km), average vegetation condition (AVC) and fraction of high condition vegetation (HCV), density of *Eucalyptus*, *Casuarina* and *Allocasuarina* spp. trees (T, number of stems > 10 cm dbh per 100 m of transect), density (number per 100 m of transect) of three host plants, *Daviesia divaricata* (Dd), *Jacksonia sternbergiana* (Js), and *Patersonia occidentalis* (Po), and number of resident (S_{res}), non-resident (S_{nres}) and total (S_{tot}) species

Appendix 2

See Table 6.

Table 6 Counts of butterfly and day-flying moth species in 46 urban habitat fragments, south-west Western Australia

Site number	1	2	3	4	5	6	7	8	9	10	11	12	13	14	15	16	17	18	19	20	21	22	23
Resident butterflies																							
Trapezites sciron												19									9	1	
Trapezites argenteoornatus							7																
Anisynta sphenosema	3	4			2					4										18			
Hesperilla donnysa																							
Hesperilla chrysotricha			1																				
Motasingha trimaculata									1		1										6		
Croitana croites																							
Mesodina cyanophracta	1		2	1	1			2	4		1	9	1								41	3	
Taractrocera papyria	2	1		7	2	2	14			2	2	2	6		2	3				3	7		

Table 6 continued

Site number	1	2	3	4	5	6	7	8	9	10	11	12	13	14	15	16	17	18	19	20	21	22	23
Geitoneura klugii	63	22		20	53	10	180	34	25	3	82	12	8	17	27	21				192	104		1
Geitoneura minyas	60	1			42	49	435	4		1	1				55	7				145			11
Heteronympha merope	22			2	2	1	30	11		1	2				1	3				8	4		
Hypochrysops halyaetus									10			8									41		
Ogyris amaryllis																	4						
Candalides acastus		5	16	1			2						2	1									
Nacaduba biocellata		1			1		8			1		2						3			9	2	
Theclinesthes miskini		1					3																
Theclinesthes serpentata							2																
Neolucia agricola	1	3			5			7	10		1	3				1				9	231	6	
Zizina labrudus	1		1	4			4	3				8	2	2	2		1				4		4
Resident day-flying moths																							
Synemon sp. (Perth)					2			13	2			7		1							66	5	
Synemon gratiosa												1									1		
Pollanisus cupreus	38	15	1	1	52			35	3			10	16	14			1	1			24	33	
Hecatesia thyridion																							
Periscepta polysticta	3	2						2			2												
Non-resident butterflies																							
Delias aganippe							3										1						
Pieris rapae	2	12	2		3	8	16	2	16	2	4	20	6	3	10		25	3	7	7	20	26	14
Junonia villida			2									4							5		1		
Vanessa kershawi	54	8	22	22	29	8	19	18	24	6	32	31	10	13	2	10	4	3	3	192	181	28	2
Vanessa itea							1														3		
Danaus chrysippus																							
Danaus plexippus						2							1							1			1
Lampides boeticus		10			1		2	1	2			7	2		1		1			1	20	1	1

Site number	24	25	26	27	28	29	30	31	32	33	34	35	36	37	38	39	40	41	42	43	44	45	46
Resident butterflies																							
Trapezites sciron	3																		1	2	1		
Trapezites argenteoornatus														19		7	5						
Anisynta sphenosema				2					5			6											1
Hesperilla donnysa			1			1					4												
Hesperilla chrysotricha																							
Motasingha trimaculata	2	1									1									1			
Croitana croites														3		2	1						
Mesodina cyanophracta	6	1																	1	5			
Taractrocera papyria			14	1		9	1	1	1		4	2			2				1	3	1		1
Geitoneura klugii	32	19	25	19		100	11	20	8		21	44	16		5	3	1		40	36	2	11	2
Geitoneura minyas	17		31			97		19	16		187	531	35										63
Heteronympha merope				1		12			2		1	8			2				2	3	2		
Hypochrysops halyaetus	14																		7	11	13		
Ogyris amaryllis			12																				
Candalides acastus									3					1			3						
Nacaduba biocellata	1	1	1	1		2			4						1				2				1
Theclinesthes miskini																1							
Theclinesthes serpentata																							
Neolucia agricola	3	5		1		7		2			9	1	10		2				2	6		1	

Table 6 continued

Site number	24	25	26	27	28	29	30	31	32	33	34	35	36	37	38	39	40	41	42	43	44	45	46
Zizina labrudus						3			44			1	2	33	1	1						1	2
Resident day-flying moths																							
Synemon sp. (Perth)	37	12				1		11			3				9				5	11	13	8	
Synemon gratiosa	3						1												1	12			
Pollanisus cupreus	6	5	2	1		1		1				4	1						7	16	2	1	
Hecatesia thyridion									1														
Periscepta polysticta						2																	
Non-resident butterflies																							
Delias aganippe																1							
Pieris rapae	20	1	18		24	3	1	11	8	21	32	7	1	152	42	17	35	7	10	19	13	12	33
Junonia villida	1														2	1							
Vanessa kershawi	26	16	20	4		27	17	4	15	4	13	12	13	2	4	2	2	1	14	31	19	12	9
Vanessa itea										1													
Danaus chrysippus	1		2			1					1				1	1							
Danaus plexippus		1				2			1			1	3		1							1	
Lampides boeticus	4					3		1		1	14	1	7		5					1		4	2

Appendix 3

See Table 7.

Table 7 Akaike weights (w) for the best-fitting models of species incidence, multi-model averaged parameter estimates, and estimates of the relative importance (w_+) of each predictor variable

	w	Parameter estimates											
		Intercept	A	D	N	NL	AVC	HCV	T	Dd	Js	Dd+Js	Po
Trapezites sciron	0.87	−3.393	−0.002	−0.002	0.000	0.000	0.012	**2.423**	0.001				
Anisynta sphenosema	0.38	−2.673	**0.179**	0.134	−0.004	0.003	0.016	0.031	−0.003				
Motasingha trimaculata	0.54	−1.865	0.030	0.008	0.015	**0.295**	0.006	0.008	−0.011				
Mesodina cyanophracta	0.88	−9.773	0.003	**2.337**	0.000	0.000	0.050	0.186	−0.005				**0.997**
Taractrocera papyria	0.13	−0.907	**0.146**	0.069	−0.023	−0.013	−0.039	0.090	**0.084**				
Geitoneura klugii	0.64	−2.270	**0.546**	0.099	−0.005	−0.002	−0.020	−0.005	0.448				
Geitoneura minyas	0.71	−1.049	**0.701**	0.152	−0.001	−0.001	**−0.464**	−0.413	0.010				
Heteronympha merope	0.27	−2.402	**0.434**	0.027	0.001	0.001	−0.002	0.070	**0.105**				
Hypochrysops halyaetus	0.95	−3.851	0.000	0.000	0.000	0.000	0.000	0.040	0.000	**1.373**	0.000	0.014	
Candalides acastus	0.60	−1.205	0.058	0.035	−0.006	−0.008	0.022	0.216	**−0.536**				
Nacaduba biocellata	0.19	−1.847	−0.001	0.098	0.040	−0.002	0.057	**1.014**	0.007				
Neolucia agricola	0.18	−2.523	**0.331**	0.119	0.005	0.023	0.043	0.022	0.091	**0.114**	0.023	0.068	
Zizina labrudus	0.16	−1.894	**0.203**	0.222	−0.009	−0.003	0.003	0.075	−0.001				
Synemon sp. (Perth)	0.20	−2.909	0.000	0.118	0.067	0.027	**0.275**	0.416	**0.162**				
Synemon gratiosa	0.27	−2.049	−0.008	−0.051	0.002	0.004	0.009	**0.300**	0.013				
Pollanisus cupreus	0.23	−3.208	−0.002	0.247	0.002	0.003	**0.478**	0.175	−0.002				
Pieris rapae													
Junonia villida	0.30	−2.139	0.005	−0.031	0.016	0.035	0.032	**0.420**	0.016				
Vanessa kershawi													
Danaus chrysippus	0.24	−2.219	0.016	0.000	−0.012	−0.006	−0.013	−0.040	0.105				
Danaus plexippus	0.24	−2.413	**0.318**	0.038	−0.012	−0.009	−0.052	**−0.523**	0.130				

Table 7 continued

	w	Parameter estimates											
		Intercept	A	D	N	NL	AVC	HCV	T	Dd	Js	Dd+Js	Po
Lampides boeticus	0.38	−1.398	0.006	0.005	0.034	0.084	−0.035	−0.022	**0.222**		**0.629**		
Rare resident species	0.14	−4.249	**0.196**	−0.003	−0.067	−0.006	−0.034	0.923	−0.686				

	w_+										
	A	D	N	NL	AVC	HCV	T	Dd	Js	Dd+Js	Po
Trapezites sciron	0.02	0.01	0.01	0.01	0.03	**0.87**	0.01				
Anisynta sphenosema	**0.38**	0.14	0.06	0.06	0.08	0.06	0.06				
Motasingha trimaculata	0.09	0.04	0.07	**0.54**	0.04	0.05	0.05				
Mesodina cyanophracta	0.01	**0.88**	0.00	0.00	0.06	0.06	0.01				**0.98**
Taractrocera papyria	**0.51**	0.17	0.18	0.15	0.20	0.14	**0.29**				
Geitoneura klugii	**0.90**	0.12	0.04	0.04	0.08	0.03	0.69				
Geitoneura minyas	**0.91**	0.09	0.01	0.01	**0.75**	0.16	0.03				
Heteronympha merope	**0.90**	0.11	0.09	0.09	0.10	0.14	**0.30**				
Hypochrysops halyaetus	0.00	0.00	0.00	0.00	0.00	0.02	0.00	**0.95**	0.00	0.02	
Candalides acastus	0.12	0.04	0.03	0.03	0.06	0.07	**0.60**				
Nacaduba biocellata	0.12	0.18	0.21	0.12	0.22	**0.65**	0.12				
Neolucia agricola	**0.72**	0.13	0.05	0.10	0.13	0.05	0.21	0.21	0.10	**0.23**	
Zizina labrudus	**0.58**	0.34	0.14	0.13	0.13	0.15	0.12				
Synemon sp. (Perth)	0.08	0.17	0.25	0.14	**0.53**	0.26	**0.34**				
Synemon gratiosa	0.08	0.11	0.08	0.08	0.08	**0.27**	0.08				
Pollanisus cupreus	0.10	0.32	0.10	0.10	**0.81**	0.19	0.10				
Pieris rapae											
Junonia villida	0.06	0.08	0.09	0.13	0.11	**0.30**	0.07				
Vanessa kershawi											
Danaus chrysippus	0.10	0.08	0.10	0.09	0.09	0.09	0.21				
Danaus plexippus	**0.62**	0.10	0.10	0.09	0.16	**0.38**	0.26				
Lampides boeticus	0.04	0.03	0.13	0.26	0.11	0.03	**0.42**		**0.96**		
Rare resident species	**0.51**	0.08	0.29	0.10	0.14	0.43	0.20				

The most important variables, which also corresponded with those in the best fitting models, are bolded. Variables as in Appendix 1. A, D, N and NL were log transformed; density of trees and host plants were square root transformed

Appendix 4

See Table 8.

Table 8 Akaike weights (*w*) for the best models of species abundance, multi−model averaged parameter estimates, and estimates of the relative importance (w_+) for each predictor variable

	w	Parameter estimates											
		Intercept	A	D	N	NL	AVC	HCV	T	Dd	Js	Dd+Js	Po
Trapezites sciron	1.00	2.444	0.000	0.000	0.000	0.000	0.000	0.000	**−0.936**				
Anisynta sphenosema	1.00	0.502	0.000	**−0.098**	0.000	0.000	0.000	0.000	0.000				
Motasingha trimaculata	0.59	−0.015	0.013	0.000	0.000	0.000	**0.014**	0.000	0.000				
Mesodina cyanophracta	0.85	−0.045	0.000	0.000	**0.051**	0.000	0.000	−0.002	0.000				**0.138**
Taractrocera papyria	0.39	0.207	−0.001	−0.001	0.001	−0.002	−0.001	0.008	−0.001				
Geitoneura klugii	0.14	0.844	0.354	−0.092	0.036	0.124	0.025	−0.010	0.001				

Table 8 continued

	w	Parameter estimates											
		Intercept	A	D	N	NL	AVC	HCV	T	Dd	Js	Dd+Js	Po
Geitoneura minyas	1.00	2.101	0.000	0.000	**−2.940**	**3.929**	0.000	0.000	0.000				
Heteronympha merope	0.49	−0.326	**0.129**	0.007	−0.001	0.001	0.003	0.015	0.002				
Hypochrysops halyaetus	0.83	0.595	0.000	0.000	0.000	0.000	0.000	0.000	0.000	0.000	0.005	**0.089**	
Candalides acastus	1.00	−1.157	0.000	0.000	0.000	0.000	**0.257**	0.000	0.000				
Nacaduba biocellata	0.83	1.187	−0.019	**−0.259**	0.001	0.011	0.000	0.000	**−0.102**				
Neolucia agricola	0.45	−1.400	**0.197**	0.008	0.002	0.001	0.007	0.039	−0.061	0.002	0.085	**0.474**	
Zizina labrudus	0.34	0.776	−0.016	−0.028	0.005	−0.020	0.004	0.033	**−0.152**				
Synemon sp. (Perth)	0.51	0.743	0.011	−0.005	0.008	0.004	0.007	−0.037	0.008				
Synemon gratiosa	0.69	0.169	0.000	0.000	−0.004	0.030	0.001	0.000	0.000				
Pollanisus cupreus	0.30	−0.891	−0.026	**0.291**	**0.118**	0.006	0.144	**0.475**	−0.001				
Pieris rapae	0.30	2.958	**−0.437**	−0.064	−0.009	−0.034	−0.003	0.072	−0.002				
Junonia villida	1.00	0.693	**−0.129**	0.000	0.000	0.000	0.000	0.000	0.000				
Vanessa kershawi	0.09	1.534	**0.181**	**−0.497**	0.060	0.009	0.083	0.224	−0.029				
Danaus chrysippus	1.00	0.170	**−0.026**	0.000	0.000	0.000	0.000	0.000	0.000				
Danaus plexippus	0.86	0.026	0.000	**0.024**	0.000	0.000	0.000	0.000	0.000				
Lampides boeticus	0.27	0.078	0.001	0.001	0.000	0.000	−0.002	−0.001	−0.002		**0.072**		
Rare resident species	0.79	0.142	0.000	**−0.045**	**−0.007**	−0.001	0.000	0.000	−0.001				

	w_+										
	A	D	N	NL	AVC	HCV	T	Dd	Js	Dd+Js	Po
Trapezites sciron	0.00	0.00	0.00	0.00	0.00	0.00	**1.00**				
Anisynta sphenosema	0.00	**1.00**	0.00	0.00	0.00	0.00	0.00				
Motasingha trimaculata	0.39	0.00	0.01	0.00	**0.59**	0.00	0.00				
Mesodina cyanophracta	0.01	0.01	**0.85**	0.01	0.01	0.01	0.01				**1.00**
Taractrocera papyria	0.12	0.11	0.12	0.16	0.12	0.13	0.11				
Geitoneura klugii	**0.69**	0.21	0.23	0.46	0.16	0.13	0.13				
Geitoneura minyas	0.00	0.00	**1.00**	**1.00**	0.00	0.00	0.00				
Heteronympha merope	**1.00**	0.12	0.06	0.05	0.10	0.12	0.06				
Hypochrysops halyaetus	0.00	0.00	0.00	0.00	0.00	0.00	0.00	0.00	0.07	**0.83**	
Candalides acastus	0.00	0.00	0.00	0.00	**1.00**	0.00	0.00				
Nacaduba biocellata	0.17	**0.83**	0.01	0.16	0.00	0.00	**0.83**				
Neolucia agricola	**0.47**	0.03	0.03	0.02	0.04	0.05	0.17	0.00	0.17	**0.83**	
Zizina labrudus	0.15	0.10	0.08	0.19	0.06	0.09	**0.72**				
Synemon sp. (Perth)	0.10	0.06	0.10	0.08	0.08	0.09	0.09				
Synemon gratiosa	0.00	0.00	0.12	**0.69**	0.01	0.00	0.00				
Pollanisus cupreus	0.15	**0.51**	**0.79**	0.08	**0.50**	0.49	0.04				
Pieris rapae	**0.91**	0.18	0.16	0.23	0.14	0.16	0.14				
Junonia villida	**1.00**	0.00	0.00	0.00	0.00	0.00	0.00				
Vanessa kershawi	**0.66**	**0.69**	0.37	0.16	0.34	0.28	0.18				
Danaus chrysippus	**1.00**	0.00	0.00	0.00	0.00	0.00	0.00				
Danaus plexippus	0.00	**0.86**	0.02	0.01	0.01	0.01	0.01				
Lampides boeticus	0.06	0.06	0.06	0.06	0.08	0.06	0.08		**0.98**		
Rare resident species	0.01	**1.00**	0.79	0.15	0.00	0.01	0.04				

The most important variables, which also corresponded with those in the best fitting models, are bolded. Variables as in Appendix 3

References

Akite P (2008) Effects of anthropogenic disturbances on the diversity and composition of the butterfly fauna of sites in the Sango Bay and Iriiri areas, Uganda: implications for conservation. Afr J Ecol 26:3–13

Atmar W, Patterson BD (1993) The measure of order and disorder in the distribution of species in fragmented habitat. Oecologia 96:373–382

Barry S, Elith J (2006) Error and uncertainty in habitat models. J Appl Ecol 43:413–423

Baz A, Garcia-Boyero A (1995) The effects of forest fragmentation on butterfly communities in central Spain. J Biogeogr 22:129–140

Beard JS (1990) Plant life of Western Australia. Kangaroo Press, Kenthurst

Beaumont L, Hughes L (2002) Potential changes in the distribution of latitudinally restricted Australian butterflies in response to climate change. Glob Change Biol 8:954–971

Blair RB (1999) Birds and butterflies along an urban gradient: surrogate taxa for assessing biodiversity? Ecol Appl 9:164–170

Braby MF (2000) The butterflies of Australia: their identification, biology and distribution. CSIRO publishing, Collingwood

Braby MF, Edwards ED (2006) The butterfly fauna of the Griffith district, a fragmented semi-arid landscape in inland southern New South Wales. Pac Conserv Biol 12:130–154

Brown KS, Freitas AVL (2004) Butterfly communities of urban forest fragments in Campinas, São Paulo, Brazil: structure, instability, environmental correlates, and conservation. J Insect Conserv 6:217–231

Burbidge AA (2004) Threatened animals of Western Australia. Department of Conservation and Land Management, Kensington

Burnham KP, Anderson DR (1998) Model selection and inference: a practical information-theoretic approach. Springer, New York

Burnham KP, Anderson DR (2002) Model selection and multimodel inference: a practical information-theoretic approach, 2nd edn. Springer, New York

Common IFB (1990) Moths of Australia. Melbourne University Press, Carlton

Connor EF, Hafernik J, Levy J, Moore VL, Rickman JK (2002) Insect conservation in an urban biodiversity hotspot: the San Francisco Bay area. J Insect Conserv 6:247–259

Dapporto L, Dennis RLH (2008) Species richness, rarity and endemicity on Italian offshore islands: complementary signals from island-focused and species-focused analyses. J Biogeogr 35:664–674

Dennis RLH (2004a) Butterfly habitats, broad-scale biotype affiliations, and structural exploitation of vegetation at finer scales: the matrix revisited. Ecol Entomol 29:744–752

Dennis RLH (2004b) Just how important are structural elements as habitat components? Indications from a declining lycaenid butterfly with priority conservation status. J Insect Conserv 8:37–45

Dennis RLH (2010) A resource-based habitat view for conservation. Butterflies in the British Landscape. Wiley-Blackwell, Chichester

Dennis RLH, Eales HT (1997) Patch occupancy in *Coenonympha tullia* (Muller, 1764) (Lepidoptera: Satyrinae): habitat quality matters as much as patch size and isolation. J Insect Conserv 1:167–176

Dennis RLH, Eales HT (1999) Probability of patch site occupancy in *Coenonympha tullia* (Müller) (Lepidoptera: Satryinae) determined from geographical and ecological data. Biol Conserv 87:295–301

Dennis RLH, Hardy PB (2004) Loss rates of butterfly species with urban development. A test of atlas data and sampling artefacts at a fine scale. Biodivers Conserv 10:1831–1837

Dennis RLH, Hardy PB (2007) Support for mending the matrix: resource seeking by butterflies in apparent non-resource zones. J Insect Conserv 11:157–168

Dennis RLH, Shreeve TG, Olivier A, Coutsis JG (2000) Contemporary geography dominates butterfly diversity gradients within the *Aegean archipelago* (Lepidoptera: Papilionoidea, Hesperioidea). J Biogeogr 27:1365–1383

Dennis RLH, Shreeve TG, Van Dyck H (2003) Towards a functional resource-based concept for habitat: a butterfly biology viewpoint. Oikos 102:417–426

Dennis RLH, Hodgson JG, Grenyer R, Shreeve TG, Roy DB (2004) Host plants and butterfly biology. Do host-plant strategies drive butterfly status? Ecol Entomol 29:12–26

Dennis RLH, Shreeve TG, Isaac NJB, Roy DB, Hardy PB, Fox R, Asher J (2006a) The effects of visual apparency on bias in butterfly recording and monitoring. Biol Conserv 128:486–492

Dennis RLH, Shreeve TG, Van Dyck H (2006b) Habitats and resources: the need for a resource-based definition to conserve butterflies. Biodivers Conserv 15:1943–1946

Dennis RLH, Dapporto L, Shreeve TG, John E, Coutsis JG, Kudrna O, Saarinen K, Ryrholm N, Williams WR (2008) Butterflies of European islands: the implications of the geography and ecology of rarity and endemicity for conservation. J Insect Conserv 12:205–236

Dover JW, Rowlingson B (2005) The western jewel butterfly (*Hypochrysops halyaetus*): factors affecting adult butterfly distribution within native *Banksia* bushland in an urban setting. Biol Conserv 122:599–609

Dover JW, Sparks TH, Greatorex-Davies JN (1997) The importance of shelter for butterflies in open landscapes. J Insect Conserv 1:89–97

Dover JW, Dennis RLH, Atkins L (2008) The western jewel butterfly (*Hypochrysops halyaetus*) II: factors affecting oviposition and egg distribution within native Banksia bushland in an urban setting. J Insect Conserv 13:487–503

Edwards ED (1997a) Castniidae and the republic. ANIC News 11:October 1997

Edwards ED (1997b) Moths in the sun. ANIC News 10:April 1997

Ewers RM, Didham RK (2006) Confounding factors in the detection of species responses to habitat fragmentation. Biol Rev 81:117–142

Fartmann T (2006) Oviposition preferences, adjacency of old woodland and isolation explain the distribution of the Duke of Bugundy butterfly (*Hamearis lucina*) in calcareous grasslands in central Germany. Institute of Landscape Ecology, Muenster

Fattorini S (2006) To fit or not to fit? A poorly fitting procedure produces inconsistent results when the species-area relationship is used to locate hotspots. Biodivers Conserv 16:2531–2538

Fawcett T (2006) An introduction to ROC analysis. Pattern Recog Lett 27:861–874

Fletcher D, Mackenzie D, Villouta E (2005) Modelling skewed data with many zeros: a simple approach combining ordinary and logistic regression. Environ Ecol Stat 12:45–54

Fred MS, Brommer JE (2003) Influence of habitat quality and patch size on occupancy and persistence in two populations of the Apollo Butterfly (*Parnassius apollo*). J Insect Conserv 7:85–98

Fry JC (ed) (1993) Biological data analysis: a practical approach. The Practical Approach Series. Oxford University Press, Oxford

Garden J, McAlpine C, Peterson A, Jones D, Possingham H (2006) Review of the ecology of Australian urban fauna: a focus on spatially explicit processes. Austral Ecology 31:126–148

Gibb H, Hochuli DF (2002) Habitat fragmentation in an urban environment: large and small fragments support different anthropod assemblages. Biol Conserv 106:91–100

Government of Western Australia (2000) Bush forever. Western Australian Planning Commission, Perth

Gripenberg S, Roslin T (2005) Host plants as islands: resource quality and spatial setting as determinants of insect distribution. Ann Zool Fenn 42:335–345

Grundel R, Pavlovic NB (2007) Resource availability, matrix quality, microclimate, and spatial pattern as predictors of patch use by the Karner blue butterfly. Biol Conserv 135:135–144

Hardy PB, Dennis RLH (1999) The impact of urban development on butterflies within a city region. Biodivers Conserv 8:1261–1279

Hill CJ (1987) Butterflies. In: Catterall CP, Wallace CJ (eds) An island in suburbia: the natural and social history of Toohey forest. Institute of Applied Environmental Research, Griffith University, Brisbane, Queensland, pp 122–125

Hogsden KL, Hutchinson TC (2004) Butterfly assemblages along a human disturbance gradient in Ontario, Canada. Can J Zool 82:739–748

Hosmer DW, Lemeshow S (2000) Applied logistic regression. Wiley series in probability and regression, 2nd edn. Wiley, New York

Kadlec T, Benes J, Jarosik V, Konvicka M (2008) Revisiting urban refuges: changes of butterfly and burnet fauna in Prague reserves over three decades. Landsc Urban Plan 85:1–11

Keighery BJ (1994) Bushland plant survey. A guide to Plant Community Survey for the Community. Wildflower Society of WA, Nedlands

Kennewell C, Shaw BJ (2008) Perth, Western Australia. Cities 25:243–255

Kitching RL, Scheermeyer E, Jones RE, Pierce NE (eds) (1999) Biology of Australian butterflies, vol 6. Monographs on Australian Lepidoptera. CSIRO Publishing, Collingwood

Koh LP, Sodhi NS (2004) Importance of reserves, fragments and parks for butterfly conservation in a tropical urban landscape. Ecol Appl 14:1695–1708

Krauss J, Steffan-Dewenter I, Tscharntke T (2003) Local species immigration, extinction, and turnover of butterflies in relation to habitat area and habitat isolation. Oecologia 137:591–602

Krauss J, Steffan-Dewenter I, Tscharntke T (2004) Landscape occupancy and local population size depends on host plant distribution in the butterfly *Cupido minimus*. Biol Conserv 120:355–361

Krauss J, Steffan-Dewenter I, Muller CB, Tscharntke T (2005) Relative importance of resource quantity, isolation and habitat quality for landscape distribution of a monophagous butterfly. Ecography 28:465–474

Lamont BB, Klinkhamer PG, Witkowski ETF (1993) Population fragmentation may reduce fertility to zero in *Banksia goodii*—a demonstration of the Allee effect. Oecologia 94:446–450

MacKenzie DI, Nichols JD, Hines JE, Knutson MG, Franklin AD (2003) Estimating site occupancy, colonization and local extinction probabilities when a species is detected imperfectly. Ecology 84:2200–2207

Maes D, Van Dyck H (2001) Butterfly diversity loss in Flanders (north Belgium): Europe's worst case scenario? Biol Conserv 99:263–276

McCaw L, Hanstrum B (2003) Fire environment of Mediterranean south-west Western Australia. In: Abbott I, Burrows N (eds) Fire in ecosystems of south-west Western Australia: impacts and management. Backhuys, Leiden, pp 87–106

McKinney ML (2008) Effects of urbanization on species richness: a review of plants and animals. Urban Ecosyst 11:161–176

Nelson GS, Nelson SM (2001) Bird and butterfly communities associated with two types of urban riparian areas. Urban Ecosyst 5:95–108

New TR (1984) Insect conservation: an Australian perspective. Dr W. Junk, Netherlands

New TR (1991) Butterfly conservation. Oxford University Press, Melbourne

New TR (ed) (1993) Conservation biology of Lycaenidae (Butterflies). Occasional Paper of the IUCN Species Survival Commission No. 8. IUCN, Gland

New TR, Sands DPA (2002a) Conservation concerns for butterflies in urban areas of Australia. J Insect Conserv 6:207–215

New TR, Sands DPA (2002b) Narrow-range endemicity and conservation status: interpretations for Australian butterflies. Invertebr Syst 16:665–670

Newland G (2003) Literature review—urban areas and butterfly conservation. Vic Entomol 33(4):60–64

Pollard E, Yates TJ (1993) Monitoring butterflies for ecology and conservation. Chapman and Hall, London

Rosenzweig ML (1995) Species diversity in space and time. Press Syndicate of the University of Cambridge, Cambridge

Ruszczyk A, De Araujo AM (1992) Gradients in butterfly species diversity in an urban area in Brazil. J Lepid Soc 46:255–264

Samways MJ (2007) Insect conservation: a synthetic management approach. Annu Rev Entomol 52:465–487

Sands DPA, New TR (2002) The action plan for Australian butterflies. Environment Australia, Canberra

Shapiro AM (2002) The Californian urban butterfly fauna is dependent on alien plants. Divers Distrib 8:31–40

Stevens J (1992) Applied multivariate statistics for the social sciences, 2nd edn. Lawrence Erlbaum, Hillsdale

Swets K (1988) Measuring the accuracy of diagnostic systems. Science 240:1285–1293

Thomas JA, Bourn NA, Clarke RT, Stewart KE, Simcox DJ, Pearman GS, Curtis R, Goodger B (2001) The quality and isolation of habitat patches both determine where butterflies persist in fragmented landscapes. Proc R Soc B 268:1791–1796

Tscharntke T, Steffan-Dewenter I, Kruess A, Thies C (2002) Contribution of small habitat fragments to conservation of insect communities of grassland-cropland landscapes. Ecol Appl 12(2):354–363

Ulrich W (2003) Species-area relationships of butterflies in Europe and species richness forecasting. Ecography 26:365–373

Ulrich W, Buszko J (2005) Detecting biodiversity hotspots using species-area and endemics-area relationships: the case of butterflies. Biodivers Conserv 14:1977–1988

Veech JA (2000) Choice of species-area function affects identification of hotspots. Conserv Biol 14(1):140–147

Vittinghoff E, Glidden DV, Shiboski SC, McCulloch CE (2005) Regression methods in biostatistics: linear, logistic, survival and repeated measures models. Springer, San Fransisco

Wahlberg N, Klemetti T, Hanski I (2002) Dynamic populations in a dynamic landscape: the metapopulation structure of the marsh fritillary butterfly. Ecography 25:224–232

Warren MS, Hill JK, Thomas JA, Asher J, Fox R, Huntley B, Roy DB, Telfer MG, Jeffcoate S, Harding P, Jeffcoate G, Willis SG, Greatorex-Davies JN, Moss D, Thomas CD (2001) Rapid responses of British butterflies to opposing forces of climate and habitat change. Nature 414:65–69

Whittingham MJ, Stephens PA, Bradbury RB, Freckleton RP (2006) Why do we still use stepwise modelling in ecology and behaviour? J Anim Ecol 75:1182–1189

Williams MR (1995) An extreme-value function model of the species incidence and species-area relations. Ecology 76:2607–2616

Williams MR (2008) Assessing biodiversity of diurnal Lepidoptera in habitat fragments: testing the efficiency of strip transects. Environ Entomol 37:1313–1322

Williams MR (2009) Butterflies and day-flying moths in a fragmented urban landscape, south-west Western Australia: patterns of species richness. Pac Conserv Biol 15:32–46

J Insect Conserv (2011) 15:165–178
DOI 10.1007/s10841-010-9335-x

ORIGINAL PAPER

High and dry or sunk and dunked: lessons for tallgrass prairies from quaking bogs

Ann B. Swengel · Scott R. Swengel

Received: 4 March 2010 / Accepted: 31 August 2010 / Published online: 14 September 2010

Abstract Northern Wisconsin bogs provide a natural experiment on butterfly population occurrence in a naturally highly fragmented vegetation type, which may provide insight on conserving butterflies in anthropogenically fragmented and degraded landscapes. We surveyed butterflies in bogs (about as unaffected by humans as possible, but naturally occurring over <1% of northern Wisconsin) primarily during 2002–2009, with additional observations from 1986 to 2001. Different bog types had different bog-specialist butterfly faunas, but bog butterfly abundance also differed in similar vegetations among subregions. Some small isolated bogs held very high densities of specialist butterflies. Summer but not spring specialists frequented adjacent lowland roadsides and utilized a variety of non-native as well as native nectar sources. Paleo-entomology indicates that insects don't evolve out of trouble; instead they move out of trouble. Given the low dispersal apparent today for species restricted to bogs, "move" might be better understood as "hunkering" within their vegetation as it expands and shrinks and moves around the landscape. Although bogs appeared to have more intact specialist butterfly faunas than tallgrass prairies (99.9% destroyed by human activities), bog butterflies do not live in average sites even in a relatively natural landscape. Just as bog butterflies are "sunk and dunked" in isolation, specialist butterflies elsewhere may have been left "high and dry" naturally, or are now due to human activities. Numerous studies have demonstrated that presence and abundance of specialist butterflies increase with increasing size and connectedness of habitat patches. But with long-term consistent vegetation, populations with high abundances in small isolated sites and with low numbers thinly occurring in large sites can be secure, as shown by bog butterflies.

Keywords Bog · Specialist butterfly · Population persistence · Dispersal · Co-evolution · Stasis · Habitat quality · Landscape configuration · Habitat fragmentation · Vegetative consistency

A. B. Swengel (✉) · S. R. Swengel
909 Birch St., Baraboo, WI 53913, USA
e-mail: swengel@naba.org

Introduction

A major threat to biodiversity is human-caused habitat loss and fragmentation, with degradation of extant patches by edge effects and disruption of ecological processes (Brown 1997; van Swaay et al. 2006; Forister et al. 2010). Butterfly populations fare more poorly in isolated or small sites, with nearer and larger sites more likely to remain occupied or become re-colonized (Bulman et al. 2007; Hanski and Pöyry 2007; Dover and Settele 2009). Both habitat quality and landscape configuration are important for maintaining butterfly populations (Dennis and Eales 1997; Thomas et al. 2001; Hanski and Pöyry 2007; Dennis 2010), and both are under more threat in a human-degraded, fragmented landscape. As a result, much research in conservation biology concerns how to counteract the effects of habitat loss, degradation, and fragmentation on vulnerable species (e.g., Mattoni et al. 2001; Thomas et al. 2001; Dennis and Hardy 2007; Swengel and Swengel 2007; Spencer and Collins 2008; Dennis 2010).

In the midwestern USA, tallgrass prairie (predominately herbaceous flora) and oak savanna (trees and brush mixed in with herbaceous patches) have been about 99% destroyed primarily by conversion to agriculture (Curtis 1959; Nuzzo 1986; Samson and Knopf 1994). Pine barrens (a savanna on very sandy soil) have also declined but not to the same degree (Curtis 1959; Riegler 1995). Fragmentation and degradation greatly affect remaining examples of these vegetations and their associated butterflies (Samson and Knopf 1994; Riegler 1995; Swengel and Swengel 1999a, 2001).

In temperate areas of North America and Europe, bog (peatland) vegetation is also rare, being naturally isolated and forming a low proportion of the natural landscape. Although often viewed as a long-lived successional stage between open water and forest in glaciated landscapes, peatlands can get reset to an earlier successional stage (Curtis 1959). Since bogs are well known for their relatively stable vegetations and insect faunas over the long term, they can also be viewed as a climax community (Spitzer et al. 1999; Spitzer and Danks 2006; Whitehouse 2006; Whitehouse et al. 2008). While often considered relatively uniform floristically both within and among sites, bogs actually contain many microhabitats (Väisänen 1992; Spitzer and Danks 2006; Turlure et al. 2009, 2010). In Wisconsin, bogs occur primarily in central and northern areas (Curtis 1959). Prior to European settlement, peatlands occurred in <1% of the Wisconsin landscape (even counting only the northern third of the state), and most of that vegetation is still extant, with only 9% loss (Hoffman 2002), more lost in central than northern Wisconsin. Much of what is left, especially in northern Wisconsin, is relatively undegraded. Primary human impacts are roads and ditches; adjacent lands are more affected by timber harvesting, agriculture, and urbanization (pers. obs.). Conversion to cranberry agriculture and peat harvesting has occurred more in central Wisconsin bogs (Curtis 1959). By contrast, in Europe bog vegetation is much destroyed and degraded by human activities, along with the associated butterfly species of high conservation concern (Vandewoestijne and Baguette 2004; Schtickzelle et al. 2006; Spencer and Collins 2008; Turlure et al. 2009). The four bog-related vegetation types ranked highest in proportion of threatened butterfly species of their typical faunas (van Swaay et al. 2006).

In addition to observations by a few other lepidopterists, Nekola (1998) conducted a systematic survey of northern Wisconsin peatlands and their associated butterflies in 1996. He described three peatland types: muskeg (black spruce-cottongrass-wiregrass savanna similar in elevation to surrounding uplands), kettlehole (sphagnum-leatherleaf mats, often floating on lakes or sunk in depressions much lower than the surrounding landscape), and coastal peatland (tamarack-sedge mats with ridges of muskeg-like vegetation in estuaries along the Lake Superior coast). Many aspects of the flora are similar among these three types (Nekola and Kraft 2002), echoing Curtis's (1959) description of remarkably uniform bog structure and composition throughout the circumboreal region. Nekola (1998) nevertheless found significant differences in bog-obligate butterfly occurrence among these three bog types, and noted variation in flora amongst sites, especially kettleholes.

We have recorded butterflies in Wisconsin bogs since 1986. In this paper, we analyze these results to expand and extend Nekola's study in order to describe patterns of bog butterfly occurrence and abundance. We examine butterfly occurrence and abundance in bogs in relation to bog type, size, and isolation, and in adjacent lowland roadside ditches and nearby upland roadsides. This provides evidence on the relative importance to specialist butterfly population persistence of dispersal tendency, habitat quality, and landscape configuration. During the same period, we conducted surveys of butterflies in prairies in seven midwestern states (Swengel 1996; Swengel and Swengel 1999b, Swengel and Swengel 2007) and Wisconsin pine barrens (Swengel 1998; Swengel and Swengel 2005, 2007). Based on this field work and others' studies, we contrast the occurrence of specialist butterflies between vegetations altered and fragmented by humans (prairie, barrens) and naturally fragmented ones (bogs). These results should be useful for application to conservation of bog butterflies where they are vulnerable, and vulnerable butterflies in other fragmented vegetations.

Methods

Study regions

The primary study region contains 73 bog sites scattered across an area 367 km east–west by 169 km north–south (45.33–46.86°N, 88.21–92.56°W) in twelve contiguous counties spanning the entire breadth of northern Wisconsin. At 20 of these sites, we also surveyed the lowland (wetland) roadside ditch through or adjacent to the bog, and at five sites, we surveyed a more upland roadside corridor 20–350 m from the bog. In three large muskeg complexes, we counted surveys in each separate area as a separate site. In central Wisconsin, the three bogs in two contiguous counties (Jackson, Wood) are in an area 29 km east–west by 4 km north–south (44.31–44.34°N, 90.19–90.56°W), which is 169 km south of the nearest study site in the northern study region. Nekola's (1998) study region comprises sites in and adjacent to the Lake Superior drainage basin in four contiguous counties (Ashland, Bayfield, Douglas, Iron) bordering the south lakeshore. This area is the north part of the west half of our northern study region. All our sites in those counties fall within his study region.

Within each study region, we biased toward high-quality examples of bog vegetation that were open to public visitation and efficient to gain access to and travel between due to clustering of sites and/or efficient routing among sites, including proximity to other kinds of sites of interest for other species. Nonetheless, peatlands often present difficulties of access both to them and across them, which reduces efficiency and amount of transect distance surveyed in a day. Roadside survey areas were selected because we noticed bog butterflies using them, they were en route to or from a bog survey route, or they appeared potentially of interest for either bog or other butterfly species.

Surveys

On 114 informal visits during 1986–2001 in both study regions (widely in the northern one), we recorded number of individuals by species per site, but did not standardize a route or record weather and effort (time and distance spent surveying). We began formal transect surveys in bogs in 1990, with most conducted during 2002–2009 (Table 1). In those last 8 years, we surveyed in a rotation through the western, central, and eastern sections of the northern study region, trying to cover one section per weekend, or more if a section was missed the previous weekend and/or if time allowed. But we missed an occasional weekend per year due to weather or another commitment. Surveys occurred between 23 April and 12 September, usually early/mid May through early/mid August in most years. We also continued to record bog specialists informally observed in uplands and roadsides as we accessed bogs for formal surveys.

Our peatland transect surveys were like those in prairie and barrens (similar to Pollard 1977 and as described in Swengel 1996, 1998; Swengel and Swengel 1997). We walked along a similar route per visit to a prairie, barrens, or bog at a slow pace (about 2 km/h) on parallel routes 5-10 m apart. We counted all adult butterflies observed ahead and to the sides, to the limit at which an individual could be identified, possibly with the aid of binoculars after detection, and tracked. A new sampling unit was designated whenever the vegetation along the route varied by management (type and/or years since last treatment), type (wet, mesic, dry), quality based on type of brush and diversity and abundance of native and exotic flora (undegraded, semi-degraded, highly degraded), and/or estimated macrosite canopy (grassland or open bog < 10%, open savanna 10-24%, closed savanna 25-49%, forest opening 50-75%). Routes crossed rather than followed ecotones and management boundaries to reduce edge effects, and were designed to minimize number of unit changes while covering representative areas of the site. Temperature (starting, ending, low, high), wind speed, percent cloud cover, percent time sun was shining, route distance, and time spent surveying were recorded for each unit. Data from each unit were kept separate. Surveys occurred during a wide range of times of day and weather, occasionally in intermittent light drizzle so long as butterfly activity was apparent, but not in continuous rain. All butterfly species found were counted, but survey times and locations were selected to study butterflies specialized to that vegetation.

In prairie and barrens, we categorized the species by habitat niche breadth (Swengel 1996, 1998): (1) specialist (restricted or nearly so to herbaceous flora in prairie and/or savanna; sensitive to vegetative quality); (2) grassland species (widely inhabiting both native and degraded herbaceous flora); (3) generalist (inhabiting grassland and other vegetation types); and (4) immigrant (occurring in the study region during the growing season but unlikely to overwinter). In bogs, we used an analogous categorization applicable to this study region only, and these categories correspond approximately to those (in parentheses) described by Spitzer and Danks (2006) (Table 2): (1) bog specialist (tyrphobiontic)—restricted or nearly so to peatlands; (2) bog affiliate (tyrphophilic)—breeding in bogs as well as other vegetations (limited to species of north temperate or boreal affinity); (3) generalist (tyrphoneutral)—year-round resident primarily using vegetation other than bogs (if the species also breeds in bogs, its range includes non-montane areas well south of Wisconsin); and (4) immigrant (tyrphoxenous)—not a year-round resident of the region and unlikely to breed in bogs. In Wisconsin, the bog specialists are all at the southern end of their eastern North American range, with their known range not extending into the state immediately south of Wisconsin, but further east *L. epixanthe* and *L. dorcas* may occur in areas more southerly than Wisconsin (Opler 1992; Glassberg 1999; Nielsen 1999).

Especially in bogs, we surveyed in a wider range of weather conditions than prescribed in Pollard (1977) and Wikström et al. (2009). We experienced severe constraints

Table 1 *N* surveys of units and survey effort (km, h) in central and northern Wisconsin at 76 bog sites, 20 lowland roadsides, and 5 upland roadsides, from 23 April to 12 September

	N	Years	Km	Hours
1987–2001				
All sites	50	1987–2001	44.0	25.8
Bog	27	1990–2001	21.5	13.1
Lowland	5	1999–2001	3.1	2.1
Upland	18	1987–1996, 1998–2001	19.5	10.7
2002–2009				
All sites	1,973	2002–2009	921.9	377.2
Bog	1,699	2002–2009	806.5	321.3
Lowland	223	2002–2009	80.5	42.5
Upland	51	2002–2009	34.9	13.5

Table 2 Total individuals of each specialist and affiliate species in bogs, lowland roadsides, and upland roadsides during 1987–2009 on formal surveys (informal observations in lowland and upland roadsides in parentheses)

Species			Bog	Lowl	Upl	Notes
Bog specialists						
L	MCK	Bog copper *Lycaena epixanthe*	3,986	674	2	23 Jun–15 Aug
N	MCK	Bog fritillary *Boloria eunomia*	1,084	0	1(1)	9 Jun–14 Jul
N	MCK	Jutta arctic *Oeneis jutta*	962	1	0	12 May–10 Jul
N	MCK	Freija fritillary *Boloria freija*	722	0	0	7 May–15 Jun
N	M	Frigga fritillary *Boloria frigga*	264	0(1)	0	19 May–20 Jun
N	M	Purplish fritillary *Boloria montinus*	192	1,093	0	27 Jul–23 Aug
L	MC	Dorcas copper *Lycaena dorcas*	163	379	0	26 Jun–15 Aug
N	M	Red-disked alpine *Erebia discoidalis*	128	1	0	10 May–14 Jun
Bog affiliates						
L	Brown elfin *Callophrys augustinus*		4,304	6	0	
N	Inornate ringlet *Coenonympha tullia*		607	167	10	
Pi	Pink-edged sulphur *Colias interior*		302	229	12	
N	Eyed brown *Satyrodes eurydice*		245	251	81	
H	Dreamy duskywing *Erynnis icelus*		168	7	2	
N	Silver-bordered fritillary *Boloria selene*		123	95	19	
N	Atlantis fritillary *Speyeria atlantis*		46	299	134	
N	Harris' checkerspot *Chlosyne harrisii*		32	29	0	
N	Green comma *Polygonia faunus*		1	0	0	

Family is abbreviated as *P* Papilionidae, *Pi* Pieridae, *L* Lycaenidae, *N* Nymphalidae, *H* Hesperiidae. Abbreviations for site types each specialist was recorded in: *M* muskeg, *C* coastal peatland, *K* kettlehole. Butterfly names follow Cassie et al. (2001)

on obtaining suitable weather especially in spring. The bog-specialist species varied greatly in detectability relative to weather. For example, temperatures of 13–14°C (if sunny) were excellent for surveying *Boloria freija*, which is active but not too active then, while *Erebia discoidalis* is most effectively surveyed at mid-day in the warmest sunniest conditions possible (at least 16°C but preferably 18–19°C or warmer as described in Swengel 2009). Both these species have similar spring flight periods (Table 2). The tan and brown color of the vegetation in spring results in warmer microclimates in and between peat hummocks relative to overall air temperature, which may facilitate butterfly activity in cooler conditions than usually recommended for butterfly surveying. The negative effect of subpar weather on butterfly detection in our survey results is not a systematic bias, as we did not (and could not) bias which sites were surveyed in better weather, so that this factor confounds statistical power, but doesn't systematically misdirect it.

Analyses

As in Nekola and Kraft (2002), we identified the flight period per species by the first and last date per year we detected it, separately for the two study regions, and in northern Wisconsin, separately for coastal peatlands, which were much later in phenology than the other sites (Swengel 2009). We also determined flight period spans separately for different areas of inland northern Wisconsin when phenological differences were evident for a particular species (e.g., *L. epixanthe* in Swengel 2009). In the case of single outlier individuals observed, we did not consider it flight period at sites where not seen. Our population index is the peak survey count per site per brood, standardized to survey time, to create an observation rate (relative abundance) per hour per unit survey. Since *Oeneis jutta* may have a two-year life cycle, with one cohort larger than the other (Opler and Krizek 1984), we tested for a skewed distribution by even and odd years. Although significantly different (Table 3), even-year cohorts were 73% the abundance of odd-year cohorts, well within the range of typical variation from annual fluctuation reported for specialist butterflies (Pollard and Yates 1993; Swengel and Swengel 2007). For this reason, we pooled all available years for this species' analyses.

Analysis was done with ABstat 7.20 software (1994 Anderson-Bell Corp., Parker, Colorado). Statistical significance was set at two-tailed $P < 0.05$. Since significant results occurred at a frequency well above that expected due to spurious Type I statistical error, the critical P value was not lowered further, as more Type II errors (biologically meaningful patterns lacking statistical significance)

Table 3 Distribution of *Oeneis jutta* individuals on peak count per year at sites ever reporting the species, by even and odd years during 2002–2009, significantly skewed from expected based on survey effort (h) in each group of years (chi square goodness of fit test $P = 0.0000$)

	Observed individuals	Survey time	Expected individuals	Observation rate (h)
Odd years	503	47.19 (54.6%)	442	10.7
Even years	307	39.26 (45.4%)	368	7.8
Total	810	86.45	810	

would be created than Type I errors eliminated. All correlations were done with the Spearman rank correlation. To test for significant skewing compared to distribution proportional to survey effort (h), we used the Chi Square Goodness of Fit test, and to test for significant differences in relative abundance among sites, we used the Mann-Whitney *U* test. All statistical tests in this study are non-parametric, which do not require any assumptions about how the data are distributed (e.g., normality).

Results

We recorded the same specialists in muskegs as in Nekola's (1998) study (Table 2). The few additional species we recorded in kettleholes and coastal peatlands within Nekola's (1998) study region were infrequently encountered in only one or two sites per bog type (in coastal peatlands, one *B. freija* at one site, five *O. jutta* at one site, and seven *L. dorcas* at two sites; in kettleholes, three *B. freija* at one site).

In analysis of the peak surveys per year of the four most frequently recorded specialists (Figs. 1, 2; Appendix 1), significant differences in abundances occurred among bog types. Significantly highest abundances of *L. epixanthe* and *B. eunomia* were in kettleholes and coastal peatlands,

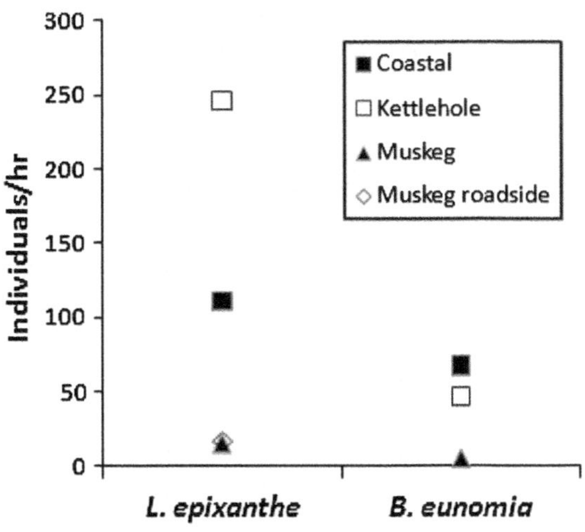

Fig. 1 Grand means of average relative abundance (peak individuals/hr per site per year) of *L. epixanthe* and *B. eunomia* in sets of sites grouped by bog type, as in "Appendix 1"

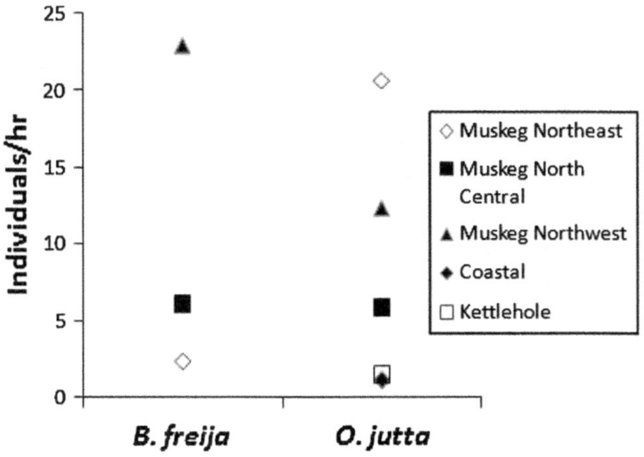

Fig. 2 Grand means of average relative abundance (peak individuals/hr per site per year) of *B. freija* and *O. jutta* in sets of sites grouped by bog type, and for muskeg, by subregion, as in "Appendix 1" (coastal peatlands and kettleholes were not plotted for *B. freija*, since we so rarely recorded it there)

despite their small size. Within each bog type, abundance of some species also significantly differed among sites (Appendix 1); for muskegs (for which we have the largest sample of sites over the greatest geographical spread), these differences occurred both within and among subregions. The percentage of years we found the species covaried significantly with mean relative abundance for each species (Table 4). Nonetheless, at muskegs with longer survey routes (Armstrong, Glidden), we had a fairly reliable detection of presence each year (50–100% of years surveyed) for species occurring in low abundance (*B. freija* at both sites, *B. eunomia* at Glidden) (Appendix 1), despite few visits (1–3) during the flight period per year.

Specialists with adequate formal sampling in nearby upland roadsides were significantly under-represented there (Fig. 3; Appendix 2). Flight period timing explained

Table 4 Coefficients (r) and significance (P) of Spearman rank correlations of proportion of years recorded as present in N sites to mean abundance in each site as in "Appendix 1", by species

	N	r	P
Lycaena epixanthe	27	+0.760	<0.01
Boloria eunomia	18	+0.536	<0.05
Boloria freija	15	+0.849	<0.01
Oeneis jutta	23	+0.633	<0.01

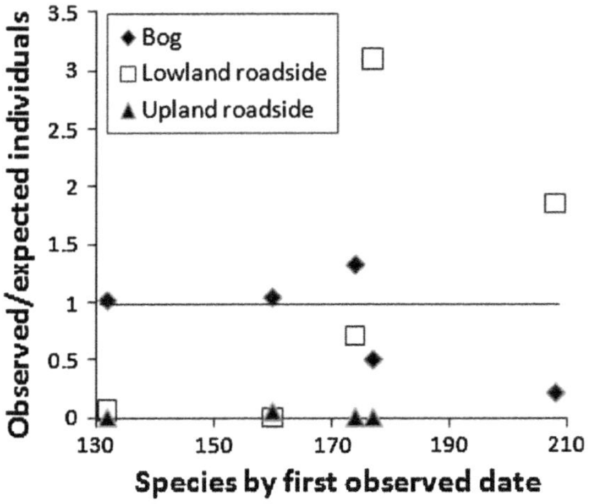

Fig. 3 Ratio of observed/expected individuals based on survey effort in bog, adjacent lowland roadside, and nearby upland roadside, by species (see "Appendix 2"), in order of first observed date (as in Table 2) expressed as sequentially numbered date. Random distribution = 1 (observed = expected)

whether bog specialists occurred in any numbers in adjacent lowland roadsides (Fig. 3; Appendix 2). Spring species were significantly under-represented there. Two summer species were significantly over-represented in these lowland roadsides compared to bogs, and the third (*L. epixanthe*) occurred in substantial numbers (Appendix 2), although still under-represented there. Bogs were relatively nectar-rich in spring, more so than the roadsides, but nectar-poor in summer, when the roadsides were nectar-rich.

L. epixanthe was remarkably detectable in cool, cloudy weather in coastal peatlands as well as inland kettleholes in the same county (Table 5). This was especially noticeable in the coastal peatlands due to the lower mean temperature and percent sunshine on surveys there. Some surveys in cool, cloudy weather produced "good" counts (Table 6), defined as abundances at least half the mean relative abundance of that site in "Appendix 1". This contrasted with inland surveys in Douglas County muskegs and adjacent lowland roadsides, which had the one significant weather effect (positive with percent sunshine) and no "good" counts in poor weather (Tables 5, 6). However, in 2009 in central Wisconsin (the hottest climate in this study), it was in poor weather that we obtained our highest count ever (82 individuals) of *L. epixanthe* there (Table 6).

Table 6 Surveys with low temperature <24°C and ≤30% sunshine with *Lycaena epixanthe* observation rate half or more the mean observation rate on peak surveys in "Appendix 1", at sites (grouped by type, subregion, and county) analyzed in the table

Site	Date	Temperature	% Sunshine	Obs rate
Coastal peatlands (Northwest: Bayfield coastal)				
Bark Bay	27 Jul 93	17	0%	64.9
Bark Bay	24 Jul 08	19	5	112.3
Bibon Lake	24 Jul 08	22	2	96.9
Port Wing Boreal Forest	24 Jul 08	19	30	228.0
Bibon Lake	23 Jul 09	17	15	114.3
Bark Bay	7 Aug 09	22	20	109.4
Kettleholes (Northwest: Bayfield interior)				
East Roger Lake	9 Jul 06	20	1	240.0
Pine Lake	24 Jul 08	22	0	525.0
East Wishbone Lake	24 Jul 09	23	25	369.2
East Wishbone Lake	9 Aug 09	21	10	200.0
Muskeg (Central: Jackson)				
Highway 54	4 Jul 09	20	0	213.9

Table 5 Spearman rank correlation coefficients (*r*) (boldfaced if significant) of peak *Lycaena epixanthe* observation rates as analyzed in "Appendix 1" with low temperature and percent sunshine on each unit survey, excluding sites (Lost Creek, East Crane Lake) significantly lower in abundance within that group of sites (see "Appendix 1")

Bog site type	Coastal	Kettlehole	Muskeg	Muskeg lowland rdside
County	Bayfield	Bayfield	Douglas	Douglas
N unit surveys	24	24	45	41
Mean temperature (°C)	18.6	24.6	25.6	25.3
Range of temperature (°C)	17–32	19–31	17–32	18–32
Mean % sunshine (%)	38.9	71.7	69.0	72.4
Range of % sunshine (%)	0–100	0–100	0–100	0–100
Correlation (*r*) with				
Low temperature	−0.224	−0.047	+0.129	−0.020
Percent sunshine	+0.255	−0.363	**+0.359**	+0.282

Discussion

Cautions

Frequent surveys under strict weather and timing protocols (e.g., Pollard 1977; Pollard and Yates 1993; Wikström et al. 2009) produce more robust results than our less frequent surveys in more variable weather and times of day. As a result, our abundance indices are more confounded by phenology and variation in detection due to weather. Therefore, the strongest findings in this study are relatively high abundances (despite these confounding factors), previously unreported presence of species in coastal peatlands and kettleholes, and results that are most consistent with others' independent findings, as discussed below.

Characterization of bog butterfly fauna

Nekola's (1998) report of *L. epixanthe* and *B. eunomia* as the dominant species in kettleholes, and the latter dominant in coastal peatlands, is consistent with the high abundances we recorded for both species in both bog types (Fig. 1) despite small bog sizes (Appendix 1). Agreeing with Nekola's (1998) report of significantly different peatland butterfly faunas among the three bog types, we found some significant differences in species' abundances among bog types (Appendix 1). Furthermore, we found significant differences in relative abundance within type among sites. Extraordinarily high abundances of a few species (*L. epixanthe*, *B. eunomia*) occurred in small kettleholes, consistent with a few reports of high numbers for *L. epixanthe* in the 4th of July butterfly counts (Opler and Powell 1984: 2000 individuals at Union City, Pennsylvania; Wander 2009: 1509 individuals at Central Franklin County, Massachusetts). High densities of *B. aquilonaris* have also been reported in European bogs (Baguette and Schtickzelle 2006). Furthermore, at two bogs of the same vegetative classification with similar host plant abundance, *B. aquilonaris* abundance varied six-fold in relation to sphagnum structure (carpet or hummock), which affected larval microclimate (Turlure et al. 2010). Small prairies can also have very high specialist numbers, e.g., 906 regal fritillaries (*Speyeria idalia*) in 1.25 h of surveying at an occasionally hayed 12 ha prairie on the Northern Loess Hills, Iowa count (Swengel and Opler 1996). Viable populations of localized butterfly have been reported for small British sites (Thomas 1984).

As Nekola (1998) noted, the North American literature indicated bog butterflies' preference for peatland margins, but that was generally not the case in his experience or ours. He noted an exception that *B. montinus* adults were primarily found at adjacent upland nectar, a description that includes flowers in the gravel edge of roads by lowland ditches adjacent to bogs. We found a few summer species, including that one, frequenting adjacent lowland roadsides but virtually none occurring in nearby upland roadsides (Fig. 3; Appendix 2). Furthermore, the use of adjacent lowland roadsides should be interpreted with caution. We sampled relatively few sites and they were biased toward opener, less brushy ditches with more of a wetland characteristic, which may exaggerate specialist use of roadsides. Some bogs, including small kettleholes and coastal peatlands, are entirely surrounded by formidable barriers (dense forest, Lake Superior). But others have places where they border more open vegetations, such as roadsides, grasslands, and marshes, so that it is not an obvious barrier (water, forest) that prevents specialist movement out of these bogs.

Movement vs. dispersal limitations

Nekola and Kraft (2002) reported significant clustering of both bog butterfly presences and absences. This may be attributable to metapopulation dynamics (both dispersal and dispersal limits). Although virtually undetected in this study (Table 2; Fig. 3; Appendix 2), some movement of specialists out of bogs and between them is still possible. As mapped in Nekola and Kraft (2002), many bogs occur in clusters with at least one other bog within 5 km. This is within or near the maximum dispersal distance reported by Baguette and Schtickzelle (2006) for two bog specialists (13.5 km for *B. aquilonaris* and 4.6 km for *B. eunomia*) and within the effective colonization distance documented for other localized butterflies (e.g., Davies et al. 2005). However, in more fragmented landscapes, bog butterflies may be particularly disinclined to disperse out of a patch (Vandewoestijne and Baguette 2004, Schtickzelle et al. 2006), with maximum dispersal usually <2 km (Mennechez et al. 2003). Many species of butterflies show variability in dispersal behavior among populations (Stevens et al. 2010), so that mobility observed in one context may not apply to another. Furthermore, examples of each bog type are mapped as >10–15 km from the next nearest bog; e.g., Pine Lake (Appendix 1) is >15 km from the next nearest bog, except for a few tiny patches much smaller than Pine Lake that we found. Thus, other factors besides metapopulation dispersal may contribute to this spatial patterning, such as subtle habitat quality factors and subdetectable numbers, leading to false negatives (a species may not be found every year in lower density sites; cf. Table 4; Appendix 1). Spatially correlated patterns today could also result from non-dispersal and non-colonization following extirpation from local catastrophes (e.g., wildfire, protracted flood, extreme drought) that may have occurred a very long time ago. It would, however, be difficult to distinguish outlier colonization events from outlier retention events.

Over geologic time, these bog-specialist butterflies had to move around the landscape because the bog vegetation they occupy now developed on topography under an ice sheet in the last glaciation (Curtis 1959, Wright 1983). By contrast, limits on their movement now are prominently apparent. Many bog specialists do not even frequent adjacent lowland roadside ditches, while all specialists seem to penetrate virtually no further than that (Tables 2; Fig. 3; Appendix 2). Bogs are scattered widely around northern Wisconsin but are <1% of the landscape as a whole (Hoffman 2002), approximately as isolated as habitat patches in highly fragmented anthropogenic landscapes, such as tallgrass prairie (Samson and Knopf 1994). Layberry et al. (1998) report additional evidence for post-glacial dispersal barriers. Tundra-adapted butterflies surviving the last Ice Age in the "Beringian Refugium" (Northern Yukon and areas westward) "dispersed northward and eastward with varying degrees of success": 32 tundra species in northern Yukon, only 11 as far east as northern Quebec. At the end of the last Ice Age, the vegetations occupied by these study species were likely not as disjunct then as they are now, given that most of these bog butterflies occupy a wider array of vegetation types in Canada now (Layberry et al. 1998). The exception is *L. epixanthe*, rather consistently restricted to bogs throughout its relatively more limited and more southerly range. More dispersal tendency may not have occurred then than now. Rather these butterflies may have occurred in more vegetation types in more of the landscape then, subsequently becoming more restricted to bogs, "high and dry" (or perhaps more appropriately, "sunk and dunked") as this vegetation became isolated patches.

Habitat quality vs. landscape configuration

Some kettleholes and coastal peatlands were relatively small and isolated (see map in Nekola and Kraft 2002) yet supported remarkably high densities of *L. epixanthe* and *B. eunomia*, as well as low but predictable numbers of *O. jutta* (Figs. 1, 2; Appendix 1), and occasionally even *B. freija* (Table 2). High habitat quality in the species-specific sense of consistently available required resources (as construed in Dennis and Eales 1997; Thomas et al. 2001; Dennis et al. 2007; Dennis 2010) appears more effective at explaining this than spatial reinforcement either from patch size or from the surrounding landscape. These dense populations of *L. epixanthe* and *B. eunomia* probably represent long-term population persistence in situ with little chance for immigration. However, some species occured only (e.g., *E. discoidalis*) or primarily (*B. freija*) in muskegs (Table 2; see "Results"), which were often larger (Appendix 1) and more connected in complexes (per map in Nekola and Kraft 2002), suggesting the value of large and connected habitat patches. Densities of some species were relatively high in some muskegs: e.g., *B. freija* (Fig. 2; Appendix 1), and cf. high numbers of *B. montinus* in the few bogs and adjacent lowland roadsides (Table 2; Fig. 3; Appendix 2) in the one county where found at all. However, while widely findable, bog specialists sometimes occurred only in small numbers: e.g., numerous sites with mean observation rates < 5/h in "Appendix 1", and these were the four most frequently recorded specialists (cf. Table 2). Pristine as these bogs are (a kind of high quality), the habitat can't be called all that good quality for those small populations–otherwise, they should be more abundant. Large size may be necessary to amass an adequate amount of required resources to sustain these populations (cf. Turlure et al. 2009, 2010), suggesting that habitat quality may be more important than it first seems in explaining the better occurrence of some specialists in bigger bogs.

Co-evolution vs. stasis

As Elias (1994) summarized, insects don't evolve out of trouble from climatic cycles; they move out of trouble. That is, remarkable stasis (species stability) is evident over millennia in the insect fossil record, but the location of these species has moved tremendously in association with vegetative changes responding to climatic oscillations (Ashworth 2001; Whitehouse 2006; Whitehouse et al. 2008). Given the low dispersal apparent today for species restricted to bogs (Spitzer et al. 1999; Spitzer and Danks 2006; see "Results"), "move" might be better understood as "hunkering" within their vegetation as it expands and shrinks and moves around the landscape. While evolution is often portrayed as an either-or choice ("adapt or die"—discussed further below), it's actually a three-way possibility: "adapt or hold out or die". On the one hand, bog butterflies evidence ongoing evolutionary change. The "*B. titania*" species complex has a European taxon and in North America, a tundra taxon and a boreal taxon (*B. montinus*, the taxon studied here); many species or species complexes have tundra (sub)species and boreal (sub)species (Layberry et al. 1998). Local races or subspecies are evident among isolated bog regions for many bog insects (Wright 1983; Spitzer and Danks 2006). On the other hand, temperate bog butterflies appear not to have co-evolved at all; they're hunkering in with the way of living they've been doing with remarkable long-term faunal stability, even though they can do this only in a very small part of the landscape (<1%) in Wisconsin (per Hoffman 2002).

Even what might look like evolution may not be after all. *L. epixanthe* in coastal peatlands had seemed more detectable in poorer weather than in sites further inland (such as Douglas County muskegs) (Table 5), until we encountered the high density in central Wisconsin (Table 6) immediately after a drenching thunderstorm. This could instead be explained as pre-existing behavioral plasticity expressed differently in different circumstances. Likewise, dispersal tendency may evolve in response to varying degrees of habitat fragmentation (as described in Baguette et al. 2003; Hanski and Pöyry 2007) or the same pre-existing behavioral repertoire may instead be expressed differently depending on varying landscape circumstances as well as individual variation.

Some European butterflies have an affinity (even apparent modern-day restriction) to some forms of old-fashioned, unintensive agriculture and forest harvesting (van Swaay and Warren 2006). Although this apparent dependence may be attributed at least in part to co-evolution, this is often explained as butterflies expanding or contracting in these semi-natural landscapes based on the species' predispositions for these resources and conditions, without necessarily co-evolving (Erhardt and Thomas 1991; Dennis 1992; Dennis 2010). This latter is more consistent with the apparent lack of ongoing co-evolution occurring in the modern landscape (why would it stop a century or two ago?), and with the present-day "biodiversity crisis" most marked for more specialized species, but even affecting numerous widely occurring species (Maes and Van Dyck 2001; van Swaay et al. 2006; Dennis 2010).

Many assert co-evolution of prairie biota with fire (e.g., Curtis 1959; Wisconsin DNR 1995; Packard and Mutel 1997). Apart from controversy over just how much fire actually occurred prehistorically (cf. Higgins 1986; Umbanhowar 1996; Russell 1997), this assumes aboriginal inhabitants as a single culture applied fire in an inescapable and frequent way everywhere to become an unequivocal and unavoidable evolutionary pressure (forcing the outcome down to "adapt or die"). Even modern civilization hasn't fully and uniformly transformed the entire landscape, witness the continued existence of never-plowed botanically "high quality" prairie as well as temperate-zone bogs in very small proportions of the landscape (see "Introduction"). If instead Native Americans varied among clans and tribes in activity and culture, with variable influence on the landscape (Russell 1997), including no-man's zones between antagonistic tribes as reported in Martin and Szuter (1999), then butterflies could sort themselves out along such gradients as fire frequency and native herbivory pressure mediated by variable aboriginal hunting pressure. Some butterflies may have co-evolved in some form, but others may have successfully hunkered in, possibly in very small parts of the landscape, just as bog butterflies have reliably persisted in the <1% of Wisconsin that is bog.

Whether it's co-evolution or not is less important when management proceeds based on conservation evidence for the species affected, as promulgated in Pullin and Knight (2001). But when conservation implements ecosystem management based on theoretical assumptions instead, then picking the right theory (e.g., co-evolution or stasis) becomes critical.

Conclusions for conservation

Dispersal, colonization, and landscape configuration matter (Dennis and Eales 1997; Thomas et al. 2001; Hanski and Pöyry 2007; Dover and Settele 2009; Dennis 2010). If metapopulation dynamics and/or the ecosystem approach are successfully conserving a particular site's insects (e.g., Davies et al. 2005), continue that. If not (e.g., specialist declines in relatively large preserves as in Schlicht et al. 2009), then it may be appropriate to imitate how nature maintains regionally appropriate numbers of highly localized bog butterflies in extremely fragmented patches via remarkable vegetative stability, as well as high habitat quality in the species-specific sense on small sites. Instead of relying on colonization, focus instead on retention because the otherwise inevitable long-term outcome for rare species is to have many more extirpations than colonizations (Bulman et al. 2007). This approach can be applied via detailed species-specific knowledge, but where that is lacking, the following multi-species/ecosystem approaches are indicated (in addition to vigorous effort to acquire species-specific knowledge). Be retrospective (not prospective) in continuing site stability (maintain consistency with past site management before investigating subtle, gradual approaches to restoration). Do not conflate vegetative variation among locations with change within a location (vegetative heterogeneity can be compatible with location-specific consistency).

Acknowledgments We greatly appreciate Mrs. Sandra McKibben and Drs. William and Elsa Boyce for funding our bog surveys. We also thank them, Jed Bromfield and Henya Rachmiel, U.S. Fish and Wildlife Service, and Wisconsin Department of Natural Resources for funding some barrens surveys. We thank Jeff Nekola for generously sharing tips, site locations, and patch sizes. Michel Baguette, an anonymous reviewer, and the editor John Dover provided numerous helpful comments on the submitted manuscript.

Appendix 1

See Table 7.

Table 7 Mean relative abundance (individuals/h) of four most abundant specialists on unit surveys of peak survey per site per year during 2002–2009, by groups of sites, sorted by bog type (with subregion and counties in parentheses)

	Years	Mean	SD	Size	Significance within among
Lycaena epixanthe					
Kettleholes (Northwest: Bayfield interior)					A
East Crane Lake	4/4	23.7	8.1	7.82	B
East Roger Lake	7/7	222.3	194.9	1.84	A
East Wishbone Lake	7/7	335.5	181.2	4.12	A
Pine Lake	4/4	404.8	374.9	1.96	AB
Coastal Peatlands (Northwest: Bayfield coastal)					B
Bark Bay	8/8	101.1	61.9	9.23	A
Bibon Lake	7/7	109.9	65.0	16.56	A
Lost Creek	6/6	28.5	19.7	2.93	B
Port Wing Boreal Forest	4/4	204.3	142.6	2.40	A
Muskegs (Northwest: Douglas)					C
Bear Lake	6/7	17.5	11.8	72.02[a]	A
Bear Lake North	4/6	12.4	21.7	72.02[a]	AB
Lyman Lake	7/8	9.8	17.4	50.55	B
Milchesky Road	7/8	29.5	37.3	114.80	AB
Muskeg lowland roadsides (Northwest: Douglas)					C
Bear Lake	8/8	35.3	50.6		A
Bear Lake North	6/8	32.9	59.4		A
Lyman Lake	3/8	4.1	6.0		A
Milchesky Road	3/8	45.9	93.5		A
Muskegs (North Central: Ashland, Iron, Price)					C
Caroline Lake	5/5	28.7	29.7	52.58	A
Glidden	8/8	20.0	31.8	80.16	A
Highway 70/137	3/4	5.8	5.3		A
Muskeg lowland roadsides (North Central: Ashland, Iron)					C
Caroline Lake	1/4	2.7	5.5		A
Caroline Lake east 0.2	0/3	0.0	0.0		–
Caroline Lake east 1.0	0/3	0.0	0.0		–
Glidden	6/8	20.7	31.8		A
Muskegs (Northeast: Forest)					C
Armstrong	8/8	20.1	31.8		A
Forest Road 2182 West	4/6	14.2	16.7		A
Forest Road 2182/2414	4/7	5.0	6.3		A
Forest Road 2414	4/7	4.6	5.7		A
Boloria eunomia					
Kettleholes (Northwest: Bayfield interior)					A
East Crane Lake	4/6	36.7	57.6	7.82	A
East Roger Lake	6/6	85.5	66.1	1.84	A
East Wishbone Lake	6/6	48.4	56.6	4.12	A
Pine Lake	5/6	22.6	16.5	1.96	A
Coastal Peatlands (Northwest: Bayfield coastal)					A
Bark Bay	7/7	43.0	39.5	9.23	BC
Bibon Lake	7/7	59.6	39.8	16.56	AB
Lost Creek	6/7	24.0	17.4	2.93	C
Port Wing Boreal Forest	7/7	144.0	124.5	2.40	A

Table 7 continued

	Years	Mean	SD	Size	Significance within among
Muskegs (Northwest: Douglas)					B
Bear Lake	7/7	5.2	6.4	72.02[a]	A
Bear Lake North	6/7	8.2	9.3	72.02[a]	A
Lyman Lake	5/8	2.8	4.4	50.55	A
Milchesky Road	5/7	4.6	5.0	114.80	A
Muskegs (North Central: Ashland, Price, Washburn)					B
Glidden	8/8	1.8	4.8	80.16	C
Highway 77	5/6	5.9	4.7		AB
Forest Road 137 1.9	2/5	1.2	2.9		C
Forest Road 137 2.3	2/5	2.3	3.3		BC
Highway 70/137	4/4	15.8	8.9		A
Muskegs (Northeast: Forest)					B
Armstrong	6/6	5.7	10.8	–	
Boloria freija					
Muskegs (Northwest: Douglas)					A
Bear Lake	8/8	31.0	14.7	72.02[a]	A
Bear Lake North	7/7	25.9	15.3	72.02[a]	A
Lyman Lake	8/8	17.8	17.5	50.55	A
Milchesky Road	8/8	16.7	13.6	114.80	A
Muskegs (North Central: Ashland, Iron, Price)					B
Caroline Lake	5/5	16.5	13.3	52.58	A
Glidden	4/7	0.9	2.6	80.16	CD
Forest Road 137 1.2	4/8	2.8	5.0		C
Forest Road 137 1.9	7/8	6.4	7.6		AB
Forest Road 137 2.3	6/8	6.2	9.8		AB
Forest Road 505 1.6	7/7	4.0	6.4		BC
Muskegs (Northeast: Forest)					C
Forest Road 2182 0.5	7/8	8.4	7.0		A
Forest Road 2182 1.4	1/8	1.0	2.9		B
Forest Road 2182/2414	2/7	0.9	1.6		B
Forest Road 2414	2/8	0.8	1.5		B
Armstrong	4/8	0.6	2.5		B
Oeneis jutta					
Kettleholes (Northwest: Bayfield interior)					CD
East Roger Lake	1/5	0.7	1.6	1.84	AB
East Wishbone Lake	1/6	0.6	1.5	4.12	B
East Crane Lake	4/5	4.6	4.4	7.82	A
Pine Lake	0/4	0.0	0.0	1.96	AB
Coastal Peatlands (Northwest: Bayfield coastal)					D
Bark Bay	0/6	0.0	0.0	9.23	B
Bibon Lake	5/7	3.6	2.6	2.40	A
Lost Creek	0/5	0.0	0.0	2.93	AB
Port Wing Boreal Forest	0/6	0.0	0.0	16.56	B
Muskegs (Northwest: Douglas)					B
Bear Lake	8/8	11.2	15.5	72.02[a]	AB
Bear Lake North	8/8	19.3	12.8	72.02[a]	A
Lyman Lake	8/8	11.3	12.2	50.55	B
Milchesky Road	5/7	7.9	6.9	114.80	B

Table 7 continued

	Years	Mean	SD	Size	Significance within among
Muskegs (North Central: Ashland, Iron, Price, Washburn)					AC
Glidden	7/8	6.0	10.9	80.16	A
Highway 77	5/7	7.4	6.4		A
Forest Road 137 1.9	6/8	10.1	15.9		A
Forest Road 137 2.3	6/7	4.6	7.3		A
Highway 70/137	1/4	1.2	2.4		A
Caroline Lake	4/5	6.0	6.0	52.58	A
Muskegs (Northeast: Forest)					A
Forest Road 2182 0.5	4/8	14.4	18.5		AB
Forest Road 2182 1.4	5/7	32.3	31.7		A
Armstrong	6/7	10.2	22.7		B
Forest Road 2182/2414	5/7	17.9	16.0		AB
Forest Road 2414	5/7	28.5	31.3		AB

"Yrs" = *N* years found/*N* years surveyed. Patch size (ha) provided by J. Nekola (personal communication), except we estimated patch size for Port Wing Boreal Forest. For each species, within each group of sites, sites sharing no letters are significantly different, and among groups, groups sharing no letters are significantly different: A, significantly highest; B, significantly lower than A; C, significantly lower than B, and so on

[a] Treated as one site by J. Nekola (personal communication)

Appendix 2

See Table 8.

Table 8 Distribution of specialist individuals in bog, adjacent lowland roadsides, and nearby upland roadsides during the entire study, in order of first observed date in Table 2

	Bog	Lowland roadside	Upland roadside
B. freija			
Observed	722	0	0
Survey time	100.0		
E. discoidalis			
Observed	128	1	0
Survey time	71.7		
Oeneis jutta			
Observed	962 (**99.9%**)	1 (0.1%)	0 (0.0%)
Expected	940.9 (97.7%)	14.9 (1.6%)	7.1 (0.7%)
Survey time	159.6	2.5	1.2
B. frigga			
Observed	264	0 (1)	0
Survey time	107.9		
B. eunomia			
Observed	1084 (**99.9%**)	0 (0.0%)	1 (1) (0.1%)
Expected	1035.3 (95.4%)	30.6 (2.8%)	19.1 (1.8%)
Survey time	82.4	2.4	1.5
L. epixanthe			
Observed	3986 (**85.5%**)	674 (14.5%)	2 (0.0%)
Expected	2995.7 (64.3%)	946.0 (20.3%)	720.2 (15.4%)
Survey time	91.7	29.0	22.0
L. dorcas			
Observed	161 (29.4%)	386 (**70.6%**)	0 (0.0%)
Expected	319.7 (58.5%)	124.5 (22.8%)	102.8 (18.8%)
Survey time	61.0	23.8	19.6
B. montinus			
Observed	192 (11.1%)	1538 (**88.9%**)	(0)
Expected	898.7 (52.0%)	831.3 (48.0%)	
Survey time	23.4	21.6	

Observed individuals for each species are significantly skewed relative to expected based on proportion of survey time (h) in each site type (chi square goodness of fit $P = 0.0000$ for each species). Most over-represented is boldfaced; most under-represented is underlined. If roadside survey time was <1 h, that site type was excluded from statistical analysis but *N* individuals observed on those surveys is provided. *N* individuals on informal observations is in parentheses but excluded from analysis

References

Ashworth AC (2001) Chapter 8: Perspectives on quaternary beetles and climate change. In Gerhard LC, Harrison WE, Hanson BM (eds) Geological perspectives of global climate change. American Association of Petroleum Geologists Studies in Geology #47, Tulsa, pp 153–168

Baguette M, Schtickzelle N (2006) Negative relationship between dispersal distance and demography in butterfly metapopulations. Ecology 87:648–654

Baguette M, Mennechez G, Petit S, Schtickzelle N (2003) Effect of habitat fragmentation on dispersal in the butterfly *Proclossiana eunomia*. C R Biol 326:S200–S209

Brown KS (1997) Diversity, disturbance, and sustainable use of neotropical forests: insects as indicators for conservation monitoring. J Insect Conserv 1:25–42

Bulman CR, Wilson RJ, Holt AR, Galvez Brávo L, Early RI, Warren MS, Thomas CD (2007) Minimum viable metapopulation size, extinction debt, and the conservation of declining species. Ecol Appl 17:1460–1473

Cassie B, Glassberg J, Swengel A, Tudor G (2001) North American Butterfly Association (NABA) checklist & English names of North American butterflies, 2nd edn. N. Am. Butterfly Assoc, Morristown

Curtis JT (1959) The vegetation of Wisconsin: an ordination of plant communities. Univ of Wisconsin Press, Madison

Davies ZG, Wilson RJ, Brereton TM, Thomas CD (2005) The re-expansion and improving status of the silver-spotted skipper butterfly (*Hesperia comma*) in Britain: a metapopulation success story. Biol Conserv 124:189–198

Dennis RLH (1992) The ecology of butterflies in Britain. Oxford Univ Press, Oxford

Dennis RLH (2010) A resource-based habitat view for conservation: butterflies in the British landscape. Wiley-Blackwell, Oxford

Dennis RLH, Eales HT (1997) Patch occupancy in *Coenonympha tullia* (Müller, 1764) (Lepidoptera: Satyrinae): habitat quality matters as much as patch size and isolation. J Insect Conserv 1:167–176

Dennis RLH, Hardy PB (2007) Support for mending the matrix: resource seeking by butterflies in apparent non-resource zones. J Insect Conserv 11:157–168

Dennis RLH, Shreeve TG, Sheppard DA (2007) Species conservation and landscape management: a habitat perspective. In: Stewart AJA, New TR, Lewis OT (eds) Insect conservation biology: proceedings of the Royal Entomological Society's 23rd symposium. CABI, Oxfordshire, pp 92–126

Dover J, Settele J (2009) The influences of landscape structure on butterfly distribution and movement: a review. J Insect Conserv 13:3–28. doi:10.1007/s10841-008-9135-8

Elias SA (1994) Quaternary insects and their environments. Smithsonian Institution, Washington

Erhardt A, Thomas JA (1991) Lepidoptera as indicators of change in the semi-natural grasslands of lowland and upland Europe. In: Collins NM, Thomas JA (eds) The conservation of insects and their habitats. Academic Press, London, pp 213–236

Forister ML, McCall AC, Sanders NJ et al (2010) Compounded effects of climate change and habitat shift patterns of butterfly diversity. Proc Nat Acad Sci (Philadelphia) 107:2088–2092

Glassberg J (1999) Butterflies through binoculars: the East. Oxford Univ Press, New York

Hanski I, Pöyry J (2007) Insect populations in fragmented habitats. In: Stewart AJA, New TR, Lewis OT (eds) Insect conservation biology: proceedings of the Royal Entomological Society's 23rd Symposium. CABI, Oxfordshire, pp 175–202

Higgins KF (1986) Interpretation and compendium of historical fire accounts in the northern Great Plains. US Fish and Wildlife Service Resource Publication 151, Washington

Hoffman RM (2002) Wisconsin's natural communities: how to recognize them, where to find them. Univ of Wisconsin Press, Madison

Layberry RA, Hall PW, LaFontaine JD (1998) The butterflies of Canada. Univ of Toronto Press, Toronto

Maes D, Van Dyck H (2001) Butterfly diversity loss in Flanders (north Belgium): Europe's worst case scenario? Biol Conserv 99:263–276

Martin PS, Szuter CR (1999) War zones and game sinks in Lewis & Clark's West. Conserv Biol 13:36–45

Mattoni R, Longcore T, Zonneveld C, Novotny V (2001) Analysis of transect counts to monitor population size in endangered insects: the case of the El Segundo blue butterfly (*Euphilotes bernardino allyni*). J Insect Conserv 5:197–206. doi:10.1023/A:1017987929824

Mennechez G, Schtickzelle N, Baguette M (2003) Metapopulation dynamics of the bog fritillary butterfly: comparison of demographic parameters and dispersal between a continuous and a highly fragmented landscape. Landsc Ecol 18:279–291

Nekola JC (1998) Butterfly (Lepidoptera: Lycaenidae, Nymphalidae, and Satyridae) faunas of three peatland habitat types in the Lake Superior drainage basin of Wisconsin. Great Lakes Entomol 31:27–38

Nekola JC, Kraft CE (2002) Spatial constraint of peatland butterfly occurrences within a heterogeneous landscape. Oecologia 130:53–61

Nielsen M (1999) Michigan butterflies and skippers. Michigan State Univ Extension, East Lansing

Nuzzo VA (1986) Extent and status of Midwest oak savanna: presettlement and 1985. Nat Areas J 6:6–36

Opler PA (1992) A field guide to eastern butterflies. Houghton Mifflin, New York

Opler PA, Krizek GO (1984) Butterflies east of the great plains. Johns Hopkins Univ Press, Baltimore

Opler PA, Powell JA (eds) (1984) Butterfly counts 1982 & 1983. Xerces Society, New Haven

Packard S, Mutel CF (1997) The tallgrass restoration handbook: for prairies, savannas and woodlands. Island Press, Washington

Pollard E (1977) A method for assessing changes in abundance of butterflies. Biol Conserv 12:115–133

Pollard E, Yates TJ (1993) Monitoring butterflies for ecology and conservation. Chapman & Hall, London

Pullin AS, Knight TM (2001) Effectiveness in conservation practice: pointers from public health and medicine. Conserv Biol 15:50–54

Riegler M (1995) Development of a pine barrens recovery plan. In: Borgerding, EA, Bartelt GA, McCown MA (eds) The future of pine barrens in northwest Wisconsin: a workshop summary. Wisconsin Dept Nat Res PUBL-RS-913-94, pp 28–33

Russell EWB (1997) People and the land through time: linking ecology and history. Yale Univ Press, New Haven

Samson F, Knopf F (1994) Prairie conservation in North America. Bioscience 44:418–421

Schlicht D, Swengel A, Swengel S (2009) Meta-analysis of survey data to assess trends of prairie butterflies in Minnesota, USA during 1979–2005. J Insect Conserv 13:429–447

Schtickzelle N, Mennechez G, Baguette M (2006) Dispersal depression with habitat fragmentation in the bog fritillary butterfly. Ecology 87:1057–1065

Spencer S, Collins S (2008) Reversing the decline in butterflies and moths across Europe—the importance of particular farming practices and the implications for CAP reform. www.birdlife.eu/eu/pdfs/BCEurope_CAPreformpaperFeb08.pdf viewed 15 Jan 2010

Spitzer K, Danks HV (2006) Insect biodiversity of boreal peat bogs. Annu Rev Entomol 51:137–161

Spitzer K, Bezděk A, Jaroš J (1999) Ecological succession of a relict Central European peat bog and variability of its insect biodiversity. J Insect Conserv 3:97–106

Stevens VM, Pavoine S, Baguette M (2010) Variation within and between closely related species uncovers high intra-specific variability in dispersal. PLoS ONE 5(6):e11123. doi:10.1371/journal.pone.0011123

Swengel AB (1996) Effects of fire and hay management on abundance of prairie butterflies. Biol Conserv 76:73–85

Swengel AB (1998) Effects of management on butterfly abundance in tallgrass prairie and pine barrens. Biol Conserv 83:77–89

Swengel A (2009) The fascinating butterflies of northwestern Wisconsin bogs. Southern Wisconsin Butterfly Assocation, Madison, Wisconsin. http://www.naba.org/chapters/nabawba/watching.html viewed 11 Feb 2010

Swengel AB, Opler PA (1996) NABA-Xerces Fourth of July Butterfly Counts 1995 Report. North American Butterfly Association, Morristown

Swengel AB, Swengel SR (1997) Co-occurrence of prairie and barrens butterflies: applications to ecosystem conservation. J Insect Conserv 1:131–144

Swengel A, Swengel S (1999a) Observations on prairie skippers (*Oarisma poweshiek*, *Hesperia dacotae*, *H. ottoe*, *H. leonardus pawnee*, and *Atrytone arogos iowa*) (Lepidoptera: Hesperiidae) in Iowa, Minnesota, and North Dakota during 1988–1997. Great Lakes Entomol 32:267–292

Swengel SR, Swengel AB (1999b) Correlations in abundance of grassland songbirds and prairie butterflies. Biol Conserv 90:1–11

Swengel AB, Swengel SR (2001) Effects of prairie and barrens management on butterfly faunal compostion. Biodivers Conserv 10:1757–1785

Swengel AB, Swengel SR (2005) Long-term population monitoring of the Karner Blue (Lepidoptera: Lycaenidae) in Wisconsin, 1990–2004. Great Lakes Entomol 38:107–134

Swengel AB, Swengel SR (2007) Benefit of permanent non-fire refugia for Lepidoptera conservation in fire managed sites. J Insect Conserv 11:263–279

Thomas JA (1984) The conservation of butterflies in temperate countries: past efforts and lessons for the future. In: Vane-Wright RI, Ackery PR (eds) The biology of butterflies. Princeton Univ Press, Princeton, pp 333–353

Thomas JA, Bourn NAD, Clarke RT et al (2001) The quality and isolation of habitat patches both determine where butterflies persist in fragmented landscapes. Proc R Soc Lond B 268:1791–1796

Turlure C, Van Dyck H, Schtickzelle N, Baguette M (2009) Resource-based habitat definition, niche overlap and conservation of two sympatric glacial relict butterflies. Oikos 118:950–960

Turlure C, Choutt J, Baguette M, Van Dyck H (2010) Microclimatic buffering and resource-based habitat in a glacial relict butterfly: significance for conservation under climate change. Glob Change Biol 16:1883–1893

Umbanhowar CE Jr (1996) Recent fire history of the northern Great Plains. Am Midl Nat 135:115–121

Väisänen R (1992) Distribution and abundance of diurnal Lepidoptera on a raised bog in southern Finland. Ann Zool Fennici 29:75–92

van Swaay CAM, Warren MS, Loïs G (2006) Biotope use and trends of European butterflies. J Insect Conserv 10:189–209

van Swaay CAM, Warren MS (2006) Prime butterfly areas of Europe: an initial selection of priority sites for conservation. J Insect Conserv 10:5–11

Vandewoestijne S, Baguette M (2004) Genetic population structure of the vulnerable bog fritillary butterfly. Hereditas 141:199–206

Wander S (ed) (2009) NABA butterfly counts (2008). North American Butterfly Assoc, Morristown

Whitehouse NJ (2006) What can forest managers learn from research on fossil insects? Linking forest ecological history, biodiversity and management. In: Grove SJ, Janula JL (eds) Insect biodiversity and dead wood: proceedings of a symposium for the 22nd international congress of entomology. Gen. Tech. Rep. SRS-93. USDA Forest Service, Southern Research Station, Asheville, pp 30–41

Whitehouse NJ, Langdon PG, Bustin R, Galsworthy S (2008) Fossil insects and ecosystem dynamics in wetlands: implications for biodiversity and conservation. Biodivers Conserv 17:2055–2078

Wikström L, Milberg P, Bergman K-O (2009) Monitoring of butterflies in semi-natural grasslands: diurnal variation and weather effects. J Insect Conserv 13:203–211

Wisconsin Department of Natural Resources (1995) Wisconsin's biodiversity as a management issue. Dept of Nat Res, Madison

Wright DM (1983) Life history and morphology of the immature stages of the bog copper butterfly *Lycaena epixanthe* (Bsd. & Le C.) (Lepidoptera: Lycaenidae). J Res Lepid 22:47–100

J Insect Conserv (2011) 15:69–81
DOI 10.1007/s10841-010-9324-0

ORIGINAL PAPER

Landscape-scale conservation in practice: lessons from northern England, UK

Sam Ellis · Dave Wainwright · Frank Berney · Caroline Bulman · Nigel Bourn

Received: 23 March 2010 / Accepted: 3 August 2010 / Published online: 22 August 2010

Abstract Landscape-scale conservation projects are now Butterfly Conservation's main delivery mechanism to conserve the UK's threatened Lepidoptera. Butterfly Conservation currently works or has recently worked in 79 different landscapes targeted at key areas for threatened species, all of which involve partnerships with government agencies, other conservation organisations and landowners. Broadly we utilise two approaches firstly by providing advice to landowners and encouraging or assisting with the uptake of agri-environment or woodland grant schemes and secondly by securing external funding to directly undertake habitat management under the guidance of Butterfly Conservation project officers. These are not mutually exclusive with most projects having elements of both approaches. We describe two projects from northern England which utilise the direct funding approach and provide evidence for some of the lessons learnt by Butterfly Conservation's 10 years experience of landscape-scale conservation delivery. From County Durham we demonstrate that for the Small Pearl-bordered Fritillary *Boloria selene* Denis & Schiff. it is possible to restore even very small fragmented landscapes. On the North York Moors we show that a non-equilibrium declining Duke of Burgundy *Hamearis lucina* L. metapopulation can be stabilised by carefully targeting restoration management, demonstrating the cost benefits of early intervention.

Keywords Lepidoptera · Landscape-scale conservation · Habitat restoration · *Boloria selene* · *Hamearis lucina*

S. Ellis (✉) · D. Wainwright · F. Berney · C. Bulman · N. Bourn
Butterfly Conservation, Manor Yard, East Lulworth, Wareham, Dorset BH20 5QP, UK
e-mail: sellis@butterfly-conservation.org

Introduction

Butterfly Conservation defines landscape-scale conservation as the coordinated conservation and management of habitats for a range of species across a large natural area, often made up of a network of sites (Bourn and Bulman 2005). Metapopulation theory has re-orientated conservation priorities to the landscape scale by emphasising the importance of area and isolation (e.g. Hanski 1999). However, research suggests that because rare species are restricted to very specific habitats or niches, it is just as important to maintain high habitat quality by targeted management within individual sites, as maintaining the site network (Thomas et al. 2001). This concept is central to Butterfly Conservation's approach to landscape-scale conservation delivery. Moreover, in the context of climate change, a landscape-scale approach would appear to be the best option for creating the habitat heterogeneity likely to be needed for species with changing ecological requirements as well as providing the opportunities for them to move through the landscape.

Since the turn of the millennium, Butterfly Conservation is or has been involved to a greater or lesser extent with 79 landscape-scale projects across the UK (Fig. 1). These projects have targeted key areas for some of the UK's most threatened species, as identified through Butterfly Conservation's Regional Action Plans (see http://www.butterfly-conservation.org/downloads/74/Regional_Action_Plans.html). Nearly all have received some external funding, directly or indirectly, to enable delivery and all involve partnerships with government agencies, other conservation organisations and landowners. Broadly we utilise two approaches firstly by providing advice to landowners and encouraging or assisting with the uptake of agri-environment or woodland grant schemes and secondly by securing funding to

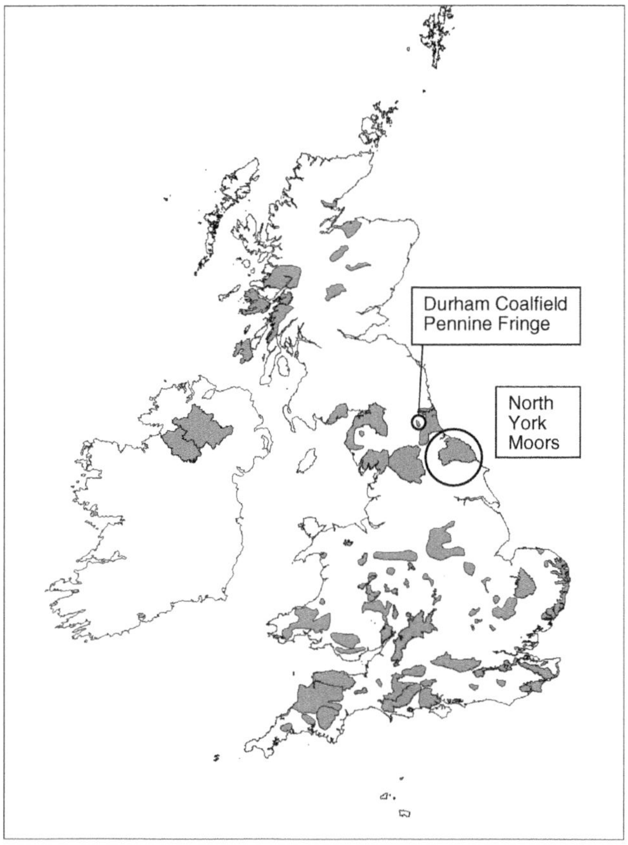

Fig. 1 Location of Butterfly Conservation's 79 landscape target areas in the UK. See Bulman et al. (2008) for further details of landscapes. Locations of the landscapes for the two case studies are *highlighted*

directly undertake habitat management under the guidance of Butterfly Conservation project officers. These are not mutually exclusive with most projects having elements of both approaches.

The need for evidence-based as opposed to experience-based conservation is now well recognised (Pullin and Knight 2001). We describe here two case studies which provide evidence for some of the lessons learnt from delivering landscape-scale conservation over the last ten years. Most landscape-scale projects operate over hundreds or thousands of hectares. In County Durham we demonstrate that for the Small Pearl-bordered Fritillary *Boloria selene* Denis & Schiff. it is possible to restore relatively small fragmented landscapes. We also demonstrate that even in small landscapes, the multi-site approach enshrined in the landscape-scale approach is essential to conserve our threatened Lepidoptera.

Where the local extinction rate exceeds the rate of colonisation in a metapopulation then species eventually become extinct in that landscape. On the North York Moors we show that with limited resources, a Duke of Burgundy *Hamearis lucina* L. non-equilibrium declining metapopulation can be stabilised by carefully targeting restoration management and demonstrating the cost benefits of early intervention.

Case study 1: restoring very small fragmented landscapes: *Boloria selene* in County Durham, UK

Introduction

Boloria selene has undergone major declines in both distribution and abundance in Britain in recent years but especially in central and eastern England (Fox et al. 2006). In the Durham Coalfield Pennine Fringe landscape of County Durham, north-east England, it was reduced to four known colonies in 2000 (Ellis 2000). In northern and western Britain, *B. selene* is found in two main habitats, (1) damp grassland/moorland, especially around wet flushes, and (2) grassland with Bracken *Pteridium aquilinum* Kuhn. and/or scrub. Breeding occurs in grassy vegetation, where one or both of the two main larval foodplants, Marsh Violet *Viola palustris* L. and Common Dog-violet *Viola riviniana* Rchb. are abundant. Between 2000 and 2002 we undertook ecological research (Ellis 2000, 2001; Berney 2002) to identify the constraints for the butterfly in this upland landscape. Research focussed on mapping occupied habitat patches, assessing population sizes, measuring dispersal and identifying ovipositing preferences. The latter, together with larval feeding preferences, are the key determinants of habitat quality for conservation of butterflies in temperate countries (Thomas 1991). We used these data to plan and implement a conservation programme from 2002 onwards to restore occupied and unoccupied habitat as well as improve connectivity between patches.

Ecological research

Methods

Occupied habitat patches were defined by the presence of the adult butterfly and the presence of one or both larval foodplants. Patches dominated by either *V. palustris* or *V. riviniana* were mapped separately on GIS. Estimates of absolute adult population size and other population parameters were made using mark, release and recapture (MRR), at the four sites: Waskerley Way (NZ 065445), Greenfield Farm (NZ 073447), Emma's Wood (NZ 084442) and Stuartfield Plantation (NZ 085455). Each site was sampled between five and nine times during the main flight period in 2000–2002 (16th to 27th June 2000; 25th June to 4th July 2001; 20th June to 13th July 2002). Population parameters were estimated using frequency of capture techniques based on a Poisson distribution which best fitted the observed data (Craig 1953). Dispersal was measured by recording the direct distance (m) between both same-day and different-day captures.

Ovipositing preferences were studied at a mainly damp grassland/wet flush site (Emma's Wood) and a mainly grassland/Bracken and scrub site (Greenfield Farm) between 6th and 16th July 2002, by following females as they searched for suitable ovipositing sites. When an egg was located, a 0.25 m^{-2} quadrat was placed over the plant, centring on the egg. Variables at the egg locations recorded were the egg-laying material, the height at which the egg was laid (cm) and the distance to nearest violet leaf (cm). A series of habitat variables were recorded from the egg-laying quadrats and 30 stratified random quadrats (allocated proportionally according to the relative patch size of the two main habitat types) at each site: number of leaves on the *Viola* spp. plant nearest the quadrat centre, cover (%) of *V. palustris* and/or *V. riviniana*, moss cover (%), non-Viola spp. herb cover (%), grass cover (%), rush/sedge cover (%), *P. aquilinum*/scrub cover (%), litter cover (%), vegetation height (cm, measured using the direct method in Stewart et al. 2001), aspect (1 = north, 2 = north-east and north-west, 3 = east and west, 4 = south-east and south-west, 5 = south, 6 = no aspect) and slope (1 = flat to 4 = steepest slope). Data were analysed by both parametric (*t*-test) and non-parametric tests (Mann–Whitney *U*-test) comparing variables in egg-laying quadrats with random quadrats on the same site.

Results

All four sites comprised very small areas of breeding habitat between 0.11 and 0.45 ha and totaling only 0.99 ha for the entire system (Fig. 2). All four sites supported small populations with absolute estimates varying from 10 to 70 individuals on the 27 days of marking during three years of study. The mean direct distance between recaptures was 34.7 m. Analysis of variance on log-transformed data showed no significant differences between male and female mobility or between the four sites (Two-way ANOVA: effect of sex: $F_{1,264} = 0.78$, $P = 0.377$; effect of site: $F_{3,264} = 1.44$, $P = 0.229$). Over 80% of recaptures were of males. Only 23% of recaptured individuals had moved over 50 m and only 4% over 100 m (Fig. 3). The maximum distance for a male was 177 m and for a female 166 m. Movements between patches of habitat on the same site were detected but not between sites, which were between 0.9 and 1.3 km apart.

Eggs were laid on a variety of substrates but 69% were on grass or litter and only 9% on *Viola* spp. (Fig. 4) and 81% of eggs were laid at medium vegetation heights between 5 and 15 cm (Fig. 5). Although 40% of eggs were laid within 5 cm of the nearest violet leaf, 21% were laid between 25 and 100 cm from the nearest violet leaf. In both habitats a preference for ovipositing in shorter than average vegetation with a high moss cover was demonstrated (Table 1). At the mainly damp grassland/wet flush site, egg quadrats also had significantly higher cover of litter; steeper slopes with more northerly aspects were avoided. In egg quadrats the cover of *Viola* spp. was similar to, or lower than in random quadrats. Moreover *Viola* spp. were entirely absent from 33% of egg quadrats at the mainly damp grassland/wet flush site and from 18% of egg quadrats at the mainly grassland/Bracken and scrub site.

Discussion

All four sites were shown to support only small populations and all were likely to be vulnerable to local extinction. Furthermore, although MRR underestimates dispersal

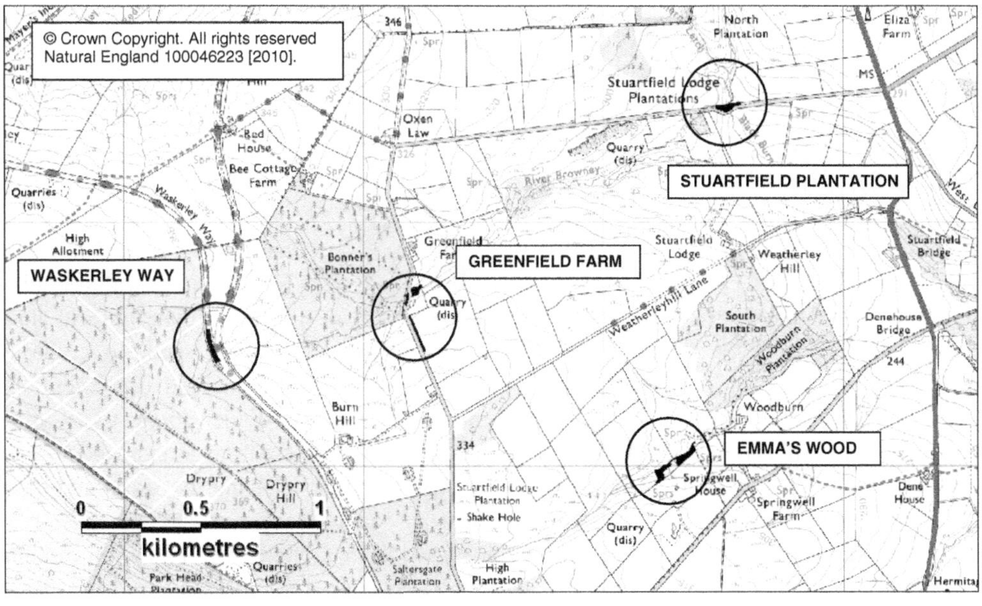

Fig. 2 Location of four *Boloria selene* sites in the Durham Coalfield Pennine Fringe in 2002. *Solid patches* indicate the extent of breeding habitat prior to management

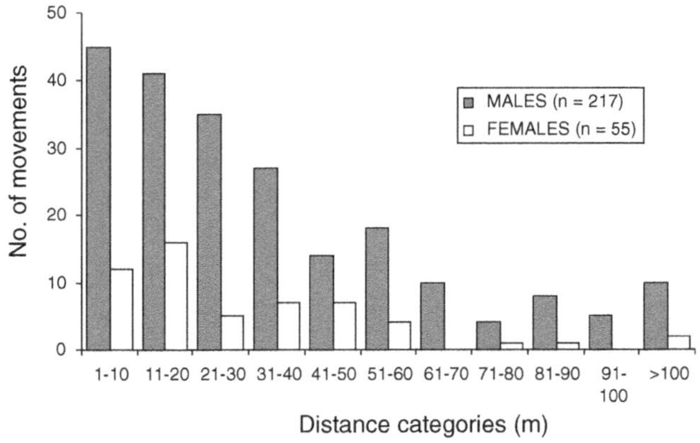

Fig. 3 Number of adult *Boloria selene* movements in distance categories on four Durham Coalfield Pennine Fringe sites during 2000–2002

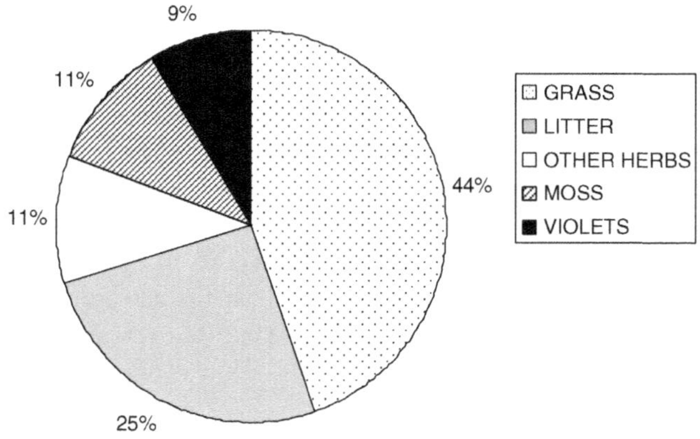

Fig. 4 Location of *Boloria selene* ovipositing sites ($n = 47$) on two Durham Coalfield Pennine Fringe sites in 2002

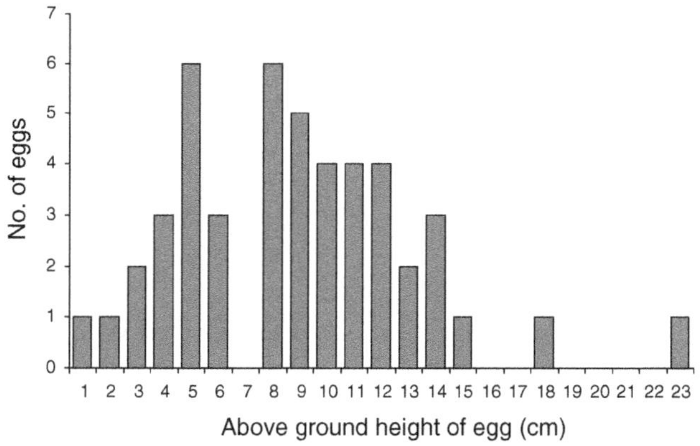

Fig. 5 Height of *Boloria selene* ovipositing sites ($n = 47$) on two Durham Coalfield Pennine Fringe sites in 2002

ability and it is often difficult to detect movement between sites (Wilson and Thomas 2002), our data suggested that in this landscape *B. selene* lives in discrete colonies, where dispersal may be limited. Ovipositing preference data indicates that *B. selene* selects egg-laying sites based more on habitat characteristics rather than the species, size and cover of the larval foodplant. In this landscape *B. selene* was able to utilise both main habitat types for breeding, but there was a clear preference in both for medium height swards over longer vegetation. This may well reflect a microclimatic requirement because shorter swards provide warmer conditions for the egg and larval stages (Thomas 1983a). This hypothesis is supported by the preference for high litter cover, which could provide warmer larval basking sites, and the avoidance of sloping ground with more northerly aspects.

Conservation management

Our research suggested that in order to restore this small fragmented landscape, conservation effort should be targeted at both improving habitat quality on currently occupied patches, restoring currently unoccupied patches and improving connectivity between them. In 2002 three of the four extant populations were located along roadside verges or paths and were ungrazed with the fourth only very lightly grazed. Succession to rank vegetation, dense Bracken and scrub was evident on all four sites. Improvements in habitat quality on occupied patches could therefore be best achieved by reversing succession to establish medium height swards, providing a more suitable microclimate but also encouraging violet regeneration. Conversely, unoccupied potential patches identified were mainly located within pastures overgrazed by sheep and characterized by short swards with relatively few violets. Unoccupied patches were all relatively close to or between occupied patches and mostly located along existing watercourses. Restoration of such unoccupied patches would therefore not only increase the overall habitat resource but also improve connectivity between them enabling more rapid colonisation by *B. selene*.

From 2002, scrub control, Bracken management and mowing rank vegetation were undertaken on occupied and some potential patches (Table 2). On both extant and potential patches we established small grazing compartments (14 compartments; mean area = 1.20 ha) by fencing actual or potential breeding habitat. On more exposed patches we established small shelterbelts (6 shelterbelts; mean area = 0.05 ha). On two patches where there were few or no larval foodplants we planted both species of violet, grown by local authority horticulturists from locally sourced seed, to create potential breeding habitat. We investigated the potential to restore breeding habitat and provide connectivity along plantation watercourses by creating streamside rides, but were unsuccessful in purchasing the woodland to implement this strategy. Much of the practical management was undertaken by volunteers but over £40 k of funding was secured from a range of sources and paid for the fencing.

Table 1 Mean values for *Boloria selene* ovipositing habitat variables on two Durham Coalfield Pennine Fringe sites in 2002

Habitat variable	Egg quadrats ($n = 27$)	Random quadrats ($n = 30$)	Test statistic[a]	P
a) Damp grassland/wet flush: Emma's Wood				
Viola palustris cover (%)	8.70	8.33		
Viola riviniana cover (%)	0	1.17		
No. *Viola palustris* leaves	2.89	2.90		
No. *Viola riviniana* leaves	0	1.50		
Moss cover (%)	39.44	18.17	$U = 235.5$	0.006
Other herb cover (%)	13.33	11.50		
Grass cover (%)	39.63	38.83		
Rush/sedge cover (%)	5.00	20.33		
Bracken/scrub cover (%)	0	7.33		
Litter cover (%)	22.96	6.50	$U = 172$	<0.001
Vegetation height (cm)	15.96	22.47	$t = -2.89$	0.006
Aspect[b]	4.26	2.93	$U = 71.5$	0.001
Slope[c]	1.59	2.00	$U = 284$	0.030
Habitat variable	**Egg quadrats ($n = 28$)**	**Random quadrats ($n = 30$)**	**Test statistic[a]**	**P**
b) Grassland/Bracken and scrub: Greenfield Farm				
Viola palustris cover (%)	1.25	8.00		
Viola riviniana cover (%)	9.82	12.83		
No. *Viola palustris* leaves	0.21	0.90		
No. *Viola riviniana* leaves	10.00	6.63		
Moss cover (%)	43.57	18.50	$U = 227$	0.003
Other herb cover (%)	14.29	16.50		
Grass cover (%)	45.89	39.00		
Rush/sedge cover (%)	1.79	5.17		
Bracken/scrub cover (%)	1.79	8.00		
Litter cover (%)	17.68	13.17		
Vegetation height (cm)	15.11	22.23	$t = -2.60$	0.019
Aspect[b]	3.93	3.87		
Slope[c]	1.82	1.77		

[a] Test statistics only shown for significant variables

[b] Aspect categories: 1 = north, 2 = north-east and north-west, 3 = east and west, 4 = south-east and south-west, 5 = south, 6 = no aspect

[c] Slope categories: 1 = flat, 2 = gentle, 3 = steep, 4 = very steep

The preferred grazing regime proposed for the grazing compartments was light cattle grazing as some poaching encourages violet regeneration in grassland and trampling of Bracken habitats helps break up dense standing trash. However, farming beef cattle in the upland fringe is increasingly uneconomic and only one site was cattle grazed. The new grazing compartments either remained ungrazed or were lightly sheep grazed for several years. However, from 2009 the compartments have been grazed on rotation by three fell ponies.

Between 2002 and 2009 the habitat resource in this network increased from 0.99 to 5.88 ha occupied habitat, nearly six times the 2002 resource (Table 2). This increase was partly attributable (2.95 ha) to the discovery of previously unrecorded patches on existing sites and two new sites at Cat Back (NZ 055434) and at High Plantation (NZ 077437). Nevertheless, the landscape restoration project has already resulted in an additional 1.93 ha occupied habitat, tripling the 2002 resource and this figure is likely to increase further as more potential patches come into suitable condition.

Monitoring the impact of landscape restoration

All sites were monitored annually from 2003 either by single transect counts (Thomas 1983b), or in the case of

Table 2 Management implemented and changes in patch occupancy on *Boloria selene* sites in the Durham Coalfield Pennine Fringe 2002–2009

	Cat Back	Waskerley Way	Greenfield Farm	High Plantation	Emma's Wood	Stuartfield Plantation	Total
Vegetation management							
Total area managed (ha)		0.24	0.39		1.35	0.23	2.21
Scrub control (ha)		0.13	0.01		1.35	0.23	1.72
Bracken management (ha)			0.19			0.20	0.39
Grassland mowing and raking (ha)		0.24	0.20			0.20	0.64
Controlled grazing							
Fencing installed (km)		2.03	0.56	0.48	3.03	0.56	6.66
Grazing compartments (ha)		5.42	1.89	0.55	8.30	0.66	16.82
Grazing compartments (no.)		5	2	1	4	2	14
Mean compartment size (ha)		1.08	0.95	0.55	2.08	0.33	1.20
Planting							
Shelterbelts (ha)		0.31					0.31
Shelterbelts (no.)		6					6
Mean shelterbelt size (ha)		0.05					0.05
Violet planting (area)		0.06	0.09				0.15
Changes in patch occupancy							
Occupied patch area in 2000 (ha)		0.21	0.11		0.45	0.22	0.99
Occupied patch area in 2009 (ha)	1.47	1.61	0.21	1.15	0.49	0.99	5.88
Previously unrecorded patches (ha)	1.47	0	0	0.71	0	0.77	2.12
Previously unoccupied patches (ha)	0	1.39	0.10	0.44	0	0	1.93

Waskerley Way and Greenfield Farm from 2004, by a weekly transect (Pollard 1977) encompassing both populations. All restored and created patches were separated from occupied patches by a maximum of 0.1 km and all those supporting appropriate breeding habitat have been recolonised by *B. selene* (Fig. 6). Some patches now support larger populations than the original occupied patches. Annual indices on the Waskerley Way and Greenfield Farm sites have remained more or less stable (Fig. 7a), despite the impact of the poor weather in 2007–2008 and in contrast to the significant 66% national UK decline between 1976–2008 (Botham et al. 2008). These two populations are now distributed over more patches (2002: 4 patches; 2009: 11 patches), but numbers have declined on the original occupied patches, where habitat quality remains below optimum. Connectivity has been increased with minimum distances between sites reduced from 0.81 to 0.34 km.

At Greenfield Farm a grazing compartment was established in 2002 to enclose an unoccupied heavily sheep grazed mire. *B. selene* colonised the patch in 2003 with an overall positive transect trend thereafter (Fig. 7b). At Waskerley Way a grazing compartment was established on a large area of heavily sheep grazed mire in 2005. *B. selene* colonised this patch later that year and the trend thereafter on the transect has also been positive (Fig. 7c). Likewise colonisation of a grazing compartment planted with about 800 violets occurred the year after (Fig. 7d), although numbers remained low until 2008–2009 when a breeding population appears to have finally established. In 2009 this small patch accounted for 25% of the transect records for *B. selene*.

We believe this case study demonstrates that restoration of small fragmented landscapes is not only achievable, but essential to enable the dynamic changes in habitat quality to occur and still maintain the species in the area. Without intervention on both occupied and unoccupied patches further local extinctions may well have occurred in this landscape. However, this only represents the first project phase, much more extensive habitat restoration both within, between and beyond the existing sites is required to secure the long-term future of this metapopulation.

Case study 2: targeting restoration management to stabilize metapopulations: *Hamearis lucina* on the North York Moors, UK

Introduction

Hamearis lucina has also undergone major declines in both distribution and abundance in Britain in recent years (Fox et al. 2006). The butterfly is mainly restricted to the

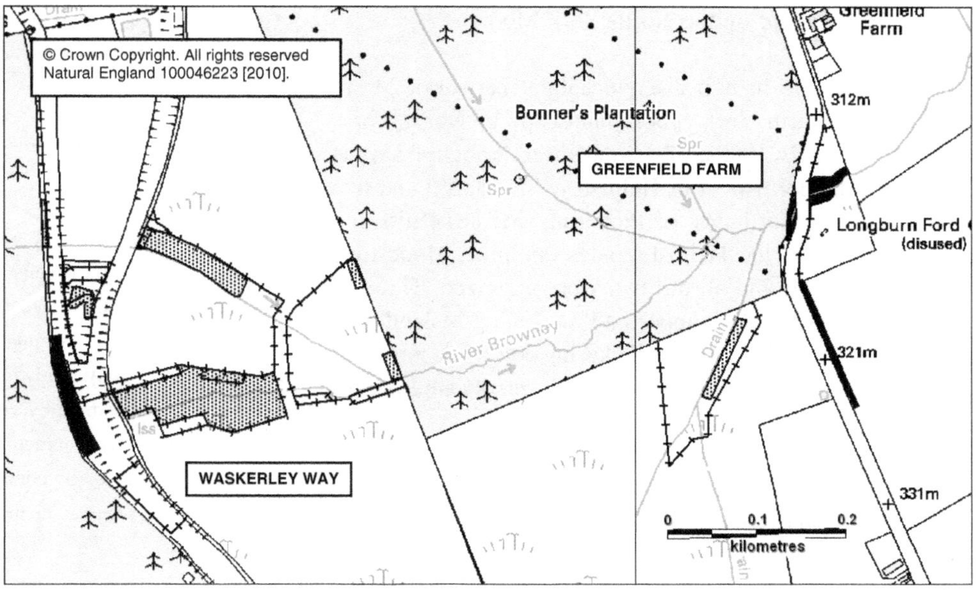

Fig. 6 Changes in *Boloria selene* patch occupancy at the Waskerley Way and Greenfield Farm sites in the Durham Coalfield Pennine Fringe following landscape-scale management. *Black habitat patches* occupied in 2002, *dotted patches* colonized following management 2002 onwards. New grazing units and shelterbelts indicated by *crossed lines*

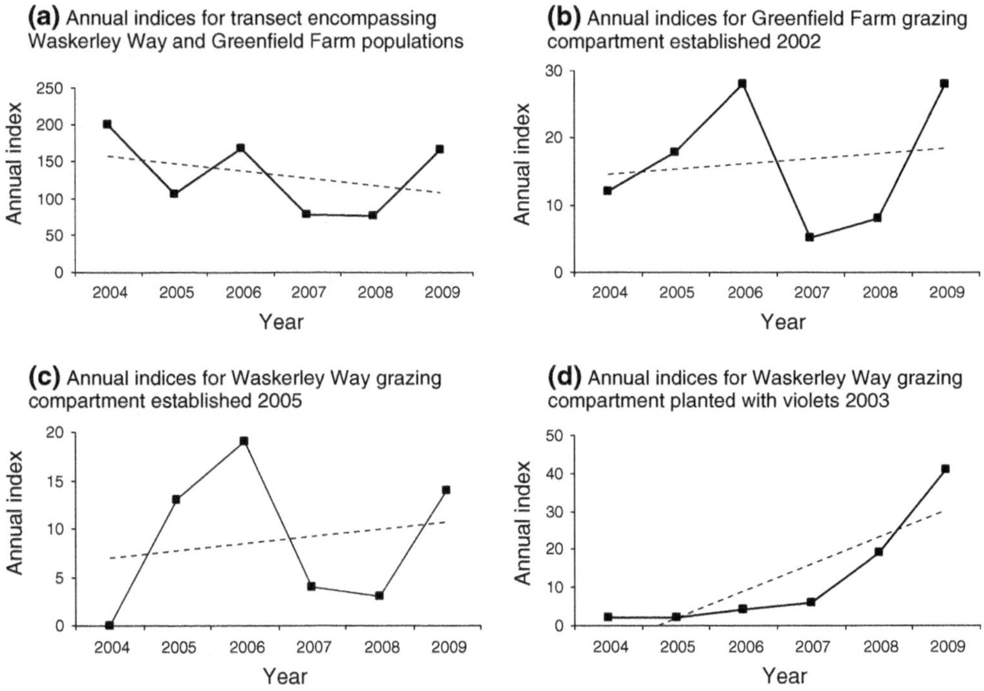

Fig. 7 *Boloria selene* weekly transect counts on two Durham Coalfield Pennine Fringe sites 2004–2009 following habitat restoration. *Dotted lines* indicate trends

limestone and chalk of southern England but more isolated colonies occurs on the Morecambe Bay Limestones and North York Moors of northern England. *H. lucina* utilises two principal habitats, (1) chalk or limestone grassland with either extensive areas of scrub or topographical shelter and (2) clearings on ancient woodland sites, either regenerating coppice, young plantations, sizeable glades or wide rides. Breeding occurs in tussocky vegetation when one or both of the two main larval foodplants, Cowslip *Primula veris* L. or Primrose *Primula vulgaris* Huds. are abundant. Larger-leaved plants, unlikely to drought during the larval feeding period are preferred (e.g. Turner et al. 2009). In the North York Moors *H. lucina* is restricted to valleys bisecting its southern edge with remaining colonies representing approximately 10% of the national resource (Butterfly Conservation, unpublished data). We implemented a conservation programme from 2003 onwards to restore occupied and unoccupied habitat.

Status of *H. lucina* on the North York Moors

H. lucina occurs in two discrete habitat networks 24 km apart in the North York Moors landscape of North Yorkshire. In the early 1990s, 17 colonies were known from the valleys to the north-west of Helmsley (SE 58, 59) and five colonies to the north-east of Pickering (SE 88) (Ellis and Parks 2003). All the Helmsley sites comprised limestone grassland habitat and all but two were ungrazed. Three of the Pickering sites also supported limestone grassland, two of which were ungrazed. One Pickering colony occurred on deciduous woodland rides and another was located on rail verge ballast.

Data on historical management were limited but nearly all sites were susceptible to succession to dense scrub and woodland. Consequently limited efforts to reverse population declines on extant sites were undertaken by volunteers through small-scale scrub management. However, two extinctions occurred in the Helmsley network in the mid 1990s. The metapopulation structure of the Helmsley colonies was studied in 2000 and suggested that occupied sites supported the largest breeding areas in close proximity to other occupied sites whereas unoccupied sites tended to be small and isolated (McAndrew 2002). The study concluded that this was a non-equilibrium metapopulation and predicted further extinctions. By 2002 only nine Helmsley and four Pickering sites were occupied, with respectively eight and one extinctions since 1993. However, at the commencement of the restoration project many of these losses were unconfirmed because Butterfly Conservation uses a five year recorded absence criterion to verify extinction.

Conservation management

On calcareous grasslands, the aim of management is to maintain a mosaic of open, sunny grassland with abundant *Primula* spp. in medium height swards (5–20 cm), with scrub edges or patches. Taller vegetation is utilised for breeding and shorter vegetation ensures continuity of foodplant supply by provision of bare ground for germinating *Primula* spp. In woodlands the aim is to ensure a continuous supply of clearings with abundant *Primula* spp. in open, sunny conditions.

From 2003 both occupied and unoccupied sites in the Helmsley network were targeted to restore suitable breeding habitat (Table 3, Fig. 8). On some sites small-scale scrub management was undertaken by removing small groups of shrubs and trees, but on most sites larger areas of scrub were removed to either create glades within woodland developing over grassland or restore more extensive areas of grassland. Wherever larger-scale management was implemented, scrub margins were retained but pushed back from the original line, invariably creating bare ground to be recolonised by *Primula* spp. In total, scrub management was implemented over a six-year period on 15 of 18 Helmsley sites. Mowing of a very rank sward was also undertaken on one site where grazing was impractical and bruising of dense Bracken followed by *P. vulgaris* planting on another. Connectivity between two sites was also improved by creating a ride through dense scrub. No management was undertaken on two Helmsley sites which supported only very small areas of habitat and where little potential to restore was identified. An occupied site which may be a recent colonisation was only discovered in 2008 and management requirements have yet to be determined.

Table 3 Management implemented on *Hamearis lucina* sites in two North York Moors networks 2003–2008

	Helmsley network	Pickering network
No. sites in landscape	18	5
No. sites managed	15	2
No. sites managed by scrub control	15	2
No. sites managed by Bracken bruising	1	0
No. sites managed by mowing	1	0
No. sites managed by Primula planting	1	0
No. sites managed by ride widening	1	1
No. of work parties held by volunteers	16	2
No. days worked by contractors	>116	0
Habitat area managed by contractors (ha)	>6	0
Project funding secured for management	£20,680	£0

In contrast only two Pickering sites were targeted by positive management (Fig. 9). Scrub control was undertaken on one and woodland ride widening on another although not specifically for *H. lucina*. Unfavourable management for *H. lucina* was implemented on two sites; one was heavily sheep grazed to maintain suitable habitat condition for the flora and the railway verge site was sprayed with herbicide to control weeds.

Management was undertaken on some sites by North York Moors National Park volunteers but mainly by contractors. Ideally intervention in both networks would have begun a decade earlier before extinctions occurred. However, nature conservation often depends on the right opportunities arising. The Helmsley network was not specifically targeted in preference to Pickering, but reflects more established relationships with landowners to enable access for management and the availability of funding to support the work undertaken by contractors (CAN DO project funding, a partnership between Natural England, the National Park and English Heritage, was only available in the Helmsley network).

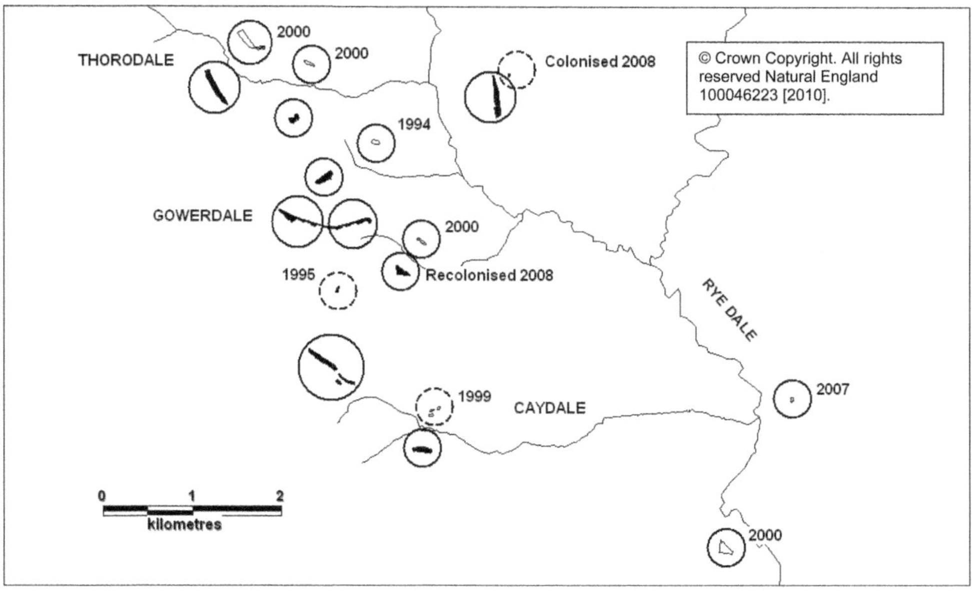

Fig. 8 Location and 2009 status of 18 *Hamearis lucina* sites in the Hemsley network on the North York Moors. Sites occur within four valleys (rivers and streams shown as solid lines) and *solid* and *empty patches* indicate extent of respectively occupied and unoccupied breeding habitat. Dates are year of last records (or re/colonisations). Sites managed during 2003–2008 indicated by *closed circles* and unmanaged sites by *dashed circles*

Monitoring the impact of landscape restoration

Monitoring of *H. lucina* populations has been undertaken annually since 1993, but not all sites were visited each year. Adult numbers have always been low on most sites and therefore thorough searches of the entire breeding areas were undertaken, recording the total number of adults present. Where several visits were made in the same year the highest recorded count for that season was utilised.

The Pickering network has continued to decline with three further extinctions since 2003 (Figs. 9 and 10a) and is now reduced to a single population occupying <1.5 ha habitat (Fig. 10b). As a consequence sites have become progressively more isolated, with the remaining colony now >28 km from its nearest occupied site in the Helmsley network, compared to a mean of ~0.4 km in 1993 and 2002 (Fig. 7c).

Conversely the decline in the Helmsley network appears to have been stabilized (Figs. 8, 10a and 10b). Only a single extinction has occurred since management commenced and this on the smallest (0.12 ha) and most isolated site (4 km from nearest occupied site). Another occupied site was discovered in 2008 and is presumed to be a colonization from a site only 0.15 km distant and a colonisation may have occurred in 2008 following a 1997 extinction, probably from a site 0.56 km distant. Due to the extinction of small outlying populations, the occupied Helmsley sites are now concentrated in a core area with a mean distance to other occupied sites of 0.46 km in comparison to 1.14 km in 2002 (Fig. 10c). However, unoccupied sites are now more isolated with a mean distance to an occupied site of 1.29 km compared to 0.68 km in 2002, but still less than that recorded for the Pickering network.

Count data from 1999, when a greater proportion of sites were monitored, were analysed by TRIM and show that the population in the Helmsley network has increased by 317% ($P < 0.05$) during that period (Fig. 11). Conversely the population index for the Pickering network declined by 81% ($P < 0.01$) during the same period and across the UK by 33% ($P < 0.01$).

These data demonstrate that habitat restoration for *H. lucina* can be achieved at the landscape-scale. However, responses have been variable between sites and unpredictable with the greatest increases on sites that appeared least promising. At one site for example, only one adult was recorded in 2002 preceding management in 2003 and none were seen in 2003–2005. However, counts of between 10 and 24 adults were made in 2006–2009 presumably as habitat quality reached an optimum following restoration management. We conclude that, providing they are not isolated, careful targeting of all potential sites within a network is important, regardless of whether they are currently occupied.

The contrasting fortunes of the two networks demonstrate the cost benefits of early intervention. The work undertaken in the Helmsley network is the first phase in the restoration of this landscape. We have identified a further 10 sites in the Helmsley network and eight in the Pickering network (Fig. 9), including several woodlands, as well as further patches on existing sites which could with appropriate management, support the butterfly. We estimate direct management costs of this second phase to be in the region of £110 k in comparison to about £21 k for the first phase. Significant funding will need to be directed at the smaller Pickering network as these sites have become progressively more unsuitable through lack of appropriate

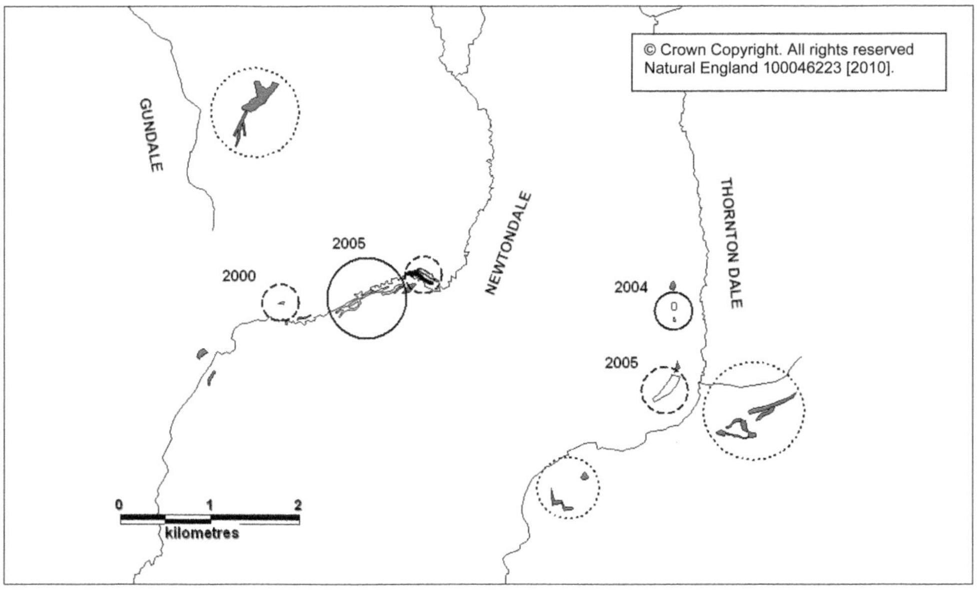

Fig. 9 Location and 2009 status of 5 *Hamearis lucina* sites in the Pickering network on the North York Moors. Sites occur within three valleys (streams shown as solid lines) and *solid* and *empty patches* indicate extent of respectively occupied and unoccupied breeding habitat. Dates are year of last records. Sites managed during 2003–2008 indicated by *closed circles* and unmanaged sites by *dashed circles*. *Grey patches* show potential breeding habitat awaiting restoration and three potential *H. lucina* reintroduction sites are indicated by *dotted circles*

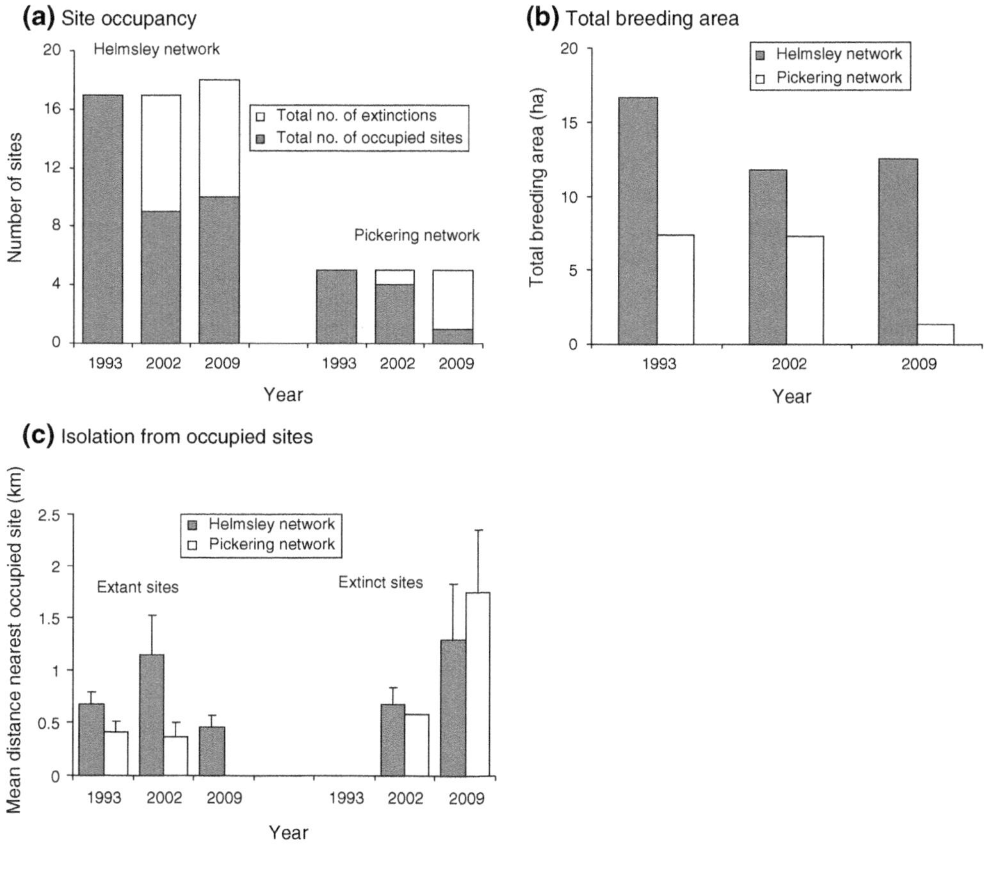

Fig. 10 Impact of habitat restoration on *Hamearis lucina* sites in two North York Moors landscapes

management and the butterfly has declined to the point of extinction. Strategic re/introductions will almost certainly be required to fully restore the Pickering network (Fig. 9), adding further to the cost.

Whilst some habitat improvements can be maintained for some time by the slow rate of succession on steep grassland slopes, in the longer-term consideration will need to be given to establishing maintenance regimes. One of the largest sites in the Helmsley network has been fenced with the aim of introducing stock grazing. However, grazing to maintain *H. lucina* habitat at a later successional stage than the shorter, open swards established under many

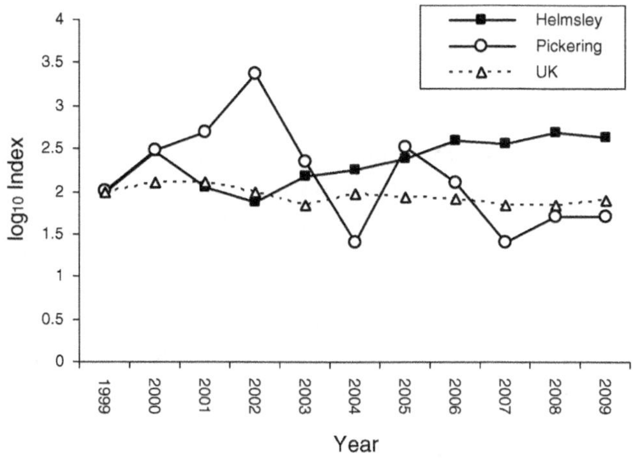

Fig. 11 Impact of habitat restoration on *Hamearis lucina* population trends in two North York Moors landscapes. Data analysed by TRIM; UK national trend included for comparison

conservation grazing regimes is a potential conflict that remains to be resolved.

Discussion

Although our case studies do not assess the relative importance of area, isolation and habitat quality in metapopulation dynamics, they do provide evidence that rare species conservation can be very effective at a landscape-scale if all three are incorporated into conservation strategies as proposed by Thomas et al. (2001). Without improvements in habitat quality which raised the carrying capacity of occupied patches and restored the potential of unoccupied patches, it is unlikely the overall increases in breeding area occupied and improvements in connectivity would have been achieved.

Local extinctions and colonisations occur naturally within metapopulations, but our two case studies both demonstrate that the extinction of species on small, isolated sites within a landscape need not be inevitable if they are properly managed. Furthermore, as land management is rarely applied across whole landscapes, budgetary and other constraints (e.g. land ownership, current land use) mean that careful targeting is required to maximize benefits. In practice networks of sites in close enough proximity are managed to enable natural colonization of unoccupied patches to occur. Improvements in connectivity are often achieved along linear features (e.g. woodland rides) or by removing barriers to dispersal (e.g. plantation forestry) but usually where there is some prospect of restoring breeding habitat as well. By working on a number of sites with different owners and circumstances it is possible to 'fail' on some sites or patches without compromising the overall success of a project restoring a functioning metapopulation. In other words, dynamic changes in habitat quality can still occur whilst still maintaining the target species in the landscape.

Our two case studies illustrate that successful landscape-scale conservation can operate at a relatively local scale and with relatively limited resources. However, these apparent successes should still be treated with a degree of caution. Although the declines experienced over several decades have been halted and some progress made in reversing them, larger-scale projects are being developed in both landscapes to increase the number of sites, the area of occupied habitat and to improve connectivity. Only when these second phase projects are successfully implemented could these landscapes be considered fully restored. However, larger-scale projects can be very expensive partly because more sites need to be managed but also because per capita management costs are much greater when intervention occurs later rather than sooner. Earlier intervention for non-equilibrium declining metapopulations is clearly more cost effective.

A further advantage of larger-scale projects is greater opportunity to ensure continuity on an economic scale of a traditional low intensity or an innovative management system that might otherwise not be commercially viable. The emergence of woodfuel for example, offers opportunities to restore small woodlands in different ownerships. On more open habitats, flying flocks or herds can be used to graze a suite of sites that individually would be impractical to manage. This is being utilized to an extent in the *B. selene* case study but the pony grazing has little commercial value, whereas a larger landscape project might have the potential to support a beef cattle grazing enterprise.

The development and implementation of landscape-scale projects is underpinned by ecological research. The *B. selene* case study illustrates that limited autecological research, which identifies the species' habitat requirements, can be invaluable in identifying management prescriptions. Similarly an understanding of a species' mobility enables connectivity improvements to be factored into management of unoccupied sites isolated from existing populations.

Project planning requires good quality spatial data, both of the distribution and abundance of the target species, and the habitats they utilise. For projects which aim to secure funding to deliver management directly, mapping proposals agreed with landowners on sites enables projects to be costed effectively. Suitable monitoring systems, principally of the target species populations, also need to be established to gauge the effectiveness of the management implemented. Habitat condition assessments can be undertaken where the target species may not respond in the timeframe of a funded project to provide some measure of success. In both our case studies demonstrating a response from the target species was critical in the development of larger-scale second phase projects.

Both our case studies could only be delivered with adequate funding. As well as the costs of land management, funding is required to pay for staff to coordinate project delivery, to manage the finances, to coordinate and train volunteers and to report back to funders. For projects which are principally advisory, the delivery mechanisms are well designed agri-environment or woodland grant schemes, with realistic payments for management options. Schemes can last for 10 years, a more realistic timescale than the typical three or four year funded project. Where habitat restoration is undertaken through a funded project then these schemes provide an opportunity to maintain the project gains, and therefore potentially a very effective exit strategy. However in the UK funding project officers to facilitate the uptake of schemes is far more difficult than funding direct management or the development of a volunteer network.

Our two case studies describe projects implemented by skilled project officers who are essential, not only to manage often complex land management and budgets, but to build up long-term relationships with landowners, contractors and volunteers. In general landowners have been very receptive to landscape-scale projects and especially if funding covers the cost of capital expenditure to enable them to implement more effective land management. However, it can take time to develop relationships with landowners who can be cautious when dealing with what appear to be innovative management techniques. In the North York Moors for example, most of the Helmsley sites are located on shooting estates and it took many years to persuade the estate managers that scrub management for *H. lucina* was compatible with management for game birds. Since both traditional and innovative habitat management techniques may be utilised in landscape-scale projects, experienced and sympathetic contractors with a good knowledge of nature conservation are also crucial.

Local volunteers can play a significant role in landscape-scale conservation and their recruitment and training is an important component of most such projects. They can contribute to practical conservation where a work party is a needed to undertake sensitive management, a regular feature of both our case studies, but are mainly most active in survey and monitoring on a scale that could not be achieved by project staff alone. In the North York Moors for example, much of the *H. lucina* monitoring has been undertaken by Butterfly Conservation volunteers. Publicising landscape-scale projects is important for both partner organisations and funders, but also for local communities. Engagement with the local community can be critical where a project proposes what may appear to be significant changes to the landscape and is also a potentially valuable source of new volunteers.

Butterfly Conservation staff developed and led both the case studies, as they do on 86% of the landscape-scale projects with which we are or have been recently engaged. However, none would be delivered effectively without the development of partnerships. This arises partly because targeted management for Lepidoptera often takes place on land owned by others, but often simply reflects the size and complexity of such projects which would be beyond the means of most conservation organisations to deliver. Even in the relatively small landscapes of our two case studies, partnership working was essential and these projects would not have progressed as far without their invaluable input. For advisory projects a close working relationship with the government agencies administering agri-environment or woodland grant schemes is essential.

There are a number of threatened moth species which are restricted nationally to one or a few sites, where a landscape-scale approach would be inappropriate (e.g. *Coloephora wockeella* Zeller). The potential to conserve some species on single sites should not be ignored if targeted management can maximise a site's carrying capacity and minimise the chances of local extinction occurring. However, there is little doubt that large, well-funded landscape-scale conservation projects will be the most successful in conserving our most threatened Lepidoptera species (e.g. Bulman et al. 2008). For most threatened and declining butterflies and moths, Butterfly Conservation has and will continue to focus resources on landscapes where there is a good chance of restoring larger networks of occupied and unoccupied habitat as well improving connectivity between breeding patches.

Acknowledgments We would like to thank all our partners and volunteers who contribute to the delivery of our 79 landscape-scale projects, but especially Durham County Council (Dave Liddle), Durham Biodiversity Partnership (Andy Lees), North York Moors National Park Authority (Rona Charles) and Natural England (Andrew Windrum) for their support delivering the two landscape-scale conservation projects described in this paper. Dave Liddle also walks the *Boloria selene* transect in County Durham. The *Hamearis lucina* data in Fig. 7 was analysed by Ian Middlebrook (Butterfly Conservation). Butterfly Conservation would also like to acknowledge the continuing support of Natural England for many of our national landscape-scale projects.

References

Berney F (2002) The status, mobility and habitat requirements of the Small Pearl-bordered fritillary *Boloria selene* in County Durham. MSc thesis, University of Sunderland

Botham MS, Brereton TM, Middlebrook I, Cruickshank KL, Harrower C, Beckmann B, Roy DB (2008) United Kingdom butterfly monitoring scheme report for 2008. CEH Wallingford

Bourn NAD, Bulman CR (2005) Landscape scale conservation, theory into practice. In: Kuhn E, Feldmann R, Thomas JA, Settele J (eds) Studies on the ecology and conservation of butterflies in Europe. Vol 1: general concepts and case studies. Butterfly Conservation, Wareham, Dorset, pp 111–112

Bulman CR, Bourn NAD, Warren MS (eds) (2008) Conservation review, 2000–2008. Butterfly conservation, Wareham, Dorset
Craig CC (1953) On the utilisation of marked specimens in estimating populations of flying insects. Biometrika 40:170–176
Ellis S (2000) The small pearl-bordered fritillary *Boloria selene* in County Durham. Contract report to English Nature, Durham County Council and Northumbrian Water
Ellis S (2001) Population ecology and conservation management of the Small Pearl-bordered fritillary *Boloria selene* in County Durham. Contract report to English Nature
Ellis S, Parks R (2003) The conservation of the Duke of Burgundy *Hamearis lucina* on the North York Moors. Butterfly Conservation, Wareham
Fox R, Asher J, Brereton T, Roy D, Warren M (2006) The state of butterflies in Britain and Ireland. Pisces Publications, Berkshire
Hanski I (1999) Metapopulation ecology. Oxford University Press,
McAndrew D (2002) Habitat fragmentation and survival of the Duke of Burgundy butterfly around Helmsley in North Yorkshire. MSc thesis, University of Leeds
Pollard E (1977) A method for assessing change in the abundance of butterflies. Biol Conserv 12:115–134
Pullin AS, Knight TM (2001) Effectiveness in conservation practice: pointers from medicine and public health. Conserv Biol 15: 50–54
Stewart KEJ, Bourn NAD, Thomas JA (2001) An evaluation of three quick methods commonly used to assess sward height in ecology. J Appl Ecol 38:1148–1154
Thomas JA (1983a) The ecology and conservation of *Lysandra bellargus* (Lepidoptera: Lycaenidae) in Britain. J Appl Ecol 20: 59–83
Thomas JA (1983b) A quick method of estimating butterfly numbers during surveys. Biol Conserv 27:195–211
Thomas JA (1991) Rare species conservation: case studies of European butterflies. In: Spellerbeg IF, Goldsmith FB, Morris MG (eds) The scientific management of temperate communities for conservation. Blackwell Scientific Publications, Oxford, pp 149–197
Thomas JA, Bourn NAD, Clarke RT, Stewart KE, Simcox DJ, Pearman GS, Curtis R, Goodger B (2001) The quality and isolation of habitat patches both determine where butterflies persist in fragmented landscapes. Proc R Son Lond 268:1791–1796
Turner EC, Granroth HMV, Johnson HR, Lucas CBH, Thompson AM, Froy H, German RN, Holdgate R (2009) Habitat preference and dispersal of the Duke of Burgundy butterfly (*Hamearis lucina*) on an abandoned chalk quarry in Bedfordshire, UK. J Insect Conserv 13:475–486
Wilson RJ, Thomas CD (2002) Dispersal and the spatial dynamics of butterfly populations. In: Bullock JM, Kenward RE, Hails RS (eds) Dispersal ecology. Blackwell Publishing, Oxford, pp 257–278

J Insect Conserv (2011) 15:111–119
DOI 10.1007/s10841-010-9328-9

ORIGINAL PAPER

Securing viable metapopulations of the Marsh Fritillary butterfly, *Euphydryas aurinia*, (Lepidoptera: Nymphalidae) in Northern England

Keith Porter · Sam Ellis

Received: 25 March 2010 / Accepted: 3 August 2010 / Published online: 15 August 2010

Abstract The Marsh Fritillary butterfly, *Euphydryas aurinia* (Rott.), is regarded as the classic species that requires landscape-scale management in order to maintain viable populations. This paper describes the background to an extinction of this species in Northern England, United Kingdom, and the action taken to reintroduce and monitor the progress of sub-populations between 2007 and 2009. A captive breeding stock was secured using the last 155 known wild caterpillars in 2004 and supplemented by 95 caterpillars collected under licence from the nearest extant populations in western Scotland. Through captive breeding the source population for subsequent reintroduction was increased dramatically and 42,400 caterpillars were used in 2007 to reintroduce the butterfly to four locations in North and West Cumbria. Each of the release sites is within known historical network areas where the butterfly had thrived in past decades. The intention was to restore multiple habitat patches across each network area and eventually establish viable metapopulations across these landscapes. Adults and larval webs have been monitored since 2007 and show that introduction has been successful at three out of the four locations. The failure at the fourth site appears to be unrelated to habitat condition and the causes are currently being investigated. Work is continuing to secure appropriate management of suitable habitat within each landscape area and the number of sub-populations in one area will be increased in the 2010 season.

K. Porter (✉)
Natural England, Northminster House, Peterborough
PE7 3SU, UK
e-mail: Keith.porter@naturalengland.org.uk

S. Ellis
Butterfly Conservation, Manor Yard, East Lulworth, Wareham, Dorset BH20 5QP, UK

Keywords Lepidoptera · *Euphydryas aurinia* · Reintroduction · Conservation · Landscape · United Kingdom

Introduction

The Marsh Fritillary butterfly, *Euphydryas aurinia* (Rott.), is known to have declined dramatically in the United Kingdom (Warren 1994; Asher et al. 2001; Fox et al. 2006) and more widely throughout Europe (Van Sway and Warren 1999). The causes of decline are attributed to land use change, especially agricultural improvement or afforestation (Schtickzelle et al. 2005). As a habitat specialist species with a metapopulation structure, the loss of habitat patches within a diverse landscape is especially damaging for Marsh Fritillary (Wahlberg et al. 2002a; Betzholtz et al. 2007). The pressure on agricultural production, especially from the early 1950s to the present, has resulted in the loss of semi-natural grassland habitats used by the Marsh Fritillary. This pressure has increasingly isolated colonies of the butterfly until local extinction leads to wholesale regional extinction. This is well documented in England using the wealth of historical biological records (Fig. 1) that show the continuing decline east to west and the isolation and ultimate loss of populations. One of the most telling statistics is the loss of 66% of colonies between 1990 and 2000 (Asher et al. 2001) showing that the Marsh Fritillary is still critically endangered in the UK. In Western Scotland and Ireland the status of the Marsh Fritillary is reasonably strong, with significant numbers of new populations being found in recent years (unpublished data).

Throughout most of its range in the British Isles, the Marsh Fritillary occurs on wet grasslands that are described as purple moor grass–rush pasture, especially the plant

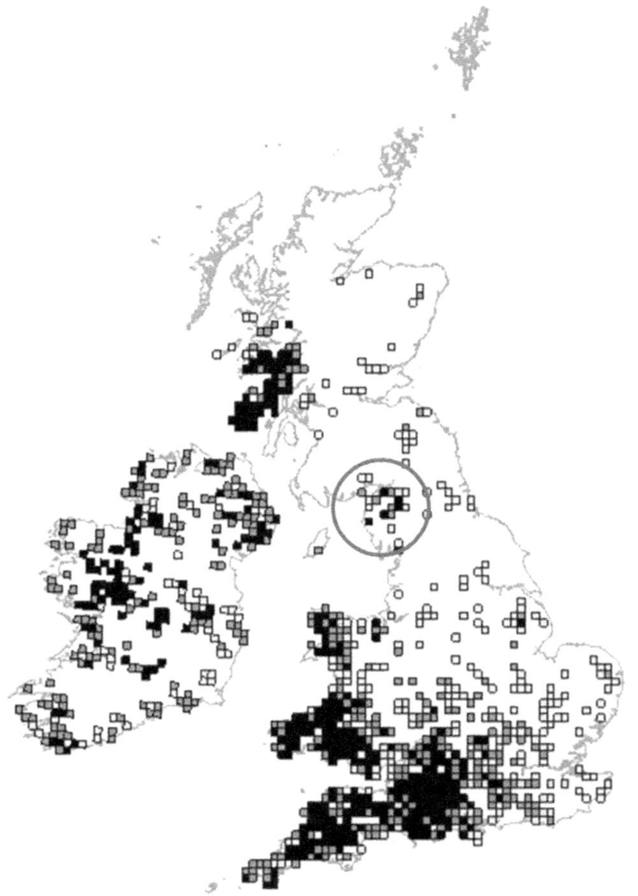

Fig. 1 Ten kilometre square distribution trend of Marsh Fritillary in the British Isles, showing three time periods; *open squares* 1600–1950, *grey squares* 1951–2000, and *filled squares* 2001–present. This shows the westward drift in distribution as land uses pressures in the east lead to local and regional extinctions. The *circled area* highlights the Cumbrian part of the natural range. *Source* http://data.nbn.org.uk/

communties characterised by purple moor grass–tormentil, *Molinia caerulae–Potentilla erecta,* and purple moor grass–meadow thistle, *M. caerulea–Cirsium dissectum,* mire communities.[1] This habitat is typically a complex of plant communities ranging from bogs and wet heath to drier, species-rich meadow. The common link is the presence of the larval foodplant Devil's-bit Scabious, *Succisa pratensis*. In areas of southern England the butterfly also occurs on calcareous, dry grasslands where *S. pratensis* also grows. This is especially true of the large areas on Salisbury Plain Military Training Area where the landscape has been untouched by agricultural improvement; this area probably contains the largest, and most viable, populations of the Marsh Fritillary in England.

The female butterfly emerges with a fully formed egg batch and once mated will lay the first and largest batch of over 300 eggs on the underside of *S. pratensis* leaves. The larvae from each egg batch remain as discrete feeding groups within silken webs for the first three instars, prior to hibernation in early autumn. The third instar larval webs may split if the entire original plant host is eaten, but will still maintain a loose association of satellite webs. The fourth instar enters hibernation without feeding and larvae spin a stronger communal web at the base of the sward. In early spring the larvae remain gregarious until the moult to the fifth instar. Thereafter, larvae feed in smaller groups and by the final, sixth, instar they feed as highly mobile, solitary larvae.

The pattern of habitat patches within a landscape is believed to be critical for viable, long-term population occurrence. Some research studies have shown the link between habitat pattern and persistence of Marsh Fritillary; it has been suggested that this is driven by the interaction between the butterfly and its specific Braconid parasitoids, *Cotesia melitaearum* Wilk. and *C. bignellii* Marsh. (Wahlberg et al. 2002b; Bulman et al. 2007; Junker and Schmitt 2010). The host–parasitoid interaction is clearly a major driver of fluctuations in population size characteristic of Marsh Fritillary. These fluctuations in population size are well documented for other *Euphydryas spp*, notably *E. editha* in North America (Ehrlich 1984). Thus the metapopulation dynamics attributed to the Marsh Fritillary, and the Glanville Fritillary, *Melitaea cinxia*, in northern Europe (Lei and Hanski 1998) are an integral part of any conservation planning for this species. The interaction between the parasitoid and Marsh Fritillary has been recently studied (Klapwijk et al. 2010) using experimental manipulation of temperature and shading of both Marsh Fritillary caterpillars and cocoons of *C. bignellii*. This study has shown that whilst caterpillars do grow faster in warmer and/or sunnier conditions, the development time of the parasitoid cocoon did not appear to be correlated to air temperature or shading over the range of temperatures pertaining in the two experimental years. Thus the relationship between the parasitoid, host and weather conditions is more complex than previously believed (Porter 1982). The experimental finding of Klapwijk were also supported by analysis of weather variables and population size on a data set derived from a 20 year monitoring exercise on a single population of Marsh Fritillary in Wales where no significant correlation was found between population size and weather. Thus the other drivers of population fluctuation, including habitat quality, patch size and patch distribution, must be taken into account when considering causes of local extinction.

Historical context

Cumbria, in North West England, represents a key link in natural range between populations in Wales and those in

[1] Respectively M25 and M24 in the British National Vegetation Classification (Rodwell 1991) .

Western Scotland. Due to its endangered status, the Marsh Fritillary is protected under UK national and European legislation, and is listed as Priority Species under the UK Biodiversity Action Plan (UK Biodiversity Group 1999). The Habitats Directive (CEC 1992) requires that the Marsh Fritillary is included in a representative series of Special Areas of Conservation (SACs) in order to maintain or restore the Favourable Conservation Status of this species. This legislation requires EU member states to maintain viable populations of the Marsh Fritillary within its natural range.

Cumbrian populations of the Marsh Fritillary have been well documented, with records going back to the early 1800s. The documented decline shows how the species became restricted to a few localities in the last few decades, with a single colony remaining in 2004 (Fig. 2). The strong populations accessible to local entomologists around Carlisle were regularly visited by collectors who noted changes in population size. This is the area where E.B. Ford and his father documented the fate of one particular colony that went through a genetic 'bottle neck' and gave rise to many abnormal colour varieties of adult (Ford and Ford 1930). The last Marsh Fritillary recorded from the site of E.B. Ford's colony was in 1975 when the final wet field in that area was ploughed in an attempt to improve the grassland by re-seeding. The sward has since recovered to a semi-natural state, but *S. pratensis* has been lost from the sward.

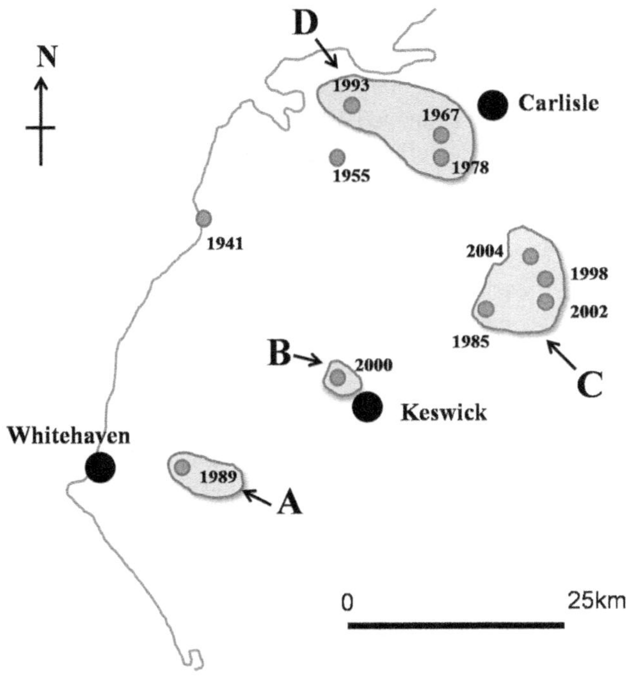

Fig. 2 This shows the extinction pattern of Marsh Fritillary Northern Cumbria as known from records held by the Tullie House Museum. The gradual loss of populations began in the mid 20th Century and by 2004 the final colony had become extinct. The four target networks (A–D) are indicated on this *map*

The decline of the species in Cumbria has been attributed to loss and fragmentation of habitat through agricultural improvement with rapid decline in the 1970s when drainage and improvement to re-seeded leys was encouraged by government funding. The four broad metapopulations recognised in Cumbria differ in the date of discovery by entomologists, but all became extinct between 1967 and 2004 (Fig. 3). Populations within a fragmented landscape are at risk of extinction, especially where the habitat extent and quality continue to decline; this has been the situation in Cumbria over past decades. The added pressures of parasitoids, adverse weather and lack of suitable management are likely to increase the chance of local extinctions and this is believed to explain the ultimate fate of the Cumbrian populations.

Reintroduction programme

The proposal

Four geographic areas, from which the Marsh Fritillary has disappeared in the last 20 years, have been defined in Cumbria. These broadly equate to distinct metapopulations and are taken as target areas for restoration of the Marsh Fritillary. The first phase of the reintroduction programme was to secure viable core populations on the best managed habitat patches within a given network. If successful, the second phase is to encourage natural colonisation of suitable nearby habitat patches, with further introductions to more remote patches within the network. The overall aim was to re-establish dispersal between patches and thereby create dynamic metapopulations that will persist within a network area. Parasitoids have been shown to play a powerful role in local fluctuations of size of sub-populations of Melitaeid butterflies (Lei and Hanski 1997, 1998; Lei and Camara 1999) and the Cumbrian populations of Marsh Fritillary were known to hold the specific parasitoid *C. melitaearum* (Ford and Ford 1930; Porter 1981). The final larvae found in the wild in Cumbria were carefully monitored in captivity to check for *Cotesia,* but none were found. In situations of rapidly declining populations it is inevitable that the parasitoid will become extinct before its host. Evidence from the above European study shows that the parasitoid-host relationship, in conjunction with hyper-parasitoids of *Cotesia*, dampens population fluctuations and, provided that the landscape is sufficiently rich in suitable habitat patches, drives the dynamics of the metapopulation. In the Cumbrian situation, with the absence of the parasitoid, it is anticipated that populations will expand to the point where food availability becomes limiting and local extinctions or severe declines will result. Once viable populations across networks are established in Cumbria the

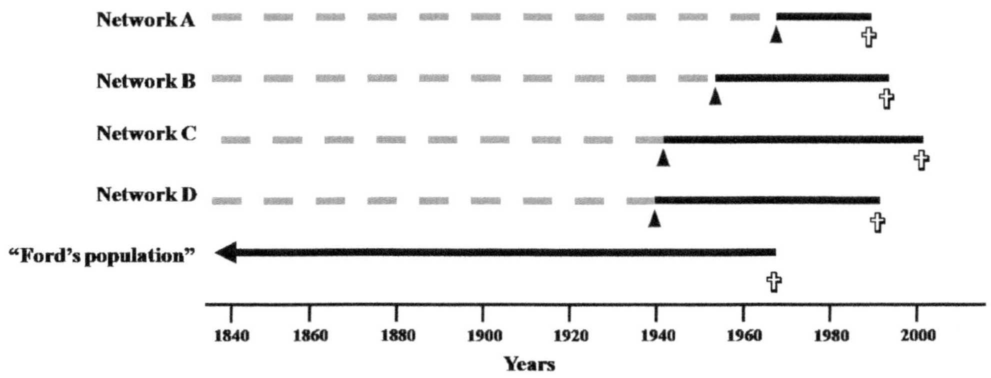

Fig. 3 Discovery and extinction records show that the ultimate extinction of all defined metapopulations occurred between 1960 and 2004, irrespective of the date when conservationists were aware of the existence of the butterfly in each area. This *diagram* assumes that Marsh Fritillary has existed within each metapopulation for centuries and by the time of discovery the sustainable network of habitat patches over a landscape were already breaking down. This assumed past occurrence is shown by the *dashed lines*, known duration of the population by *solid lines*; the discovery point by *filled triangles* and; the date of recorded extinction by a *cross*. The colony where Ford and Ford (1930) undertook their observations on fluctuation in population size is also shown for comparison. This site sits within metapopulation area D

parasitoid will be reintroduced to allow natural dynamics to re-establish.

Recent history

Concern was raised over the decline of the Marsh Fritillary in Cumbria in the 1980s with extinctions being recorded on well known sites, including National Nature Reserves and other protected areas. The final colonies in Cumbria were extensively monitored by volunteers and staff of Butterfly Conservation from the mid 1980s, with additional effort to locate any previously unrecorded colonies and improve the management of the remaining few sites. Two separate introductions were attempted in 1997 and 1998 by the Cumbrian Marsh Fritillary Action Group (CMFAG), but both failed in their first season (unpublished data). The cause of failure is uncertain but very small numbers were released and no significant management was undertaken at the release site.

By the autumn of 2004 only a single female and a few male adult butterflies were seen on the last extant site; after thorough search over August and September two small larval webs were found adjacent to one another. The decision was made by the CMFAG to take the 155 remaining wild larvae, under licence, into a captive breeding programme. This decision was made on the basis that this small number of wild larvae, likely to be the offspring of a single female, was unlikely to restore a viable population on the site and extinction in the short term was probable.

Given the likely restricted genetic heterogeneity of the rescued larvae, captive rearing commenced with a pure Cumbrian line and a sample 95 larvae taken under licence from western Scotland. The Scottish populations are genetically similar to the Cumbrian colonies (Joyce 2001) and the licence allowed removal of up to five larvae from discrete sub-populations in the wild. In total 19 Scottish sites were sampled ranging from the island of Mull in the North to Islay in the South. It is assumed that the genetic diversity of the overall sample was great, taken from so many sites over such a large geographic area. These two breeding lines were used to establish a hybrid line in case the Cumbrian stock proved to be too inbred. The cause of loss remains unproven, but within two generations, the pure Cumbrian line females failed to produce fertile eggs. From a starting point of 155 Cumbrian larvae and 95 Scottish larvae in spring 2005, the hybrid captive stock was increased to an estimated 50,000 larvae in captivity by the spring of 2007. Rearing was done under natural conditions using large netted cages and pot grown plants of *S. pratensis*, supplemented by foliage of garden varieties of evergreen shrub Honeysuckle, *Lonicera spp*, Snowberry, *Symphoricarpos albus* and wild Honeysuckle, *Lonicera periclymenum*. The captive stock was distributed across six volunteer breeders to insure against accidental loss of stock or disease outbreaks.

Between March and April 2007 42,000 larvae were released onto four sites, each site being at the core of a network of existing or potential habitat patches (Table 1). The larvae were released in small groups of approximately 100, equating to the average size of wild feeding larval groups based on extensive fieldwork elsewhere (Porter 1981). Larval groups were placed on clusters of *S. pratensis* plants to ensure that the IV and V instar larvae had access to sufficient foodplant. The final, VI instar larvae are extremely mobile and forage over larger areas in search of *S. pratensis* or occasionally *Lonicera periclymenum*. Releases were restricted to a single core site within each network because these were the best managed and where reintroduction most likely to succeed. Prior to this project no successful reintroductions had been made in Cumbria. Four sites across four networks were chosen to increase the chances of success.

Site preparation

Leading up to, and after local populations became extinct in Cumbria, many attempts were made by the CMFAG to improve site management on existing or potential sites in Cumbria. The introduction of agri-environment schemes has now enabled all the core sites to be brought back into suitable condition. These schemes enable farmers to be funded for both capital work and ongoing management to favour semi-natural habitats and appropriate stocking levels. On all sites an active programme of scrub control has restored the former extent of open grassland and on Site B removal of overgrown tree lines has removed adverse shading impacts from fields. On all sites, grazing regimes using cattle (three sites) or New Forest ponies (one site) have been reinstated with extensive grazing regimes of around 0.4 livestock units per hectare. Water level management is being addressed on two sites where inundation or drying was found to be adversely affecting the condition of the swards. On Site E, which lies on Common Land, management restoration has required two applications to UK Secretary of State to enable grazing to be reintroduced by enclosing the land and installing cattle grids. The final approval was given during 2010. Due to difficulties of securing grazing, this site has been subject to remedial cutting regimes, with cut and removal of rank vegetation, particularly *Molinia* and other coarse grasses, to remove the build up of leaf litter and encourage recruitment of herbs into the sward. This has been very successful at stimulating the growth of *S. pratensis*, but this technique cannot be used once colonies are re-introduced. In Eastern Europe the rotational cutting of litter meadows is a standard approach to the management of Marsh Fritillary habitats, but the overall site sizes facilitate the use of rotational patch cutting. At the scale of English sites this approach is best applied as a restoration technique. On all sites, the expansion of suitable habitat, using these approaches is continuing in order to increase overall patch size and further reduce the risk of local extinction. The estimated area of available habitat, as assessed using aerial photographs, is given for each site in Table 1. Modelling work to estimate the minimum viable habitat area needed to support Marsh Fritillary (Bulman et al. 2007) suggest that at least 70 ha of habitat within a 4 × 4 km area is needed to give a 95% probability of persistence for 100 years. The figures given in Table 1 show the estimated area within single sites that represent the first patch within an overall network. At present only Site B appears to sit within an area of existing or potential habitat that meets this threshold. The project is focussed on restoring habitat patches within each network area using the minimum threshold figure above as a target.

Table 1 Area of suitable habitat for Marsh Fritillary breeding on four core release sites and number of larvae released in 2007

Site code	Area of suitable habitat (ha)	No. of larvae released
Site M	7.38	11,000
Site F	12.24	5,000
Site B	20.04	22,900
Site E	7.79	3,500

Monitoring methods

All monitoring was undertaken by the local volunteers of Butterfly Conservation, with the assistance and guidance of professional staff of both Butterfly Conservation and Natural England. The basic methods are taken from standard approaches recommended by Butterfly Conservation.

Adults

The four re-introduction sites were allocated to lead volunteers and visited from mid-May onwards to assess adult numbers using a timed count method (Warren, et al. 1984). All adult butterflies are counted during suitable weather conditions along a zigzag route which representatively samples the flight area. The search time was recorded and counts are expressed as number of sightings per hour per hectare of flight area to provide a Population index (P):

$$\text{Population index (P)} = \text{Adults counted per hour} \times \text{Flight Area (ha)}$$

Counts are made over the flight period and the peak day (maximum) count is used to estimate total population size using the formula taken from Warren (1994) and based upon field calibration of timed counts compared with absolute population estimates made using mark recapture techniques:

$$\text{P} = 6.039 + 0.198 \times \text{Absolute Population estimate}$$

This method has to assume that adults are likely to occur across the population breeding area, in reality high densities adults are often restricted to certain parts of the overall area, thus this formula gives an index of adult numbers rather than true absolute estimates.

Larvae

The gregarious feeding behaviour of pre-hibernation Marsh Fritillary larvae within silken webs provides a convenient monitoring stage. Each web, or closely associated webs, originates from at least one egg batch. The largest individual host plants may hold more than one egg batch, laid by different females (Porter 1981). The webs of actively feeding third instar larvae are conspicuous within the vegetation and can be counted in most weather conditions from mid-August onwards and are the most efficient measure of population size.

Each of the re-introduction sites is monitored by volunteers and all the patches of host plant within the site are systematically searched and the total number of webs present is recorded. The hostplant patches are easily seen in late summer when the blue flowers of *S. pratensis* are held high above the general vegetation and all available patches are checked for larval webs.

Results

Adult numbers

The flight period of Marsh Fritillary in Cumbria ranges from early May until the end of June and variability is related to the weather conditions in the spring; sunny, warm conditions in March and April allow accelerated growth of larvae and their foodplants. The larvae are adapted to utilise sunlight to raise their body temperature above ambient by basking (Porter 1982) and are able to maintain a body temperature of 35–37°C when ambient temperature can be as low as 4°C. This increases the rate of metabolism of food, and thus accelerates larval growth (Heinrich 1977).

After a rapid increase in the first part of the flight period, adult numbers tail away as the flight period continues. Variation between sample days is partly an artefact of weather differences and partly due to observer differences, with two volunteers sharing the task. A profile of adult numbers taken from a detailed population assessment over an entire flight period in southern England (Fig. 4) shows that peak adult numbers occur 2–8 days from first emergence.

The population estimates for the four release sites from 2007 to 2009 (Fig. 5) suggests that each of the reintroduction sites is responding differently. Site M is failing to show any significant response, with adult numbers remaining low; site F shows a steady increase in population size over the 3 years of monitoring, whilst site B shows consistent decline over the same period. The final site is maintaining numbers with some fluctuation year to year.

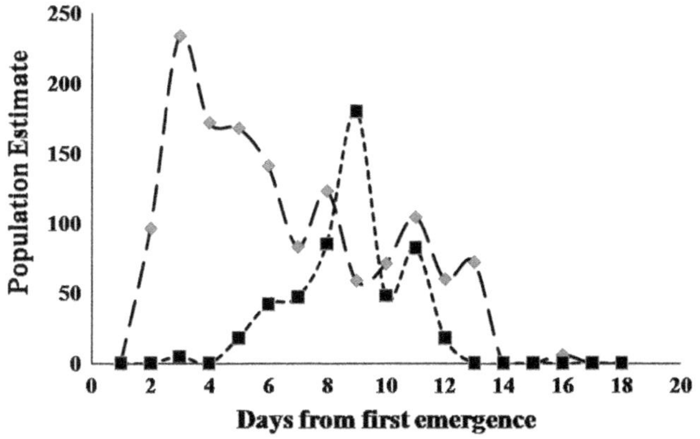

Fig. 4 *Example data* showing difference between male *grey diamonds* and female *black squares* emergence from detailed mark-recapture study of an isolated population in Oxfordshire in 1979, based on field data collected by Porter (1981)

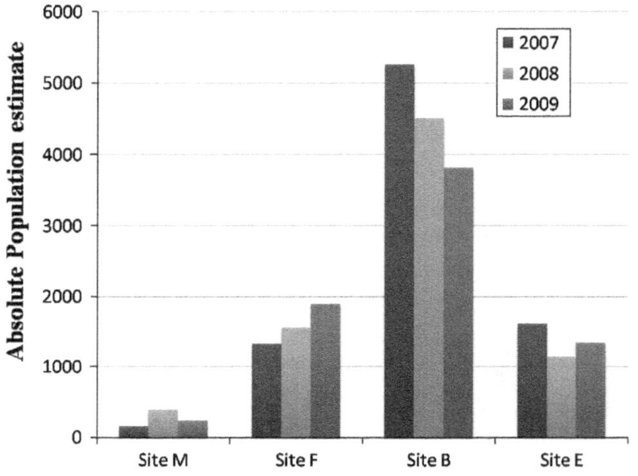

Fig. 5 Adult absolute population estimates for the four reintroduction sites over the 3 years since first release

These conclusions are challenged by the assessment of population performance as judged by larval web counts below.

Larval web monitoring

The monitoring process for larval webs provides far greater confidence in the population status for each of the release sites and, as shown in Fig. 6, gives a different conclusion to that provided by adult counts. Site M remains a relative failure, in terms of population viability, with indications of decline in the already small population size. Site F shows very strong growth over the 3 years of monitoring with a 500% increase in population size, as assessed by web number, over the 3 years. Site B, with a slight decrease in 2008, still shows a 144% increase, with the final site showing 282% increase over the same period.

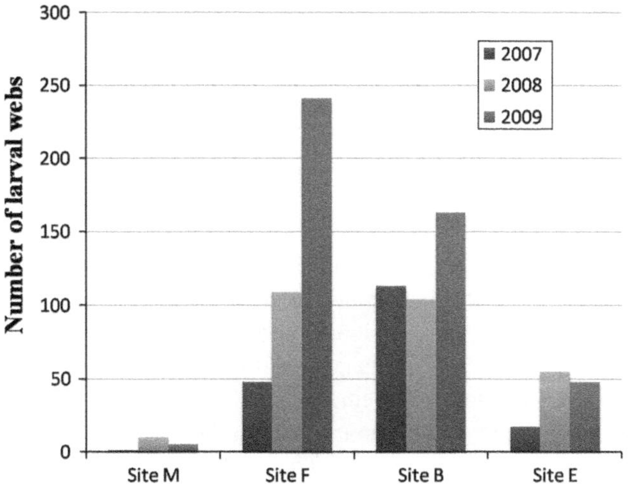

Fig. 6 Larval counts as a measure of population size for the four reintroduction sites over 3 years since first release

The location of webs is increasingly being mapped using GPS to record locations and enable spatial comparisons of habitat use between years. This data provides reasonably accurate data that can be related to sub-compartments within a site and detect changes in breeding location in response to site management or improvement in the quality patches.

Adult dispersal

At site F, where the population is showing greatest growth, adult butterflies have been seen to move away from the core area in 2009. The most significant sighting was a mated female[2] found over one kilometre in a straight line from the core population. This individual can only have flown from the core population and its condition and body size suggested that it had already laid its main egg batch; presumably before dispersal. The terrain between the core population and where the female was found contains what were believed to be significant barriers to dispersal. In European studies (Betzholtz et al. 2007) adult Marsh Fritillary are presumed to have moved over high forest in order to disperse between meadows. Mark-recapture studies at a population in central England showed that movement across a thick, four metre high hedgerow was restricted (Porter 1981). The evidence gained from the project to date support the hypothesis that adults can disperse across non-grassland habitats.

[2] This can readily be detected by the basal plate of the female genitalia showing as a bare plate, devoid of white scales.

Discussion

Factors affecting decline and local extinction

Records from Cumbrian colonies show that populations can exist on isolated sites for decades. The Marsh Fritillary was first recorded on Site M in 1971 and was recorded regularly by observers, and monitored closely from the late 1980s. The specific parasitoid *C. melitaearum* was present on the site in 1979 (Porter 1981) and was also confirmed from wild collected larvae in 1997 when an unsuccessful attempt was made to introduce the species elsewhere in Cumbria. The last known larvae from Site M were taken from the wild in 2004 and were free of *Cotesia* parasitoids. It is assumed that the parasitoid had become extinct before its host. Given the potential for high rates of parasitism by *Cotesia* spp. in Marsh Fritillary larvae, isolated populations may be at risk from severe population reduction when weather conditions favour the parasitoid (Porter 1983). However the interaction between host and parasitoid is complex and the potential impact of climate change on interaction clearly requires further study (Klapwijk 2010). Experience gained in captive breeding of the last remaining Cumbrian larvae suggests that inbreeding is one of the possible, but unproven, causes of extinction in very small, isolated colonies of Marsh Fritillary.

Reintroduction success

The restoration of Marsh Fritillary to historical networks in Cumbria is showing success on three out of four sites, with encouraging increases in population size, as judged by larval web counts. The results show that this approach can succeed where habitats are being maintained in good condition by appropriate grazing levels or restoration mowing prior to re-introduction. The ultimate causes of local extinction will always remain unknown as monitoring focussed purely on the numbers of adults, or larval webs. No assessment was made of parasitoid loading or genetic heterozygosity.

Releasing post-hibernation larvae enables the caterpillars to disperse and utilise the most suitable patches of *S. pratensis* within the site. This is especially true of the final instar larvae that are highly mobile and will forage for fresh foodplants; individual larvae have been recorded moving over 30 metres in 2 h when accidently released from captive rearing cages (unpublished observation).

The reason for the very poor response of site M, despite intensive effort to improve the sward structure by restoring grazing and density of foodplant, is not yet understood. This was the location of the last wild colony in Cumbria and from where the larvae were taken to form the captive stock. This is also the weakest network in terms of area and distribution of potential habitat. The only factor that distinguishes site M from other reintroduction sites is the proximity of pheasant rearing pens in adjacent woodland. Some basic exclusion experiments are planned to test the impact of the game rearing on the Marsh Fritillary on this site.

Efficacy of monitoring methods

The comparison of adult versus larval web counts show that, despite standardisation of approach, adult counts can be misleading as a measure of population trend from year to year. The adult Marsh Fritillary is relatively easy to see in the wild as both sexes are relatively sedentary and rest or bask openly in the vegetation. The number seen will be affected by weather conditions and observer skill; basking or resting individuals can be overlooked in taller swards. Larval web counts are more reliable as measures of population strength, aided by the long time period when they can be found and counted. Detecting the presence of webs is made easier when *S. pratensis* is in flower and can be readily located and searched for webs.

Response of reintroduced populations to individual sites

Each restored population shows a different trend over the 3 years from first release. Site F shows a steady increase over the 3 years at a site with a contiguous block of suitable habitat that is being actively managed by pony grazing. Site E shows a levelling out of larval web numbers on a site that is about to have cattle grazing restored after initial habitat improvement by restoration mowing. The suitable habitat patches are distributed across Site E as small pockets rather than extensive areas with *S. pratensis.* Site B consists of several fields, each with different densities of *S. pratensis.* In 2009 78% of webs were located in a single 0.41 ha field where the sward has high densities of *S. pratensis* within a *Molinia* sward. Distribution of webs across the remainder of the site is directly influenced by the occurrence of swards with moderate densities of *S. pratensis.* The pattern of use at Site M is atypical as the restoration management using cattle grazing and scrub removal has created large areas of good quality habitat on areas where the butterfly previously existed for decades. The reasons are under investigation and are not believed to be related to habitat quality.

Future potential for expansion

The best recorded site is Site M, where over 200 webs were recorded in the 1990s, prior to the population decline. This suggest that even the smallest of the release sites has a yet to reach its carrying capacity. Each site reported here lies at the heart of discrete potential habitat networks; some contain existing patches of unoccupied habitat that are currently being restored.

Partnership approach

The success of the first phase of reintroduction of the Marsh Fritillary to Cumbria is a result of a strong partnership between professional conservation staff, volunteers and landowners. The management of sites and restoration of habitat patches is being secured through use of government funded agri-environment schemes and the collaboration of landowners. The maintenance of captive stock is a result of the hard work and skill of volunteer breeders and plant growers. The monitoring programme has been coordinated and undertaken by local volunteers to provide some high quality consistent data that allows the success of this project to be assessed and reported. This partnership has established a strong foundation of core sites from which the more ambitious phase of landscape scale restoration can build.

Ongoing work

Work is continuing to increase the amount and condition of habitat within and adjacent to the reintroduction sites using a wide range of approaches. On Site B each field within the site has been affected to a different degree by past agricultural management and past attempts at improvement through re-seeding, using more productive fodder grasses or intensive sheep grazing that eradicates *S. pratensis* from the sward. Unlike most other herbs in the *Molinia-Potentilla* plant community the seed of *S. pratensis* does not persist in the soil seed bank. Thus, destruction of semi-natural swards means that other species will recover if the disturbance is short-lived, but *S. pratensis* will require restoration through seeding or transplants. Experimental work is underway, using pot grown transplants, to restore *S. pratensis* to many of the fields at Site B.

Within network areas, opportunities to restore or create new habitat patches are being explored and in 2010 further releases of larvae are underway on two further sites within one of the network areas. This increase in source patches within a landscape is intended to encourage dispersal from and between sub-populations. Given the existing pattern of small habitat patches within the wider network, and assuming success of this additional reintroduction, success will be measured by assessing dispersal from the source populations in coming years. This provides the exciting prospect of securing a true metapopulation operating in a landscape rather than focussing in discrete habitat patches.

Monitoring has so far focussed on the reintroduction sites as a direct assessment of the success or otherwise of the restoration programme. As more habitat patches are restored, and core populations become established, a wider monitoring approach is needed to detect range expansion and colonisation of new patches. Success will ultimately be measured by the maintenance of a number of sub-populations across a landscape, with both local extinction and colonisation occurring. Each network area has been mapped to identify existing and potential habitat. Future monitoring will extend to sample unoccupied habitat patches in a network, with sampling intensity related to distance from known core sub-populations. Larval web counts will be used as the measure of successful colonisation or local extinction.

Acknowledgments We wish to thank: the enthusiastic volunteers of the Cumbria Branch of Butterfly Conservation for their dedication and hard work in rearing livestock and monitoring the adults and larval webs; the staff in Natural England and farmers who we have worked with to restore and manage habitat for the Marsh Fritillary; staff of the National Trust, National Park and Environment Agency who have provided support and funding to secure sites; and finally to the horticultural skills of Messrs R and J Bennett who have supplied rearers and farmers with many thousands of *S. pratensis* plants.

References

Asher J, Warren M, Fox R, Harding P, Jeffcoate G, Jeffcoate S (2001) The millennium atlas of butterflies in Britain and Ireland. Oxford University Press, Oxford

Betzholtz PE, Ehrig A, Lindeborg M, Dinnétz P (2007) Food plant density, patch isolation and vegetation height determine occurrence in a Swedish metapopulation of the Marsh Fritillary *Euphydryas aurinia* (Rottemburg 1775) (Lepidoptera. Nymphalidae). J Insect Conserv 11:343–350

Bulman CR, Wilson RJ, Holt AR, Bravo LG, Early RI, Warren MS, Thomas CD (2007) Minimum viable metapopulation size, extinction debt, and the conservation of a declining species. Ecol Appl 17:1460–1473

CEC (1992) The conservation of natural habitats and wild fauna and flora (Council Directive 92/43/EEC). CEC, Brussels

Ehrlich PR (1984) The structure and dynamics of butterfly populations. In: Vane-Wright RI, Ackery PR (eds) The biology of butterflies. Academic Press, London, pp 25–40

Ford HD, Ford EB (1930) Fluctuations in numbers and its influence on variation in *Melitaea aurinia* Rott. (Lepidoptera). Trans R Ent Soc Lond 78:345–352

Fox R, Asher J, Brereton T, Roy D, Warren M (2006) The state of butterflies in Britain and Ireland. Nature Bureau, Newbury, England

Heinrich B (1977) Why have some animals evolved to regulate a high body temperature? Am Nat 111:623–640

Joyce DA (2001) The use of molecular genetics in the formulation of conservation strategies for Lepidoptera. Ph.D. Thesis. University of Birmingham, UK

Junker M, Schmitt T (2010) Demography, dispersal and movement pattern of *Euphydryas aurinia* (Lepidoptera: Nymphalidae) at the Iberian Peninsula: an alarming example in an increasingly fragmented landscape ? J Insect Conserv 14:237–246

Lei GCN, Camara MKD (1999) Behaviour of a specialist parasitoid, *Cotesia melitaearum*: from individual behaviour to metapopulation processes. Ecol Entomol 24:59–72

Lei G, Hanski I (1997) Metapopulation structure of Cotesia melitaearum, a specialist parasitoid of the butterfly Melitaea cinxia. Oikos 78:91–100

Lei G, Hanski I (1998) Spatial dynamics of two competing parasitoids in a host metapopulation. J Anim Ecol 67:422–433

Porter K (1981) The population dynamics of small colonies of the butterfly *Euphydryas aurinia*. D.Phil. Thesis. Oxford University, UK

Porter K (1982) Basking behaviour in larvae of the butterfly *Euphydryas aurinia*. Oikos 38:308–312

Porter K (1983) Multivoltinism in *Apanteles bignellii* and the influence of weather on synchronisation with its host *Euphydryas aurinia*. Ent Exp Appl 34:155–162

Rodwell J (1991) British plant communities, vol 2. Mires and Heaths, Cambridge, England

Schtickzelle N, Choutt J, Goffart P, Fichefet V, Baguette M (2005) Metapopulation dynamics and conservation of the Marsh Fritillary butterfly: population viability analysis and management options for a critically endangered species in Western Europe. Biol Conserv 126:569–581

UK Biodiversity Group (1999) Tranche 2 action plans. Volume IV–Invertebrates. English Nature, Peterborough

Van Sway CAM, Warren MS (1999) Red data book of European Butterflies (Rhopalocera). Nature and environment series No. 99, Council of Europe, Strasbourg

Wahlberg N, Klemetti T, Hanski I (2002a) Dynamic populations in a dynamic landscape: the metapopulation structure of the Marsh Fritillary butterfly. Ecography 25:224–232

Wahlberg N, Klemetti T, Selonen V, Hanski I (2002b) Metapopulation structure and movements in five species of checkerspot butterflies. Oecologia 130:33–43

Warren MS (1994) The UK status and suspected metapopulation structure of a threatened European butterfly, the Marsh Fritillary, *Eurodryas aurinia*. Biol Conserv 67:239–249

Warren MS, Thomas CD, Thomas JA (1984) The status of the heath fritillary butterfly *Mellicta athalia* Rott. Biol Conserv 29: 287–305 (In Britain)

J Insect Conserv (2011) 15:345–350
DOI 10.1007/s10841-010-9308-0

SHORT COMMUNICATION

Is the Natura 2000 network sufficient for conservation of butterfly diversity? A case study in Slovenia

Rudi Verovnik · Marijan Govedič · Ali Šalamun

Received: 23 March 2010 / Accepted: 30 June 2010 / Published online: 31 August 2010

Abstract Slovenia has one of the most extensive Natura 2000 networks in Europe with 259 SAC's covering 31.4% of the country. To determine how well does the current network cover the areas of high butterfly diversity and/or aggregation of the butterfly species of conservation concern, the data from the recent survey for a distribution atlas were used. Altogether 99,423 records of 173 species collated after 1979 were used. The data distribution is slightly biased towards SAC's, with 44.8% of localities within them, most likely due to sparsely sampled urban areas and intensive farmland areas which are found only outside SAC's. The diversity and distribution of red listed species was evaluated at a 5 × 5 km grid square level. Additionally the importance of the size of the SAC's was compared to their butterfly species diversity. In general the high diversity areas also hold the largest aggregation of red listed species with core areas concentrated in SW Slovenia. The SAC's cover majority of areas with high diversity and the distribution of all but one threatened butterfly species. That species is *Colias myrmidone*, which is now considered extinct in Slovenia with no records after 1993. The most prominent areas with high conservation value in Slovenia not included in the SAC's network are the Koroška region, Goriška Brda region, lower Sava River valley and Slovenske Gorice region. The butterfly diversity in small SAC's is relatively high with increases in size only gradually increasing the species numbers, thus emphasizing the importance and conservation value of small SAC's for sustaining high butterfly diversity in Slovenia.

Keywords Lepidoptera · Butterflies · Distribution · SAC network · Habitat directive

Electronic supplementary material The online version of this article (doi:10.1007/978-94-007-1442-7_29) contains supplementary material, which is available to authorized users.

R. Verovnik (✉)
Biotechnical Faculty, Department of Biology, University of Ljubljana, Večna pot 111, 1000 Ljubljana, Slovenia
e-mail: rudi.verovnik@bf.uni-lj.si

M. Govedič · A. Šalamun
Centre for Cartography of Fauna and Flora, Antoličičeva 1, 2204 Miklavž na Dravskem polju, Slovenia

Introduction

The Natura 2000 network is the most important conservation effort being implemented Europe-wide. Its main aim is to conserve an extensive range of habitat types and wildlife species throughout Europe, maintaining listed habitat and species at "favorable conservation status" (European Commission 2000). It is also an important tool for improvement of existing networks of conservation areas in EU and the main strategy of the EU to halt, or at least significantly reduce, biodiversity loss by 2010 (Balmford et al. 2005). In Slovenia the implementation of the legislation associated with Natura 2000 network has vastly improved the finances for the nature conservation. Also it shifted from basic scientific research to the management of protected areas. However, the biggest leap forward was the rise in the quantity of the conservation areas, which were before limited to small forest reserves and a single national park covering only 4% of the Slovenian territory. Currently there are 27 Special Protection Areas (SPA's) designated for protection of threatened bird species with total coverage of 23.0% of the land. Additionally, there are 259 designated Special Areas of Conservation (SAC's) protecting threatened habitats and species other than birds covering 31.4% of the Slovenian territory, both being approved by the Commission in 2008 (European Commission 2009). Due to wide overlap of the SAC's and SPA's the total coverage of the current Natura 2000 network is now approximately 36%. The current network is based on available historical distribution data of qualifying species and habitats and limited mapping of current distribution for some selected groups of taxa. Following from the outcomes of the biogeographical seminars, a meeting of government nature conservation agencies and nongovernmental organisations, where sufficiency of coverage of the current SAC's network of the threatened species and habitats range was evaluated, some additions to the Natura 2000 network is required (Zagmajster 2005; Zagmajster and Skaberne 2006). With expected additions, Slovenia will be firmly at the top of the EU member states regarding the percent of land cover of the Natura 2000 network. The effectiveness of such a large network will however remain to be tested by the establishment of regular monitoring.

In order to test how well does the current network of SAC's cover the areas of high biodiversity the distribution data for butterflies collated during last three decades in Slovenia were used. Butterflies are an important component of biodiversity, and are highly sensitive to environmental changes (Maes and Van Dyck 2001; Roy et al. 2001; Franco et al. 2006; van Swaay et al. 2006). Therefore, they are useful indicators of the health of the environment (Thomas and Clarke 2004; Thomas 2005) and can thus be considered as suitable bioindicators of biodiversity (Fleishman et al. 2005; Thomas 2005). Slovenia has a rich butterfly fauna owing to its geographic position at the junction of the Alpine, Mediterranean and Pannonian regions. Around 180 butterfly species are known from Slovenia with historical presence of some additional species still under debate (Carnelutti 1992). Nine species are listed in Annex II of the Habitats directive requiring establishment of SAC's. Among these *Nymphalis vaualbum* (Dennis & Schiffermüller, 1775) with no records in the last 60 years and *Colias myrmidone* (Esper, 1781) with no records despite deliberate surveys in the last 17 years (Predovnik and Verovnik 2004) could be considered regionally extinct. Although available, butterfly distribution data were only marginally used to establish SPA network in Slovenia. Therefore, it seems reasonable to state that the butterfly distribution data were collated independently from the current Natura 2000 network and in most cases the records predate its establishment.

The main aims of our research are: (1) to identify areas of high butterfly diversity in Slovenia; (2) to evaluate the efficiency of the current SAC's network in conservation of butterfly diversity, (3) the coverage of the current SAC's network of the distribution of the butterfly species of conservation concern, and (4) to compare the importance of small versus large SAC's for butterfly diversity conservation.

Methods

With an area of 20,273 square kilometers, Slovenia is among the smallest European countries. However, its position on junction of different biogeographical zones, mountainous relief preventing large scale intensive farming and large forest cover make it one of the centers of biodiversity in Europe. Climatic factors are equally complex with Alpine climate dominating the northern and central part of the country, continental climate in the northwest and Mediterranean climate along the coast and Vipava valley. The main geographic features of Slovenia are the south eastern part of the Alps connected in a wide belt with Dinaric Alps covering most of the western and southern part of the country. The only flatlands of Slovenia are alluvial plans along major rivers Sava, Drava, and Mura (Fig. 1).

The butterfly database maintained by the Centre for Cartography of Fauna and Flora collated mostly by the members of the Society for the Conservation and Study of Lepidoptera in Slovenia was used for present research. Data were collected from literature sources, national and private collections, but mostly by field work using standard datasheet forms for field recording. All the records were georefferenced at the level sufficient to asign them to the 5×5 km grid, less precise data were discarded. In order to identify gaps in data coverage a 5×5 km grid representing ¼ of the UTM grid squares was used. The total count of 5×5 km grid squares covering Slovenia is 873. The records were mapped on the grid and squares or clusters of squares with no or insufficient records were identified and appointed to volunteers for surveys. To ensure wide coverage of the survey with limited number of volunteers available, a minimum of 15 butterfly species recorded per square was required to consider it sufficiently studied. This meant that in most cases only the potentially best butterfly habitats were surveyed, so the recording within a certain square was not random. The data collection was also biased towards oversampling in regions with known high butterfly diversity and areas in proximity of the residence of the volunteers. To limit that bias the standard Jackknife1 estimate of species richness (Heltshe and Forrester 1983) was calculated for each square with sites considered as sampling units. Within each square all the species recorded over all sites were counted and singletons were the species present only on a single site.

In order to identify concentrations of butterfly species of conservation importance a weighting scheme similar to the one used by Early and Thomas (2007) was used with addition of the nationally important species (see appendix I in Electronic Supplementary Material). The main rationale of this method is to give higher value to the species of higher conservation priority. We assigned a weight of 1 to

Fig. 1 Map of SAC's network in Slovenia with main toponyms mentioned in the text

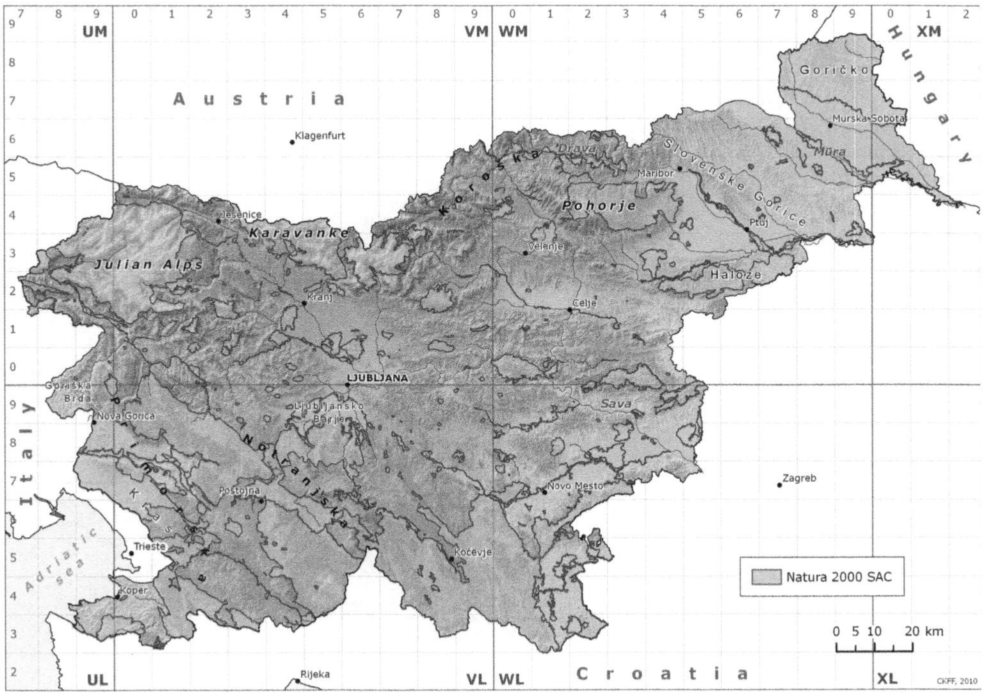

species considered near threatened, a weight of 3 to species classified as vulnerable, and a weight of 5 to species classified as endangered in both the European red list of butterflies (van Swaay et al. 2010) and the national red list (Uradni list RS 2002). Additionally the species endemic to Europe were assigned weight 1 regardless of the threat status.

To check how does the current SAC's network cover the distribution of the species of conservation concern conservation value of each square was calculated by adding up the values of the species present in the square. Values of all squares with records within SAC's were than added up and compared to sums gained from random positioning of the SAC's that was generated 1000×. The actual value was than compared to null distribution by single distribution Student's *t*-test.

To test the importance of small versus large SAC's for the butterfly diversity conservation the number of observed species was plotted against the size of the SAC's. Only those SAC's with more than 15 recorded species were included as they were considered sufficiently surveyed. Simple Pearson correlation coefficient between log transformed size of the SAC's and species recorded was calculated. All statistics were performed with Microsoft Office Excel 2003.

Results

The butterfly database includes 135,766 records, 22.1% literature records and 73.2% records collated in the last 30 years, which were used for this analysis. The vast majority of the squares had targeted surveys in the last three decades with only 0.6% of the squares remaining with no records and 2.8% squares with no observations in the last 30 years. The proportion of records from last three decades within SAC's is 46.9%. Butterflies were recorded at 6,061 sites with 2,718 (44.8%) localities within SAC's and 3,343 (55.2%) outside them.

Coverage of butterfly diversity

As expected the areas of highest butterfly diversity are concentrated in SW part of the country (Fig. 2), where the mixture of Mediterranean, Alpine and Continental fauna is most apparent. Other centers of butterfly diversity are scattered throughout the country with notable concentrations in lower part of the Sava valley, NE part of the Alps in Koroška region, Haloze, Slovenske Gorice and Goričko, the last three in NE Slovenia. The Julian Alps and Karavanke Mts. despite large altitudinal spans in some of the squares and the presence of rare mountainous species did not figure among the high biodiversity regions.

The current network of SAC's does well match the high butterfly diversity areas in Slovenia covering 72.7% of the top 5% high biodiversity squares (Fig. 2). The only noticeable exceptions are the Goriška brda region in northern Primorska region (W Slovenia), lower part of the Sava valley (E Slovenia), and Koroška region (N Slovenia). These three regions are characterized by abundance of seminatural grasslands which host high butterfly diversity. At species level all but one of the species recorded in last

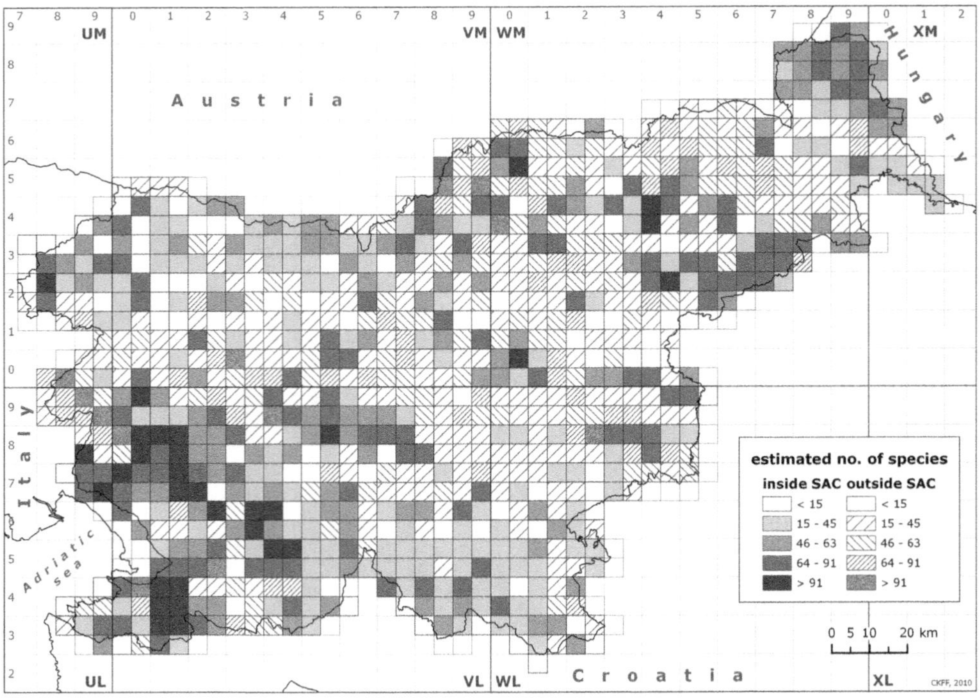

Fig. 2 Distribution of high butterfly diversity areas in Slovenia. *Squares* with records within SAC's are *colored* in *different shades* of *grey*, those with records outside SAC's are marked with *line pattern* of different densities

30 years were found in SAC's, which additionally illustrates a high SAC's total coverage of the butterfly diversity in Slovenia. The only species not found is the Habitats directive listed species *Colias myrmidone*, for which there are only few records available after 1979 and is now considered regionally extinct in Slovenia.

Coverage of the species of conservation concern

The SAC's network does well correspond to the distribution of the butterfly species of conservation concern with 63% of the total conservation value distributed in squares with SAC's. This value was statistically significantly different from random distribution ($P = 0{,}001$).

The concentrations of the butterfly species of conservation concern closely follow the distribution of the general high butterfly diversity areas (Fig. 3). The difference is only evident in areas with the presence of specialists of humid grasslands, which are among the most severely threatened butterfly species in Slovenia, some also at the European level. The most prominent example is the Ljubljansko Barje region in central Slovenia, most of which is covered by one large SAC. Among the areas with concentrations of the threatened species not covered by SAC's again Goriška brda, Koroška region, and additionally Slovenske Gorice (NE Slovenia) are the most obvious exceptions.

Size of SAC's and diversity

Among 259 SAC's in Slovenia only one is larger than 100,000 ha and 126 of them smaller than 100 ha. Many of these small SAC's represent cave entrances with a buffer zone for breeding colonies of bats, special habitat type (bogs) or streams for crayfish. Therefore, it is no surprise that 135 SAC's have no butterfly records and additionally 35 SAC's have only sporadic records (<15) and were not included in comparison of size and butterfly species counts. There is an obvious positive correlation between the size and the number of species recorded in SAC's confirmed also by Pearson correlation coefficient ($R = 0.75$). However, the butterfly diversity in small SAC's is comparatively large if one considers their size (Fig. 4). Some of the most prominent outliers are: Žejna dolina—60 species on 55 ha, Vipava Valley—78 species on 1,460 ha, Haloze—94 species on 6,300 ha, and Notranjska—125 species on 15,200 ha. Kras SAC with 47,490 ha has the highest species count of 137 butterfly species.

Discussion

Currently there are only few studies available on effectiveness of the Natura 2000 network in reaching its main objective of halting the erosion of biodiversity in Europe, two on plants (Dimitrakipoulos et al. 2004; Chiarucci et al. 2008) and one on vertebrates (Maiorano et al. 2007). This study thus represents the first evaluation of the SAC's conservation value for a group of insects, which in general represent the largest portion of biodiversity in any given area of terrestrial habitats. The establishment of the SAC network was certainly the most important step forward in nature conservation in Slovenia with an almost tenfold

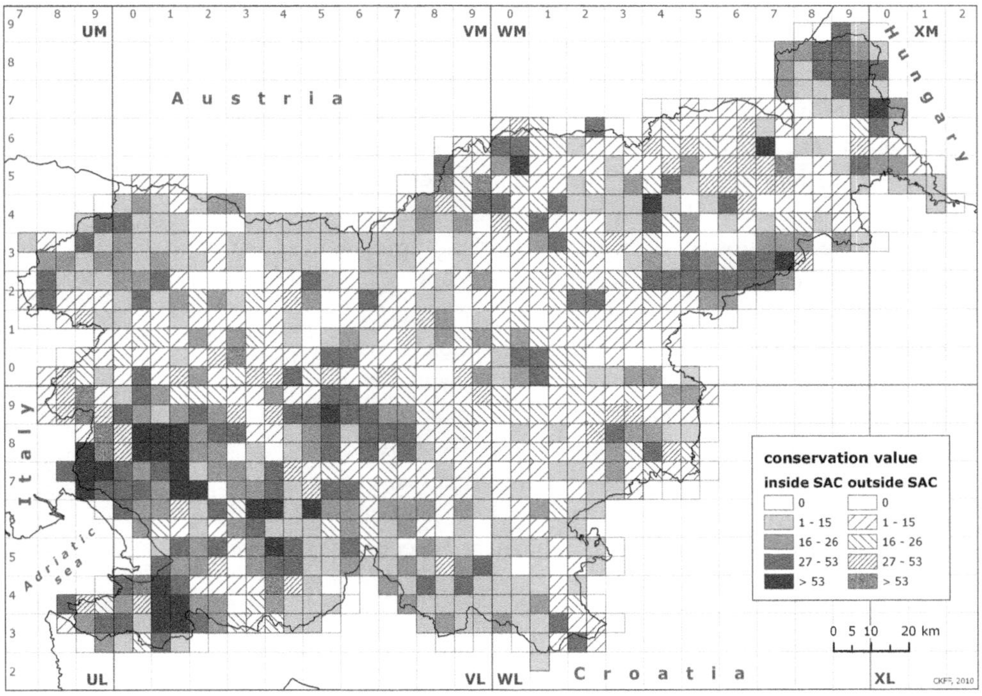

Fig. 3 Conservation value species distribution in Slovenia (see the "Methods" for the weighting system). *Squares* with records within SAC's are colored in *different shades* of *grey*, those with records outside SAC's are marked with *line pattern* of different densities

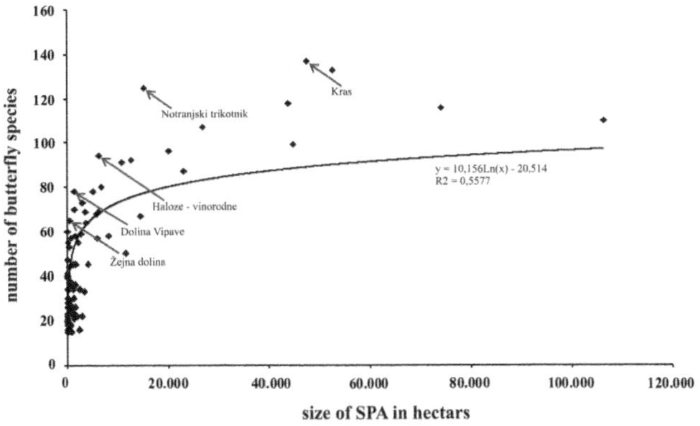

Fig. 4 The relation between number of species in a given SAC and its size. Outliers are named according to the list of accepted SAC's in Slovenia

increase of the size of the protected areas. However, it is important to note that several large SAC's were established only for protection of large carnivores with large proportion of land covered by closed canopy forests. Therefore, it is not straightforward to presume that these areas are suitable also for conservation of other target species, in particular grassland butterflies. As we have shown the small SAC's can host high butterfly diversity (Fig. 4) and could play an important role as source populations of many species in the wider countryside. However, the SAC's should be large enough to host viable metapopulations of threatened species, to ensure their long term survivor.

Butterflies are a well accepted biodiversity indicator group (Thomas 2005) and are among the best studied invertebrates in Slovenia. Although the number and density of records is not comparable with the Netherlands (Bos et al. 2006) or Britain (Asher et al. 2001), with 6.7 records per square kilometer, it is still higher than in most other atlases published in recent years. Due to targeted surveys in the last decade the number of squares with no records represents only a negligible 0.6% of the 873 squares. The surveys were not entirely systematic as some remote areas were visited only once while some were repeatedly visited due to easier access or known high butterfly diversity. Therefore, the distribution of the butterfly diversity hot spots (Fig. 2) is likely to be slightly different with additional recordings in under surveyed squares, but certainly the vast majority of the high butterfly diversity areas were already identified.

It is surprising how well the current SAC network covers the high butterfly diversity areas and those with concentration of the butterfly species of conservation concern in Slovenia (Figs. 2, 3), especially if one takes into consideration the lack of input from the butterfly distribution data available in establishment of this network. This is well emphasized by the results of the biogeographical seminars where all qualifying butterfly species were given the status of 'insufficient moderate' for Slovenia (Zagmajster 2005; Zagmajster and Skaberne 2006) requiring inclusion of further populations into SAC network. Our comparison shows that some areas with high conservation value for butterflies, in particularly Koroška region, Goriška brda, Slovenske Gorice, and lower part of the Sava valley, were not included in the current SAC network. Fortunately in the additions of the SAC network, that are currently under evaluation, all four mentioned areas are now partially covered with one large or several small proposed SAC's. With these additions approved we could consider the network of SAC's in Slovenia well suited for the long term conservation of butterfly diversity and to halt the decline of the species of conservation concern. The SAC's network in Slovenia has all the foundations for reaching the main objective of halting the biodiversity loss, however more deliberate action will need to follow. Currently almost no species protection oriented management is present in the SAC's and the monitoring has only been established for the butterflies and some other groups. In general, keeping traditional small scale farming with low intensity grazing, haying, and minute use of pesticides would be ideal management for majority of the target butterfly species. Hopefully in 10 or 20 years we will be able to report positive effects of our large SAC network on maintaining the high butterfly diversity and reversing the declines of threatened species.

Conclusions

High butterfly diversity areas are concentrated in Primorska region and in low intensity farmland areas like lower part of the Sava valley, Koroška region, Haloze, Slovenske Gorice and Goričko. Most of these areas as well as the areas with concentrations of the species of the conservation concern are covered by the existing SAC network. The major gaps in the coverage will be removed by the planned extensions following the results of the biogeographical seminars. Small SAC's with high butterfly diversity play an important role in sustaining high butterfly diversity in Slovenia.

Acknowledgments The authors would wish to thank all the volunteers for their attentive field work and data provided from their private collections. We are especially grateful to Chris van Swaay on the comments and improvements of the early version of the manuscript. The Centre for Cartography of Fauna and Flora provided all necessary resources for digitalization and maintenance of the records database.

References

Asher J, Warren M, Fox R, Harding P, Jeffcoate G, Jeffcoate S (2001) The millenium atlas of butterflies in Britain and Ireland. Oxford University Press, Oxford

Balmford A et al (2005) The convention on biological diversity's 2010 target. Science 307:212–213

Bos F, Bosveld M, Groenendijk D, van Swaay C, Wynhoff I (2006) De dagvlinders van Nederland: verspreiding en bescherming (Lepidoptera: Hesperoidea, Papilionoidea). Nationaal Natuurhistorisch Museum Naturalis, Leiden, KNNV Uitgeverij, Urecht, European Invertebrate Survey—Nederland, Leiden

Carnelutti J (1992) Red list of Lepidoptera (Macrolepidoptera) in Slovenia (in Slovene). Varstvo narave 17:61–104

Chiarucci A, Bacaro G, Rocchini D (2008) Quantifying plant species diversity in a Natura 2000 network: old ideas and new proposals. Biol Conserv 141:2608–2618

European Commission (EU) (2000) Managing Natura 2000 sites. The provisions of article 6 of the 'Habitats' Directive 92/43/CEE. EU, Luxembourg

Dimitrakipoulos PG, Memtsas S, Troumbis AY (2004) Questioning the effectiveness of the Natura 2000 special areas of conservation strategy: the case of Crete. Global Ecol Biogeogr 13:199–207

Early R, Thomas CD (2007) Multispecies conservation planning: identifying landscapes for the conservation of viable populations using local and continental species priorities. J Appl Ecol 44: 253–262

European Commission (EU) (2009) Natura 2000 barometer. EU, Luxembourg [http://ec.europa.eu/environment/nature/natura 2000/barometer/index.en.htm]. Accessed Jan 2010

Fleishman E, Thomson JR, Mac Nally R, Murphy DD, Fay JP (2005) Using indicator species to predict species richness of multiple taxonomic groups. Conserv Biol 19:1125–1137

Franco AMA, Hill JK, Kitschke C, Collingham YC, Roy DB, Fox R, Huntley B, Thomas CD (2006) Impacts of climate warming and habitat loss on extinctions at species' low-latitude range boundaries. Glob Change Biol 12:1545–1553

Heltshe J, Forrester NE (1983) Estimating species richness using the jackknife procedure. Biometrics 39:1–11

Maes D, Van Dyck H (2001) Butterfly diversity loss in Flanders (north Belgium): Europe's worst case scenario? Biol Conserv 99:263–276

Maiorano L, Falcucci A, Garton EO, Boitani L (2007) Contribution of the Natura 2000 network to biodiversity conservation in Italy. Conserv Biol 21:1433–1444

Predovnik Ž, Verovnik R (2004) New records of rare pierids (Lepidoptera: Pieridae) in Slovenia. Nat Slov 6(2):39–47

Roy DB, Rothery P, Moss D, Pollard E, Thomas JA (2001) Butterfly numbers and weather: predicting historical trends in abundance and the future effects of climate change. J Anim Ecol 70: 201–217

Thomas JA (2005) Monitoring change in the abundance and distribution of insects using butterflies and other indicator groups. Philos Trans Roy Soc B Biol Sci 360:339–357

Thomas JA, Clarke RT (2004) Extinction rates and butterflies—response. Science 305:1563–1564

Uradni list RS (2002) Regulations on the classification of endangered plant and animal species in the Red list (in Slovene). Ur L Rep Slov 82:8893–8975

van Swaay C, Warren M, Lois G (2006) Biotope use and trends of European butterflies. J Insect Conserv 10:189–209

van Swaay C, Cuttelod A, Collins S, Maes D, Munguira ML, Šašić M, Settele J, Verovnik R, Verstrael T, Warren M, Wiemers M, Wynhoff I (2010) European red list of butterflies. EU, Luxembourg

Zagmajster M (2005) Overview of the final decisions of the Biogeographical seminars—Alpine region (in Slovene). Kranjska Gora (version 7.6.2005)

Zagmajster M, Skaberne B (2006) Overview of the final decisions of the Biogeographical seminars—continental region (in Slovene). Darova (CZ) (version 28.5.2006)

J Insect Conserv (2011) 15:361–365
DOI 10.1007/s10841-010-9354-7

SHORT COMMUNICATION

Status and conservation of *Asclepiadaceae* and *Danaus* in southern Spain

Juan Fernández Haeger · Diego Jordano · Mateo León Meléndez

Received: 23 March 2010 / Accepted: 9 September 2010 / Published online: 22 September 2010

Abstract A 900 km^2 coastal area in southern Spain was searched for both *Danaus plexippus* and *D. chrysippus* and their foodplants. A set of 63 single or mixed patches of *Asclepias curassavica* and *Gomphocarpus fruticosus* was detected and visited regularly from June 2008 to December 2009. *Danaus plexippus* was on the wing throughout our study period, but it was never detected in 21 patches while in 4 patches it was observed on more than 80% of our visits. *Danaus chrysippus* was on the wing only during summer and autumn of 2009 and it was never detected in 49 patches. The proportion of plant patches occupied by butterflies varies seasonally, increasing during summer and autumn. These results are discussed in relation to the origin, persistence and conservation of both species in Europe.

Keywords *Danaus plexippus* · *Danaus chrysippus* · *Gomphocarpus fruticosus* · *Asclepias curassavica* · Spain · Patchy distribution · Conservation

Introduction

Habitats are being increasingly fragmented by human activities. The result of this process is an ensemble of habitat units or patches isolated across landscapes. This fragmentation has been frequently identified as one of the main factors leading to the extinction of species (Wilcox and Murphy 1985; Thomas 2000; Debinsky and Holt 2000).

J. F. Haeger (✉) · D. Jordano · M. L. Meléndez
Área de Ecología, Universidad de Córdoba,
14071 Córdoba, Spain
e-mail: bv1fehaj@uco.es

Species persistence will depend on the quality and size of the remaining patches, their distribution in, and nature of, the matrix or intervening non-habitat, and the frequency at which they are created or destroyed, combined with movements of the focal species amongst them (Zalucki and Lammers 2010). Human intervention in an area may also promote the presence of some species which grow preferentially on these disturbed habitats. For these species their persistence depends on the intensity and quality of the anthropogenic management of the habitat. In both cases, the resulting ensemble of habitat units is usually described as a metapopulation for the focal species. The persistence of a species in this kind of landscape depends on a combination of dynamics between remnant habitat patches and dynamics across such patches (Hanski 1989; Hanski and Thomas 1994; Dover and Settele 2009).

Butterflies have frequently been used as model systems when exploring population persistence and landscape structure. Their habitat requirements are relatively well known and include: foodplants for oviposition, microclimatic conditions for immature stages and for thermoregulation for adults, mating sites and nectar sources for adults, hibernation sites, escape from predators, etc. (Dover et al. 1997; Walter and Zalucki 1999; Shreeve et al. 2001; Dennis et al. 2003; Zalucki and Lammers 2010).

In southern Spain, *Danaus plexippus* (*L.*) (Lepidoptera: Danaidae) is a nonmigratory, non-diapausing species, frequently cited by local entomologists since the sixties (see Fernández Haeger and Jordano 2009 and references therein). In most cases its presence in Spain has been attributed to immigrant individuals arriving from North America or Atlantic archipelagos, colonising patches of its foodplants and establish ephemeral colonies. *Danaus chrysippus* (L., 1758) has been also frequently observed in southern Spain (Fernández Haeger 1999) and along the

Mediterranean coast to Turkey (van der Heyden 2009) occasionally extending to inland areas. For this species, immigrants are believed to come from northern Africa. Both species are frequently found in the same patches, ovipositing on the same foodplants and maintaining ephemeral local colonies. The main foodplants for both species in our study area are *Gomphocarpus fruticosus* and *Asclepias curassavica* (Fernández Haeger 1999). *Cynanchum acutum*, a native species that has been cited as alternative host plant in Granada and Almeria (Gil-T 2006), is rarely found in our study area.

Asclepiadaceae (milkweeds) are very rich nectar producing species (Wyatt and Broyles 1994). During the summer drought most Mediterranean pastureland species dry off and the flowering period for shrub species is over, whereas both *A. curassavica* and *G. fruticosus* are in full bloom. Therefore, those patches not only contain milkweed foodplant for oviposition and immature development, but are also the main adult nectar sources and both *Danaus* species may aggregate around them. Human management has created a landscape in which patches of milkweeds are associated with farmlands with high herbivore pressure around damp locations, but the area in between is almost empty, a situation analogous to a classic metapopulation (Zalucki and Lammers 2010). Both *Danaus plexippus* and *D. chrysippus* are very vagile and can move long distances between patches of various sizes.

Our objective here is to address the status of both species in an area of southern Spain where the butterflies have been frequently detected. We regularly sampled a set of foodplant patches to analyse patch use, butterfly abundance and persistence within and between years. These data provide new insights for the understanding of the origin of these species in Europe, and their persistence under current management practices.

Materials and methods

During 2008 and 2009 we studied an approximately 900 km^2 coastal area in the proximity of the Strait of Gibraltar (southern Spain) by traversing all available roads and forest tracks for patches of milkweed. Central point coordinates of each plant patch were taken with a Garmin GPS with a precision of $\pm$3 m. These data were used to establish the geographical location of the patches using Arc GIS. Each patch was visited regularly from June 2008 until December 2009 and any new patches added. Total number of visits to all patches detected reached 1,067. On each visit, butterflies were counted by standing and counting for 10 min. For 10–15 min, plants were sampled to detect eggs, larvae or pupae. The presence of immature stages was used as an indicator that those patches had been used by butterflies, although adults may have not been present during our visit.

Results

A total of 63 plant patches were located in the study area (Fig. 1). The maximum distance between patches was 49 km. In 30 patches the only species found was *G. fruticosus* and in 13 only *A. curassavica*, with both in 20 patches. Thus *G.fruticosus* was present in 50 patches and *A. curassavica* in 33. Plant numbers in patches varied from a single individual up to about 4,500 plants.

Considering all patches together, *D. plexippus* was on the wing throughout our study period (June 2008–December 2009). *D. chrysippus* appeared in the patches only during the summer and autumn of 2009. Patch use by both species was very variable. Of the 63 patches studied, 21 were never occupied (we did not find adults or immatures) and 35 were occupied on up 20% of our visits. *D. plexippus* was detected on more than 80% of the visits in only four patches (Fig. 2). *D. chrysippus* was detected in 14 patches (absent in 49) and only during the summer and autumn of 2009.

The proportion of patches used by *D. plexippus* increased throughout the seasons in both years (Fig. 3). During 2009, in which the greatest number patches were visited, the number of patches utilised by increased from 16% during the winter, to 49% during autumn. This result suggests that the butterfly expands patch occupancy during the most favourable months and remained in a few favourable patches during the winter, expanding again during the next year. The proportion of patches occupied by *D. chrysippus* doubled from 8 to 17% from summer to autumn during 2009.

Discussion

To our knowledge, no equivalent work has been undertaken with these butterfly species in Europe. Our data provides a different point of view to the one proposed previously, namely that *D. plexippus* individuals detected in ephemeral colonies along coastal areas of Spain and Portugal are the result of immigrant butterflies arriving to continental Europe from Atlantic archipelagos (Canary Islands, Madeira, Azores) or even from North America (Fernández-Vidal 2002; Maravalhas 2003). Although this possibility cannot be totally excluded and these kind of arrivals are periodically registered in Ireland and the United Kingdom (Asher et al. 2001) or northern Spain (Fernández-Vidal 2002), our data suggests that successful colonisation in southern Spain could have happened much

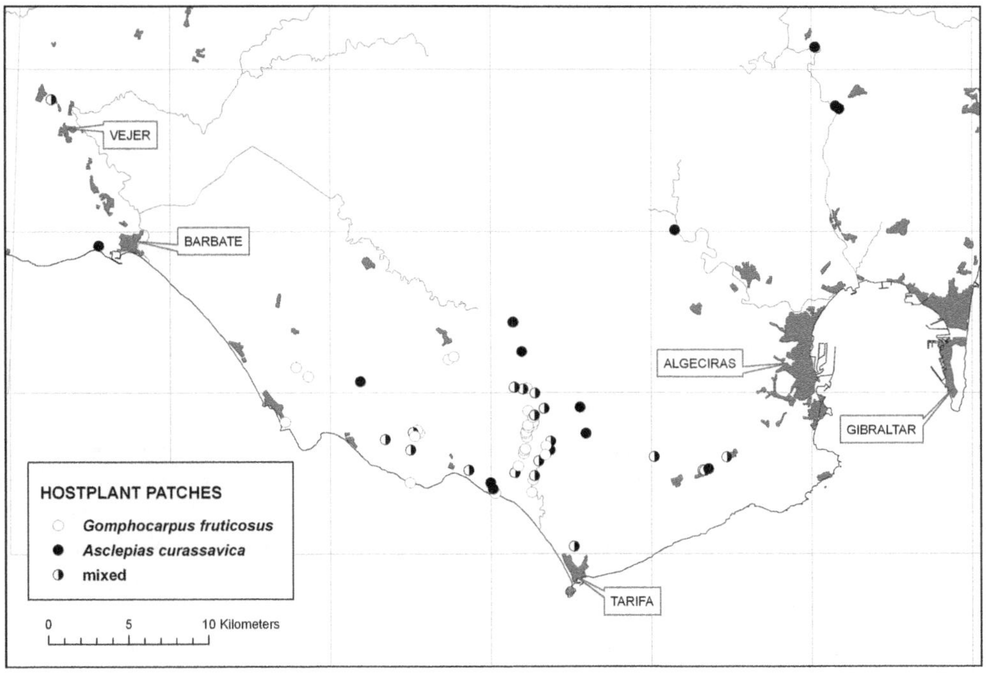

Fig. 1 Plant patches location ($n = 63$) in the study area around the Gibraltar strait (southern Spain)

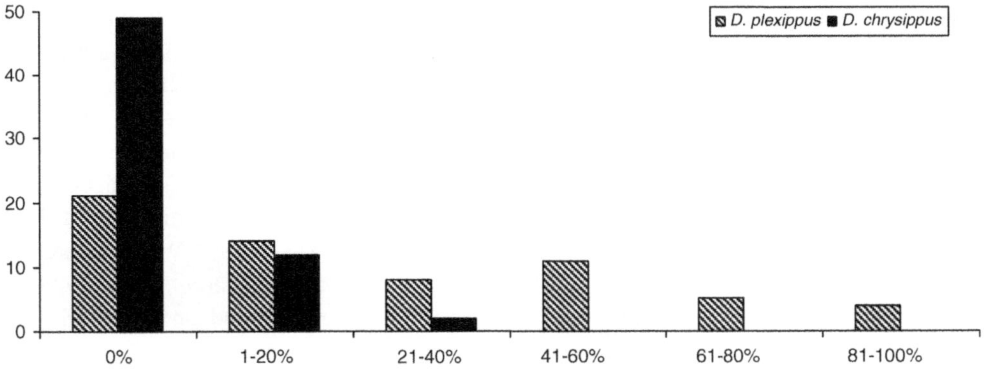

Fig. 2 Frequency distribution of patch occupancy for both species based on observations of adults and immatures in each patch

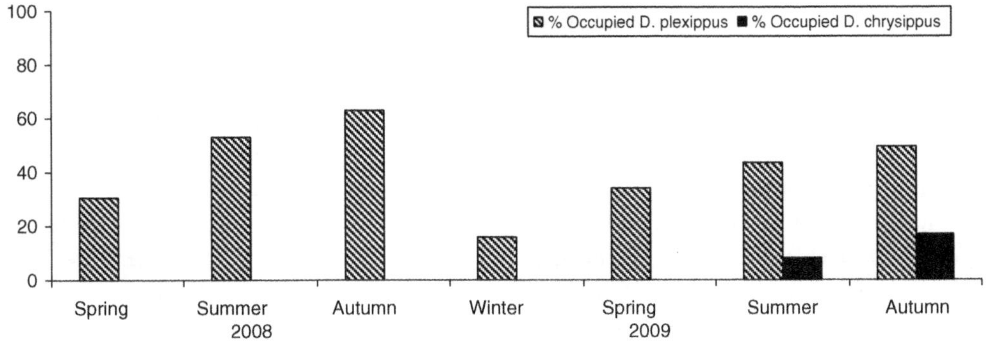

Fig. 3 Proportion of patches in which both species of butterfly were detected. In both years, the proportion of patches occupied increased throughout the year, as a consequence of dispersal during the favorable season

earlier. In fact, *D. plexippus* was registered at Gibraltar in the nineteenth century (Walker 1886). During this earler period *D. plexippus* expanded through the Pacific Ocean, arriving in Australia (Vane-Wright 1993; Zalucki and Clarke 2004) and was detected in the above mentioned archipelagos. Elderly local people in Spain to whom pictures of the butterflies have been shown, remembered that this butterfly was very common 50 years ago, before they were recorded by entomologists.

For a successful expansion of a butterfly three basic requirements are necessary: opportunities for transport or flying potential, favourable climatic conditions in the area of arrival and foodplants for immature development. In Asher et al. (2001) frequent arrivals of *D. plexippus,* beginning in 1876, are recorded in the British Islands. Similar arrivals could have happened in southern Spain, but were not formally recorded until the 1960s except by Walker (1886) in Gibraltar. In this area climatic conditions are appropriate (Zalucki and Rochester 1999) and foodplants were introduced some centuries before (Quer 1762) which subsequently expanded into the managed landscape. *D. plexippus* has been consistently recorded in the study area since 1997 (Fernández Haeger and Jordano 2009). Our data suggests that *D. plexippus* is a permanent resident in the area and colonizes a variable number of patches and that the number of patches used by the butterfly varies seasonally and between years. In a few patches adults and/or immature stages are present during the whole year, but in others they occur sporadically or have never been detected. We suggest that a metapopulation has probably persisted in the area since the ninetieth century.

The status of D. chrysippus is different. It also frequents the area (it was registered in 1998, 1999, 2000, 2007 and 2009) and has similar requirements to *D. plexippus,* but its density and patch use is lower. During our study (June 2008–December 2009) it was only detected during the last summer and autumn in one year. Butterflies were in a very good condition, suggesting that they emerged in the same area and they were not immigrants from northern Africa. Notwithstanding this, movements across the Gibraltar strait (14 km wide at its narrowest point) cannot be excluded. It is a very vagile species and has expanded through the Mediterranean to Turkey (van der Heyden 2009). During the autumn of 2009 it was also detected in some inland localities (Seville and Madrid).

During the summer drought, almost no grasses and ephemeral plants persist on the ground except both species of *Asclepiadaceae*, which are well defended by their cardenolide content (Agrawal et al. 2008; Agrawal and Konno 2009). Chemically well defended foodplants,with a long flowering period and high nectar production are an ideal resource for an adapted host specialist. Both *Danaus* species are concentrated on the patches where they find the essential resources they need for surviving; reproducing: foodplants for oviposition and development of immature stages, mating opportunities and nectar sources.

For plant patches to persist a high herbivore pressure on less well defended plants seems to be necessary. Along river-banks and irrigation ditches other hygrophitic plants (e.g. *Rubus ulmifolius, Scirpus holoschoenus*) are competitors of both species of milkweed. If the density of ungulates decreases, those species can drive milkweeds to local extinction. Eventual clearing of those species is followed by regrowth of both milkweeds.

Our data suggests that *D. plexippus* is maintained by a low number of high quality patches from which butterflies disperse to other patches. During our study period two of those high quality patches (in which butterflies were present in more than 80% of visits) were the ones in which density was remarkably high: more than 100 individuals on the wing were counted in each during the summer of 2008. This high density of adults resulted in an extremely high number of larvae, which completely defoliated the plants (only *A. curassavica* was present in these patches) and, as a consequence, high mortality by starvation of the developing larvae was observed. Most of the adult butterflies abandoned the patch after defoliation, probably in search of higher quality patches.

Since the work of Granados Corona and García Novo (1983) on the status of *G. fruticosus* in the National Park of Doñana, both species of milkweed are included on the checklist of invasive plants of Southern Spain (Dana et al. 2005). Therefore, in the "Natural Park of the Alcornocales" which was part of our study area, control of these plants is occasionally undertaken and one of the biggest mixed patches of *G. fruticosus* and *A. curassavica* was cleared in 2007. In this patch *D. plexippus* was only sporadically seen, but we registered up to 45 *D. chrysippus* flying during the summer of 2009. In the National Park of Doñana (150 km to the NW of the area) both plants have been systematically uprooted. In 1983 the monarch butterfly was not included on the checklist of this National Park, but both species of *Danaus* have been detected in past years. At least one flourishing colony was eradicated in 2004 (Fernández Haeger and Jordano 2009). Nevertheless the total extinction of plants is not easy. Patches recovered in a few months after being cleared, because both species resprout easily from roots, from seed already in the soil and the arrival of seeds from surrounding patches might be frequent and germination rates of seeds are high (unpublished data). Therefore, if herbivore density and water availability does not change, recovery of patches occurs in a short period of time. In any case, there is a conflict between the conservation of these specialist butterflies and the eradication of their foodplants considered as invasive species.

Acknowledgments We wish to thank to the "Fundación Migres" for providing financial support for this research. Charo Rivas and Carlos Camacho helped in searching for patches and counting plants and butterflies and their immature stages in the field. We wish to thank M. Zalucki for commenting on an earlier draft of the manuscript.

References

Agrawal AA, Konno K (2009) Latex: a model for understanding mechanisms, ecology and evolution of plant defense against herbivory. Annu Rev Ecol Evol S 40:311–331

Agrawal AA, Lajeunesse MJ, Fishbein M (2008) Evolution of latex and its constituent defensive chemistry in milkweeds (*Asclepias*): a phylogenetic test of plant defense escalation. Entomol Exp Appl 128:126–138

Asher J, Warren M, Fox R, Harding P, Jeffcoate G, Jeffcoate S (2001) The millennium atlas of butterflies in Britain and Ireland. Oxford Univ Press, Oxford

Dana ED, Sanz M, Vivas S, Sobrino E (2005) Especies vegetales invasoras de Andalucía. Dirección General de la RENPA. C.M.A. Junta de Andalucía

Debinsky DM, Holt RD (2000) A survey and overview of habitat fragmentation experiments. Conserv Biol 14(2):255–342

Dennis RLH, Shreeve TG, Van Dick H (2003) Towards a functional resource-based concept for habitat: a butterfly biology point of view. Oikos 102:417–426

Dover J, Settele J (2009) The influence of landscape structure on butterfly distribution and movement: a review. J Insect Conserv 13:3–27

Dover JW, Sparks TH, Greatorex-Davis JN (1997) The importance of shelter for butterflies in open landscapes. J Insect Conserv 1:89–97

Fernández-Haeger J (1758) La mariposa monarca *Danaus plexippus* (L. 1758) en el estrecho de Gibraltar (*Lepidoptera: Danaidae*). Shilap Rev Lepidopt 37(148):421–438

Fernández-Haeger J (1999) *Danaus chrysippus* (Linnaeus, 1758) en la Península Ibérica: ¿migraciones o dinámica de metapoblaciones? (*Lepidoptera: Nymphalidae, Danainae*). Shilap Rev Lepidopt 27(107):423–430

Fernández-Vidal EH (2002) La *Danaus plexippus* (Linnaeus, 1758) en Galicia (España). Nuevos datos y noticias sobre la "Operación Monarca" (Lepidoptera: Danaidae). Bol S.E.A. 31: 243–246

Gil-T F (2006) A new hostplant for *Danaus plexippus* (Linnaeus, 1758) in Europe. A study of cryptic preimaginal polymorphism within *Danaus chrysippus* (Linnaeus, 1758) in southern Spain (Andalusia). Atalanta 37(1/2):143–149

Granados Corona M, García Novo F (1983) Ensayo de control de la mata de la seda *Gomphocarpus fruticosus* (L.) en el Parque Nacional de Doñana. Bol Est Ecol 15(29):73–80

Hanski I (1989) Metapopulation dynamics: does it help to have more on the same? Trends Ecol Evol 4:113–114

Hanski I, Thomas CD (1994) Metapopulation dynamics and conservation: a spatially explicit model applied to butterflies. Biol Conserv 68:167–180

Maravalhas E (2003) As borboletas de Portugal. Multipunto S.A, Oporto

Quer J (1762) Flora española, vol 3. Imprenta Joachim Ibarra, Madrid, pp 74–77

Shreeve TG, Dennis RLH, Roy DB, Moss D (2001) An ecological classification of British butterflies: ecological attributes and biotope occupancy. J Insect Conserv 5:145–161

Thomas CD (2000) Dispersal and extinction in fragmented landscapes. Proc R Soc Lond B 267:139–145

Van der Heyden T (2009) Bemerkungen zur Biologie, zur Ökologie und Verbreitung von Danaus chrysippus Linnaeus, 1758 im Mittelmeerraum, insbesondere in der Türkey (Lepidoptera: Nymphalidae, Danainae). Nachr Entomol Ver Apollo N.F. 30(3):173–176

Vane-Wright RI (1993) The Columbus hypothesis: an explanation for the dramatic 19th century range expansion of the monarch butterfly. In: Malcolm SB, Zalucki MP (eds) Biology and conservation of the monarch butterfly. Natural History Museum, Los Angeles, pp 179–187

Walker JJ (1886) *Anosia plexippus* L. (*Danaus plexippus*, F.) at Gibraltar. Entomologists Monthly Magazine, 23:162

Walter GH, Zalucki MP (1999) Rare butterflies and theories of evolution and ecology. In: Kitching et al (eds) Biology of Australian butterflies. CSIRO Publishing, Melbourne, pp 349–368

Wilcox BA, Murphy DD (1985) Conservation strategy: the effects of fragmentation on extinction. Am Nat 125(6):879–887

Wyatt R, Broyles SB (1994) Ecology and evolution of reproduction in milkweeds. Annu Rev Ecol Syst 25:423–441

Zalucki MP, Rochester WA (1999) Estimating the effect of climate on the distribution and abundance of the monarch butterfly, *Danaus plexippus* (L.): a tale of two continents. In: Hoth J, Merino L, Oberhauser K, Pisanty I, Price S, Wilkinson T (eds) The 1997 North American conference on the Monarch butterfly. Commission for Environmental Cooperation, Montreal, pp 151–163

Zalucki MP, Clarke AR (2004) Monarchs across the Pacific: the Columbus hypothesis revisited. Biol J Linn Soc 82:111–121

Zalucki MP, Lammers JH (2010) Dispersal and egg shortfall in Monarch butterflies: what happens when the matrix is cleaned up? Ecol Entomol 35:84–91

J Insect Conserv (2011) 15:129–137
DOI 10.1007/s10841-010-9332-0

ORIGINAL PAPER

Grassland butterflies and low intensity farming in Europe

**J. W. Dover · S. Spencer · S. Collins ·
I. Hadjigeorgiou · A. Rescia**

Received: 24 March 2010 / Accepted: 18 August 2010 / Published online: 4 September 2010

Abstract In this paper we describe the impact of the abandonment of traditional farming practices on butterflies and their habitats in traditional, often montane, pastoral systems. We link these declines to socioeconomic factors: illustrating how the failure of the CAP to support traditional farming leads to structural changes in farming enterprises—features which may be obscured by crude statistics on stock. We then call for the scheduled CAP reforms in 2013 to be radically realigned to support rather than destroy biodiversity so that any new EU agri-biodiversity commitments have an effective funding stream to support them.

Keywords Common Agricultural Policy · Grassland · Butterflies · Land-use change · Abandonment · Europe · Traditional management

J. W. Dover (✉)
Institute for Environment, Sustainability and Regeneration, Staffordshire University, College Road, Stoke-on-Trent ST4 2DE, UK
e-mail: j.w.dover@staffs.ac.uk

S. Spencer
Butterfly Conservation European Interests Group, Rhoslan, Brithdir, Llanfyllin, Powys SY22 5HB, UK

S. Collins
Butterfly Conservation, Manor Yard, Wareham BH20 5QP, UK

I. Hadjigeorgiou
Department of Nutritional Physiology and Feeding, Agricultural University of Athens, 75 Iera Odos, 118 55 Athens, Greece

A. Rescia
Departamento de Ecología, Complutense University, Madrid, Spain

Introduction

The European red list of butterflies lists a third of the 482 European butterfly species as having had population declines since 2000; 9% of species are considered threatened (van Swaay et al. 2010). The main processes are loss and degradation of habitat through agricultural intensification and land abandonment (and consequent fragmentation), much of it on grasslands (van Swaay 2002; van Swaay and Warren 2006; van Swaay et al. 2006, 2010). It is sobering to consider that much work in spatial ecology (see refs in Dover and Settele 2009) is not concerned with understanding processes in naturally fragmented systems, but with understanding and ameliorating anthropogenically mediated fragmentation that threatens biodiversity. It is perhaps ironic that many of the biodiverse systems in Europe which are now threatened by intensification and abandonment (Plieninger et al. 2006; Stoate et al. 2009; van Swaay et al. 2010) were created by human action in the first place as a side-effect of traditional farming practices (MacDonald et al. 2000; Plieninger et al. 2006). It is also easy, without data, to underestimate the scale of land-use change (and its impact on biodiversity) as people are adept at forgetting what landscapes were like when they were younger and hence unreliable witnesses when questioned. Younger members of a population are also likely to assume that there has been little change in biodiversity in their landscape because of a lack of communication between generations. Both phenomena are summarised by the term 'shifting baseline syndrome' (Papworth et al. 2009).

The Common Agricultural Policy (CAP) is the consumer of the lion's share of the European Union's budget and directly impacts on biodiversity. Wenzel et al. (2006) describe a shocking situation where seven remnant high quality fragments of calcareous grasslands, managed as nature reserves, lost more than 50% of their habitat specialist species whilst the generalists remained stable over the period 1972–2001. The results were interpreted as changes in the wider landscape surrounding the reserves impacting on the reserves either by reduced connectivity/loss of habitat or by affecting dynamics within the reserves. Designating nature reserves alone is not an answer to biodiversity conservation, the quality of the wider countryside is important too. Whilst the CAP does support agri-environment measures, it is clear that it is currently ineffective in protecting some of the most valuable biodiversity remaining on (high nature value) farmed land as well as in adjacent reserves (see van Swaay et al. 2010).

High nature value farmland

The term high nature value farmland (HNV) was coined in the 1990s in an attempt to focus attention on traditional farming systems which support high levels of biodiversity, but which were likely to decline without additional support through the CAP (Beaufoy, undated). Projections made by Klijn (2004) suggest that marginal lands, including HNV, are threatened by wide-scale abandonment whilst Kleijn et al. (2009) point out that conservation is more cost-effective in "extensively farmed areas that still support high levels of biodiversity".

Existing EU policy explicitly promotes the support of low-intensity farming systems, primarily through Rural Development Programmes funded via Axis 2 of the CAP, but little progress has been made in the rebalancing of support from intensive farmland to HNV farmland (Beaufoy, undated) and there are conflicts within rural policy itself (Beaufoy et al. 2008). Paracchini et al. (2008) have refined initial attempts to map HNV farmlands (which used CORINE land cover data and information on farm systems) by incorporating biodiversity data from the Natura 2000 network, butterflies (Prime Butterfly Areas; van Swaay and Warren (2003)), birds (Important Bird Areas; Heath et al. (2000)) and national biodiversity datasets along with land cover and national-specific environmental data. Only Prime Butterfly Areas that included at least one of 27 species that use alpine, humid or dry grassland were used in the analyses (Paracchini et al. 2008).

Three criteria were used to identify HNV land

1. Farmland with a high proportion of semi-natural vegetation
2. Farmland with a mosaic of low- intensity agriculture and natural and structural elements, such as field margins, hedgerows, stone walls, patches of woodland or scrub, etc.
3. Farmland supporting rare species or a high proportion of European or World populations

The results gave an outline estimate of 32% of agricultural land as potentially of High Nature Value (using data from 26 of the EU-27 states) (Paracchini et al. 2008). This mapping approach has been criticised by Beaufoy et al. (2008) as not truly representing HNV farmland as it does not 'define' the HNV farming systems in member states, but is an "exercise in spatial targeting of support".

Abandonment

Land-use change is a catch-all and can represent abandonment of management, change in timing or intensity of management, change in cropping, and destruction of habitat. HNV farmland is susceptible to two processes: intensification (conversion to intensive arable or increased grazing pressure in livestock systems) or, as much of it is on marginal land, abandonment. Such impacts can have catastrophic effects on grassland composition; for example, Foley (1987) identified eight separate factors accounting for the loss of 83 sites of *Orchis ustulata* L. in northern England: agricultural improvement (42%), building (22%), ploughing (8%), overgrazing (20%), scrubbing up (5%), afforestation (2%) and quarrying (1%). A further 20 sites were shown to have lost the species, but the causal factor could not be identified. Whilst the abandonment of farmed land can also be the end result of a complex of factors, the two main reasons are: (1) that it is no longer economic to farm (MacDonald et al. 2000) or (2) for social reasons (Stenseke 2006). The less accessible (Mottet et al. 2006) and least productive fields (MacDonald et al. 2000) are those most likely to be abandoned; this generality holds for whether the fields are in Europe (e.g. Sweden, Stenseke 2006) or Japan (Uematsu et al. 2010). Of 24 montane study areas in Europe, MacDonald et al. (2000) estimated 21 were suffering from abandonment. Few young people entering farming and rural depopulation are also common causes of abandonment (Rescia et al. 2008; Uematsu et al. 2010).

Impacts of abandonment on the sward

Whilst the vegetation of lightly grazed meadows and traditional mown (hay) meadows is different (Erhardt 1985), continuity in management is essential in both cases for the maintenance of their characteristic flora. Abandonment of management leads to substantial declines of characteristic hay meadow species and succession through invasion by coarse grasses, shrubs and eventually trees (Erhardt 1985;

see also Jensen et al. 2001). Losvik (1999) found that overall plant species richness could be maintained, or even increase over a 10-30 year period, although in most cases succession to woodland would ultimately result. However, cessation of mowing for just 2–3 years reduces hay meadow plant species richness (Kull and Zobel 1991). Abandoned open meadows can become very nutrient poor, whilst those colonized by woody vegetation can become more fertile (Losvik 1999). The cessation of mowing, with its associated raking-off of the cut sward, will necessarily increase litter accumulation (Mašková et al. 2009) and may reduce seedling survival rates (Fowler 1988).

The abandonment of mowing need not immediately necessarily lead to conversion of an area to one dominated by woody vegetation; in the Pyrennes such meadows are often used, if only temporarily, as grazing meadows (Mottet et al. 2006). A switch from hay meadow to grazing meadow will maintain the open character of the field, but will have significant impacts on plant structure through the growing season and on plant composition and richness. For example, hay meadow and grazing management has been reported to result in different routes for the principal modes of reproduction by the toxic perennial false helleborine (*Veratrum album*) in subalpine Swiss grasslands (Kleijn and Steinger 2002). Essentially, hay meadow populations are mainly made up of small plants resulting from seed; whereas pastures have a higher proportion of large plants resulting from vegetative reproduction. The avoidance of grazing on the alkaloid-containing tissues is thought to result in increased vegetative growth whilst also promoting increased seed production (Kleijn and Steinger 2002). If the results of a study of *Scabiosa columbaria* by Reisch and Poschlod (2009) can be generalised, it also appears that plants in mown meadows flower earlier than in grazed meadows, and show genetic differentiation. Some meadows, such as those in the Picos de Europa, Spain, are artificially irrigated partly to prevent desiccation, but also to distribute waterborne macronutrients. Drainage channels are cut by hand through the meadows and water diverted from springs or streams as required. The abandonment of irrigation may precede abandonment of mowing, and is likely to modify plant communities in a similar way to the installation of drainage (Grootjans et al. 1996)

Better to maintain than restore

Mottet et al. (2006) noted that abandonment of meadows was sometimes temporary, with management reinstated after 10–20 years. Restarting management may not be as good as maintaining it however: Stampfli and Zeiter (1999) reinstated mowing regimes in two hay meadow plots on the slope of Monte Generoso in Switzerland. The abandoned meadows had become dominated by *Brachypodium pinnatum* within 8 years of the cessation of mowing and 26% of the herb species lost. Over a 10-year period of restored mowing biomass increased, but growth was strongly affected by drought and was strongly correlated with relative humidity. Reinstating mowing did not lead to the recovery to pre-abandonment levels of species richness. Many of the meadow species in their study area had low persistence in the seed-bank (<5 years), and they concluded that if viable seed sources were not within a few metres of restoration management recolonisation would take a very long time and would also be affected by the microsite quality experienced by seeds. Eriksson and Ehrlen (1992) considered seed limitation to be an important, and underestimated, factor in recruitment in plant populations. Zobel et al. (1996) considered that mowing reduced shoot competition for light, allowing less-competitive species to survive, and removal of litter by raking improved seedling production. In effect, mowing constrains the process of competitive exclusion. The message of Kull and Zobel (1991) is clear: maintenance is simpler than restoration; when fully abandoned, estimates of timescales for restoration are of the order of 50 years due to impoverished seed banks and seed-rain.

Impacts of abandonment on butterflies

Traditional grazing and hay management (mowing) are both valuable management techniques for butterflies and perhaps comparable: cessation of mowing may be compensated for by low intensity grazing, at least for many species (Dolek and Geyer 1997; Saarinen and Jantunen 2005). The expected trajectory for species richness of butterflies in abandoned traditional grasslands was outlined by Erhardt (1985): essentially, newly abandoned grassland increases in butterfly species richness as grazing/mowing pressure is relaxed (see Söderström et al. 2001), mainly a result of improved survival of juvenile stages, whilst plant species richness, which tends to correlate with butterfly richness (Cremene et al. 2005; Marini et al. 2009a, b), is maintained. As shrubs invade, plant diversity declines and shading increases. In a later paper on butterflies in the Swiss Jura Mountains Balmer and Erhardt (2000) indicated that the later successional stages of abandoned ('fallow') land, provided that it did not succeed to forest, was more valuable than extensively managed grassland for butterfly richness and called for the incorporation of the seral stage in conservation management as a rotation. This 'mosaic' approach to management intensity was also considered useful by Cremene et al. (2005), Kruess and Tscharntke (2002a), Marini et al. (2009a), and Schwarzwälder et al. (1997); the latter also considered that later seral stages in adjacent meadows were useful for complementation/supplementation (Dunning et al. 1992; Ouin et al. 2004). Stefanescu et al. (2009) reported changes in butterfly fauna in abandoned hay meadows over a 7-year period in northeast Spain, following a baseline year with hay management; a single continuously managed hay meadow acted as a control. The sward responded to abandonment typically through increased turf height, invasion by coarse grasses, brambles (*Rubus* spp.) and tree seedlings, loss of leguminous plants, colonisation by *Cirsium* spp. The principal results were that changes in the butterfly fauna were rapid with much change in species composition in the first 4 years, including the extinction of *Plebejus argus* along with its host plant *Lotus corniculatus*. Species richness did not change significantly in abandoned meadows, contrary to expectations, but species with specialist habitat requirements were replaced by generalists. The nearby, continuously managed control gained species; the assumption being that species relocated from the abandoned meadows.

With variations, elements of abandonment, e.g. succession following abandonment, leading to (often) initial increases in species richness, followed by decline as trees invade and displacement of grassland specialists by generalists, have been identified in many studies: e.g. Pöyry et al. (2005, 2006, 2009) in Finland, Kruess and Tscharntke (2002a, b) in Germany, Cremene et al. (2005) in Romania, Öckinger et al. (2006) and Nilsson et al. (2008) in Sweden. Whilst community trends may be similar, individual species' responses may differ in different geographic areas as Pöyry et al. (2005) caution.

The above studies were concerned, largely, with community effects; individual species studies reveal impacts of abandonment as well. A detailed study of *Mellicta athalia* ssp. *celadussa* in hay meadows, abandoned grasslands, and newly restored grasslands on Monte Generoso in Switzerland (Schwarzwälder et al. 1997) found distinct partitioning of activities by adult males between the different management regimes (mating in hay meadows and recently abandoned meadows, feeding in mature abandoned meadows); host plants and larvae were only found in the active hay meadows. Restoration of hay cutting in mature abandoned meadows did not result in colonisation by host plants after 2 years indicating that restoration management was a long-term process; Pöyry et al. (2005) found 'old-pasture' butterfly species had not recolonised restored pasture even after 5 years in Finland. Bergman and Kindvall (2004) considered that abandonment of grazing or mowing, and consequent succession, in meadows threatened the long-term survival of *Lopinga achine* in Sweden.

An example from Spain (hay and grazing meadows)

MacDonald et al. (2000) included in their paper a snapshot of the Picos de Europa, a prime butterfly area (van Swaay and Warren 2003): predominant farming: sheep and cattle; 26% of land area (180,000 ha total) used for agriculture; farm sizes 2–30 ha, % of farmers in the working population = 40% (45% of population in employment), 10 people per km^2, decreasing population with a "High dependence on farming for employment". They also considered that traditional land management practices and the land itself was being abandoned, with consequent impacts on the landscape. Rescia et al. (2008) defined the socio-ecological system in the Deva Valley of the Picos de Europa to be one of extensive mountain livestock farming. Cheese is a speciality product of the area (Álvarez 2006).

Rescia et al. (2008) assessed loss of open grassland in a study area at the head of the Deva Valley in the 'county' of Liébana, Picos de Europa: 51 meadow parcels were examined and 31 (61%) were shown, over the period 1957–2002, to have undergone changes in vegetative cover associated with succession (shrinkage and interspersion) whilst two were replaced by infrastructure. Dover et al. (in revision) working in the same general area (the municipality of Cameleño) estimated that most of the loss was associated principally with a decline in grazing meadows rather than hay meadows, though loss and shrinkage was evident in these too. In 1900 there were approximately 2,700 people in Cameleño, a decline set in around 1940, and by 1998 the population was about 1,100 (Gómez et al., undated). Within Liébana as a whole only the tourist town of Potes experienced a population increase over the same period. Between 1960 and 2003 Potes increased in population by 17.8% with all the other municipalities declining by between 49.6 and 83.6% (Gómez et al., undated) (Fig. 1). In Liébana as a whole, milk production declined by 43% in the 5 years 1995/1996 to 2001/2002 with only Tresviso increasing production (from 0 to 6,320 kg, equivalent to 0.18% of the total quota for Liébana in 2001/2002) (Fig. 2). Over the same period, the total number of dairy farmers in Liébana declined from 312 to 113 a reduction of 64%—only Tresviso had an increase in the number of farmers: up 1 from zero (Fig. 3) (Anonymous 2005; Gómez et al., undated). The number of livestock (meat) farmers in 2003 stood at 447 with a total headage (number of stock) for Liébana of 12,469 but with herd sizes varying between 10 and 35 per farm, depending on municipality (Gómez et al., undated). Over the period 1984–2003 livestock farms declined by about 50% (Fig. 4), with beef cattle numbers increasing by about 3% (see Rescia et al. 2008). The age structure of the farmers shows a top-heavy age structure compared with the population of Liébana (Rescia et al. 2008). Sheep farmers declined by

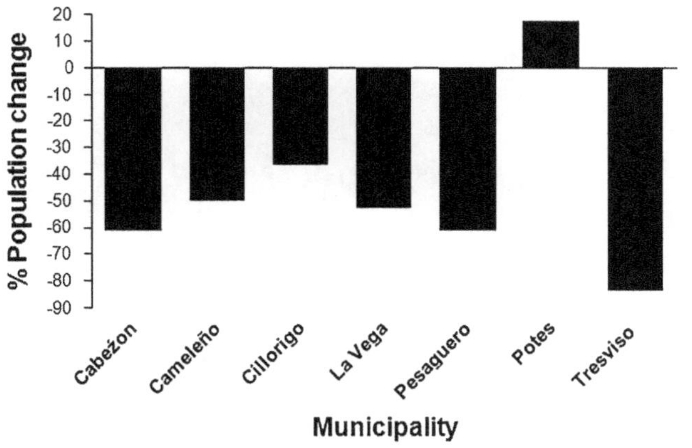

Fig. 1 Percentage human population change in the seven municipalities of Liébana, Picos de Europa, Spain 1960–2003. Data from Padrón municipal cited in Gomez et al. (undated)

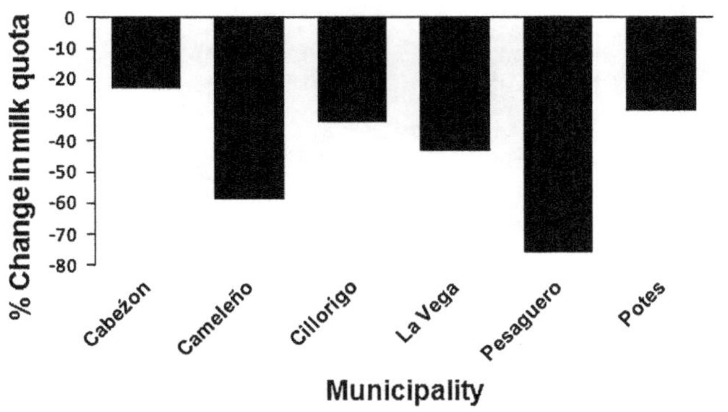

Fig. 2 Percentage change in milk quota in six of the seven municipalities of Liébana, Picos de Europa, Spain between 1995/1996 and 2001/2002. Data from Consejeria de Ganaferia, Agricultura y Pesca cited in Gomez et al. (undated)

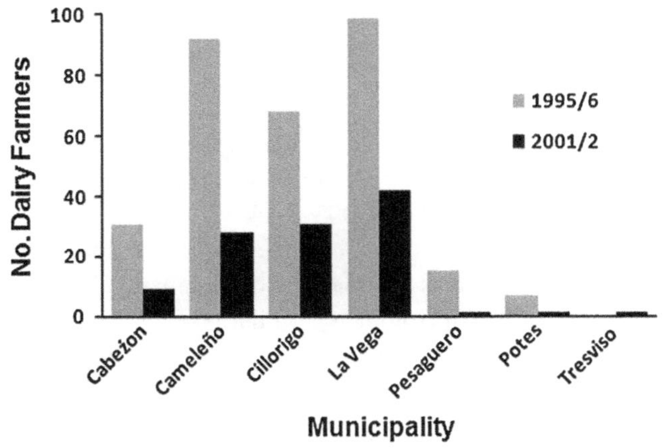

Fig. 3 Number of dairy farmers in the seven municipalities of Liébana, Picos de Europa, Spain between 1995/1996 and 2001/2002. Data from Consejeria de Ganaferia, Agricultura y Pesca cited in Gomez et al. (undated)

65% to 161 over the period 1992–2003, and goat farmers by 45% to 63, sheep numbers declined by 79% over the same period (to 7,575 animals) whilst goats declined by

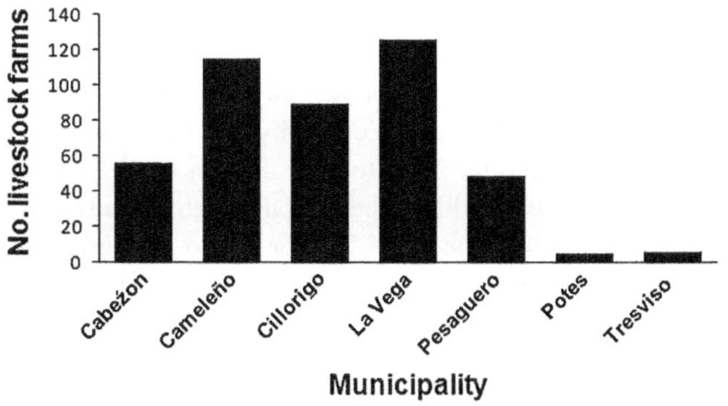

Fig. 4 Number of livestock farms in the seven municipalities of Liébana, Picos de Europa, Spain in 2003. Data from Extension Agraria. Datos de la campaña de saneamiento 2003 cited in Gomez et al. (undated)

just 8% (to 4,972 animals) (data in Gómez et al., undated). Whilst all municipalities lost goat farmers, headage losses did not occur in Cameleño or Cillorigo (up 8.5 and 25% respectively) and this may reflect the needs of the cheese industry which is concentrated in these municipalities (Gómez et al., undated).

The results collectively demonstrate a collapse of traditional farming: the dairy industry has declined, remaining farmers have moved over to meat production. Massive depopulation of the rural areas has occurred, with tourism replacing farming as an economic activity. Shrinkage and loss of open grassland has been identified, with changes associated with the early stages of abandonment being picked up in the responses of the butterfly fauna (Dover et al., in revision, Dover et al., submitted). Whilst hay meadows appear relatively unaffected at present, the economics of cattle farming may result in an increase in barn-reared cattle, with hay imported from outside the valley—hay meadows are then likely to suffer the fate of grazing meadows.

The example of Greece (shepherded grazing)

Traditionally, in Greece, grazing was unfenced and communal; sheep and goats were folded during the night, shepherded on mountain pastures during the day, and milked in the afternoon. However, observations suggest shepherded grazing in mountain regions has gone into a decline (SS, personal observation) and Zervas (1998) warned of problems of abandonment and overgrazing in Greek mountains. Hadjigeorgiou et al. (2005) cite some data on the depopulation and abandonment of mountain areas, where in the study area of Pindus Mountains a 46% decrease in cultivated land and a 43% decline in rough grazing was observed between years 1945 and 1992. The following data come from the Hellenic Statistical

Authority in Athens. National statistics (classified by altitude 0–300 m = lowland, 301–600 m = semi-mountainous, >600 m = mountainous) actually indicate an increase in the national flocks of sheep and goats over the period 1981–2000 (sheep up 2.3% to 8.75 million head; goats up 12.3% to 5.30 million head) although trends vary depending on location (Fig. 5). For sheep and goats the semi-mountainous areas have seen substantial headage increases (44.4 and 57.8% respectively) but declines are evident in both the lowlands and mountains for sheep (−12.1 and −14.7%), whilst goats have increased in the lowlands but decreased in the mountains (27.5 and −22.7%). Whilst goats have declined more than sheep and the declines in mountain areas are relatively modest, these figures do not reflect the *extent* of grazing declines observed in some areas. Despite the rise in sheep and goat numbers, there is a clear and substantial decrease in the number of farms over the same time period with sheep (−41.8% to 128,551) and goats (−60.6% to 138,251) (Fig. 5); the declines are less in the semi-mountainous areas for sheep, and goats (Fig. 6). A decrease in farms coupled with an increase in headage (Fig. 5) would suggest that enterprises would be intensifying stock production by feeding in barns rather than by shepherding or transhumance, especially as much of a sheep and goat farm's income derives from milk production requiring high hygiene standards and easy access to dairies (see Anonymous 2009; Kalantzopoulos et al. 2002). However, the national statistics do not demonstrate any increase in such 'home feeding' of animals; in the case of sheep and goats there are declines in both 'home-feeding' and transhumance, but large increases in the proportion of animals that were shepherded—especially for goats (Fig. 7). This seems extraordinary given that shepherded flocks made up 85.1 and 82.8% of the national sheep and goat flocks respectively. Recent legislative changes in Greece now require all dairy sheep and goat farms to have permanent sheds approved by the state authorities (Law N° 1579 of 1985 as amended by 3698 of 2008), in order to ensure milk and stock hygiene, animal welfare and manure management. However, this implies significantly higher costs for shed construction, restriction of shed locations near electricity and water supply networks, etc., therefore a greater emphasis on 'home-fed' or sedentarised flocks. It is possible that the historic method of data collection (registration of flocks) does not reveal the changes that are happening on the ground; the latest date that we have for comparisons is 2000 and much can happen in 10 years. The lack of up-to-date information on trends in husbandry, national flocks, their location relative to topography, and grazing pressure may hamper efforts to provide support for HNV farming (de Rancourt et al. 2006). For example, EU Sheep and Goat support has been '100% decoupled' for

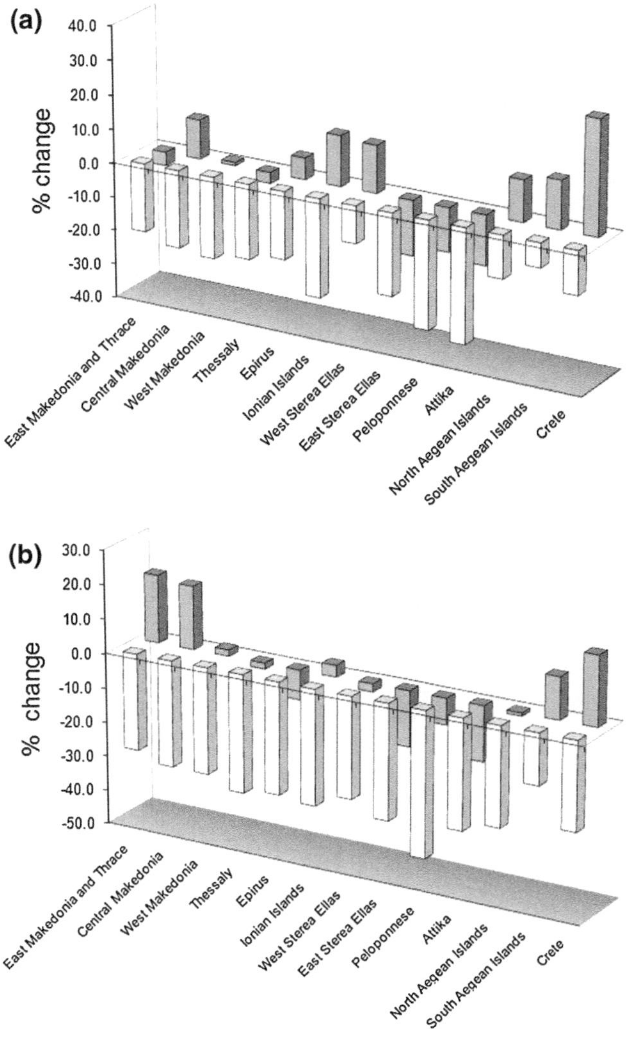

Fig. 5 Percentage change in the number of farms (*white*) and headage (*grey*) for **a** sheep and **b** goats at different locations in Greece over the period 1991–1990. *Source*: Hellenic Statistical Authority, Athens

Greece i.e. converted to a "headage payment" based on historic rights (Stoate et al. 2009) so shepherds receive a payment based on their 2003 flock size. Whether these shepherded flocks still exist in some places is a moot point. Some mountain areas occupied by transhumant Vlachs are overgrazed (e.g Mt Smolikas), whilst other areas such as Northern Epirus, the Pindus Mountains and the Central Peloponnese are largely abandoned (SS and IH, personal observation). Anecdotally, shepherds on Mt. Chelmos indicate a ×10 decline in free ranging sheep and goats, but an increase in cattle.

Why is this important for butterflies? Pamperis (2009) gives the latest trend for 235 butterfly species found in Greece. Over the period 1983–2009, 24 species have declined by 15–25%, 7 by 25–50% and 1 by 50–75%, no species have been shown to increase although the geranium bronze *Cacyreus marshalil* has recently colonised. For

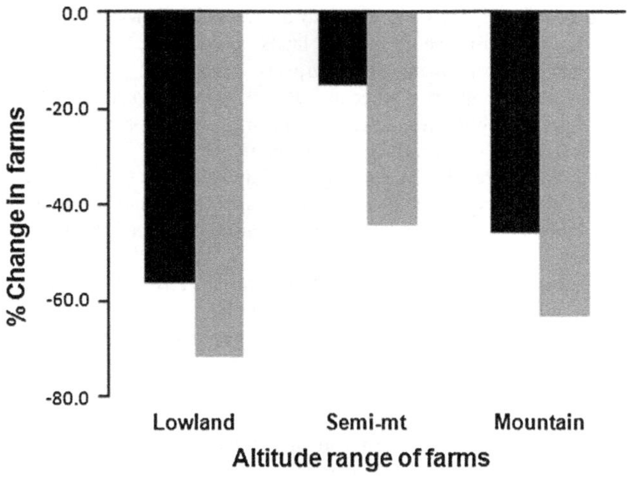

Fig. 6 Percentage change in the number of Greek sheep and goat farms over the period 1981–2000 by altitude range. Hellenic Statistical Authority, Athens. *Black* sheep, *grey* goats, lowland = 0–300 m, semi-mountainous = 301–600 m, mountainous = >600 m

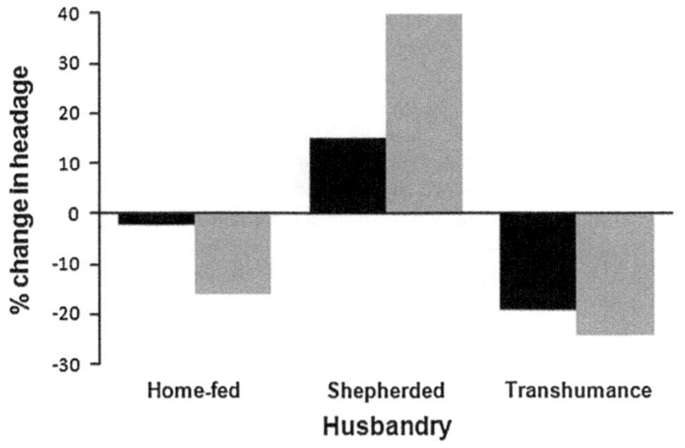

Fig. 7 Percentage change in the number of Greek sheep and goats under different husbandry regimes 1981–2000. Hellenic Statistical Authority, Athens. *Black* sheep, *grey* goats

Greece ten sites are given as Prime Butterfly Areas in van Swaay and Warren (2003); most are mountain areas.

Policy

Farm size is negatively correlated with both plant and butterfly species richness in Italian hay meadows; larger farms also utilise flatter fields compared with small farms (Marini et al. 2009b). The decoupling of farmer support under the CAP in 2005/2008 disadvantages small farmers and effectively promotes production in covered production areas (Stoate et al. 2009). There are clearly problems in maintaining traditional, extensive, sheep and goat grazing in montane areas (Anonymous 2009).

The entry of new states into the EU exposes them to the impact of the CAP; inflexible agri-environmental programmes can have disastrous impacts, including causing the extinction of some species, as happened to *Colias myrmidone* in the Czech Republic (Konvicka et al. 2008). Schmitt and Rákosy (2007) examined the impact of accession to the EU on Romanian agriculture and used butterflies as an indicator group. Their conclusion was that accession would result, as elsewhere, in intensification and abandonment of semi-natural grasslands with consequent negative effects on the country's butterfly fauna unless policy instruments directed funds towards their conservation and traditional management. Current agri-environment schemes are inadequate to prevent further biodiversity loss in the EU (Rounsevell et al. 2006). Giupponi et al. (2006) coupled the loss of small dairy farms in montane grasslands in Italy with biodiversity loss. They suggested that small dairy units might be retained by encouraging on-site/nearby processing units so that the monetary rewards from added-value products could be retained by local producers. Just such a system is evident in the Picos de Europa (Álvarez 2006; JWD, personal observation), and whilst it does seem to be halting the overall decline in goats in two of the seven municipalities in the county of Liébana, production is declining elsewhere. The implication is that any positive effect of processing plants would be very local and, whilst real, may be difficult to roll out across whole regions. Maintaining or even increasing milk production will not, of itself, maintain traditional grasslands as dairy stock can be housed in covered sheds—as happens in Pido in Camaleño (JD, personal observation)—support is needed for the traditional husbandry practices as well as for adding value locally.

Clearly changes are needed to the main support system in Europe (the CAP), and there does seem to be acknowledgement that reform is needed to support some producer systems, not least for their contribution to maintaining biodiversity (Anonymous 2009). Whilst agri-environment schemes designed to soften the impact of intensive agriculture are welcome and valuable, it is imperative that CAP reform in 2013 provides support for truly HNV farmland, and we suggest that the recommendations contained in (Anonymous, undated) by BirdLife International is a good starting point.

References

Álvarez P (2006) Los quesos de Liébana. De la tradición a las denominaciones de origen protegidas. Gobiniero de Cantabria, Santander

Anonymous (2005) Anuario de Estadística Agraria Pesquera Cantabria 2003. Gobiernero de Cantabria, Santander

Anonymous (2009) European parliament resolution of June 2008 on the future of the sheep/lamb and goat sector in Europe (2007/2192(INI)). Off J Eur Union 286:41–45

Anonymous (undated) New challenges, new CAP: BirdLife International's vision for the future of the EU Common Agricultural Policy. BirdLife International, Brussels

Balmer O, Erhardt A (2000) Consequences of succession on extensively grazed grasslands for central European butterfly communities: rethinking conservation practices. Conserv Biol 14:746–757

Beaufoy G (undated) HNV farming—explaining the concept and interpreting EU and national policy commitments. European Forum on Nature Conservation and Pastoralism available at: www.efncp.org/download/EFNCP-HNV-farming-concept.pdf

Beaufoy G, Jones G, De Rijck K, Kazakova Y (2008) High nature value farmlands: recognising the importance of south east European landscapes. Final summary report (Bulgaria & Romania) WWF Danube-Carpathian programme and the European forum on nature conservation and pastoralism

Bergman KO, Kindvall O (2004) Population viability analysis of the butterfly *Lopinga achine* in a changing landscape in Sweden. Ecography 27:49–58

Cremene C, Groza G, Rákosy L, Schileyko AA, Baur A, Erhardt A, Baur B (2005) Alterations of steppe-like grasslands in Eastern Europe: a threat to regional biodiversity hotspots. Conserv Biol 19:1606–1618

de Rancourt M, Fois N, Lavin MP, Tchakerian E, Vallerand F (2006) Mediterranean sheep and goats production: an uncertain future. Small Rumin Res 62:167–179

Dolek M, Geyer A (1997) Influence of management on butterflies of rare grassland ecosystems in Germany. J Insect Conserv 1:125–130

Dover J, Settele J (2009) The influences of landscape structure on butterfly distribution and movement: a review. J Insect Conserv 13:3–27

Dover JW, Rescia A, Fungariño S, Fairburn J, Carey P, Lunt P, Arnot C, Dennis RLH, Dover CJ (in revision) Land-use, environment, and their impact on butterfly populations in a mountainous pastoral landscape: species richness and family-level abundance. J Insect Conserv

Dover JW, Rescia A, Fungariño S, Fairburn J, Carey P, Lunt P, Arnot C, Dennis RLH, Dover CJ (submitted) Land-use, environment, and their impact on butterfly populations in a mountainous pastoral landscape: individual species distribution and abundance. J Insect Conserv

Dunning J, Danielson B, Pulliam H (1992) Ecological processes that affect populations in complex habitats. Oikos 65:169–175

Erhardt A (1985) Diurnal Lepidoptera—sensitive indicators of cultivated and abandoned grassland. J Appl Ecol 22: 849–861

Eriksson O, Ehrlen J (1992) Seed and microsite limitation of recruitment in plant-populations. Oecologia 91:360–364

Foley MJY (1987) The current distribution and abundance of *Orchis ustulata* L. in northern England. Watsonia 16:409–415

Fowler NL (1988) What is a safe site—neighbor, litter, germination date, and patch effects. Ecology 69:947–961

Gómez AF, López MCI, Suero ÁA, Serna FG (undated) Análysis diagnóstico territorial de la comarca de Liébana. Agencia desarrollo local Liébana, Potes

Giupponi C, Ramanzin M, Sturaro E, Fuser S (2006) Climate and land use changes, biodiversity and agri-environmental measures in the Belluno province, Italy. Environ Sci Policy 9:163–173

Grootjans AP, Fresco LFM, deLeeuw CC, Schipper PC (1996) Degeneration of species-rich *Calthion palustris* hay meadows: some considerations on the community concept. J Veg Sci 7:185–194

Hadjigeorgiou I, Osoro K, Fragoso de Almeida JP, Molle G (2005) Southern European grazing lands: production, environmental and landscape management aspects. Livest Prod Sci 96:51–59

Heath MF, Evans MI, Hoccom DG, Payne AJ, Peet NB (2000) Important bird areas in Europe: priority sites for conservation (2 vols). BirdLife International, Cambridge

Jensen C, Vorren KD, Eilertsen SM, Samuelsen R (2001) Successionary stages of formerly cultivated grassland in northern Norway, abandoned for 10, 20 and 35 years. Nord J Bot 21:305–320

Kalantzopoulos G, Dubeuf JP, Vallerand F, Pirisi A, Casalta E, Trujillo T (2002) Characteristics of the sheep and goat milks: quality and hygienic stakes for the sheep and goat dairy sectors. Standing committee on microbiological hygiene meeting 28th Sep 2002, International Dairy Federation, Brussels. www.sagarpa.gob.mx/ganaderia/Publicaciones/Lists/Sistemas%20Productos%20Pecuarios/Attachments/77/lech_cap_1.pdf

Kleijn D, Steinger T (2002) Contrasting effects of grazing and hay cutting on the spatial and genetic population structure of *Veratrum album*, an unpalatable, long-lived, clonal plant species. J Ecol 90:360–370

Kleijn D, Kohler F, Baldi A, Batary P, Concepcion ED, Clough Y, Diaz M, Gabriel D, Holzschuh A, Knop E, Kovacs A, Marshall EJP, Tscharntke T, Verhulst J (2009) On the relationship between farmland biodiversity and land-use intensity in Europe. Proc R Soc Lond B Biol 276:903–909

Klijn JA (2004) Driving forces behind landscape transformation in Europe, from a conceptual approach to policy options. In: Jongman R (ed) The new dimensions of the European landscape. Springer, Dordrecht, pp 201–218

Konvicka M, Benes J, Cizek O, Kopecek F, Konvicka O, Vitaz L (2008) How too much care kills species: grassland reserves, agri-environmental schemes and extinction of *Colias myrmidone* (Lepidoptera: Pieridae) from its former stronghold. J Insect Conserv 12:519–525

Kruess A, Tscharntke T (2002a) Grazing intensity and the diversity of grasshoppers, butterflies, and trap-nesting bees and wasps. Conserv Biol 16:1570–1580

Kruess A, Tscharntke T (2002b) Contrasting responses of plant and insect diversity to variation in grazing intensity. Biol Conserv 106:293–302

Kull K, Zobel M (1991) High species richness in an Estonian wooded meadow. J Veg Sci 2:715–718

Losvik MH (1999) Plant species diversity in an old, traditionally managed hay meadow compared to abandoned hay meadows in southwest Norway. Nordic J Bot 19:473–487

MacDonald D, Crabtree JR, Wiesinger G, Dax T, Stamou N, Fleury P, Lazpita JG, Gibon A (2000) Agricultural abandonment in mountain areas of Europe: environmental consequences and policy response. J Environ Manage 59:47–69

Marini L, Fontana P, Battisti A, Gaston KJ (2009a) Agricultural management, vegetation traits and landscape drive orthopteran and butterfly diversity in a grassland-forest mosaic: a multi-scale approach. Insect Conserv Divers 2:213–220

Marini L, Fontana P, Klimek S, Battisti A, Gaston KJ (2009b) Impact of farm size and topography on plant and insect diversity of managed grasslands in the Alps. Biol Conserv 142:394–403

Mašková Z, Doležal J, Květ J, Zemek F (2009) Long-term functioning of a species-rich mountain meadow under different management regimes. Agric Ecosyst Environ 132:192–202

Mottet A, Ladet S, Coque N, Gibon A (2006) Agricultural land-use change and its drivers in mountain landscapes: a case study in the Pyrenees. Agric Ecosyst Environ 114:296–310

Nilsson SG, Franzen M, Jonsson E (2008) Long-term land-use changes and extinction of specialised butterflies. Insect Conserv Divers 1:197–207

Öckinger E, Eriksson AK, Smith HG (2006) Effects of grassland abandonment, restoration and management on butterflies and vascular plants. Biol Conserv 133:291–300

Ouin A, Aviron S, Dover J, Burel F (2004) Complementation/supplementation of resources for butterflies in agricultural landscapes. Agric Ecosyst Environ 103:473–479

Pamperis LN (2009) The butterflies of Greece 2010. Hellenic Society for the Protection of Nature/KOAN, Athens

Papworth SK, Rist J, Coad L, Milner-Gulland EJ (2009) Evidence for shifting baseline syndrome in conservation. Conserv Lett 2:93–100

Paracchini ML, Petersen J-E, Hoogeveen Y, Bamps C, Burfield I, van Swaay C (2008) High nature value farmland in Europe: an estimate of the distribution patterns on the basis of land cover and biodiversity data. Office for Official Publications of the European Communities, Luxembourg

Plieninger T, Hochtl F, Spek T (2006) Traditional land-use and nature conservation in European rural landscapes. Environ Sci Policy 9:317–321

Pöyry J, Lindgren S, Salminen J, Kuussaari M (2005) Responses of butterfly and moth species to restored cattle grazing in semi-natural grasslands. Biol Conserv 122:465–478

Pöyry J, Luoto M, Paukkunen J, Pykälä J, Raatikainen K, Kuussaari M (2006) Different responses of plants and herbivore insects to a gradient of vegetation height: an indicator of the vertebrate grazing intensity and successional age. Oikos 115:401–412

Pöyry J, Paukkunen J, Heliölä J, Kuussaari M (2009) Relative contributions of local and regional factors to species richness and total density of butterflies and moths in semi-natural grasslands. Oecologia 160:577–587

Reisch C, Poschlod P (2009) Land use affects flowering time: seasonal and genetic differentiation in the grassland plant *Scabiosa columbaria*. Evol Ecol 23:753–764

Rescia A, Pons A, Lomba I, Esteban C, Dover JW (2008) Reformulating the social–ecological system in a cultural rural mountain landscape in the Picos de Europa region (northern Spain). Landsc Urban Plan 88:23–33

Rounsevell MDA, Berry PM, Harrison PA (2006) Future environmental change impacts on rural land use and biodiversity: a synthesis of the ACCELERATES project. Environ Sci Policy 9:93–100

Saarinen K, Jantunen J (2005) Grassland butterfly fauna under traditional animal husbandry: contrasts in diversity in mown meadows and grazed pastures. Biodivers Conserv 14:3201–3213

Schmitt T, Rákosy L (2007) Changes of traditional agrarian landscapes and their conservation implications: a case study of butterflies in Romania. Divers Distrib 13:855–862

Schwarzwälder B, Lörtscher M, Erhardt A, Zettel J (1997) Habitat utilization by the heath fritillary butterfly, *Mellicta athalia* ssp. celadussa (Rott) (Lepidoptera: Nymphalidae) in montane grasslands of different management. Biol Conserv 82:157–165

Söderström B, Svensson B, Vessby K, Glimskär A (2001) Plants, insects and birds in semi-natural pastures in relation to local habitat and landscape factors. Biodivers Conserv 10:1839–1863

Stampfli A, Zeiter M (1999) Plant species decline due to abandonment of meadows cannot easily be reversed by mowing. A case study from the Southern Alps. J Veg Sci 10:151–164

Stefanescu C, Penuelas J, Filella I (2009) Rapid changes in butterfly communities following the abandonment of grasslands: a case study. Insect Conserv Divers 2:261–269

Stenseke M (2006) Biodiversity and the local context: linking seminatural grasslands and their future use to social aspects. Environ Sci Policy 9:350–359

Stoate C, Baldi A, Beja P, Boatman ND, Herzon I, van Doorn A, de Snoo GR, Rakosy L, Ramwell C (2009) Ecological impacts of early 21st century agricultural change in Europe—a review. J Environ Manage 91:22–46

Uematsu Y, Koga T, Mitsuhashi H, Ushimaru A (2010) Abandonment and intensified use of agricultural land decrease habitats of rare herbs in semi-natural grasslands. Agric Ecosyst Environ 135:304–309

van Swaay CAM (2002) The importance of calcareous grasslands for butterflies in Europe. Biol Conserv 104:315–318

van Swaay CAM, Warren MS (2003) Prime butterfly areas of Europe: priority sites for conservation. National Reference Centre for Agriculture, Nature and Fisheries; Ministry of Agriculture, Nature Management and Fisheries, The Netherlands

van Swaay CAM, Warren MS (2006) Prime butterfly areas of Europe: an initial selection of priority sites for conservation. J Insect Conserv 10:5–11

van Swaay C, Warren M, Lois G (2006) Biotope use and trends of European butterflies. J Insect Conserv 10:305–306

van Swaay C, Cuttelod A, Collins S, Maes D, Munguira ML, Šŝić M, Verovnik R, Verstrael T, Warren MS, Wiemers M, Wynhoff I (2010) European red list of butterflies. Publications Office of the European Union, Luxembourg

Wenzel M, Schmitt T, Weitzel M, Seitz A (2006) The severe decline of butterflies on western German calcareous grasslands during the last 30 years: a conservation problem. Biol Conserv 128:542–552

Zervas G (1998) Quantifying and optimising grazing regimes in Greek mountain systems. J Appl Ecol 35:983–986

Zobel M, Suurkask M, Rosén E, Pärtel M (1996) The dynamics of species richness in an experimentally restored calcareous grassland. J Veg Sci 7:203–210

J Insect Conserv (2011) 15:279–290
DOI 10.1007/s10841-010-9345-8

ORIGINAL PAPER

Developing and launching a wider countryside butterfly survey across the United Kingdom

T. M. Brereton · K. L. Cruickshanks · K. Risely · D. G. Noble · D. B. Roy

Received: 25 May 2010 / Accepted: 3 September 2010 / Published online: 30 September 2010

Abstract Many butterfly species that were historically common throughout the UK are thought to have declined due to land use intensification, whilst others have increased regionally due to climate change. Population trends in these species are inadequately assessed by current monitoring programmes: the UK Butterfly Monitoring Scheme (UKBMS) mainly samples semi-natural habitats; coarse-scale distribution recording under-estimates declines in population abundance. We designed a scheme (the wider countryside butterfly survey—WCBS) to provide unbiased abundance trends for widespread butterfly species across the whole countryside. The WCBS has been developed over a 3 year pilot study, culminating in a UK-wide roll out in 2009. After testing a range of survey methods, the scheme involves visiting randomly selected 1-km squares at least twice during July–August and walking a linear transect route, counting the number of individual butterflies seen within 5 m. A method for volunteer recorders to collect abundance data on non-lepidopteran insects was also tested. The results of the first year of the WCBS and future plans for the scheme will be described.

T. M. Brereton (✉) · K. L. Cruickshanks
Butterfly Conservation, Manor Yard, East Lulworth, Wareham, Dorset BH20 5QP, UK
e-mail: tbrereton@butterfly-conservation.org

K. Risely · D. G. Noble
British Trust for Ornithology, The Nunnery, Thetford, Norfolk IP24 2PU, UK

D. B. Roy
NERC Centre for Ecology and Hydrology, Maclean Building, Benson Lane, Crowmarsh Gifford, Wallingford, Oxfordshire OX10 8BB, UK

Keywords Biodiversity · Generalist species · Volunteer monitoring schemes · Occupancy · Survey design · Invertebrate monitoring methods

Introduction

Due to the efforts of thousands of amateur and professional recorders, more is probably known of the butterfly fauna of the UK than any country on the planet (Fox et al. 2006). Data come from two main sources. The Butterflies for the New Millennium (BNM) project collects data from 'casual' sightings to document changes in butterfly distribution; the dataset currently comprises more than 6.25 million records (Butterfly Conservation unpublished data). The UK Butterfly Monitoring scheme (UKBMS) collects data on butterfly abundance recorded at a network of sites monitored each year using a peer-reviewed methodology (Pollard 1977; Pollard and Yates 1993); the dataset currently comprises 3.85 million counts from 1,726 sites recorded since 1973 (Brereton et al. this volume).

In spite of this wealth of data, we still lack an unbiased assessment of abundance trends for many wider countryside butterflies, defined as species with relatively high mobility that occur in a range of habitat types across the landscape including those heavily modified by human activity (Asher et al. 2001). Wider countryside species comprise half of the UK's butterfly species and have their main breeding populations in the general countryside, where relatively few UKBMS sample sites are located (Pollard et al. 1995; Greatorex-Davies and Roy 2001; Pywell et al. 2004; Brereton et al. this volume). These data deficiencies are of potential concern because recent research indicates that many of these species underwent substantial declines in local and regional abundance during the 20th century, that were largely undetected by monitoring and mapping schemes (Cowley et al. 1999; Leon-Cortes et al. 1999; León-Cortés et al. 2000). Agricultural intensification has been highlighted as a major cause of generalist and specialist species declines across the whole of Europe (van Swaay 1990) and continues to be a factor in the decline of widespread species of other well monitored taxa, including birds (Donald et al. 2001). The decline of once common taxa associated with lowland farmland is one of the most pressing issues in UK biodiversity conservation (Krebs et al. 1999; Aebischer et al. 2000; Vickery et al. 2004; Firbank et al. 2008). Habitat loss, including through urban development and afforestation has also been an important major factor in the decline of common butterflies (Asher et al. 2001). In the latest review of priority species (UKBAP 2007), two wider countryside butterfly species Small Heath (*Coenonympha pamphilus*) and Wall Brown (*Lasiomata megera*) were added to the list for the first time, owing to recent rapid declines in both range and abundance (Fox et al. 2006, Fox et al. 2010).

Currently, abundance trends in wider countryside butterflies are derived from the UKBMS and largely inferred from data derived from sampling higher quality semi-natural habitat and protected areas in the lowland areas of central and southern UK. Sites are selected by recorders rather than through a random sampling framework. Provisional analysis suggests that abundance trends in wider countryside species from UKBMS sample sites may be representative of the wider population (Brereton et al. this volume). However, it can be argued that species trends are being assessed from too small a proportion of the population and from a fundamentally inappropriate survey design (Buckland et al. 2005). For example, the Meadow Brown *Maniola jurtina* is widespread on intensively managed farmland (Asher et al. 2001) which extends over nearly half of the UK landscape (Carey et al. 2008) but these habitats account for less than 6% of UKBMS monitoring sites (UKBMS unpublished data). A bigger concern is that ongoing species declines may currently be under-estimated, and that wider countryside butterflies may have undergone more extreme declines as has been seen in other north European countries (e.g. the Netherlands, van Swaay et al. 2006; van Dyck et al. 2009) and has been demonstrated for common moths in the UK (Conrad et al. 2006). Due to deficiencies in sample coverage of intensively managed/modified habitats, stable or positive trends and northward range expansions shown in some wider countryside species attributable to climate warming (Roy et al. 2001; Fox et al. 2006) may possibly be masking regional declines due to land management change.

A previous obstacle to improved monitoring coverage of wider countryside habitats in the UK (Pollard and Greatorex-Davies 1997) has been the likely low diversity of butterflies encountered for the a high level of recording effort. Up to 26 surveys per annum are required for Pollard Walk methods used in the UKBMS (Pollard 1977; Pollard and Yates 1993). A consensus has developed amongst the recording and scientific community which started in the mid-1990s (Pollard and Greatorex-Davies, 1997), that a more efficient scheme based on random sampling is required to generate more accurate abundance trends for widespread butterfly species. Given that butterfly population trends are regarded as a good indicator of the state of the environment by the UK Government (Defra 2006), data from such a scheme have a number of potential policy applications related to priority issues in the wider countryside, including assessment of climate change impacts and mitigation measures, impacts of new urban development and the potential benefits of agri-environment schemes on biodiversity (UK Biodiversity Partnership 2007).

In 2004, Butterfly Conservation (BC) and the Centre for Ecology & Hydrology (CEH) secured resources from a consortium of government agencies led by the Department of the Environment food and Rural Affairs (Defra) to develop a wider countryside butterfly monitoring method. The development of a new method requires consideration of a number of fundamental design and practical issues including (1) which sampling framework to adopt, (2) what field methods to employ, (3) when and how often to sample, (4) assessment of scheme power (5) identifying who will do the sampling, (6) how will data and recorders be co-ordinated and (7) data analysis and reporting methods. In this paper we will describe how these issues were considered in relation to a new component of the UKBMS—the wider countryside butterfly survey (WCBS). We will describe the design issues investigated, field testing and other method and scheme validation work carried out over 4 years, approaches to co-ordination and partnerships and preliminary results from the launch of the WCBS in 2009. We conclude by outlining the future plans for the scheme.

Designing the scheme

Methodological issues were investigated through consultation with a number of experts, field testing by volunteers and research projects targeted at specific design issues.

Sampling framework

The remit of the WCBS was to produce unbiased abundance indices and trends for wider countryside species that were representative of the whole of the countryside. Primary design considerations were the need for a random sampling framework and practical consideration of the suitability of field methods for the volunteer recorders who would complete the surveys. For a combination of scientific and practical reasons, it was decided to adopt the broad sampling framework used for the Breeding Bird Survey (BBS) co-ordinated by the British Trust for Ornithology (BTO). The BBS uses 1-km grid squares of the Ordnance Survey national grid system as the unit of sampling. This was assessed as a convenient sampling unit for volunteer recorders, is readily identifiable in the field, is a scale at which climate and environmental data are readily available and is a unit that is commonly used by other national monitoring programmes in the UK in addition to the BBS (e.g. Countryside Survey, Common Plants Survey, National Bat Monitoring Programme). In the BBS, 1-km squares are selected by stratified random sampling. Regions receive a differing number of randomly selected squares, with the number of allocated squares being proportional to regional recorder density (a measure of likely participation levels). During the consultation process, no major scientific concerns were identified in using the BBS design to sample butterflies, but a number of practical advantages were identified, including the potential to involve BTO recorders in recording their squares and the potential to analyse changes in bird and butterfly abundance from the same sample areas. Note, however, that a decision was made that the WCBS would use a range of other randomly selected 1-km squares (again stratified by recorder density) in addition to BBS squares, to enable participation by butterfly recorders not involved in the BBS.

Field methods

A number of sampling methods within 1-km squares were considered, including point counts (transects; Buckland 2006), practical routes (e.g. following only footpaths) and different types of linear transects (e.g. zig zag walks versus evenly spaced parallel lines). For both scientific and practical reasons, it was decided to follow the approach used by bird recorders in the BBS by recording along two linear transects (‘survey lines’), but making adaptations for butterflies using Pollard Walk recording criteria (Pollard 1977; Pollard and Yates 1993). The BBS method involves counting all animals detected along two evenly spaced (ideally 500 m apart) parallel 1-km long survey lines subdivided into sections (Fig. 1), with birds seen/heard assigned to one of three distance bands. Adaptation of the BBS method for butterflies included recording (1) in a more restricted area than for birds (within an imaginary 5 m box), (2) at different times of the day (between 10:45 and 15:45) (3) at different times of the year (predominantly summer rather than spring) and (4) under set weather conditions proven as suitable for butterfly activity. The similarity of the method to the BBS provided the added benefit that BTO recorders could potentially take part on their existing squares by making separate ‘butterfly visits’ in the summer months.

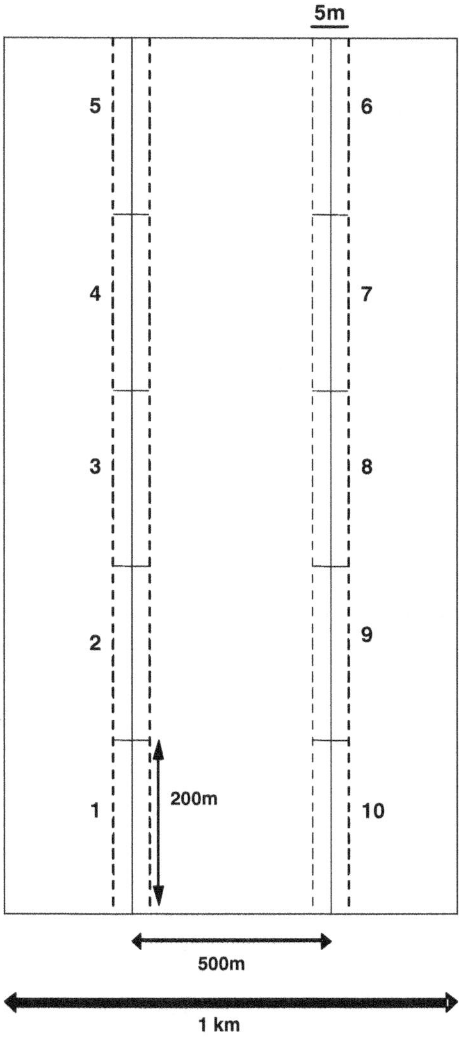

Fig. 1 Sampling design within a 1-km square. *Parallel survey lines* can run north–south or east–west through the square. Deviations from this design following paths and linear features are accepted for ease of surveying

A number of more detailed refinements were considered and tested in an effort to make the scheme more attractive to potential recorders and thus increase participation levels. These included assessing (1) whether transect width could be widened to increase the number of butterflies counted (2) whether the time of day recording period could be extended to give greater flexibility in when recording could be carried out and (3) if other insect groups (e.g. syrphid flies, day-flying moths (lepidoptera), *Bombus* species, odonata, orthoptera) could be monitored as part of the survey, to increase returns from survey effort.

The application of distance sampling (Buckland et al. 2001) along survey lines was tested as an approach to widening the recording area. Distance sampling has further benefits in that it enables (1) detectability (which diminishes with distance) to be accounted for and (2) absolute density to be estimated as done for birds in the BBS

(Newson et al. 2008). Pilot studies in 2005 showed that increasing the transect width from 2.5 to 5 m, increased the number of butterflies seen by a third. However, distance sampling proved unworkable on practical grounds (1) because of the difficulties of having to maintain recording quality in a 'worse cast scenario' with potentially large numbers of fast flying individuals of different species over short distances and time spaces and (2) because butterflies have a biased distribution towards linear features—to effectively sample this through distance sampling would be impractical (i.e. would involve lots of survey lines running perpendicular to linear features i.e. into crops and closed canopy forest) (Isaac et al. in review).

For time of day criteria, field studies were carried out to determine if butterfly abundance across wider countryside butterfly species and groups varied over the day, including immediately before and after conventional BMS recording times (10:45–15:45). Log-linear modelling (Poisson regression) was used to test and estimate differences between time periods and interaction effects. The results showed that for sites largely unshaded during the morning or late afternoon periods and when weather conditions were within BMS criteria (Pollard 1977), the recording window could be extended by up to 2 h—with additional recording permissible from 09:30–10:45 and 15:45–16:30 (Cruickshanks et al. in prep).

During the design phase there was considerable interest amongst the recording community and policy makers in using the WCBS to generate abundance data on a broader range of insect groups. The pilot studies indicated that only day-flying moths and dragonflies could be recorded at the same time as butterflies, due to them having similar search images. In terms of other insect groups, it was considered best to have optional, supplementary recording following completion of butterfly recording, with the assumption that a sufficient number of recorders would take part to make the surveys scientifically worthwhile. These could take a number of forms, including repeat annual surveys for specific species/groups or periodic surveys for different species/groups on a rolling programme. For this, a range of survey methods were tested for individual species and groupings over the pilot studies, with an Insect Flower Search method tested widely during 2009 (see below).

Timing of visits and number of visits required within a season and across sites

Seasonal variation in butterfly abundance on UKBMS transect routes demonstrates that the optimal time to concentrate sampling for almost all wider countryside species is during July and August (70% of the flight period; Fig. 2). Only one wider countryside butterfly species, the Orange-tip *Anthocharis cardamines* has its main flight period outside of July and August.

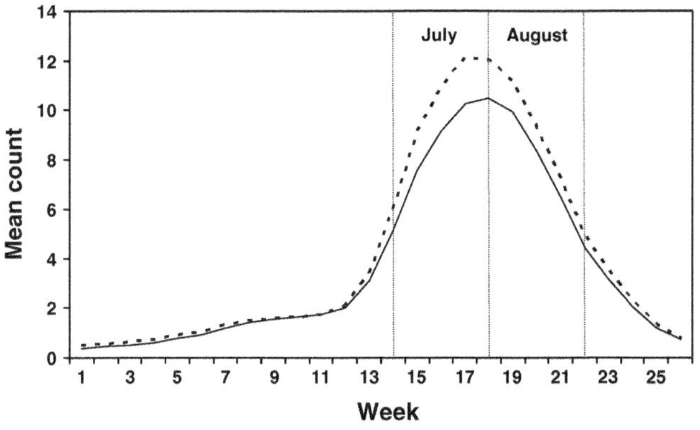

Fig. 2 Seasonal pattern of mean abundance of butterflies recorded on UKBMS transects. Mean number of individuals per transect locations are given for all species (*solid line*) and 20 species classified as wider countryside species (*dashed line*). Week 1 = April 1st and week 26 = the end of September

A power analysis was used to determine the sample size that was likely to be required to detect moderate declines (25%) in 20 of the target wider countryside species (all except canopy-dwelling species) over a short time period (10 year) with reasonable precision. The power analysis used existing UKBMS transect data to test the efficiency of reduced sampling (e.g. only three to four visits) versus a more intensive, 26-week scheme for detecting trends. The power analysis is reported in detail elsewhere (Roy et al. 2007) and is only briefly summarised here. Overall, 80% power (5% significance level) for detecting a 25% decline in abundance over 10 years could be achieved with 430 sites visited three times in July–August and 630 sites with two visits in the same period. These figures highlight substantial efficiency savings, with only a doubling of the number of sites required in the full 26-week scheme (based on an analysis of 180 sites) to achieve the same power to detect trends.

For practical reasons a mandatory two visit scheme in July and August was proposed. Three visits was not considered to be practical, particularly for established BBS recorders, as it would require more effort by bird recorders than completing bird surveys on their squares. Furthermore, reducing the number of required visits to a minimum would again make the scheme more appealing to a wider range of potential recorders.

Assessing potential sampling bias of the scheme

Bias through sampling along actual rather than ideal routes

Butterflies show a clumped distribution towards linear features in the wider countryside (Asher et al. 2001). This was demonstrated in the pilot studies, where in samples squares (n = 10) there were significantly fewer butterflies in open habitats compared with linear habitats ($F_{(1,78)} = 41.24$, $P < 0.001$). 'Actual' WCBS routes are more likely to oversample linear features such as footpaths along hedgerows, field margins and woodland edges compared with 'ideal' routes (randomly available habitat—see Fig. 1) due to ease of access. Again, this was supported by results from a sample (n = 10 squares) in the pilot studies, where for example, open habitats comprised over 80% of available habitat, but accounted for only 9% of habitat along actual routes. To assess the potential impact of oversampling of linear features, research was carried out on arable farms in southern England in 2006 to investigate the extent to which butterfly diversity differed between actual and ideal survey routes. The study showed that, on average approximately, 40% fewer species (mean n = 7 random route and n = 12 survey lines; $t = 3.176$, $P = 0.011$) and less than half the number of butterflies were present on 'actual' compared to 'ideal' routes (mean n = 50 individuals per square ideal route and n = 120 actual route; $t = 2.9$, $P = 0.017$). Therefore actual transects may overestimate butterfly species diversity and abundance.

Do the survey lines effectively sample/represent what is in the square?

Birds are generally detectable at far greater distances than butterflies, consequently a much larger proportion of bird diversity present in the square is likely to be detected through the sampling effort. This is illustrated by the differences in strip widths and areas sampled. In the BBS, relative bird density is estimated over a minimum of 20% of the square, whilst for butterflies only 1% of the surface area of the square is sampled in the WCBS. Surveys were carried out to investigate what was seen along survey lines in relation to more extensive surveys of squares (up to 6% of the surface area of squares surveyed). Four to fivefold increases in coverage within individual squares yielded two to four times more species and two to six times more individuals. However, across squares, only four additional species (18% of species) were detected by the more extensive sampling, with butterfly abundance often extremely low or absent in intensively managed agricultural fields. Thus it was concluded that the survey lines adequately sampled the species complement present across squares.

Testing the WCBS with volunteer recorders

The WCBS was widely tested by volunteer recorders between 2005 and 2008. There were two main objectives to the pilot studies (1) assessing the likely levels of recorder participation and survey coverage that would be achieved if the scheme was launched nationally and (2) determining whether the method is likely to detect the target wider countryside species in sufficient quantity to be able to general national indices and trends.

Volunteer participation

The most extensive testing was undertaken in 2008, with 310 squares sampled in selected regions through the combined efforts of BC and BTO volunteers. The level of participation by both recorder types approximately doubled expectations and feedback was very positive. A questionnaire indicated that three quarters of participants would be prepared to take part in a WCBS on an annual basis. Moreover, by extrapolating to all areas of the UK, it was predicted that in the region of 1,400 squares (approximately half each covered by BC and BTO recorders) could potentially be sampled in a fully-funded WCBS. The results gave real encouragement to the possibility of rolling out the scheme nationally.

Species seen in the pilot studies

On average, nearly 100 individuals of ten species were seen per square based on two visits, indicating that typical WCBS visits were likely to be of sufficient interest to recorders to merit their participation. Thirty-eight species were seen across squares receiving two summer visits. Nearly all wider countryside species (24 of 25) were present in more than 3% of squares surveyed. Therefore a scheme achieving the predicted uptake of more than 1,000 squares would likely yield abundance data for each of these species in more than 30 squares—a statistically valid sample size for generating UK-wide indices and trends.

Launching the WCBS in 2009

Sampling strategy and co-ordination

To capitalise on the success of the pilot studies, a decision was made by BC/BTO/CEH to launch the WCBS in 2009, albeit on a limited budget. The scheme was operated by extending two existing co-ordination mechanisms. Butterfly recorders were organised through a network of local champions within each BC branch (local areas) overseen by a central co-ordinator. Volunteers contributing to the BTO BBS were co-ordinated centrally by the BBS National Organiser. Resources provided for recorders included downloadable recording forms, guidance notes, identification guides and online data entry facilities. WCBS

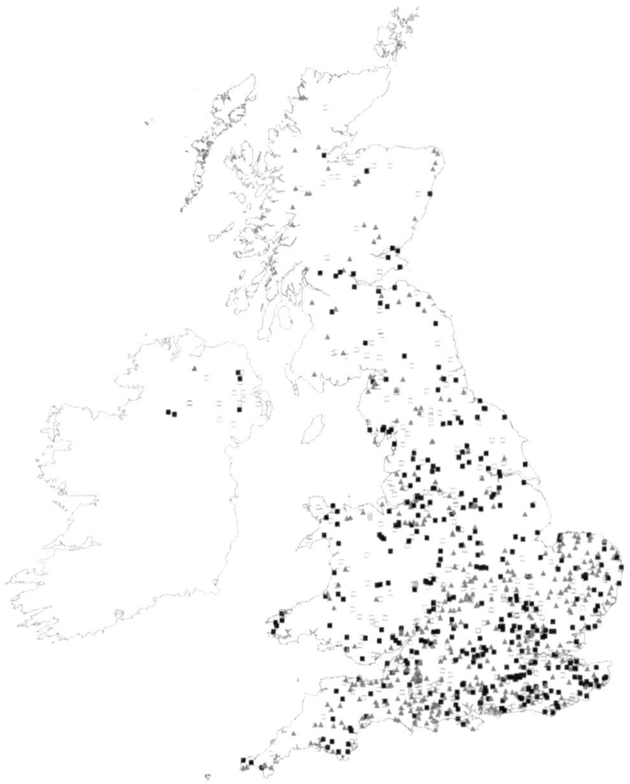

Fig. 3 Location of WCBS sampling squares in 2009. *Filled black squares* (n = 360) represent squares sampled by BTO BBS recorders, *gray triangles* (n = 403) by BC recorders, whilst *open gray squares* (n = 343) represent squares allocated to BC or BTO recorders that subsequently were not sampled

survey squares were selected from two sources. Squares were selected at random, stratified by recorder density for each BC Branch (as measured by the number of active members per branch) and made available to BC recorders. BTO recorders were asked to sample their BBS squares, which had previously been determined by stratified random sampling (pool of 3,211 surveyed squares in 2008). In total, recorders signed up to survey 1,001 squares (Fig. 3), including 11% of BBS squares.

Field methods

The WCBS method used was the same as described above for the pilot study, with refinements including dragonfly and day-flying moth recording during the butterfly survey and an extended time period from 09:30–16:30. Surveyors were requested to undertake a minimum of two visits to squares over July and August, with visits spaced at least 10 days apart. Optional visits were encouraged between May and September, but targeted over the May to June period to pick up Orange-tip and other spring flying species. The importance of following the ideal route as closely as possible to limit bias in habitat sampling was emphasised.

An additional optional component to the WCBS was a rapid survey for selected other insect species, to be completed at the end of the survey. The primary purpose of this was to further evaluate the extent to which a large body of recorders would be prepared to record other insect groups for which there are currently no national data on abundance trends. The *Insect Flower Search* involved counting the numbers of 12 selected insect species on flower heads in a 1 m wide strip to one side of the butterfly transect route over a 10 min period. The twelve species were : Common Carder Bumblebee (*Bombus pascuourum*), Marmalade Hoverfly (*Episyrphus balteatus*), Common Red Soldier Beetle (*Rhagonycha fulva*), Red-tailed Bumblebee (Bombus lapidarius), Hoverfly (*Sericomyia silentis*), the Hoverfly (*Chrysotoxum bicinctum*), Great Pied Hoverfly (*Volucella pellucens*), Thick-legged Flower Beetle (*Oedemera nobilis*), Duck-billed Hoverfly (*Rhingia campestris*), Common Malachite Beetle (*Malacius bipustulatus*), Black and Yellow Longhorn Beetle (*Rutpela maculata*) and the wasp beetle (*Bombus pascuourum*). Recorders were encouraged to note other insect species present identifiable to species level, whilst basic habitat recording (nectar abundance on a scale from 1 to 5) was also completed as part of the survey (see www.ukbms.org.uk).

Participation levels

In total, 1,642 visits were made to 763 1-km squares by 611 recorders, who collectively walked 3,300 km of survey line. BC volunteers sampled 53% of squares, whilst BBS recorders sampled 47%, with regional differences apparent between the two groups (Fig. 3). During July and August, 597 squares (80%) received the required two visits. Optional, extra visits included ~200 in the spring. Recording standards and data quality were thought to be high, with only 0.2% of records obvious misidentifications based on species known distributions and flight times. Of those who participated, more than 70% used the online system for submitting their data.

Butterfly species

Over the 1,642 visits, recorders counted in excess of 119,000 individual butterflies, with an average visit during July and August yielding approximately 80 individuals of 8 species. In total, more than three-quarters (47 of 58) of the UK's regularly occurring (breeding species and annual migrants) butterfly species were recorded (Table 1). Sufficient coverage to compile national relative abundance indices (i.e. present in at least 30 squares) was achieved for all of the target wider countryside species, with the exception of Scotch Argus, *Erebia aethiops* (found in only 17 squares).

Table 1 Abundance and occupancy figures for 46 butterfly species based on two summer visits to 593 squares

Species	Occupancy			Abundance		
	No. squares	%	Rank	Total counted	% of all butterflies counted	Rank
Large White *Pieris brassicae*[a]	534	89.4	1	14,817	15.71	2
Meadow Brown *Maniola jurtina*[a]	521	87.3	2	15,281	16.20	1
Small White *Pieris rapae*[a]	518	86.8	3	13,489	14.30	3
Painted Lady *Vanessa cardui*[b]	493	82.6	4	6,803	7.21	6
Gatekeeper *Pyronia tithonus*[a]	450	75.4	5	10,520	11.15	4
Speckled Wood *Parage aegeria*[a]	441	73.9	6	4,825	5.11	8
Green-veined White *Pieris napi*[a]	429	71.9	7	8,505	9.02	5
Peacock *Inachis io*[a]	408	68.3	8	2,632	2.79	10
Small Tortoiseshell *Aglais urticae*[a]	358	60.0	9	2,642	2.80	9
Comma *Polygobia c-album*[a]	319	53.4	10	1,058	1.12	12
Red Admiral *Vanessa atalanta*[b]	303	50.8	11	910	0.96	14
Ringlet *Aphantopus hyperantus*[a]	278	46.6	12	5,446	5.77	7
Common Blue *Polyommatus icarus*[a]	233	39.0	13	1,643	1.74	11
Small Copper *Lycaena phlaeas*[a]	156	26.1	14	379	0.40	20
Small Skipper *Thymelicus sylvestris*[a]	139	23.3	15	897	0.95	15
Brimstone *Gonepteryx rhamni*[a]	109	18.3	16	277	0.29	22
Large Skipper *Ochlodes sylvanus*[a]	106	17.8	17	403	0.43	19
Small/Essex Skipper[a]	98	16.4	18	960	1.02	13
Small Heath *Coenonympha pamphilus*[a]	81	13.6	19	422	0.45	17
Wall Brown *Lasiommata megera*[a]	66	11.1	20	405	0.43	18
Holly Blue *Celastrina argiolus*[a]	64	10.7	21	95	0.10	27
Essex Skipper *Thymelicus lineola*[a]	62	10.4	22	261	0.28	23
Brown Argus *Plebeius agestis*[a]	58	9.7	23	194	0.21	24
Marbled White *Melanargia galathea*[a]	51	8.5	24	476	0.50	16
Silver-washed Fritillary *Argynnis paphia*	41	6.9	25	139	0.15	26
Clouded Yellow *Coleus croceus*[b]	31	5.2	26	88	0.09	28
Purple Hairstreak *Neozephyrus quercus*[a]	19	3.2	27	50	0.05	29
Orange Tip *Anthocharis cardamines*[a]	11	1.8	28	13	0.01	35
White Admiral *Limenitis camilla*	10	1.7	29	49	0.05	30
Scotch Argus *Erebia aethiops*[a]	9	1.5	30	379	0.40	20
Grayling *Hipparchia semele*	9	1.5	30	169	0.18	25
Dark Green Fritillary *Argynnis aglaja*	9	1.5	30	18	0.02	33
Chalk-hill Blue *Polyommatus coridon*	7	1.2	33	21	0.02	32
Small Pearl-bordered Fritillary *Boloria selene*	3	0.5	34	16	0.02	34
Small Blue *Cupido minimus*	3	0.5	34	6	0.01	36
White-letter Hairstreak *Satyrium w-album*[a]	3	0.5	34	3	<0.01	38
Adonis Blue *Polyommatus bellargus*	2	0.3	37	25	0.03	31
Brown Hairstreak *Thecla betulae*	2	0.3	37	3	<0.01	38
Green Hairstreak *Callophrys rubi*	2	0.3	37	3	<0.01	38
Silver-studded Blue *Plebeius argus*	2	0.3	37	2	<0.01	42
Large Heath *Coenonympha tullia*	1	0.2	41	6	0.01	36
Lulworth Skipper *Thymelicus acteon*	1	0.2	41	3	<0.01	38
Dingy Skipper *Erynnis tages*	1	0.2	41	2	<0.01	42
Pearl-bordered Fritillary *Boloria euphrosyne*	1	0.2	41	1	<0.01	44
Purple Emperor *Apatura iris*	1	0.2	41	1	<0.01	44
Wood White *Leptidea sinapis*	1	0.2	41	1	<0.01	44

[a] Wider countryside species

[b] Regular migrant species

Optional spring visits improved coverage for several species, including Brown Argus (*Plebeius agestis*), Wall Brown (*Lasiommata megera*) but especially Orange-tip which was found in 65 squares. The most widely recorded species over the core July and August period was the Large White (*Pieris brassicae*) occurring in nearly 90% of squares, whilst the most abundant species was Meadow Brown (*Maniola jurtina*) with over 15,000 individuals counted (Table 1). Two satyrids (Meadow Brown and Gatekeeper, *Pyronia tithonus*) and three Whites (Large White; Small White, *Pieris rapae*; Green-veined White, *Pieris napi*) accounted for two-thirds of all butterflies counted. During July and August the three regular migrants species Red Admiral (*Vanessa atalanta*), Painted Lady (*Vanessa cardui*) and Clouded Yellow (*Coleus croceus*) comprised nearly one in ten of all butterflies counted, chiefly because 2009 was one of the best years since 1996 for the Painted Lady (UKBMS unpublished data).

Although habitat specialists (low mobility species chiefly restricted to semi-natural habitats) comprise more than half of UK butterfly species (Asher et al. 2001), they accounted for less than 0.5% of the total number of butterflies counted on WCBS squares in July and August. The most widely recorded habitat specialist was the Silver-washed Fritillary (*Argynnis paphia*), occurring in 52 squares.

Through the pilot studies, 310 squares were surveyed in 2007 and 192 in 2008, whilst 100 squares have been surveyed in all 3 years from 2007 to 2009, enabling comparisons of abundance between years. On average, butterfly abundance on WCBS squares in 2009 was nearly a third higher than 2008 (a similar change to that detected on transect sites, UKBMS unpublished data), with one to two more species seen per visit. 2009 was a particularly good year for the Large White and as previously mentioned the Painted Lady.

Other insects

Dragonflies and day-flying moths were recorded in 383 squares (50% of the total). Fifty-two species of moth (439 individuals) were counted in 233 squares. A migrant species, the Silver-Y (*Autographa gamma*) was the most widespread and abundant species, with 194 individuals counted from 152 squares. The only other moderately widespread species was the Six-spot Burnet (*Zygaena filipendulae*) with 49 individuals counted in 40 squares. Twenty-eight species (plus 3 'Unidentified' groups) of dragonfly (2,370 individuals) were counted in 285 squares. Common Darter (*Sympetrum striolatum*) was the most widespread species (24% of squares), closely followed by Brown Hawker (*Aeshna grandis*) (21%) and Southern Hawker (*Aeshna cycnea*) (21%). The most abundant species was the Common Blue Damselfly (*Enallagma cyathigerum*) accounting for 31% of all individuals seen.

The Insect Flower Search was completed on 40% of squares, with one or more target species detected on 81% of surveys. Across squares, all of the 12 target species were detected, with four species occurring in more than half of the squares sampled. These were Common Carder Bumblebee, Marmalade Hoverfly, Common Red Soldier Beetle and Red-tailed Bumblebee. In contrast, six species occurred in fewer than 10% of squares. The Common Carder Bumblebee was the most widespread species being present in 71% of squares, whilst Marmalade Hoverfly was the most abundant species totalling 34% of all individuals seen.

Habitat coverage

To assess whether the squares sampled in 2009 were a representative sample of the UK landscape, broad habitats (Jackson 2000) sampled in the WCBS squares were compared to their occurrence across the UK as a whole using the national Land Cover Map dataset (Fuller et al. 2002). No difference was found between the proportions of land cover types within the WCBS squares sampled in 2009 ($N = 757$ squares, excluding Northern Ireland) and the proportions found across the UK (*t*-test, $t = 0.00$, $P = 1$; Table 2).

Discussion

Overall evaluation of scheme design

The WCBS borrows heavily from the BBS survey method developed by the BTO, with fine-tuning made for butterflies. In addition to copying a sound scientific design, there are a number of benefits in using the BBS survey design for butterflies. The BBS is a widely regarded scheme (e.g. data used to compile a governmental biodiversity indicator of sustainable development), that has been peer reviewed and thoroughly tested (Greenwood et al. 1995). During the first year of the WCBS, encouraging BBS bird recorders to participate by surveying their squares effectively doubled the coverage that would have been achieved using only established (UKBMS) butterfly recorders. There is considerable further potential to improve coverage with bird recorders, as there are three times more bird recorders contributing to the BBS than there are butterfly recorders in the UKBMS. The combined efforts of bird and butterfly recorders (accounting for overlap), could potentially yield a WCBS sampling 4,000 or more squares. There are also efficiency savings in using BBS squares, because survey routes have already been established.

Table 2 Broad habitats sampled by WCBS squares in 2009 in comparison to the proportions found across the whole UK (CEH Land Cover Map)

Broad habitat	Land cover map (N = 242,308 squares)	WCBS in 2009 (N = 757 squares)
Broadleaved, mixed and yew woodland	6.31	10.24
Coniferous woodland	5.35	3.78
Arable and horticulture	23.45	33.92
Improved grassland	20.15	22.97
Neutral grassland	5.28	4.30
Calcareous grassland	4.40	5.88
Acid grassland	6.04	2.04
Bracken	0.78	0.36
Dwarf shrub heath	10.83	2.65
Fen, marsh and swamp	0.08	0.11
Bogs	2.12	0.72
Standing open water and canals	0.87	0.40
Montane habitats	1.64	0.00
Inland rock	0.90	0.68
Built-up areas and gardens	6.69	10.81
Supra-littoral rock	0.02	0.00
Supra-littoral sediment	0.09	0.14
Littoral rock	0.06	0.02
Littoral sediment	1.19	0.31
Inshore sublittoral	3.73	0.67

However, use of the BBS and bird recorders imposes a number of constraints. A mandatory limit of two summer visits was imposed, because it was thought unrealistic to request bird recorders (chiefly volunteering for a bird monitoring organisation) to put more effort into sampling butterflies than birds. Although efficiency increases with fewer visits (Roy et al. 2007), if occupancy is low then an unrealistic number of sample squares may be required to detect short-term trends with a two visit scheme. A further constraint in the BBS is that section lengths are fixed at 200 m long, rather than variable according to habitat type, making fine-scale habitat analysis more complicated.

There is a tendency for BBS routes to oversample linear features and WCBS routes established by butterfly recorders showed a similar bias. This sampling bias is not a major concern in bird recording as sampling is likely to detect a good proportion of birds present in squares from sampling routes, due to birds being apparent (seen or heard) well beyond survey lines. However, this is not the case for butterflies. Although WCBS results from 2009 indicates that overall sample squares are representative of the UK in habitat terms, within square sampling bias towards linear habitats needs to be accounted for to avoid overestimating absolute butterfly abundance. We suggest it is preferable to account for sampling bias post hoc through a weighting procedure (van Swaay et al. 2002), because eliminating sampling bias in the field would likely involve considerable access problems for volunteer recorders including physical obstacles, potential crop damage and negotiating access permissions as well as spending more time in areas with few or no butterflies; all these factors would be unpopular and threaten the long-term viability of the scheme.

Distance sampling methods were considered in the survey design to enable absolute density estimates to be derived, but rejected on practical grounds. As a consequence, the WCBS generates only relative density estimates and an assumption is made that a constant proportion of butterflies are detected at sample sites. There is a need to test this assumption using long term changes in detection probabilities using UKBMS data and methods to analyse multiple visits as applied for butterflies (Kéry and Plattner 2007, Pellet 2008).

Only limited testing of habitat survey methods was undertaken in the pilot studies. Further research is required as the WCBS would be improved through the incorporation of a habitat survey component specific to butterflies, to aid interpretation of butterfly distribution and trends.

Data quality is a potential issue in the WCBS which relies heavily on bird recorders whose primary interest and recording skills are in another taxonomic group. However, few obvious errors were evident in the survey results in terms of unusual counts, dates or distributions (0.2% of records) and a proportion of these may have been data input errors. The provision of free identification sheets to recorders may have helped to reduce misidentifications. A preliminary analysis of field validation data (squares repeat sampled by both volunteers and professionals) showed that results were extremely similar, again indicating that error rates were low and data quality was high. However, sample sizes were low and more research is required (and planned) in this area.

Evaluation of results from the first year of the survey

The WCBS represents the first UK-wide assessment of the distribution and abundance of common and widespread butterflies using a scientific sampling framework. Moreover it is the first scheme in the UK to assess multi-species invertebrate abundance through random sampling. In total, 763 1-km squares were sampled primarily by volunteer recorders, which we believe is a considerable achievement for a first year especially given limited publicity, a late announcement to the roll out, poor weather in some regions and the scheme running on minimal resources. To put this into context, it took nearly 30 years to achieve a similar level of site coverage through the UKBMS. The involvement of bird recorders was particularly beneficial in improving geographical coverage in regions with few BC recorders.

Volunteer recorders signed up to surveying ~1,000 squares, but only ~75% of these were ultimately surveyed, with final coverage ~15% less than expected. Whilst poor weather over the summer was undoubtedly a contributory factor, these figures highlight the need to better predict coverage in scheme planning. To achieve coverage of more than 1,000 squares, a fully resourced scheme is required to improve promotion, facilitate training and to allow a longer lead in time for the survey to gain momentum.

There is a potential concern due to not all allocated BBS and BC squares being sampled. With surplus squares, recorders may choose those squares where butterfly diversity is likely to be higher, which could lead to overestimates of occupancy and abundance at a national scale. Sampling bias is currently being investigated and will be accounted for when abundance indices and trends are estimated.

More than three-quarters (47 of 58) of the UK's regularly occurring butterfly species were recorded. The scheme successfully detected patterns in butterfly distribution and abundance evident in results from other butterfly recording and monitoring schemes, including a substantial Painted Lady (*Vanessa cardui*) migration (Fox et al. in press), highly restricted distributions in the majority of habitat specialist species (Cowley et al. 1999) and an absence of the Wall Brown across much of central Great Britain (Fox et al. 2006). In particular, abundance of the majority of habitat specialist was extremely low, because these species have highly restricted distributions within semi-natural habitats, confirming the need for other sampling approaches.

All of the target wider countryside species (except Scotch Argus) were detected in a sufficient number of squares (more than 30) to enable a UK-wide abundance index to be compiled, showing that the scheme successfully delivered on its primary purpose. However, no individual species was found in 630 or more squares (maximum 534 for Large White), which is the average number of occupied squares required to detect a 25% change over 10 years. These results suggest that scheme power may be low to detect trends over the short term with the current emphasis on two visits and that optional visits should be encouraged, especially amongst BC recorders. Moreover, optional visits proved valuable in 2009, particularly in the spring, and enabled a baseline to be established for Orange-tip.

On average, recorders counted approximately 80 individual butterflies of eight species per single survey made over the July and August period, confirming that an 'average WCBS' visit is likely to yield a reasonable butterfly return. This was especially so given that it was a below average year for butterflies (UKBMS unpublished data). Over the same period, an average of 130 butterflies from ten species was seen on UKBMS butterfly transects of similar survey length (UKBMS unpublished data). It is encouraging that butterfly diversity for an average visit is not vastly different to transects, as it gives hope that volunteers will find visits sufficiently rewarding to repeat them regularly in future years.

Through a combination of the UKBMS site-based monitoring network and the WCBS, we believe suitable survey methods and adequate sampling strategies are now in place to enable UK-wide assessments of abundance for 53 of the 58 regularly occurring species. Exceptions include montane specialist species such as Mountain Ringlet and canopy-dwelling forest butterflies such as Purple Emperor. The WCBS has highlighted the problems in achieving sampling coverage in some regions, particularly in upland areas were recorded densities are low and the combination of remoteness, steep terrain and a cool climate present considerable recording challenges. This is a wider recording problem affecting other taxa (JNCC unpublished data) and differing approaches may be required to fill coverage gaps in remote areas, including employing contractors or recruiting/training new types of volunteers (e.g. hill walkers). We recommend periodic surveys as a more realistic tool for monitoring montane species, whilst research is required to develop improved monitoring methods for canopy species.

The primary purpose of the Insect Flower Search was to help assess interest levels amongst volunteers in counting numbers of less well studied insect groups and for which national abundance trends are unavailable. The results were encouraging, with 40% of recorders taking part, whilst participation might have been even higher if this component had been made a mandatory part of the WCBS. Moth and dragonfly recording was also well supported during butterfly surveys. However, as the survey routes inevitably included little aquatic habitat, the dragonfly results will have been biased in favour of the more mobile, less habitat-specific species. Nevertheless, these results indicate that there is considerable potential to use the WCBS to collate national abundance data on other insect species. This is important in biodiversity monitoring terms, as insects comprise 56% of known species (Groombridge 1992).

Future plans and prospects

BC, CEH and the BTO aim to run the WCBS on an annual basis, at least for the next 5 years. This will enable a fuller validation of the scheme and an assessment of whether trends in common and widespread butterfly species from WCBS sites are similar to trends obtained from UKBMS transects. The WCBS will be repeated in 2010 with similar levels of resourcing, with a target of maintaining or possibly exceeding coverage levels achieved in the first survey year. Extra visits will be encouraged, as will new recording in areas where coverage was poor in 2009, whilst more validation surveys will be carried out.

The UK was the first country to develop a national scheme to monitor butterfly abundance. The Butterfly Monitoring Scheme, established in 1976 (now the UKBMS) has inspired many others and 17 schemes now operate across Europe, with more planned (van Swaay et al. 2008). The WCBS represents another step forward, by incorporating a random sampling framework into the survey design that should enable unbiased indices to be generated that are representative of the whole landscape. Therefore, the WCBS provides a new monitoring protocol that could help other countries establish efficient and effective butterfly monitoring schemes. This is increasingly important given that butterflies have now been adopted as biodiversity indicators at a European level.

Acknowledgments Funding was provided to develop the WCBS through the UKBMS project (WCO2014 and BD1453) by a consortium of government agencies including Defra, the Joint Nature Conservation Committee, Natural England, Countryside Council for Wales, Natural Environment Research Council, Scottish Natural Heritage, Scottish Executive Environment and Rural Affairs Department, Northern Ireland Environment Agency and the Forestry Commission. We are grateful to the continued support from BC and BTO/JNCC/RSPB Breeding Bird Survey recorders and co-ordinators. We are grateful to many for helpful discussions of field and analytical methods including Valerie Brown, Nick Greatorex-Davies, Roger Dennis, Keith Porter, Richard Pywell, Peter Rothery, Helen Roy, David Sheppard, Tim Shreeve, Mark Stevenson, Chris Thomas, Lawrence Way and Rob Wolton. Help with fieldwork in the pilot studies was given by Julian Hughes, Poppy Lakeman-Fraser, Ian Middlebrook, Thomas Mills, Helena Romo, Silvia Ruggieri, and Duncan Stewart and Sarah Walley, with permission from the Lulworth Estate and the tenant farmer, Paul Simpson. Finally, thanks to Martin Warren for commenting on an earlier version of this manuscript.

References

Aebischer NJ, Evans AD, Grice PV, Vickery JA (2000) Ecology and conservation of lowland farmland birds. British Ornithologists Union, Tring

Asher J, Warren MS, Fox R, Harding P, Jeffcoate G, Jeffcoate S (2001) The millennium atlas of butterflies in Britain and Ireland. Oxford University Press, Oxford

Buckland ST (2006) Point transect surveys for songbirds: robust methodologies. Auk 123:345–357

Buckland ST, Anderson DR, Burnham KP, Laake JL, Borchers DL, Thomas L (2001) Introduction to distance sampling estimating abundance of biological populations, Oxford University Press, Oxford

Buckland ST, Magurran AE, Green RE, Fewster RM (2005) Monitoring change in biodiversity through composite indices. Philos Trans R Soc B 360:243–254

Carey PD, Wallis S, Chamberlain PM, Cooper A, Emmett BA, Maskell LC, McCann T, Murphy J, Norton LR, Reynolds B, Scott WA, Simpson IC, Smart SM, Ullyett JM (2008) Countryside survey: UK results from 2007. NERC/Centre for Ecology & Hydrology, 105 pp

Conrad KF, Warren MS, Fox R, Parsons MS, Woiwod IP (2006) Rapid declines of common, widespread British moths provide evidence of an insect biodiversity crisis. Biol Conserv 132:279–291

Cowley MJR, Thomas CD, Thomas JA, Warren MS (1999) Flight areas of British butterflies: assessing species status and decline. Proc R Soc B 266:1587–1592

Defra (2006) Working with the grain of nature–taking it forward: volume II. Measuring progress on the England Biodiversity Strategy: 2006 assessment. Defra, London

Donald PF, Green RE, Heath MF (2001) Agricultural intensification and the collapse of Europe's farmland bird populations. Proc R Soc B 268:25–29

Firbank LG, Petit S, Smart S, Blain A, Fuller RJ (2008) Assessing the impacts of agricultural intensification on biodiversity: a British perspective. Phil Trans R Soc B 363:777–787

Fox R, Asher J, Brereton T, Roy DB, Warren MS (2006) The state of butterflies in Britain and Ireland. Pisces, Newbury

Fox R, Warren MS, Brereton TM (2010) A new red list of British butterflies, species status, vol 12. Joint Nature Conservation Committee, Peterborough, pp 1–32

Fuller RM, Smith GM, Hill RA, Thomson AG (2002) The UK Land Cover Map 2000: construction of a parcel-based vector map from satellite images. Cartograph J 39:15–25

Greatorex-Davies JN, Roy DB (2001) The butterfly monitoring scheme progress report for 2000/2001. Joint nature conservation committee contract report. NERC Centre for Ecology and Hydrology, Huntingdon

Greenwood JJD, Baillie SR, Gregory RD, Peach WJ, Fuller RJ (1995) Some new approaches to conservation monitoring of British breeding birds. Ibis 137:S16–S28

Groombridge B (1992) Global biodiversity: status of the Earth's living resources. Chapman & Hall, London

Jackson DL (2000) Guidance on the interpretation of the biodiversity broad habitat classification (terrestrial and freshwater types): definitions and the relationship with other habitat classifications. JNCC Report, No. 307. JNCC, Peterborough

Kéry M, Plattner M (2007) Species richness estimation and determinants of species detectability in butterfly monitoring programmes. Ecol Entomol 32:53–61

Krebs JR, Wilson JD, Bradbury RB, Siriwardena GM (1999) The second silent spring? Nature 400:611–612

Leon-Cortes JL, Cowley MJR, Thomas CD (1999) Detecting decline in a formerly widespread species: how common is the common blue butterfly *Polyommatus icarus*? Ecography 22:643–650

León-Cortés JL, Cowley MJR, Thomas CD (2000) The distribution and decline of a widespread butterfly *Lycaena phlaeas* in a pastoral landscape. Ecol Entomol 25:285–294

Newson SE, Evans KL, Noble DG, Greenwood JJD, Gaston KJ (2008) Use of distance sampling to improve estimates of national population sizes for common and widespread breeding birds in the UK. J Appl Ecol 45:1330–1338

Pellet J (2008) Seasonal variation in detectability of butterflies surveyed with Pollard walks. J Insect Conserv 12:155–162

Pollard E (1977) A method of assessing changes in the abundance of butterflies. Biol Conserv 12:115–134

Pollard E, Greatorex-Davies JN (1997) Butterfly monitoring scheme: review and suggestions for the future. CEH, Huntingdon

Pollard E, Yates TJ (1993) Monitoring butterflies for ecology and conservation. Chapman and Hall, London

Pollard E, Moss D, Yates TJ (1995) Population trends of common British butterflies at monitored sites. J Appl Ecol 32:9–16

Pywell R, Warman EA, Sparks TH, Greatorex-Davies NJ, Walker K, Meek WR, Carvell C, Petit S, Firbank LG (2004) Assessing habitat quality for butterflies on intensively managed arable farmland. Biol Conserv 118(3):313–325

Roy DB, Rothery P, Moss D, Pollard E, Thomas JA (2001) Butterfly numbers and weather: predicting historical trends in abundance and the future effects of climate change. J Anim Ecol 70(2): 201–217

Roy DB, Rothery P, Brereton TM (2007) Reduced-effort schemes for monitoring butterfly populations. J Appl Ecol 44:993–1000

UK BAP (2007) UK list of priority species and habitats. Available at http://www.ukbap.org.uk/newprioritylist.aspx

UK Biodiversity Partnership (2007) Conserving biodiversity—the UK approach. Defra, London

van Dyck H, van Strien AJ, Maes D, van Swaay CAM (2009) Declines in common, Widespread Butterflies in a Landscape under Intense Human Use. Conserv Biol 23(4):957–965

van Swaay CAM (1990) An assessment of the changes in butterfly abundance in the Netherlands during the 20th century. Biol Conserv 52:287–302

van Swaay CAM, Plate CL, Van Strien A (2002) Monitoring butterflies in the Netherlands: how to get unbiased indices. Proc Sect Exp Appl Entomol Neth Entomol Soc 13:21–27

van Swaay CAM, Warren MS, Loıs G (2006) Biotope use and trends of European butterflies. J Insect Conserv 10:189–209

van Swaay CAM, Nowicki P, Settele J, van Strien AJ (2008) Butterfly monitoring in Europe: methods, applications and perspectives. Biodivers Conserv 17:3455–3469

Vickery JA, Evans AD, Grice P, Brand-Hardy R, Aebischer NA (2004) Ecology and conservation of lowland farmland birds II: the road to recovery. Ibis 146(Suppl. 2):1–258

J Insect Conserv (2011) 15:321–326
DOI 10.1007/s10841-010-9356-5

ORIGINAL PAPER

Ten challenges for 2010 and beyond to conserve Lepidoptera in Europe

Martin S. Warren · Nigel A. D. Bourn

Received: 23 March 2010 / Accepted: 15 September 2010 / Published online: 9 October 2010

Abstract Butterflies and moths have undergone a serious decline in most European countries following rapid changes in land use in recent decades. The main drivers of loss have been agricultural and forestry intensification, abandonment of marginal land (especially in mountainous regions), loss of traditional management of grasslands and woodlands, and urban spread. Over the same period the science and practice of Lepidoptera conservation has developed considerably and concerted action to save biodiversity has been taken in many countries, with vast areas designated as nature reserves or national parks. Despite this effort, Lepidoptera are still declining at an alarming rate and it is clear that the 2010 target of halting biodiversity loss will not be met. We suggest ten challenges that conservationists in Europe need to address if we are to be successful in halting these losses over coming decades. In this continent, Lepidoptera and their habitats often rely on traditional farming and forestry systems. How can these be brought together in harmony to create a healthier environment in which both humans and wildlife can thrive? The ten challenges include reform of agricultural support, identifying and supporting beneficial forestry systems, managing the matrix between habitats, managing habitats on a landscape scale, mitigating for climate change, creating a robust planning system that protects key sites, developing a comprehensive monitoring programme for Europe, securing long term funding for nature conservation, and ensuring both political and public support.

M. S. Warren (✉) · N. A. D. Bourn
Butterfly Conservation, Manor Yard,
East Lulworth, Dorset BH20 5QP, UK
e-mail: mwarren@butterfly-conservation.org

Keywords Conservation · Butterflies · Moths · 2010 target · Biodiversity · Agriculture

Introduction

Butterflies and moths have undergone a serious decline in most European countries following rapid changes in land use in recent decades. The main drivers of loss have been agricultural and forestry intensification, abandonment of marginal land (especially in mountainous regions), loss of traditional management of grasslands and woodlands, and urban spread. These have led to extensive loss and fragmentation of habitats (Van Swaay et al. 2010). Climate change is expected to be a major additional threat in coming decades and will exacerbate existing problems (Settele et al. 2008).

While butterfly and moth populations have dwindled, the science and practice of Lepidoptera conservation has developed and our knowledge of species life cycles and ecological issues has grown immeasurably. The papers presented in this special issue of Journal of Insect Conservation highlight the progress that has been made in our understanding of the species and the conservation problems they face.

Over the same period, concerted conservation action has been taken in many countries, with large areas designated as nature reserves or national parks. In Europe, the EU Habitats Directive has led to the creation of a network of sites, known as Natura 2000. This includes 26,000 protected areas covering a total area of more than 850,000 km^2, representing approximately 18% of total EU terrestrial area (http://ec.europa.eu/environment/nature/index_en.htm). In the UK, over 10% of the land area has been protected as Sites of Special Scientific Interest and over 8% of the country is designated as National Parks (9.3% England, 19.9% Wales and 7.2% Scotland).

In 1992, heads of state from around the world came together at the Earth Summit in Rio de Janeiro where they signed the Convention of Biological Diversity (CBD) which aimed to achieve a significant reduction of the current rate of biodiversity loss. At EU level, Ministers agreed a target at Goteborg in 2001 to go even further and set an objective to *halt* biodiversity loss by 2010.

Despite this effort, Lepidoptera are still declining at an alarming rate in countries across the world (e.g. Van Swaay et al. 2010 and other papers in this issue). Moreover, where good comparative data exist, there is evidence that butterflies (and possibly moths) are declining faster than other well known taxa (Thomas et al. 2004) and we must conclude that the 2010 target will not be reached. The challenge ahead for biodiversity conservation remains daunting.

Conservationists are rightly re-examining their strategy to halt biodiversity loss and decide whether they should continue with the same approach, but try to implement it more widely after 2010, or develop a new more radical strategy that chimes with a broader agenda aimed at maintaining human health and ecosystem services. We believe that the two are not mutually exclusive and that we need all the arguments at our disposal if we are to halt biodiversity loss in the decades to come.

Within the broader framework of a post 2010 target covered in other papers, we propose ten key challenges for Lepidoptera ecologists and conservationists in the decades ahead. The challenges focus on the situation in Europe where the future of Lepidoptera is closely interwoven with the survival of traditional farming and forestry systems (Potts 1998; Van Swaay et al. 2010). A key issue is whether these can be brought together in harmony to create a healthier environment in which both humans and wildlife can thrive?

(1) *How do we reform agricultural policy to foster sustainable production and reverse the loss of biodiversity*? Agriculture is probably the biggest single driver of land use change affecting Lepidoptera in Europe. The European Red List of Butterflies identifies agricultural intensification and abandonment as by far the two biggest threats to butterflies, being the main threat for nearly 30% of species (Van Swaay et al. 2010). The EU Common Agricultural Policy has subsidised European agriculture for many decades and encouraged intensification, leading to habitat loss. It has also encouraged unsustainable systems that have high carbon emissions and rely on cheap oil (Leng 2009). Under recent CAP reform, a welcome new element has been introduced (known as pillar 2) which pays for public goods such as biodiversity and clean water. However, this shift still only accounts for 6% of CAP expenditure, while the rest still props up intensive systems (1910 million Euro of a total budget of 29,737 Million Euro in 2007) (Institute for European Environmental Policy 2007). Moreover, the 6% expenditure on pillar 2 is often poorly applied in practice and can still drive species loss through inappropriate and uniform actions (Konvicka pers.comm.).

A better system of agricultural payments is needed urgently to maintain wildlife habitats on farmland and encourage environmentally sustainable production. A key concept to target support is the concept of High Nature Value (HNV) farmland (Dover et al. 2010; EEA 2004; Paracchini et al. 2008). Typically these are low intensity farming systems in marginal hill land where abandonment is the major threat. Without such support, these systems, which support over half of Europe's butterflies, could vanish in a generation. Conservationists need to identify which systems are most valuable and why, and identify support payments which work for the farming communities on the ground. We advocate a new alliance between conservationists and farmers to maintain biodiversity and livelihoods in key landscapes. This would bring benefits for wildlife, rural communities and tourism as well as preserving an important part of our cultural heritage. It would also create more sustainable systems that are less reliant on cheap oil and have lower carbon emissions.

(2) *How do we ensure suitable management of woodlands*? Woodland is the second most important biotope for Lepidoptera in Europe and changing woodland management has been identified as an important threat to many species (Van Swaay et al. 2010). Several threatened species are associated with traditional systems of management and have declined seriously when this has ceased (e.g. Warren and Key 1991; Bergman 2001; Benes et al. 2003, 2006; Konvicka et al. 2008; Spitzer et al. 2008).

The challenge is to develop a policy framework so that forestry systems known to be beneficial to Lepidoptera are encouraged in the right regions of Europe. The need for timber and timber products will undoubtedly grow in coming decades and with it the pressure to intensify production, often through plantation woodland. However, there are opportunities to harvest natural woodlands and turn them into thriving providers of resources that we need. One important new potential market is for wood fuel, which provides a renewable resource to replace the use of fossil fuel. In England for example, 60% of the woodland resource is currently unused and woodland biodiversity is declining through neglect (Forestry Commission 2008). The usable wood fuel resource in England alone is estimated at 2 million tones per year, enough to heat 250,000 homes (FC 2008). Using this resource would help the rural economy, combat climate change and help biodiversity. The challenge is making sure it happens in the right way and in the right places across Europe.

(3) *How do we manage the matrix between semi-natural areas*? Many recent papers have suggested that the matrix of the countryside between core semi-natural areas plays a key role in Lepidoptera population survival (e.g. Hanski 1999; Dover and Fry 2004; Merckx et al. 2008, 2009; Dennis 2010). The structure and composition of the matrix may be crucial in governing dispersal between sub-populations and may also provide essential resources for species survival. However, comparatively little is known about the importance of the matrix for different species and further research is needed on the factors that determine a good versus bad matrix. This information is vital to get a correct balance between efforts to maintain the matrix vs effort to conserve semi-natural habitats. In England, the improved management of the matrix is being encouraged by an Entry Level Environment Stewardship scheme (ELS) which now covers almost 60% of farmland (Natural England 2010). However, this takes up as much of the resources as the Higher Level Scheme (HLS) which is targeted at semi-natural habitats. Is this the correct balance and how do we ensure that ELS type schemes deliver what is needed in the matrix?

(4) *How do we ensure habitat management is conducted on a landscape scale to ensure the long term survival of metapopulations*? Historic habitat loss has left many populations stranded in isolated habitats scattered through the landscape. According to the Millennium Ecosystem Assessment (2005), Europe has faced more habitat and ecosystem fragmentation than any other continent. Many species thus rely on networks of fragmented habitats and their long term survival requires the protection and appropriate management of these networks to ensure that they survive as viable metapopulations (e.g. Hanski 1999). For many European species, this requires maintenance of traditional agricultural systems on a vast scale. The only existing mechanism is a major reform of the Common Agricultural Policy (see above). However, the challenge for conservationists is to identify the critical elements of traditional farming system and ensure these are supported in any reforms, while still maintaining innovation in farming.

Even in the most optimistic scenarios, there will not be sufficient funding to manage every semi-natural habitat in every country. Thus we will have to prioritise where scarce resources are focussed to ensure the survival of all threatened Lepidoptera species and conserve butterfly hotspots of diversity. An urgent task is to identify all important butterfly areas of Europe and ensure these are used to target resources. A start has been made by Van Swaay and Warren (2003) who identified an initial selection of Prime Butterfly Areas of Europe. This needs to be built on using up-to-date information on distribution of threatened species.

Using distribution data, Moilanen et al. (2005), have mapped 20 key landscapes that would allow the survival of every resident butterfly species in the UK. Such an approach could be applied elsewhere to identify important landscapes in Europe and identify High Nature Value (HNV) farming systems in Europe to target agricultural support mechanisms in a reformed CAP.

(5) *How do we mitigate for climate change*? (*Can we mitigate for climate change*?) Climate change is expected to have a profound effect on butterfly populations. The Climatic Risk Atlas of European butterflies (Settele et al. 2008) shows that in the worst case scenario 24% of the species modelled will lose 95% of their current climate space by 2080 and even in the moderate scenario over half will lose over 50% of their space. Key factors for species survival are whether they will be able to move into the new climate space that becomes available and whether that space is suitable for breeding when they get there (e.g. will the food-plants be present in the right conditions). Fundamental questions for conservationists are whether we can slow down the rate that species have to shift or can we adapt habitats so that they may be able to support populations even within a radically changed climate. Current advice is to ensure protected habitats are larger and more heterogeneous so that they accommodate a broader range of niches and may allow species possibility of adaptation (Settele et al. 2008; Wilson and Maclean 2010). However, we need far more research to learn whether such an approach works and whether special measures can be taken to maximise chances of in situ population survival.

A good example from the UK is the Large Blue, *Maculinea arion*, which became extinct in 1979 but has since been successfully re-established at over 30 sites (Thomas et al. 2009). The species relies on a single species of ant that require short turf on freely drained soils, making it highly vulnerable to droughts. It is thought that the extreme drought of 1976 knocked the last remaining colony down so far that it could not recover. Since its inception, the recovery programme has always aimed to restore the butterfly to a wide range of sites, with a variety of different aspects and topographical conditions. This should maximise its chances of survival when the next extreme drought comes.

(6) *A robust planning system throughout Europe that protects our best sites and fosters wildlife conservation within urban areas.* Urban spread and built developments are a major threat to many European butterflies, including over half of all threatened species (Van Swaay et al. 2010). Tourism development and associated infrastructure are a particular threat to endemic island species, for example on Madeira and the Canary Islands. The last remaining site of the endemic Chelmos Blue *Polyommatus iphegenia* is also directly threatened by quarrying for aggregate to construct a road to a new ski station in Greece (S Spencer pers. comm.) and several important sites on the edge of the Black Sea in Bulgaria are threatened by golf course and other tourist developments (Abdejeiv and Beshkov 2007). Conservationists need to ensure a robust planning system across Europe that protects key sites from development and compensates for any losses. The network of Prime Butterfly Areas of Europe needs to be embedded as a first priority within the planning systems of each member states and by expanding the Natura 2000 network. A comprehensive European mapping scheme is also needed urgently to document the location of all important sites and make use of naturalists' records (see below).

(7) *A comprehensive system for recording and monitoring European Lepidoptera.* Two systems need to be developed. Firstly a pan European recording system is needed to collate distribution data and make it available to planners, conservationists and other users. Dutch Butterfly Conservation has developed an online recording system (www.telmee.nl) that could be used to capture records where no country scheme exists but a system is also needed to bring together records from all existing schemes.

The second element is a system to monitor butterflies at a population level in order to detect changes at an early stage and assess the impact of changing land use policies. This would contribute to Article 17 of EH Habitats and Species Directive, which requires an overview of the species on the annexes every 6 years. Robust systems of monitoring both butterflies and moths have now been developed, tested, and applied in several countries. The method of counting butterflies along standard transect routes is now used in over 14 countries in the first ever continental scale network (Van Swaay and Van Strien 2008; Brereton et al. 2010). The network has been used to produce an immensely valuable European Grassland Butterfly Indicator as well as a European Climate Change Indicator (Van Swaay et al. 2008). At a country level, the results are used to develop national and regional butterfly indicators as well as assessing the impact of land use policies such as agri-environment schemes (e.g. Brereton et al. 2007). However, many countries still lack a butterfly monitoring programme and have no objective system of assessing change or identifying conservation priorities. A priority is therefore to develop a monitoring network in every country both to assess national trends and contribute to a representative European trend.

An objective monitoring system for moths has also been developed in the UK, known as Rothamsted Light trap network (Conrad et al. 2006; Fox et al. 2006). This has run since 1968 and is based on a network of around 100 traps. A similar system in other countries is needed to gather information on this important group, and establish trends in a similar way to butterflies. As moths are important parts of the food-chain, they would be a valuable indicator of ecosystem health and functionality.

(8) *Long term funding for nature conservation.* Most nature conservation and research is funded as short-term projects, lasting 3–5 years at most. Yet, effective conservation requires long term research and monitoring, and long term effort to conserve species and habitats at a landscape scale. This mismatch leads to waste of effort and a grossly inefficient system of conservation. Conservationists must argue for a better system of funding which allows for continuity of effort, focused in core landscapes and on key research topics. In effect nature conservation needs to be positioned as part of the core fabric of society and not an adjunct that is good to do when convenient. Two steps to help us get to this are our final two challenges.

(9) *How do we make wildlife matter to politicians*? All the above challenges would come to nothing if we cannot persuade politicians that it is worth spending government funds on conserving nature and natural resources. Politicians face many demands on the use of government funds, so we have to make a strong case as to why a significant proportion must be spent on conserving the natural world. Many powerful

arguments have been proposed by the Millennium Ecosystem Assessment (2005) including provisioning services like food and fibre, regulating services like controlling climate change and fresh water, cultural services such as spiritual enrichment and aesthetic values, and supporting services such as biomass production and water cycling. More recently Pavan Sukdev and his colleagues have tried to put a value on ecosystem services so that society puts a proper value on natural resources, including wildlife, in the planets support services (TEEB 2009).

However, despite these compelling arguments, markets and governments are slow to respond, and spending on the environment remains small in most countries and could well lose out as the human population increases and politics becomes driven by concerns about food security and human health. Conservationists need to make a continued effort to win the hearts and minds and budgets of politicians. This type of lobbying comes hard for most scientists, and indeed many conservationists, who would far rather be out in the field or managing their favourite nature reserve. However, we need to make a concerted effort to make sure our efforts are not in vain. A firm target to halt biodiversity loss post 2010 is vital, but this must be backed by strong political leadership if we are to stand a chance of achieving this goal in coming decades.

(10) *How do we make wildlife matter to people*? We will not achieve political support for conservation unless the public care. Although there is a huge interest in wildlife amongst the public, we live in an increasingly urbanised society, where people are becoming more and more removed from direct contact with nature and wildlife.

Butterflies and moths have a special role to play in popularising nature. They are beautiful, fascinating and relatively easy to study. Many are easy to identify, even by the most inexperienced amateur. Many can also be found in urban areas, close to where people live. Although moths may not seem as popular as butterflies, they have the great advantage that they come to you if you run a moth trap. They can thus be a valuable educational tool to teach children about the world on their doorstep.

There are many good examples from around Europe of good educational programmes and citizen science projects that can raise awareness (Fox et al. 2010). We need to capitalise on these examples and use the popularity of butterflies to develop a wider interest in the public and raise awareness of the fundamental role of biodiversity in maintaining a healthy planet.

Conclusions

Biodiversity loss is one of the biggest challenges facing the world: heads of state from around the word made a commitment to slow this loss at the Rio Convention in 1992. In Europe, EU Ministers agreed to go even further and set an objective to *halt* biodiversity loss by 2010. The decline of butterflies and moths in Europe suggest that this objective will not be met and we will have to renew our efforts in the coming decade. The EU Council has already discussed a new target and has adopted a headline target of halting the loss of biodiversity and the degradation of ecosystem services in the EU by 2020, and restoring them in so far as feasible (Council of the European Union 2010). They are also considering several more specific sub-targets which will help assess progress.

Although the ten challenges we propose are ambitious, we believe that they are essential if we are to stand a chance of meeting a new target of halting biodiversity loss and maintaining Europe's rich wildlife heritage.

References

Abdejeiv S, Beshkov S (2007) Prime butterfly areas in Bulgaria. Pensoft, Sofia

Benes J, Konvicka M, Dvorak J, Fric Z, Havelda Z, Pavlicko A, Vrabec V, Weidenhoffer Z (eds) (2003) Butterflies of Czech Republic: distribution and conservation. SOM, Prague

Benes J, Cizek O, Dovala J, Konvicka M (2006) Intensive game keeping, coppicing and butterflies: the story of Milovicky Wood, Czech Republic. Forest Ecol Manag 237:353–365

Bergman K-O (2001) Population dynamics and the importance of habitat management for conservation of the butterfly *Lopinga achine*. J Appl Ecol 38:1303–1313

Brereton TM, Warren MS, Roy DB, Stewart K (2007) The changing status of the Chalkhill Blue butterfly *Polyommatus coridon* in the UK: the impacts of conservation policies and environmental factors. J Insect Conserv 12:629–638

Brereton T, Roy DB, Middlebrook I, Botham M, Warren M (2010) The development of butterfly indicators in the United Kingdom and assessments in 2010. J Insect Conserv. doi:10.1007/s10841-010-9333-z

Conrad KF, Warren MS, Fox R, Parsons MS, Woiwod IP (2006) Rapid declines of common, widespread British moths provide evidence of an insect biodiversity crisis. Biol Conserv 132: 279–291

Council of the European Union (2010) Council conclusions on biodiversity post-2010—EU and global vision and targets and international access and burden sharing regime. 3002nd Environment Council meeting, Brussels, 15 March 2010

Dennis RLH (2010) A resourced based habitat view for conservation: butterflies in the British landscape. Wiley, Oxford

Dover JW, Fry GLA (2004) Experimental simulation of some visual and physical components of a hedge and the effects on butterfly behaviour in an agricultural landscape. Ent Exp Appl 100: 221–233

Dover JW, Spencer S, Collins S, Hadjigeorgiou I, Rescia A (2010) Grassland butterflies and low intensity farming in Europe. J Insect Conserv. doi:10.1007/s10841-010-9332-0

European Environment Agency (EEA) (2004) High nature value farmland: characteristics, trends and policy challenges. EEA, Copenhagen

Forestry Commission (2008) A wood fuel strategy for England. Forestry Commission England, Cambridge

Fox R, Conrad KF, Parsons M, Warren MS, Woiwod IP (2006) The state of Britain's larger moths. Butterfly Conservation and Rothamsted Research, Dorset

Fox R, Hill L, Randle Z, Anders S, Wiffen L, Parsons MS (2010) Moths count: recording moths for conservation in the UK. J Insect Conserv. doi:10.1007/s10841-010-9309-z

Hanski I (1999) Metapopulation ecology. Oxford University Press, New York

Institute for European Environmental Policy (2007) Towards the CAP health check and the European budget review. IEEP, London

Konvicka M, Novak J, Benes J, Fric Z, Bradley J, Keil P, Hrcek J, Chobot K, Marhoul P (2008) The last population of the Woodland Brown butterfly (*Lopinga achine*) in the Czech Republic: habitat use, demography and site management. J Insect Conserv 12:1366–1638

Leng RA (2009) Peak Oil, resource depletion, global warming, financial stress and future world food and feed production. In: Preston R, Ogle B (eds) International Conference on Livestock, Climate Change and the Environment. An Giang University, Vietnam

Merckx T, Feber RE, Dulieu RL, Townsend MC, Parsons MS, Bourn NAD, Riordan P, Macdonald DW (2008) Effect of field margins on moths depends on species mobility: field-based evidence for landscape-scale conservation. Agric Ecosyst Environ 129:302–309

Merckx T, Feber RE, Riordan P, Townsend MC, Bourn NAD, Parsons MS, Macdonald DW (2009) Optimizing the biodiversity gain from agri-environment schemes. Agric Ecosyst Environ 130:177–182

Millennium Ecosystem Assessment (2005) Ecosystems and human well-being: current state and trends. Island press, Washington

Moilanen A, Franco AMA, Early RI, Fox R, Wintle B, Thomas CD (2005) Prioritizing multiple-use landscapes for conservation: methods for large multi-species planning problems. Proc R Soc B 272:1885–1891

Natural England (2010) Environmental Stewardship Update. www.naturalengland.org.uk. February 2010

Paracchini ML, Petersen J-E, Hoogeveen Y, Bamps C, Burfield I, Swaay C van (2008) High nature value farmland in Europe: an estimate of the distribution patterns on the basis of land cover and biodiversity data. EUR—Scientific and Technical Research series, JRC Scientific and Technical Reports. Office for Official Publications of the European Communities, Luxembourg

Potts R (1998) Effects of human interference on the landscape with special reference to the role of grazing livestock. In: Wallis DeVries MF, Van Bakker JP, Wieren SE (eds) Grazing and conservation management. Kluwer, Dordrecht, pp 107–134

Settele J, Kudrna O, Harpke A, Kühn I, van Swaay C, Verovnik R, Warren M, Wiemers M, Hanspach J, Hickler T, Kühn E, van Halder I, Veling K, Vliegenthart A, Wynhoff I, Schweiger O (2008) Climatic risk atlas of European butterflies. Pensoft, Sofia. www.pensoftonline.net/biorisk

Spitzer L, Konvicka M, Benes J, Tropek R, Tuf IH, Tufova J (2008) Does closure of traditionally managed open woodlands threaten epigeic invertebrates? Effects of coppicing and high deer densities. Biol Conserv 141:827–837

TEEB (2009) The economics of ecosystems and biodiversity for national and international policy makers—summary: responding to the value of nature. United Nations Environment Programme. www.teebweb.org

Thomas JA, Telfer MG, Roy DB, Preston CD, Greenwood JJD, Asher J, Fox R, Clarke RT, Lawton JH (2004) Comparative losses of British butterflies, birds and plants and the global extinction crisis. Science 303:1879–1881

Thomas JA, Simcox DJ, Clarke RT (2009) Successful conservation of a threatened Maculinea butterfly. Science 325:80–83

Van Swaay CAM, Van Strien AJ (2008) The European butterfly indicator for grassland species 1990–2007. Report VS2008.022, De Vlinderstichting, Wageningen

Van Swaay CAM, Warren MS (eds) (2003) Prime butterfly areas of Europe: priority sites for conservation. National reference centre for agriculture, nature conservation and fisheries, The Netherlands

Van Swaay CAM, Van Strien AJ, Julliard R, Schweiger O, Brereton T, Heliölä J, Kuussaari M, Roy D, Settele J, Stefanescu C (2008) Developing a methodology for a European butterfly climate change indicator. Report VS2008.040, De Vlinderstichting, Wageningen

Van Swaay C, Cuttelod A, Collins S, Maes D, López Munguira M, Šašić M, Settele J, Verovnik R, Verstrael T, Warren M, Wiemers M, Wynhof I (2010) European red list of butterflies. IUCN and butterfly conservation Europe. Publications Office of the European Union, Luxembourg

Warren MS, Key RS (1991) Woodlands: past, present and potential for insects. In: Collins NM, Thomas JA (eds) The conservation of insects and their habitats. Academic press, London, pp 155–211

Wilson RJ, Maclean IMD (2010) Recent evidence for the climate change threat to Lepidoptera and other insects. J Insect Conserv. doi:10.1007/s10841-010-9342-y

SUBJECT INDEX

2010 target, 1, 2, 5, 12, 15, 16, 20, 21, 24, 25, 40, 73, 332, 361, 362
abandonment, 22, 58, 86, 91, 92, 97, 189, 190, 247, 248, 250, 256–259, 339–343, 345–347, 361, 362
abiotic, 126, 137, 201, 208, 247–258
aboriginal, 297
abundance, 3, 15–18, 20–23, 25, 26, 29, 30, 35, 39, 40–47, 49, 52–55, 57–59, 63, 67, 69, 74, 75, 77–81, 83, 97, 111, 120, 123–126, 129, 130, 133, 139, 141–146, 148, 150, 152, 154, 155, 158, 160, 161, 163–169, 172–180, 183–187, 196, 202, 207, 213–215, 220, 225–227, 229, 231–241, 244, 247–249, 251, 256, 258, 259, 261, 265, 267, 271, 272, 274–277, 279, 280, 285, 289–295, 298, 301, 302, 304, 308, 313, 315, 329, 332, 334, 337, 346, 349, 350–360
abundances, 54, 67, 150, 159, 164–166, 174, 175, 177, 198, 204, 207, 208, 237, 238, 240, 241–243, 256, 267, 274, 289, 293, 294, 295
abundance-temperature relationship, 142
Acherontia atropos, 33
Achillea millefolium, 228, 244
adaptation, 1, 109, 110, 120, 121, 147, 148, 351, 363
adaptations, 119
adapted, 23, 64, 108, 109, 120, 125, 142, 157, 169, 280, 322, 336
Adonis blue, 355
adult resources, 123, 124, 129, 130, 222, 234
Aeromachus inachus, 87
afforestation, 89, 91, 92, 96, 155, 157, 317, 340, 350
Africa, 67, 70, 74, 75, 80, 82, 83, 104, 334, 336
afrotropical, 70, 71–73, 79, 83
afrotropics, 80
age, 138, 160, 236, 238, 269, 296, 342, 347
aggregate, 334, 364
Aglais io, 228, 229, 234, 244
Aglais urticae, 55, 56, 58, 65, 130, 138, 144, 146, 228, 229, 244, 250, 355
agricultural improvement, 39, 125, 189, 317–319, 340
agricultural landscape, , 3, 101, 120, 187, 234, 237, 260, 261, 263, 268, 365
agricultural systems, 2, 267, 363
agri-environment, 226, 259, 303, 314, 340,
agri-environment schemes (AES), 2, 3, 15, 17, 18, 21, 24, 41, 85, 225, 226, 234, 235, 244, 245, 248, 259–262, 266–269, 321, 324, 345, 350, 364, 366
Agrochola lychnidis, 38
Agrostemma githago, 244
Agrotis puta, 35, 36
Agrynnis adippe, 127, 130
Agrynnis aglaja, 130
air quality, 39
Alcea rosea, 240
Alcis jubata, 38
algae, 39, 47, 115
alien species, 87–89, 91, 93, 94, 97, 99, 102
allee effect, 198, 199, 288
Allocasuarina, 274, 276, 282
alps, 49, 99, 101, 110, 112, 163, 164, 168–170, 238, 267, 328, 329, 346, 347
altitude, 112, 139, 165, 170, 172, 198, 204, 247, 248, 250–255, 257, 344, 345
American Indians, 149
Andorra, 51, 55, 56, 58, 63–66
Andorran Butterfly Monitoring Scheme, 2, 52, 63
Anisynta sphenosema, 277, 280, 282–286
Anolis carolinensis, 97, 100, 102
anomalous, 38
ant brood, 132
ant host, 112, 123, 132
Anthemis tinctoria, 244
Anthocharis cardamines, 56, 88, 91, 144, 225, 228, 251, 352, 355
Antigius butleri kurinodakensis, 86, 87
ants, 109, 124, 163–165, 168–180, 187, 188
Apatura ilia, 241, 242
Apatura iris, 355
Aphantopus hyperantus, 8, 144, 229, 244, 251, 355
Aphrodite fritillary, 152
Aporia crataegi, 56, 251
Aporia hippia, 88, 95, 96, 100
Aporophyla lutulenta, 38
arable, 26, 52, 92, 134, 137, 138, 179, 222, 226, 227, 232, 235, 236, 238, 244, 245, 259, 340, 353, 357, 359
arable plants, 226
Araschnia levana, 244
Arctia caja, 39, 40
Arctiidae, 30
area, 2, 3, 9, 17, 18, 25, 31, 39, 40, 51–53, 59, 64, 68, 69, 71–73, 75–80, 83, 86, 89–98, 100, 101, 107–109, 114, 127, 128, 135, 136, 139, 151, 160, 164–169, 172–175, 178, 187, 190–193, 198, 202, 203, 207, 209, 211–213, 215–219, 222, 227, 238, 239–244, 247–251, 254–257, 262–264, 270–282, 287, 288, 290, 303, 306, 308, 310–313, 317–321, 323, 325, 328, 330, 333–336, 341–343, 351, 353, 357, 361
area of occupancy (AOO), 71, 72
Argynnis adippe, 7, 38, 130, 140, 144
Argynnis aglaja, 144, 355
Argynnis paphia, 54–56, 125, 130, 144, 242, 251, 355, 356
Argyreus hyperbius, 94
Aricia agestis, 127, 130, 144
Aricia eumedon, 65
Aricia nicias, 65
Aristolochia pistolochia, 64
Arogos skipper, 152, 156, 160
Asclepiadaceae, 333, 334, 336

Asclepias curassavica, 333, 334
aspect, 65, 120, 126, 173, 203, 216, 227, 247, 250, 252–257, 305, 307
Aspitates gilvaria, 33
assemblages, 25, 40, 41, 60, 101, 111, 112, 124, 134, 166, 236, 244, 245, 271, 287, 288
Atewa range, 76, 77, 83
Athetis pallustris, 33
Atlantic archipelagos, 333, 334
Atlantis fritillary, 292
Atolmis rubricollis, 35–37, 39
Atrytone arogos, 149, 152, 156, 160, 161, 302
Australia, 3, 104, 160, 271–274, 277–279, 281, 282, 287, 288, 335
autumn cut, 232
autumnal rustic, 38
Azores, 334

Balearic Islands, 52
banana bait, 77
Banksia, 273, 287, 288
bare ground, 109, 130, 190, 201, 203–208, 211, 213–216, 221, 227, 310
barley, 227
barrens, 26, 150–152, 155, 159–161, 289–291, 297, 301, 302
barriers, 138, 207, 208, 211, 212, 222, 262, 264, 265, 280, 295, 296, 313, 323
basking, 190, 199, 306, 322, 324, 325
beaded chestnut, 38
beans, 227
bees, 170, 187, 225, 226, 235, 259, 346
behaviour, 3, 26, 37, 73, 92, 112, 123, 124, 139, 169, 173, 187, 199, 212–214, 218, 219, 222, 223, 261, 264, 265, 268–270, 288, 322, 325, 365
behavioural, 261, 264–268
behavioural response, 264
Belgium, 25, 39, 41, 58, 74, 288, 301, 332
Bern Convention, 65, 103–110, 170, 189, 201
Bicyclus istaris, 81
Bicyclus larseni, 77, 81
Bicyclus safitza, 81
Bicyclus sylvicolus, 77, 81
Bicyclus uniformis, 81
bioclimate models, 103–105
biodiversity action plan, 15, 33, 38, 41, 134, 189, 211, 223, 319
biodiversity hotspot, 58, 76, 287
biodiversity hotspots, 60, 75, 83, 238, 259, 272, 288, 346
biodiversity indicators, 15–17, 20, 23, 25, 26, 67, 69, 359
biodiversity monitoring programme, 237, 239, 241–243
biotic homogenisation, 141, 145
biotope, 42, 60, 161, 236, 245, 259, 263, 268, 270, 302, 332, 337, 347, 360, 362
biotope association, 267
biotopes, 5, 30, 52, 60, 125, 130, 137, 146, 262
birds, 6, 12, 13, 16, 18, 24–26, 30, 40–43, 48, 49, 54, 58–60, 68, 69, 72–74, 76, 102, 170, 235, 236, 287, 314, 327, 340, 347, 350, 351, 353, 357, 359, 360, 366
Bischofia javanica, 97
black and yellow longhorn beetle, 354
Black sea, 364
Blair's shoulder-knot, 32, 35, 36, 38
blue-bordered carpet, 38
BMS, 17, 20–22, 125, 126, 135, 352
BNM, 5, 6, 7, 349
bog copper, 292, 302
bog fritillary, 269, 292, 301, 302
bogs, 3, 172, 289–292, 294–297, 301, 302, 318, 330, 357
Boloria bellona, 152
Boloria dia, 56, 251
Boloria eunomia, 292, 293, 298
Boloria euphrosyne, 7, 125, 130, 135, 144, 251, 355
Boloria freija, 292, 293, 299
Boloria frigga, 292
Boloria montinus, 292
Boloria pales, 65
Boloria selene, 65, 127, 130, 144, 250, 292, 303–309, 314, 315, 355
Bombus lapidarius, 354
Bombus pascuourum, 354
Borneo, 39, 106
botanical diversity, 48, 235–267
Brachypodium, 130, 341
Brachypodium pinnatum, 130, 341
bracken, 12, 304–308, 310, 357
braconid, 318
breeding, 12, 16, 41, 49, 60, 94, 100, 111, 114, 120, 128, 129, 136, 147, 191, 197, 271, 291, 304–306, 308–314, 317, 320, 321, 323, 330, 349, 351, 354, 359, 363
Breeding Bird Survey (BBS), 351–354, 356–358
breeding habitat, 197, 305, 306, 308, 310–313
Brenthis daphne, 89, 242
brimstone, 355
brindled ochre, 38
British Isles, 139, 147, 223, 236, 259, 267, 317, 318
broad-bordered yellow underwing, 38
brown argus, 137, 355, 356
brown elfin, 292
brown hairstreak, 113, 114, 355
budburst, 2, 113–121
buddleja, 129
buff footman, 35, 36, 38, 39
buffer strips, 225–227, 232–234
building, 15, 71, 98, 119, 141, 142, 201, 225, 261, 340, 349
Bulgaria, 101, 346, 364, 365
bumblebees, 235–237, 242–245
burning, 92, 97, 101, 149–152, 154–161, 190, 193
burnt, 193, 273
Bursaphelenchus xylophilus, 94
bushland, 274, 287, 288
Butterflies for the New Millennium (BNM), 5, 349
butterfly collectors, 126
butterfly communities, 51, 55, 57–60, 75–79, 92, 98, 100, 145, 237, 242, 259, 269, 287, 288, 346, 347
butterfly community recovery, 3, 75
Butterfly Conservation Europe, 20, 26, 171, 366
Butterfly Monitoring Scheme (BMS), 17, 125

cacyreus marshalli, 56, 344
calcareous, 2, 7, 19, 20, 60, 125, 132, 134, 201, 202, 205–208, 287, 310, 318, 339, 347, 357
calcareous grassland, 7, 19, 20, 60, 134, 201, 202, 205–208, 347, 357
Callophrys augustinus, 292
Callophrys rubi, 55, 56, 144, 355
Calophasia lunula, 32, 35, 36, 39
camberwell beauty, 54
Canary Islands, 334, 364
Candalides acastus, 277, 283–286
Canopy (forest), 75–77, 79, 80, 157, 158, 213, 217, 291, 331, 352, 358
captive breeding, 111, 317, 320, 323
Carabus olympiae, 105, 110, 112
Carcharodus alceae, 56, 240, 244, 245
carrying capacity, 123, 125, 131–133, 135, 136, 146, 262, 263, 313, 314, 324
Carterocephalus palaemon, 87, 88, 124, 127, 130, 138, 139
Carterocephalus palaemon akaishianus, 87
cassava, 76
Castniidae, 273, 277, 287
Casuarina, 274, 276, 282
Catalan Butterfly Monitoring Scheme, 52, 59, 63
Catalonia, 51–56, 58, 59, 63
caterpillars, 49, 113, 114, 120, 121, 139, 170, 172, 173, 178, 317, 318, 323
Catocala nupta, 35, 36
cattle, 178, 189, 190, 194, 199, 206, 307, 313, 321, 324, 342–344, 347
cattle grazing, 189, 190, 199, 206, 307, 313, 324, 347
CEH, 12, 17, 20, 24, 59, 123, 198, 247, 314, 350, 353, 357–359
Ceiba pentandra, 76
Celastrina argiolus, 55, 56, 129, 144, 242, 251, 355
Celastrina ogasawaraensis, 86, 87, 93, 95, 96, 99, 101, 102
Centaurea cyanus, 244
Centaurea jacea, 241, 242, 244
Centaurea nigra, 228, 234
Centaurea spp., 250, 252–255
Central Europe, 52, 110, 120, 187
Centre for Ecology and Hydrology (CEH), 5, 141, 160, 208, 209, 235, 349, 350, 359
Cephrenes augiades, 275
Cephrenes trichopepla, 275
Cerambyx cerdo, 105, 110
Cercyonis pegala, 152
cereal, 238
chalk carpet, 33
chalk-hill blue, 355
challenges, 1–3, 25, 60, 67, 71, 72, 74, 100, 147, 151, 160, 346, 358, 361, 362, 364–366
Channel Islands, 5, 31, 32
Chapman's blue, 52
Charaxes candiope, 77, 81
Charaxes fulvescens, 80, 81
Charaxes pollux, 81
Charaxes zelica, 81
Charaxinae, 77, 81
charcoal, 93
Chelmos blue, 364
chestnut heath, 52
China, 86, 89, 94
Chloroclysta siterata, 38
Chlorophora excelsa, 76
Chlosyne harrisii, 292
Chrysotoxum bicinctum, 354
Chrysozephyrus hisamatsusanus, 92, 102
Cichorium intybus, 244
Cirsium, 233, 234, 318, 342
Cirsium dissectum, 318
citizen science, 31, 365
clearing, 79, 91, 125, 221, 273, 274, 281, 336
clearings, 89, 90, 130, 220, 309, 310
climate, 1, 3, 5–7, 11–13, 16, 22–27, 30, 32, 39–43, 46, 48, 49, 51, 52, 59, 60, 61, 67, 75, 91, 94, 99, 100, 102–113, 115, 118, 120, 121, 125, 126, 128, 132, 135, 137–150, 158–160, 163, 164, 170, 201, 209, 223, 261, 262, 265, 268, 272, 273, 287, 288, 294, 300–303, 323, 328, 332, 337, 346, 349–351, 358, 360, 361, 363–366
climate change, 1, 3, 5, 6, 11, 12, 22–26, 30, 32, 39, 41, 42, 46, 49, 51, 52, 59, 60, 61, 67, 75, 94, 100, 102–113, 120, 121, 125, 128, 137–139, 141–148, 158, 160, 201, 209, 223, 261, 265, 268, 272, 287, 300–303, 323, 332, 349, 350, 360, 361, 363–366
climate envelope, 135, 141
climate space, 105, 106, 108, 363
climatic envelope, 59
climax community, 290
CLIMIT, 137, 147, 169
close-cropped, 131
closed population, 126, 251
Clossiana dia, 242
clouded yellow, 240, 355, 356
coastal peatland, 290, 292
cocksfoot, 226
cocoa, 76
Coenonympha arcania, 55, 56, 250, 251
Coenonympha dorus, 54–56, 59
Coenonympha glycerion, 52, 56, 58
Coenonympha hero, 89, 105, 107–109, 111
Coenonympha oedippus, 86, 87, 92, 105, 107, 108
Coenonympha pamphilus, 55, 56, 125, 144, 146, 228, 240, 243, 244, 350, 355
Coenonympha tullia, 128, 130, 137, 144, 156, 268, 287, 292, 301, 355
co-evolution, 158, 289, 296, 297
Coleus croceus, 355, 356
Colias alfacariensis, 55, 56, 239
Colias crocea, 55, 56, 65, 251
Colias croceus, 7, 144, 228, 244
Colias hyale, 239, 240, 242, 244
Colias interior, 292
Colias myrmidone, 3, 245, 327, 328, 330, 345, 346
collecting, 40, 44, 72–74, 96, 99, 102
collectors, 92, 126–129, 319
Coloephora wockeella, 314
colonise, 126, 226, 232
colonization, 63, 109, 111, 112, 145, 179, 180, 187, 199, 271, 288, 295, 297, 311, 313

colonized, 171, 262, 309, 341
Colorado, 153, 267, 292
Colotis evagore, 59
comma, 7, 8, 10–12, 25, 35, 55, 56, 106, 111, 124, 125, 127, 130, 132, 134, 137–139, 144, 262, 268, 269, 292, 301, 355
Common Agricultural Policy (CAP), 2, 151, 152, 259, 301, 339, 340, 345, 346, 362, 363, 366
common blue, 229, 240, 269, 355, 356, 359
common carder bumblebee, 354, 356
common darter dragonfly, 35
common dog-violet, 304
common malachite beetle, 354
common red soldier beetle, 354, 356
common sorrel, 230
common species, 24, 38, 43, 45, 46, 51, 141, 174, 214, 237, 242, 262, 267
common toadflax, 32
common wood-nymph, 152
communities, 25, 26, 31, 32, 40, 51, 54, 55, 57–60, 75–79, 83, 92, 98, 100, 129, 139, 141, 145, 147, 158, 159, 166, 169, 170, 172, 174, 179, 188, 203, 233, 235–237, 242, 243, 258, 259, 262, 269–271, 287, 288, 301, 314, 315, 318, 325, 341, 346, 347, 362, 366
community change, 346, 347, 362, 366
competition, 48, 91, 93, 94, 159, 164, 172, 208, 233, 341
competitive grasses, 225, 233
connectedness, 289
connectivity, 48, 190, 201, 208, 209, 238, 243, 244, 262–265, 267, 268, 270–272, 274, 276, 279, 280, 304, 306, 308, 310, 313, 314, 340
conservation priorities, 60, 69, 83, 103, 105, 109, 145, 272, 303, 364
conservation priority, 5, 125, 328
conspecifics, 264, 265
consumables, 262, 265, 266, 268
Convention on Biological Diversity (CBD), 1, 2, 15, 16, 22–24, 26, 362
core, 23, 25, 92, 141, 142, 147, 148, 151, 156, 159, 189, 192, 199, 204–209, 239, 245, 267, 272, 311, 319–321, 323–325, 327, 356, 363, 364
Coreana raphaelis, 88, 95, 100, 101
CORINE, 53, 54, 60, 340
correlation, 18, 32, 39, 57, 129–131, 136, 141, 142, 146, 153, 154, 166, 167, 204, 217, 234, 249, 275, 293, 294, 318, 329, 330
correspondence analysis, 240, 241, 243
corridors, 76, 160, 264, 268, 273
Cossidae, 30
Cotesia bignellii, 318
Cotesia melitaearum, 318, 325
courtship, 219, 222, 223
cowslip, 309
cranberry agriculture, 290
Crematogaster, 73
critically endangered, 38, 68, 69, 72, 73, 86, 89, 90, 105–107, 199, 240, 244, 317, 325
Croitana croites, 277, 282, 283
Cryphia domestica, 38
Ctenuchidae, 30
cultural heritage, 362
cultural history, 247
Cupido minimus, 56, 125, 144, 241, 251, 288, 355
Cupressaceae, 32
currants, 40
cutting, 77, 92, 93, 100, 151, 152, 157–160, 171, 173, 175, 177–180, 196, 199, 227, 235, 321, 342, 346
Cyaniris semiargus, 240, 243, 244
cycles, 16, 60, 103, 112, 129, 146, 150, 296, 361
Cynthia cardui, 55, 56, 60, 64, 65
cypress pug, 32, 35
Cyrestis thyodamas, 94

Dactylis glomerata, 226
Dakota skipper, 152, 156, 160, 161
dams, 91
Danaus chrysippus, 59, 277, 280, 283–286, 333, 337
Danaus plexippus, 59, 152, 277, 283–286, 333, 334, 337
dark green fritillary, 355
dark spinach, 38, 45, 47
dark-barred twin-spot carpet, 38
Dasypolia templi, 38
data book, 26, 63, 65, 66, 70, 74, 85, 86, 89, 90, 95, 99, 101, 102, 133, 140, 170, 199, 209, 240, 244, 270, 325
data deficient, 68, 69, 71–73
Daucus carota, 243, 244
Daviesia divaricata, 274, 276, 282
day-flying moths, 3, 41, 49, 271–273, 275, 277, 278, 281, 283, 284, 288, 351, 352, 356
death's head hawk-moth, 33
decline, 1, 2, 7–12, 15, 16, 19–22, 24, 38–43, 45–48, 51, 56, 57, 60, 68, 69, 71, 72, 85–89, 91–94, 96, 97, 99–102, 104–107, 111, 112, 114, 124, 125, 134, 141, 144–146, 148, 156, 157, 160, 163, 189, 194, 197, 201, 202, 211, 225, 226, 234–237, 242, 248, 261, 269, 271, 272, 301, 308, 311, 317, 319, 320, 322–324, 332, 340, 342–345, 347, 350, 352, 359, 361, 365
declining, 7, 11, 15, 16, 22, 43, 46–49, 51, 57, 69, 90, 114, 123, 129, 131, 134, 137, 139, 150, 153, 154, 157, 189, 198, 208, 262, 271, 280, 287, 301, 303, 304, 313, 314, 319, 325, 342, 345, 361–363
deep-brown dart, 38
deer, 22, 91, 93, 94, 258
degraded, 3, 75, 80, 83, 149, 158, 276, 279–281, 289–291
Deileptenia ribeata, 38, 39
Delias aganippe, 283, 284
density-dependent, 124, 129, 132, 136, 138, 223, 264, 268
Department of the Environment Food and Rural Affairs (Defra), 350
desert orange tip, 59
developmental history, 265
devil's-bit scabious, 189, 190, 318
Devon carpet, 38
Dianthus barbatus, 32
Diloba caeruleocephala, 38, 45, 47
dingy footman, 35, 36, 38, 39
dingy skipper, 355
Dipsacus fullonum, 244
dispersal, 2, 26, 75, 76, 83, 94, 105–112, 119, 123–125, 127, 128, 134, 138, 147, 172, 180, 198, 199, 207, 208, 211, 243, 258, 263, 268–272, 274, 279, 280, 289, 290, 295–297, 300, 301, 304–306, 313, 315, 319, 323–325, 335, 337, 363

dispersive, 202, 264, 265, 274
distance decay function, 263
distance sampling, 351, 352, 357, 359
distribution, 3, 5–8, 10–12, 16, 22, 23, 26, 29–35, 37–42, 44–47, 49, 52, 60, 64, 65, 67–69, 71–74, 86–90, 93, 94, 97, 99, 100, 102, 106–112, 114, 115, 117, 119–121, 131, 135, 138, 139, 141, 142, 144, 146–148, 152, 153, 159, 163–165, 168, 171–174, 180, 187, 189, 191, 198, 199, 201–204, 206, 207, 209, 211, 213, 216, 217, 219, 221, 226, 235, 236, 240, 243, 247, 248, 251, 258, 259, 261, 262, 264, 265, 268, 269, 271, 272, 277, 281, 287, 288, 292–294, 297, 300–302, 304, 308, 313, 318, 324, 327–333, 335, 337, 346, 347, 349, 352, 357, 358, 359, 363–366
distribution change, 6, 11, 30, 32–34, 37, 38, 90, 93, 106
distribution modelling, 42, 111, 141, 147
distribution patterns, 3, 41, 49, 109, 271, 272, 281, 347, 366
distribution trends, 29, 32, 38, 39
disturbance, 2, 59, 60, 75–77, 80, 83, 111, 172, 177, 190, 207, 209, 234, 235, 247, 251, 258, 261, 267, 271–274, 279–281, 288, 301, 324
disturbed, 75–80, 172, 174, 179, 207, 258, 266, 272, 273, 280, 333
ditches, 92, 173, 179, 212, 290, 295, 296, 336
diversity, 1, 2, 5–8, 12, 13, 15–17, 23–26, 35, 39, 41, 42, 48, 49, 63, 67, 68, 73–80, 83, 92, 93, 97, 100, 101, 107, 148, 159, 160, 163, 164, 166, 167, 169, 170, 187, 199, 202, 207–209, 225–227, 229, 231–238, 240, 242–245, 259, 260, 267–269, 271, 279, 281, 287, 288, 291, 297, 301, 320, 327–332, 341, 346, 347, 350, 353, 356, 358, 362, 363
Doñana National Park, 336
Dorcas copper, 292
dormancy, 234
dotted carpet, 38
double dart, 35, 38, 39
dragonflies, 30, 71, 97, 100, 352, 356
drainage, 91, 125, 290, 301, 319, 341
dreamy duskywing, 292
drepanidae, 30
drought, 22, 113, 132, 133, 273, 295, 309, 334, 336, 341, 364
dry meadows, 59
duck-billed hoverfly, 354
Duke of Burgundy, 7, 9, 248, 303, 304, 315
dusky heath, 54, 59
dusky thorn, 38
dusky-lemon sallow, 38
Dutch Butterfly Conservation, 43, 44, 113, 114, 120, 171, 364
dwarf cream wave, 32, 35, 38

early seral, 124, 130
Earth Summit, 1, 362
Echium vulgare, 244
ecological compensation areas, 238, 243
ecological traits, 11, 12, 41
economics, 2, 343, 366
ecosystem services, 1, 16, 41, 103, 111, 147, 225, 362, 365
ecosystems, 3, 26, 29, 35, 41, 43, 85, 92, 98, 111, 149, 157, 161, 198, 226, 235, 245, 269, 288, 346, 366
eco-tourism, 80
Ectopistes migratorius, 150
edge, 83, 114, 130, 169, 179, 187, 213, 223, 250, 257, 264, 269, 280, 289, 291, 295, 309, 364
edge effect, 77, 136, 187, 227, 289, 291
egg, 2, 97, 113–120, 125, 129, 130, 133, 190, 211–216, 219–222, 245, 287, 305–307, 318, 320, 322, 323, 337
egg counts, 114, 120, 125
egg hatch, 2, 113–120
eggs, 46, 92, 113–120, 124, 130–133, 163, 171, 172, 179, 180, 190, 205, 207, 208, 211–216, 220, 221, 223, 243, 305, 306, 318, 320, 334
Eilema complana, 38
Eilema depressa, 35, 36, 38, 39
Eilema griseola, 35, 36, 38
elevational gradients, 106
embankment, 173
emigrants, 123, 136, 137
emigration, 130, 136, 138, 198, 265
endangered, 1, 25, 38, 68, 69, 72– 74, 85–87, 89, 90, 92, 94–96, 98–100, 105–107, 111, 112, 134, 138–140, 150, 159–161, 163, 170, 189, 199, 209, 240, 241, 244, 269, 273, 301, 317, 319, 325, 329, 332
Endromidae, 30
England, 3, 7, 9, 11, 12, 15, 17–25, 40, 49, 106, 124, 134, 137, 138, 141, 142, 144–146, 168, 169, 187, 189, 191, 198, 199, 201, 202, 211, 212, 226, 235, 259, 303–305, 309, 310, 314, 317, 318, 321–323, 325, 340, 346, 353, 359, 362, 363, 366
England Biodiversity Strategy, 20
Ennomos erosaria, 38
Ennomos fuscantaria, 38
entomological, 25, 40, 41, 60, 92, 95, 112, 125, 268, 301
Entry Level Stewardship (ELS), 226, 227, 235, 363
Environmental Impact Assessment (EIA), 91, 92
environmental stewardship, 18, 226, 366
Epilobium ciliatum, 241, 242
Episyrphus balteatus, 354
equilibrium, 123, 129, 132, 281, 303, 304, 310, 313
Erebia aethiops, 144, 354, 355
Erebia discoidalis, 292
Erebia sudetica, 105, 109
Erynnis icelus, 292
Erynnis tages, 55, 56, 134, 144, 355
Essex skipper, 355
Eucalyptus, 75, 273, 274, 276, 282
Euchloe belemia, 59
Euchloe simplonia, 65
Eugnorisma glareosa, 38
Eulithis mellinata, 38, 40
Euphaedra, 78–80, 82
Euphydryas aurinia, 2, 3, 17, 38, 56, 106–108, 127, 130, 137, 138, 144, 189, 198, 199, 201, 208, 209, 317, 325
Euphydryas editha, 318
Eupithecia phoeniceata, 32, 35, 39
Eurema laeta, 88, 91
Euriphene, 78–80, 82
Europe, 2, 3, 11, 20, 22, 24–26, 30, 40–43, 47, 48, 51, 52, 58–61, 65–67, 70, 74, 85, 86, 92, 94, 99, 102, 103, 105–107, 110, 111, 114, 120, 123, 130–132, 137–140, 148, 160, 161, 163, 164, 170–172, 179, 187, 189, 190, 199, 201, 202, 209, 226, 234–236, 245, 247, 258–260, 269, 270, 288, 290, 301, 302, 314, 317, 318, 321, 325, 327–330, 333, 334, 337, 339, 340, 345–347, 350, 359–366
European climate change indicator, 364

European Environment Agency, 20, 23, 366
European grassland butterfly indicator, 26, 161, 270, 364
European red list, 26, 42, 74, 102, 163, 170, 329, 332, 339, 347, 362, 366
European Union (EU), 1, 2, 5, 6, 12, 16, 20, 26, 42, 74, 102, 163, 170, 198, 234, 237, 248, 259, 267, 319, 327, 328, 332, 339, 344–347, 361, 362, 365, 366
Euxoa nigricans, 38
Euxoa tritici, 38
evenness, 77, 78
Everes lacturnus, 86, 87, 95
evergreen forest, 77, 83, 93
EVF, 275, 278
evidence-based, 125, 268, 304
evolve, 289, 296, 297
extensive, 2, 9, 51, 52, 71, 86, 92, 94, 97, 163, 190, 194, 196, 201, 237, 247, 258, 308–310, 320, 321, 324, 327, 342, 345, 353, 361
extensively used, 237, 238, 241–243
extent of occurrence (EOO), 71, 72
extinct, 68, 69, 72, 73, 85, 86, 90, 99, 123, 124, 126, 150, 164, 170, 189, 205, 226, 272, 281, 304, 312, 319, 321, 323, 327, 328, 330, 363
extinction, 1, 3, 11–13, 23, 26, 29, 40–42, 60, 63, 67–71, 73, 74, 76, 85, 86, 89, 93, 95, 97–99, 101, 102, 104–107, 110–112, 125, 131, 132, 135–137, 139, 148, 161, 170, 189, 190, 197–199, 202, 208, 245, 261, 262, 271, 272, 281, 288, 301, 304, 305, 310–314, 317–321, 323, 325, 332, 333, 336, 337, 342, 345, 346, 366
extinction debt, 99, 137, 198, 199, 208, 281, 301, 325
extinction risk, 11, 12, 41, 42, 68, 71, 74, 104–106, 110, 112, 148
extreme-value function model, 288
eyed brown, 152, 292

Fabaceae, 228, 243
Fabriciana nerippe, 86, 87, 93, 94, 100, 102
Fagopyrum esculentum, 244
Fagus, 94, 102
farmers, 76, 131, 226, 227, 235, 243, 248, 258, 321, 325, 342, 343, 345, 362
farmland, 2, 15, 18–23, 26, 39, 48, 49, 52, 75–77, 79, 80, 82, 83, 90–94, 97, 98, 128, 134, 138, 203, 225, 226, 228, 234–236, 245, 247, 248, 259, 268, 327, 332, 340, 345–347, 350, 359, 360, 362, 363, 366
farmland bird, 18, 48, 49, 225, 359, 360
farmland butterflies, 19, 21, 225, 226
favourable condition, 16, 18, 80, 113, 196, 197
favourable conservation status, 319
feathered gothic, 38
fecundity, 138, 234
fertiliser drift, 226
fertilisers, 226
fertilizer application, 243
Festuca ovina, 130
Festuca rubra, 226
field boundaries, 248
field boundary, 226, 227
field margins, 39, 160, 235, 236, 238, 244, 245, 267, 269, 340, 353, 366
field scabious, 202
figure of eight, 38, 45, 47
Finland, 38, 302, 342
fire, 149, 150, 151, 153, 154, 156–161, 199, 271, 273, 288, 297, 301, 302
fire management, 150, 158–160
fixed-point monitoring, 30
flagship species, 96, 103, 267
flagships, 29, 35, 86
flame shoulder, 32, 38
flight, 12, 25, 41, 42, 44, 59, 138, 146, 152, 153, 165, 166, 173, 190, 191, 196, 197, 199, 213, 216, 220–222, 227, 264, 265, 268, 273, 292, 293, 304, 321, 322, 352, 354, 359
flight direction, 265
flight period, 44, 152, 153, 165, 166, 173, 190, 191, 196, 197, 213, 216, 273, 292, 293, 304, 321, 322, 352
flood control, 91
flower abundance, 238, 239, 241
flower visits, 237, 241, 242
flower-poor, 130
fluazifop-P-butyl, 225–227
fluctuation, 47, 69, 198, 292, 318, 320, 322
focal species, 262, 269, 333
food density, 127
food security, 112, 365
food webs, 225, 237
foodplant, 123–126, 130, 131, 135, 137, 139, 163–165, 169, 202, 211, 220, 221, 223, 306, 310, 318, 320, 323, 334
forest, 3, 24, 25, 49, 58, 59, 72–80, 82, 83, 86–91, 93, 94, 97, 101, 102, 110, 111, 121, 137, 158, 160, 170, 179, 211, 212, 219, 220, 223, 237–239, 241, 248, 259, 266, 273, 287, 288, 290, 291, 294, 295, 297–300, 302, 321, 323, 327, 328, 334, 341, 346, 352, 358, 365
forestry, 15, 20, 24, 74, 81, 83, 87–89, 91, 97, 98, 101, 120, 124, 155, 180, 211, 212, 223, 313, 359, 361–363, 366
Forestry Commission (FC), 24, 211, 212, 223, 359, 363, 366
Formica fusca, 174
Formica rufibarbis, 174
fossil record, 296
four-spotted, 33
fragmentation, 2, 5, 12, 16, 22, 23, 51, 59, 60, 68, 69, 72, 74, 76, 94, 97, 109, 112, 137, 138, 140, 157, 187, 189, 201, 202, 207, 209, 226, 237, 250, 257, 259, 261, 262, 268, 269, 272, 273, 280, 281, 287–290, 297, 301, 315, 319, 333, 337, 339, 361, 363
fragments, 22, 24, 72, 83, 157, 158, 202, 207, 208, 268, 270–273, 275, 277–282, 287, 288, 339
France, 109, 111, 123, 139, 187, 267
Freija fritillary, 292
Frigga fritillary, 292
fritillaries, 129, 135, 139, 242, 295
fruit-feeding, 75–77, 83
fundamental niche, 134
funding, 18, 31, 68, 70, 99, 159, 237, 258, 297, 303, 306, 310, 311, 313, 314, 319, 325, 339, 359, 361, 363, 364
fungal pathogens, 273, 274
fusilade, 226, 236

Galium verum, 228
game birds, 314
gap effect, 80

garden dart, 38
garden tiger, 39, 40
gatekeeper, 59, 145, 229, 355, 356
Geitoneura klugii, 275, 277, 283–286
Geitoneura minyas, 271, 276, 277, 280, 283–286
general linear mixed models, 229
generalist, 5, 7, 8, 11–13, 15, 18–24, 51, 53–55, 57–60, 78, 80, 107, 145, 149, 150, 158, 223, 226, 291, 340, 342, 349, 350
generation, 123, 129, 131, 137, 161, 197, 199, 362
generations, 52, 69, 126, 129, 131, 136, 157, 197, 208, 320, 339
genetics, 198, 265, 325
genotype, 265
geographical variability, 51
geometric decay, 104
geometridae, 30, 32, 106
germination niches, 225–227, 232, 258
Ghana, 3, 75, 76, 80, 81, 83, 85
glaciated landscapes, 290
glanville fritillary, 199, 318
global, 1, 2, 13, 15, 16, 24–26, 29, 30, 40, 42, 60, 67–71, 73–75, 85, 87, 91, 94, 98, 100–104, 107, 111, 112, 142, 148, 170, 209, 300, 332, 359, 365, 366
global warming, 26, 85, 87, 91, 94, 98, 100, 111, 112, 148, 366
goats, 343–346
Gomphocarpus fruticosus, 333, 334, 337
Gonepteryx maxima, 88
Gonepteryx rhamni, 55, 56, 144, 228, 242, 244, 250, 355
graminicide, 225–227, 229–235
grant schemes, 16, 18, 22, 303, 314
Graphiphora augur, 35, 38, 39
grass, 38, 92, 93, 98, 102, 156, 157, 159, 160, 167, 169, 171, 173, 190, 191, 214, 220, 225–227, 229–235, 238, 258, 269, 305–307
grass margins, 235, 238, 269
grass rivulet, 38
grassland, 3, 7, 19, 20, 26, 49, 51, 54, 55, 57, 58, 60, 85–94, 97–102, 111, 125, 126, 130, 132–134, 137–139, 151, 152, 160, 161, 169, 171, 177, 187, 191, 196, 198, 199, 201, 202, 205–208, 232, 235–238, 243–245, 247, 257–259, 261, 266, 269, 270, 288, 291, 302, 304, 305, 307–310, 312, 317, 319, 321, 323, 331, 339–343, 346, 347, 357, 364–366
grassland management, 97, 169, 171, 237, 269
grassland species, 26, 58, 86, 90, 257, 270, 291, 366
grayling, 355
grazed, 7, 169, 172, 193, 194, 196, 198, 205, 206, 248, 257–259, 306–308, 310, 340, 341, 346, 347
grazing, 18, 58, 92, 102, 131–133, 150–152, 155–159, 161, 163, 164, 169, 170, 178, 187, 189, 190, 192–196, 199, 206, 207, 247, 248, 250–253, 256–259, 274, 306–310, 312, 313, 321, 323, 324, 332, 340–347, 366
grazing management, 102, 132, 206, 247, 341
Great Britain, 32, 35, 39, 40–42, 49, 208, 235, 236, 358
great pied hoverfly, 354
Greece, 339, 343–345, 347, 364
green anole, 97, 102
green comma, 292
green hairstreak, 355
green-striped white, 59
green-veined white, 355, 356
grey shoulder-knot, 38
grizzled skipper, 52, 137
ground beetles, 226, 233
growth rate, 132, 133, 136, 142, 168, 263
growth-form, 123, 131
guild, 124, 134

habitat area, 139, 217, 238, 270, 272, 288, 310, 321
habitat availability, 12, 26, 41, 125, 139, 148
habitat destruction, 29, 51, 67, 149, 177, 199, 272
Habitat Directive, 163, 327
habitat disturbance, 80
habitat indicators, 51, 54, 55, 57, 58
habitat loss, 6, 12, 39, 60, 71, 76, 94, 95, 111, 157, 187, 237, 261, 270, 289, 332, 350, 362, 363
habitat networks, 109, 262, 310, 324
habitat patches, 2, 77, 97, 123, 124, 129, 134–136, 139, 163, 164, 189, 190, 198, 202, 208, 270, 272, 274, 288, 289, 296, 302, 304, 309, 315, 317–321, 324, 325, 333
habitat preferences, 54, 57, 58, 83, 163, 164
habitat quality, 23, 26, 41, 48, 78, 80, 123, 124, 130, 131, 134, 135, 137, 139, 163, 172, 180, 192, 198, 201, 208, 225, 236, 247, 257, 259, 268, 272, 279, 280, 287–290, 295, 296, 297, 301, 303, 304, 306, 308, 311, 313, 318, 324, 359
habitat requirements, 51, 108, 109, 110, 139, 158, 163, 171, 187, 189, 190, 201–203, 207, 208, 212, 242, 248, 313, 314, 333, 342
habitat resources, 3, 271, 272
habitat specialist, 17, 110, 317, 340, 356, 358
habitat structure, 22, 109, 172
habitat type, 15, 53, 54, 58, 237, 241, 330, 357
Habitats and Species Directive, 189, 201, 364
Habitats Directive, 170, 319, 328, 330, 361
Hadena compta, 32, 35, 36, 39
Hamearis lucina, 7, 133, 144, 248, 250, 287, 303, 304, 308, 310–315
harassment, 221, 264
Harris checkerspot, 292
hay meadows, 172, 179, 248, 257, 258, 270, 342, 343, 345, 346
heath fritillary, 7, 12, 27, 140, 259, 325, 347
Hecatesia thyridion, 277, 283, 284
hedge rustic, 38
hedgerows, 114, 120, 197, 226, 257, 259, 268, 340, 353
Helianthemum nummularium, 203
Hemaris tityus, 33
Hepialidae, 30
herb, 158, 169, 305, 307, 341
herbicides, 226, 235, 272
Hesperia comma, 7, 12, 25, 55, 56, 106, 111, 124, 127, 130, 137–139, 144, 262, 268, 269, 301
Hesperia dacotae, 149, 151, 156, 159–161, 302
Hesperia florinda, 87
Hesperia ottoe, 149, 152, 156, 159
Hesperiidae, 82, 83, 87, 88, 102, 135, 139, 161, 253, 256, 277, 292, 302
Hesperilla chrysotricha, 282, 283
Hesperilla donnysa, 277, 282, 283
Hestina assimilis, 83, 93, 94
heterogeneity, 2, 11, 109, 111, 112, 146, 199, 208, 248, 260, 261, 267–269, 297, 303, 320

Heterogeneous, 41, 110, 112, 141, 142, 190, 196, 206–209, 248, 264, 269, 301, 363
Heteronympha merope, 277, 283–286
heterozygosity, 323
Hibernation, 46, 257, 318, 322, 323, 333
High Brown Fritillary, 7, 9, 38, 140
High Nature Value (HNV), 2, 247, 340, 346, 347, 362, 363, 366
Higher Level Environmental Stewardship (HLES), 18
Higher Level Stewardship (HLS), 226, 235, 363
high-quality, 123, 126, 137, 141, 149, 245, 290
Hipparchia semele, 55, 56, 125, 128, 130, 144, 355
Hipparchia statilinus, 56, 65
Hippocrepis comosa, 126, 131, 135, 203
Historical records, 5, 33, 156
HNV, 247, 259, 340, 344–346, 362, 363
holly blue, 139, 148, 355
honeydew, 99, 127, 128
honeysuckle, 320
Hoplodrina ambigua, 35, 36, 38
hotspots, 7, 60, 75, 83, 131, 238, 259, 268, 272, 287, 288, 346, 363
human, 1, 13, 39–42, 52, 59, 60, 64, 91, 94, 98, 111, 149, 150, 158, 163, 212, 245, 247, 270, 271, 288–290, 333, 339, 343, 344, 349, 360, 362, 365, 366
human health, 1, 94, 362, 365
humming-bird hawk-moth, 33
Hymenoptera, 106, 169, 170, 187, 245, 267
Hypericum perforatum, 244
Hypochrysops halyaetus, 274, 277, 283–287
Hypodryas maturna, 105–108

ice age, 160, 296
Idaea fuscovenosa, 32, 35, 38
Idaea rusticata, 35, 38, 39
Idaea trigeminata, 38
immigrant, 136, 291, 333, 334
impacts, 1, 2, 11, 12, 18, 19, 21–25, 30, 41, 52, 60, 83, 85, 104, 111, 112, 125, 126, 128, 133, 137, 147, 155, 158, 170, 192, 202, 245, 248, 257, 280, 288, 290, 321, 332, 339–342, 345, 347, 350, 359, 365
Inachis io, 8, 56, 144, 355
inbreeding, 110, 197–199, 202, 207, 323
inbreeding depression, 110, 198
Incidence Function Model, 198, 263
index of habitat specialisation, 53
indicator groups, 26, 42, 74, 75, 77, 83, 139, 332
Individual Based Models, 263–265
inornate ringlet, 292
insect diversity, 16, 41, 107, 169, 170, 238, 346
insecticides, 94, 126, 129, 272
instars, 114, 118, 119, 163, 318
intensification, 22, 42, 49, 58, 85, 171, 202, 226, 237, 248, 339, 340, 345, 349, 350, 359, 361, 362
intensive, 2, 5, 7, 19, 20, 30, 67, 90, 93, 98, 124, 128, 132–134, 142, 158, 169, 190, 194, 237, 238, 242, 243, 247, 266, 267, 271, 274, 323, 324, 327, 328, 340, 345, 352, 362, 365
intensive agriculture, 2, 158, 345
interaction, 124, 206, 207, 226, 229–232, 277, 279, 281, 318, 323, 352
interior, 80, 158, 280, 292, 294, 298, 299
inter-patch movements, 264
International Union for the Conservation of Nature and Natural Resources (IUCN), 68
interspecific compelition, 91, 93, 94, 172
intraspecific, 26, 72, 125, 139, 148, 172
intrinsic growth rate, 132, 133, 263
Iowa, 149–154, 156, 157, 159–161, 295, 302
IPCC scenario, 105
Iphiclides spp, 251, 255–257
Ireland, 5, 6, 7, 12, 15, 24, 25, 30, 32, 33, 34, 37, 40–42, 49, 137–139, 147, 160, 198, 208, 223, 235, 268, 315, 317, 325, 332, 334, 337, 356, 359
island, 12, 40, 41, 86, 89, 94, 97, 155, 157, 160, 274, 279, 287, 288, 301, 320, 364, 366
Isle of Man, 5, 31, 32
isolated, 11, 92, 134, 151, 153, 158, 160, 190, 199, 201, 202, 205, 207–209, 212, 257, 267, 271, 272, 275, 276, 289, 290, 296, 309–311, 313, 317, 322, 323, 333, 363
isolation, 12, 76, 94, 96, 97, 109, 111, 123, 135, 137, 139, 168, 198, 204, 207, 208, 223, 268–272, 274, 279, 287–290, 301–303, 312, 313, 315, 317, 325
Issoria lathonia, 55, 56, 244, 251
IUCN, 26, 40, 68–74, 83, 99, 101, 104, 106, 288, 366
IUCN red list categories, 68, 71, 74, 104

Jacksonia sternbergiana, 274, 276, 282
Japan, 3, 85–87, 89–96, 98–102, 123, 340
Japan Butterfly Conservation Society (JBCS), 85, 96
Japonica onoi mizobei, 86
Japonica onoi onoi, 88
JBCS, 85, 89, 96–98, 100
juniper carpet, 38
Junonia villida, 277, 280, 283–286
Jura, 238, 341
Jutta arctic, 292

Kakum Forest, 3, 75, 76
Kakum National Park, 76, 77, 83
kettlehole, 290, 292–296, 298, 299
key factor analysis, 124
key factors, 1, 125, 363
Khaya ivoriensi, 76
knapweed fritillary, 54
Knautia arvensis, 207, 240
Korea, 201

lackey, 38
Lampides boeticus, 56, 274, 277, 283–286
Lampropteryx otregiata, 38
landowners, 192, 198, 303, 310, 313, 314, 324
landscape, 2, 3, 13, 23–26, 30, 33, 37, 39, 42, 51, 52, 58, 60, 85, 92, 93, 96, 98, 99, 101, 102, 111, 120, 126, 128, 135–137, 139, 140, 147, 149, 152, 157–159, 170–172, 177, 187, 190, 198, 199, 201–205, 207–209, 222, 234, 237, 238, 243–245, 247, 248, 257–272, 287–290, 296, 297, 301–304, 306–308, 310, 311, 313, 314, 317–320, 324, 325, 333, 334, 336, 337, 339, 340, 342, 346, 347, 349, 350, 356, 359–366
landscape change, 262, 267
landscape configuration, 261–264, 266, 289, 290, 296, 297
landscape context, 101, 238, 245, 259, 262, 270

landscape features, 39, 248, 262–267
landscape heterogeneity, 2, 260, 261, 267
landscape planning, 262
landscape simplification, 248, 261
landscape structure, 3, 52, 60, 102, 257, 259, 265, 268, 301, 333, 337, 346
landscapes, 2, 11, 41, 48, 60, 91, 98, 109, 110, 112, 123, 124, 126, 134, 135, 137–139, 147, 158, 190, 199, 201, 202, 204–209, 226, 236–238, 242–245, 248, 257, 259, 261–271, 287– 290, 295–297, 302–304, 308, 312–315, 317, 332, 333, 337, 339, 346, 347, 363, 364, 366
landscape-scale, 1, 2, 3, 109, 124, 149, 190, 259, 261, 303, 304, 309, 311, 314, 317, 366
land-use, 3, 25, 51, 52, 111, 134, 187, 247, 248, 259, 263, 267, 339, 340, 346, 347
large blue, 52, 109, 138, 139, 169, 170–172, 187, 270, 363
large heath, 137, 355
large skipper, 230, 355
large white, 229, 355, 356, 358
large yellow underwing, 33, 38
larvae, 43, 47, 48, 97, 109, 112, 114, 115, 117–120, 123, 130–132, 135, 163–165, 168, 169, 172, 190, 196, 198, 199, 202, 205, 207, 209, 211, 225, 226, 234, 242, 243, 272, 318–325, 334, 336, 342
larval development, 132
larval feeding, 118, 304, 309
larval food plant, 113, 114, 189, 237, 242, 243
larval growth rate, 142
larval habitat, 123, 126, 128, 130–135, 198, 208, 237, 243
larval nests, 97
larval niche, 123, 124, 128, 130–133, 170
larval resource, 123, 124, 130
larval webs, 191–196, 201–208, 317, 318, 320, 322, 323, 325
Lasiocampidae, 30
Lasiommata maera, 251
Lasiommata megera, 7, 55, 56, 138, 144, 146, 228, 244, 355, 356
Lasius flavus, 172, 174, 176, 178, 180, 187
Lasius niger, 171, 172, 175, 176, 178, 180, 183–185, 187
Lasius umbratus, 174
Lathyrus pratensis, 211, 214, 215, 220, 221
least black arches, 38
least carpet, 32, 35, 38, 39
Legousia speculum-veneris, 244
Leontodon autumnalis, 240
Leontodon spp, 203
Lepidoptera, 1, 3, 24–26, 29, 41–44, 49, 67, 70, 73, 74, 82, 83, 99, 101–103, 105–114, 120, 121, 137–139, 147, 149, 159–161, 170, 187, 190, 198, 199, 208, 209, 212, 223, 225, 235, 245, 259, 268, 269, 287, 288, 301–304, 314, 315, 317, 325, 327, 328, 332, 333, 337, 346, 347, 351, 361, 362–364, 366
Lepindex, 70, 71, 73
Leptidea amurensis, 88
Leptidea sinapis, 55, 56, 127, 130, 133, 140, 144, 211, 223, 250, 355
Leptothorax acervorum, 174
Leptothorax nylanderi, 174
Lethe marginalis, 88
Leucanthemum vulgare, 228, 234, 240, 244
lichens, 23, 39, 47, 49, 115
life, 191, 198, 199
life cycle, 46, 146, 292
life-history traits, 58
life-stage, 124
life-tables, 124, 126, 128–130
light, 37, 40, 48, 51, 104, 130, 150, 151, 155, 156, 159, 195, 291, 307, 364
light trap, 364
lightning, 149, 160
Limacodidae, 30
Limenitinae, 77, 82
Limenitis camilla, 52, 127, 144, 242, 355
Limenitis populi, 88, 91
Linaria purpurea, 32
Linaria repens, 32
Linaria vulgaris, 32
linear features, 262, 267, 269, 313, 351–353, 357
linear regression, 18, 45–47, 53, 55, 240, 241, 274, 275
Liptena ilaro, 72
Listrodomus nycthemerus, 129
Lithophane leautieri, 32, 35–39
Lithophane ornitopus, 38
litter, 109, 149, 305–307, 321, 341, 346
litter meadows, 321
livestock, 102, 151, 158, 321, 325, 340, 342, 343, 366
local extinction, 189, 202, 208, 261, 271, 281, 288, 304, 305, 314, 317, 318, 321, 323, 325, 336
logging, 72, 79, 91, 96, 274
logistic regression, 247, 249–252, 256, 257, 274, 275, 287, 288
long dash, 152
Lonicera, 209, 320
Lonicera periclymenum, 320
Lopinga achine, 105, 107, 108, 342, 346, 365, 366
Lotus corniculatus, 203, 211, 214, 215, 220, 228, 342
Lotus pedunculatus, 211, 214, 215, 220, 221
low-density, 125, 273
lowland, 3, 25, 55, 83, 111, 237–239, 241, 242, 247, 261, 263, 289–292, 294–296, 298, 300, 301, 344, 345, 350, 359, 360
Luehdorfia japonica, 86, 87, 91, 92, 94–96, 99, 100
Luehdorfia puziloi, 88, 95, 96, 101
Lulworth skipper, 355
Lycaeides melissa, 149, 150, 152, 156, 161
Lycaena alciphron, 52, 55, 56
Lycaena bellargus, 253
Lycaena coridan, 250
Lycaena dispar, 105, 107, 108, 124, 134, 138
Lycaena dorcas, 292
Lycaena epixanthe, 292–294, 298, 302
Lycaena phlaeas, 55, 56, 127, 137, 144, 228, 229, 244, 269, 355, 359
Lycaena tityrus, 65, 242, 250
Lycaena virgaureae, 251
Lycaenidae, 71, 74, 81, 83, 87, 88, 101, 114, 120, 121, 135, 139, 161, 170, 187, 252, 254, 277, 288, 292, 301, 302, 315
Lymantriidae, 30
Lysionotus pauciflorus, 97

Macaria wauaria, 35, 38, 39, 40
Macroglossum stellatarum, 33
macro-moths, 29–35, 37–39, 45, 47
Macrothele calpeiana, 110, 111

Maculinea, 74, 109, 112, 130, 137–140, 148, 163, 164, 170–173, 175, 178, 187, 264, 268, 270, 363, 366
Maculinea alcon, 125, 138
Maculinea arion, 52, 56, 58, 65, 88, 105, 107–109, 112, 124, 127, 130, 132, 133, 138, 139, 163, 170, 270, 363
Maculinea nausithous, 105, 107–109, 111, 125, 139, 187
Maculinea rebeli, 124, 138, 169, 170
Maculinea teleius, 86, 88, 92, 105, 107, 108, 125, 140, 188
Madeira, 334, 364
Malacius bipustulatus, 354
Malacosoma neustria, 38
mallow skipper, 240, 242, 243, 245
Malva alcea, 240
Malva moschata, 244
Malva neglecta, 240
Malva sylvestris, 244
management, 1–3, 7, 12, 16–18, 21, 22, 24–26, 60, 75, 83, 86, 91, 92, 95–97, 99, 102–104, 109, 110, 114, 124, 126, 128, 129, 131–134, 137–140, 142, 146, 149, 150–161, 163, 169, 171–180, 187, 189–199, 201–203, 206–208, 211, 212, 221, 223, 225–227, 230–238, 243–245, 247–251, 253, 256–261, 263, 266–274, 281, 287, 288, 291, 297, 301–306, 308–315, 317, 319–321, 323–325, 327, 332–334, 339–342, 344–347, 350, 361–363, 365, 366
management costs, 311, 313
Maniola jurtina, 55, 56, 125, 141, 144, 145, 228, 229, 243, 244, 251, 264, 269, 350, 355, 356
marbled beauty, 38
marbled white, 58, 240, 355
marginal, 51, 57, 212, 222, 248, 259, 263, 267, 340, 361, 362
mark-recapture, 211–213, 216, 222, 223, 304, 323
marmalade hoverfly, 354, 356
marsh, 2, 3, 17, 25, 33, 38, 106, 137, 138, 140, 189, 198, 199, 201, 208, 209, 270, 288, 304, 317–325, 357
marsh fritillary, 2, 3, 17, 318
marsh moth, 33
marsh violet, 304
mating, 127–130, 212, 218, 219, 221, 222, 333, 336, 342
matrix, 2, 3, 80, 129, 166, 207, 248, 257, 259, 261–265, 268, 269, 271, 273, 274, 279, 280, 287, 288, 301, 333, 337, 361, 363
matrix occupying, 274
matrix sensitive, 274
mazarine blue, 240
meadow, 94, 97, 171, 180, 187, 188, 211, 234, 239–241, 243, 248–258, 268, 270, 318, 340–342, 346
meadow brown, 145, 229, 350, 355, 356
meadow fritillary, 152
meadows, 59, 140, 157, 159, 165, 168, 171, 172, 177, 179, 180, 187, 199, 209, 226, 237–244, 247–258, 270, 321, 323, 340–343, 345–347
mechanistic, 59, 124, 132, 133, 141, 147
mechanistic models, 124
Mediterranean basin, 51, 52, 57, 58, 60, 112, 270
Mediterranean endemics, 59
Mediterranean scrub, 51, 58
Melanargia galathea, 125, 144, 228, 229, 240, 243, 244, 251, 355
Melanargia occitanica, 55, 56, 58, 59
Melanitis phedima, 94
Melilotus alba, 244
Melinaea spp, .72
Melitaea ambigua, 88, 93
Melitaea athalia, 7, 124, 134, 144, 242
Melitaea aurinia, 135, 198, 325
Melitaea cinxia, 55, 56, 124, 125, 127, 130, 198, 262, 318, 325
Melitaea phoebe, 54–56, 250
Melitaea protomedia, 87, 95, 97, 99–102
Melitaea scotosia, 86, 87, 90, 93, 97, 99–101
Melitaeninae, 129, 135
Mellicta athalia, 12, 27, 140, 259, 325, 342, 347
Mesodina cyanophracta, 274, 277, 282–286
Mesophilous forest, 78, 80
meta-analysis, 126–130, 161, 301
metapopulation, 12, 25, 97, 111, 112, 123, 134–140, 161, 165, 187, 189, 190, 197–199, 201, 202, 208, 209, 223, 261–263, 265, 268–272, 279, 281, 288, 295, 301, 303, 304, 308, 310, 313, 315, 317, 319, 320, 324, 325, 333, 334, 336, 366
metapopulation dynamics, 138, 189, 198, 199, 209, 269, 295, 297, 301, 313, 318, 325, 337
microclimate, 109, 128, 171, 172, 179, 180, 288, 295, 306
microclimatic conditions, 173, 190, 206, 333
Microdon analis, 169
microhabitat, 7, 109, 123, 138, 168, 198
mid-successional, 130
migration mortality, 263
milk production, 342, 344, 345
minimum area, 280
Ministry of Defence, 202
Minnesota, 150–154, 156, 159–161, 301, 302
mitigating, 105, 126, 361
mitigation, 1, 350
mobility, 18, 22, 75, 118, 137, 140, 147, 174, 216, 217, 222, 261, 263, 267, 295, 305, 313, 314, 349, 356, 366
model, 17, 18, 25, 35, 42, 59, 95, 107, 110, 111, 118, 119, 124, 132, 133, 136, 138, 139, 141–147, 159, 168, 170, 174–176, 192, 193, 198, 203–206, 216, 229–232, 239, 244, 249, 252–256, 262, 263, 265, 266, 268, 269, 274, 275, 277–279, 284, 285, 287, 288, 333, 337
Molinia caerulae, 318
monarch, 59, 152, 336, 337
monitoring, 2, 5, 12, 15–18, 23–26, 29–32, 35, 40–42, 44, 49, 51–53, 59, 60, 63, 66–68, 71, 73, 74, 94–100, 109, 114–117, 120, 121, 123, 125, 131, 134, 138, 139, 141, 142, 150, 152, 158–161, 173–175, 177, 178, 180, 184, 189, 191, 192, 197–199, 207–209, 225, 235, 237, 239, 241–243, 245, 270, 281, 287, 288, 301, 302, 307, 311, 313, 314, 318, 321–325, 328, 332, 349, 350, 351, 357–361, 364
Monochamus alternatus, 94
montane, 26, 223, 259, 261, 339, 340, 345, 347, 357, 358
morphology, 111, 138, 147, 265, 302
mortalities, 129, 132
mortality, 69, 113, 118, 124, 136, 137, 146, 156, 159, 164, 190, 207, 263, 267, 336
mosaic, 76, 92, 93, 101, 180, 190, 191, 199, 202, 209, 244, 259, 268, 310, 340, 341, 346
moss, 13, 26, 27, 42, 53, 60, 61, 139, 140, 148, 270, 288, 305–307, 332, 337, 359, 360
Motasingha dirphia, 273
Motasingha trimaculata, 277, 282–286
moth, 3, 23, 26, 29–35, 38–49, 111, 112, 114, 117, 120, 121, 199, 243, 272, 275, 279, 282, 314, 347, 354, 356, 358, 361, 365

moths, 3, 12, 23, 25, 29–35, 37–49, 60, 73, 92, 100, 101, 137, 270–275, 277, 278, 281, 283, 284, 287, 288, 301, 314, 347, 350–352, 356, 359, 361, 362, 364–366
Moths Count, 3, 29–33, 35, 40, 366
mountainous, 3, 164, 247, 248, 259, 328, 329, 344–346, 361
mowing, 58, 89, 92, 97, 102, 150–152, 158, 159, 171, 173, 177–180, 187, 196, 220, 238, 243, 306, 308, 310, 323, 324, 341, 342, 347
MRR, 151, 165, 304, 305
muskeg, 290, 292–294, 298
myrmecophage, 275
myrmecophilic, 171
myrmecophily, 71, 74, 124
Myrmica hellenica, 168
Myrmica lobulicornis, 166
Myrmica lonae, 163, 166, 170
Myrmica rubra, 166, 172, 174, 176, 178, 180, 187
Myrmica ruginodis, 166
Myrmica rugulosa, 168, 174
Myrmica sabuleti, 109, 126, 133, 164, 170, 187
Myrmica scabrinodis, 164, 171, 172, 175, 176, 180, 187
Myrmica schencki, 170
Myrmica sulcinodis, 163, 166
myxomatosis, 131

Nacaduba biocellata, 277, 283–286
Narathura bazalus, 94
narrow-bordered bee hawk-moth, 33
natality, 124, 127–130, 132, 136
National Moth Recording Scheme, 29, 31, 33, 34, 40
National Nature Reserves (NNR), 191, 320
National Parks, 95, 99, 361, 362
National Vegetation Classification, 197, 318
Natura 2000, 164, 327, 328, 330, 332, 340, 361, 364
natural enemies, 103, 120, 146, 147, 197, 209, 245
natural regeneration, 238
natural resources, 74, 91, 92, 159, 160, 161, 169, 297, 302, 364, 365
natural woodlands, 362
nature reserves, 19, 114, 124, 129, 136, 172–177, 179, 180, 320, 340, 361
nectar, 49, 111, 127–130, 132, 135, 138, 190, 196, 203, 222, 226, 228, 234, 237, 251, 257, 269, 289, 294, 295, 333, 334, 336, 354
nectar resources, 234, 237, 257
Neolucia agricola, 274, 277, 283–286
neotropical, 24, 70, 71, 79, 301
Neozephyrus quercus, 55, 56, 355
Netherlands, 23, 25, 26, 30, 38, 39, 41, 43–49, 60, 85, 106, 113–115, 121, 133, 140, 171, 173–176, 178, 180, 184, 188, 260, 270, 288, 331, 347, 350, 360, 366
nettles, 234
network approach, 243
networks, 31, 109, 190, 198, 208, 262, 281, 310, 311, 313, 314, 319, 321, 323, 324, 327, 344, 363
niche, 2, 7, 11, 106–108, 123–125, 128, 130–134, 139, 141, 142, 147, 148, 168–170, 187, 201, 223, 225–227, 232, 235, 258, 302, 303, 363
niche breadth, 142, 146, 291
Niphanda fusca, 86, 87
nitrogen, 49, 52, 101, 112, 130, 148
Noctua fimbriata, 38
Noctua pronuba, 33, 38
Noctuidae, 30, 41, 277
Nola confusalis, 38
Nolidae, 30
non-hibernating, 129
non-resident, 274–280, 282–284
North America, 67, 94, 97, 103, 106, 123, 150, 158, 160, 290, 296, 301, 318, 333, 334
Northern Europe, 41, 42, 58, 59, 106, 107, 130–132, 160, 235, 236, 269, 318
northward shifts, 32, 34–36
Notodontidae, 30
nuptial gifts, 130, 137
nutrient, 49, 134, 139, 179, 195, 341
Nymphalinae, 77, 81, 209, 277
Nymphalis antiopa, 54–56, 242
Nymphalis vaualbum, 328

oak savanna, 289, 301
Oarisma poweshiek, 149, 152, 156, 160, 161, 302
oats, 227
occupancy, 6–8, 10–12, 69, 71, 72, 123, 124, 132, 134, 135, 137, 145, 146, 202, 207, 208, 250, 268, 270–272, 274, 276, 277, 279, 280, 287, 288, 301, 308, 309, 312, 334, 335, 337, 349, 355, 357, 358
Ochlodes subhyalina, 87
Ochlodes sylvanus, 144, 228–230, 244, 251, 355
Ochropleura plecta, 32, 38
Oedemera nobilis, 354
Oeneis jutta, 292, 293, 299, 300
Oeneis norna asamana, 88
Oeneis norna sugitanii, 86
Ogasawara Islands, 87, 93, 100, 102
Ogyris amaryllis, 277, 283
Ogyris idmo, 275
oilseed rape, 227
Onobrychis, 243
Onobrychis viciifolia, 244
Operophtera brumata, 114, 120, 121
optimum, 123, 125, 126, 130–132, 135–137, 189, 190, 192–196, 212, 308, 311
orange tip, 59, 225, 354–356, 358
orchards, 92, 238
Orchis ustulata, 340, 346
ordination, 159, 240, 249, 301
Origanum, 132, 163, 237, 241, 242, 244
Origanum vulgare, 132, 163, 241, 242, 244
orthophotographs, 238
Osmoderma eremita, 105, 110, 112
Ottoe skipper, 152, 156
overcollecting, 87, 88, 97, 272
overgrazing, 258, 340, 343
overwintering habitat, 238
ovipositing, 114, 120, 129–131, 139, 172, 173, 187, 188, 199, 205, 208, 209, 287, 304–307, 333, 334, 336

paddy fields, 93, 97
painted lady, 55, 229, 355, 356, 358
pale clouded yellow, 240
pale toadflax, 32
paleo-entomology, 289
Panchala ganesa, 88, 89, 95
Panolis flammea, 38, 39
Papaver rhoeas, 244
Papilio machaon, 56, 127, 134, 244
Papilio memnon, 94
Pararge aegeria, 8, 55, 56, 59, 65, 139, 144, 229, 244, 257, 262, 269
parasitism, 97, 146, 165, 323
parasitoids, 23, 124, 127, 128, 197, 201, 207, 208, 318, 319, 323, 325
Parnara ogasawarensis, 87
Parnassius apollo, 65, 105, 107–109, 111, 169, 287
Parnassius mnemosyne, 105, 107–109, 112
parsnip, 243
partnership, 5, 15, 20, 24–26, 31, 191, 223, 303, 310, 314, 324, 350, 360
passenger pigeon, 150, 160
Pastinaca sativa, 243, 244
pastoral management, 258
pasture, 64, 82, 91, 92, 97, 110, 163, 165, 168, 169, 171, 172, 187, 189, 191, 306, 317, 341, 343, 347
patch, 2, 111, 123, 135–137, 150–152, 165, 166, 168, 187, 190, 191, 198, 204, 207, 221, 222, 238, 239, 262–267, 274, 295–297, 300, 305, 308, 309, 318, 321, 334–336, 346
patch arrangement, 263
patch connectivity, 263, 264
patch leaving threshold, 265, 266
patch occupancy, 123, 135, 137, 263, 268, 287, 301, 308, 309, 334, 335
patch-matrix, 263, 265
Patersonia occidentalis, 274, 276, 282
Patrinia scanbiosaefolia, 97
peacock, 8, 10, 11, 234, 355
pearl-bordered fritillary, 7, 9, 304, 314, 315, 355
peat harvesting, 290
peat hummocks, 292
peatland, 290–292, 295, 301
Pelurga comitata, 38, 45, 47
perception, 265
perceptual ability, 265, 267
Periscepta polysticta, 277, 283, 284
Perizoma albulata, 38
permeability, 264
pest control, 41, 125, 126, 128, 237
pesticide, 91, 94, 97, 127, 128, 226
Phengaris, 170–172, 178–180, 187, 270
phenological window, 119
phenology, 30, 41, 49, 52, 60, 110–113, 120, 121, 140, 148, 212
phenotypes, 125
Phleum pratense, 226
physiological state, 265
physiology, 124, 339
Picos de Europa, 247, 248, 252, 254, 258, 259, 341–343, 345, 347
Pieris brassicae, 55, 56, 65, 144, 146, 228, 229, 253, 355, 356
Pieris napi, 56, 65, 144, 229, 244, 355, 356
Pieris rapae, 55, 56, 59, 64, 65, 111, 126, 127, 137, 138, 144, 228, 229, 244, 251, 275, 277, 283–286, 355, 356
Pilosella officinarum, 203
pine barrens, 26, 150, 152, 155, 159, 161, 289, 290, 301, 302
pine beauty, 38
pink-edged sulphur, 292
Plagodis dolabraria, 32, 38
plain tiger, 59
plant communities, 25, 26, 54, 147, 159, 166, 169, 235, 258, 259, 271, 301, 318, 325, 341
Plantago lanceolata, 228, 234, 240, 250
plantation woodland, 26, 211, 362
Plebeius agestis, 355, 356
Plebejus argus, 88, 90, 101, 125, 127, 130, 139, 144, 342
Plebejus argyrognomon, 88, 91, 93
Plebejus subsolanus, 86, 88, 91
Plebicula dorylas, 65
Plemyria rubiginata, 38
ploughing, 178, 243, 340
Poa pratensis, 234
Poaceae, 239
poaching, 97, 307
Polites mystic, 152
politicians, 40, 364, 365
Pollanisus cupreus, 277, 283–286
pollination, 41, 111, 225
pollinators, 43, 103, 226, 237, 245
pollution, 23, 39, 40, 48, 49, 52, 85, 91, 94, 99, 272
Polygonia c-album, 355
Polygonia faunus, 292
Polyommatus amanda, 65
Polyommatus bellargus, 55, 56, 124, 127, 130, 131, 135, 144, 355
Polyommatus coridon, 24, 55, 56, 65, 125, 126, 137, 144, 355, 365
Polyommatus icarus, 55, 56, 130, 137, 144, 228, 229, 240, 243, 244, 269, 355, 359
Polyommatus iphegenia, 364
Polyommatus thersites, 52
Ponera coarctata, 174
pony grazing, 313, 324
population, 9, 11, 17, 18, 20, 24–26, 29, 30, 32–35, 38–41, 44, 45, 47–49, 51–60, 63, 65, 68, 69, 71, 73, 74, 76, 86, 97–106, 109–112, 116, 118–120, 123–126, 128–148, 150–153, 155–161, 163–165, 168–170, 187, 190, 196–199, 202, 207–209, 211, 216–218, 222–234, 249, 251–253, 256, 261, 263, 264, 266–270, 273, 280, 287–290, 292, 296, 300–302, 304, 308, 310, 311, 313, 315, 317–325, 333, 339, 342, 343, 346, 349, 350, 359, 363–366
population decline, 11, 68, 104, 106, 124, 211, 324
population density, 74, 123, 125, 126, 128, 135, 141–147, 264
population dynamics, 49, 55, 110, 112, 123–125, 135, 137–141, 147, 199, 207, 209, 263, 269, 325, 365
population estimate, 321
population growth rate, 168, 263
population index, 32, 34, 156, 292, 311, 321
population persistence, 111, 196, 289, 333
population size, 30, 68, 69, 103–106, 123, 124, 126, 128–131, 133, 136, 146, 151, 159, 196, 197, 216, 222–234, 249, 256, 261, 263, 264, 270, 280, 288, 301, 304, 318–323

population time series, 141, 198
population trends, 25, 30, 32, 33, 40, 41, 44, 47, 49, 51–59, 124, 128, 141, 146, 158, 160, 313, 349, 350, 359
Portugal, 110, 111, 334, 337
potato, 238
Potentilla erecta, 318
Potentilla reptans, 214, 220
prairie butterflies, 149, 152, 161, 301, 302
prairie management, 149, 150
prairies, 3, 149–155, 157, 160, 161, 289, 290, 295, 301
predatory invertebrates, 226
predictive modelling, 25, 267
primary forest, 75–78, 80, 82
primary rainforest, 75, 76
Prime Butterfly Areas, 258, 259, 270, 302, 340, 345, 347, 363–366
Primula veris, 309
Primula vulgaris, 309
priority species, 20, 22, 24, 26, 33, 38, 189, 211, 262, 267, 319, 350, 360
process-response model, 265, 266, 268
Proclossiana eunomia, 223, 262, 268, 301
protected areas, 15–17, 19, 21, 22, 24, 83, 98, 99, 320, 327, 331, 350, 361
Provence hairstreak, 52
Prunus spinosa, 2, 113, 114
Pseudaletis arrhon, 73
Pteridium aquilinum, 12, 250, 304
public involvement, 29
purple emperor, 100, 355, 358
purple hairstreak, 355
purple moor grass, 191, 317, 318
purple toadflax, 32
purple-shot copper, 52
purplish fritillary, 292
Pyrenees, 63, 259, 346
Pyrgus maculatus, 86, 87, 90, 92, 102
Pyrgus malvae, 87, 94, 102, 127, 130, 137, 144, 250
Pyrgus malvoides, 52
Pyronia tithonus, 55, 56, 144–146, 228, 229, 250, 264, 355, 356

quantity, 201, 203–205, 235, 244, 265, 271, 272, 288, 327, 353
quarrying, 340, 364

rabbits, 48, 131, 132
railway, 134, 173, 211, 310
rain forests, 125
rainfall, 26, 94, 142, 197
rainforest regeneration, 75
random walk, 263
range change, 5, 94
range margin, 6, 8, 11, 12, 29, 30, 32, 34–36, 39, 41, 111, 128, 139, 146, 148
range margins, 11, 12, 29, 30, 32, 34, 35, 39, 41, 111, 128, 139, 146
range shifts, 6, 13, 41, 60, 103, 105, 108, 109, 111, 112
ranges, 12, 13, 22, 25, 40, 41, 52, 59, 61, 105, 109, 110, 112, 119, 124, 125, 128, 130, 137–139, 147, 148, 207, 211, 217–220, 262, 272, 322
Ranunculus, 167, 203, 240
rare species, 26, 44, 48, 58, 60, 139, 199, 237, 279, 297, 303, 313, 315, 340
recolonisation, 135, 208, 341
recording, 3, 5–9, 11, 17, 29–34, 37, 40, 42, 44, 52, 53, 126, 135, 153, 191, 202, 203, 238, 239, 241, 287, 304, 311, 328, 331, 349–354, 357–359, 364, 366
recreation, 96, 138, 271
red admiral, 55, 60, 355, 356
Red Data Book, 63, 65, 70, 73, 85, 89, 90, 95, 133, 240, 244
red fescue, 226
Red List Index (RLI), 67, 68
red list, 3, 26, 38, 40–42, 67–74, 86–89, 95, 96, 99, 101, 102, 104, 114, 163, 170, 208, 241, 242, 327, 329, 332, 339, 347, 359, 362, 366
red underwing, 35, 36, 52
red underwing skipper, 52
red-disked alpine, 292
red-green carpet, 38
red-necked footman, 35, 36
red-tailed bumblebee, 354, 356
regal fritillary, 151, 156, 160, 161
regional species pool, 237, 238, 241, 242
reintroduction, 97, 109, 110, 140, 171, 172, 187, 188, 198, 312, 317, 319, 321–325
reintroductions, 110, 321
relictual populations, 271, 279
remnant vegetation, 3, 271–274, 277, 279
reproduction, 43, 48, 49, 120, 137, 138, 159, 180, 223, 271, 297, 337, 341
reproductive potential, 265
rescue effect, 261
residence, 217, 264–266, 328
residence times, 217, 264, 266
resident, 2, 18, 32, 39, 45, 266, 272–280, 282–286, 291, 336, 363
resource, 2, 3, 5, 6, 12, 25, 31, 39, 68, 74, 83, 91–93, 98, 99, 102, 109, 123–125, 129–133, 137, 138, 149, 157, 159–161, 163, 168, 169, 171, 187, 206, 208, 211, 212, 222, 223, 226, 229, 234, 236, 237, 243, 245, 257–259, 261–272, 274, 279–281, 287, 288, 296, 297, 301, 302, 304, 306, 307, 309, 313, 314, 332, 336, 337, 347, 350, 353, 357–359, 362–366
resource quality, 265, 271, 288
restoration management, 303, 304, 308, 311, 324, 341, 342
restoration, 26, 49, 75, 82, 85, 95, 96–98, 100, 126, 132–134, 139, 160, 180, 187, 190, 198, 232, 266, 268, 269, 271, 280, 281, 297, 301, 303, 304, 306–314, 319, 321, 323–325, 341, 342, 347
restored grasslands, 342
Rhagonycha fulva, 354
Rhinanthus, 250, 252, 253
Rhingia campestris, 354
Ribes, 40
richness estimator, 77, 78
ringlet, 8, 10, 11, 112, 156, 229, 292, 355, 358
Rio de Janeiro, 1, 362
Riodinidae, 71, 248
riparian, 54, 93, 101, 187, 288
rivers, 91, 92, 172, 311, 328

roadsides, 1, 67, 91, 97, 123, 134, 158, 171–175, 177–180, 212, 217, 221, 222, 225, 247, 261, 269, 273, 280, 289–296, 298–300, 334, 339, 360, 364
Romania, 342, 346, 347
Rothamsted Insect Survey, 23, 27, 30, 32, 35, 39, 42, 47, 49
Rubus spp, 342
Rubus ulmifolius, 336
Ruderal, 54, 59, 148, 266
Rumex, 228, 230, 239, 249–251, 254, 255
Rumex acetosa, 228, 230
Rutpela maculata, 354
rural communities, 362
rural depopulation, 60, 340
rural development, 340
Rural Development Programmes, 340
rush, 154, 305, 307, 317

Saga pedo, 105, 110, 111
Salisbury plain, 201, 202, 209, 318
Sampled Red List Index (SRLI), 69–73
Sanguisorba officinale, 130
saproxylic beetle, 110
Sasakia charonda, 89, 94, 95, 99, 100
satin beauty, 38
Satoyama, 85– 88, 91–93, 95, 96, 98–102
Saturniidae, 30
Satyrinae, 77, 81, 83, 269, 277, 287, 301
Satyrium esculi, 56, 64
Satyrium pruni, 124, 127, 130, 133, 134
Satyrium w-album, 355
Satyrodes eurydice, 152, 292
savanna, 157, 160, 161, 259, 289–291, 301
savannah, 78, 80, 83
savannah butterflies, 78
Scabiosa columbaria, 202, 207, 341, 347
scale, 1–3, 6, 22, 30, 35, 37, 39, 42, 44, 48, 58, 59, 71, 79, 80, 83, 85, 91, 94, 98, 109, 111, 123, 125, 128, 130, 136, 137, 147, 148, 165, 166, 170, 172, 209, 212, 213, 215, 227, 229, 244, 245, 250, 259, 261–263, 267, 268, 270, 274, 287, 303, 313, 314, 321, 324, 328, 332, 339, 351, 354, 358, 361, 363, 364
scarce footman, 38
scarification, 225–227, 229–234
Scirpus holoschoenus, 336
scorched wing, 32, 38
Scotch Argus, 354, 355, 358
Scotland, 7, 11, 15, 17–23, 26, 138, 139, 209, 317, 319, 320, 362
Scotopteryx bipunctaria, 33
scrub, 51, 54, 55, 57–59, 110, 129, 133, 134, 191–193, 211, 213, 215, 221, 248, 250–258, 263, 304–310, 314, 321, 324, 340
scrub control, 306, 308, 310, 321
season, 17, 52, 53, 76, 120, 149, 160, 237, 241, 242, 291, 311, 317, 320, 335, 341, 352,
secondary forest, 80, 83, 102
sedge, 290, 305, 307
seed bank, 227, 324, 341
seed mixture, 227, 228, 234, 236–238, 243, 245
selective graminicide, 225, 227
semi-natural, 16, 18, 19, 21–25, 51, 58, 91, 92, 97, 99, 100, 125, 160, 297, 301, 302, 321, 340, 345, 347, 349, 350, 356, 358, 363
semi-natural areas, 363
semi-natural grassland, 91, 92, 97
september thorn, 38
seral, 101, 124, 130, 137, 273, 341, 342
seral stages, 101, 124, 130, 273, 342
Sericinus montela, 93, 94, 99–101
Sericomyia silentis, 354
Sesiidae, 30
settlements, 238
shade, 26, 130, 140, 179, 213–216, 221, 223, 274
sheep, 48, 187, 306–308, 310, 324, 342–346
shelter, 123, 126–129, 131, 138, 287, 309, 337
shelterbelt, 308
shifting baseline syndrome, 339, 347
Shijimia moorei, 86, 87, 89, 95, 97, 99, 101
Shijimiaeoides divinus, 86, 87, 90, 92, 97, 99–102
shrubland, 52, 267
shuttle-shaped dart, 35, 36
Sierra Leone, 80, 83
Silene dioica, 228
Silene pratensis, 244
silver-bordered fritillary, 292
silver-spotted skipper, 7, 12, 25, 111, 137, 139, 268, 301
silver-studded blue, 355
silver-washed fritillary, 54, 355
Single Large or Several Small (SLOSS), 281
sink, 131, 133
site occupancy, 124, 132, 135, 137, 145, 268, 271, 272, 276, 277, 279, 280, 287, 288, 312
Sites of Special Scientific Interest (SSSI), 17–19, 22, 25, 36
slender scotch burnet, 33
slope, 45–47, 56, 114, 144, 173, 174, 176, 177, 179–181, 183–187, 192, 215, 248, 250, 252–255, 305, 307, 341
Slovenia, 85, 109, 327–332
small blue, 355
small copper, 137, 355
small heath, 240, 350, 355
small pearl-bordered fritillary, 304, 314, 315, 355
small scabious, 202
small skipper, 225, 229, 355
small tortoiseshell, 355
small white, 59, 111, 229, 355, 356
small-scale agriculture, 238
smooth meadow grass, 234
snowberry, 320
social justice, 2
socioeconomic, 248, 258, 339
socio-economic, 259, 263, 267
soil depth, 126
soil temperature, 164, 179
solar radiation, 265, 266
source habitat, 124, 135, 136
source patch, 135, 136
sources, 17, 18, 31, 43, 44, 70, 96, 104–106, 128, 129, 131, 135, 142, 156, 158, 196, 207, 228, 234, 265, 267, 289, 306, 328, 333, 334, 336, 341, 349, 354
south east Asia, 97
southern gatekeeper, 59
sowing, 226, 228, 243
sown, 3, 225–227, 229–235, 237, 238, 240, 242–245

spain, 3, 51, 52, 59, 63, 106, 109, 148, 247, 248, 259, 287, 333–337, 339, 341–343, 347
spatial, 2, 5, 12, 23, 26, 32, 33, 35, 42, 44, 60, 71, 79, 83, 99, 111, 112, 115, 119, 137, 138, 141, 142, 144–147, 163, 168, 187, 188, 192, 204, 219, 220, 239, 243, 245, 257, 261–264, 266–270, 272, 288, 295, 296, 301, 313, 315, 323, 325, 337, 339, 340, 346
spatial data, 147, 313
spatial heterogeneity, 267, 269
spatial patterning, 295
spatially explicit, 124, 198, 287, 337
Spatially Realistic Metapopulation Theory, 263
Special Areas of Conservation (SAC), 119, 319, 327–332
Special Protection Areas (SPA), 164, 327, 328
specialist, 11–13, 15, 17–24, 54, 94, 95, 108, 110, 132, 138, 149, 150–153, 155–159, 209, 223, 257, 269, 289–292, 295–297, 300, 317, 325, 336, 340, 342, 350, 356, 358
specialists, 11, 18, 20–22, 44, 51, 53–55, 57–60, 80, 101, 124, 134, 149, 150, 152, 153, 155, 157–159, 289, 291–296, 298, 330, 342, 356
species diversity, 5–8, 12, 97, 148, 160, 225, 229, 232, 233, 271, 281, 288, 327, 332, 346, 353
species occupancy, 6, 274, 279, 355
species richness, 7, 9, 12, 23–25, 29, 30, 41, 42, 59, 60, 61, 67, 71, 75, 77–80, 92, 112, 142, 146, 157, 166, 169, 172, 174, 177, 187, 196, 203, 225–227, 229–235, 237–242, 248, 259, 267–270, 280, 287, 288, 328, 332, 341, 342, 345–347, 359
species-area relationships, 272, 278, 288
species-rich calcareous grassland, 20
species-specific, 55, 134, 142–144, 149–156, 158, 159, 164, 187, 261, 268, 272, 281, 297
species temperature index, 59
speckled wood, 8, 10, 59, 139, 269, 355
Speyeria aphrodite, 152, 153, 155, 157
Speyeria atlantis, 292
Speyeria idalia, 149, 151, 156, 160, 161, 295
sphagnum-leatherleaf mats, 290
Sphingidae, 30, 41
Spialia sertorius, 52
spiders, 49, 226, 233
spinach, 38, 45, 47
spruce carpet, 38
spruce-cottongrass-wiregrass savanna, 290
Starvation (larval), 109, 113, 117, 132, 190, 205, 336
stasis, 289, 296, 297
sterility, 130
Stilbia anomala, 38
stochastic patch occupancy model, 263
stochastic processes, 124
stratification, 76, 83
straw belle, 33
sub-populations, 317, 319, 320, 324, 325, 363
suburbs, 273
succession, 49, 97, 102, 149, 190, 221, 240, 244, 247, 248, 258, 259, 301, 306, 310, 312, 340–342, 346
successional stage, 158, 290, 312
Succisa pratensis, 189, 198, 201, 202, 318
summer grazing, 247, 248, 250–253, 256–258
sun-moths, 273
survey design, 155, 349, 350, 356, 357, 359
survival, 43, 74, 99, 109–111, 113, 114, 124, 129, 130, 132, 136–138, 151, 163, 164, 168, 170, 179, 196, 206–208, 213, 221, 223, 272, 280, 288, 315, 341, 342, 362–364
sustainable production, 362
swallowtail, 70, 74, 99, 120, 137, 243
sward, 109, 131, 132, 166, 169, 189, 190, 192–196, 199, 201, 203–208, 227, 229, 233–235, 243, 247, 250–258, 310, 315, 318, 319, 321, 323, 324, 340–342
sward condition, 247
sward cuts, 243
swards, 130, 132, 135, 232, 257, 258, 306, 310, 312, 321, 324
Sweden, 133, 212, 222, 235, 237, 244, 259, 340, 342, 346
sweet William, 32
Swiss Alps, 49, 267
Switzerland, 74, 85, 99, 101, 237–239, 242, 244, 245, 268, 341, 342
Sympetrum striolatum, 35, 356
Symphoricarpos albus, 320
Synemon gratiosa, 273, 277, 280, 283–286

tall grassland, 266
tallgrass prairies, 3, 289
tamarack-sedge mats, 290
Tanacetum vulgare, 244
Taractrocera papyri, 277, 282–286
temperate, 16, 26, 29, 60, 67, 93, 100, 103, 107, 129, 136, 139, 161, 201, 225, 280, 290, 291, 296, 297, 302, 304, 315
temperature, 26, 42, 51, 55, 58, 59, 94, 103, 106, 108, 110, 111, 113–119, 128, 138, 139, 141–147, 164, 169, 179, 220, 221, 229, 263, 265, 266, 291, 292, 294, 318, 322, 325
temporal, 18, 30, 31, 32, 39, 44, 55, 57, 80, 83, 89, 111, 128, 138, 141, 147, 192, 195, 197, 262–265
Terminalia ivoriensis, 76
Tetramorium caespitum, 174
Thaumetopoeidae, 30
The Netherlands, 23, 26, 30, 38, 39, 41, 43–49, 60, 85, 106, 113, 114, 121, 133, 140, 171, 173–176, 178, 180, 184, 188, 260, 270, 331, 347, 350, 360, 366
Thecla, 2, 113, 114, 120, 121, 125, 127, 130, 355
Thecla betulae, 113, 114, 125, 127, 130, 355
Theclinesthes miskini, 277, 283
Theclinesthes serpentata, 283
Thera britannica, 38, 39
Thera juniperata, 38
Thermal environment, 265
thermophilous, 131, 132, 139, 148, 164, 172, 179
thermoregulatory, 129
thick-legged flower beetle, 354
thistles, 233
Tholera cespitis, 38
Tholera decimalis, 38
Thyatiridae, 30
Thymelicus, 55, 56, 65, 88, 124, 127, 130, 135, 139, 143, 225, 228, 229, 239, 242, 244, 251, 355
Thymelicus acteon, 55, 56, 65, 130, 135, 139, 143, 355
Thymelicus lineola, 228, 229, 242, 244, 355
Thymelicus sylvestris, 55, 56, 225, 228, 229, 251, 355
Thymus, 132, 163–168
Thymus pulegioides, 163, 164
timber harvesting, 290

time of day, 17, 215, 351, 352
timed count, 17, 321
Timothy, 159, 226
toadflax brocade, 32, 35, 36
Tomares ballus, 52, 55, 56
tourism, 88, 91, 96, 258, 343, 362, 364
trade, 138, 272
traditional farming, 190, 339, 340, 343, 361–363
traditional management, 86, 177, 261, 339, 345, 361
training, 31, 33, 93, 202, 207, 209, 314, 318, 358
transect, 17, 18, 26, 52–54, 56, 57, 63–65, 71, 77, 79, 96, 125, 135, 142–144, 150, 151, 191–196, 203, 216, 227, 229, 237–243, 273, 274, 276, 282, 291, 301, 307–309, 314, 349, 351, 352, 354, 356, 359, 364
transects, 17, 19, 21, 52, 53, 76–80, 125, 141, 145, 150, 152, 153, 189, 191, 192, 194, 195, 237, 238, 240, 242, 247, 249, 258, 273, 274, 280, 288, 351–353, 358
transhumance, 344
trap, 77, 364, 365
Trapezites argenteoornatus, 277, 282, 283
Trapezites sciron, 277, 282–286
treble brown spot, 38
trees, 32, 39, 73, 76, 77, 90, 92, 93, 110, 112, 115, 121, 150, 174, 177, 179, 221, 242, 258, 274, 276, 277, 282, 285, 310, 340, 342
trend, 1, 10, 15, 16, 18, 20, 21, 23, 25, 26, 30, 34, 38, 40, 43, 45–48, 51, 56–59, 71, 79, 86, 93, 126, 137, 150, 153, 155, 161, 175, 192, 197, 226, 241, 308, 313, 318, 324, 344, 364
Trifolium, 167, 226, 228, 240
Trifolium pratense, 167, 228
Trifolium repens, 167
TRIM, 25, 53, 54, 56, 60, 311, 313
tropical forest, 75, 79, 83, 137
two-year life cycle, 292
tyrphobiontic, 291
tyrphoneutral, 291
tyrphophilic, 291
tyrphoxenous, 291
Tyta luctuosa, 33

ubiquitous species, 33, 78, 80
UK, 1–3, 5, 6, 12, 15–24, 29–35, 38–40, 49, 51, 58, 67, 73, 74, 77, 86, 99, 103, 109, 112, 123–126, 128–138, 140–143, 145, 146, 169, 170, 189, 190, 201, 202, 207–209, 211, 222, 223, 225, 226, 235, 236, 238, 243, 245, 261, 266, 269, 303, 304, 308, 311, 313, 314, 317–319, 321, 325, 339, 349, 350, 351, 353, 356–361, 363, 364, 366
UK Biodiversity Action Plan (UKBAP), 15, 16, 18, 350, 360
UK Butterfly Monitoring Scheme (UKBMS), 5, 15, 17–20, 22, 24, 125, 136, 141–147, 349, 350, 352, 354, 356–359
umbrella species, 163, 169, 262, 268, 269
unfertilised, 130, 131
unoccupied patches, 139, 264, 306, 308, 313
unsown annual, 229, 230, 231
unsown perennial, 231, 234
upland, 25, 52, 54, 55, 58, 76, 83, 261, 263, 266, 290–295, 300, 301, 304, 307, 358
Upper Guinean Forest Zone, 3, 75
urban, 3, 24, 87, 88, 91–93, 111, 201, 259, 268, 270–274, 276–282, 287, 288, 327, 347, 350, 361, 364, 365
urban adapter, 274
urban avoider, 274
urban spread, 361, 364
urbanization, 58, 272–274, 280, 288, 290
Urtica dioica, 138, 239
utilities, 262, 264–266, 268

vagility, 134, 271
Vanessa atalanta, 7, 55, 56, 60, 144, 146, 229, 244, 250, 355, 356
Vanessa cardui, 144, 228, 229, 244, 250, 355, 356, 358
Vanessa itea, 283, 284
Vanessa kershawi, 275, 277, 283–286
varied coronet, 32, 35, 36
vegetation condition, 3, 271, 272, 274, 276, 277, 279–282
vegetation height, 137, 166, 167, 198, 208, 214, 215, 221, 223, 305, 307
Veratrum album, 341
Verbascum lychnitis, 244
Verbascum thapsus, 244
verges, 134, 171–175, 177–180, 184, 214, 215, 267, 269, 306, 310
Vicia cracca, 214, 215, 220
Vicia sativa, 215
Vicia sepium, 215
Vine's rustic, 35, 36, 38
viola, 130, 239, 250, 251, 254, 255, 304, 305, 307
Viola palustris, 304, 307
Viola riviniana, 304, 307
v-moth, 35, 38, 39
voltinism, 137
Volucella pellucens, 354
volunteers, 5, 6, 12, 31, 32, 35, 44, 120, 147, 258, 273, 306, 310, 314, 320–322, 324, 325, 328, 332, 350, 353, 354, 357, 358
vulnerable, 2, 38, 47, 68, 69, 72, 86, 89, 90, 106, 107, 113, 118, 136, 189, 202, 204, 208, 241, 262, 289, 290, 305, 329, 364

Wales, 9, 12, 15, 17, 18, 20, 24, 40, 211, 318, 359, 362
wall brown, 138, 350, 355, 356, 358
warming climate, 132, 135, 141
water courses, 226
water vole, 258
weather, 17, 20, 22, 43, 48, 91, 94, 123, 127–129, 131, 159, 173, 189, 191, 197, 213, 221, 239, 261–265, 273, 291, 292, 294–296, 308, 318, 319, 321–324, 351, 352, 357, 358
weeds, 157, 158, 271, 273, 274, 280, 310
Western Australia, 3, 271–274, 277–279, 281, 282, 287, 288
western marbled white, 58
wet grasslands, 196, 317
wetlands, 54, 125, 130, 134, 268, 302
wheat, 227, 238
white admiral, 52, 355
white-letter hairstreak, 355
white-line dart, 38
wider countryside, 2, 15, 18, 20–22, 24, 242, 331, 340, 349, 350, 352–355, 358
Wider Countryside Butterfly Survey (WCBS), 349–354, 356–359
widespread species, 6, 7, 11, 12, 154, 262, 269, 350, 356

wild boar, 258
wild carrot, 243
wildflower, 3, 225–235, 237–245, 288
wildlife, 16, 18, 23, 41, 42, 59, 75, 81, 83, 85, 92, 94–101, 103–105, 111, 134, 139, 148, 151, 152, 159, 160, 189, 191, 198, 259, 297, 301, 327, 361, 362, 364, 365
winter diapause, 109, 113, 190
winter grazing, 247, 248, 250, 256–258
Wisconsin, 149–161, 289–292, 294, 296, 297, 301, 302
within-patch, 123, 135, 263, 267
wood fuel, 93, 362, 363, 366
wood white, 140, 211, 223, 355
woodland, 12, 15, 18–22, 26, 51, 54, 55, 57, 58, 85–88, 91–93, 96, 98–100, 110, 124, 125, 129, 130, 133, 137, 140, 147, 160, 211, 212, 216, 217, 221–223, 238, 257, 261, 263, 268, 273, 287, 301, 303, 306, 309–311, 313, 314, 324, 340, 341, 353, 357, 361–363, 366
Woodland Grant Scheme, 18, 22, 303, 314
woodland management, 12, 22, 129, 362

Xanthia gilvago, 38
Xanthorhoe ferrugata, 38
xeric, 266
xero-thermophilous, 164

Ypthima multistriata niphonica, 88

Zerynthia rumina, 56, 64
Zizeeria karsandra, 88
Zizina emelina, 87, 91, 93, 100
Zizina labrudus, 277, 283–286
Zygaena loti, 33
Zygaenidae, 30, 277

GPSR Compliance

The European Union's (EU) General Product Safety Regulation (GPSR) is a set of rules that requires consumer products to be safe and our obligations to ensure this.

If you have any concerns about our products, you can contact us on ProductSafety@springernature.com

In case Publisher is established outside the EU, the EU authorized representative is:

Springer Nature Customer Service Center GmbH
Europaplatz 3
69115 Heidelberg, Germany

Batch number: 10755939

Printed by Printforce, the Netherlands